STRATEGIES FOR PALLADIUM-CATALYZED NON-DIRECTED AND DIRECTED C—H BOND FUNCTIONALIZATION

CONTENTS

LIST OF CONTRIBUTORS

Bhalchandra M. Bhanage
Institute of Chemical Technology, Mumbai, Maharashtra, India

Aniruddha Dey
Indian Institute of Technology Bombay, Mumbai, Maharashtra, India

Uttam Dhawa
Indian Institute of Technology Bombay, Mumbai, Maharashtra, India

Ian J.S. Fairlamb
University of York, York, United Kingdom

Prashant Gautam
Institute of Chemical Technology, Mumbai, Maharashtra, India

Vijay Gayakhe
Institute of Chemical Technology, Mumbai, Maharashtra, India

Aniket Gholap
Institute of Chemical Technology, Mumbai, Maharashtra, India

Ye-Qiang Han
Zhejiang University, Hangzhou, China

Fang Hu
Zhejiang University, Hangzhou, China

Lin-Yu Jiao
Northwest University, Xi'an, Shaanxi, PR China

Anant R. Kapdi
Institute of Chemical Technology, Mumbai, Maharashtra, India

Fuk-Yee Kwong
The Hong Kong Polytechnic University, Kowloon, Hong Kong; The Chinese University of Hong Kong, Shatin, New Territories, Hong Kong

Gabriele Laudadio
Eindhoven University of Technology, Eindhoven, The Netherlands

Gang Li
Fujian Institute of Research on the Structure of Matter, Chinese Academy of Sciences, Fuzhou, Fujian, China

Bin Liu
Zhejiang University, Hangzhou, China

Debabrata Maiti
Indian Institute of Technology Bombay, Mumbai, Maharashtra, India

Timothy Noël
Eindhoven University of Technology, Eindhoven, The Netherlands

Martin Oestreich
Technical University of Berlin, Berlin, Germany

Benudhar Punji
CSIR-National Chemical Laboratory (CSIR-NCL), Pune, Maharashtra, India

Bing-Feng Shi
Zhejiang University, Hangzhou, China

Vineeta Soni
CSIR-National Chemical Laboratory (CSIR-NCL), Pune, Maharashtra, India

Neetipalli Thrimurtulu
Indian Institute of Technology Bombay, Mumbai, Maharashtra, India

Chandra M.R. Volla
Indian Institute of Technology Bombay, Mumbai, Maharashtra, India

Shun-Man Wong
The Hong Kong Polytechnic University, Kowloon, Hong Kong

FOREWORD

Methods for the functionalization of otherwise unreactive C—H bonds have emerged as an increasingly viable platform in molecular synthesis, with enabling applications in various research areas ranging from material sciences and natural product chemistry to crop protection and pharmaceutical industries. While various transition metals have been identified as viable catalysts for step-economical C—H activations, palladium catalysis has arguably thus far proven among the most versatile tools. Thus, pioneering studies by Chatt on stoichiometric cyclopalladation reactions indicated, among others, the potential of this approach for the positional selective cleavage of inert C—H bonds. Hence, Lewis-basic functionalities proved instrumental for the site-selective C—H palladation on arenes by chelation assistance. The strategy of proximity-induced C—H activation set the stage for a plethora of protocols for directing group-guided C—H functionalizations. However, from a historical perspective it is interesting to note that the first palladium-catalyzed C—H functionalizations were actually accomplished in the absence of directing groups—an observation that holds true for most of the initial catalyzed C—H functionalizations, irrespective of the transition metal. Specifically, the oxidative alkenylation of simple benzene derivatives proved viable by the action of palladium (II/0) catalysis in a nondirected fashion. The practical importance of the cross-dehydrogenative olefination was quickly realized and is, for instance, reflected by this chemical transformation being recognized as a named reaction. Yet, the synthetic utility of palladium catalysis in C—H activation chemistry is not limited to the palladium(II/0)-catalyzed Fujiwara – Moritani reaction. Indeed, the palladium(II/IV) reaction manifold was viable through oxidation-induced reductive eliminations as the key elementary step. Arguably, the most powerful approach continues to be represented by the palladium(0/II) catalysis regime. Here, the oxidative addition of organic electrophiles onto catalytically competent palladium (0) complexes set the stage for various C—H transformations, including C—H arylations and difficult C—H alkylations. The unique power of the palladium(0) catalysis approach became evident by ligand-accelerated C—H functionalizations employing challenging organic chlorides, tosylates, and even mesylates. Since the substrates of interest in palladium-catalyzed C—H functionalizations usually display various C—H bonds with close dissociation energies, achieving positional selectivity in intermolecular reactions continues to be the key challenge. While the early studies by Chatt and Fujiwara/Moritani (*vide supra*) highlighted the power of directed and nondirected C—H transformations, respectively, more recent advances exploited strategies to control site-selectivity by means of remote arene functionalizations. However, the palladium-catalyzed C—H activation approach was not restricted to the functionalizations of $C(sp^2)$—H bonds, but also allowed for proximity-induced functionalizations of aliphatic $C(sp^3)$—H bonds. Particularly, in the arena of challenging $C(sp^3)$—H activation, palladium catalysis displayed a yet unparalleled broad substrate scope.

With a remarkable toolbox for palladium-catalyzed C—H functionalization protocols at hand, it does not come as a surprise that transformative applications to the preparation of structurally complex target molecules have appeared in the past few years. Thus, the power of palladium-catalyzed C—H activation was illustrated by efficient syntheses of natural products, novel polymerization techniques, and the late-stage modification of bioactive compounds, such as nucleobases or peptides in a bioorthogonal fashion.

This book summarizes all facets of palladium-catalyzed C—H activation of relevance to the assembly of organic molecules and the design of molecular architectures, which should prove invaluable for scientists in academia, as well as practitioners in applied areas, such as chemical and pharmaceutical industries. In consideration of the increasing practical importance of C—H functionalizations, along with the outstanding versatility of palladium catalysis, further exciting applications are expected. The full potential of catalytic C—H activation chemistry will only be unleashed through the transition from academic research to industrial applications on scale.

Lutz Ackermann
Göttingen, August 2016

Traditionally, the assembly of heterocycles largely involved the interconversion of preexisting functional groups, the installation of which often required lengthy synthetic operations. Since these approaches, thus, produce major amounts of undesired by-products and waste, the focus has shifted during the past decade towards the development of methods that capitalize upon the activation of otherwise inert C—H bonds. Hence, metal-catalyzed C—H functionalizations have revolutionized the efficiency of heterocycle syntheses, leading to an overall streamlining of the construction of increasingly complex target structures. Particularly, oxidative transformations were identified as versatile tools for the effective de novo synthesis of, among others, oxygen-, nitrogen-, and phosphorous-containing heterocycles.

The synthetic utility of the C—H activation approach is, however, not limited to the assembly of heterocycles. Indeed, the past decade has gained enormous momentum in establishing tools for the step-economical functionalization of heterocycles. Furthermore, in a more general context, heterocycles are not only of relevance as substrates for C—H functionalizations, but are also mandatory as ligands and/or as site-selectivity-ensuring directing groups in C—H activation chemistry.

Organic substrates of interest for direct functionalizations usually bear numerous C—H bonds of comparable dissociation energies. Therefore, a major obstacle in achieving synthetically useful C—H functionalizations is constituted by fully controlling the site-selectivity in intermolecular transformations. While this represents a formidable challenge in direct arene functionalizations, heteroarenes often display C—H bonds with increased kinetic acidity, mostly proximal to an electronegative heteroatom. Hence, recent years have witnessed a major growth in protocols for metal-catalyzed C—H functionalizations that proceed in a (concerted metalation) deprotonation fashion. For instance, earth abundant base metals, such as copper, nickel, or cobalt catalysts, allow for site-selective C—H functionalizations of various five-membered heteroarenes, provided rather strong bases were utilized. Since the functional group tolerance of such reactions is often compromised, a powerful alternative was established by carboxylate-assisted C—H metalations, largely with the aid of 4d transition metal catalysts based on palladium, rhodium, or ruthenium. Thereby, challenging heteroarene functionalizations proved ultimately viable under exceedingly mild reaction conditions.

From a historical perspective, it is noteworthy that several C—H functionalization protocols were in fact initially developed for heterocyclic substrates. For instance, one of the arguably most powerful tools in C—H functionalization chemistry, palladium(0)-catalyzed direct arylation with organic electrophiles, was originally devised for the site-selective diversification of indoles. This pioneering report nicely illustrated the power of controlling the site-selectivity in heteroarene C—H functionalizations.

Thus far, most of the advances in heterocyclic C—H activation were accomplished through the cleavage of $C(sp^2)$—H bonds. However, major recent progress was constituted by the functionalization of thermodynamically less stable, but frequently more difficult to activate, $C(sp^3)$—H bonds.

CHAPTER 1

Introduction

Aniruddha Dey[1], Anant R. Kapdi[2] and Debabrata Maiti[1]
[1]Indian Institute of Technology Bombay, Mumbai, Maharashtra, India
[2]Institute of Chemical Technology, Mumbai, Maharashtra, India

Organic chemistry has often been nominated as collateral to a celestial divinity bestowing Earth with the art of creation of life. Since the realization of a tangible cognizance about the prowess of this wonderful repository, mankind began their venture into developing deeper comprehension about the same. Theories akin to "*Vitalism*" were dubbed as outmoded clichés, as burgeoning scientific tendencies separated abstract convictions into perceptible rationale. Credulous minds gradually evolved into inquisitive intellects, and the convergence between beliefs and truth was thus laid, resulting in the foundation of unfettered knowledge and newfound discoveries. This had been proactive in carving the existence and subsequent development of the human race since its genesis. This saga remained persistent throughout centuries, leading to the creation of rewarding inventions time and again whenever cogent queries hit covert speculations. This is what defines science in its purest form, and organic chemistry, being an inseparable and indispensable subset, fulfills the criteria in more ways than are actually expected. Organic chemistry serves to be an omnipotent benefactor for human civilization beside other subsidiaries of chemical science. Beginning with the discovery of the first organic compound, i.e., urea in 1828, chemists continued to seek startling revelations about the chemical structure of organic components, and develop synthetic procedures to mimic the formation of natural products, produce life-saving pharmaceuticals, and address solutions to product manufacture in complex industrial settings. The late 19th century witnessed the first ever synthesis of aspirin and since then, the pharmaceutical industry has undergone a boom in drug production and development, thereby emerging to occupy a voluminous portion of the world's economy in recent times. Classical organic synthetic procedures paved the way for chemists to find a path for the creation of important bio-relevant components that were otherwise preeminent in their occurrence through natural chemical processes. Importantly, cross-coupling reactions, typical

Strategies for Palladium-Catalyzed Non-Directed and Directed C−H Bond Functionalization
DOI: http://dx.doi.org/10.1016/B978-0-12-805254-9.00001-3

of the Wurtz reaction, appealed to the minds of erudite chemists. However, expectations rose further in the search for strategies that would render high yields of final products while utilizing raw materials at low cost. Implementation of transition metal-based catalytic procedures were determined as perfunctory measures to address this conjecture. Consequently, famous reactions like those devised by Fujiwara – Moritani, Heck, Suzuki, and others rose to the zenith of popularity, as these seemed to assure cost-effective and productive routes for synthesis of complex moieties from simpler ones. Advances were made in transforming various types of carbon bonds into the desired ones, using an assortment of functionalization techniques. The advent of catalysis imbued synthetic chemistry with the right panache to provide remedies for several limitations that had been prevalent since preexisting times. The most vibrant of these was the modification of a target site in a molecule beside the concurrent presence of several competitive sites with identical reactivities. This has been addressed through a plethora of metal-mediated transformations that ensured accentuated levels of site-selectivity within a complex molecular cluster containing bonds with innately similar chemical properties. This has been found to be prevalent during functionalization of heterocycles, aromatic-rings, and several other members of the aliphatic class. Recognition of directing groups in performing remote functionalization was another feather in the crown that rose to attain distinctive stature at the turn of the 21st century. Leading the fleet of innovations and ushering in interest in the betterment of science and research, directing group assisted C—H activation was established as the modus operandi for complicated functionalization in organic molecules. This has enabled transformations to be executed at remote sites, be it in aromatic or in long aliphatic chains. Discovery of detachable nucleophilic anchors for directing remote sp^2 C—H activation in recent years has further advanced the chain of development in regioselective functionalization. Additionally, curious quests to blend computational tools with synthetic methodology have unlocked a new field of inquiry and élan.

This book summarizes the recent advances in synthetic chemistry through the eyes of C—H functionalization-based strategies. Renowned research scholars have penned a collective treatise from their distinctive fields of interest, and pooled a coherent discussion about the fundamental concepts. Insightful and detailed address of the intricacies for the various methodologies further mark the cogent and analytical comprehension of the ideal mindsets of the 21st century chemists. However, the variety of metals used for catalytic conversions stands legion, and a vivid discourse encompassing all of them is beyond a corporeal attempt. Undoubtedly, the countable occasions in which palladium was used to perform such historical transformations are innumerable. This could be attributed to its greater catalytic efficiency, ease of procurement, and a multitude of advantages, coupled with the cost-effectiveness. Subsequent chapters in this book therefore revolve around chemical strategies which discern palladium as a catalyst, and thus rightfully recognize it as the protagonist in the book's context (Fig. 1.1).

The opening chapter introduces readers to the fascinating diorama of ligand supported functionalization of C—H bonds in aromatic cores. Owing to the synthetic significance of metal-mediated C—H activation strategies, the benefit of using ligands are well-precedented in the literature

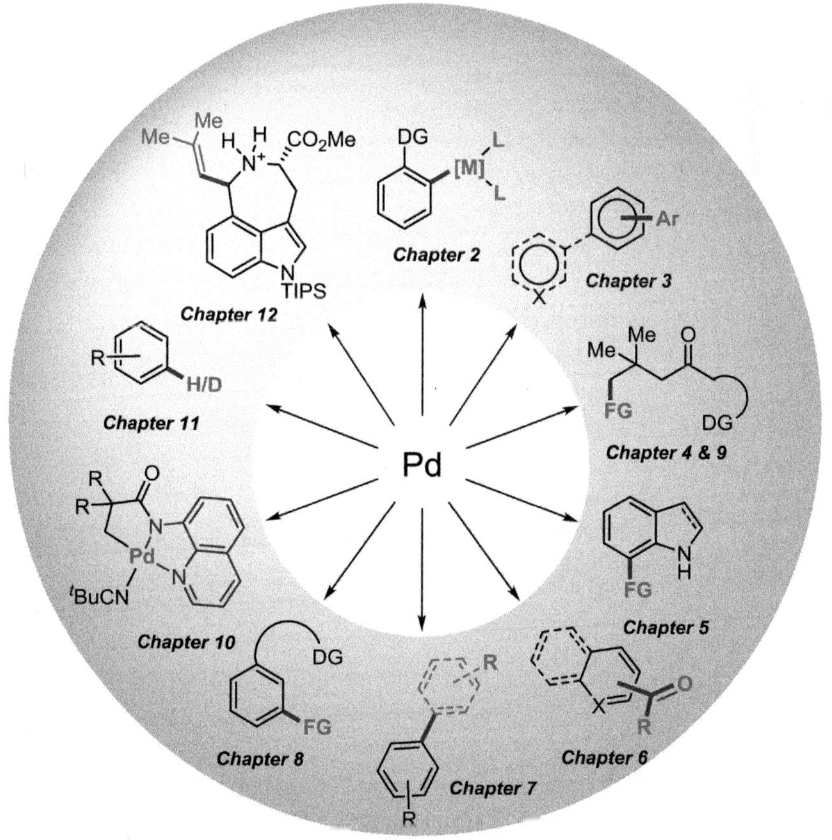

Figure 1.1 Schematic representation of the various aspects of palladium-catalyzed protocols covered in this book.

of organometallic chemistry. With respect to the modes of coordination, ligands can either form a strong or weak coordination bond to the metal precatalyst. This in turn compensates for lowering the energy of intermediates formed in the chemical transformation. A low energy requiring pathway is thereby established which renders the transformation effective on a catalytic scale. Strongly coordinating ligands are believed to yield a more stable intermediate compared to the weakly binding ligands. In this chapter, Fairlamb and Kapdi present a vivid exemplar of C—H activation-based transformation based on the effective and logical use of ligand control-based strategies.

In the chapter about nondirected C—H bond functionalization, Wong et al. discuss viable strategies undertaken for synthesis of biheteroarenes. The latter class of compounds enjoy a celebrated mention among a multitude of applicative fields like biological and pharmaceutical materials, and natural products, as well as in synthetic research. Correspondingly, a deep insight into the contemporary synthetic procedures which allow their production is significant. Wong et al. underline nondirected catalytic routes used in the last decade for effectuating the synthesis of biheteroarenes from simple starting precursors. These ensure cost-effective and greener alternatives to the directing group-based functionalization phenomenon. The chapter talks about strategies for palladium-catalyzed cross-coupling reactions between arenes/heteroarenes and arylhalides. Owing to the harsh synthetic procedures required for preparation of non-commercially available arylhalides, application of pseudohalides for biheteroarene synthesis is discussed as well.

The discussion on biheteroarene synthesis is then succeeded by an account of the activation of sp^3 C—H bonds. The high bond dissociation energy of an aliphatic sp^3 C—H bond renders these types of systems extremely inert. Moreover, an absolute absence of an easily polarizable electron cloud renders them unreactive. In this chapter, Liu et al. talk about a highly challenging aspect of incorporating arene rings at these unactivated centres. An assortment of novel strategies encompassing the efficacy of using directing groups and other chelation assisted strategies to perform a highly selective arylation in aliphatic carbon motifs have been explored. These include a lucid imagery of the benign directing groups which have been designed by Daugulis, Carratero, Ma, and others. More about this and about other types of functionalizations at unactivated sp^3 C—H centers have been discussed by Dey et al. in a later chapter of this book.

In the following chapter, Jiao et al. concentrate on the synthetic approaches for C—H bond functionalization at the benzene core of heterocycles, namely indole and indolines. The prominent importance of the indole motif in natural products and biologically significant materials recognizes the divulgence of their synthetic strategies. Jiao and Oestreich conscientiously portray recent advances in functionalization at the C-4 and C-7 positions of indole, and C-6 and C-7 positions of indolines under the broad umbrella of palladium catalysis. Preferable reactivity at the C-2 and C-3 positions of indoles render selective functionalization at the other positions challenging. The chapter describes measures that circumvent such a possibility to ensure the desired functionalization. Drawing analogies from the traditional Fujiwara – Moritani reaction, selective functionalization at the C-4 position of indole has been effected. Interestingly, chemists have employed indolines as plausible precursors for C-7 transformation in indoles. Execution of such phenomenon through the use of directing groups is also tangible. Numerous reports on palladium-catalyzed C—C bond formation reactions are detailed in this chapter.

Subsequently, Gautam et al. present highlights on carbonylative and carboxylative reactions in C—H functionalization. The latter two categories of functionalization are recognized to be some of the oldest examples of chemical transformations to be executed. In this chapter, the authors describe an account of palladium-catalyzed carbonylative and carboxylative methodologies. Additionally, the importance of directing groups like those of amidines, oximes, sulfonamides, etc., in attaining sp^2 C—H carbonylation has been emphasized. Further, the same stretches over a discussion on carbonylation reactions with heterocyclic systems containing single or multiple heteroatoms. Carboxylative reactions which typify the use of CO_2 as a reacting agent are also mentioned towards the close of the chapter.

While the sojourn through the metal-catalyzed organic transformations continues, Laudadio et al. introduce us to the riveting applications of "Flow Chemistry" in organic synthesis. A boon for the chemists of the modern era, flow chemistry has created a completely new dimension for synthetic processes by allowing faster reaction rates at cheaper production costs. Laudadio and Noël draw a contour emphasizing the potential of microflow reactors in performing catalytic transformations in microliter scale. The hardware setup for flow chemistry is meticulously designed to cause self-degeneration of the hazardous chemical wastes that are formed as by-products in a reaction mixture. Eminent research groups like those of Davies, Jones, Lapkin, and Noël have showcased the efficiency of this technique in performing transition metal-catalyzed C—H activation reactions that proceed with shorter time scales and precise selectivity. Integrated multistep synthesis coupled with effective scalability augment the attractiveness of this approach in a practical laboratory setup.

As Noël builds the stage for presenting the new-fangled advances of recent eons, Gang Li presents the subsequent chapter describing a phenomenon occurring in the area of directed C—H functionalization chemistry: *meta*-C—H functionalization of aromatic cores. In this chapter, Li summarizes the significant reports on sp^2 C—H functionalization at the distal *meta*- position of an arene. Pioneered by Yu in 2012, directed *meta*-C—H activation has emerged currently as a hot area of research in the arena of organometallic catalysis. Regioselective formation of C—C/C—O bonds continues to make way at the *meta*-position of unactivated arenes, thereby overcoming a problem persistent since the onset of metal-mediated C—H bond activation in an arene core. Gang Li pens a chronological sequence of the gradual evolution of *meta*-functionalization. The latter popularizes an emerging class of cleavable directors which were subsequently brought to operational existence by the Yu, Tan, Maiti, and Li groups to perform distinguished *meta*-C—H olefination, acetoxylation, arylation, hydroxylation, and iodination.

In the succeeding chapter, Dey et al. discuss transformation of sp^3 C—H bonds catalyzed by palladium precatalysts. As discussed in Chapter 4, activation of an sp^3 C—H bond is quite challenging, owing to its low reactivity and relative inertness compared to the sp^2 and sp congeners. In such cases, complexity is further amplified by the attempts at functionalization at remote positions with respect to the preinstalled functional groups. Functionalization at sites which are close to the preinstalled functionality (say, at α/β position) has been addressed mostly through classical or in some cases, by application of traditional cross-coupling methods. Functionalization at sites remote to these positions requires skilled construction of the catalytic operations. A suitable requirement of directing groups in promoting such remote aliphatic C—H activation was realized. This chapter therefore highlights some recent advances made in the field of aliphatic distal C—H functionalization which have been made possible through the employment of an assortment of strategies.

In Chapter 10, Palladacycles for Directed and Nondirected C—H Bond Functionalization of (Hetero)arenes, Soni et al. focus on the use of palladacycle precatalysts in C—H bond functionalization, as well as providing an overview of palladacycle intermediates in directed C—H bond functionalization reactions. The described palladacycle precatalysts are more efficient and more selective than the traditional palladium complexes for the C—H bond functionalization processes. The second half of the chapter summarizes the involvement of palladacycle intermediates in various directed C—H bond functionalizations that are proposed with experimental evidence. Some of the described palladacycle intermediates are shown to be competent as catalyst precursors, coupled with their stability during the functionalization phenomenon.

The penultimate chapter deals with the mechanistic aspects of metal-mediated C—H bond functionalization as part of its context. Gayakhe et al. present an account of the fundamental and key concepts behind the operation of organometallic catalysis. With the earlier reports about oxidative addition to H_2, the authors discuss the basic mechanistic pathways involving electrophilic aromatic substitution and oxidative additions. Further, owing to the recent reports on the popularity of cyclometallation – demetallation (CMD)-based reaction schemes, an insight into the CMD mechanism is provided. Finally, the discussion concludes with mention of the Heck-type carbometallation pathways that lessen the requirement for any preactivation step. The latter is particularly popular in the case of palladium-catalyzed cross-coupling reactions.

In the closing chapter, Thrimurtulu et al. collectively summarize the application of C—H functionalization in the total synthesis of organic molecules. This chapter discusses the synthetic routes to form a host of natural products and other complex substrates, starting from simple and readily available starting precursors. All of these involve C—H activation as a key step in achieving the desired transformation. The applicability of sp^3 and sp^2 C—H functionalization has been showcased as an effective method of ensuring functional group installation in a highly site-selective manner. Examples from recent literature reports are detailed with substantial analysis of the reaction pathways. This chapter is therefore important from the point of view of developing a sound concept about simplifying retrosynthetic disconnections in the course of synthesizing any complex molecules as the target motif.

An intuitive mind is always a sacred gift for humanity, and the curiosity to know opens up boundless possibilities for evolution and advancement. This has been the sole truth behind every piece of advancement made throughout history since bygone ages, which continues to be in existence and shall remain the same for eternity. Attempts to bring modernization into chemistry led to the development of C—H activation-based protocols which have significantly reshaped scientific temperament and created a novel dimension for rethinking synthetic transformations. This book, besides summarizing the notion of scientific progress made in the field of palladium-catalyzed C—H functionalization in recent times, aims to inculcate cognizance among the scientific fraternity about the same, and wishes to encourage chemists of both the present and future generations to contribute their ideas towards the welfare of the scientific community, as well as for all and sundry.

CHAPTER 2

Directed C—H Bond Functionalization Strategies for Synthesis

Ian J.S. Fairlamb[1] and Anant R. Kapdi[2]
[1]University of York, York, United Kingdom
[2]Institute of Chemical Technology, Mumbai, Maharashtra, India

Contents

2.1 INTRODUCTION

Organic synthesis in recent decades has undergone a tremendous change with the advent of catalytic processes providing an efficient alternative to non-catalyzed reactions. The unparalleled success obtained by these catalytic processes was based on the rapid development of newer and more efficient catalytic systems functioning under milder reaction conditions. Selective functionalization of a variety of unactivated bonds having applications across a wide spectrum of fields has contributed further to the popularity of these processes. The prerequisite for achieving the required success involves the pre-functionalization of the substrate leading to further activation of other bonds. Although the installation of C—O, C—X, C—N, and C—C bonds could be achieved via these processes, cost escalation for the multi-step reactions renders the overall process commercially less attractive. Circumventing this problem would involve the possibility of employing an unfunctionalized substrate that could be directly functionalized without the requirement for a prefunctionalization step.

Direct functionalization via the activation of unreactive C—H bonds could provide researchers with the required sustainable solution that could

Strategies for Palladium-Catalyzed Non-Directed and Directed C—H Bond Functionalization
DOI: http://dx.doi.org/10.1016/B978-0-12-805254-9.00002-5

also be cost-effective.[1] This is achieved via the insertion of transition metals into the unreactive C—H bond, resulting in a chemically more reactive C—[TM] bond being available for further functionalization. Although an attractive alternative to the multi-step coupling processes,[2] selective functionalization of chemically equivalent C—H bonds presents researchers with a unique synthetic challenge. Initial reports on the stoichiometric activation of C—H bonds have now been replaced in recent years with efficient catalytic activation processes employing transition metals such as Ru, Cu, Fe, Ni, Rh, and Pd.[3] Palladium, with its unique reactivity, has been at the forefront in these processes, allowing the functionalizations to be performed under milder conditions.

To address the problem of selective C—H activation in substrates, researchers have strategized several methods based on factors that could assist metal insertion into a solitary C—H bond. One of the strategies that has found extensive application in palladium-catalyzed C—H bond activation takes advantage of the coordinating ability of ligands (functional groups) on substrates that can direct the metal into C—H bonds allowing selective functionalization (Fig. 2.1). Such a strategy would not just allow a cost-effective method for functionalization, but also would make it possible to install a variety of bonds such as carbon—oxygen, carbon—nitrogen, carbon—sulfur, and carbon—carbon, etc.

The focus of this chapter involves the classification of the coordinating ligands based on their coordinating ability and the type of bond formation they have been employed for. Some discussion on the effect of base for the abstraction of proton, as well as the role of acids (as solvents) for promoting the directed C—H functionalization, is also undertaken. Most of the examples discussed herein are related to *ortho*-functionalization of aromatic substrates proceeding commonly via a cyclopalladated intermediate that governs the selectivity in these reactions. Functionalization of other sites on the benzene ring or on aliphatic substrates falls outside the scope

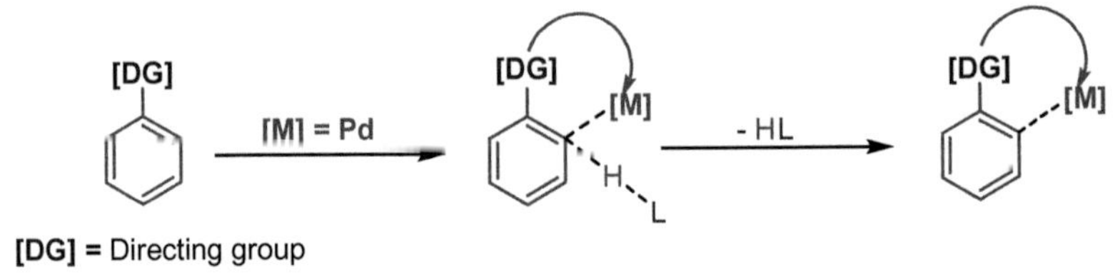

Figure 2.1 Schematic representation of directing group effect on C—H bond activation.

of this chapter, and readers are advised to refer to some of the recent literature on this topic,[4] as well as the further chapters of this book where these topics are discussed in depth by Li and Maiti.

2.2 ORTHO-PALLADATED C—H BOND FUNCTIONALIZATION VIA DIRECTING GROUP EFFECT

The innate nature of the C—H bond makes the process of selective activation all the more challenging. Researchers have developed various methods to circumvent this problem, with the use of directing groups (coordinating ligands) on the substrate to be functionalized the most popular of all. Depending on the coordinating capacity of the directing groups a lot of chemistry has been developed in the past decade that utilizes palladium precursors for catalyzing these transformations. These coordinating ligands could be divided into two types (only for those functional groups that have been employed in this chemistry) as strongly coordinating or weakly coordinating with regards to their interaction with the palladium center. Some of the commonly employed coordinating ligands in palladium-catalyzed C—H bond functionalization are listed below:

(a) Strongly coordinating ligands: pyridine, anilide, urea, etc.

(b) Weakly coordinating ligands: ketone, hydroxyl, amides, phosphate, esters, sulfonamides, carbamates, formate esters, alkynes, alkenes, carboxylic acids, etc.

There are many other functionalities that have also found applications in selective C—H bond functionalization reactions; however, based on the number of examples in literature, only the most applicable ligand systems (listed above) will be discussed in detail in this chapter.

2.2.1 Effect of Coordination Capacity on Functionalization

At the outset it is important to understand the effect of the coordinating power of ligand on the effectiveness of selective C—H bond activation, and the functionalization possibility presented by such types of coordination. A glance through the literature suggests that the number of examples based on the employment of strongly coordinating ligands as directing groups far exceeds that of their weakly coordinating counterparts. On a fundamental level, this difference in coordination with the metal center directly affects the stability of the resultant cyclopalladated intermediate that would translate into a more stable intermediate for the strongly coordinating ligands, and conversely a weaker intermediate for the weaker

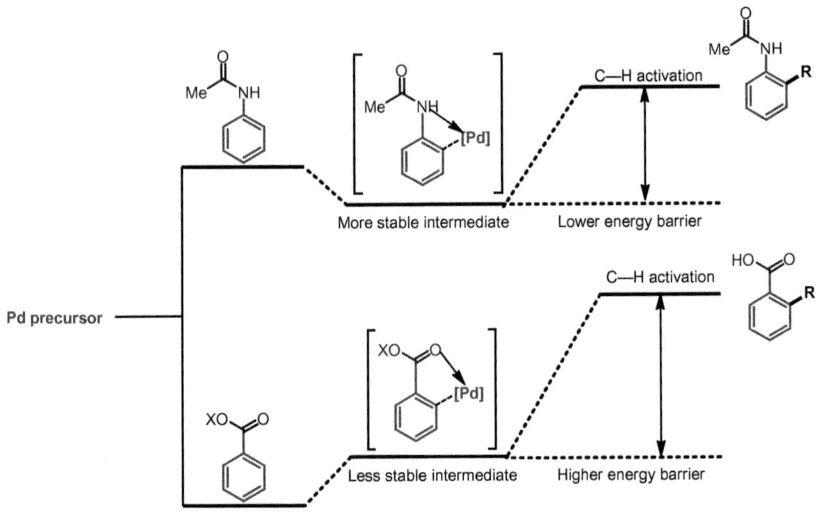

Figure 2.2 Energy representation for strong and weakly coordinating ligands in C—H bond activation processes.

ligands. In terms of the energetics of the reactions, it could be observed that complexation of the metal precursor with strongly coordinating ligands could lead to a lowering of energy, thus making the resultant intermediate more stable (Fig. 2.2). Conversely, binding the metal with a weakly coordinating ligand increases the energy of the intermediate, which is less stable and has a higher possibility of dissociation.

Based on these details, it would be fair to say that the effectiveness of selective C—H bond functionalization could be achieved in the presence of strongly coordinating ligands on the substrate. However, pioneering work by Yu and co-workers[5] in recent years demonstrating the potential of weakly coordinating ligands for the functionalization of a diverse range of substrates also holds a lot of promise, and could help promote further development in this area. From the myriad of functional groups that have found major applications in palladium-catalyzed C—H bond functionalization, we will discuss in detail examples involving heterocyclic rings, amides, anilides, and carboxylic acids as the directing groups.

2.2.1.1 Heterocycles as Directing Groups for Palladium-Catalyzed C—H Bond Functionalization

One of the most common and applicable heterocyclic directing groups coordinating with the palladium center through the lone pairs available on the nitrogen atom is pyridine.[6] Although recent examples have also incorporated other heterocyclic ring systems (quinolines, benzoquinolines, pyrrolidinones, oxazolidinones, etc.), pyridine has led the way due to its better coordinating capacity, and the ease of synthesis of the required substrates for carrying out such transformations.

One of the earliest examples for the employment of pyridine as a directing group in palladium-catalyzed C—H bond functionalization was reported by Sanford in 2005[7] (Scheme 2.1A), with diphenyliodonium trifluoroborate used as the coupling reagent. Subsequently, an oxidative homocoupling protocol was also reported by the same group[8] (Scheme 2.1B).

The possibility of using simple aryl iodides as coupling partners was first put forth by Daugulis and co-workers,[9] making the process synthetically more attractive. This was further extended to benzylic substrates rather than the general arenes, allowing the generalization of substrate combination for effective C—H bond functionalization (Scheme 2.2A). Despite all the advances involving the development of highly active catalytic systems for efficient C—X activation (especially C—Br or C—Cl bond activation) in cross-coupling processes, their application in directed arylations has been limited. A major synthetic challenge therefore remains to be addressed for using commercially viable aryl bromides, or preferably chlorides, as the arylating reagents. However, recently Sun and co-workers[10] have reported on aryltrimethoxysilanes as coupling partners for the arylation of phenylpyridines (Scheme 2.2B).

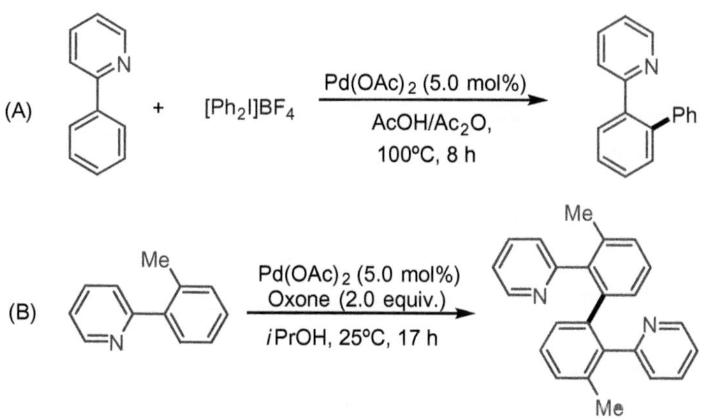

Scheme 2.1 (A) Functionalization using diphenyliodonium salt; (B) oxidative homocoupling of phenylpyridines using oxone as the oxidant.

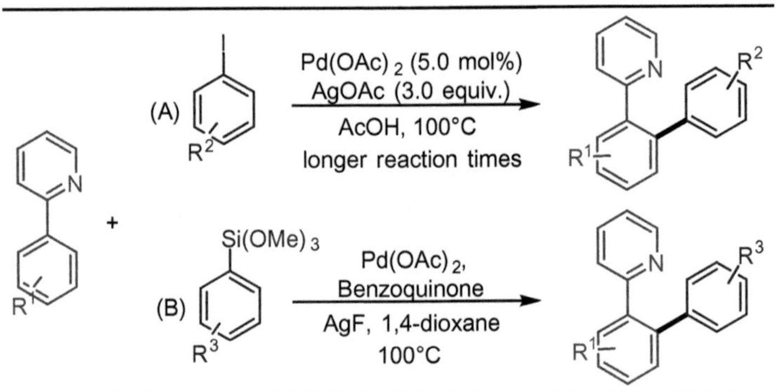

Scheme 2.2 (A) Aryl iodides as the arylating agents; (B) aryltrimethoxysilanes as coupling partners.

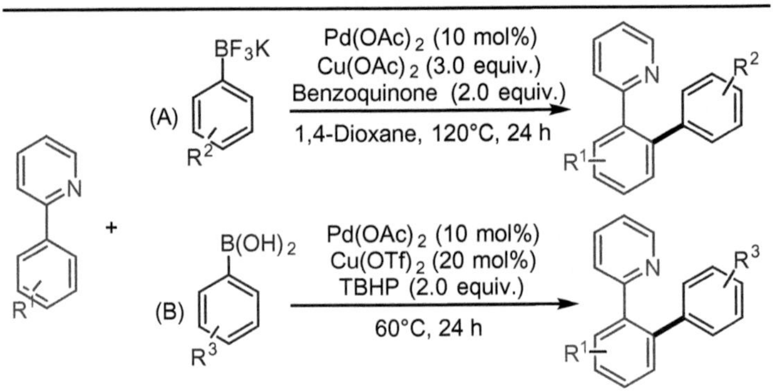

Scheme 2.3 Boron reagents for arylation: (A) potassium aryltrifluoroborates; (B) arylboronic acids.

The employment of boron compounds as arylating reagents has also been explored, with the first example reported in 2009 for the arylation of phenylpyridine using potassium aryltrifluoroborates, although the catalyst loading of $Pd(OAc)_2$ was found to be very high (Scheme 2.3A).[11] Arylboronic acids were also tried as a possible cheaper coupling partner, but this failed to furnish the arylated product in good yields. A few years later, Cai and co-workers, in 2013, employed the economical arylboronic acids for the directed arylation reactions in the presence of TBHP and catalytic copper salt as oxidants, with the amount of copper restricted to 20 mol% as compared to a large excess used in the earlier protocol (Scheme 2.3B).[12]

Besides the examples mentioned above for the arylation of phenylpyridines, another protocol involving a lesser known arylating agent, acyl peroxides, was also demonstrated to provide the *ortho*-arylated product in good yields.

In comparison to the directed arylation protocols involving the use of a variety of arylating reagents (*vide infra*), alkenylation or alkylation of pyridines yielding molecules of synthetic importance due to the possibility of further manipulation are less pronounced. Pyridine as a directing group allowed the olefination to take place in the presence of a Pd(II) catalyst and a catalytic amount of a heteropolyacid in the presence of air as the oxidant.[13] Such a synthetic strategy allows easy access to 6,5-*N*-fused bicyclic cores of further biological relevance (Scheme 2.4). The protocol, however, suffers from a limited reaction scope attributed to the dependence of catalytic activity on the reactivity of the olefin with activated olefins being preferred over the unactivated. Surprisingly, this is the only report for olefination being carried out utilizing the directing property of pyridine.

Another synthetic challenge that has intrigued researchers in the past decade has been the alkylation of substrates using highly reactive alkylboron reagents that are known to dissociate rapidly under the conditions via the deboronation pathway. The employment of these alkylboron reagents as alkylating agents in palladium-catalyzed C—H bond functionalization of phenylpyridines was first reported by Yu and co-workers[14] (Scheme 2.5). Given the reactive nature of the alkylboron reagents, rather than using methylboronic acid (for methylation) that is known to deboronate at a rapid rate, its cyclic analogue methylboroxine (stable over a longer period of time) was employed, providing good yields of the methylated product. Based on these results, the authors developed an asymmetric variant involving the use of butylboronic acid in the presence of $Pd(OAc)_2$/chiral amino acid ligand.

Methylation of phenylpyridine could also be carried out by the employment of a unique methylating agent in the form of *tert*-butyl peroxides.

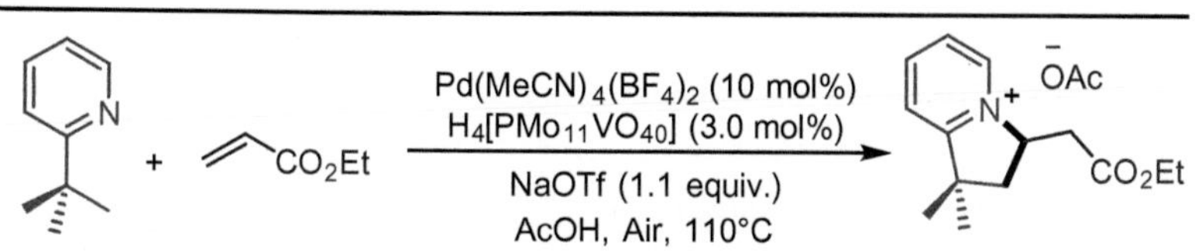

Scheme 2.4 Pyridine as directing group for olefination leading to 6,5-*N*-fused bicyclic cores.

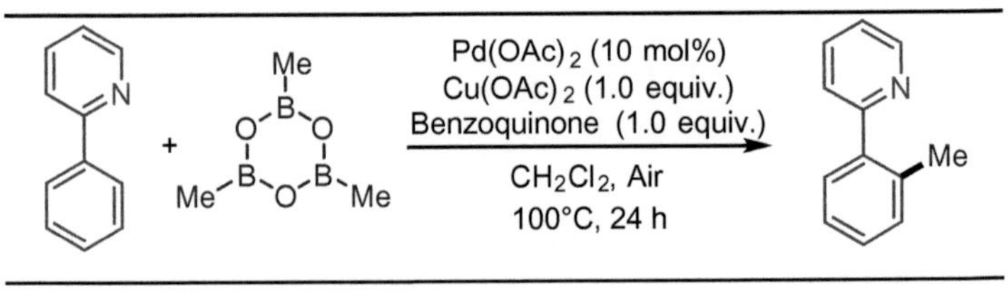

Scheme 2.5 Methylation of phenylpyridine using boroxine as an alkylating agent.

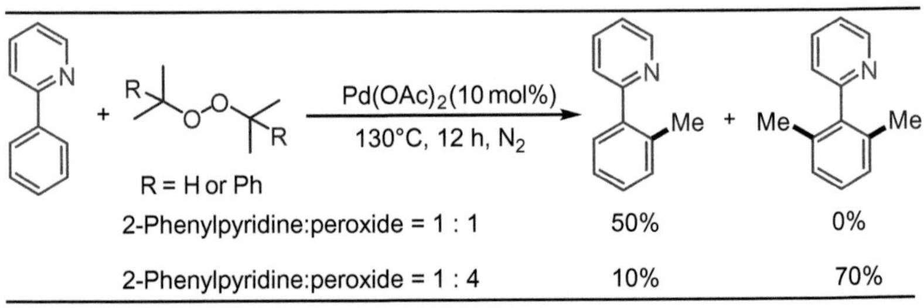

Scheme 2.6 Methylation of phenylpyridine using peroxide as methylating agent.

This unusual methylation protocol was recently disclosed by Li and co-workers,[15,16] with the peroxide serving as the possible methylating agent (Scheme 2.6). An interesting aspect of the *ortho*-arylation process is the selectivity achieved for the *mono*-methylated product over *di*-methylation. Support for the selectivity observed, as well as the possible pathway followed, was provided by Sunoj and co-workers,[17] based on computational studies that were performed. A combination of radical and non-radical pathways involving a palladium—methyl intermediate was found to be operative.

Trifluoromethylation of phenylpyridine under palladium-catalyzed C—H bond functionalization conditions was performed using 5-(trifluoromethyl)-dibenzothiophenium tetrafluoroborate as the trifluoromethylating agent by Yu and co-workers[18] (Scheme 2.7). It is the only literature report on phenylpyridines with trifluoroacetic acid as the solvent. The protocol showed remarkable functional group tolerance, with a variety of functional groups showing compatibility to the conditions employed.

Aryl nitriles are important synthetic intermediates that have found wide application due to their ease of conversion to other functional groups. The introduction of a cyanide group via palladium-catalyzed C—H bond functionalization would therefore provide synthetic chemists with a handle for further manipulation, leading to molecules of synthetic

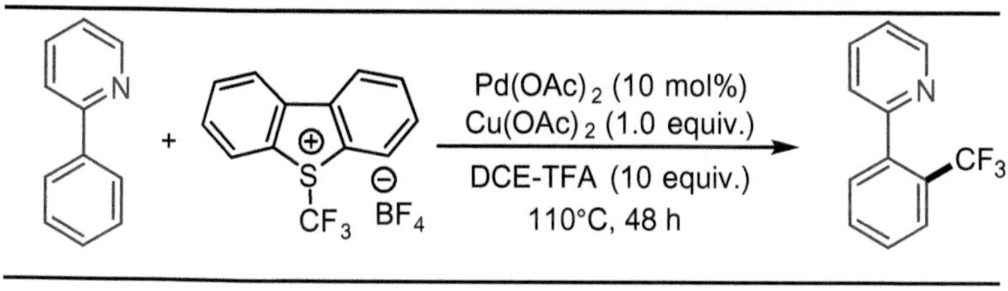

Scheme 2.7 Trifluoromethylation via *ortho*-C—H bond functionalization of phenylpyridine.

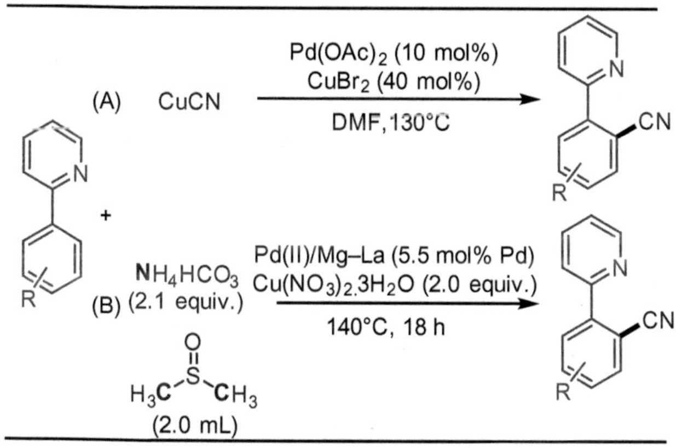

Scheme 2.8 Palladium-catalyzed C—H cyanation of phenylpyridines using cyanating agent: (A) copper cyanide; (B) a combination of ammonium hydrogen carbonate and dimethyl sulfoxide.

relevance. The first example of C—H cyanation using CuCN as the cyanide source was reported in 2009 by Cheng and co-workers[19] (Scheme 2.8A). The cyanation protocol thus developed was applied to the synthesis of a key intermediate of an alkaloid, Menispermum dauricum DC, which is known to exhibit promising analgesic and antipyretic properties. An important advanceinvolving a palladium(II)/magnesium—lanthanum mixed oxide as a heterogeneous catalyst for palladium-catalyzed C—H cyanation of phenylpyridines was put forth by Kantam and co-workers in 2015 (Scheme 2.8B).[17] In situ generation of CN^- was made possible from the combination of NH_4HCO_3 and DMSO, rather than the commonly applicable inorganic cyanide sources. Besides achieving excellent regioselectivity for cyanation, the catalytic solution was also found to be recyclable for up to three cycles. The active nature of the heterogeneous catalyst allowed a tandem Suzuki coupling/C—H cyanation reaction of 2-halopyridines.

Another biologically and chemically important process that is known to define numerous cellular processes is acylation. A functional group that has defined a large number of industrial processes due to its unique reactivity demands a special place in the synthetic chemist's arsenal. To access these important structural motifs, a cross-dehydrogenative coupling strategy between phenylpyridine and arylaldehyde (electron-rich or electron-deficient) in air as an oxidant was envisaged by Cheng and co-workers (Scheme 2.9A).[20] This transformation was made possible through the coordination-assistance provided by the pyridine nitrogen. The protocol was further extended to other heterocyclic ring systems such as quinolines. On similar lines, Li and co-workers[21] also reported an efficient acylation protocol for such heterocyclic ring systems allowing aliphatic, alicyclic aldehydes to be used as acylating agents (Scheme 2.9B).

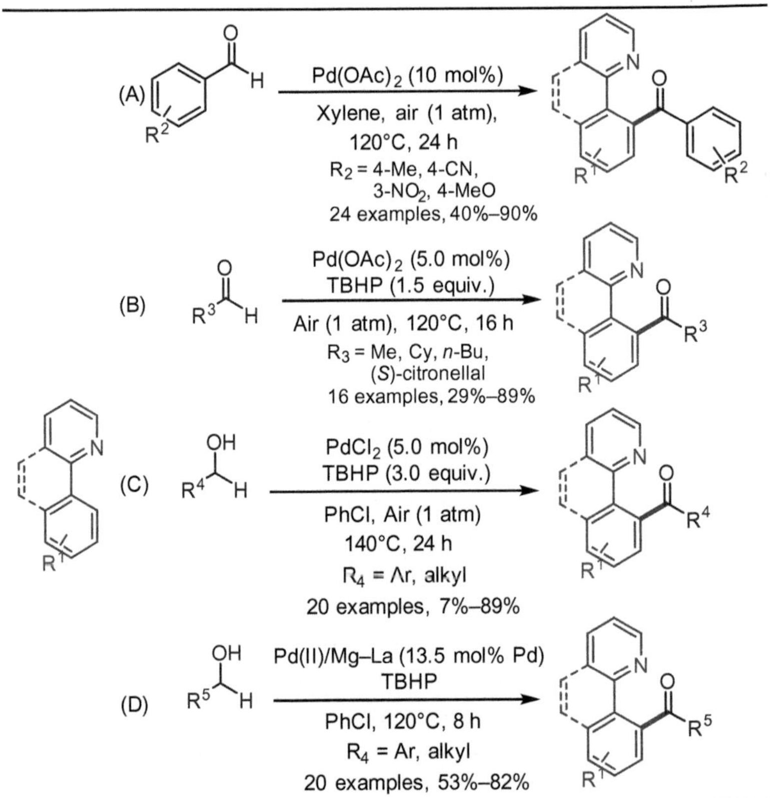

Scheme 2.9 Acylation strategies via palladium-catalyzed C—H bond functionalization. A) Arylaldehyde as acylating reagents B) Alkyl aldehydes as acylating reagents C) Alcohols as acylating reagents D) Alcohols as acylating reagents with heterogeneous catalyst.

Further developments in this area allowed the C—H bond functionalization to be performed with primary alcohols rather than aldehydes as the acylating reagents, which was shown independently by Deng and Li to be efficiently oxidized in situ using TBHP as the oxidant (Scheme 2.9C).[22] Recently, a heterogeneous catalytic version (mixed metal catalyst) developed by Kantam and co-workers,[23] allowed the transformation to be carried out with good selectivity and with some level of recyclability also achieved (Scheme 2.9D). Besides using aldehydes or alcohols as the acylating agents, Yang, Fu, and co-workers,[24] recently revealed the possibility of utilizing a cheap and readily available acylating reagent in the form of arylmethyl amines. In the presence of a palladium precursor, $PdCl_2$, and with TBHP as the oxidant, benzylamines were efficiently converted into acyl functionalities via the chelation-assisted C—H activation protocol.

Halogenation of substrates in a selective manner has important synthetic applications, as it could provide researchers with a useful handle for further manipulation. Pioneering work by Sanford and co-workers deserves special mention in this regard, with a large variety of functional groups including halogens (I, Cl, Br, and F) as well as pseudohalides (OAc) being introduced on the aryl moiety of the phenylpyridine via the directing capabilities of the pyridine nitrogen. These types of functionalizations could be brought about by the employment of mild and selective halogenating reagents. One of the earliest examples of the functionalization of phenylpyridine and related structural features was reported by Sanford and co-workers in 2004[25] for the acetoxylation using $PhI(OAc)_2$ as a stoichiometric oxidant (Scheme 2.10).

The resultant *ortho*-acetoxylated product was obtained with very good yields. The protocol was further extended towards the functionalization of other heterocycles in an efficient way.[26] Mechanistic studies performed on the catalytic acetoxylation reaction revealed the initiation of such transformations taking place via a chelation-assisted cyclometalation pathway resulting in a possible Pd(IV) intermediate **[A]**, although no direct evidence for such a proposal has been provided.[27] Given the synthetic utility of this procedure, Ritter and co-workers conducted a thorough investigation into the scope of the acetoxylation reaction, and were successful in extending the protocol to a vast variety of heterocycles and other substrates.[28] Rather than employing the hypervalent $PhI(OAc)_2$ reagent for acetoxylation, Yu and co-workers, revealed the capability of IOAc for effecting the same transformation also in good yields.[29]

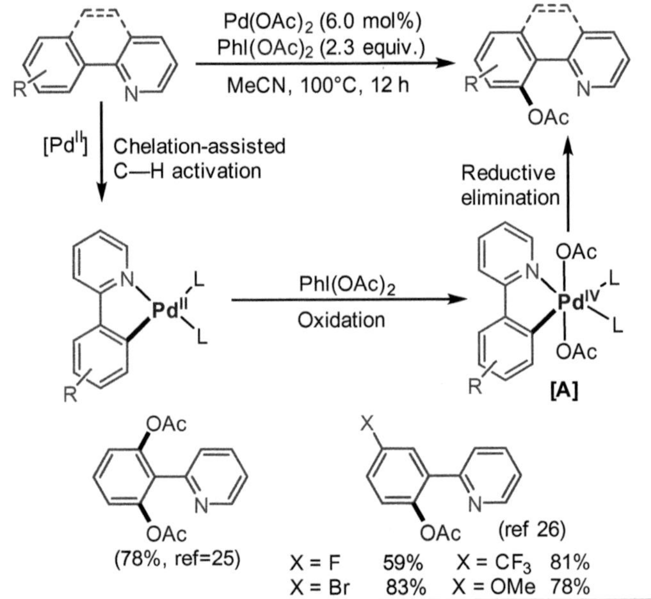

Scheme 2.10 Selective acetoxylation of phenylpyridines.

Directed halogenation of substrates has been known in the literature since 1970,[30] however, the problems associated with this transformation in terms of the selectivity, reactivity, and the employment of substoichiometric amounts of the palladium catalyst, were addressed more recently. In 2004, a catalytic version was first reported by Sanford and co-workers as part of their study into the functionalization of heteroarenes (*vide infra*).[25] *N*-Halosuccinimides for the first time were demonstrated to promote the selective halogenation of benzo[*h*]quinoline,[25] which was later extended to phenylpyridines[28c] (Scheme 2.11). These results triggered a flurry of activity leading to the identification of a large variety of halogenating (bromination or chlorination) reagents that have now found application in the palladium-catalyzed C–H activation of heteroarenes. Some of the commonly applicable reagents are copper halides (CuX_2; X = Cl, Br),[31] $PhICl_2$,[28c] and electrochemical oxidation using HX (X = Cl or Br).[32]

Iodination of heteroarenes in a selective manner was also achieved by Martin-Matute and co-workers,[33] using a heterogeneous (Pd@MIL-88B-NH_2(Fe)) catalyst and NIS as the iodinating agent (Scheme 2.12). Selectivity was switched between mono-*ortho*-iodination to di-*ortho*-iodination by simply varying the solvent from AcOH (providing majority mono-selectivity)

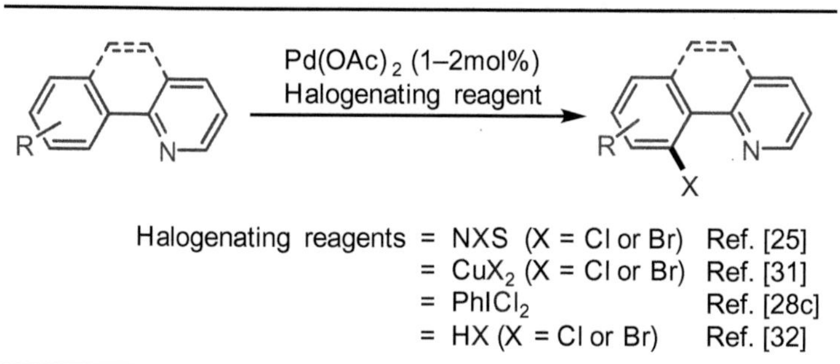

Scheme 2.11 Palladium-catalyzed C—H halogenation of phenylpyridines.

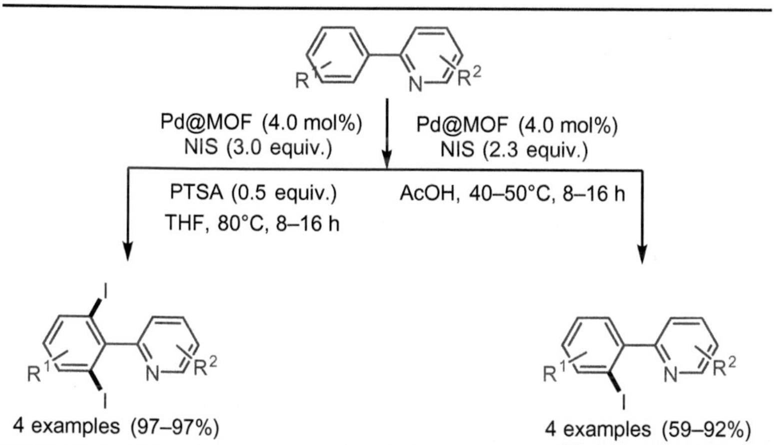

Scheme 2.12 Selective iodination of heteroarenes using Pd@MOF heterogeneous catalyst.

to THF (complete di-iodination). The addition of PTSA was helpful in promoting the di-iodination to take place in a quantitative manner, and no traces of mono-iodination were observed.

Finally, selective fluorination of phenylpyridines has also been performed using *N*-fluoro-2,4,6-trimethylpyridinium tetrafluoroborate as an electrophilic fluorinating reagent (Scheme 2.13).[34] Although a variety of fluorinating reagents (Selectfluor, *N*-fluoro-2,4,6-trimethylpyridinium tetrafluoroborate, *N*-fluoropyridinium tetrafluoroborate, 4-iodotoluenedifluoride, and *N*-fluorodisulfonamide) have been reported in literature[35] and have shown excellent selectivity for the fluorination of organic substrates, the best selectivity in the case of phenylpyridines was achieved with *N*-fluoro-2,4,6-trimethylpyridinium tetrafluoroborate.

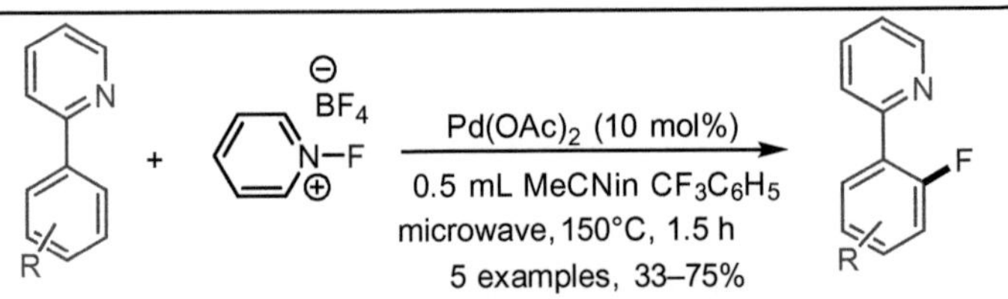

Scheme 2.13 Selective fluorination of phenylpyridines using electrophilic fluorinating agents.

Beside the examples discussed so far highlighting the directing influence of pyridine as an important heteroarene for palladium-catalyzed C—H functionalization, several other examples including those for other heteroarenes are known, and readers are encouraged to refer to the recent reviews that have given a comprehensive account of these types of transformations.

2.2.1.2 Anilides and Ureas as Directing Group for Palladium-Catalyzed C—H Bond Functionalization

Another example of strongly coordinating ligands on organic substrates exhibiting superior directing influences and assisting efficient palladium-catalyzed C—H bond functionalization of innumerable substrates are the anilides or ureas. The versatility of anilides as a functional group could be understood from the myriad of catalytic functionalizations, ranging from olefination, arylation, acylation, amination, halogenation, trifluoromethylation, alkynylation, and many more. In this section, we will be discussing some of these examples in further detail.

C—H bond functionalization reactions performed on substrates with anilides as the directing group have been well documented over the past decades; however, the possibility of utilizing the strong coordinating power of the anilide group was first put forth by Leeuwen and co-workers[36] in 2002 (Scheme 2.14). Olefination of acetanilides via palladium-catalyzed C—H bond activation with the employment of benzoquinone as the stoichiometric oxidant was performed with high *ortho*-selectivity for the first time.

The above transformation was recently revisited by Brown and co-workers[37a] to understand the pathway followed, and to ascertain the role of the added *p*-toluenesulfonic acid (TsOH). The reaction was found to be catalyzed by an electrophilic palladium species that is generated from the combination of $Pd(OAc)_2$ and TsOH.[37b] The process proceeded

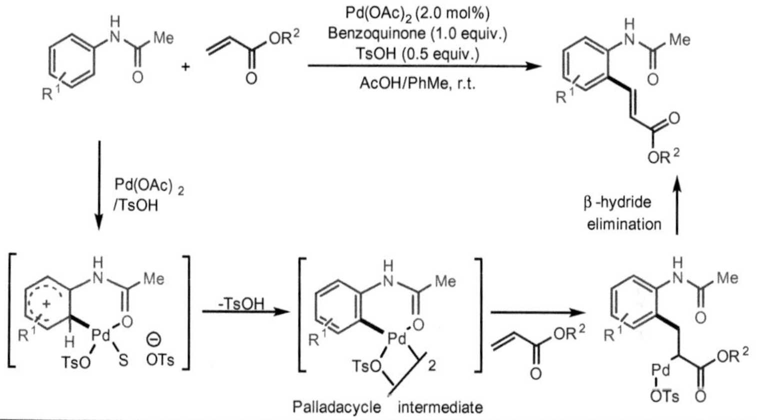

Scheme 2.14 Olefination of acetanilides via palladium-catalyzed C—H bond activation.

through the formation of a palladacyclic intermediate that was isolated and shown to perform the catalytic functionalization efficiently. Using NMR, single crystal X-ray diffraction, and kinetic studies, conclusive evidence on the pathway for such a transformation was put forth. This has laid the foundation for the rapid development of C—H bond activation processes for the functionalization of a vast variety of substrates.

One of the synthetically useful catalytic reactions involving the selective introduction of an aryl group on different substrates has commonly been achieved via the cross-coupling pathway which suffers from a number of drawbacks, including the requirement for pre-functionalized substrates. A more efficient and sustainable solution to this problem was provided in the form of a useful directed *ortho*-arylation protocol, first developed by Daugulis and co-workers[38] in 2005 (Scheme 2.15A).

The potential of an anilide functionality as a strongly directing group for effecting selective *ortho*-arylation was achieved for the first time using a combination of aryl iodides and silver acetate. Daugulis also demonstrated the effectiveness of the developed protocol with diphenyliodonium salts. Daugulis, later in 2007, followed up this result with a useful synthetic procedure for the preparation of phenanthridine derivatives via a palladium-catalyzed anilide directed-arylation/cyclization strategy in the presence of trifluoroacetic anhydride (Scheme 2.15B).[39] A similar strategy, taking into account the directing capability of protected *N*-methylbenzylamines (with

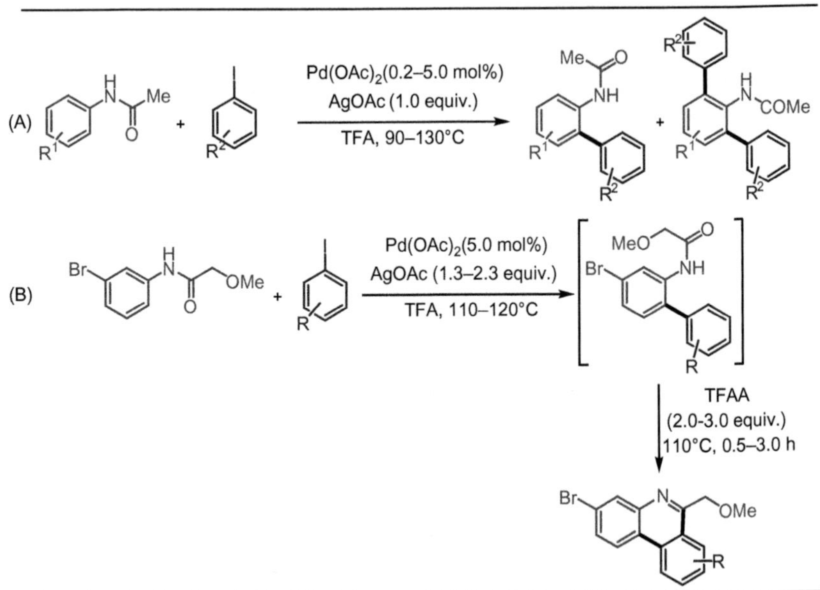

Scheme 2.15 Directed arylations using anilide as directing group. A) ortho-arylation or diarylation B) ortho-arylation/cyclization

trifluoroacetic anhydride), was shown to provide *ortho*-arylated product via palladium-catalyzed C—H bond activation.[40]

One more method for the arylation of aromatic ring systems with arylboronic acids as organometallic coupling reagents via the directing group effect of the anilide functionality was first explored by Shi and co-workers (Scheme 2.16A).[41] The insertion of the palladium species into the *ortho*-position of the arene is facilitated by the strongly coordinating power of the anilide group bringing about the desired oxidative arylation, with the arylboronic acid acting as the nucleophilic reagent. Further developments in this area have been restricted due to the lower reactivity of palladium species in effecting such transformations compared to its counterpart, ruthenium, that has shown promising results for a wide variety of functionalization reactions. Another class of organometallic reagents that was demonstrated to provide the *ortho*-arylation product in good yields is the arylsilanes[42] (Scheme 2.16B), however, this is the only example present in literature for palladium-catalyzed *ortho*-arylation employing such reagents.[43]

In 2014, Wang and co-workers[44] reported a unique palladium-catalyzed decarboxylative, anilide-assisted, *ortho*-arylation with acyl peroxides as the arylating agent (Scheme 2.17A). The reaction proceeded through the release of CO_2 to afford the *orth*-arylated products; however,

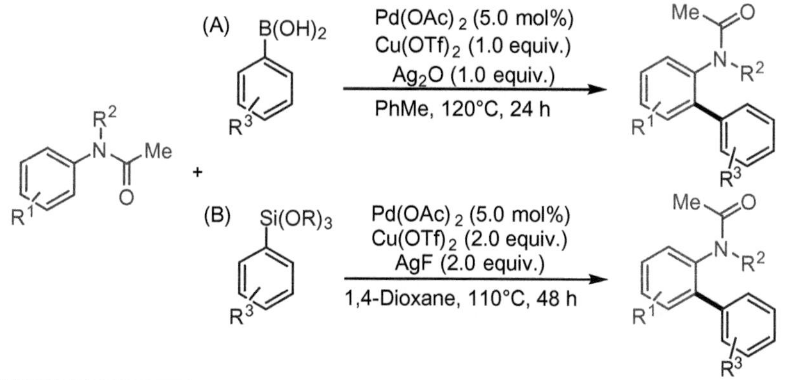

Scheme 2.16 *Ortho*-arylation via anilide-assistance using organometallic reagents. A) Aryl boronic acids for arylation B) Arylsilanes for arylation

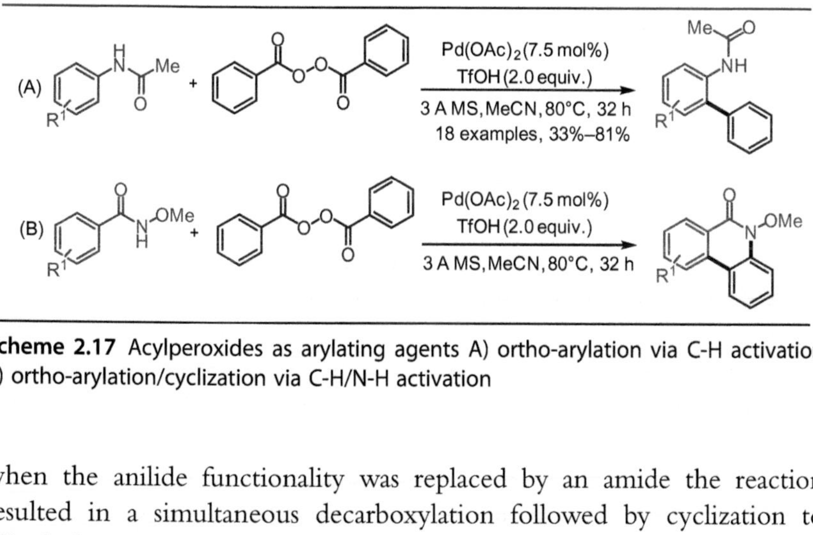

Scheme 2.17 Acylperoxides as arylating agents A) ortho-arylation via C-H activation B) ortho-arylation/cyclization via C-H/N-H activation

when the anilide functionality was replaced by an amide the reaction resulted in a simultaneous decarboxylation followed by cyclization to afford phenanthridinones in good yields (Scheme 2.17B).

An affordable and more desirable way of arylating substrates would be the employment of simple unfunctionalized arenes as coupling partners. The attractiveness of such a methodology is dwarfed by the unreactive nature of the arenes and problems related to the selective activation of a C—H bond over others. Development of efficient dehydrogenative coupling protocols are therefore highly desirable, and given this possibility Shi and co-workers[45] first reported a double C—H bond activation strategy involving directed-assistance from the anilide group providing excellent regioselectivity for the arylation reaction (Scheme 2.18A). Due to

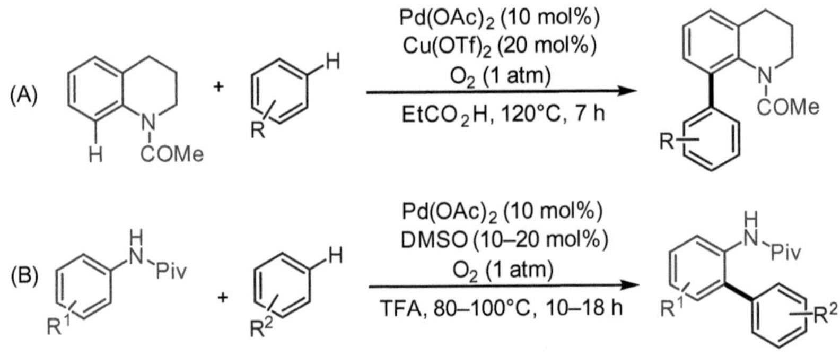

Scheme 2.18 Unsubstituted arenes as arylating agents for dehydrogenative coupling.

the lower reactivity of the arenes, an excess amount was commonly required to drive the reaction in the forward direction.

Advances in this regard were reported by Buchwald and co-workers,[46] allowing the requirement for 4 – 11 equivalents of the arene rather than excess of it (Scheme 2.18B). This was brought about by the subtle modification in the protocol by incorporating TFA as the solvent under an atmosphere of pure oxygen as the oxidant. An unprecedented result for the arylation of anilides at room temperature was achieved by You and co-workers,[47] demonstrating the power of a highly electrophilic palladium species ($Pd[TFA]^+$) obtained from the reaction of $Pd(OAc)_2$ and TFA to efficiently insert into the *ortho*-position assisted by the anilide at ambient temperature.

Acylation strategies for the functionalization of aromatic compounds have undergone rapid development, given the synthetic and commercial importance of these substrates. Directing-group assisted C—H bond activation strategies for the installation of acyl functionality would be a promising and more effective alternative to other uncatalyzed and catalyzed processes. Kwong and co-workers first reported the strategy for *ortho*-acylation by employing aldehydes as the acylating agents in the presence of TBHP for oxidation.[48] Around the same time, Yu[49] and Wang[50] also published similar strategies for acylation with aldehydes (Scheme 2.19B). However, a synthetically attractive strategy for acylation was later reported by Yu and co-workers,[49] allowing the in situ oxidation of primary alcohols and further activation to provide the *ortho*-acylated products in good yields (Scheme 2.19B). Benzylic alcohols have also found applications as acylating agents in aqueous media under

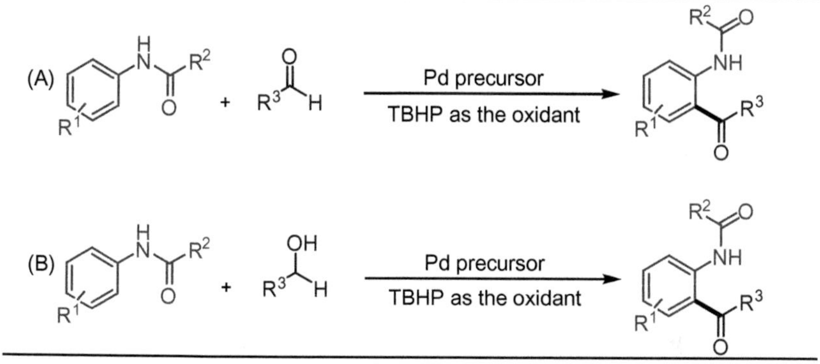

Scheme 2.19 Aldehydes and alcohols as acylating agents for anilide-directed acylation strategy. A) Aldehydes as acylating reagents B) Alcohols as acylating reagents

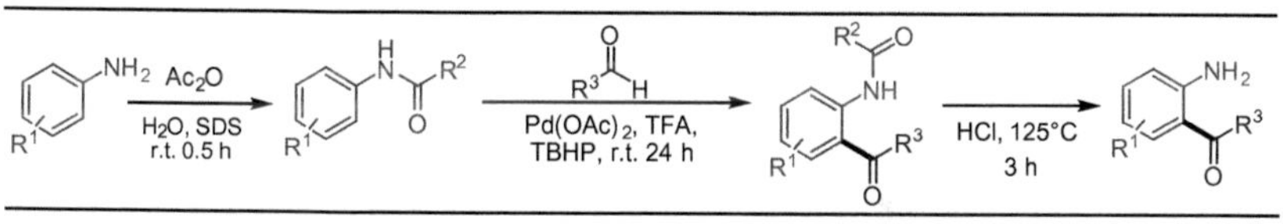

Scheme 2.20 One-pot step-economic acylation strategy for anilides.

palladium-catalyzed conditions by Zhou and co-workers in 2015 for obtaining substituted benzophenones.[51]

In continuation of the discussion on the employment of aldehydes as acylating agents, it is worth mentioning a mild, practical, and selective method reported by Novak and co-workers in 2014 for the one-pot synthesis of *ortho*-aminobenzophenones (Scheme 2.20).[52] The process involves the protection of anilines as acetanilide, followed by *ortho*-selective palladium-catalyzed C—H bond activation with TBHP as the oxidizing agent. The acylated product was finally hydrolyzed to afford *ortho*-aminobenzophenone, providing a step-economic alternative to such molecules.

Taking advantage of the strong coordination possibility of the anilide group, Ge and co-workers[53] also developed a decarboxylative *ortho*-selective acylation protocol using α-oxocarboxylic acid as the acylating reagent. One of the unique features of the protocol was the replacement of TBHP as the oxidizing agent with ammonium thiosulfate, commonly employed in most of the earlier protocols (Scheme 2.21). The attractiveness of the protocol developed was further enhanced by the possibility of performing the catalytic functionalization at ambient temperature. A similar synthetic strategy

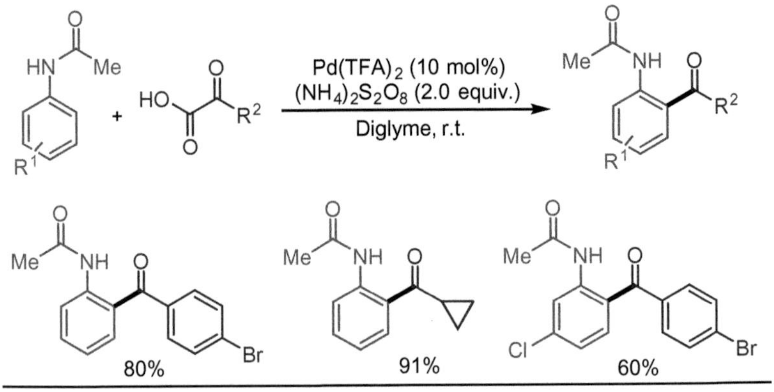

Scheme 2.21 Acylation strategy using α-oxocarboxylic acid.

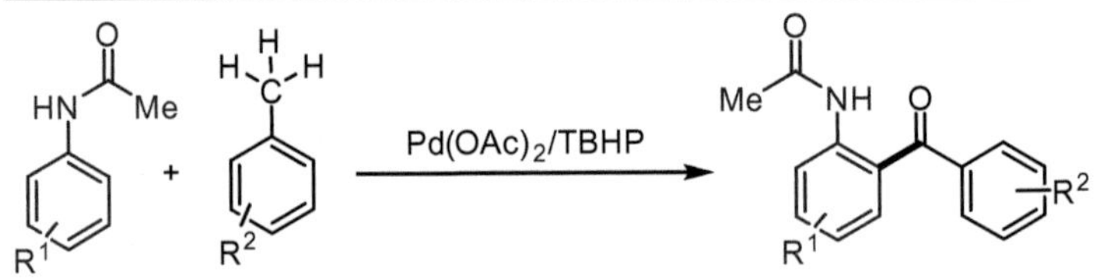

Scheme 2.22 Toluene as acylating reagent for substituted benzophenone synthesis.

was recently employed by Kim and co-workers for palladium-catalyzed decarboxylative acylation of phenylacetamides.[54]

Toluene, when employed as a substrate, provides researchers with an economic and synthetically useful alternative to alcohols and aldehydes, as it could be readily oxidized in situ with suitable oxidizing agents. Sun and co-workers,[55] first introduced this strategy towards the acylation of acetanilides with toluene as the direct acylating reagent using TBHP as the oxidizing agent, providing a highly regioselective protocol for *ortho* acylation. Subsequently, Kwong and co-workers,[56] in 2013, further extended the scope of the reaction allowing easy access to functionalized benzophenones under mild reaction conditions (Scheme 2.22).

Regioselective halogenation of substrates presents researchers with a challenge that also provides a useful handle for further functionalization. Taking into consideration the directing ability of anilides, Shi and co-workers[57] reported a highly regioselective and efficient protocol for *ortho*-halogenation using Cu(II)halide (halides = Cl or Br) salts with $Cu(OAc)_2$ as the oxidizing agent (Scheme 2.23A). Mechanistic studies performed on the catalytic system revealed the presence of a cyclopalladated complex that was proposed to be

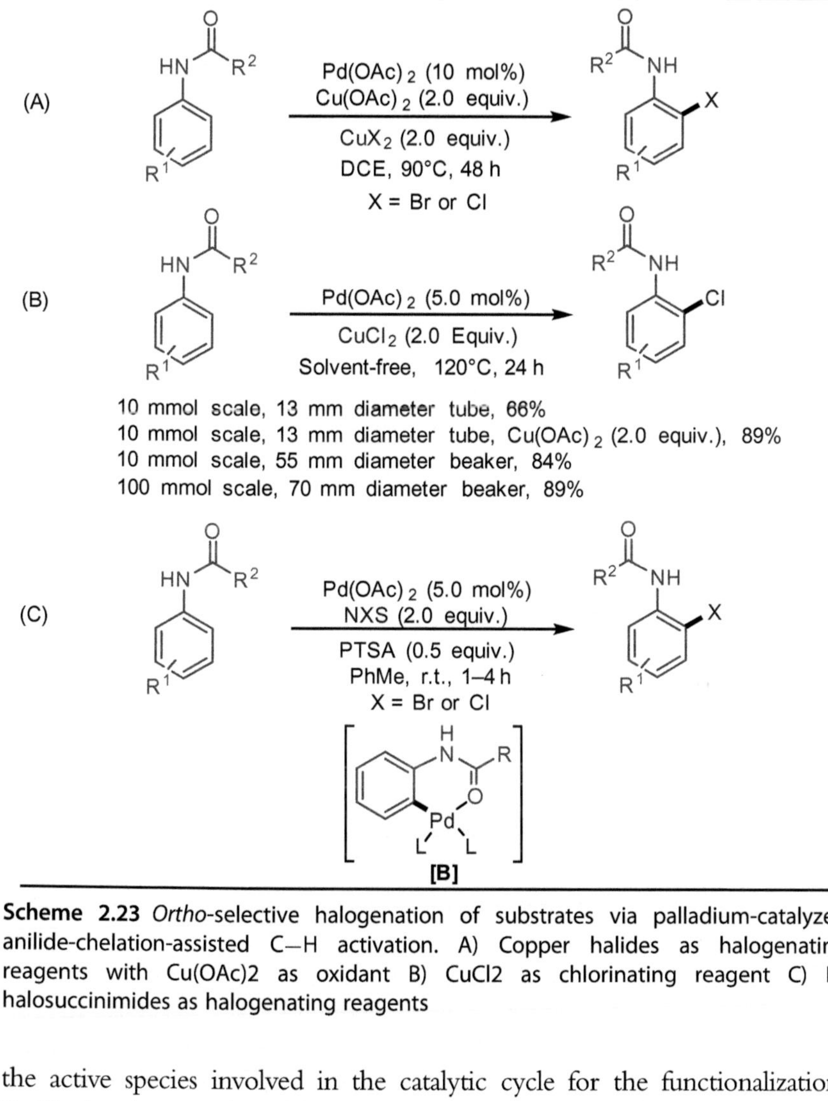

Scheme 2.23 *Ortho*-selective halogenation of substrates via palladium-catalyzed anilide-chelation-assisted C—H activation. A) Copper halides as halogenating reagents with Cu(OAc)2 as oxidant B) CuCl2 as chlorinating reagent C) N-halosuccinimides as halogenating reagents

the active species involved in the catalytic cycle for the functionalization. Bedford and co-workers followed this up with a solvent-free palladium-catalyzed C—H bond activation – halogenation protocol for scale-up possibilities (Scheme 2.23B).[58] The reaction conditions were the same as those developed by Shi, although the solvent was completely removed as part of the scale-up procedure, with the catalytic reaction performed with and without $Cu(OAc)_2$ still resulting in good to excellent yields.

Based on the initial assumptions for the presence of palladacyclic intermediates as a part of the functionalization process, Bedford and co-workers developed a mild and selective *ortho*-halogenation protocol, with

N-halosuccinimide as the halogenating reagent and p-toluenesulfonic acid as the additive (Scheme 2.23C).[59] The functionalization was found to proceed effectively under aerobic conditions at ambient temperature via an *ortho*-palladated PTSA complex **[B]** that was proposed to catalyze the reaction.

Anilide-directed C—H bond activation has been explored extensively in the past several years, and many more functionalization protocols involving C—C and C—heteroatom bond formation have been reported. Recent reviews[3] have elegantly highlighted these examples in more detail, and the reader is advised to refer to these for further reading.

Structurally identical functional groups to anilides, urea, and formamides are also known for their strong coordinating properties, which were recently exploited for the development of a variety of functionalization strategies such as arylation, alkenylation, carbonylation, etc. Most notable contributions in this regard comes from Xu and Huang's group[60] for developing a palladium-catalyzed *ortho*-arylation of formanilides (Scheme 2.24). These substructures in the past have allowed their easy conversion to important heterocyclic molecules.

Urea as a directing group has also provided promising results for the *ortho*-arylation of substrates. Ease of removal of the urea functionality also makes it highly desirable as a directing group in palladium-catalyzed C—H bond functionalization reactions. An excellent example employing urea as the directing group for direct olefination with *n*-butyl acrylate was recently disclosed by Yu and co-workers (Scheme 2.25A).[61] The unique feature of the protocol was the achievement of efficient olefination at ambient temperature, made possible by the $Pd(OAc)_2$/*p*-TsOH/AcOH combination. Another useful protocol for palladium-catalyzed fluoroalkylation of aromatic ureas was reported for the first time by Novak and co-workers[62] in 2017 (Scheme 2.25B). The fluoroalkylation was facilitated by the application of a novel trifluoroethyl(mesityl)iodonium salt that enabled the

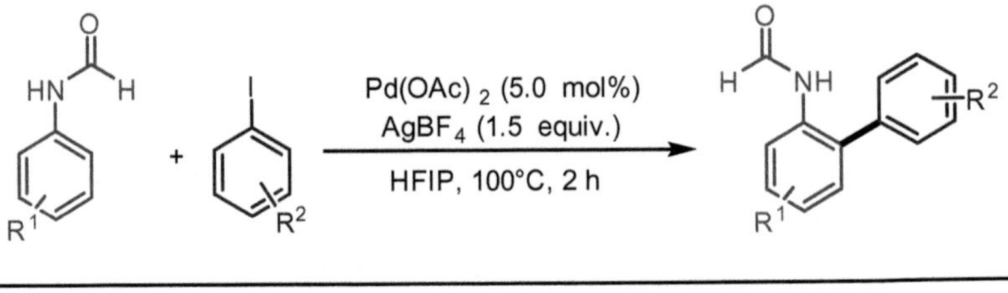

Scheme 2.24 Formanilide as a directing group for palladium-catalyzed arylation of aromatic amines.

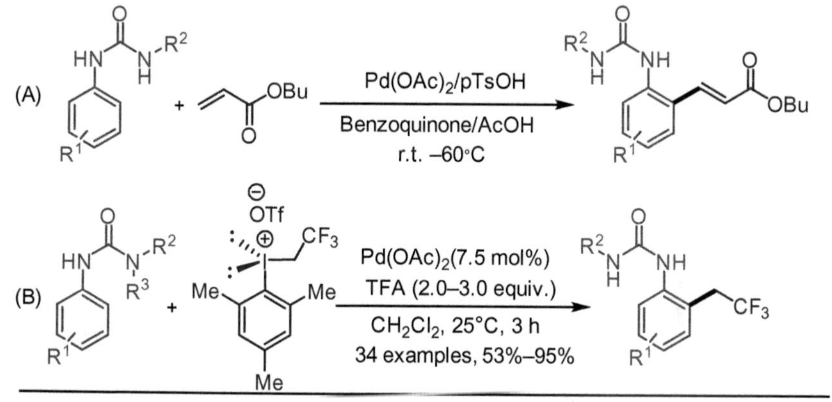

Scheme 2.25 Urea as a directing group for olefination and trifluoroalkylation. A) Olefination using acrylates B) Trifluoroalkylation using iodonium salt

installation of the trifluoroethyl moiety on the aromatic ureas at ambient temperature in high yields. Beside these examples for the application of urea as a directing group for the functionalization of substrates, various strategies for arylation and carbonylation are also known, and have contributed immensely in the further development of this area.

2.2.1.3 Amides as a Directing Group for Palladium-Catalyzed C—H Bond Functionalization

Examples in the earlier sections were representative of functional groups that exhibit strong coordination (possibly due to the availability of a lone pair of electrons on the heteroatoms) towards palladium, facilitating a wide variety of catalytic functionalization processes. In comparison to these, several functional groups such as ketone, hydroxyl, amides, phosphate, esters, sulfonamides, carbamates, formate esters, alkynes, alkenes, and carboxylic acids are known to be weakly coordinating to the metal center. Amides are one such example of weakly coordinating directing groups that have found extensive applications in the palladium-catalyzed C—H bond functionalization of substrates.

The earliest example for the use of amides (COHNOMe) as a directing group was reported by Yu and co-workers[63] in 2008 for palladium-catalyzed functionalization of *o*-methyl hydroxamic acids with organometallic boron reagents (Scheme 2.26A). The same group subsequently reported[64] the arylation protocol using aryl iodides as coupling partners and $CONHC_6F_5$ as the directing group (Scheme 2.26B).

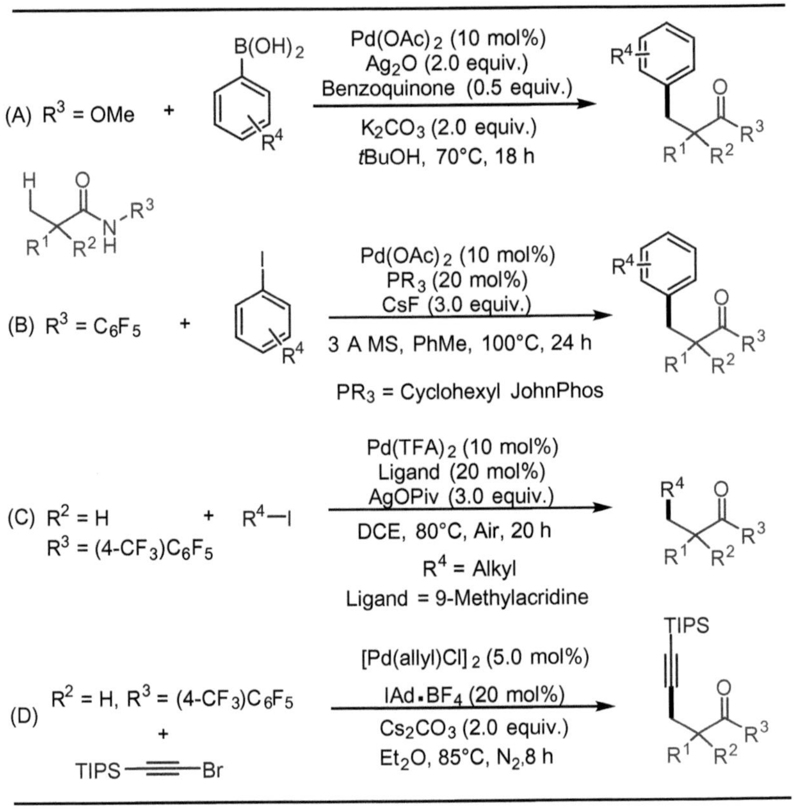

Scheme 2.26 Early examples of using amides as directing groups. A) Arylboronic acids as arylating reagents B) Aryl iodides as arylating reagents C) Alkyl iodides as alkylating reagents D) Alkynylation using halo alkynes

Another notable contribution to the functionalization of simple amides was presented by Yu and co-workers[65] in 2014, with alkyl halides as alkylating agents. The strategy was similar to that employed in Scheme 2.26B for arylation of amides. Instead of using phosphine as ligand, Yu and co-workers employed 9-methyl acridine and other pyridine based ligands to obtain the desired reactivity (Scheme 2.26C). Finally, Yu and co-workers in 2013 also reported an amide-directed palladium-catalyzed alkynylation protocol using a Pd(0)/*N*-heterocyclic carbene catalytic system (Scheme 2.26D).[66]

Functionalization of aromatic amides via palladium-catalyzed C—H bond activation takes place through the chelation-assistance provided by the weakly coordinating amides, thus allowing selective *ortho*-arylation.

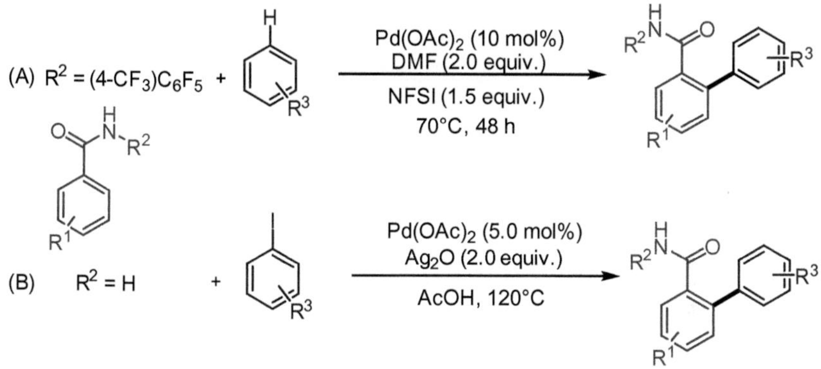

Scheme 2.27 *Ortho*-C—H functionalization using aromatic amides as directing groups. A) Arenes as arylating reagents B) Aryl iodides as arylating reagents

Yu and co-workers, in 2011,[67] first reported the dehydrogenative arylation of aromatic amides (although with a protected form of amide) using simple substituted arenes in the presence of NFSI as the oxidizing agent (Scheme 2.27A). The incoming aryl functionality showed an exceptionally high preference for para-coordination that was unprecedented at the time and readily attributed to the use of NFSI. Later, Wang and co-workers[68] reported in 2012 the arylation of aromatic amides (the first example of unprotected amide used as a directing group) using aryl iodides as the coupling partners (Scheme 2.27B). More recently, Srinivasu and co-workers in 2014 reported the *ortho*-arylation of benzamides catalyzed by palladium supported over mesoporous silica as a recyclable catalytic system.[69] The catalyst was found to retain its activity for up to four recycles.

Further developments in this area were centered around the *ortho*-activation of aromatic amides followed by oxidative cyclization of the resultant *ortho*-arylated benzamides, allowing easy access of phenanthridinones, a structural motif of relevance to synthetic organic chemists pertaining to their occurrence in a large number of naturally occurring bio-active molecules. The first of these reports highlighting the importance of the amide directing group towards cyclization was by Wang and co-workers[70] in 2011, with aryl iodides as the arylating agent (Scheme 2.28A). The simultaneous activation of two C—H bonds and one N—H in a single pot (one-pot) procedure without the isolation of intermediate *ortho*-arylated *N*-methoxybenzamides provides researchers with a

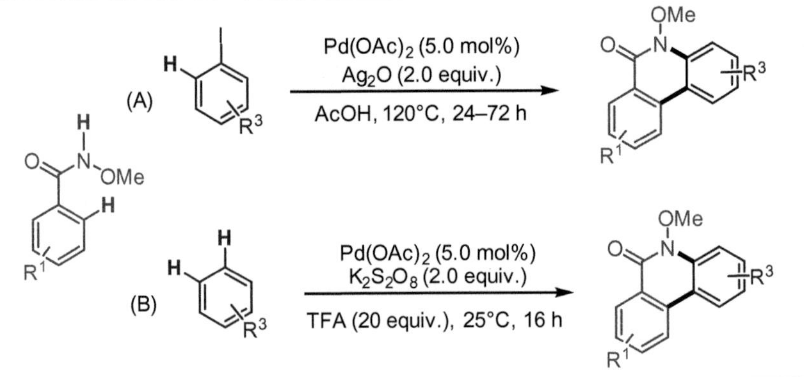

Scheme 2.28 Phenanthridinone synthesis via one-pot multi C—H bond activations. A) Using Aryl iodides B) Using arenes

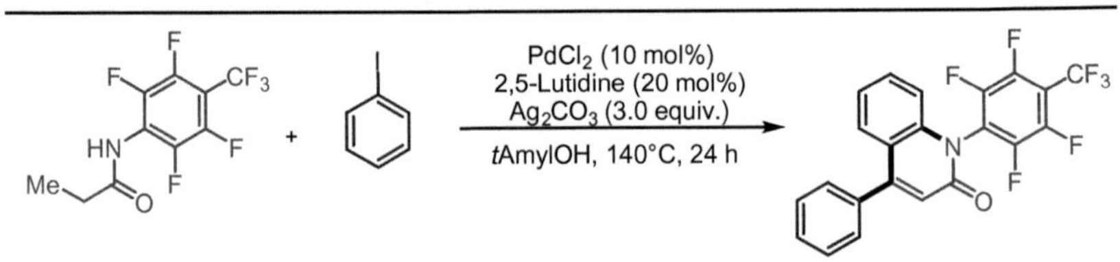

Scheme 2.29 Cascade multi-bond activation for 4-aryl-2-quinolinone synthesis.

step-economic alternative for the synthesis of phenanthridinones. Cheng and co-workers[71] later provided an attractive, atom-economic alternative involving the activation of three C—H and one N—H through the employment of simple arenes as arylating agents (Scheme 2.28B).

Following on the cascade processes (*vide infra*) allowing the activation of multiple bonds in one-pot, Yu and co-workers[72] in 2014 reported on a similar cyclization strategy for obtaining diversely substituted 4-aryl-2-quinolinones (Scheme 2.29). The uniqueness of this reaction was the simultaneous C(sp^2)—H arylation, dehydrogenative coupling, Heck reaction, followed by amidation in a single synthetic procedure.

Acylation strategies have also been developed that employ amides as weakly coordinating ligands for palladium-catalyzed functionalization using different acylating reagents. One such strategy, reported by Kim and co-workers[54] in 2013, discusses a decarboxylative methodology involving α-oxocarboxylic acids as acylating agents (Scheme 2.30A). Interestingly, the decarboxylation of the acid was performed at the relatively lower temperature of 70°C, with $(NH_4)_2S_2O_8$ as the oxidant. Another useful synthetic

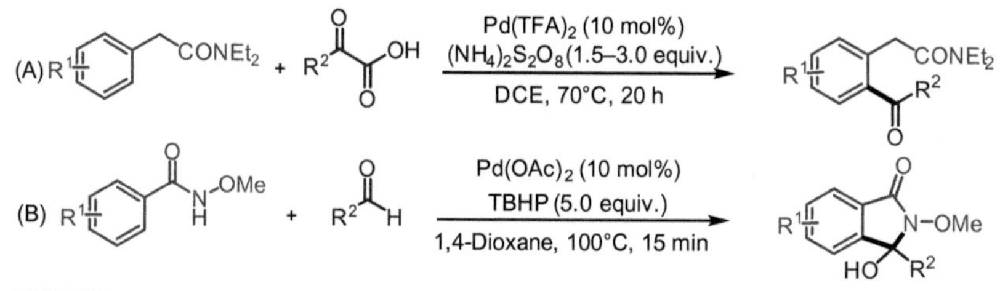

Scheme 2.30 Acylation strategies in C—H activation using benzamides as a directing group. A) Oxocarboxylic acids as acylating reagents B) Aldehydes as acylating reagents

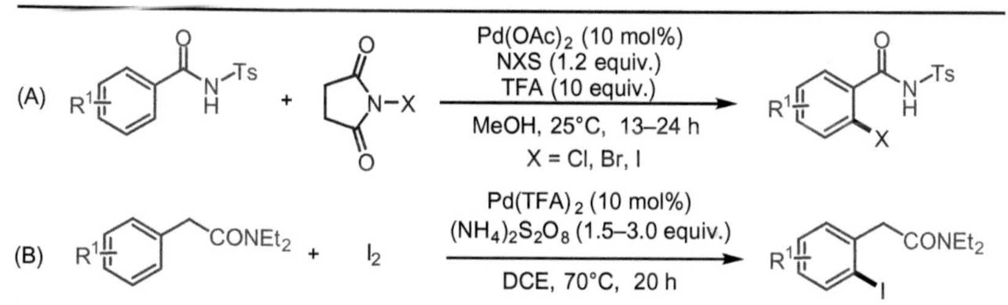

Scheme 2.31 Benzamide-directed halogenation of arenes. A) N-halosuccinimides as halogenating reagents B) Iodine as iodinating reagent

methodology using palladium-catalyzed C—H activation (acylation with benzaldehyde)/annulation for the synthesis of hydroxyl isoindolones was reported by Zhao and co-workers[73] in 2013 (Scheme 2.30B). Besides the use of a benign oxidant, TBHP, the reaction protocol proceeded with a short reaction time, excellent atom economy, and good selectivity.

Palladium-catalyzed amide-directed halogenation reactions could provide useful solutions to the mild functionalization of further exploitation. With this in mind, Fabis and co-workers[74] developed a mild and selective iodination protocol for arenes (Scheme 2.31A). The *N*-tosylbenzamide functional group helps direct the incoming halide (*N*-halosuccinimide as the halide source) to the *ortho*-position at ambient temperature. A cheaper alternative for iodination was subsequently reported by Yu and co-workers[75] that uses molecular iodine as a readily available and economic iodination reagent, and also satisfies the oxidizing requirement for the process (Scheme 2.31B).

All the above examples highlighting the application of amides as directing groups, suggest the influence of a weakly coordinating ligand on the C—H activation reactions. In comparison to the strongly coordinating phenylpyridines or anilides as directing groups, it could be observed that weakly coordinating amides have lesser applicability, although the conditions are milder and more suitable for further modification. Several other examples of benzamide-directed functionalization are known in literature as well as the ones we have discussed here.

Sulfonamides are another class of functional groups with structural similarities to the amides, that have found applications as directing groups in the C—H bond activation processes. Although the number of protocols reported in literature is more limited, this could be attributed to the rather unreactive and rigid nature of sulfonamides.

In spite of all the problems associated with the sulfonamide structural motif, researchers have successfully applied them as effective directing groups for palladium-catalyzed C—H bond activation in processes such as arylation, carbonylation, carboxylation, halogenation, methylation, and olefination. Pioneering work by Yu and co-workers[76] deserves a special mention for their contribution to the development of sulfonamides as directing groups in C—H activation reactions (Scheme 2.32). Yu and co-workers demonstrated the applicability of the developed methodology towards the late-stage modification of a non-steroidal anti-inflammatory blockbuster drug celecoxib.

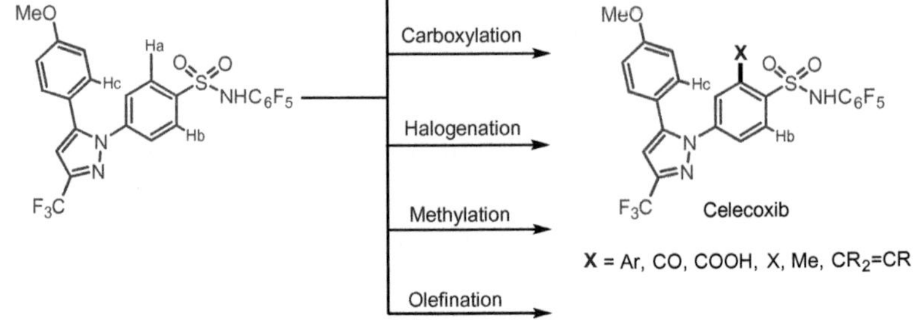

Scheme 2.32 Sulfonamides as directing groups in palladium-catalyzed C—H activation.

2.2.1.4 Carboxylic acids as a Directing Group for Palladium-Catalyzed C—H Bond Functionalization

One of the synthetically useful and highly versatile functional groups, in terms of ease of conversion to other functional groups and unique metal-coordination properties, are carboxylic acids. The ability to bind weakly to the metal center significantly improves the possibility of incorporating these structural motifs as effective directing groups for palladium-catalyzed C—H bond functionalization. A glance through the literature suggests the role played by carboxylic acid directing groups in palladium-catalyzed C—H bond activation is significant, given the number of examples that have been reported.

The role of carboxylic acids as directing groups was first realized, by Miura and co-workers in 1998, for the palladium-catalyzed C—H olefination of arenes (Scheme 2.33).[77] The catalytic functionalization was made possible by a $Pd(OAc)_2/Cu(OAc)_2$ catalyst combination as the reaction proceeded through a possible Wacker-type oxidative cyclization.

The resulting products range from simple *ortho*-olefins, to phthalides and isocoumarin derivatives, and the selectivity could be improved substantially by the placement of groups in specific positions on the benzoic acid substructure. Miura and co-workers,[77] through this pioneering work, provided researchers with a useful handle in their quest to find improved selectivity and the possibility to move away from the strongly coordinating ligands that cause metal-dissociation problems in certain catalytic processes.

Benzoic acids have since been exploited as directing groups, and a variety of functionalization reactions have been developed to highlight the potential of the synthetic methodology. An arylation strategy for

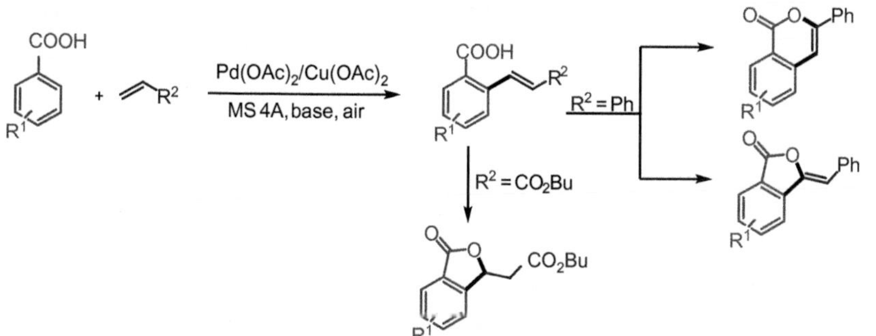

Scheme 2.33 First example of carboxylate-directed palladium-catalyzed C—H olefination.

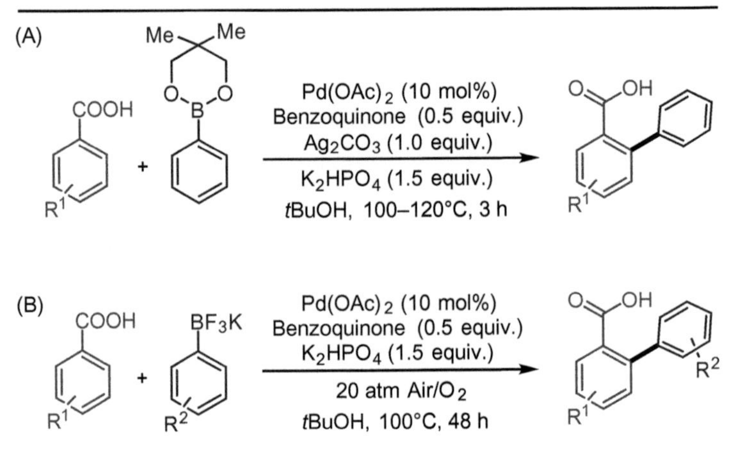

Scheme 2.34 First examples of carboxylate-directed palladium-catalyzed C—H arylation. A) Phenylboronate ester as arylating reagent B) Aryltrifluoroborates as arylating reagents

ortho-functionalization of benzoic acids was first reported by Yu and co-workers[78] as early as 2007 (Scheme 2.34A). The choice of an organometallic reagent, and phenyl boronate esters as aryl donors, allowed smooth conversion of substituted benzoic acids to their corresponding *ortho*-arylated products. Yu and co-workers, as a part of their study, proposed for the first time the involvement of carboxylate ion-assisted/directed C—H bond functionalization, which now has become an indispensable tool in organic synthesis. Yu and co-workers[79] further demonstrated the potential of the methodology by using readily prepared aryltrifluoroborates as coupling partners (Scheme 2.34B). The ease of synthesis of these boron reagents allowed expansion of the scope for the functionalization reaction, although an increase in reaction time to 48 h and transformation performed under high O_2 pressures make the reaction protocol synthetically less attractive.

Direct arylations of benzoic acids were next visited by Daugulis and co-workers[80] in 2007, using aryl halides as the coupling partners. A highly chemoselective protocol was developed with aryl iodides as the arylating agents that allowed bromo or chloro functionality to be well tolerated, highlighting the mildness of the reaction conditions (Scheme 2.35A). The role of base was performed by the stoichiometric amount of Ag salt (AgOAc) that helped scavenge the iodide released in the reaction mixture. In order to make the protocol synthetically more attractive, aryl chlorides, a cheaper alternative to their more expensive

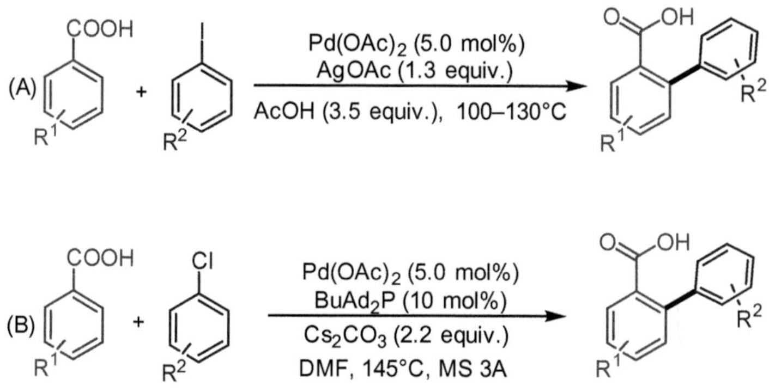

Scheme 2.35 Aryl halides as coupling partners for carboxylate-directed palladium-catalyzed C—H arylation. A) Aryl iodides as arylating reagents B) Aryl chlorides as arylating reagents

iodide counterparts, were also coupled under slightly modified conditions by using a $Pd(OAc)_2/BuAd_2P$ active catalytic system (Scheme 2.35B).[80]

A synthetically more challenging reaction with immense applicability is the introduction of alkyl functionality on aromatic or hetero-aromatic ring systems. This process has been further complicated by the reactivity of the alkylating reagents, such as alkyl boronic acids, boronates, or tri-fluoroborates, which are known to dissociate rapidly under basic conditions. This was evident from the conditions that allowed Yu and co-workers to methylate benzoic acids with methyl boronic acid (Scheme 2.36).[78] Unfortunately, this turned out to be the only alkyl boron reagent that was able to provide functionalized product in decent yields, while most others failed. Although this problem was later resolved for other carboxylic acids, up till now alkylation of benzoic acid using alkylboron reagents still represents a challenge.

Selective *ortho*-alkylation of benzoic acids, although it has several drawbacks and lacks the efficient application of organometallic boron reagents, could be achieved more effectively, as shown by Yu and co-workers[81] in 2009 through a sequential C—H alkylation/lactonization strategy (Scheme 2.37). This transformation was achieved by *ortho*-selective carboxylate-directed C—H alkylation with dihaloalkane, prompting a sequential S_N2 substitution reaction resulting in the formation of five or six-membered lactone rings having immense synthetic relevance.

Benzoic acids have also played a significant role as directing groups for the development of useful acylation strategies. Although the earliest

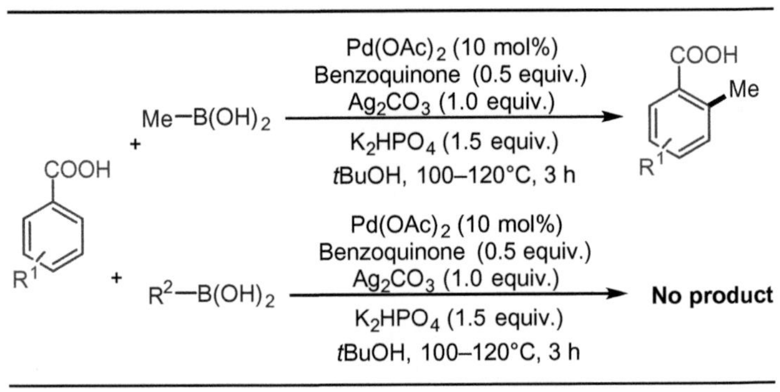

Scheme 2.36 Methylation of benzoic acids via carboxylate-assisted C—H activation.

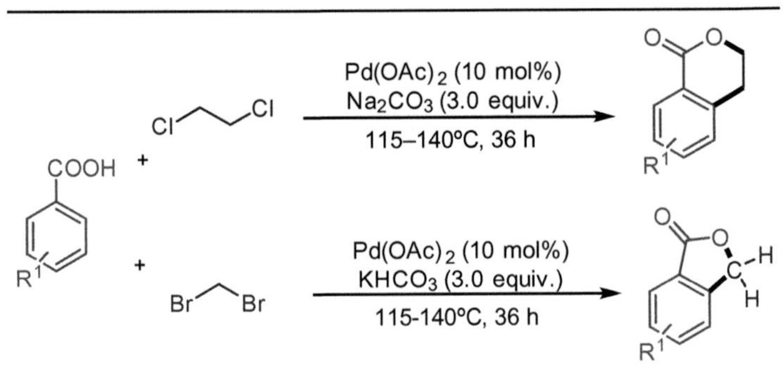

Scheme 2.37 Alkylation/lactonization strategy via carboxylate-directed C—H activation.

example for the *ortho*-acylation of benzoic acid was reported for an Rh catalytic system by Goosen and co-workers,[82] significant work in this area has been achieved with Pd as the catalyst. This was demonstrated elegantly by Ge and co workers[83] in the form of an efficient protocol allowing high chemoselectivity for the decarboxylative coupling of benzoic acid α-oxocarboxylic acids (Scheme 2.38). The synthetic potential of the developed methodology allowed the authors to achieve the synthesis of pitefenone (which has antiplasmodic properties, commonly used as a combination drug in Spasmalgon). The combination of two different carboxylic acids in the same reaction pot for performing contrasting roles is a major synthetic challenge. Ge and co-workers, with the mediation of a transition metal (palladium in this case) and subtle variation in the reaction conditions, were able to overcome this challenge with great success.

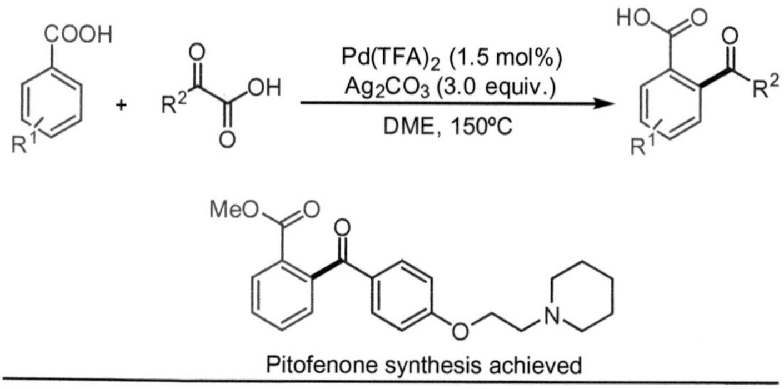

Scheme 2.38 Selective *ortho*-acylation of benzoic acids with α-oxocarboxylic acids.

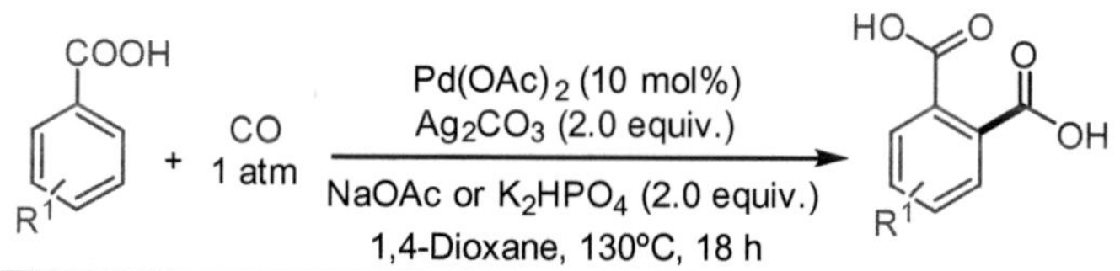

Scheme 2.39 *ortho*-Carboxylation of benzoic acids.

Carboxylation of benzoic acids for the installation of another carboxylic acid functionality at the *ortho-* position, thus resulting into the formation of phthalic acid derivatives, was achieved by Yu and co-workers[84] in 2008 (Scheme 2.39). The carboxylate-assisted C—H activation is followed by the activation of CO and subsequent insertion into the *ortho*-position, to furnish dicarboxylated products with very high selectivities.

As one could discuss the importance of benzoic acids in synthesis, arylacetic acids and 3-aryl propionic acids have found a place for themselves, due to their ease of conversion to molecules of synthetic relevance. The presence of carboxylic functionality in close proximity to the arene structural motif provides researchers with a unique opportunity for further exploitation as directing groups in palladium-catalyzed C—H bond functionalizations. Yu and co-workers[85] in 2010 first demonstrated the potential of these moieties in a C—H bond functionalization protocol for the *ortho*-selective olefination of phenylacetic acid (Scheme 2.40). With the loss of only a H_2 molecule as a part of the dehydrogenative reaction between phenylacetic acid and olefins in the presence of O_2 as oxidant, Yu and co-workers[85] effectively achieved excellent atom economy in the

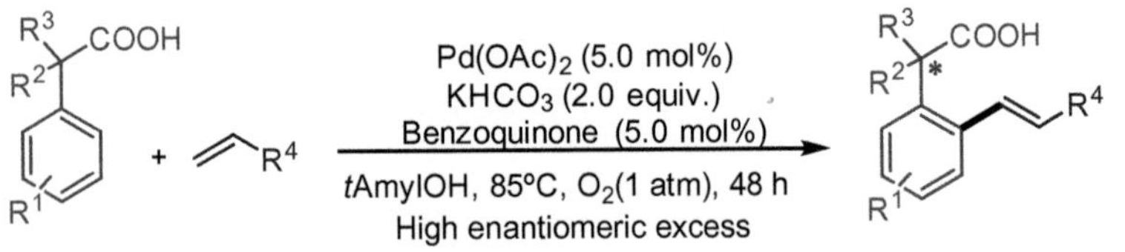

Scheme 2.40 Alkenylation of phenylacetic acid.

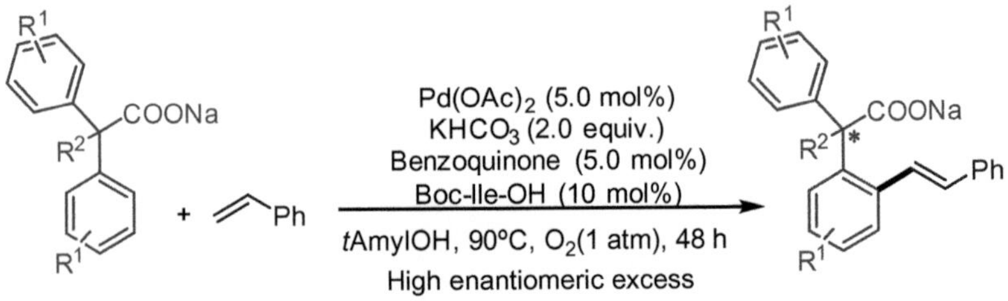

Scheme 2.41 *Mono*-N-protected amino acid ligand-assisted olefination.

whole process. The mildness of the conditions was further exploited in achieving very high enantiomeric excess.

Yu and co-workers[86] further developed another enantioselective protocol for selective C—H activation using a pioneering *mono*-N-protected amino acid ligand-assisted olefination directed by the carboxylate ion (Scheme 2.41). The incorporation of the amino acid was influential in achieving selective *ortho*-alkenylation, possibly due to steric factors. The reactivity and selectivity was governed to a certain extent by the electronics on the aromatic ring of phenylacetic acid, with electron-releasing and mildly electron-attracting substituents providing better yields of the *ortho* alkenylated product. Such a synthetic strategy has the potential to provide researchers with easy access to enantiomerically enriched starting materials for further applications.

A few examples have also emerged in literature for the carboxylate-assisted functionalization of alkyl carboxylic acids for obtaining useful synthetic building blocks. Although the methodology followed for these types of transformations is similar to the ones for benzoic acids and arylacetic acids, a recent example reported by Sasai and co-workers[87] for palladium-catalyzed enantioselective intramolecular oxidative cyclization is worth mentioning (Scheme 2.42). Enantioselective synthesis of furanones was

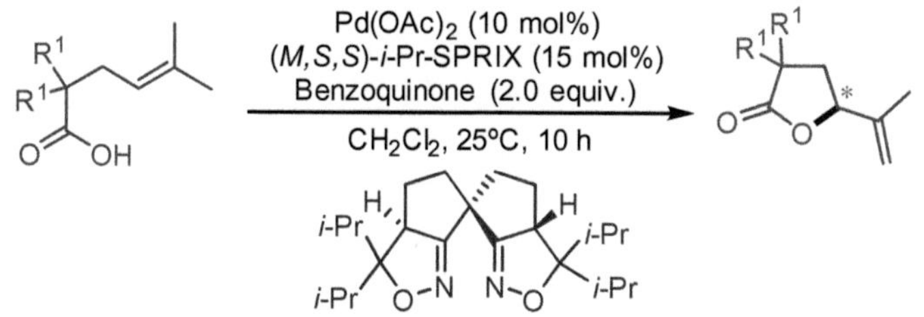

Scheme 2.42 Enantioselective carboxylate-assisted intramolecular cyclization.

made possible by the use of a chiral ligand, (*M*,*S*,*S*)-*i*-Pr-SPRIX, with the cyclization of carboxylate ion taking place on the olefin.

2.2.1.5 Carbonyls (Aldehydes and Ketones) as Directing Groups for Palladium-Catalyzed C—H Bond Functionalization

Other classes of functional groups related to the carboxylic acids are the carbonyl compounds, such as aldehydes and ketones. Although their reactivity, as well as their directing group ability, is much less pronounced than their carboxylic acid counterparts, the possibility of synthetic manipulation makes them indispensable. To date, aldehydes have seldom found applications as directing groups for palladium-catalyzed C—H bond functionalization. This could be attributed to the large number of reports on palladium-catalyzed arylation of aldehydes leading diarylmethanols.[88]

Ketones, in this regard, are less reactive and portray better coordinating ability to the metal center. One of the earliest examples for the multiple arylation of benzylphenyl ketones with aryl bromides was reported by Miura and co-workers[89] in 1999. The feasibility of ketones as efficient directing groups was further tested by Cheng and co-workers[90] for the synthesis of phenanthrones by palladium-catalyzed C—H activation followed by cyclization (Scheme 2.43A). The isolation of a palladacyclic intermediate and characterization with single crystal X-ray diffraction helped authors to confirm the coordination of a carbonyl moiety to the palladium center, and to predict the pathway for the cyclization. A dehydrogenative C—H activation/cyclization strategy was next to be reported by Cheng and co-workers[91] for the synthesis of fluorenones (Scheme 2.43B). The reaction was found to be promoted by the use of TFA as the solvent, which helps the formation of an electrophilic palladium species required for such

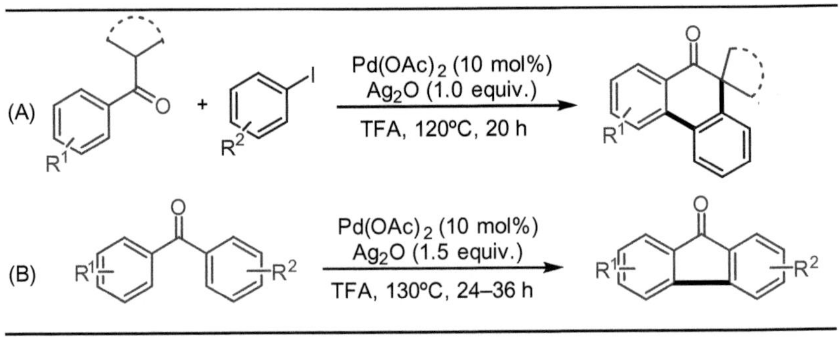

Scheme 2.43 Ketones as directing groups in palladium-catalyzed C–H functionalizations.

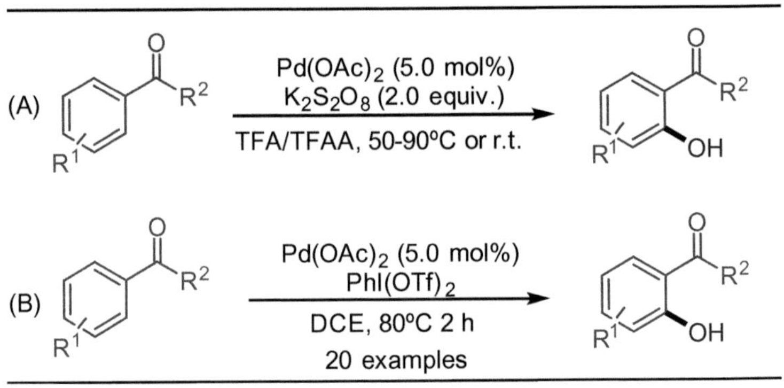

Scheme 2.44 2-Acylphenol synthesis by ketone-directed palladium-catalyzed C–H activation. A) Hydroxylation in TFA/TFAA system B) Hydroxylation using PhI(OTf)2

transformations. Around the same time, Shi and co-workers[92] also reported on a similar protocol for fluorenone synthesis.

Another useful synthetic methodology for the synthesis of 2-acyl phenols brought about by the intervention of ketone functionality involving the selective installation of a hydroxyl group *ortho*- to the acyl group was achieved by Rao and co-workers[93] in 2012 as part of their study into the functionalization of a variety of functional groups (Scheme 2.44A). Given the broad functional group tolerance ability of the developed protocol, the authors were also able to convert the synthesized *ortho*-hydroxy derivatives into important heterocycles. A more exclusive report focusing on the 2-acyl phenols was subsequently reported by Kwong and co-workers,[94] utilizing $PhI(OTf)_2$ as the oxidant (Scheme 2.44B).

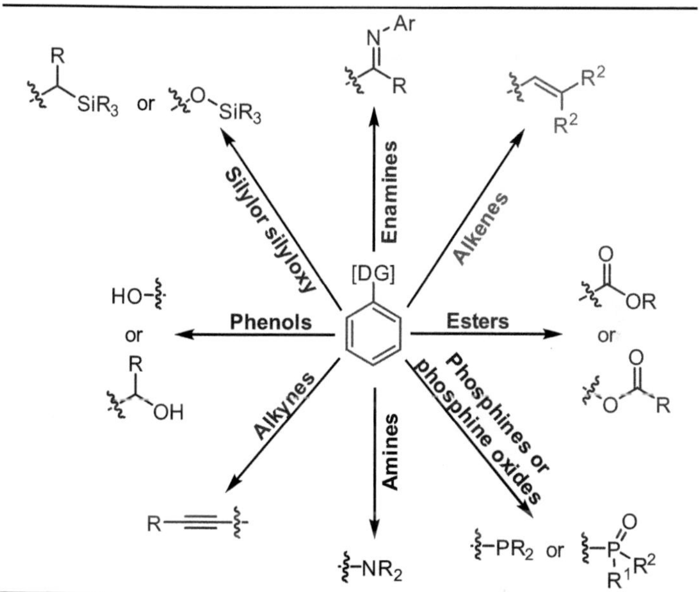

Scheme 2.45 Important directing groups for palladium-catalyzed C—H activation.

2.2.1.6 Other Directing Groups for Palladium-Catalyzed C—H Bond Functionalization

Beside the functional groups that have been extensively highlighted in this chapter (based on the quantum of applications cited in literature), a growing list of many more functionalities with promising directing abilities has provided researchers with a bigger opportunity to functionalize a large variety of substrates in the near future. The schematic provided below would give the reader a better idea about what is now available, besides the ones discussed earlier, for coordination to palladium and further functionalization of substrates (Scheme 2.45). More information on the functionalizations performed using these directing functional groups can be obtained from recent reviews[3] providing extensive discussion of the applications.

2.3 CONCLUSION

Palladium-catalyzed directed arylations, as seen from all the examples presented in this chapter, have the requirements of a functional group that could be weakly or strongly binding to the metal center. Most of the functionalities that are commonly known to provide assistance have been explored; however, applicability of this methodology has been limited owing to the requirements of such a group. The methodology certainly has found applications in the synthesis of certain bio-active molecules, but given the above-mentioned limitations, researchers have started exploring the non-directed arylation strategy which provides better flexibility, as there is no direct involvement of any functional group in the C—H activation step. This aspect of catalytic functionalization will be explored in the next chapter.

REFERENCES

1. Mousseau, J. J.; Charette, A. B. *Acc. Chem. Res.* **2013,** *46*, 412.
2. Molnar, A. *Palladium-Catalyzed Cross-Coupling Reactions: Practical Aspects and Future Developments;* Wiley-VCH, Verlag: Weinheim, 2013.
3. Recent reviews on metal-catalyzed directed C—H bond activation/functionalizations see:
   (a) Zhang, M.; Zhang, Y.; Jie, X.; Zhao, H.; Li, G.; Su, W. *Org. Chem. Front.* **2014,** *1*, 843.
   (b) Gensch, T.; Hopkinson, M. N.; Glorius, F.; Wencel-Delord, J. *Chem. Soc. Rev.* **2016,** *45*, 2900.
   (c) Chen, Z.; Wang, B.; Zhang, J.; Yu, W.; Liu, Z.; Zhang, Y. *Org. Chem. Front.* **2015,** *2*, 1107.
4. For review on meta-selective palladium-catalyzed C—H functionalization see: Dey, A.; Agasti, S.; Maiti, D. *Org. Biomol. Chem.* **2016,** *14*, 5440. For reports on para-selective palladium-catalyzed C—H functionalization see:
   (a) Wang, X.; Leow, X.; Yu, J.-Q. *J. Am. Chem. Soc.* **2011,** *133*, 13864.
   (b) Wu, Z.; Luo, F.; Chen, S.; Li, Z.; Xiang, H.; Zhou, X. *Chem. Commun.* **2013,** *49*, 7653. For a general overview of site-controlled palladium-catalyzed C—H functionalization see: Neufeldt, S. R.; Sanford, M. S. *Acc. Chem. Res.* **2012,** *45*, 936.
5. Engle, K. M.; Mei, T.-S.; Wasa, M.; Yu, J.-Q. *Acc. Chem. Res.* **2012,** *45*, 788.
6. Lyons, T. W.; Sanford, M. S. *Chem. Rev.* **2010,** *110*, 1147.
7. Kalyani, D.; Deprez, N. R.; Desai, L. V.; Sanford, M. S. *J. Am. Chem. Soc.* **2005,** *127*, 7330.
8. Hull, K. L.; Sanford, M. S. *J. Am. Chem. Soc.* **2007,** *129*, 11904.
9. Shabashov, D.; Daugulis, O. *Org. Lett.* **2005,** 7, 3657.
10. Li, W.; Yin, Z.; Jiang, X.; Sun, P. *J. Org. Chem.* **2011,** *76*, 8543.
11. Cho, J.-H.; Tsai, S.-L.; Wu, M.-J. *Synthesis* **2009,** 3757.
12. Feng, J.; Lu, G.; Lv, M.; Cai, C. *Synlett* **2013,** *24*, 2153.
13. Stowers, K. J.; Fortner, K. C.; Sanford, M. S. *J. Am. Chem. Soc.* **2011,** *133*, 6541.
14. Chen, X.; Goodhue, C. E.; Yu, J.-Q. *J. Am. Chem. Soc.* **2006,** *128*, 12634.
15. Zhang, Y.; Feng, J.; Li, C.-J. *J. Am. Chem. Soc.* **2008,** *130*, 2900.
16. Li, C.-J. *Acc. Chem. Res.* **2009,** *42*, 335.
17. Kishore, R.; Yadav, J.; Venu, B.; Venugopal, A.; Kantam, M. L. *New J. Chem.* **2015,** *39*, 5259.
18. Wang, X.; Truesdale, L.; Yu, J.-Q. *J. Am. Chem. Soc.* **2010,** *132*, 3648.
19. Jia, X.; Yang, D.; Zhang, S.; Cheng, J. *Org. Lett.* **2009,** *11*, 4716.
20. Jia, X.; Zhang, S.; Wang, W.; Luo, F.; Cheng, J. *Org. Lett.* **2009,** *11*, 3120.
21. Baslé, O.; Bidange, J.; Shuai, Q.; Li, C.-J. *Adv. Synth. Catal.* **2010,** *352*, 1145.
22. Xiao, F.; Shuai, Q.; Zhao, F.; Baslé, O.; Deng, G.; Li, C.-J. *Org. Lett.* **2011,** *13*, 1614.

23. Ramineni, K.; Kantam, M. L.; Yadav, J.; Akula, V. *J. Mol. Catal. A* **2013,** *379*, 213.
24. Zhang, Q.; Yang, F.; Wu, Y. *Chem. Commun.* **2013,** *49*, 6837.
25. Dick, A. R.; Hull, K. L.; Sanford, M. S. *J. Am. Chem. Soc.* **2004,** *126*, 2300.
26. (a) Kalyani, D.; Sanford, M. S. *Org. Lett.* **2005,** 7, 4149.
(b) Desai, L. V.; Malik, H. A.; Sanford, M. S. *Org. Lett.* **2006,** *8*, 1141.
27. Guo, L.; Xu, Y.; Wang, X.; Liu, W.; Lu, D. *Organometallics* **2013,** *32*, 3780.
28. (a) Powers, D. C.; Geibel, M. A. L.; Klein, J. E. M. N.; Ritter, T. *J. Am. Chem. Soc.* **2009,** *131*, 17050.
(b) Powers, D. C.; Ritter, T. *Nat. Chem.* **2009,** *1*, 419.
(c) Powers, D. C.; Ritter, T. *Nat. Chem.* **2009,** *1*, 302.
29. Wang, D. H.; Hao, X. S.; Wu, D. F.; Yu, J. Q. *Org. Lett.* **2006,** *8*, 3387.
30. (a) Fahey, D. R. *J. Chem. Soc. Chem. Commun.* **1970,** 417.
(b) Fahey, D. R. *J. Organomet. Chem.* **1971,** *27*, 283.
31. (a) Zhao, X.; Dimitrijevic, E.; Dong, V. M. *J. Am. Chem. Soc.* **2009,** *131*, 3466.
(b) Kalyani, D.; Dick, A. R.; Anani, W. Q.; Sanford, M. S. *Tetrahedron* **2006,** *62*, 11483.
32. Kakiuchi, F.; Kochi, T.; Mutsutani, H.; Kobayashi, N.; Urano, S.; Sato, M.; Nishiyama, S.; Tanabe, T. *J. Am. Chem. Soc.* **2009,** *131*, 11310.
33. Pascanu, V.; Carson, F.; Solano, M. V.; Su, J.; Zou, X.; Johansson, M. J.; Martin-Matute, B. *Chem. Eur. J.* **2016,** *22*, 3729.
34. Hull, K. L.; Anani, W. Q.; Sanford, M. S. *J. Am. Chem. Soc.* **2006,** *128*, 7134.
35. Sankar Lal, G.; Pez, G. P.; Syvret, R. G. *Chem. Rev.* **1996,** *96*, 1737.
36. Boele, M. D. K.; van Strijdonck, G. P. F.; de Vries, A. H. M.; Kamer, P. C. J.; de Vries, J. G.; van Leeuwen, P. W. N. M. *J. Am. Chem. Soc.* **2002,** *124*, 1586.
37. (a) Rauf, W.; Thompson, A. L.; Brown, J. M. *Dalton Trans.* **2010,** *39*, 10414.
(b) Bedford, R. B.; Haddow, M. F.; Mitchell, C. J.; Webster, R. L. *Angew. Chem., Int. Ed.* **2011,** *50*, 5524.
38. Daugulis, O.; Zaitsev, V. G. *Angew. Chem., Int. Ed.* **2005,** *44*, 4046.
39. Shabashov, D.; Daugulis, O. *J. Org. Chem.* **2007,** *72*, 7720.
40. Lazareva, A.; Daugulis, O. *Org. Lett.* **2006,** *8*, 5211.
41. Li, B.-J.; Yang, S.-D.; Shi, Z.-J. *Synlett* **2008,** 949.
42. Yang, S.; Li, B.; Wan, X.; Shi, Z.-J. *J. Am. Chem. Soc.* **2007,** *129*, 6066.
43. Sun, C.-L.; Li, B.-J.; Shi, Z.-J. *Chem. Commun.* **2010,** *46*, 677.
44. Li, D.; Xu, N.; Zhang, Y.; Wang, L. *Chem. Commun.* **2014,** *50*, 14862.
45. Li, B.-J.; Tian, S.-L.; Fang, Z.; Shi, Z.-J. *Angew. Chem., Int. Ed.* **2008,** *47*, 1115.
46. Brasche, G.; García-Fortanet, J.; Buchwald, S. L. *Org. Lett.* **2008,** *10*, 2207.
47. Yang, F.; Song, F.; Li, W.; Lan, J.; You, J. *RSC Adv.* **2013,** *3*, 9649.
48. Wu, Y.; Li, B.; Mao, F.; Li, X.; Kwong, F. Y. *Org. Lett.* **2011,** *13*, 3258.
49. Chan, C. W.; Zhou, Z.; Yu, W. Y. *Adv. Synth. Catal.* **2011,** *353*, 2999.
50. Li, C.; Wang, L.; Li, P.; Zhou, W. *Chem. Eur. J.* **2011,** *17*, 10208.
51. Luo, F.; Yang, J.; Li, Z.; Xiang, H.; Zhou, X. *Eur. J. Org. Chem.* **2015,** 2463.
52. Szabo, F.; Simko, D.; Novak, Z. *RSC Adv.* **2014,** *4*, 3883.
53. Fang, P.; Li, M.; Ge, H. *J. Am. Chem. Soc.* **2010,** *132*, 11898.
54. Park, J.; Kim, M.; Sharma, S.; Park, E.; Kim, A.; Lee, S. H.; Kwak, J. H.; Jung, Y. H.; Kim, I. S. *Chem. Commun.* **2013,** *49*, 1654.
55. Yin, Z.; Sun, P. *J. Org. Chem.* **2012,** *77*, 11339.
56. Wu, Y.; Choy, P. Y.; Mao, F.; Kwong, F. Y. *Chem. Commun.* **2013,** *49*, 689.
57. Wan, X.; Ma, Z.; Li, B.; Zhang, K.; Cao, S.; Zhang, S.; Shi, Z. *J. Am. Chem. Soc.* **2006,** *128*, 7416.
58. Bedford, R. B.; Engelhart, J. U.; Haddow, M. F.; Mitchell, C. J.; Webster, R. L. *Dalton Trans.* **2010,** *39*, 10464.

59. Bedford, R. B.; Haddow, M. F.; Mitchell, C. J.; Webster, R. L. *Angew. Chem., Int. Ed.* **2011,** *50*, 5524.
60. Wan, C.; Zhao, J.; Xu, M.; Huang, J. *J. Org. Chem.* **2014,** *79*, 4751.
61. Wang, L.; Liu, S.; Li, Z.; Yu, Y. *Org. Lett.* **2011,** *13*, 6137.
62. Kovacs, S.; Toth, B. L.; Borsik, G.; Bihari, T.; May, N. V.; Striling, A.; Novak, Z. *Adv. Synth. Catal.* **2017,** *359*, 527.
63. Wang, D.-H.; Wasa, M.; Giri, R.; Yu, J.-Q. *J. Am. Chem. Soc.* **2008,** *130*, 7190.
64. Wasa, M.; Engle, K. M.; Yu, J.-Q. *J. Am. Chem. Soc.* **2009,** *131*, 9886.
65. Zhu, R.-Y.; He, J.; Wang, X.-C.; Yu, J.-Q. *J. Am. Chem. Soc.* **2014,** *136*, 13194.
66. He, J.; Wasa, M.; Chan, K. S. L.; Yu, J.-Q. *J. Am. Chem. Soc.* **2013,** *135*, 3387.
67. Wang, X.; Leow, D.; Yu, J.-Q. *J. Am. Chem. Soc.* **2011,** *133*, 13864.
68. Li, D.-D.; Yuan, T.-T.; Wang, G. W. *J. Org. Chem.* **2012,** *77*, 3341.
69. Parsharamulu, T.; Venkanna, D.; Kantam, M. L.; Bhargava, S. K.; Srinivasu, P. *Ind. Eng. Chem. Res.* **2014,** *53*, 20075.
70. Wang, G.-W.; Yuan, T.-T.; Li, D.-D. *Angew. Chem., Int.Ed.* **2011,** *50*, 1380.
71. Karthikeyan, J.; Cheng, C.-H. *Angew. Chem., Int. Ed.* **2011,** *50*, 9880.
72. Deng, Y.; Gong, W.; He, J.; Yu, J.-Q. *Angew. Chem., Int. Ed.* **2014,** *53*, 6692.
73. Yu, Q.; Zhang, N.; Huang, J.; Lu, S.; Zhu, Y.; Yu, X.; Zhao, K. *Chem. Eur. J.* **2013,** *19*, 11184.
74. Péron, F.; Fossey, C.; Santos, J. S. O.; Cailly, T.; Fabis, F. *Chem. Eur. J.* **2014,** *20*, 7507.
75. Wang, X.-C.; Hu, Y.; Bonacorsi, S.; Hong, Y.; Burrell, R.; Yu, J.-Q. *J. Am. Chem. Soc.* **2013,** *135*, 10326.
76. Dai, H.-X.; Stepan, A. F.; Plummer, M. S.; Zhang, Y.-H.; Yu, J.-Q. *J. Am. Chem. Soc.* **2011,** *133*, 7222.
77. Miura, M.; Tsuda, T.; Satoh, T.; Pivsa-Art, S.; Nomura, M. *J. Org. Chem.* **1998,** *63*, 5211.
78. Giri, R.; Maugel, N.; Li, J.-J.; Wang, D.-H.; Breazzano, S. P.; Saunders, L. B.; Yu, J.-Q. *J. Am. Chem. Soc.* **2007,** *129*, 3510.
79. Wang, D.-H.; Mei, T.-S.; Yu, J.-Q. *J. Am. Chem. Soc.* **2008,** *130*, 17676.
80. Chiong, H. A.; Pham, Q.-N.; Daugulis, O. *J. Am. Chem. Soc.* **2007,** *129*, 9879.
81. Zhang, Y.-H.; Shi, B.-F.; Yu, J.-Q. *Angew. Chem., Int. Ed.* **2009,** *48*, 6097.
82. Mamone, P.; Danoun, G.; Gooßen, L. J. *Angew. Chem., Int. Ed.* **2013,** *52*, 6704.
83. Miao, J.; Ge, H. *Org. Lett.* **2013,** *15*, 2930.
84. Giri, R.; Yu, J. Q. *J. Am. Chem. Soc.* **2008,** *130*, 14082.
85. Wang, D.-H.; Engle, K. M.; Shi, B.-F.; Yu, J.-Q. *Science* **2010,** *327*, 315.
86. Shi, B.-F.; Zhang, Y.-H.; Lam, J. K.; Wang, D.-H.; Yu, J.-Q. *J. Am. Chem. Soc.* **2010,** *132*, 460.
87. Takenaka, K.; Akita, M.; Tanigaki, Y.; Takizawa, S.; Sasai, H. *Org. Lett.* **2011,** *13*, 3506.
88. For reports on the palladium-mediated diarylmethanol syntheses see:
   (a) He, P.; Lu, Y.; Dong, C.-G.; Hu, Q.-S. *Org. Lett.* **2007,** *9*, 343.
   (b) He, P.; Lu, Y.; Hu, Q.-S. *Tetrahedron Lett.* **2007,** *48*, 5283.
   (c) Qin, C.; Wu, H.; Cheng, J.; Chen, X.; Liu, M.; Zhang, W.; Su, W.; Ding, J. *J. Org. Chem.* **2007,** *72*, 4102.
   (d) Lin, S.; Lu, X. *J. Org. Chem.* **2007,** *72*, 9757.
   (e) Liu, T.-P.; Liao, Y. X.; Xing, C.-H.; Hu, Q.-S. *Org. Lett.* **2011,** *13*, 2452.
89. Satoh, T.; Kametani, Y.; Terao, Y.; Miura, M.; Nomura, M. *Tetrahedron Lett.* **1999,** *40*, 5345.
90. Gandeepan, P.; Parthasarathy, K.; Cheng, C.-H. *J. Am. Chem. Soc.* **2010,** *132*, 8569.
91. Gandeepan, P.; Hung, C.-H.; Cheng, C.-H. *Chem. Commun.* **2012,** *48*, 9379.
92. Li, H.; Zhu, R.-Y.; Shi, W.-J.; He, K.-H.; Shi, Z.-J. *Org. Lett.* **2012,** *14*, 4850.
93. Shan, G.; Yang, X.; Ma, L.; Rao, Y. *Angew. Chem., Int. Ed.* **2012,** *51*, 13070.
94. Choy, P. Y.; Kwong, F. Y. *Org. Lett.* **2013,** *15*, 270.

CHAPTER 3

Nondirected C−H Bond Functionalizations of (Hetero)arenes

Shun-Man Wong[1] and Fuk-Yee Kwong[1,2]
[1]The Hong Kong Polytechnic University, Kowloon, Hong Kong
[2]The Chinese University of Hong Kong, Shatin, New Territories, Hong Kong

Contents

3.1 INTRODUCTION

Bi(hetero)aryl derivatives are common structural motifs in numerous natural products, biological, pharmaceutical, and functional materials; and have attracted much attention to the development of a multitude of catalytic protocols for the C−C bond construction between (hetero)arenes.[1] Among the traditional palladium-catalyzed Hiyama,[2] Kumada,[3] Negishi,[4] Stille,[5] or Suzuki−Miyaura[6] reactions, additional synthetic steps are required to prepare prefunctionalized (hetero)arene coupling partners to link up the two (hetero)arenes by losing the introduced halide and organometallic functionalities as waste. Indeed, synthetic protocols circumventing prefunctionalization steps have attracted considerable interest due to their simplicity, generality, and mildness on resources and environment.[7] Palladium-catalyzed direct C−H bond functionalizations are one of the most powerful C−C bond forming methodologies.[8]

Strategies for Palladium-Catalyzed Non-Directed and Directed C−H Bond Functionalization
DOI: http://dx.doi.org/10.1016/B978-0-12-805254-9.00003-7

In order to have a broad synthetic value for a C—H functionalization reaction, one C—H bond must be selectively functionalized over the other C—H bonds. Undoubtedly, this issue can be circumvented by employing molecules with C—H bonds with different reactivity, such as functionalization of heteroarene that possesses at least one position of higher reactivity.[9]

However, in benzene derivatives, the difference in reactivity between the C—H bonds is generally not obvious and additional regio-controlling directing groups are required, such as amides, pyridines, or acetanilides. There are two roles of directing group: (1) directs the transition-metal into close proximity to functionalize the C—H bond resulting in high regioselectivity; (2) increases the reactivity due to the higher effective concentration of the catalyst at the site of interest. Yet, in most cases, the functionalization of the arene occurs only at the C—H bond *ortho* to the directing group, and hence limits the scope of accessible products.[10] Moreover, additional synthetic steps for installing the directing groups into the starting materials and removing the directing groups after the reactions are required.[11]

Therefore, a more appealing strategy for the C—H functionalization of benzene derivatives should be to avoid using the directing group. However, not many examples of transition metal-catalyzed regioselective C—H functionalization of benzene derivatives without using the directing groups have been disclosed.[12]

Catalytic C—H functionalization falls into three categories: (1) the reaction of a (hetero)arene C—H bond with a (pseudo)halide; (2) the reaction of a (hetero)arene C—H bond with an organometallic reagent; and (3) the reaction between two (hetero)arene C—H bonds. The first type of reaction is oxidation-neutral, while the second and third types of reaction require a sacrificial oxidant. Herein, selected representative examples for intermolecular nondirected C—H bond functionalizations of (hetero)arenes with (pseudo)halides for constructing bi(hetero)aryl derivatives, in recent decades, are described.

3.2 REACTIONS WITH HALIDES

The direct arylation of (hetero)arenes with aryl halides via palladium-catalyzed C—H bond functionalization has received huge attention, since the prefunctionalization of organometallic reagents can be avoided, thus reducing the corresponding metal waste.[7a,8a,9a,12c,13]

3.2.1 Pd-catalyzed C—H/C—X Cross-Coupling Reactions Between an Arene and an Aryl Halide

3.2.1.1 Simple Arenes

The direct arylation of simple arenes is one of the most challenging topics, due to the low acidity of the aromatic C—H bonds and lack of directing group to assist. To date, not many examples are disclosed on the direct arylation of simple arenes using aryl halides as the coupling partners.

Fagnou described the development of a combination of palladium-pivalic acid co-catalyst for the direct arylation of benzene.[14] In the presence of $Pd(OAc)_2$/**DavePhos** catalyst and a catalytic amount of pivalic acid (PivOH) as an additive, a variety of aryl bromides were compatible for this reaction, while aryl chlorides and iodides were not (Scheme 3.1).

A phosphine-free $Pd(OAc)_2$-catalyzed direct arylation of naphthalene or benzene with various aryl iodides was reported by Lu.[15] In the presence of a stoichiometric amount of CF_3CO_2Ag in trifluoroethanol (TFE), this catalytic system afforded biaryls in moderate yields (Scheme 3.2).

Doucet demonstrated the first $PdCl(C_3H_5)$(dppb)-catalyzed regioselective direct arylation of free —OH phenols at the C-2-position with an array of electron-deficient (hetero)aryl bromides (Scheme 3.3).[16] In the presence of KOAc as the base in DMA, the direct monoarylation at C-2 of

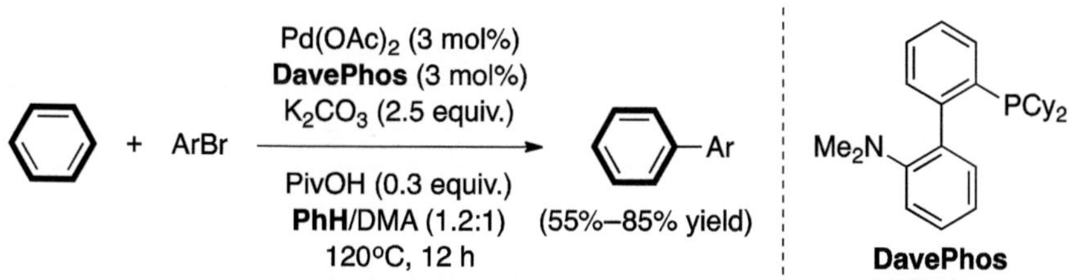

Scheme 3.1 $Pd(OAc)_2$/**DavePhos**-catalyzed direct arylation of benzene with aryl bromides.[14]

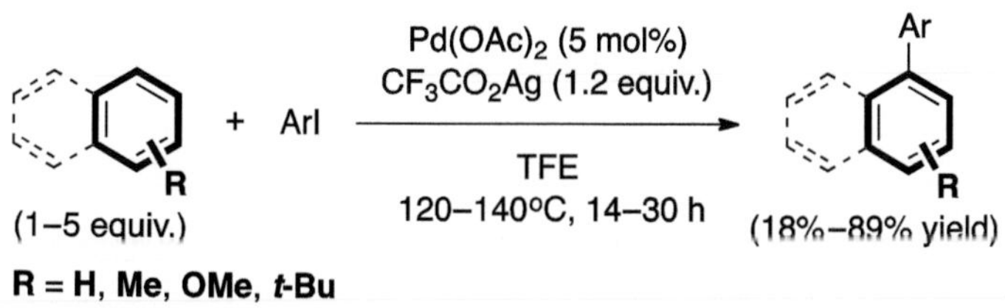

Scheme 3.2 Phosphine-free $Pd(OAc)_2$-catalyzed direct arylation of naphthalene or benzene with aryl iodides.[15]

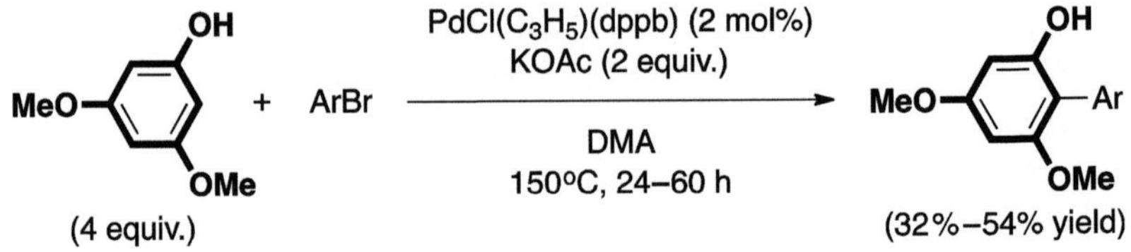

Scheme 3.3 $PdCl(C_3H_5)(dppb)$/KOAc-catalyzed regioselective direct arylation of phenols with (hetero)aryl bromides.[16]

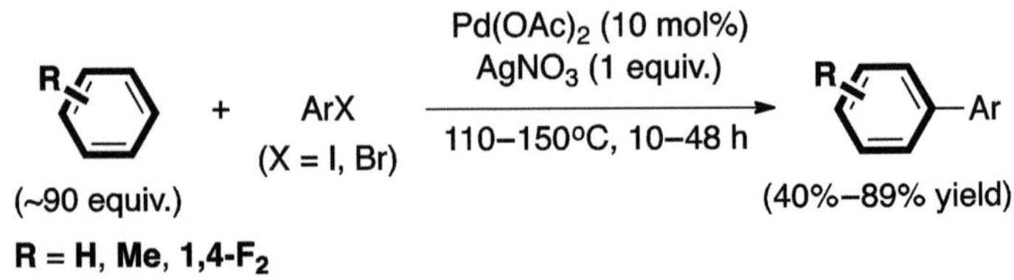

Scheme 3.4 $Pd(OAc)_2/AgNO_3$-catalyzed direct C—H arylation of unactivated arenes with aryl iodides/bromides.[17]

unprotected phenols was favored. The use of KOAc was found to be crucial to control the regioselectivity of the reaction. Particularly noteworthy was that this protocol was an environmentally attractive route, as the by-products of this protocol were AcOH associated to KBr instead of metallic salts.

Jiang discovered an efficient phosphine-free protocol for the direct C—H arylation of unactivated arenes with aryl halides (Scheme 3.4).[17] In the presence of the $Pd(OAc)_2/AgNO_3$ system, both aryl iodides and bromides were good coupling partners with arenes. Remarkably, the direct arylation of toluene was also accomplished in good yield, yet without regioselectivity. In addition, the double coupling took place smoothly and gave product in moderate yield.

Hou developed two efficient methods for the $Pd(OAc)_2$-catalyzed direct arylation of electron-deficient arenes with aryl bromides in the presence of K_2CO_3 in DMA (Scheme 3.5).[18] One protocol was ligand-free (Scheme 3.5, **Method A**) while the other (Scheme 3.5, **Method B**) employed monoprotected amino acid ligand (Ac-Ile-OH) to assist the reaction to proceed under mild conditions.

Doucet later reported the $PdCl(C_3H_5)(dppb)$-catalyzed direct arylation of guaiazulenes with a variety of aryl bromides (Scheme 3.6).[19] It was noteworthy that both sp^2 and sp^3 C—H bonds of guaiazulene could be functionalized by applying different natures of bases. In the presence of KOAc in ethylbenzene, the arylation could be controlled regioselectively at C-2- and C-3-positions.

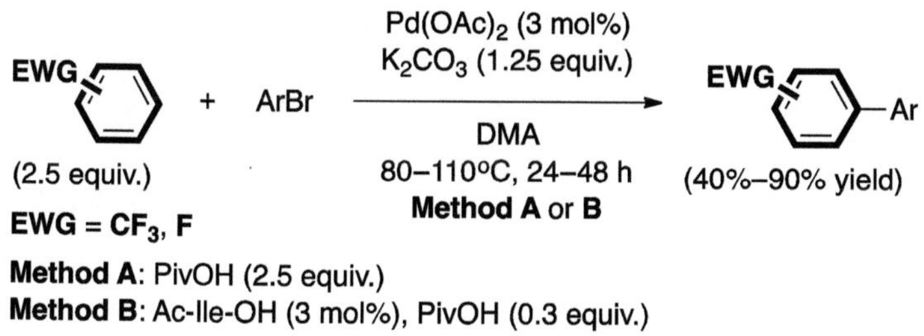

Scheme 3.5 $Pd(OAc)_2$-catalyzed electron-deficient arenes with aryl bromides.[18]

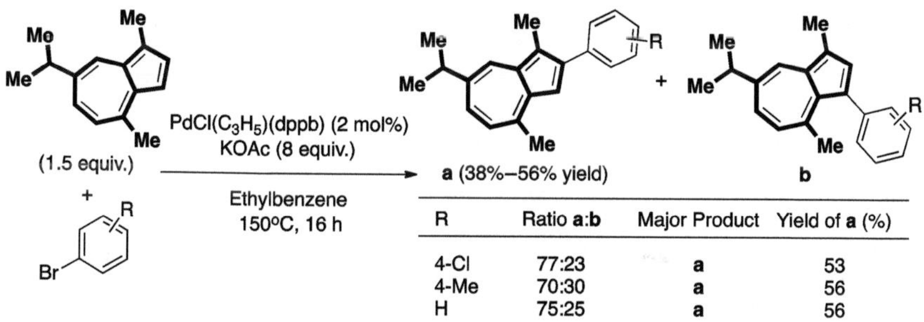

R	Ratio **a**:**b**	Major Product	Yield of **a** (%)
4-Cl	77:23	**a**	53
4-Me	70:30	**a**	56
H	75:25	**a**	56

Scheme 3.6 $PdCl(C_3H_5)(dppb)$-catalyzed direct C-2- or C-3-arylation of guaiazulenes with aryl bromides.[19]

Zhou presented the first highly *para*-selective $Pd(OAc)_2$-catalyzed direct C—H arylation of unprotected phenols with aryl iodides (Scheme 3.7).[20] It was worthy to note that inexpensive, environmental benign water was employed as a "green" solvent under mild conditions such as room temperature. Moreover, this reaction was undergone by taking the formal inverse direct arylation strategy, which envisioned that aryl iodide with an *ortho*-coordinating group would accelerate the step of oxidative addition to provide the thermodynamically stable intermediate and facilitate the activation of phenol.[21]

However, an *ortho*-coordination group was necessary for aryl iodide substrates in order to increase their activities, which would limit the appeal of the protocol in synthetic applications. Therefore, the development of new straightforward methods for C—H arylation of unfunctionalized phenols is still challenging. In continuation of the work of palladium-catalyzed C—H functionalization, Zhou later described an efficient protocol for the direct arylation of phenols with aryl iodides without an *ortho*-coordination group.[22] These arylation reactions were compatible with common functional groups and afforded highly *para*-selective arylated phenols in

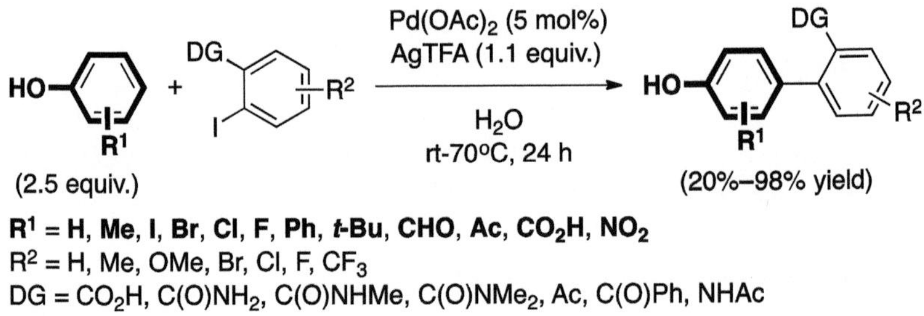

Scheme 3.7 $Pd(OAc)_2$-catalyzed *para*-selective arylation of phenols with aryl iodides in water.[20]

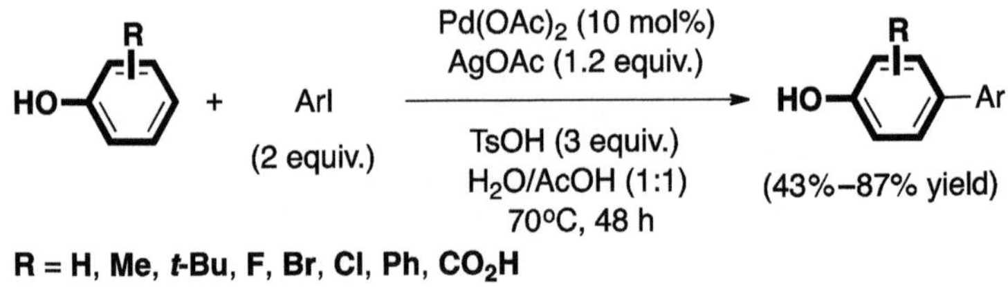

Scheme 3.8 $Pd(OAc)_2$-catalyzed *para*-selective arylation of phenols with aryl iodides.[22]

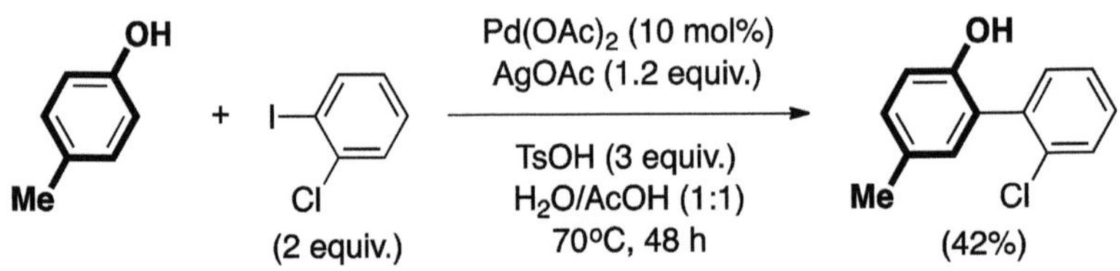

Scheme 3.9 $Pd(OAc)_2$-catalyzed *ortho*-arylation of 4-methylphenol with 2-chloro iodobenzene.[22]

moderate-to-high yields by using $Pd(OAc)_2$ as the catalyst, AgOAc and TsOH as the additives in a mixed water, and AcOH (Scheme 3.8). Interestingly, when 4-methylphenol was used as a substrate, the *ortho*-arylated product was given in a 42% yield (Scheme 3.9).

3.2.1.2 Polyfluoroarenes

During recent years, many protocols have emerged for the direct arylation of polyfluoroarenes, as a fluoro-substituted benzene ring appears to promote the C–H bond functionalization. In addition, no or a lower regioselectivity issue exists with these compounds.[23]

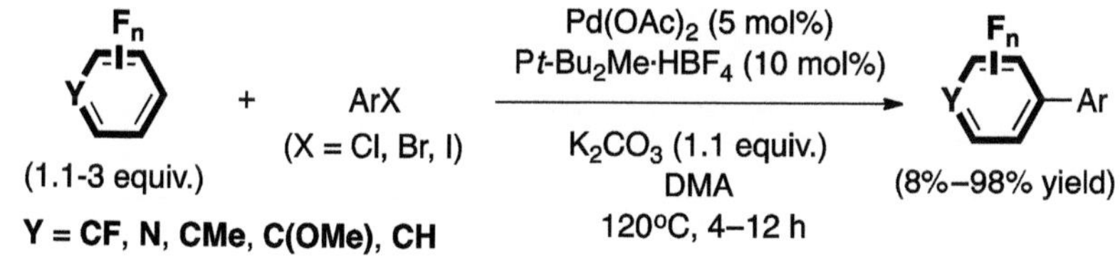

Scheme 3.10 $Pd(OAc)_2$/P*t*-$Bu_2Me \cdot HBF_4$-catalyzed direct arylation of polyfluoroarene with aryl halides.[24]

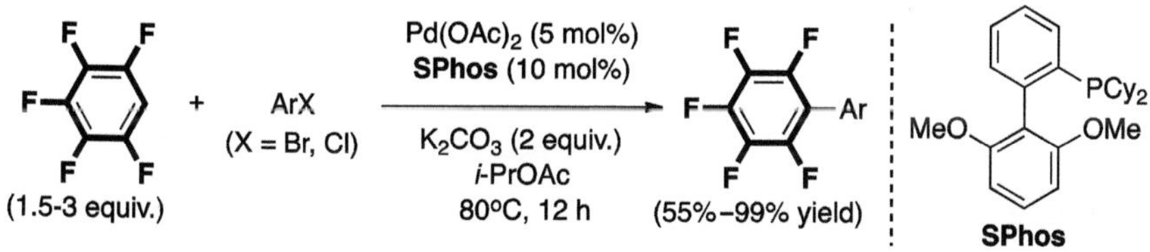

Scheme 3.11 $Pd(OAc)_2$/**SPhos**-catalyzed direct arylation of pentafluorobenzene with (hetero)aryl halides.[25]

Fagnou developed the direct arylation of electron-deficient polyfluoroarene with aryl halides.[24] By using $Pd(OAc)_2$/P*t*-$Bu_2Me \cdot HBF_4$ as the catalytic system, consisting of K_2CO_3 in DMA, the reaction proceeded smoothly and exhibited a broad scope with regard to both aryl halides and polyfluoroarenes (Scheme 3.10). Apart from (hetero)aryl bromides, aryl chlorides and iodides could also be used. Moreover, tetrafluoro-, trifluoro-, difluoro-, and even fluorobenzene could be selectively arylated, although they have more than one reaction site for mono-, di- and triarylation to occur, which could be minimized by introducing a slightly excess amount of the fluoroarene.

Fagnou subsequently described the direct arylation of pentafluorobenzene with sterically aryl halides in relatively milder reaction conditions.[25] The system of $Pd(OAc)_2$/**SPhos** was very effective for the direct arylation of pentafluorobenzene with a wide range of sterically hindered (hetero) aryl bromides and chlorides (Scheme 3.11). Later on, this protocol was applied by Matsuo's group for transformations of aryl-polyfluoro-substituted tetracenes, which is applied in organic semi-conductors.[26]

René and Fagnou developed new mild biphasic conditions for the Pd $(OAc)_2$/**MePhos**-catalyzed direct arylation of polyfluoroarenes with (hetero)aryl iodides at room temperature (Scheme 3.12).[27] With the advantage of biphasic conditions, employing an immiscible mixture of water and EtOAc allowed all components of the system to be solubilized

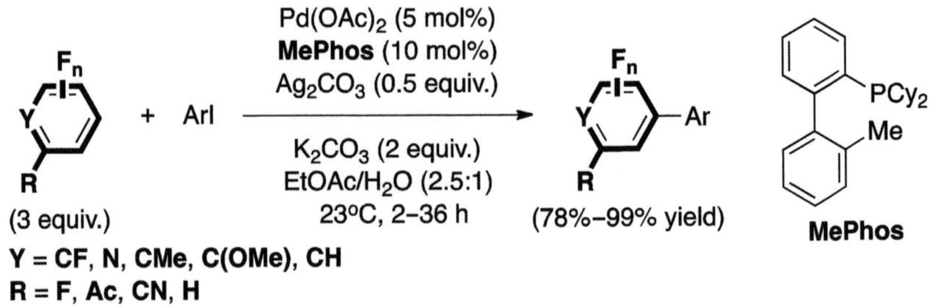

Scheme 3.12 Pd(OAc)$_2$/**MePhos**-catalyzed direct arylation of polyfluoroarenes with (hetero)aryl iodides in biphasic conditions at room temperature.[27]

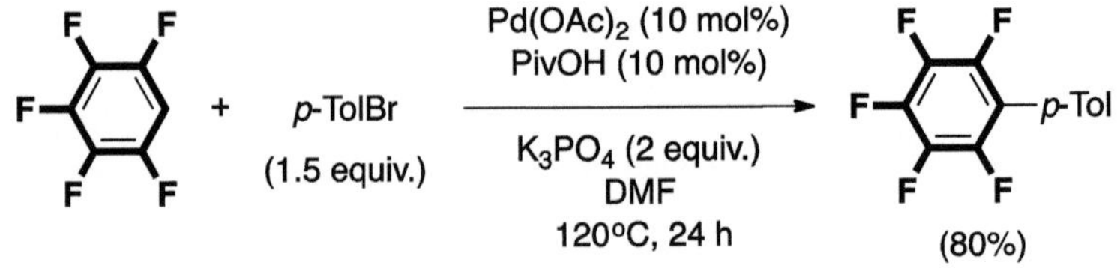

Scheme 3.13 Pd(OAc)$_2$-catalyzed direct arylation of pentafluorobenzene with *p*-bromotoluene.[28]

completely. This system was effective for a variety of polyfluoroarenes and aryl iodides as the coupling partners, and excellent yields were also obtained.

You and co-workers described an example of direct arylation of pentafluorobenzene with 4-bromotoluene via Pd(OAc)$_2$-catalyzed C–H functionalization process using PivOH as a co-catalyst (Scheme 3.13).[28] Particularly noteworthy was that this protocol was highly effective toward the direct arylation of azoles with aryl bromides (see Scheme 3.191).

Zhang reported a Pd(OAc)$_2$-catalyzed direct arylation of polyfluoroarenes with (hetero)aryl iodides.[29] In this protocol, inexpensive PPh$_3$ ligand, environmentally benign water as reaction medium, and mild reaction conditions (70°C) were applied to allow high reaction efficiency and excellent functional group compatibility (Scheme 3.14).

Zhang et al. reported a ligand-free palladium/copper bimetallic catalyst system for the direct arylation of polyfluoroarenes with aryl iodides.[30] The reaction, in the presence of Pd(PPh$_3$)$_4$/CuI, *t*-BuOLi, and DMF, exhibited good functional group compatibility with respect to polyfluoroarenes and aryl iodides (Scheme 3.15).

Huynh synthesized two new palladium(II) complexes bearing *N*-heterocyclic carbenes (NHCs), one based on 1,2,3-triazolin-5-ylidene

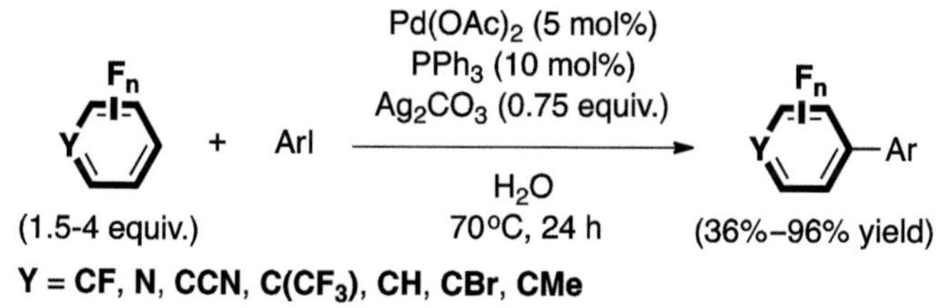

Scheme 3.14 $Pd(OAc)_2/PPh_3$-catalyzed direct arylation of polyfluoroarenes with (hetero)aryl iodides.[29]

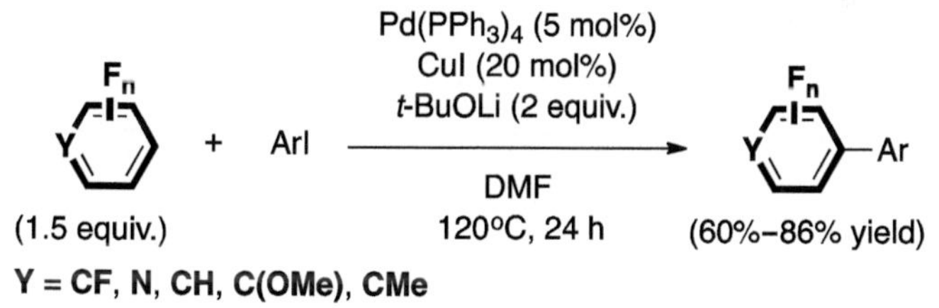

Scheme 3.15 $Pd(PPh_3)_4/CuI$-catalyzed direct arylation of polyfluoroarenes with aryl iodides.[30]

1,[31] another based on pyrazolin-5-ylidene **2**.[32] These catalysts showed good activity toward the direct arylation of pentafluorobenzene with various aryl bromides. In the presence of K_2CO_3 in DMA, this protocol gave corresponding biaryls in moderate-to-good yields (Scheme 3.16).

Later on, Huynh investigated another series of Pd(II) hetero-dicarbene complexes, in which the complex **3**, carrying the mesitylimidazolin-2-ylidene unit, proved to be the most active in the direct arylation of pentafluorobenzene.[33] The challenging 1-bromo-4-chlorobenzene was chosen to highlight the differences in the activation of Br versus Cl in the reaction (Scheme 3.17). Complex **3** promoted the Br-activated monocoupled product in a 64% yield, while the doubly coupled product was given in only a 5% yield.

Liu, Dai and co-workers developed a general approach to afford polyfluorobiphenyls by $Pd(OAc)_2/PCy_3$-catalyzed direct arylation of polyfluorobenzenes with aryl halides in the presence of Cs_2CO_3 in toluene.[34] This system gave the products from aryl bromides and chlorides in good-to-excellent yields at 80°C and 110°C, respectively (Scheme 3.18). The polyfluorobiphenyl with long alkyl substituents could also be prepared.

Metal nanocrystals (NCs) have shown excellent catalytic activities toward photocatalysis and organic synthesis because of their high surface-to-volume ratio and quantum size effects.[35] Cao synthesized palladium

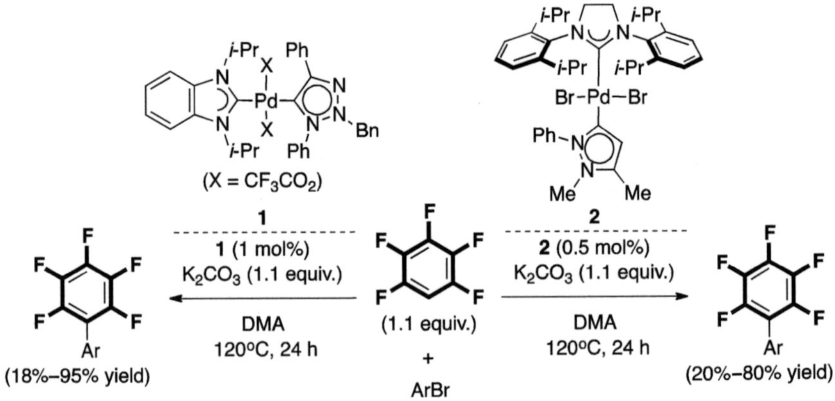

Scheme 3.16 Pd(II) complex-catalyzed direct arylation of pentafluorobenzene with aryl bromides.[31,32]

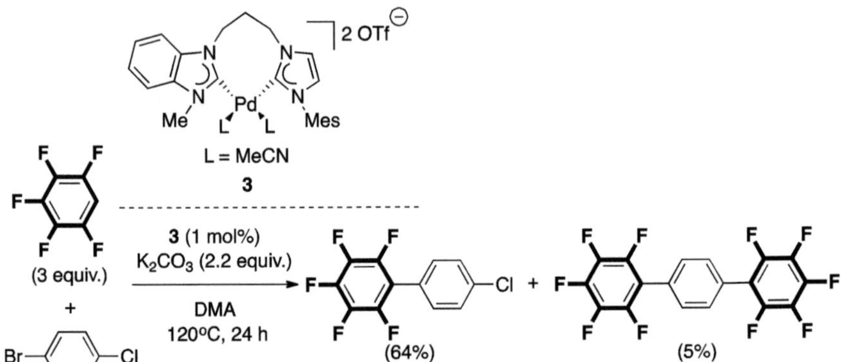

Scheme 3.17 Pd(II) hetero-dicarbene complex-catalyzed direct arylation of pentafluorobenzene with 1-bromo-4-chlorobenzene.[33]

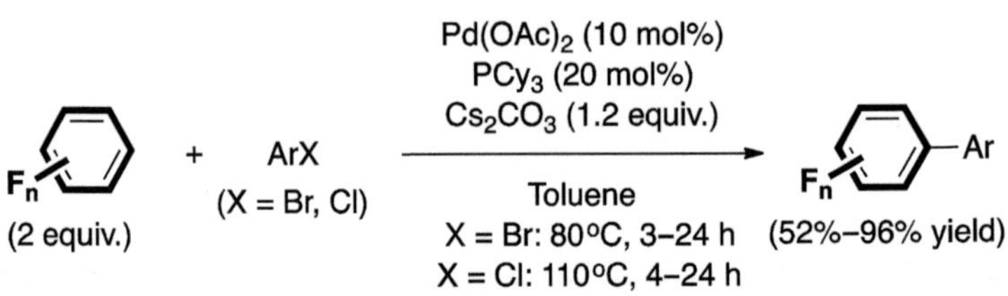

Scheme 3.18 $Pd(OAc)_2/PCy_3$-catalyzed direct arylation of polyfluorobenzenes with aryl bromides/chlorides.[34]

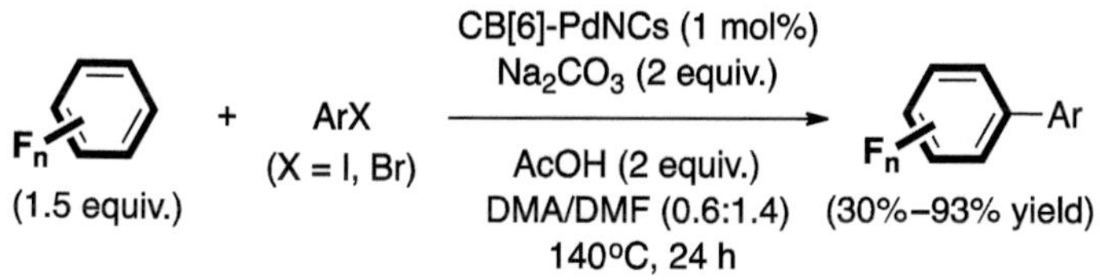

Scheme 3.19 CB[6]-PdNCs-catalyzed direct arylation of polyfluorobenzenes with aryl iodides/bromides.[36]

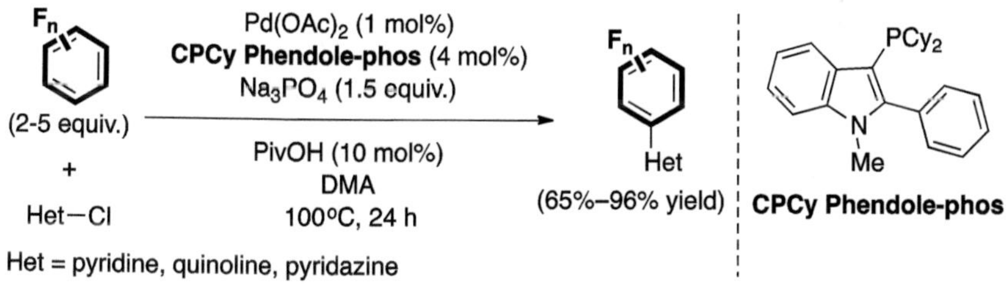

Scheme 3.20 $Pd(OAc)_2$/**CPCy Phendole-phos**-catalyzed direct arylation of polyfluoroarenes with heteroaryl chlorides.[38]

nanocrystals (PdNCs) by using cucurbit[6]uril (CB[6]) as a stabilizer.[36] By employing CB[6]-PdNCs as the catalyst, the direct arylation of polyfluorobenzenes with aryl halides proceeded efficiently (Scheme 3.19). The system exhibited excellent air and water stability, as well as good recyclability. Pentafluorobenzene could be arylated by both aryl iodides and bromides, tetra-, and trifluorobenzenes were proved to be efficient for the arylation with aryl bromides. Remarkably, the CB[6]-PdNCs catalyst could also be recycled five times without significant loss in catalytic activity.

Very recently, Kwong reported the first direct arylation of polyfluoroarenes with heteroaryl chlorides, especially 2-chloropyridine which is the difficult coupling partner,[37] by using a combination of $Pd(OAc)_2$ and indolylphosphine ligand **CPCy Phendole-phos**, as the catalytic system (Scheme 3.20).[38] A wide range of heteroaryl chlorides was an effective coupling partner and afforded corresponding products in good yields. Many new compounds were synthesized, which implied a lack of current effective methods for accessing pyridyl/polyfluoro biaryl compounds.

Apart from the direct arylation reaction with aryl iodides, bromides, and chlorides, Fu and Cai reported a $Pd(OAc)_2$-catalyzed direct arylation of polyfluoroarenes with iodobenzene diacetates.[39] This system, consisting of Ag_2CO_3 in DMSO/DMF solvent mixture, displayed a good functional

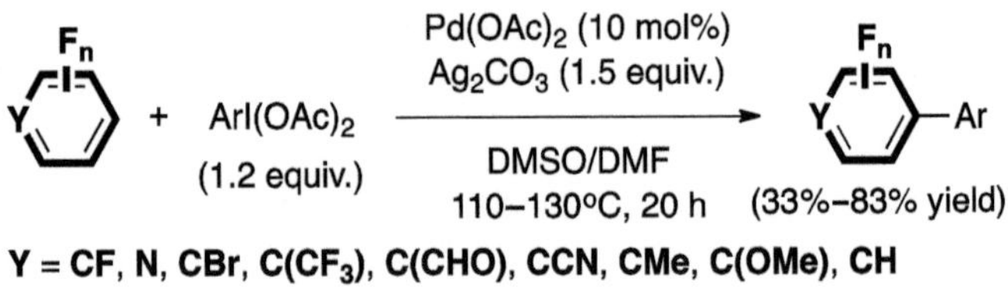

Scheme 3.21 $Pd(OAc)_2$-catalyzed direct arylation of polyfluoroarenes with iodobenzene diacetates.[39]

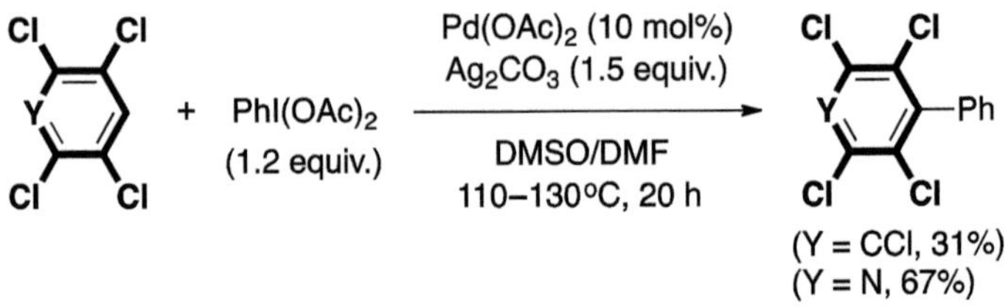

Scheme 3.22 $Pd(OAc)_2$-catalyzed direct arylation of polychloroarenes with iodobenzene diacetates.[39]

Scheme 3.23 $PdCl_2$-catalyzed direct arylation of difluorobenzenes bearing benzoxazole or benzothiazole moieties with (hetero)aryl bromides.[40]

group tolerance and provided the desired products in good yields (Scheme 3.21). Mono- and diarylation of tetrafluorobenzene were observed. Moreover, the authors also showed examples of the direct arylation of polychloroarenes (Scheme 3.22). Lower yields were observed when compared to that of polyfluoroarenes, in which it could be suggested that the C—H bond in polychloroarenes was less reactive, probably due to its lowered acidity.

Ammar, Soulé, and Doucet reported the direct arylation of difluorobenzenes and dichlorobenzenes bearing benzoxazole or benzothiazole moieties (Scheme 3.23).[40] With a variety of (hetero)aryl bromides, the reaction proceeded nicely using $PdCl_2$, and PivOK in DMA as the

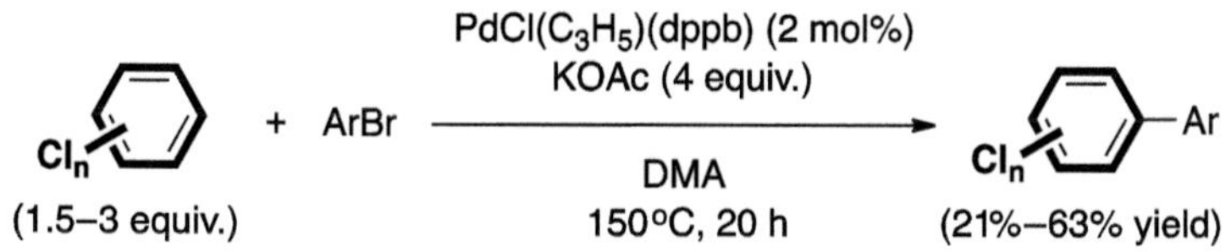

Scheme 3.24 $PdCl(C_3H_5)(dppb)/KOAc$-catalyzed direct arylation of polychlorobenzenes with (hetero)aryl bromides.[41]

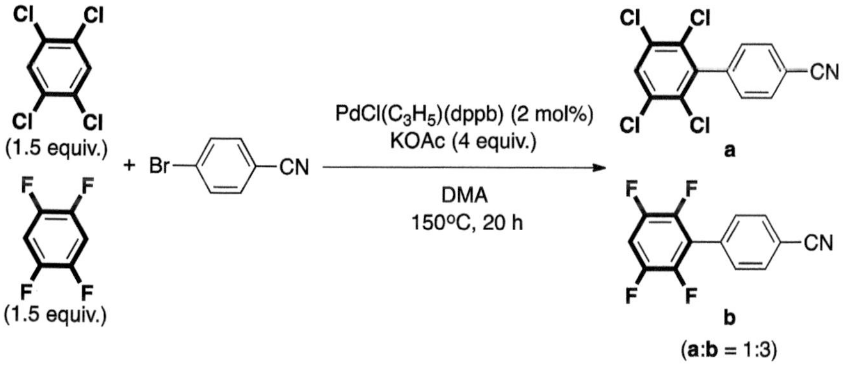

Scheme 3.25 Competitive $PdCl(C_3H_5)(dppb)/KOAc$-catalyzed direct arylation of 1,2,4,5-tetrachlorobenzene and 1,2,4,5-tetrafluorobenzene.[41]

catalytic system to afford the desired products in moderate-to-good yields. Indeed, the arylation occurred at the most acidic position between the C–H bond flanked by the two fluorine or chlorine atoms. However, no reaction occurred for the dichlorobenzene bearing benzothiazole moiety.

Besides polyfluoroarenes, Doucet studied the reactivity of polychloroarenes in direct arylation with (hetero)aryl bromides.[41] The $PdCl(C_3H_5)$(dppb)/KOAc system was found to promote the direct arylation with polychloroarenes (Scheme 3.24). In order to gain insight into the reactivity of polychloroarenes versus polyfluoroarenes, a competitive reaction was performed by using equimolar 1,2,4,5-tetrachlorobenzene and 1,2,4,5-tetrafluorobenzene with 4-bromobenzonitrile, and the result showed that the reactivity of chlorobenzene was lower than that of fluorobenzene (Scheme 3.25). This finding can be explained by Gorelsky's calculation of Gibbs free energies of activation for direct arylation via the concerted metalation deprotonation (CMD) pathway of polychloro- and polyfluorobenzenes (Fig. 3.1).[42]

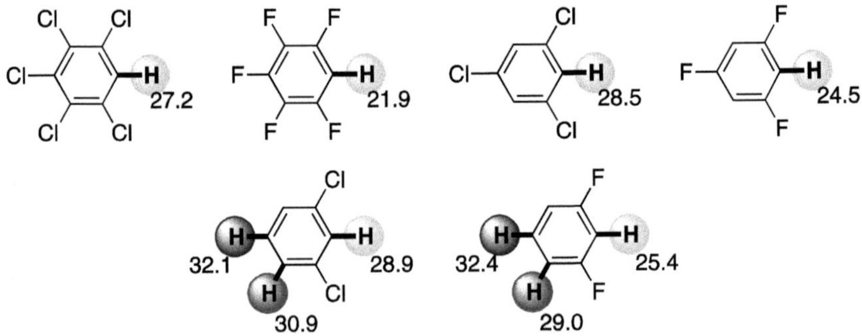

Figure 3.1 Gibbs free energies of activation ($\Delta G_{298\ K}$) of the cleavage of C–H bonds for different polyhaloarenes in the CMD pathway using $[Pd(C_6H_5)(PMe_3)(OAc)]$.[42c]

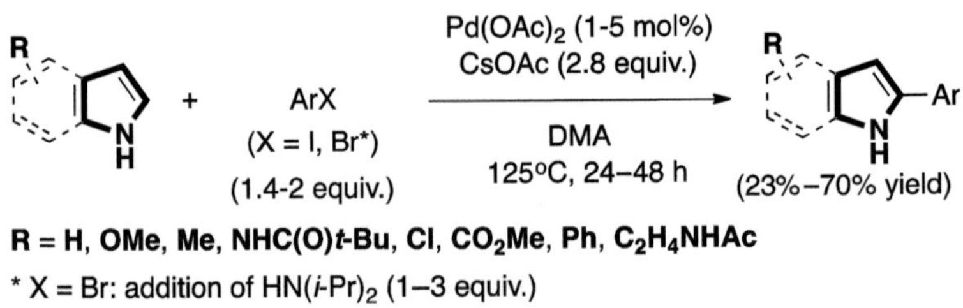

Scheme 3.26 $Pd(OAc)_2$-catalyzed direct arylation of free (NH)-indoles or pyrroles with aryl iodides/bromides.[46]

3.2.2 Pd-catalyzed C–H/C–X Cross-Coupling Reactions Between a Heteroarene and an Aryl Halide

In most cases, heteroarenes were successfully direct arylated with aryl halides, and their related reactions will herein be classified according to the type of heteroarenes.[8a]

3.2.2.1 Pyrroles and Indoles

The arylation of pyrroles or indoles is an important field of research in organic synthesis,[43] due to their biological properties and applications in medicinal agents[44] and natural products.[45]

A phosphine-free $Pd(OAc)_2$-catalyzed protocol for direct arylation of free (NH)-indoles and pyrroles with aryl iodides and bromides was disclosed by Sames and co-workers.[46] In the presence of CsOAc as the base in DMA, the arylation of indoles and pyrroles underwent regioselectively at the C-2-position. The addition of $HN(i\text{-}Pr)_2$, in the arylation with aryl bromides improved the product yields without affecting the regioselectivity (Scheme 3.26).

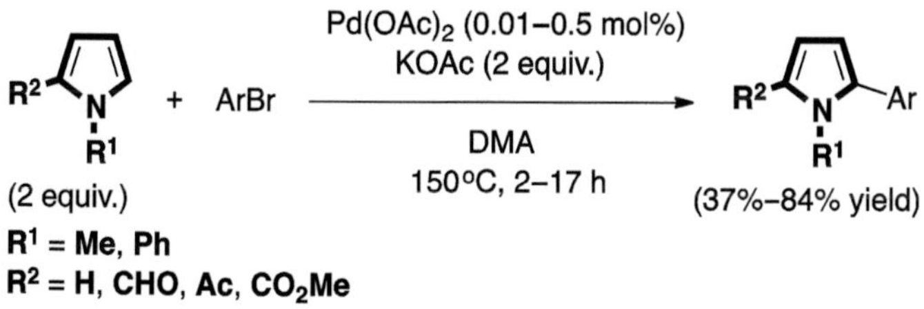

Scheme 3.27 $Pd(OAc)_2$/KOAc-catalyzed direct C-2- or C-5-arylation of pyrroles with (hetero)aryl bromides.[47]

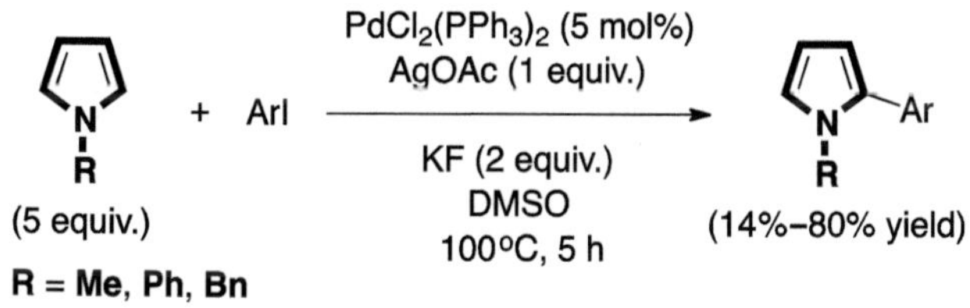

Scheme 3.28 $PdCl_2(PPh_3)_2$-catalyzed direct C-2-arylation of pyrroles with aryl iodides.[48]

Doucet reported a simple method for the regioselective C-2- or C-5-arylation of pyrroles through $Pd(OAc)_2$-catalyzed C—H bond functionalization.[47] Notably, only 0.01—0.5 mol% of $Pd(OAc)_2$ was employed in this reaction (Scheme 3.27). The presence of electron-withdrawing substituents at the C-2-position of pyrrole and broad variety functional groups on (hetero)aryl bromides were well tolerated in this catalytic system.

Meanwhile, Gryko developed a protocol for accessing 2-arylated pyrroles via $PdCl_2(PPh_3)_2$-catalyzed direct arylation from aryl iodides.[48] In the presence of AgOAc as the additive, aryl iodides with various substituents were reacted with *N*-methyl, *N*-phenyl, and *N*-benzyl pyrroles leading to desired products in moderate-to-good yields (Scheme 3.28). It should be noted that the use of anhydrous DMSO gave a crucial improvement in this protocol.

Direct functionalization of pyrrole through regioselective C—H bond functionalization in the presence of a reactive N—H functionality provides a cost-effective, environmentally attractive, and atom-economical entry because of the elimination of the need for introducing protecting groups. Jafarpour designed a straightforward route to functionalize free (NH)-pyrrole via palladium-catalyzed direct arylation with aryl iodides.[49] A number of aryl iodides could be coupled with both free (NH)-pyrrole and *N*-protected pyrrole regioselectively at the C-2-position with Pd$(OH)_2$/C in triethanolamine (TEA) (Scheme 3.29).

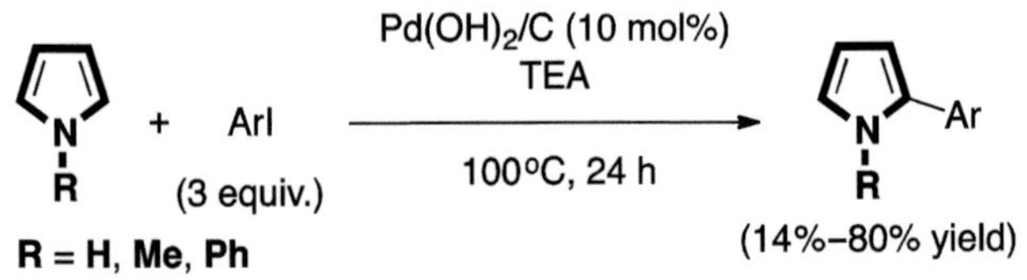

Scheme 3.29 $Pd(OH)_2/C$-catalyzed direct C-2-arylation of free (NH)-pyrrole or *N*-protected pyrrole with aryl iodides.[49]

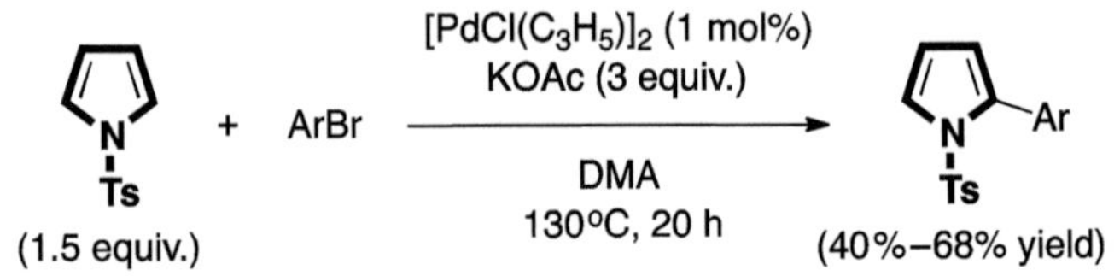

Scheme 3.30 $[PdCl(C_3H_5)]_2$-catalyzed direct C-2-arylation of *N*-tosylpyrrole with (hetero)aryl bromides.[50]

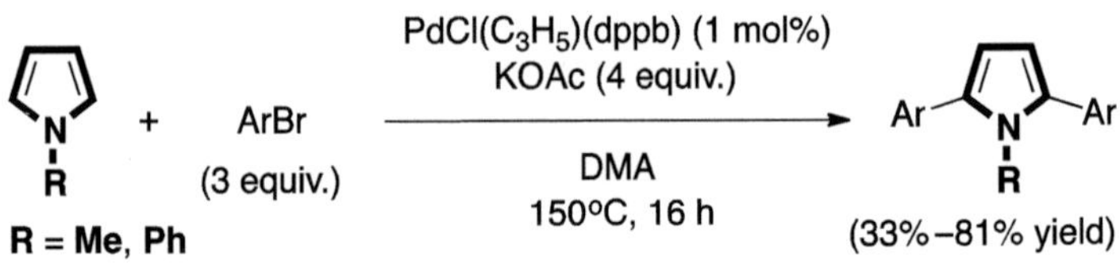

Scheme 3.31 $PdCl(C_3H_5)(dppb)/KOAc$-catalyzed direct 2,5-diarylation of *N*-methylpyrrole or *N*-phenylpyrrole with (hetero)aryl bromides.[51]

Doucet reported the $[PdCl(C_3H_5)]_2$-catalyzed direct coupling of pyrrole with (hetero)aryl bromides.[50] A range of (hetero)aryl bromides undergo regioselective coupling at the C-2-position of *N*-tosylpyrrole in moderate-to-good yields using KOAc as the base in DMA (Scheme 3.30). It was notable that the 2-arylated *N*-tosylpyrrole underwent deprotection easily in methanol using NaOH as the base, affording the free (NH)-2-arylpyrrole.

Pyrrole has four reaction sites in direct arylation reaction, the C-2- and C-5-position are generally more reactive for C—H functionalization than the C-3- and C-4-position. Doucet carried out the study of palladium-catalyzed direct polyarylation of *N*-methylpyrrole and *N*-phenylpyrrole.[51] By employing the same protocol, yet using $PdCl(C_3H_5)$ (dppb)/KOAc as the catalyst, and increasing the (hetero)aryl bromides to 3 equivalents, symmetrical 2,5-diarylated pyrroles could be obtained in moderate-to-good yields (Scheme 3.31). Moreover, tetraarylated pyrrole could also be accessed. The authors employed 4 equivalents of 3,5-bis

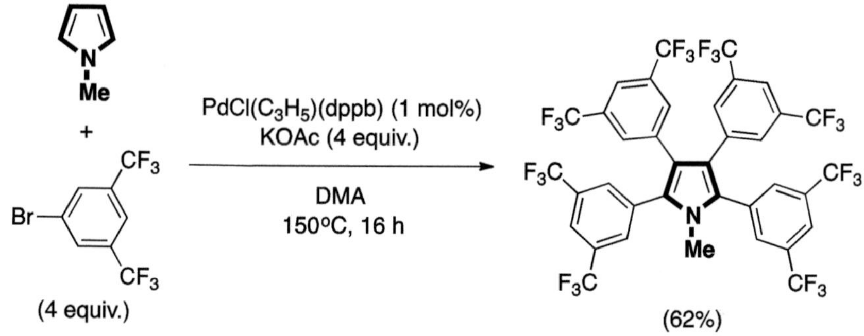

Scheme 3.32 $PdCl(C_3H_5)(dppb)$/KOAc-catalyzed direct tetraarylation of 1-methylpyrrole with 3,5-bis(trifluoromethyl)bromobenzene.[51]

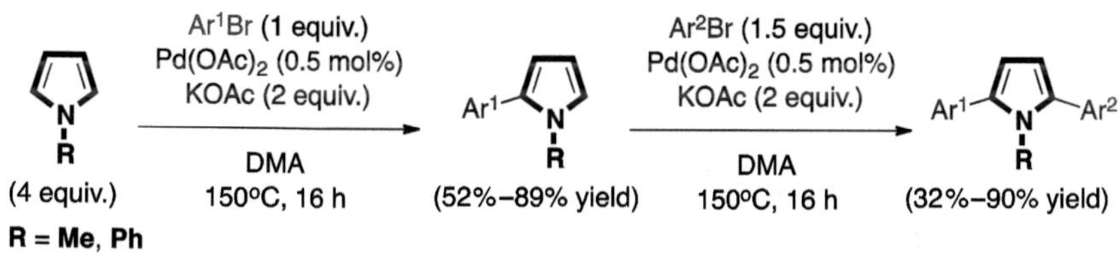

Scheme 3.33 $Pd(OAc)_2$/KOAc-catalyzed direct sequential 2,5-diarylation of 1-methylpyrrole or 1-phenylpyrrole with aryl bromides.[51]

(trifluoromethyl)bromobenzene to obtain the 2,3,4,5-tetraarylated pyrrole as the major product in 62% yield (Scheme 3.32). Particularly noteworthy was that unsymmetrically 2,5-diarylated pyrroles could also be obtained through sequential C-2-arylation followed by C-5-arylation (Scheme 3.33). For both steps, $Pd(OAc)_2$ was employed because the authors previously observed that such system allowed efficient direct monoarylation of pyrroles (see Scheme 3.27).[47]

Li and Doucet further developed the sequential direct C-2- and C-5-arylation for accessing unsymmetrical pyrroles.[52] Herein, 3-acetylpyrrole and (hetero)aryl bromides were used as the coupling partners, and it was found that the system comprising $PdCl(C_3H_5)(dppb)$/KOAc in DMA was effective for the direct regioselective C-5-arylation. Under the same conditions, the sequential direct C-5-arylation followed by C-2-arylation allowed the synthesis of unsymmetrical 2,5-diaryl-3-acetylpyrrole in high yields (Scheme 3.34).

Ammar and Doucet further explored the reactivity of a pyrazole bearing a pyrrole substituent at the C-5-position and an ester at the

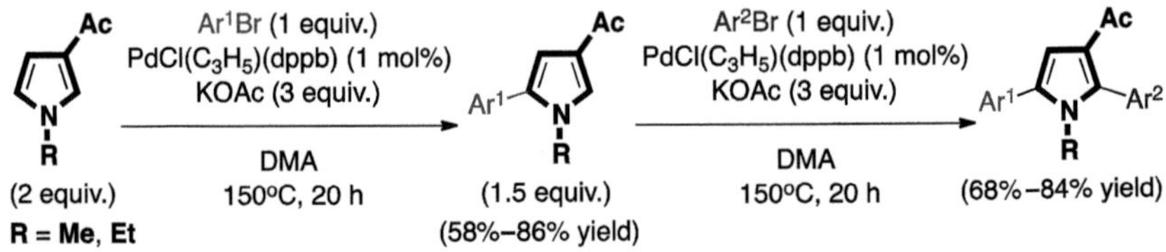

Scheme 3.34 $PdCl(C_3H_5)(dppb)/KOAc$-catalyzed direct sequential 2,5-diarylation of 3-acetylpyrrole with (hetero)aryl bromides.[52]

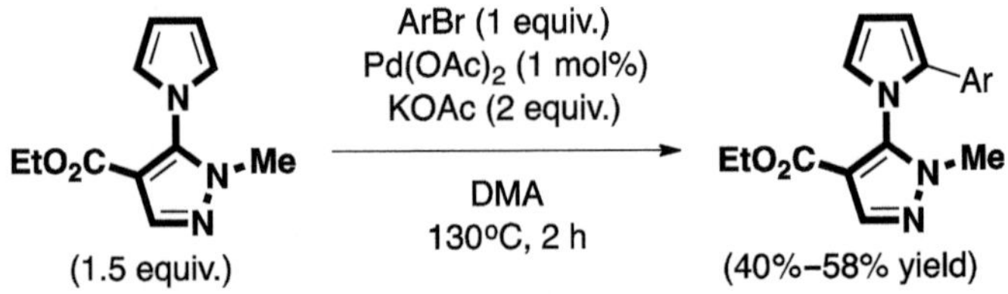

Scheme 3.35 $Pd(OAc)_2/KOAc$-catalyzed direct arylation of pyrrole bearing pyrazole moiety with (hetero)aryl bromides.[53]

C-4-position using the $Pd(OAc)_2/KOAc$ system (Scheme 3.35).[53] Since the C-3-position of pyrazole generally exhibits a poor reactivity, the arylation took place at the C-2-position of pyrrole as was expected.

A set of *N*-heterocyclic carbene-palladium catalysts (Pd-NHC), **4–12**, was synthesized (Fig. 3.2) and employed for the direct arylation of pyrrole derivatives using aryl chlorides as the coupling partners.[54] This set of Pd–NHC catalysts promoted the C-5-arylation of *N*-methylpyrrole-2-carboxaldehyde with 4-chlorobenzonitrile in good yield (Scheme 3.36). Notably, these were the first examples of direct arylation of pyrroles by using aryl chlorides as the coupling partners and Pd–NHC complexes as the catalyst.

Huynh later synthesized two series of Pd – NHC complexes, **13–24** (Fig. 3.3).[55] Under the same reaction conditions as Özdemir and Doucet's report,[54] both series of complexes were found to be catalytically active in the direct arylation of *N*-methylpyrrole with 4-bromoacetophenone to afford the corresponding product in good yield (Scheme 3.37).

Larrosa reported a phosphine-free protocol that allowed the direct C-2-arylation of indoles with aryl iodides at room temperature for the first time.[56] These mild conditions, comprising of Ag_2O, *o*-nitrobenzoic acid, and DMA, permitted a broad set of both indoles and aryl iodides to be coupled in excellent yields (Scheme 3.38).

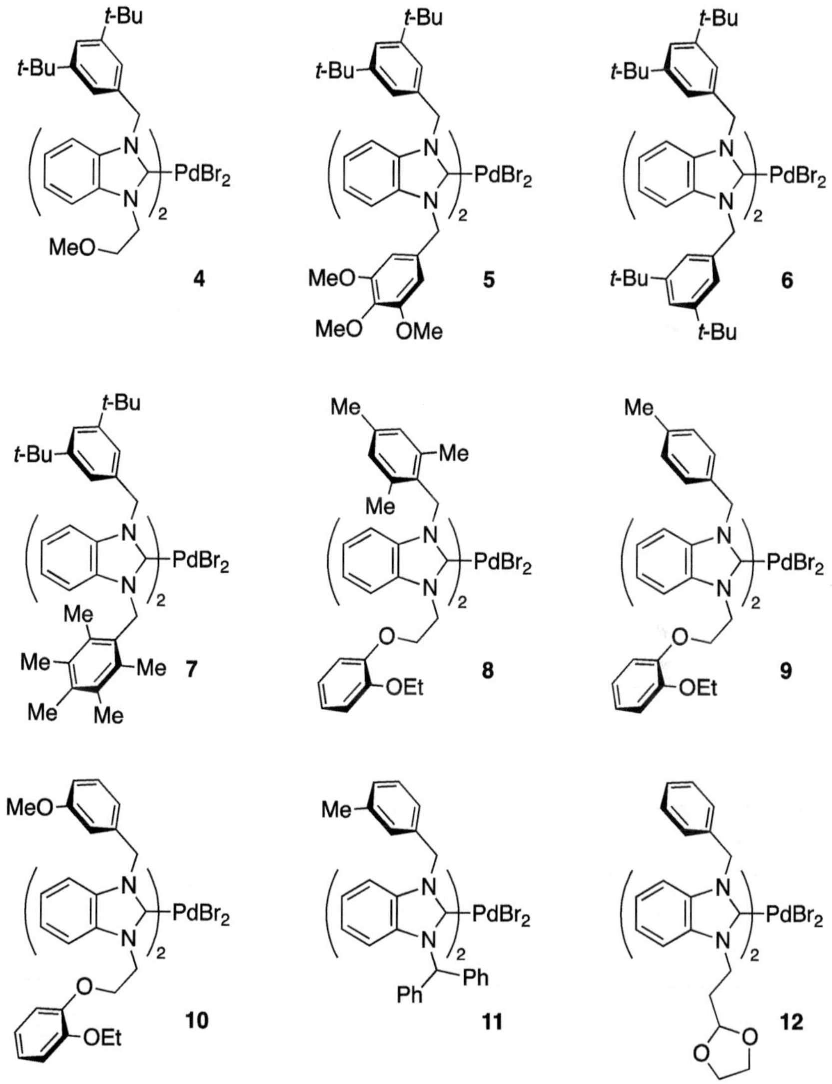

Figure 3.2 Pd—NHC complexes **4—12**.[54]

Apart from the C-2-arylation of indoles, many research groups focus their investigations on the arylation of indoles at the C-3-position. Djakovitch reported a heterogeneous palladium catalyst [Pd]/SBA-15 (mesostructured porous silica) which was applicable to the arylation of 2-phenylindole with aryl bromides, yet over long reaction times and producing low yields (Scheme 3.39).[57] Noteworthy, the C-3-arylation of 2-phenylindole could also achieved by the use of $Pd(OAc)_2/AgBF_4$ as the

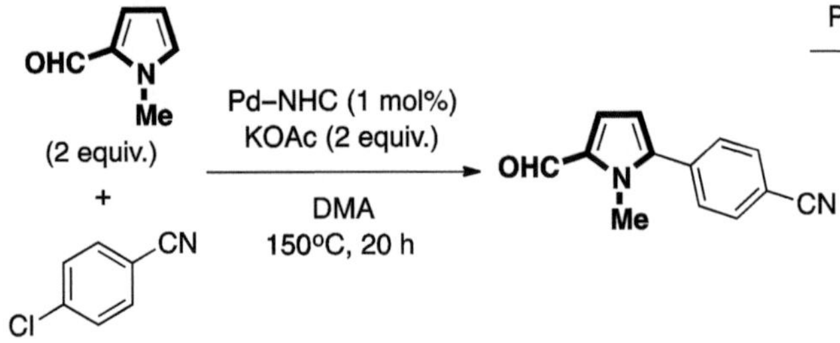

Pd–NHC	Yield (%)
4	57
5	87
6	83
7	61
8	57
9	36
10	37
11	72
12	65

Scheme 3.36 Pd–NHC-catalyzed direct C-5-arylation of *N*-methylpyrrole-2-carboxaldehyde with 4-chlorobenzonitrile.[54]

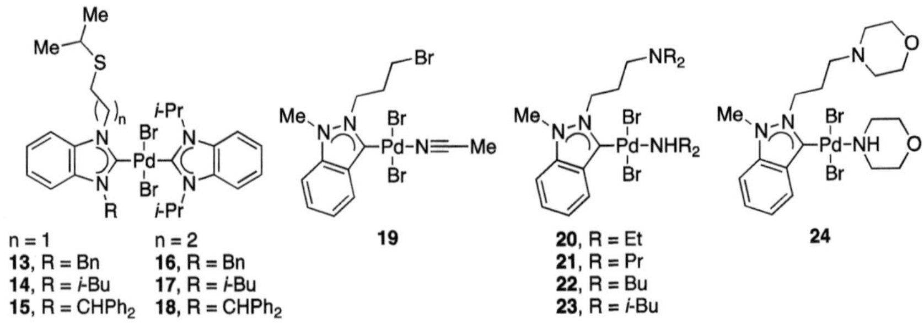

Figure 3.3 Pd–NHC complexes **13–24**.[55]

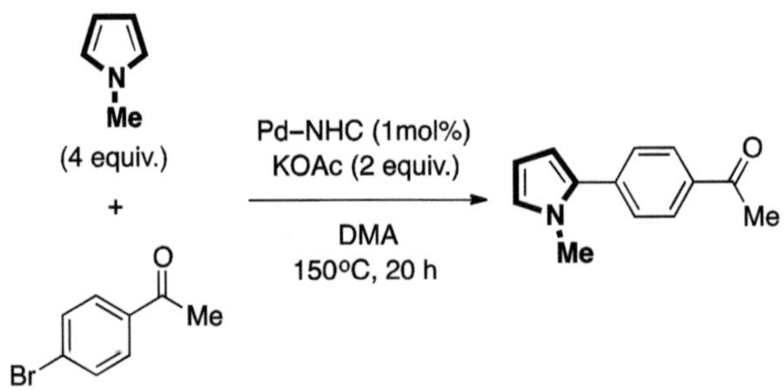

Pd–NHC	Yield (%)
13	74
14	77
15	83
16	71
17	66
18	86
19	85
20	97
21	88
22	87
23	84
24	88

Scheme 3.37 Pd–NHC-catalyzed direct arylation of *N*-methylpyrrole with 4-bromoacetophenone.[55]

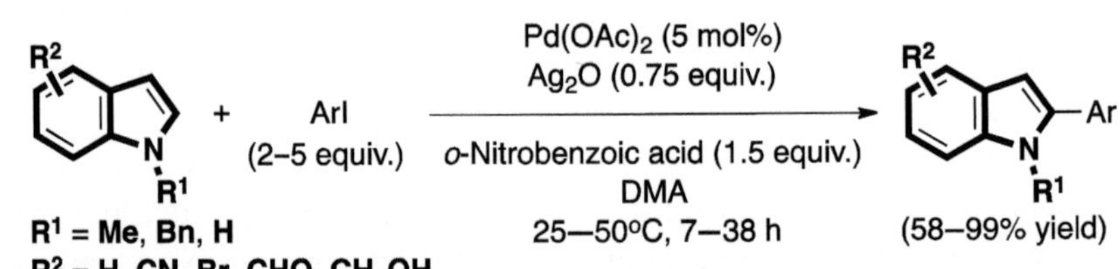

Scheme 3.38 $Pd(OAc)_2$-catalyzed direct arylation of indoles with aryl iodides at room temperature.[56]

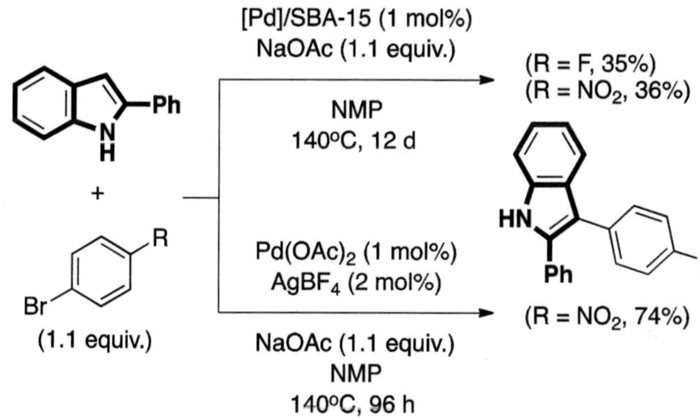

Scheme 3.39 Palladium-catalyzed direct arylation of 2-phenylindole with aryl bromides.[57,58]

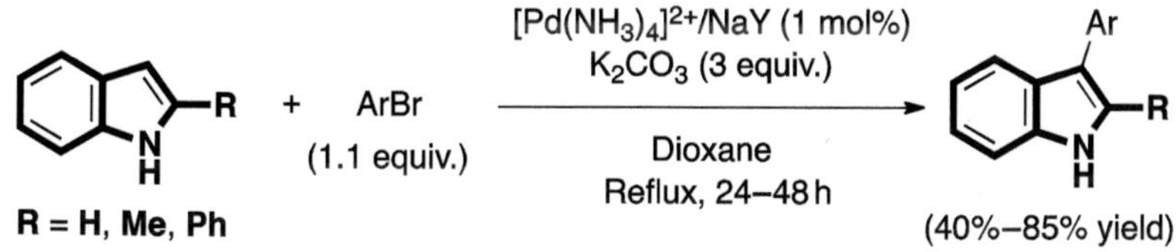

Scheme 3.40 $[Pd(NH_3)_4]^{2+}$/NaY-catalyzed direct arylation of indoles with aryl bromides.[59]

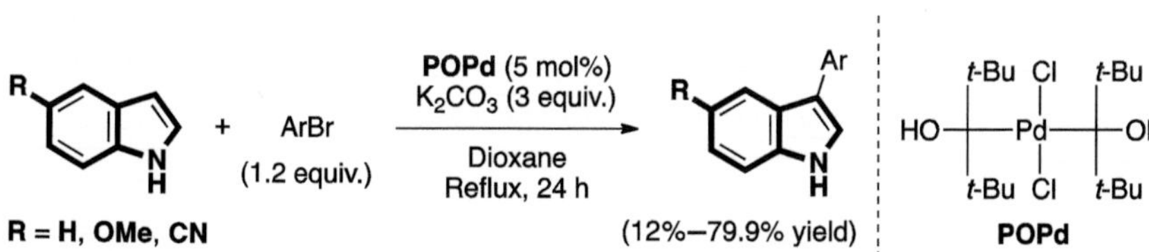

Scheme 3.41 **POPd**-catalyzed direct arylation of free (NH)-indole with aryl bromides.[60]

catalytic system (Scheme 3.39).[58] However, these protocols were limited to the use of activated aryl bromides. Djakovitch subsequently disclosed another heterogeneously palladium-catalyzed protocol for the C-3-arylation of indoles.[59] Under the $[Pd(NH_3)_4]^{2+}$/NaY catalytic system, moderate-to-good product yields were obtained (Scheme 3.40).

Meanwhile, Zhang, and He developed an efficient protocol for C-3-arylation of free (NH)-indole.[60] The air-stable palladium-phosphinous acid complex $[(t\text{-Bu})_2P(OH)]_2\text{-}PdCl_2$ (**POPd**) was investigated and found to be the most efficient catalyst for the arylation of indole with various aryl bromides (Scheme 3.41).

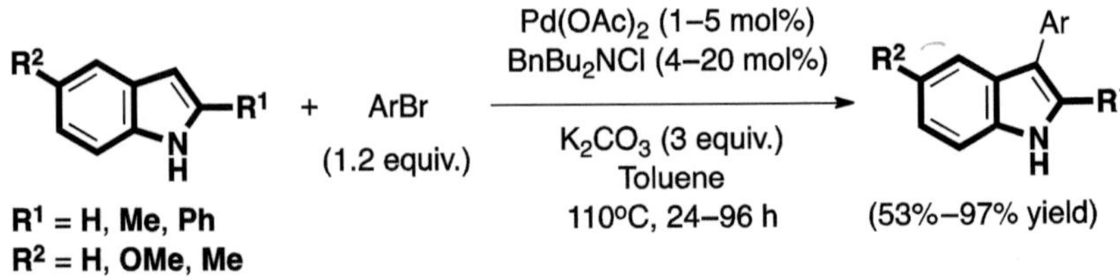

Scheme 3.42 Pd$(OAc)_2$-catalyzed direct arylation of free (NH)-indoles with aryl bromides.[61]

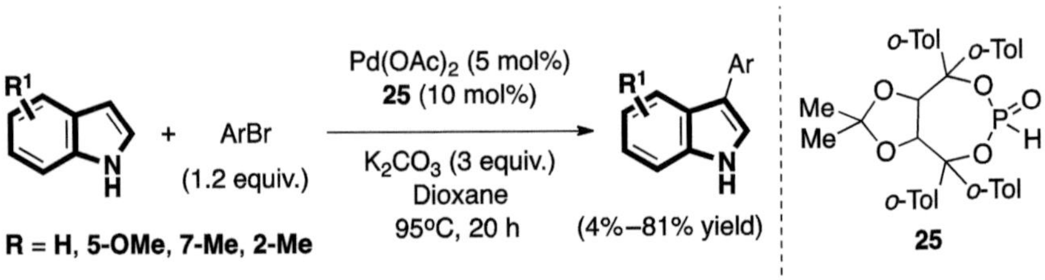

Scheme 3.43 Pd$(OAc)_2$/**25**-catalyzed direct arylation of free (NH)-indoles with aryl bromides.[63]

Since then, Bellina and Rossi started to improve the protocol developed by Zhang and He. An efficient and highly regioselective direct Pd $(OAc)_2$-catalyzed C-3-arylation of free (NH)-indoles under ligandless conditions was achieved.[61] This protocol allowed free (NH)-indoles to couple with a range of both activated and deactivated aryl bromides in refluxing toluene to furnish 3-arylindoles in good-to-excellent yields (Scheme 3.42).

Palladium complexes derived from secondary phosphine oxides (SPOs) were shown to enable highly regioselective direct C-3-arylation of indoles.[62] Ackermann et al. developed a highly efficient direct arylation of indoles which took place with high regioselectivity at the C-3-position with an in situ generated palladium complex **25** derived from an air-stable heteroatom-substituted phosphine oxide (HASPO).[63] This study revealed that a range of aryl bromides could couple with free (NH)-indoles smoothly in good yields, yet at lower yields with sterically hindered substrate (Scheme 3.43).

Lavilla described a new route for C-2-arylation of free (NH)-indoles in tryptophan core via Pd$(OAc)_2$-catalyzed C—H functionalization protocol with (hetero)aryl iodides.[64] It was found that the use of Pd$(OAc)_2$, $AgBF_4$, 2-nitrobenzoic acid in DMF resulted in an efficient arylation of the *N*-acetyltryptophan methyl ester (Ac-Trp-OMe) (Scheme 3.44).

Scheme 3.44 $Pd(OAc)_2$-catalyzed direct arylation of Ac-Trp-OMe with (hetero)aryl iodides.[64]

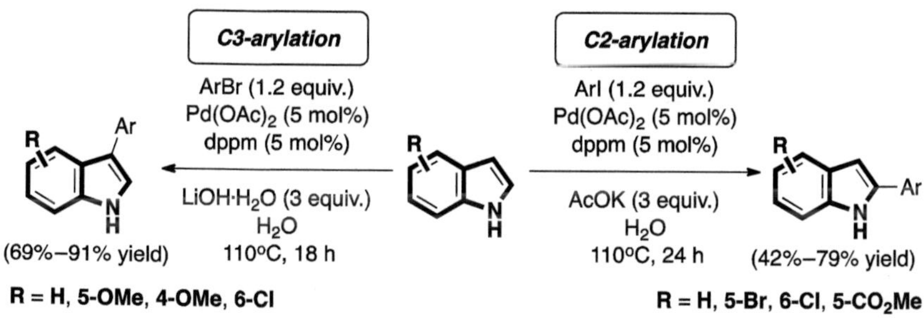

Scheme 3.45 Base/halide-controlled $Pd(OAc)_2$-catalyzed regioselective arylation of free (NH)-indoles.[65]

Remarkably, the use of microwave irradiation allowed the reaction to proceed rapidly without epimerization. This protocol guaranteed a convenient and general access to modify tryptophan peptides that provided useful applications in organic synthesis and medicinal chemistry.

Concurrently, Djakovitch developed a versatile protocol based on $Pd(OAc)_2$/dppm that worked "on water" to give site-selective C—H arylation of free (NH)-indoles.[65] This protocol allowed the preparation of either C-2- or C-3-arylindoles. The control of regioselectivity was achieved by changing the base and halide partners. When employing AcOK as the base and aryl iodides as the coupling partners, the desired C-2-arylindoles were afforded in good yields with high regioselectivity. On the other hand, having $LiOH \cdot H_2O$ as the base and aryl bromides as the coupling partners, this reaction provided C-3-arylindoles with both good yields and regioselectivity (Scheme 3.45).

Daugulis reported the use of a $Pd(OAc)_2$/P(*o*-biphenyl)Cy_2 catalytic system to promote the direct arylation of indoles, pyrroles, or furans with aryl chlorides.[66] A number of substituted indoles could be coupled with a variety of (hetero)aryl chlorides using Na_2CO_3 as the base in DMA (Scheme 3.46). 1,3-Disubstituted indoles reacted with (hetero)aryl

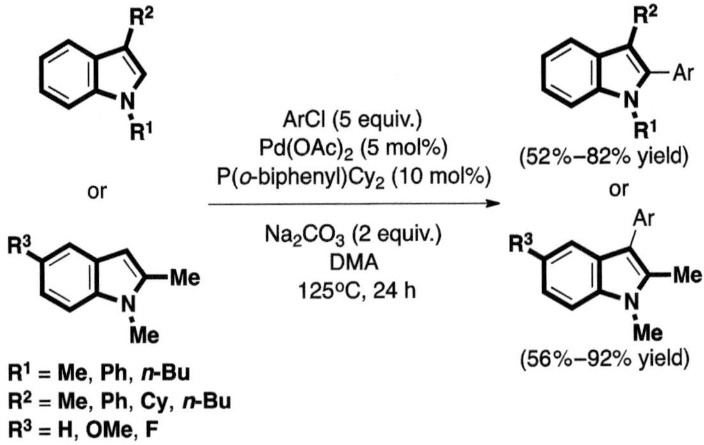

Scheme 3.46 $Pd(OAc)_2/P(o\text{-biphenyl})Cy_2$-catalyzed direct arylation of indoles with (hetero)aryl chlorides.[66]

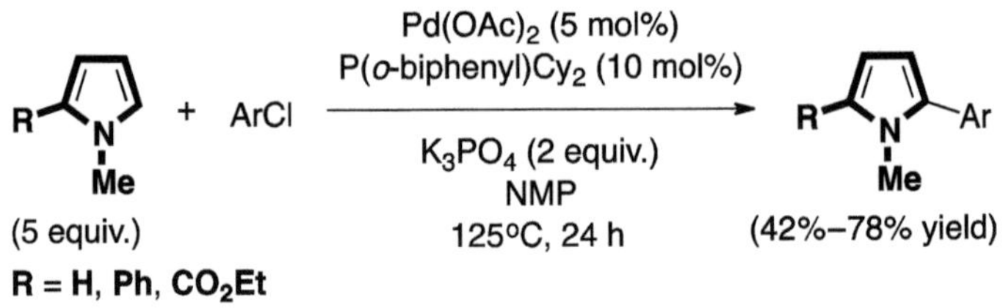

Scheme 3.47 $Pd(OAc)_2/P(o\text{-biphenyl})Cy_2$-catalyzed direct arylation of pyrroles with (hetero)aryl chlorides.[66]

chlorides to give desired C-2-arylated products, whereas 1,2-disubstituted indoles gave C-3-arylated structures. Arylation of pyrroles with (hetero)aryl chlorides was also examined by using K_3PO_4 in NMP (Scheme 3.47). This protocol allowed the arylation to take place at the α-position of pyrroles, and tolerated a range of functional groups. In addition, dichlorobenzene could react with pyrroles to form a triaryl moiety giving good yields. The arylation of furans will be discussed in the following section (see Scheme 3.125).

Lee developed a new palladium bis(carbene) complex **26** for the decarboxylative coupling of heteroaromatic carboxylic acids with aryl halides.[67] This complex could also catalyze the direct arylation of 1-methylindole with 4-bromoacetophenone that led to a mixture of C-2- and C-3-arylated indoles in 41% and 20% yields, respectively (Scheme 3.48).

Porphyrins, composing of four modified pyrrole subunits, are an important class of aromatic compounds which exhibit optical and electrochemical

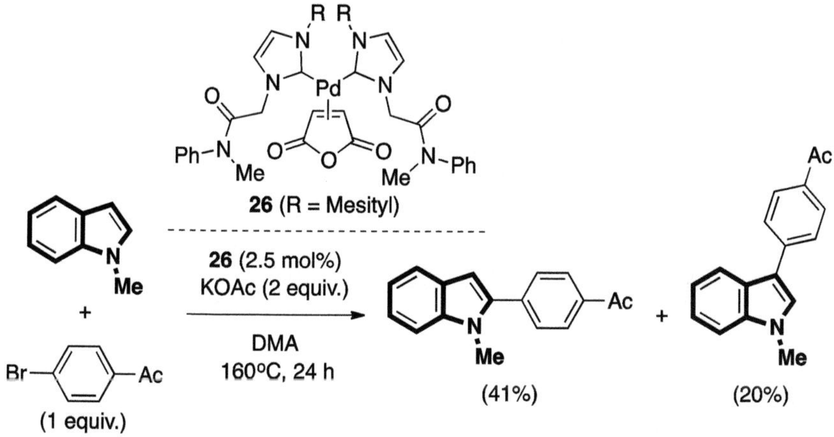

Scheme 3.48 Palladium bis(carbene) complex-catalyzed direct arylation of 1-methylindole with 4-bromoacetophenone.[67]

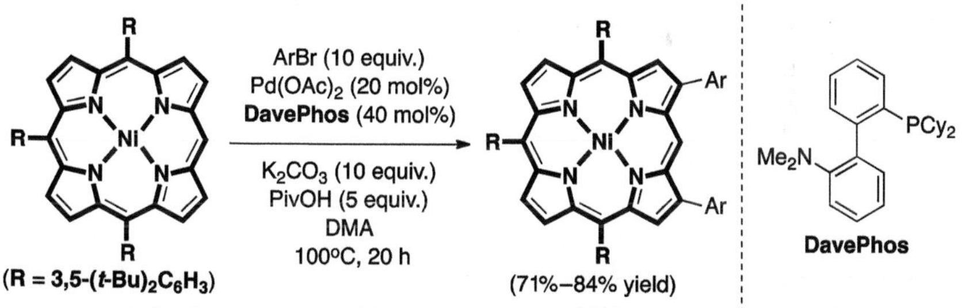

Scheme 3.49 Pd(OAc)$_2$/**DavePhos**-catalyzed dirarylation of nickel complex of 5,10,15-tris(3,5-di-*tert*-butylphenyl)porphyrin with aryl bromides.[68]

properties, and have attracted much attention as key components of functional materials and supramolecular architectures.[69] Yorimitsu and Osuka reported the use of Fagnou's Pd(OAc)$_2$/**DavePhos** catalytic system[14] (see Scheme 3.1) to catalyze the direct arylation of porphyrin with aryl bromides.[68] Employing a nickel complex of 5,10,15-tris(3,5-di-*tert*-butylphenyl)porphyrin and various aryl bromides as the coupling partners, the reaction proceeded exclusively at the β-positions with excellent regioselectivity to afford the diarylated products in high yields (Scheme 3.49).

Boron complexes of dipyrromethenes (BODIPY) are valuable fluorophores which display bright fluorescence, as well as robustness toward light and chemicals.[70] Dehaen investigated a new one-step protocol for synthesizing 3- and 3,5-diarylated BODIPY dyes via palladium-catalyzed

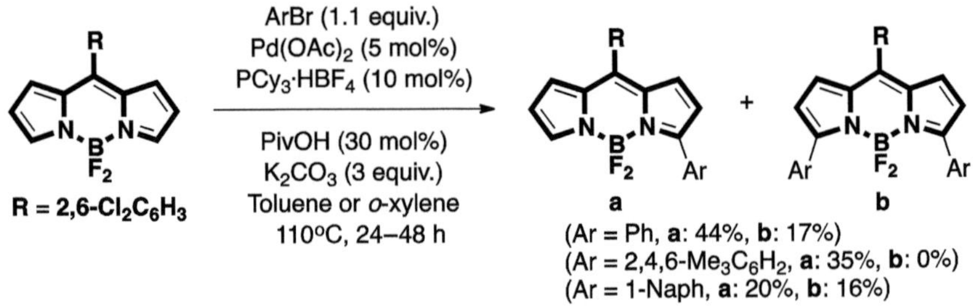

Scheme 3.50 Pd(OAc)$_2$/PCy$_3$ · HBF$_4$-catalyzed direct arylation of BODIPY dye with aryl bromides.[71]

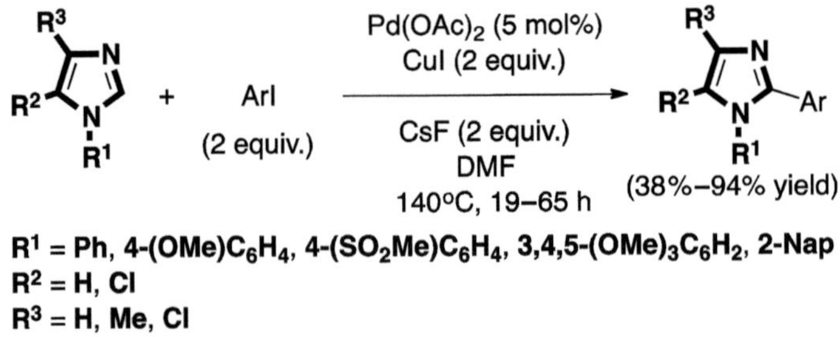

Scheme 3.51 Pd(OAc)$_2$/CuI-catalyzed direct C-2-arylation of *N*-arylimidazoles with aryl iodides/bromides.[77]

direct arylation reactions.[71] By introducing Fagnou's Pd(OAc)$_2$/PCy$_3$ · HBF$_4$ catalytic system[72] (see Schemes 3.83 and 3.178), and aryl bromides as the coupling partners, a mixture of 3- and 3,5-diarylated BODIPY dyes could be obtained (Scheme 3.50).

3.2.2.2 Imidazoles

Substituted imidazoles are important heteroarenes that are known to exhibit a broad range of biological activities such as antifungal properties,[73] analgesic activity,[74] and anti-inflammatory activity.[75] As a result, a number of methodologies for the synthesis of variously substituted imidazoles has been intensively developed.[76]

Bellina and Rossi found that 1,2-diarylimidazoles could be regioselectively synthesized in moderate-to-high yields by direct C-2-arylation of *N*-arylimidazoles with aryl iodides by using Pd(OAc)$_2$/CuI system (Scheme 3.51).[77] Notably, some imidazole derivatives prepared from this protocol have been found to exhibit significant cytotoxic activity against some human tumor cell lines.[78]

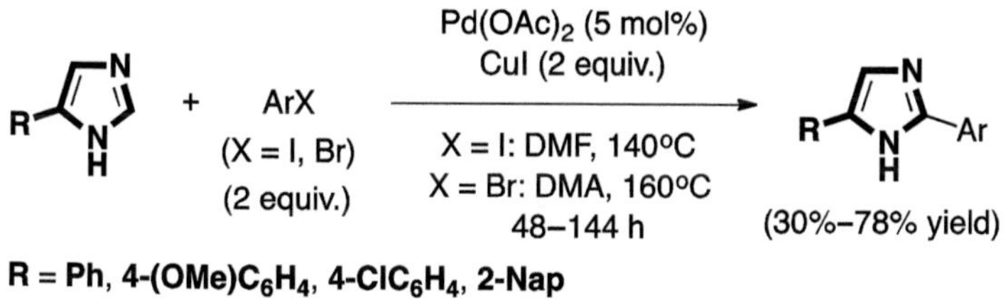

Scheme 3.52 $Pd(OAc)_2$/CuI-catalyzed direct C-2-arylation of 4(5)-aryl-1*H*-imidazoles with aryl iodides/bromides.[79]

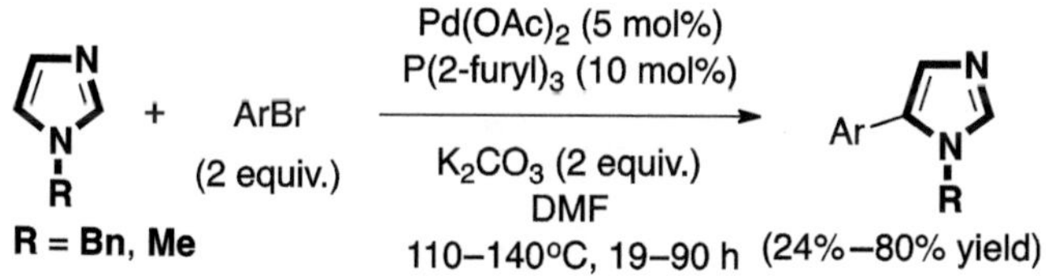

Scheme 3.53 $Pd(OAc)_2$/P(2-furyl)$_3$-catalyzed direct C-5-arylation of *N*-benzylimidazole or *N*-methylimidazole with (hetero)aryl bromides.[80,81]

The report from the same group described that 4(5)-aryl-1*H*-imidazoles could be efficiently and selectively prepared by $PdCl_2$(dppf)-catalyzed Suzuki—Miyaura reaction of 4(5)-bromo-1*H*-imidazole with arylboronic acids.[79] Then, by employing the 4(5)-aryl-1*H*-imidazoles for the direct C-2-arylation with aryl bromides or iodides under a base-free and ligandless $Pd(OAc)_2$/CuI system in DMF (for ArI) or DMA (for ArBr), 2,4(5)-diaryl-1*H*-imidazoles were regioselectively synthesized in moderate-to-good yields (Scheme 3.52). It should be noted that no *N*-arylation by-products were observed under these reaction conditions.

Subsequently, Bellina and Rossi developed the use of $Pd(OAc)_2$/P(2-furyl)$_3$ system for the direct C-5-arylation of *N*-benzylimidazole[80] or *N*-methylimidazole[81] with (hetero)aryl bromides. 5-Aryl-*N*-benzylimidazoles or 5-aryl-*N*-methylimidazoles were synthesized in moderate-to-good yields (Scheme 3.53). Again, by using the same approach as Scheme 3.52,[79] *N*-benzyl-2,5-diarylimidazoles could also be afforded. *N*-Debenzylation was successfully achieved to furnish the free (NH)-diarylated imidazoles by using palladium on charcoal.

Sames described a general and comprehensive protocol for the synthesis of complex aryl imidazoles, where all three C—H bonds of the imidazole could be arylated in a regioselective and sequential manner.[82] For the general reactivity trends of imidazole, the C-2-position bears the most acidic C—H bond, the C-5-position shows high reactivity toward

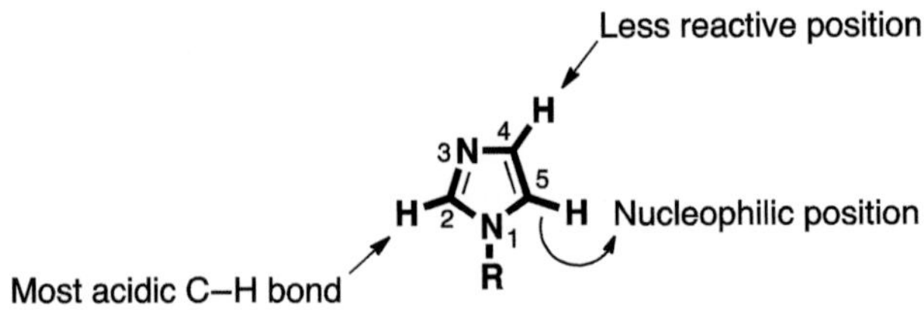

Figure 3.4 General reactivity properties of imidazole.[82]

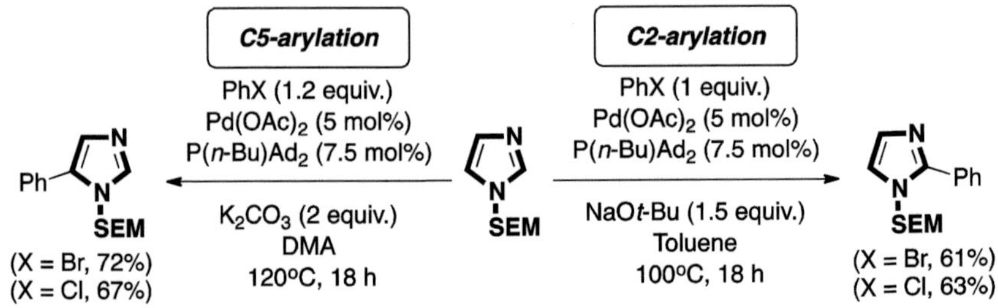

Scheme 3.54 Base/solvent-controlled selective C-5- and C-2-arylation of SEM-imidazole.[82]

electrophilic substitution, and the C-4-position is relatively less reactive (Fig. 3.4). By employing SEM-protected imidazole to the $Pd(OAc)_2$/ P(*n*-Bu)Ad_2 catalytic system, C-5-arylation product was observed when using K_2CO_3 as the base in polar DMA, whereas C-2-arylation was afforded when employing stronger NaO*t*-Bu in nonpolar toluene (Scheme 3.54). Importantly, aryl bromides and chlorides could be used as the coupling partners under these conditions. Since the C-4-position exhibited low reactivity, the authors described the SEM-group switching that transposed the SEM-group from N1 to N3 nitrogen of the imidazole and transformed the unreactive C-4-position to the reactive C-5-position. Hence, the 5-arylimidazole was converted in one step to 4-arylimidazole, followed by direct C-5-arylation again to afford the 4,5-diarylimidazole (Scheme 3.55). Remarkably, undergoing the C-2-arylation of 4,5-diarylimidazole with the $Pd(OAc)_2$/P(*n*-Bu)Ad_2 catalytic system again, the 2,4,5-triarylimidazole could be afforded in a programmable manner.

Later, Lee developed a $Pd(OAc)_2$ complex bearing phosphine and NHC **27** that was efficient in catalyzing direct C-5-arylation of imidazoles with aryl halides.[83] This catalytic system allowed a range of aryl bromides and chlorides as the coupling partners to afford the C-5-arylated imidazoles in good yields (Scheme 3.56). Notably, microwave irradiation effectively promoted the reactions with better yields in only 2 h.

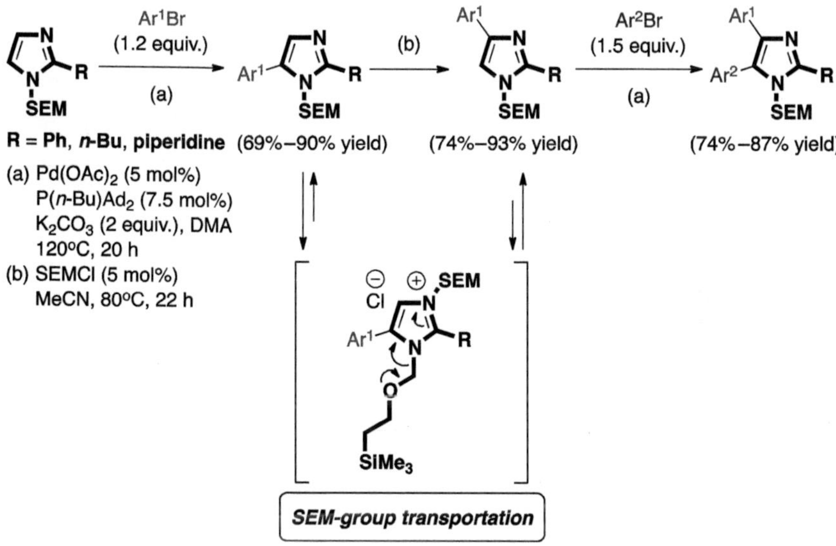

Scheme 3.55 Synthesis of 4,5-diarylimidazole via SEM-group transportation and sequential $Pd(OAc)_2$-catalyzed direct arylation with aryl bromide.[82]

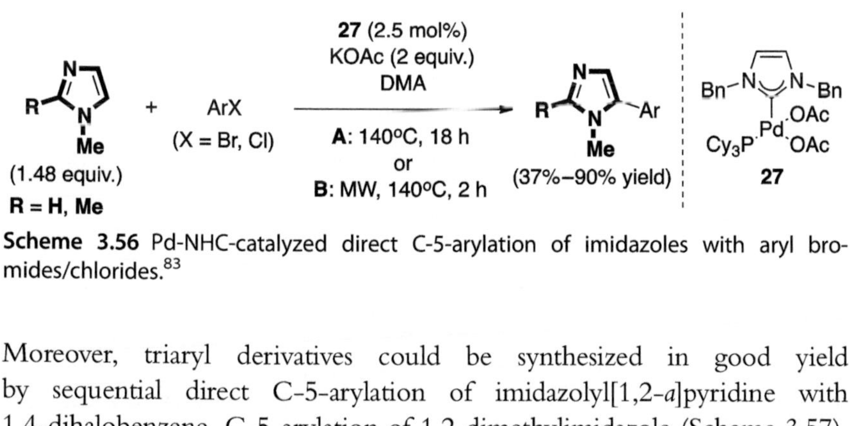

Scheme 3.56 Pd-NHC-catalyzed direct C-5-arylation of imidazoles with aryl bromides/chlorides.[83]

Moreover, triaryl derivatives could be synthesized in good yield by sequential direct C-5-arylation of imidazolyl[1,2-*a*]pyridine with 1,4-dihalobenzene, C-5-arylation of 1,2-dimethylimidazole (Scheme 3.57). It should be highlighted that in combination with the classical C–C bond formation methods, such as Suzuki–Miyaura and Heck coupling, novel imidazole derivatives bearing biaryl and styryl substituents could also be successfully obtained.

Shibahara and Murai described an efficient synthetic protocol for triarylated imidazole consisting of three different aryl groups via one-pot sequential multiple C–H bond functionalization.[84] By the combination of Rossi's $Pd(OAc)_2/P(2\text{-furyl})_3$ system[81] and the $Pd(phen)_2(PF_6)_2$ system,

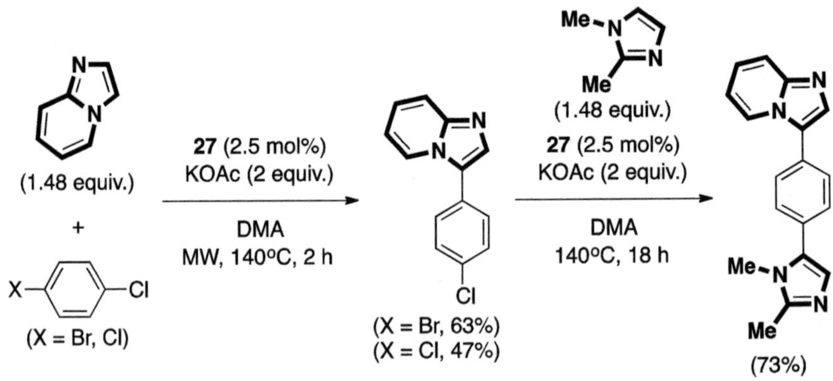

Scheme 3.57 Sequential direct C-5-arylation of imidazolyl[1,2-*a*]pyridine with 1,4-dihalobenzene, C-5-arylation of 1,2-dimethylimidazole.[83]

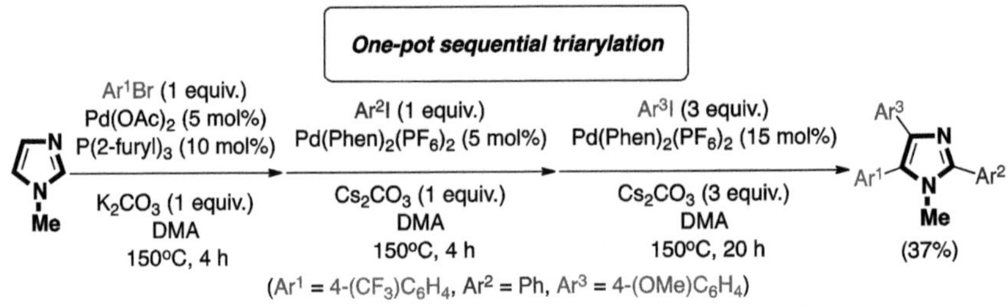

Scheme 3.58 One-pot sequential triarylation of *N*-methylimidazole.[84]

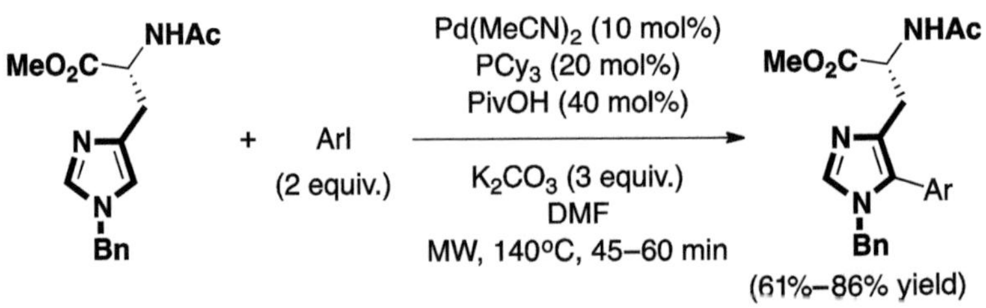

Scheme 3.59 $Pd(MeCN)_2/PCy_3$-catalyzed direct C-5-arylation of L-histidine with aryl iodides.[85]

the one-pot triarylation of *N*-methylimidazole was also successfully achieved (Scheme 3.58). Notably, the authors revealed that portionwise addition of aryl halide, base, and catalyst significantly improved the overall yield of the desired triarylated product.

Jain reported an efficient microwave-assisted direct C-5-arylation of fully protected L-histidine with aryl iodides by using a $Pd(MeCN)_2/PCy_3$ catalytic system (Scheme 3.59).[85] This system offered high functional

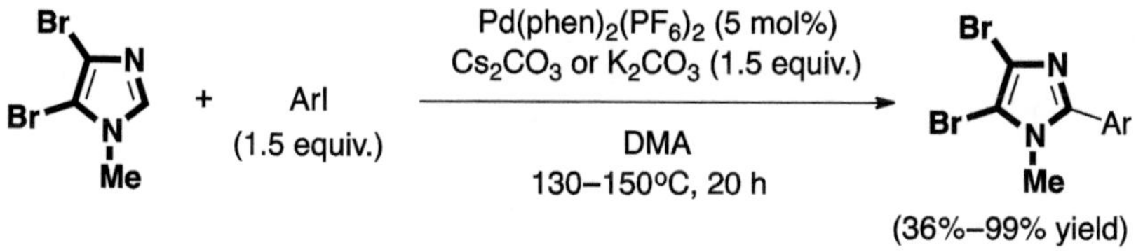

Scheme 3.60 $Pd(phen)_2(PF_6)_2$-catalyzed direct C-2-arylation of 4,5-dibromoimidazole with (hetero)aryl iodides.[86]

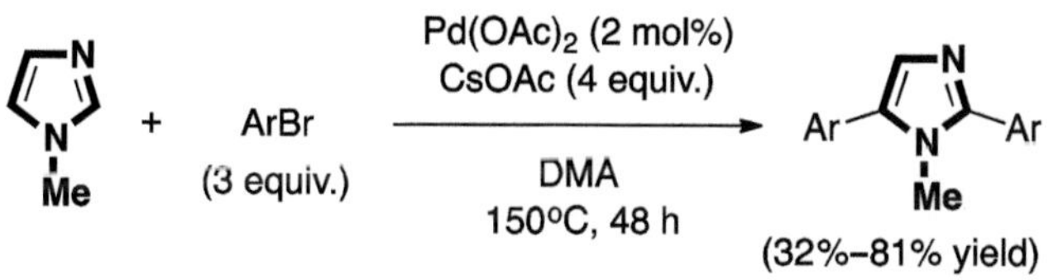

Scheme 3.61 $Pd(OAc)_2$-catalyzed direct diarylation of *N*-methylimidazole with (hetero)aryl bromides.[88]

group tolerance and gave a series of 5-aryl-L-histidines in good yields. Attempt to use aryl bromides as the coupling partners was also carried out, yet low yields were obtained.

Shibahara and Murai developed the direct C-2-arylation of 4,5-dibromoimidazole with (hetero)aryl iodides by using palladium-1,10-phenanthroline complex $Pd(phen)_2(PF_6)_2$ (Scheme 3.60).[86] Particularly noteworthy was that this protocol tolerated the presence of a variety of functional groups and prohibited the cleavage of C–Br bonds on the imidazole core to furnish the C-2-arylated 4,5-dibromoimidazoles in moderate-to-excellent yields. Importantly, taking advantage of the remaining C–Br bonds, the products formed in these reactions were successfully transformed to various polyfunctionalized imidazoles with pharmaceutical functions.[87]

Meanwhile, Doucet reported for the first time a simple one-pot protocol leading to 2,5-diarylimidazoles.[88] Using a $Pd(OAc)_2$/CsOAc system in DMA, a range of (hetero)aryl bromides underwent the direct diarylation of *N*-methylimidazole smoothly to afford 2,5-diarylimidazoles in good yields (Scheme 3.61). The authors revealed that the choice of base was found to be crucial to obtain the desired diarylated products.

Recently, Li, and Doucet reported the cascade palladium-catalyzed intermolecular and subsquent intramolecular direct arylations of 1-(2-bromobenzyl)imidazole (Scheme 3.62).[89] The authors revealed that, in

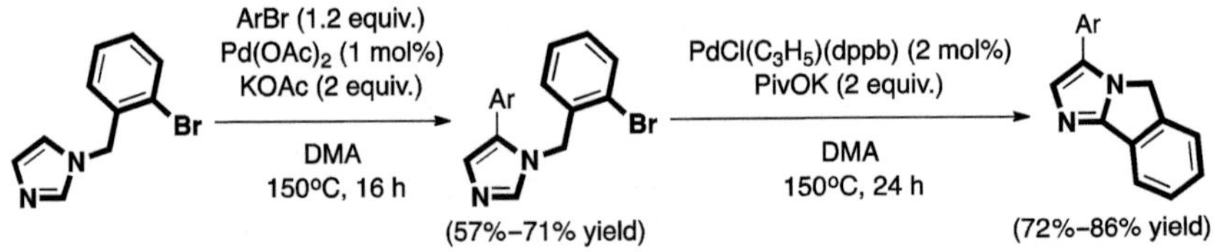

Scheme 3.62 Pd-catalyzed intermolecular followed by intramolecular direct arylations of 1-(2-bromobenzyl)imidazole.[89]

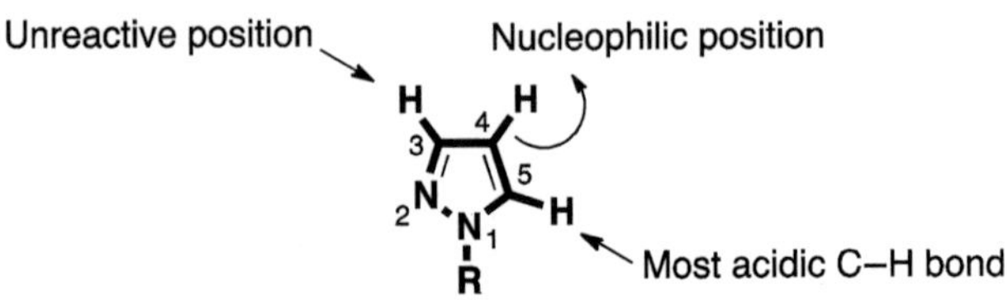

Figure 3.5 General reactivity properties of pyrazole.[91]

the presence of a $Pd(OAc)_2$/KOAc catalytic system, the intermolecular arylation between 1-(2-bromobenzyl)imidazole and electron-deficient aryl bromides proceeded faster than that of intramolecular arylation to afford the C-5-arylated products in good yields. On the other hand, by employing a $PdCl(C_3H_5)$(dppb)/PivOK system, the intramolecular arylation sequentially proceeded in good yields. It should be noted that these were the first examples of the intermolecular arylation of imidazoles substituted by a 2-bromobenzyl group at the *N*-position.

3.2.2.3 Pyrazoles and Indazoles

Arylpyrazole and arylindazole derivatives are important buildings blocks in organic synthesis due to their biological properties, including natural products, protein ligands, and pharmaceuticals.[90] Therefore, the development of versatile protocols to access highly substituted pyrazoles are always in demand.

Sames and co-workers reported the first catalytic method for arylation of pyrazoles and explored its regioselectivity trends.[91] For the rationale of regioselectivity, it is well known that the C-4-position of pyrazole is the most nucleophilic and readily undergoes electrophilic substitution, while the C-5-position has the most acidic C—H bond which can be selectively deprotonated by a strong base (Fig. 3.5).[92] The C-4-arylated pyrazoles and (hetero)aryl bromides were used to examine the reaction scope in the presence of a $Pd(OAc)_2/P(n\text{-}Bu)Ad_2$ system, where it was found that the catalytic system displayed good functional group tolerance, and afforded

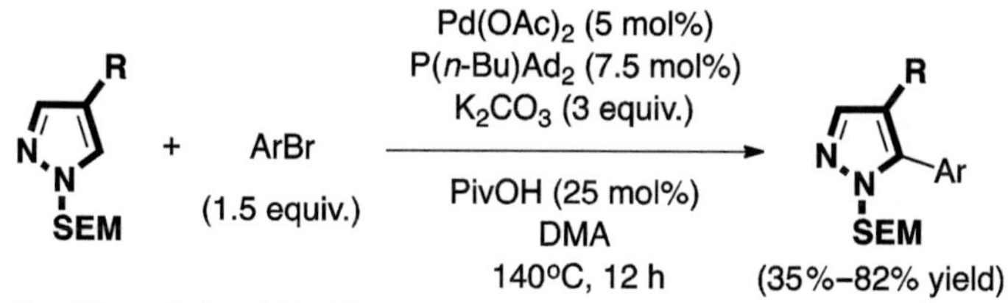

R = Ph, *o*-Tol, 4-$(CF_3)C_6H_4$, CO_2Et

Scheme 3.63 $Pd(OAc)_2$-catalyzed direct arylation of 4-aryl-SEM-pyrazoles with (hetero)aryl bromides.[91]

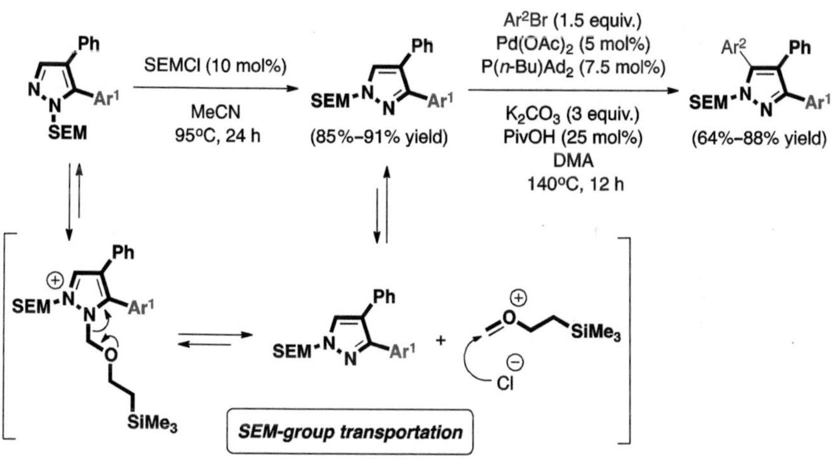

Scheme 3.64 Synthesis of triarylpyrazole via SEM-group transportation and sequential $Pd(OAc)_2$-catalyzed direct arylation with (hetero)aryl bromide.[91]

4,5-diaryl-SEM-pyrroles in good yields (Scheme 3.63). In order to synthesize triarylpyrazoles, the sequential arylation would have to take place at the C-3-position, yet the arylation of this site was not feasible because of its low reactivity. The authors considered switching the SEM-group from one nitrogen atom to another, which would transform the unreactive C-3-position to the reactive C-5-position, followed by the second arylation to give SEM-protected triarylpyrazole in high yields (Scheme 3.64). Therefore, all three arenes were introduced to the pyrazole with complete regiochemical control.

Doucet and Santelli showed 1,3,5-trisubstituted pyrazoles as suitable partners for $Pd(OAc)_2$-catalyzed direct arylation with aryl bromides.[93] The protocol used 0.5–2 mol% $Pd(OAc)_2$ with KOAc as the base in DMA and promoted regioselective C-4-arylated pyrazoles in

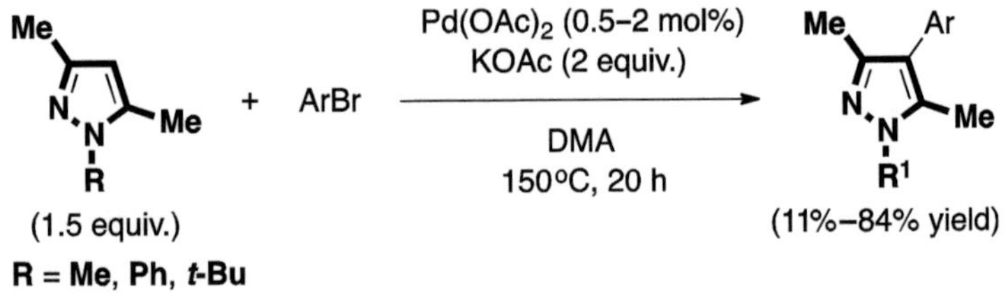

Scheme 3.65 $Pd(OAc)_2$/KOAc-catalyzed direct arylation of 1,3,5-trisubstituted pyrazoles with (hetero)aryl bromides.[93]

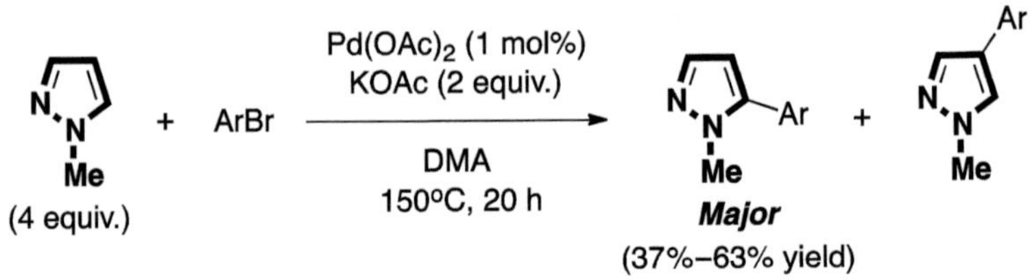

Scheme 3.66 $Pd(OAc)_2$/KOAc-catalyzed direct C-5-arylation of *N*-methylpyrazole with (hetero)aryl bromides.[94]

moderate-to-high yields (Scheme 3.65). Particularly noteworthy was that a wide variety of (hetero)aryl bromides were tolerated, and relative low catalyst loading was used.

Ammar and Doucet later applied the same catalytic system to the direct arylation of *N*-methylpyrazole with aryl bromides.[94] In all cases, C-4-arylated pyrazole was detected, associating with the desired C-5-arylated pyrazoles (Scheme 3.66). In addition, a range of (hetero)aryl bromides were found applicable.

The direct arylation of pyrazole takes place preferentially at the C-5-position rather than the C-4-position. Doucet reported the $PdCl(C_3H_5)$(dppb)-catalyzed direct C-4-arylation of nonprotected 5-aminopyrazoles with (hetero)aryl bromides and proceeded to high yields (Scheme 3.67).[95] The choice of KOAc as the base was found to be crucial to inhibit the competitive amination reaction.

Concurrently, Doucet developed the use of temporary chloro-protection at the C-5-position of pyrazoles which allowed regioselective access to the C-4-arylated pyrazoles via $Pd(OAc)_2$-catalyzed direct arylation with aryl bromides.[96] With the same protocol, the reaction proceeded in moderate-to-high yields in the presence of various (hetero)aryl bromides using down to 0.1 mol% $Pd(OAc)_2$ loading (Scheme 3.68). Furthermore,

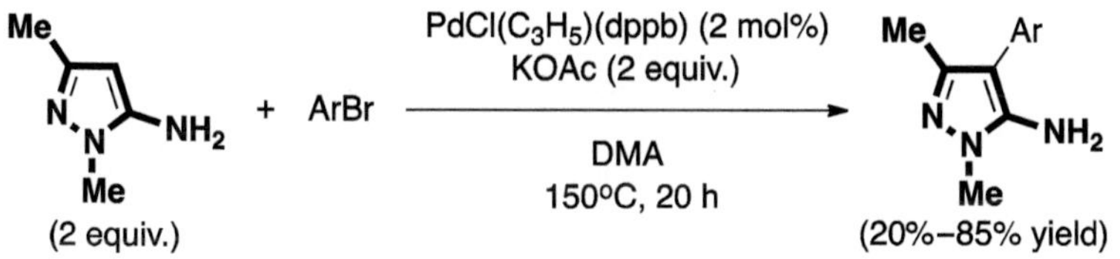

Scheme 3.67 $PdCl(C_3H_5)(dppb)$/KOAc-catalyzed direct C-4-arylation of 1,3-dimethyl-pyrazol-5-amine with (hetero)aryl bromides.[95]

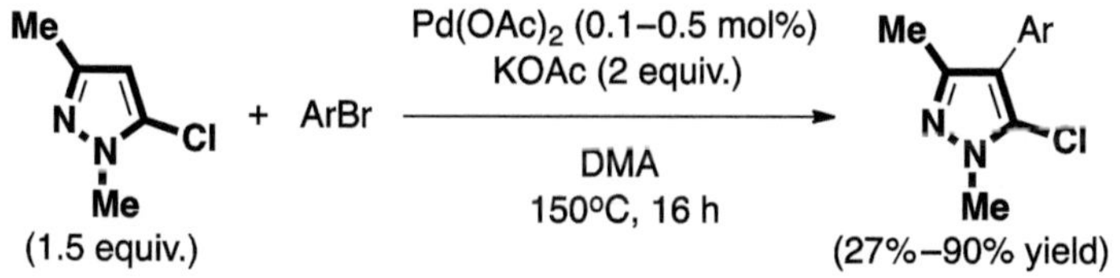

Scheme 3.68 $Pd(OAc)_2$/KOAc-catalyzed direct C-4-arylation of 5-chloro-1,3-dimethyl-pyrazole with (hetero)aryl bromides.[96]

the sequential C-4-arylation, dechlorination, and C-5-arylation was also successfully achieved and provided the controlled double arylation of pyrazole at C-4- and C-5-positions in good yields (Scheme 3.69).

By taking advantage of the relatively similar energies of activation of the C-4- and C-5-position of pyrazole for the palladium-catalyzed C—H functionalization reactions (Fig. 3.6),[42c] Touzani and Doucet developed a one-step access to 4,5-diarylpyrazoles using $PdCl(C_3H_5)(dppb)$ as a catalyst.[97] A range of (hetero)aryl bromides reacted at both C-4- and C-5-positions of *N*-methylpyrazole, yet without arylation at the C-3-position, to give 4,5-diarylpyrazoles in moderate-to-good regioselectivities and yields in only one-step; even the formation of monoarylated pyrazoles was detected in all cases (Scheme 3.70).

Mateos and Mendiola described a highly regioselective $Pd(OAc)_2$/**DavePhos**-catalyzed direct C-5-arylation of 4-chloro-1-methylpyrazole with aryl bromides.[98] The $Pd(OAc)_2$/**DavePhos** system, with Bu_4NAc as the base, isobutyric acid as the additive in NMP, allowed preparation of C-5-arylpyrazoles selectively in moderate-to-excellent yields (Scheme 3.71). In addition, this system tolerated a range of (hetero)aryl bromides with either electron-poor or electron-rich substituents.

Kumpulainen developed a highly regioselective C-5-arylation of *N*-dimethylaminosulfamoyl-protected pyrazole via $Pd(OAc)_2/PPh_3$-catalyzed direct C—H functionalization with (hetero)aryl bromides in the presence of K_2CO_3 as the base, and 2-ethylhexanoic acid (EHA) as the

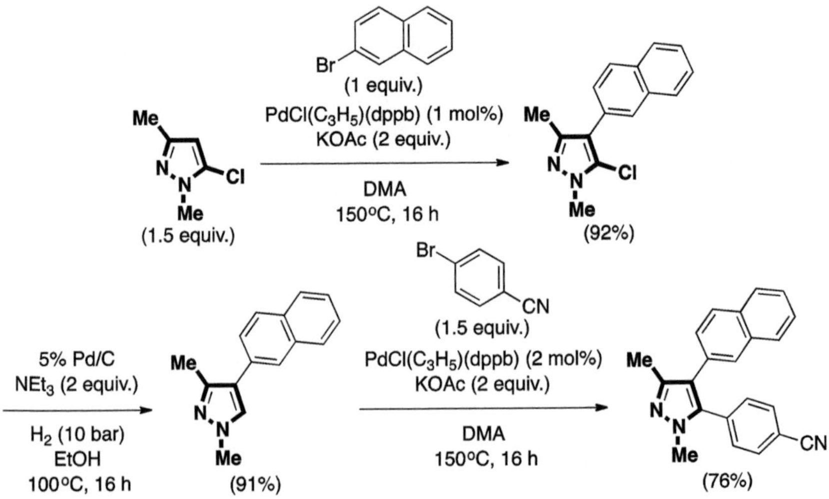

Scheme 3.69 Sequential C-4-arylation, dechlorination, and C-5-arylation of 5-chloro-1,3-dimethylpyrazole.[96]

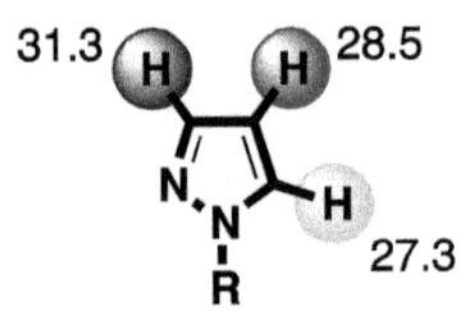

Figure 3.6 Gibbs free energies of activation ($\Delta G_{298\ K}$) of the cleavage of C—H bonds for 1-methylpyrazole in the CMD pathway using $[Pd(C_6H_5)(PMe_3)(OAc)]$.[42c]

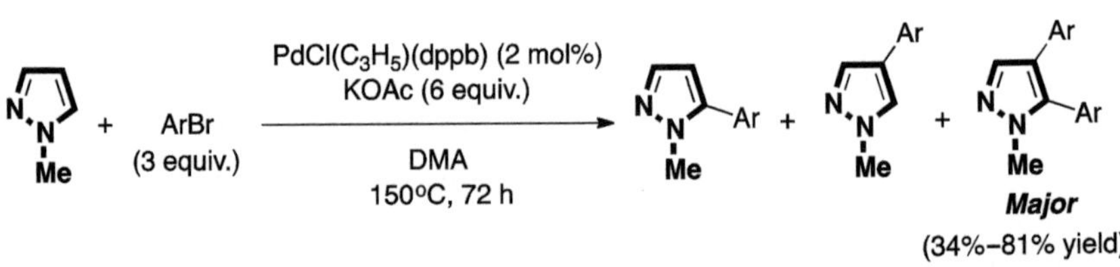

Scheme 3.70 $PdCl(C_3H_5)$(dppb)/KOAc-catalyzed direct diarylation of *N*-methylpyrazole with (hetero)aryl bromides.[97]

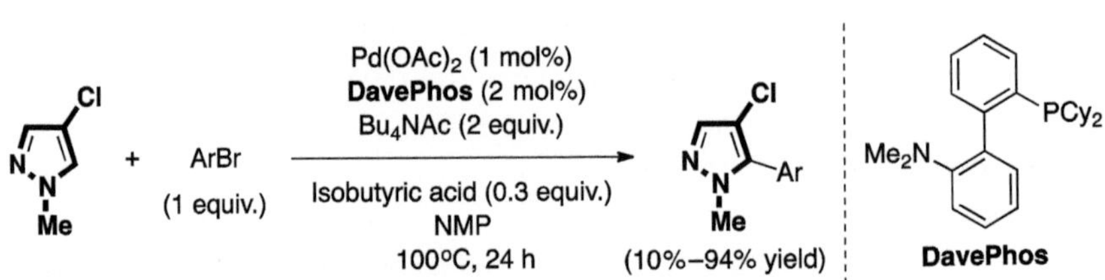

Scheme 3.71 $Pd(OAc)_2$/**DavePhos**-catalyzed direct C-5-arylation of 4-chloro-1-methylpyrazole with (hetero)aryl bromides.[98]

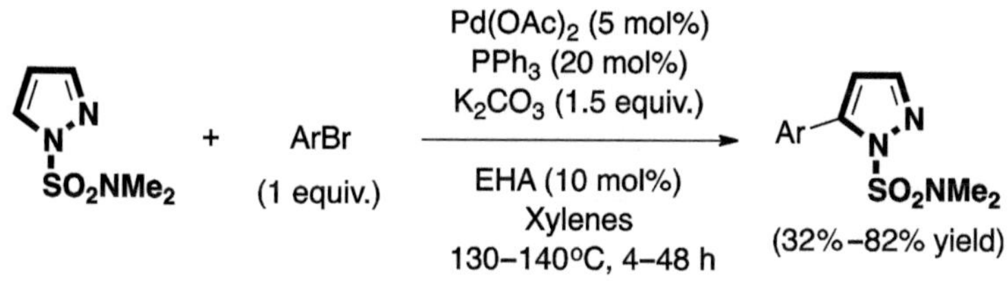

Scheme 3.72 $Pd(OAc)_2/PPh_3$-catalyzed direct C-5-arylation of *N*-SO_2NMe_2 pyrazole with (hetero)aryl bromides.[99]

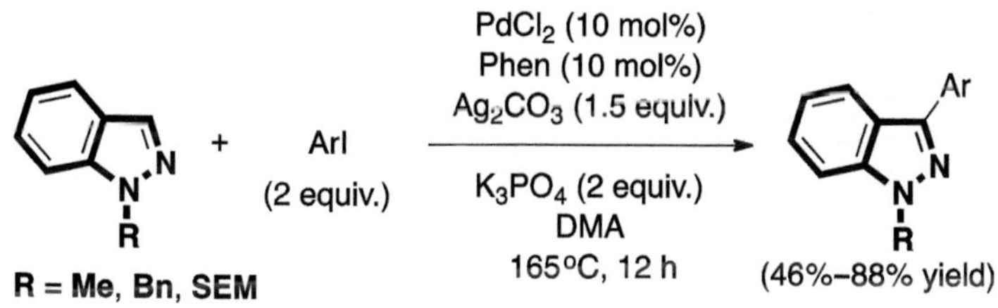

Scheme 3.73 $PdCl_2$/phen-catalyzed direct C-3-arylation of 1*H*-indazoles with (hetero) aryl iodides.[100]

additive in xylenes (Scheme 3.72).[99] This protocol showed good tolerance of functional groups and afforded the C-5-arylated pyrazoles in moderate-to-good yields. Notably, the direct arylation of *N*-SO_2NMe_2 pyrazole followed by deprotection could also be achieved. It was a viable alternative for synthesis of C-3-arylated pyrazole building blocks which have wide applications in the construction of bioactive derivatives.

Yamaguchi and Itami described the palladium-catalyzed direct arylation reactions of 1*H*- or 2*H*-indazoles with aryl halides in the presence of $PdCl_2$ catalyst.[100] The $PdCl_2$/phen catalytic system was effective for the C-3-arylation of 1*H*-indazoles with various (hetero)aryl iodides to afford the corresponding products in good yields (Scheme 3.73). Notably, the $PdCl_2$/phen catalytic system was also found to be effective for the arylation of 2*H*-indazoles to furnish C-3-arylated products (Scheme 3.74). It should be noted that this catalytic system could be employed to the application of synthesis of biologically active compounds, platelet anti-aggregation agent **YD-3**,[101] and antitumor agent **YC-1**.[102]

Concurrently, Kazzouli and Guillaumet reported the direct arylation of 1*H*-indazoles with aryl halides.[103] Various 1-substituted 1*H*-indazoles and (hetero)aryl iodides, as well as bromides, were used to afford the C-3-arylated products in good yields in the presence of $Pd(OAc)_2$/phen as the catalytic system (Scheme 3.75).

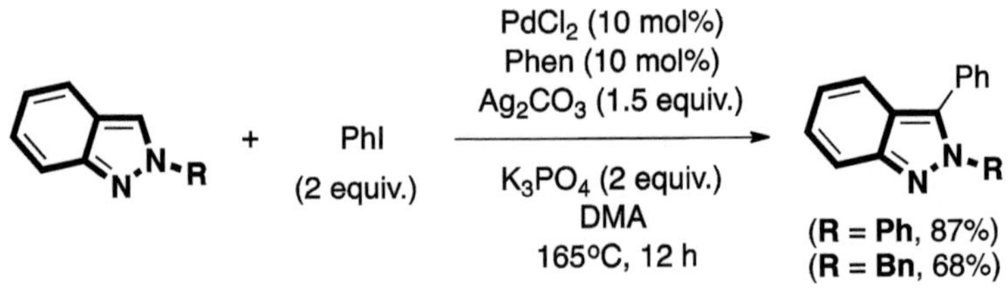

Scheme 3.74 $PdCl_2$/phen-catalyzed direct C-3-arylation of 2*H*-indazoles with iodobenzene.[100]

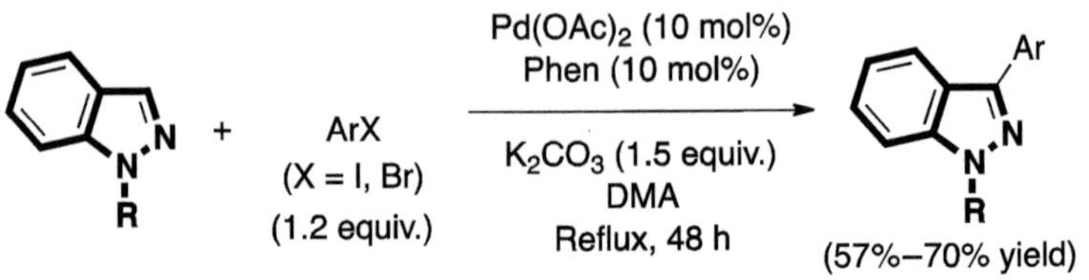

Scheme 3.75 $Pd(OAc)_2$/phen-catalyzed direct C-3-arylation of 1*H*-indazoles with (hetero)aryl iodides/bromides.[103]

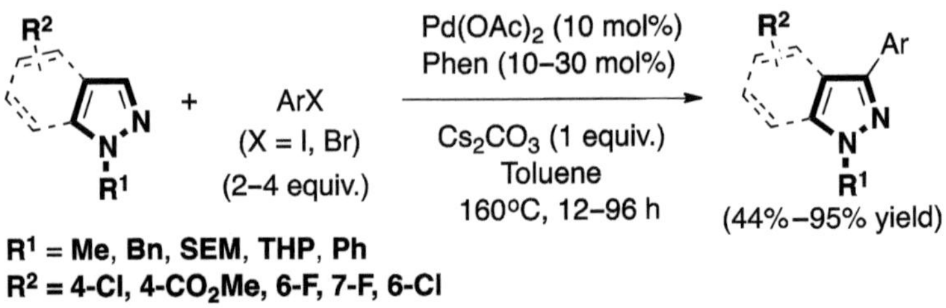

Scheme 3.76 $Pd(OAc)_2$/phen-catalyzed direct C-3-arylation of 1*H*-indazoles or pyrazoles with (hetero)aryl iodides/bromides.[104]

Yu and co-workers also used the $Pd(OAc)_2$/phen catalyst, except replacing K_2CO_3 with Cs_2CO_3 in toluene for the direct arylation of 1*H*-indazoles and pyrazoles with aryl iodides and bromides.[104] The coupling of various 1-substituted 1*H*-indazoles underwent smoothly to afford the C-3-arylated products in good yields (Scheme 3.76). The arylation of 2*H*-indazole under these reaction conditions was also achieved. In the competitive reaction between 1*H*-indazole and 2*H*-indazole with iodobenzene, C-3-arylated 2*H*-indazole was given as a major product with a 75% yield, while for C-3-arylated 1*H*-indazole the yield was 8%, confirming that the reactivity of 2*H*-indazole is superior to 1*H*-indazole (Scheme 3.77). In addition, this catalytic system was employed in the total synthesis of nigellidine hydrobromide.[105]

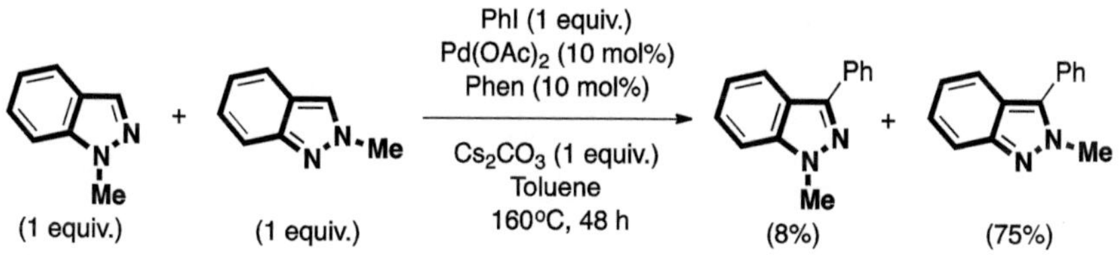

Scheme 3.77 Competitive reaction between 1*H*-indazole and 2*H*-indazole in $Pd(OAc)_2$/phen-catalyzed direct arylation with iodobenzene.[104]

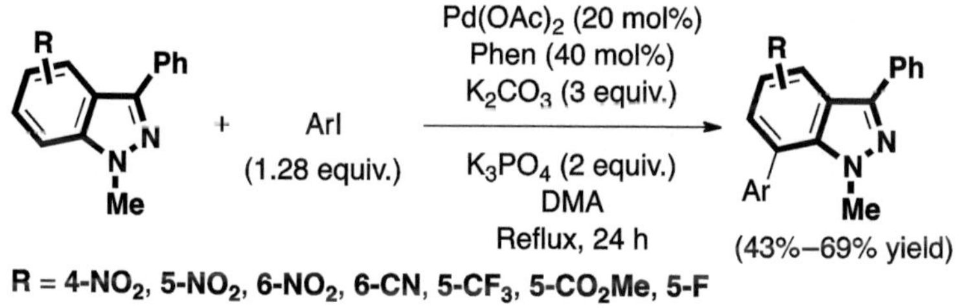

Scheme 3.78 $Pd(OAc)_2$/phen-catalyzed direct C-7-arylation of substituted indazoles with aryl iodides.[106]

Guillaumet reported a novel direct C-7-arylation of indazoles with aryl iodides also using the $Pd(OAc)_2$/phen catalytic system consisting of K_2CO_3 as the base in refluxing DMA (Scheme 3.78).[106] The authors mentioned the fact that the addition of K_3PO_4 was found to be crucial in shortening the reaction time. The direct C-7-arylation of 3-substituted indazole containing electron-withdrawing groups on the arene ring with aryl iodides gave the desired products in good yields. Yet this protocol was limited to heteroaryl iodide. In addition, a one-pot Suzuki/C-7-arylation was successfully achieved by employing 3-bromoindazoles as the coupling partners to afford 3,7-diarylated indazoles in moderate yields (Scheme 3.79).

3.2.2.4 Thiophenes and Benzothiophenes

The functionalization of thiophenes and benzothiophenes is an important and fundamental transformation in organic chemistry.[107]

Itami developed a general protocol for the programmed synthesis of tetraarylthiophenes.[108] The utilization of three catalysts, RhCl(CO){P[OCH(CF_3)$_2$]$_3$}$_2$, $PdCl_2$/P[OCH(CF_3)$_2$]$_3$, and $PdCl_2$/bipy, enabled regioselective sequential arylations at three C–H bonds of 3-methoxythiophene with aryl iodides (Scheme 3.80). Firstly, 3-methoxythiophene reacted with iodobenzene in the presence of RhCl(CO){P

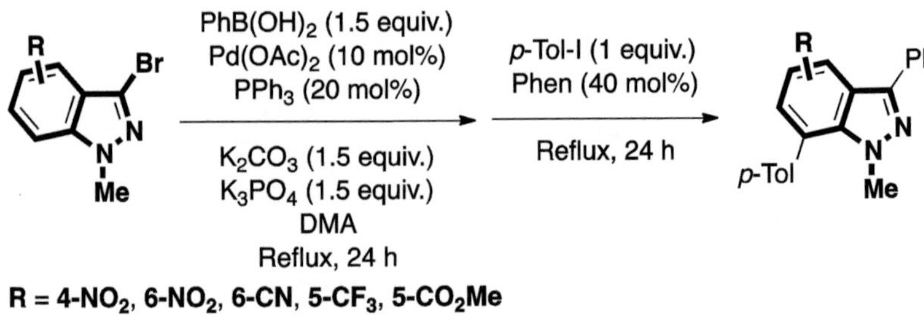

Scheme 3.79 $Pd(OAc)_2$-catalyzed one-pot Suzuki/C-7-arylation of substituted indazoles.[106]

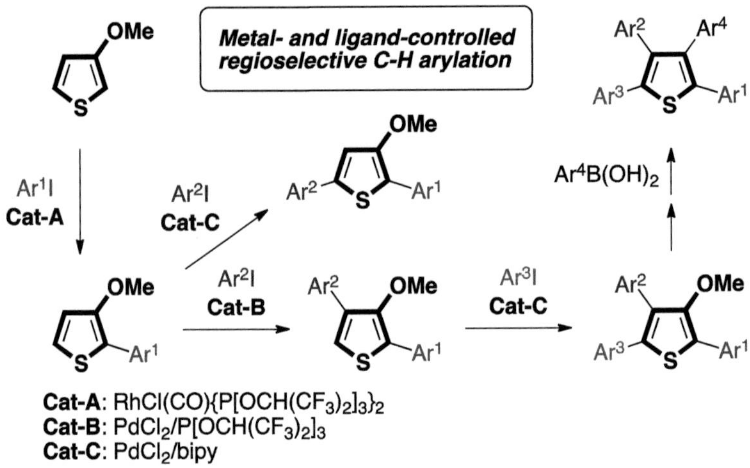

Scheme 3.80 Programmed synthesis of tetraarylthiophenes.[108]

$[OCH(CF_3)_2]_3\}_2$ catalyst to afford 2-phenyl-3-methoxythiophene in high yield with complete regioselectivity. Then, in the presence of $PdCl_2$, the site of arylation of 2-phenyl-3-methoxythiophene could be controlled by just changing the ligand, $P[OCH(CF_3)_2]_3$ for C-4-arylation and 2,2′-bipyridyl (bipy) for C-5-arylation, and the corresponding products were afforded in both high yields and regioselectivities (Scheme 3.81). Finally, the fourth arylation was carried out by demethylation/triflation, using BBr_3 for demethylation followed by treatment with Tf_2O/i-Pr_2NEt/DMAP to produce the corresponding triflates, followed by Suzuki–Miyaura coupling, in the presence of $Pd(PPh_3)_4$ catalyst, to obtain the targeted tetraarylthiophenes.

Itami then used the $PdCl_2/P[OCH(CF_3)_2]_3$ system to extend the reaction scope of β-selective arylation of thiophenes with aryl iodides.[109]

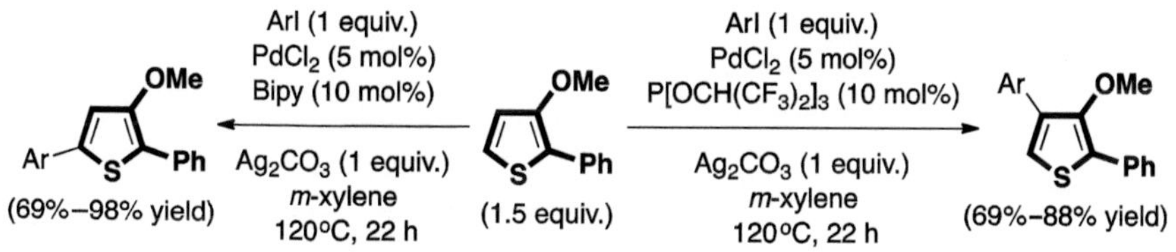

Scheme 3.81 Ligand-controlled $Pd(OAc)_2$-catalyzed regioselective arylation of 2-phenyl-3-methoxythiophene with aryl iodides.[108]

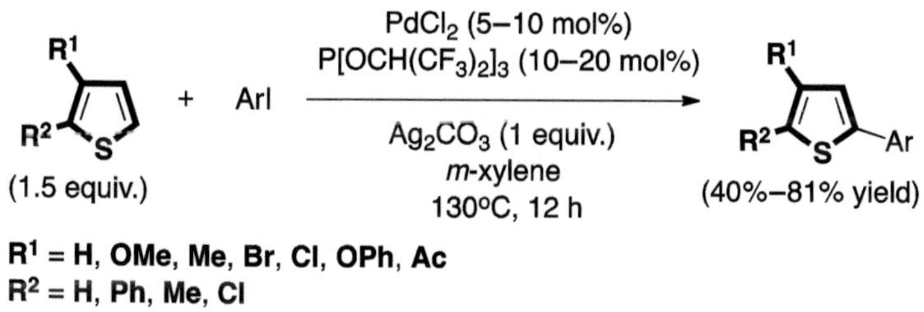

Scheme 3.82 $PdCl_2/P[OCH(CF_3)_2]_3$-catalyzed direct β-selective arylation of thiophenes with aryl iodides.[109]

Reactions of 2,3-disubstituted thiophenes did afford C-4-arylated products in moderate-to-good yields (Scheme 3.82). Aryl iodides with both electron-donating and electron-withdrawing substituents were also successfully coupled with 2-chlorothiophene in good yields. It should be noted that thiophene-containing fused aromatic compounds, such as benzothiophene and thieno[3,2-*b*]thiophene were also arylated.

The low reactivity and regioselectivity of the direct arylation process is typically observed, Liégault and Fagnou disclosed that the introduction of a chloro substituent to a heteroarene could serve to improve reactivity and regioselectivity, and provide superior outcomes.[72b] Based on the authors' previous work,[25] it was found that the chloro functional group might act as an activating group for the reactions at nearby C—H bonds. For example, C-3-subsititued thiophene could be regioselectively chlorinated at either the C-2- or C-5-position by using common organic synthetic routes (Scheme 3.83). Both C-2- and C-5-chlorinated thiophenes reacted smoothly with various aryl bromides to give single regioisomers in good yields in the presence of $Pd(OAc)_2/PCy_3 \cdot HBF_4$ as the catalytic system (Scheme 3.83). The chlorine atom on the thiophene showed a good blocking ability to facilitate other regioselective arylation. Importantly, the chlorine atom could be easily removed by reduction using Pd/C under a hydrogen atmosphere.[110] On the other hand, the

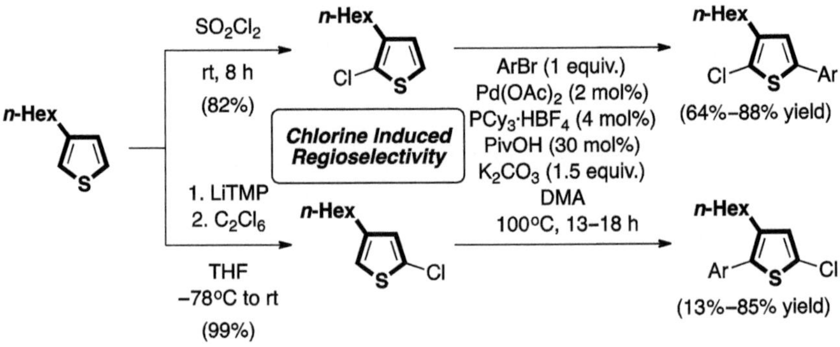

Scheme 3.83 Chlorine induced regioselective $Pd(OAc)_2/PCy_3 \cdot HBF_4$-catalyzed direct arylation of thiophene with aryl bromides.[72b]

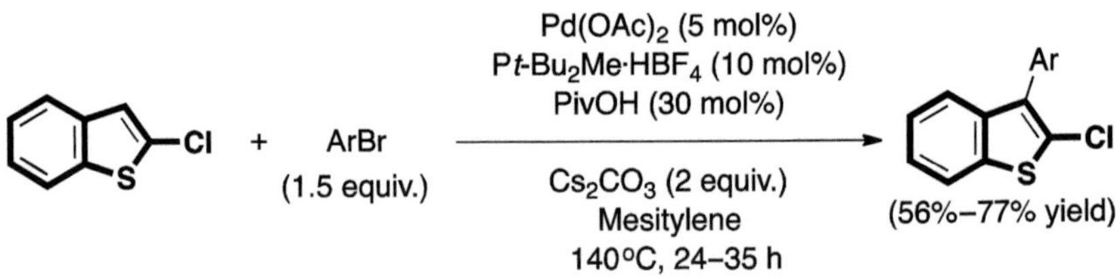

Scheme 3.84 Chlorine-diverted regioselective $Pd(OAc)_2/Pt\text{-}Bu_2Me \cdot HBF_4$-catalyzed direct arylation of benzothiophene or benzofuran with aryl bromides.[72b]

chloro functional group could be used to divert the direct arylation away from normal site selectivity. Typically, benzothiophene was an excellent substrate in palladium-catalyzed direct arylation to provide C-2-regioselectivity. By installing a C-2-chlorine atom, benzothiophene could undergo direct C-3-arylation smoothly (Scheme 3.84).

To have a better understanding of the characteristics of the direct arylation of thiophenes, Itami compared their $PdCl_2/P[OCH(CF_3)_2]_3$ catalytic system[109] with Fagnou's $Pd(OAc)_2/PCy_3 \cdot HBF_4$ catalytic system[72b,111] via intermolecular competitive experiments (Scheme 3.85). Fagnou's system favored arylation at the α-position and also reacted preferentially with acidic 2-chlorothiophene over nucleophilic 2-methylthiophene. When the competitive reaction was conducted with Itami's catalytic system, arylation occurred preferentially with nucleophilic 2-methylthiophene at the β-position. These contrasting results with respect to the preferred substrate (acidic vs nucleophilic) and the regioselectivity (α vs β) indicated that the Itami's catalytic system would react by a mechanism distinct from that proposed by Fagnou.

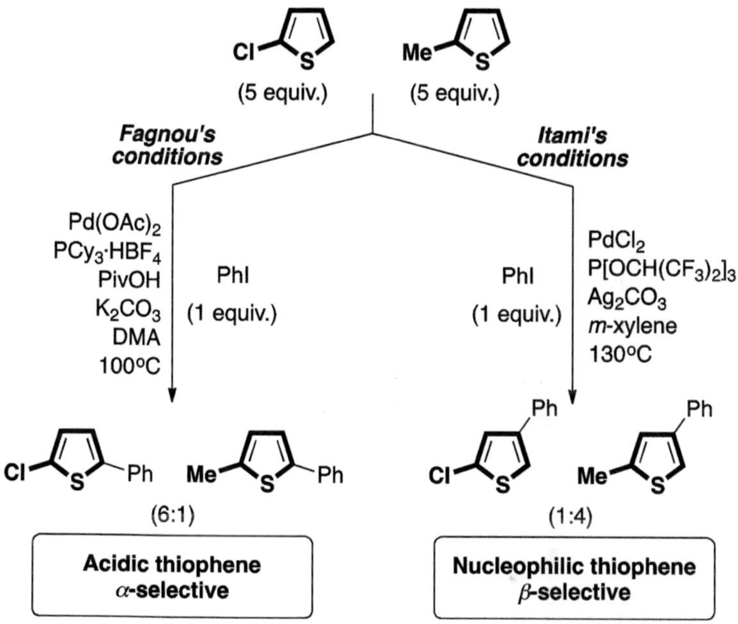

Scheme 3.85 Regioselective comparsion between Fagnou's and Itami's catalytic systems.[109]

Spirocyclic thiophenes, σ receptors, are a well established nonopioid, nonphencyclidine and haloperidol sensitive receptor family with a unique binding profile and a characteristic distribution in the central nervous-system as well as in endocrine, immune, and some peripheral tissues like kidney, liver, lung, and heart.[112] Itami and co-workers employed the $PdCl_2$/bipy system[108] (see Scheme 3.81) for the direct α-arylation of spirocyclic thiophenes with (hetero)aryl iodides to prepare the target aryl-appended spirocyclic thiophenes (Scheme 3.86).[113] This protocol allowed the preparation of a large set of diverse σ receptors for diagnosis and therapy purposes.

René and Fagnou previously described the biphasic $Pd(OAc)_2$/**MePhos** catalytic system for direct arylation of polyfluoroarenes with aryl iodides[27] (see Scheme 3.12). This system was compatible to the cross-coupling between halogenated thiophenes and aryl iodides to afford corresponding heterobiaryl products in good yields.[27] Under the EtOAc/H_2O biphasic condition, 2-chlorothiophene and 3 halogenated thiophenes reacted regioselectively at the C-5- and C-2-position, respectively (Scheme 3.87). The presence of the chloro and bromo substituents on these products could be used for subsequent synthetic transformations.

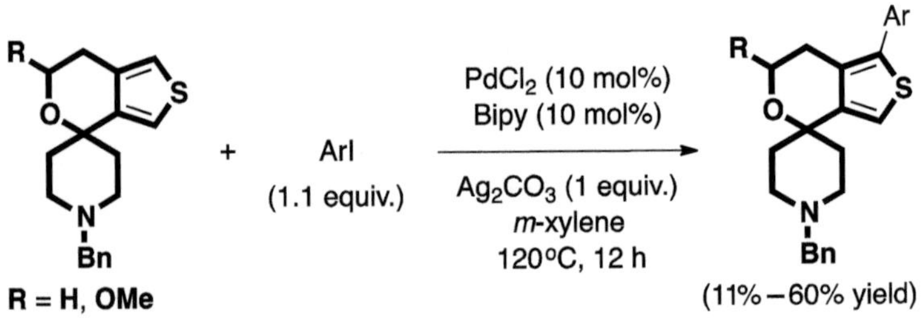

Scheme 3.86 $PdCl_2$/bipy-catalyzed direct α-arylation of spirocyclic thiophene with (hetero)aryl iodides.[113]

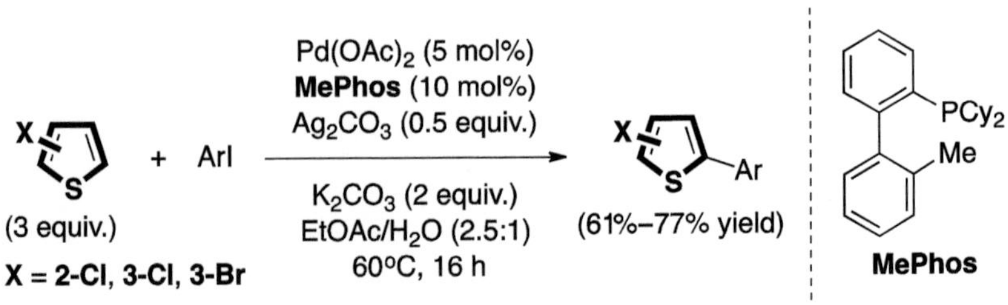

Scheme 3.87 $Pd(OAc)_2$/**MePhos**-catalyzed cross-coupling between halogenated thiophenes and aryl iodides in biphasic conditions.[27]

Doucet described, for the first time, the direct arylation of free NH_2-substituted thiophenes.[114] This simple protocol, a $PdCl(C_3H_5)$(dppb)/KOAc system, allowed the direct C-2- or C-5-arylation of free NH_2-substituted thiophenes to proceed in moderate-to-high yields using a variety of (hetero)aryl bromides. Selective C-5-arylation was observed when using 3-amino-4-methylthiophene-2-carboxylate as the coupling partners while C-2-arylation for 3-amino-4-methylthiophene (Scheme 3.88) was seen, these results revealed that the presence of an ester substituent on the C-2-position of thiophene was useful to block this highly reactive carbon. Notably, no coupling of aryl bromide with thiophene NH_2 substituent was detected.

Arylated bithiophenes are useful compounds because of their coordination and physical properties.[115] Under the same catalytic system, Doucet prepared a variety of arylated bithiophenes in good yields (Scheme 3.89).[116] The authors employed an excess of these bithiophenes to allow the formation of monoarylated products at the C-5-position. A range of functional groups on the (hetero)aryl bromides was tolerated. Noteworthy, bis(thiophen-2-yl)methanone was compatible in this system to afford the desired products in good yields, and 1,2-bis(thiophen-2-yl) ethane-1,2-dione did so yet gave a low yield.

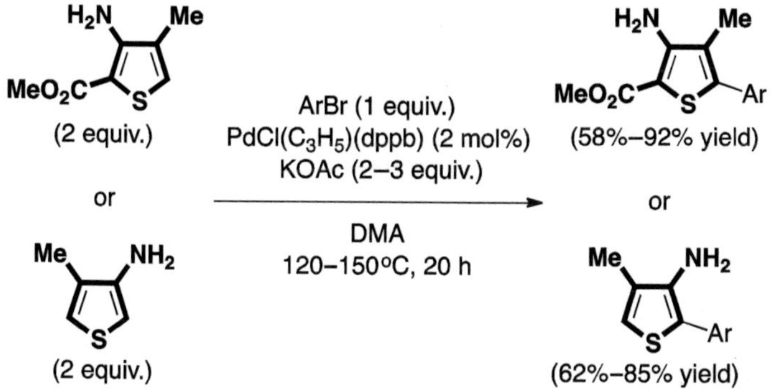

Scheme 3.88 $PdCl(C_3H_5)(dppb)$/KOAc-catalyzed direct arylation of NH_2-substituted thiophenes with (hetero)aryl bromides.[114]

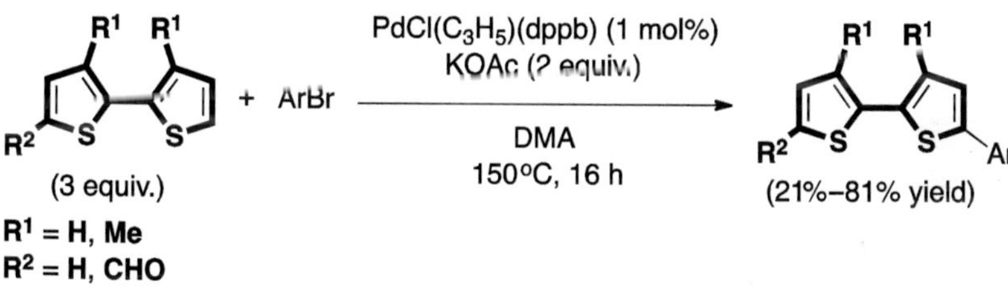

Scheme 3.89 $PdCl(C_3H_5)(dppb)$/KOAc-catalyzed direct C-5-arylation of bithiophenyl derivatives (hetero)aryl bromides.[116]

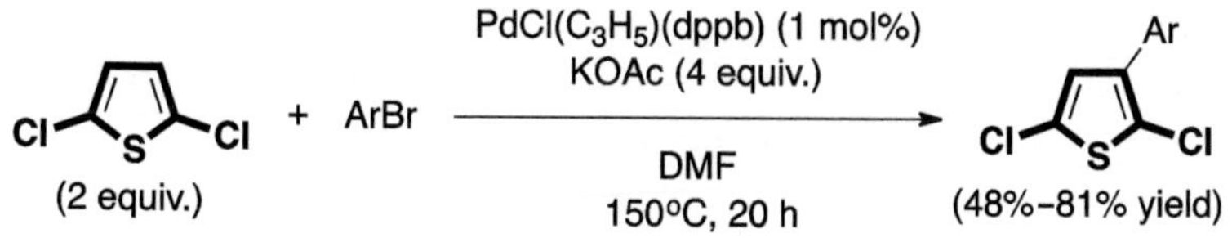

Scheme 3.90 $PdCl(C_3H_5)(dppb)$/KOAc-catalyzed direct arylation of 2,5-dichlorothiophene with (hetero)aryl bromides.[41]

Doucet revealed that the same catalytic system was also active toward the direct arylation with polychloroarenes[41] (see Scheme 3.24) and could be applied to the direct arylation of 2,5-dichlorothiophene with (hetero) aryl bromides (Scheme 3.90).[41]

Doucet continued the development on direct arylation and found that $Pd(OAc)_2$ was also an efficient catalyst to promote the direct arylation of functionalized halothiophenes.[117] Initially, the influence of the chloride position on the thiophene was examined, the reactivity of 2-acetyl-4-chlorothiophene, 2-acetyl-3-chlorothiophene (Scheme 3.91), and 2-chlorothiophene (Scheme 3.92) was found to be different, yet they

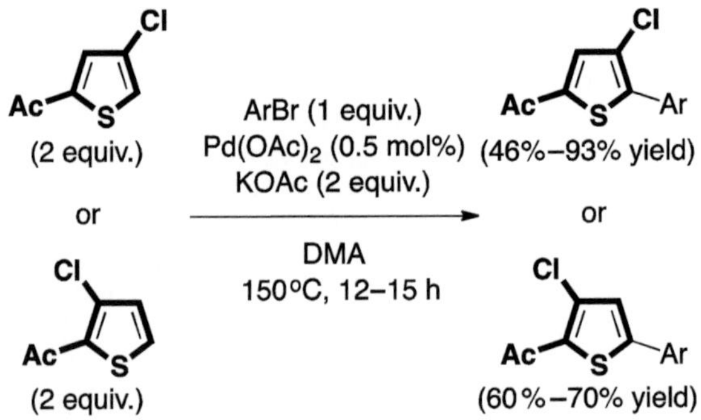

Scheme 3.91 Pd(OAc)$_2$/KOAc-catalyzed direct C-5-arylation of 2-acetyl-4-chlorothiophene or 2-acetyl-3-chlorothiophene with (hetero)aryl bromides.[117]

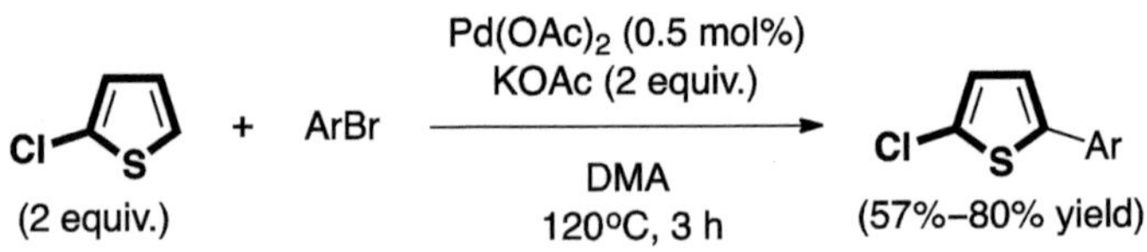

Scheme 3.92 Pd(OAc)$_2$/KOAc-catalyzed direct C-5-arylation of 2-chlorothiophene with aryl bromides.[117]

were also successfully coupled with various aryl bromides to furnish the C-5-arylated thiophenes in good yields. These reactions allowed an efficient one-step synthesis of polyfunctionalized arylated thiophenes.

The influence of the bromide position on the thiophene was then investigated.[117] However, it found that the reactivity of the direct arylation of 2-acetyl-4-bromothiophene, 2-acetyl-3-bromothiophene, and 2-formyl-3-bromothiophene were very low or failed to afford the desired products. Debromination occurred when using 2-formyl-4-bromothiophene as the coupling partner (Scheme 3.93). The coupling between 2-(4-bromothiophen-2-yl)-[1,3]dioxolane and a variety of (hetero)aryl bromides was successfully achieved to afford the desired C-5-arylated thiophenes in good yields (Scheme 3.94).

The same reaction protocol was also extended and employed to the thiophene substrate bearing sulfonyl, sulfonamide, or sulfonic ester substituents.[118] In this protocol, thiophenes bearing SO_2R-substituted at both the C-2-position[118a] and C-3-position[118b] were found to proceed regioselectively at the C-5-position, and in high yields, using a variety of (hetero)aryl bromides (Scheme 3.95). It should be noted that a catalyst loading as low as 0.1 mol% could also be successfully achieved.

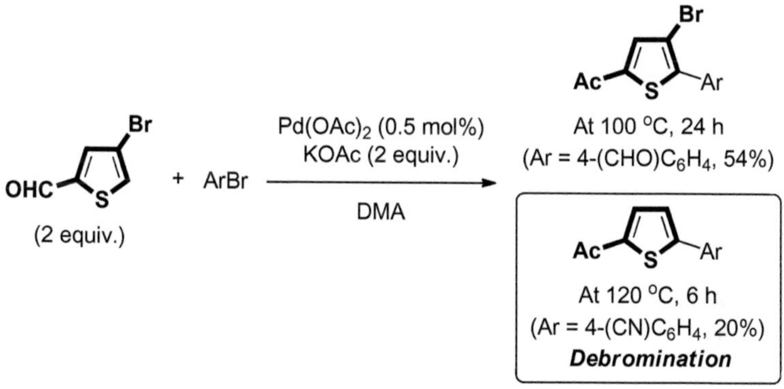

Scheme 3.93 $Pd(OAc)_2$/KOAc-catalyzed direct C-5-arylation of 2-formyl-4-bromothiophene with aryl bromides.[117]

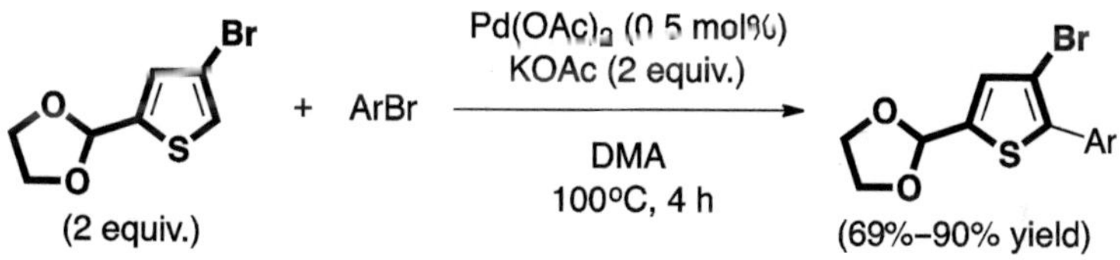

Scheme 3.94 $Pd(OAc)_2$/KOAc-catalyzed direct C-5-arylation of 2-(4-bromothiophen-2-yl)-[1,3]dioxolane with (hetero)aryl bromides.[117]

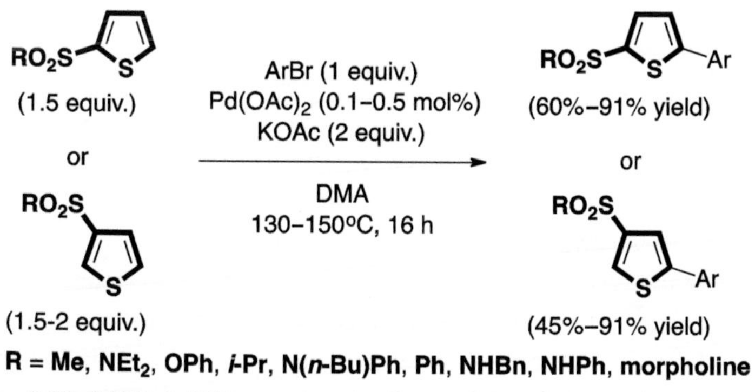

Scheme 3.95 $Pd(OAc)_2$/KOAc-catalyzed direct C-5-arylation of 2- or 3-SO_2R-substituted thiophenes with (hetero)aryl bromides.[118a]

Furthermore, thiophene bearing a cyclopropyl ketone group at the C-2-position was also compatible with the same catalytic system (Scheme 3.96).[119] The reaction proceeded regioselectively at the C-5-position without decomposition of the cyclopropyl ketone substituent. A broad range of (hetero)aryl bromides was also tolerated.

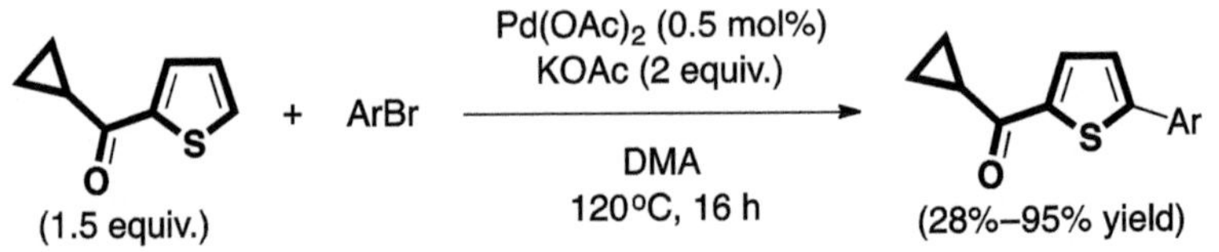

Scheme 3.96 $Pd(OAc)_2$/KOAc-catalyzed direct C-5-arylation of cyclopropyl 2-thienyl ketone with (hetero)aryl bromides.[119]

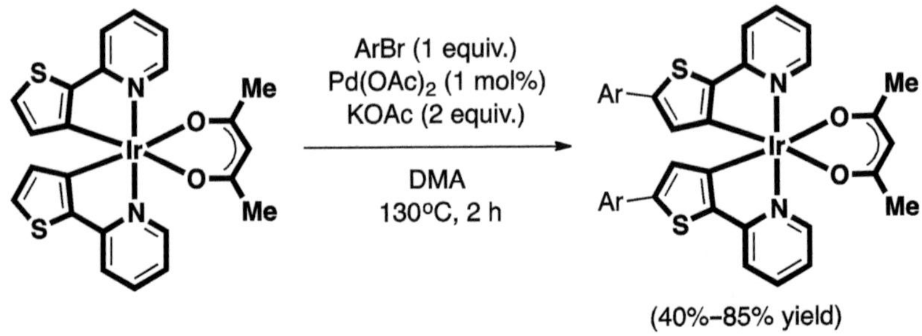

Scheme 3.97 $Pd(OAc)_2$/KOAc-catalyzed direct C-5-arylation of thiophene moiety of iridium complex.[120]

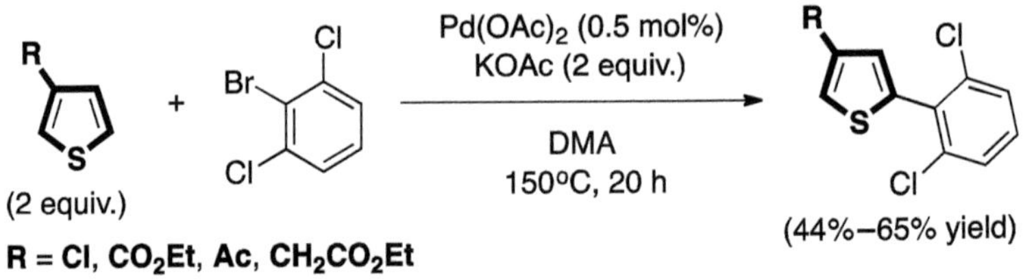

Scheme 3.98 $Pd(OAc)_2$/KOAc-catalyzed direct C-5-arylation of C-3-substituted thiophenes with 2-bromo-1,3-dichlorobenzene.[121]

This reaction protocol was also applied to the direct arylation of thiophene moieties of iridium complexes with (hetero)aryl bromides (Scheme 3.97).[120] This protocol allowed easy modification of the nature of the ligand of Ir complexes. Moreover, the photophysical properties of the new functionalized complexes were studied by means of absorption and emission spectroscopy.

The $Pd(OAc)_2$/KOAc system also allowed the coupling of C-3-substituted thiophenes with sterically hindered 2-bromo-1,3-dichlorobenzene to furnish C-5-arylated products in moderate yields with high regioselectivities (Scheme 3.98).[121]

Recently, a report from the same group described the $Pd(OAc)_2$/KOAc-catalyzed direct arylation of 3-(pyrrol-1-yl)thiophene derivatives.[53]

Scheme 3.99 Pd(OAc)$_2$/KOAc-catalyzed direct C-5-arylation of 4-methyl-3-(pyrrol-1-yl)thiophene-2-carboxylate with (hetero)aryl bromides.[53]

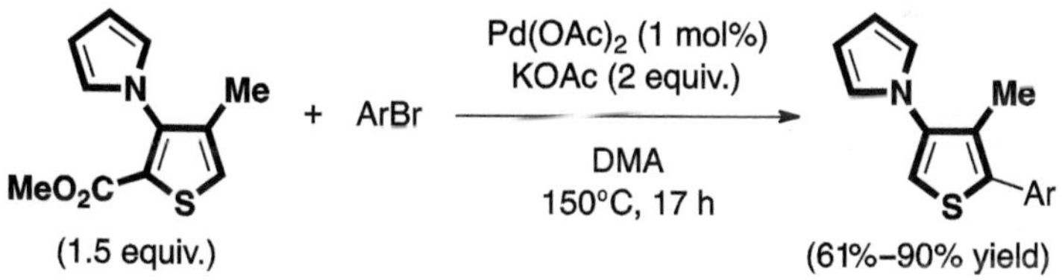

Scheme 3.100 Pd(OAc)$_2$/KOAc-catalyzed direct C-5-arylation of 4-methyl-3-(pyrrol-1-yl)thiophene-2-carboxylate with (hetero)aryl bromides with in situ decarboxylation.[53]

4-Methyl-3-(pyrrol-1-yl)thiophene-2-carboxylate could be arylated at the C-5-position with various (hetero)aryl bromides in good yields at 130°C in 2 h (Scheme 3.99). A higher temperature at 150°C and longer reaction time in 17 h led to the formation of in situ decarboxylated C-5-arylated products in good yields (Scheme 3.100). The authors revealed that the regioselectivity of the arylation of 3-(pyrrol-1-yl)thiophene derivatives was sensitive to the steric hindrance of the pyrrolyl moiety. The presence of a substituent at the C-4-position of thienyl ring was crucial to control the selectivity. Indeed, the formation of a mixture of the C-5-arylated thiophene and C-2-arylated pyrrole in a 31:35 ratio was observed when employing methyl 3-(pyrrol-1-yl)thiophene-2-carboxylate and 4-bromobenzonitrile as the coupling partner (Scheme 3.101). Moreover, the regioselectivity of the arylation of 1-(4-methylthiophen-3-yl)-pyrrole was also studied (Scheme 3.102). C-2-arylated product was obtained, which indicated that ester substituent at the C-2-position of the thienyl ring acted as a blocking group. This regioselectivity trend also matched the results that were previously reported using 3-aminothiophene derivatives (see Scheme 3.88).[114]

Doucet employed a Pd(OAc)$_2$/KOAc catalytic system for the direct arylation of benzothiophene substrates with aryl bromides.[122] This reaction took place with high regioselectivity at the C-2-position of benzothiophenes (Scheme 3.103), and it tolerated a variety of electronically

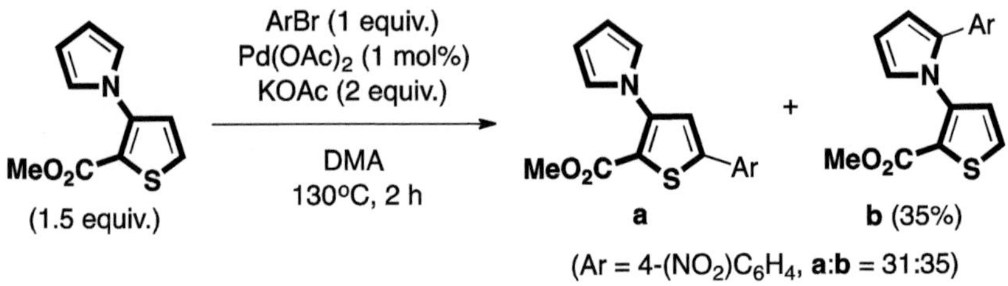

Scheme 3.101 Regioselectivity of the arylation of methyl 3-(pyrrol-1-yl)thiophene-2-carboxylate.[53]

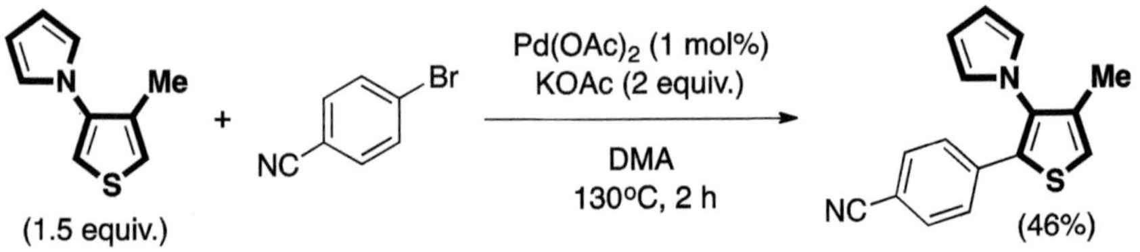

Scheme 3.102 Regioselectivity of the arylation of 1-(4-methylthiophen-3-yl)-pyrrole.[53]

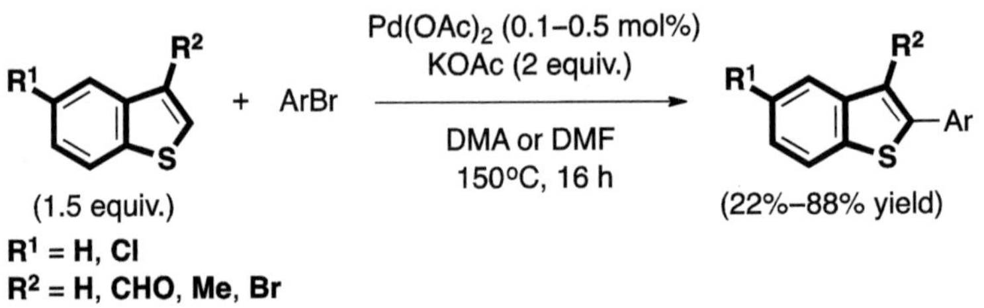

Scheme 3.103 $Pd(OAc)_2$/KOAc-catalyzed direct C-2-arylation of benzothiophenes with (hetero)aryl bromides.[122]

and sterically diverse (hetero)aryl bromides in moderate-to-good yields using low catalyst loadings down to 0.1 mol%. It should be noted that the example of the direct arylation at the C-2-position of 3-bromo-benzothiophene was also demonstrated for the first time.

For the 3-bromobenzothiophene, by employing a lower reaction temperature of 80°C and a shorter reaction time of 30 min, the reactions proceeded giving moderate-to-good yields (Scheme 3.104).[123] In order to demonstrate the synthetic potential of the prepared 2-aryl-3-bromobenzothiophenes, direct arylations with pyrrole and furan derivatives were carried out and afforded the desired products in good yields (Scheme 3.105). Furthermore, since there was a presence of C—Br bond on the thiophene, Suzuki–Miyaura coupling was also conducted and successfully achieved.

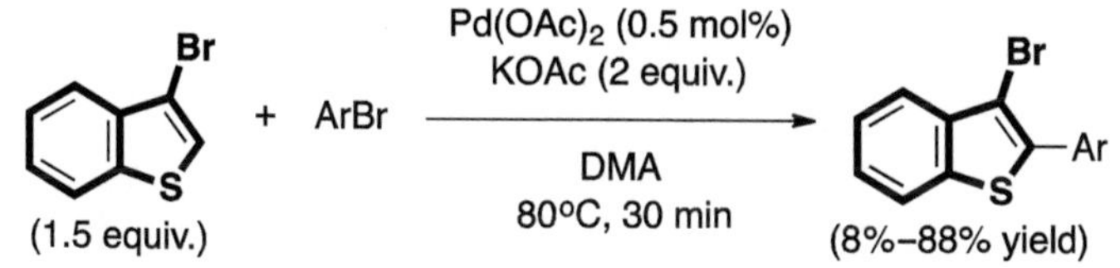

Scheme 3.104 $Pd(OAc)_2$/KOAc-catalyzed direct C-2-arylation of 3-bromobenzothiophene with (hetero)aryl bromides.[123]

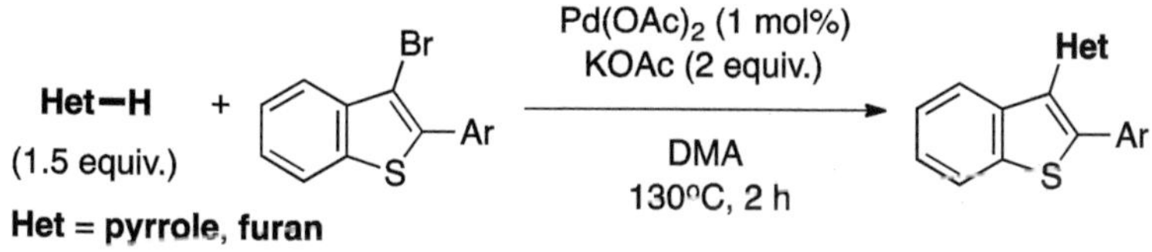

Scheme 3.105 $Pd(OAc)_2$/KOAc-catalyzed direct arylation of heteroarenes with 2-aryl-3-bromobenzothiophenes.[123]

Scheme 3.106 $Pd(OAc)_2$/dppb-catalyzed direct C-2-arylation of 3-substituted thiophenes with (hetero)aryl bromides.[124]

The $Pd(OAc)_2$/dppb system was found to be an efficient catalyst for the direct arylation of 3-substituted thiophenes.[124] The author revealed that the regioselectivity of the arylation of thiophene strongly depends on both the thiophene substituent and the nature of the aryl bromide. 3-Formyl, 3-cyano, 3-methyl, 3-hydroxymethyl, 3-bromothiophene, and ethyl 3-thiophenylacetate were examined and allowed the formation of C-2-arylated thiophenes in both good yields and regioselectivities with a variety of (hetero)bromides (Scheme 3.106). The influence of other substituents at the C-3-position of thiophene on the regioselectivity of the arylation was studied using 3-formylthiophene diethylacetal as the coupling partner (Scheme 3.107). However, the C-5-arylation was predominant and the authors suggested that this observation might be due to steric factors, as the C-2-position of 3-formylthiophene diethylacetal is more hindered than that of 3-formylthiophene. Followed by hydrolysis with hydrochloric acid, C-5-arylated 3-formylthiophenes could be obtained in moderate yields

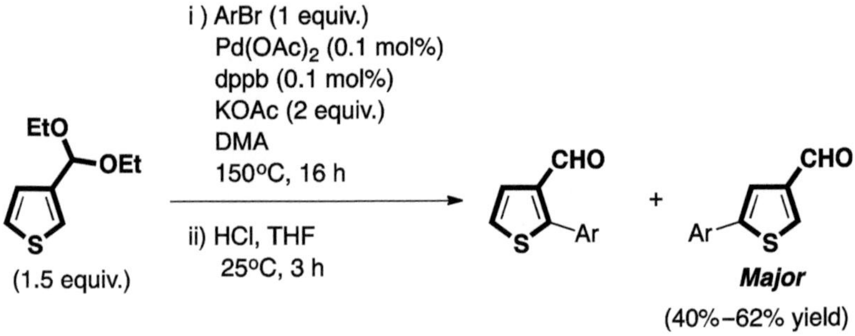

Scheme 3.107 Pd(OAc)$_2$/dppb-catalyzed direct C-5-arylation of 3-formylthiophene diethylacetal with (hetero)aryl bromides followed by hydrolysis.[124]

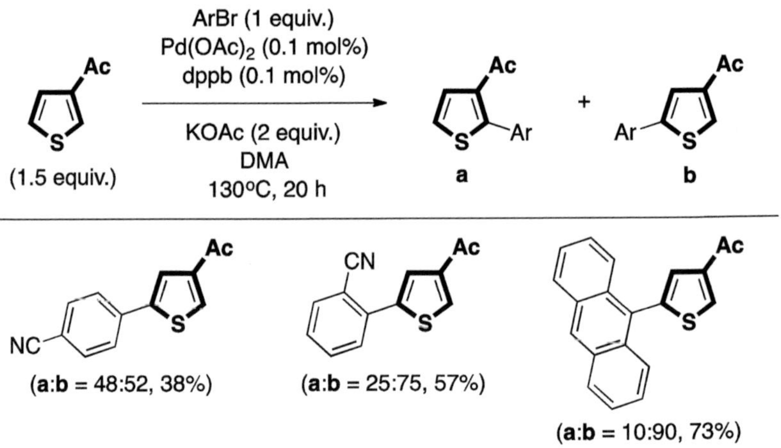

Scheme 3.108 The steric influence of aryl bromides on Pd(OAc)$_2$/dppb-catalyzed direct arylation of 3-acetylthiophene.[124]

with good regioselectivities. Thus, the regioselectivity of the arylation of thiophenes mainly depended on the steric properties of the substituents at the C-3-position rather than electronic properties. In order to confirm these observations, the slightly more congested 3-acetylthiophene was used to couple with different aryl bromides (Scheme 3.108). For 4-bromobenzonitrile, an almost equimolar amount of regioisomers was obtained. In addition, it was observed that the regioselectivity shifts to the C-5-position of thiophene in the presence of sterically more hindered 2-bromobenzonitrile. The very bulky 9-bromoanthracene led to a highly regioselective C-5-arylation. Therefore, the authors finally confirmed that the regioselectivity of the arylation mostly depended on the steric influence of both coupling partners.

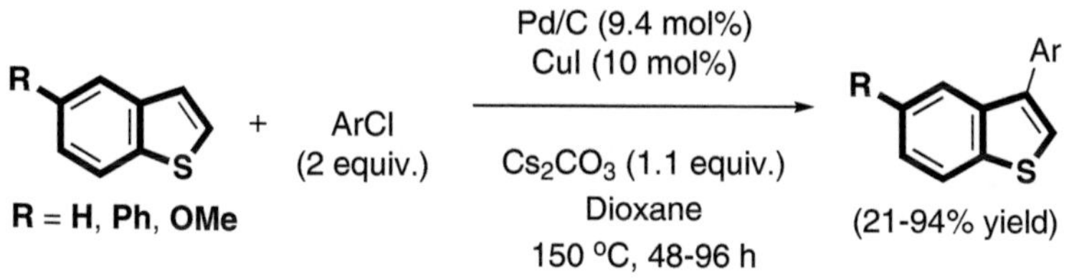

Scheme 3.109 Pd/C-catalyzed direct C-3-arylation of benzothiophenes with aryl chlorides.[125]

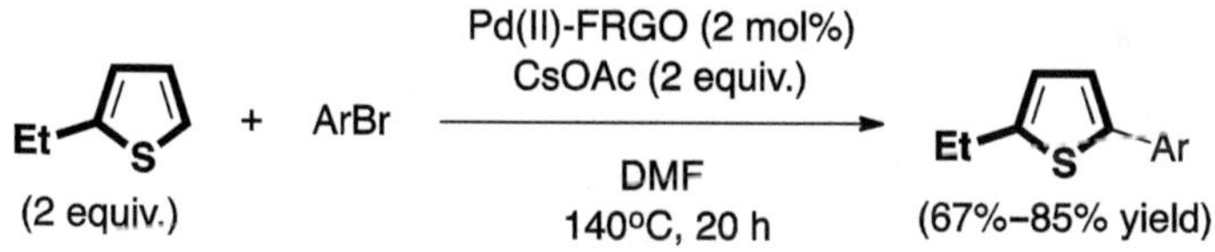

Scheme 3.110 Pd(II)-FRGO-catalyzed direct arylation of 2-ethylthiophene with aryl bromides.[126]

Glorius reported the first palladium heterogeneously-catalyzed completely regioselective C-3-arylation of benzothiophenes.[125] Benzothiophenes are coupled with various aryl chlorides using a dual catalytic system of heterogeneous palladium on charcoal and CuCl with Cs_2CO_3 in dioxane (Scheme 3.109). It should be noted that cheap and widely available catalysts and coupling partners are employed, and the reaction was both operationally simple, and tolerant of air and water.

Recently, Jia and Xi developed a novel palladium catalyst by affording Pd (II) salt on the surface of functionally reduced graphene oxide (FRGO).[126] In the combination of Pd(II)-FRGO with CsOAc in DMF, the direct arylation of 2-ethylthiophene with aryl bromides proceeded in good yields (Scheme 3.110). In addition, this catalyst also performed nicely toward Heck reaction, Suzuki–Miyaura reaction and terminal alkyne C–H functionalization.

State of the art syntheses usually rely on traditional cross-coupling reactions (Suzuki–Miyaura, Stille, Negishi, Kumada, etc.). However, these methods have drawbacks including preparation of an organometallic derivative that generates a stoichiometric amount of metal waste and is often toxic, difficult to handle, expensive, or requires a number of synthetic steps. Direct arylation has emerged as a viable alternative to traditional cross-coupling, Schipper and Fagnou reported the application of the direct arylation method (in the presence of a $Pd(OAc)_2/PCy_3 \cdot HBF_4$ catalytic system) to synthesize a broad range of thiophene-based molecules which have exhibited applications as optoelectronic materials.[127]

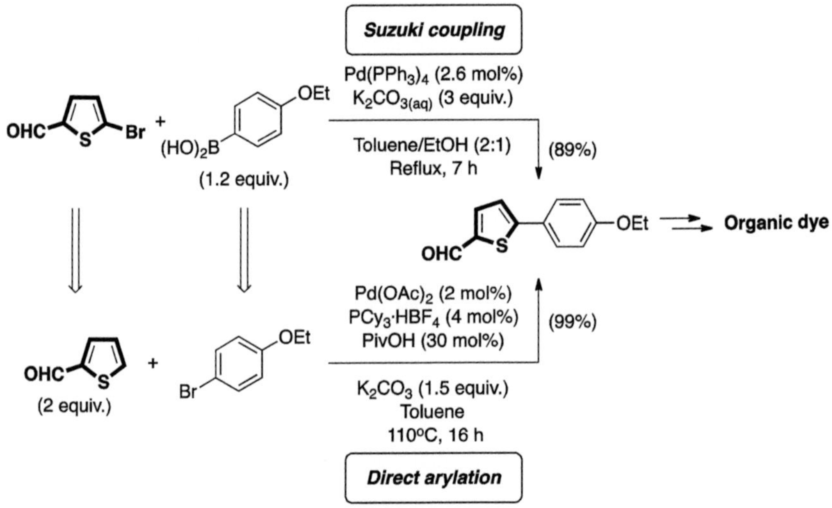

Scheme 3.111 Synthesis of organic dye via $Pd(OAc)_2/PCy_3 \cdot HBF_4$-catalyzed direct arylation pathway.[127]

For example, Park and co-workers have reported the synthesis of thiophene-based organic dye for sensitized solar cells.[128] In a previous report, the organic dye was synthesized from 5-(4-ethoxyphenyl)thiophene-2-carbaldehyde intermediate which was produced in 89% yield from a Suzuki–Miyaura coupling of 4-ethoxyphenylboronic acid and 5-bromo-2-thiophenecarboxaldehyde (Scheme 3.111). For the direct arylation, 2-thiophenecarboxaldehyde and 1-bromo-4-ethoxybenzene were used instead of bromothiophene and boronic acid, respectively, to furnish the desired intermediate in 99% yield. It would provide a more direct route using simpler and cheaper starting materials.

Melucci and co-workers reported, for the first time, that microwave-assisted double direct arylation to synthesize the target 9,10-dithienylanthracene derivatives, has particular relevance because of their suitability for light emitting[129] or charge transport based applications,[130] in only one-step and 5 min (Scheme 3.112).[131] In the presence of a $Pd(OAc)_2/PCy_3 \cdot HBF_4$ system under microwave assistance, this protocol worked more efficiently rather than a conventional multi-step Stille coupling approach.[132]

Other thiophene-based organic electronic materials were also synthesized via palladium-catalyzed direct C–H arylation reactions. Yu reported the synthesis of 3,4-ethylenedioxythiophene (EDOT)-based π-conjugated molecules in the presence of a $Pd(OAc)_2/P(m\text{-Tol})_3$ catalytic system

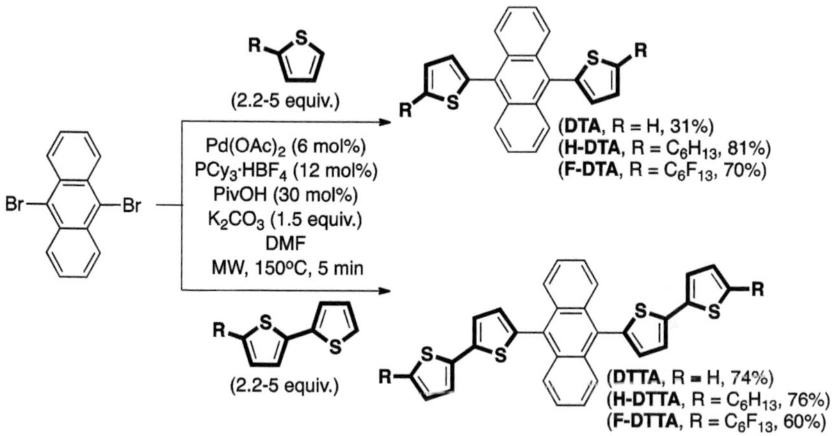

Scheme 3.112 Synthesis of 9,10-dithienylanthracene derivatives via a microwave-assisted $Pd(OAc)_2/PCy_3 \cdot HBF_4$-catalyzed direct arylation approach.[131]

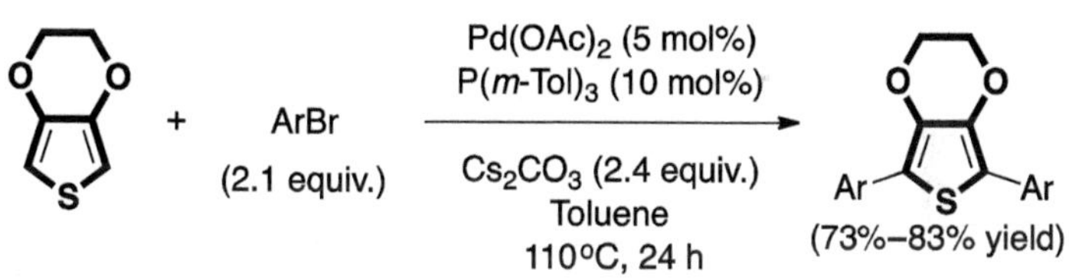

Scheme 3.113 $Pd(OAc)_2/P(m\text{-}Tol)_3$-catalyzed direct diarylation of EDOT with aryl bromides.[133]

(Scheme 3.113).[133] Notably, this protocol allowed, for the first time, the synthesis of EDOT-based π-conjugated molecules to become more straightforward and atom-economical than traditional reports.[134] The spectroscopic and electrochemical characteristics of the synthesized EDOT-based π-conjugated molecules were also examined.

Attention to the preparation of thieno[3,4-*c*]pyrrole-4,6-dione (TPD)-based derivatives has been drawn due to their unique π-conjugated system in applications as polymer,[135] organic field-effect transistors,[136] and organic photovoltaic cells (OPVCs).[137] Liu demonstrated a step-economical synthetic alternative to access a series of diarylated TPD via $Pd(OAc)_2$/$P(m\text{-}Tol)_3$-catalyzed direct C–H diarylation (Scheme 3.114).[138]

Organic photochromic compounds have attracted much attention because of the potential applications for photoswitching materials.[139] Among various photochromic systems, diarylethene is one of the best potential organic molecules over other photochromic compounds in terms of thermal stability, fatigue resistance, and intriguing photochromic

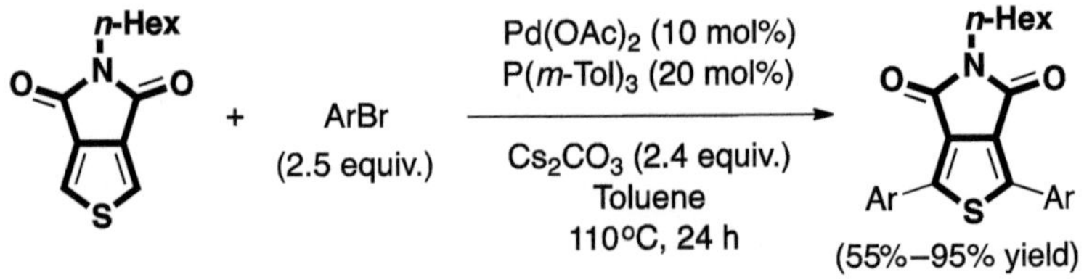

Scheme 3.114 Pd(OAc)$_2$/P(*m*-Tol)$_3$-catalyzed direct diarylation of TPD with aryl bromides.[138]

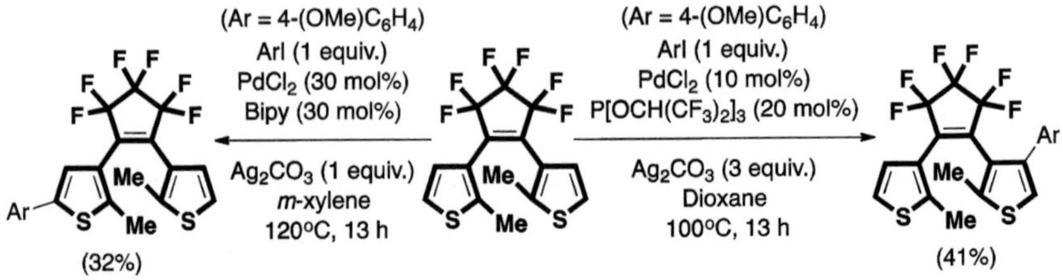

Scheme 3.115 Ligand-controlled direct regioselective arylation of DTE derivative with aryl iodides.[141]

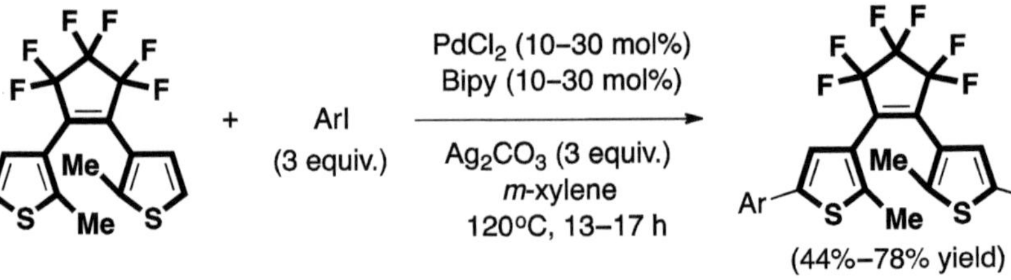

Scheme 3.116 PdCl$_2$/bipy-catalyzed direct diarylation of DTE derivative with aryl iodides.[141]

behavior.[140] Itami and Shinokubo reported ligand-controlled direct regioselective arylation on the thienyl groups of dithienylperfluorocyclopentene (DTE) derivative with various aryl iodides[141] via their previously reported conditions[108] (see Scheme 3.81). DTE could undergo arylation at the α-position by employing a $PdCl_2$/bipy system, whereas arylation at the β-position occurs using $PdCl_2/P[OCH(CF_3)_2]_3$ (Scheme 3.115). Importantly, adjusting the amount of aryl iodides that are employed could control the number of times of arylation, one equivalent of aryl iodides for monoarylation while three equivalent of aryl iodides for diarylation (Scheme 3.116). Furthermore, by combining this two-catalytic system sequentially, triarylated DTE derivatives could also be prepared.

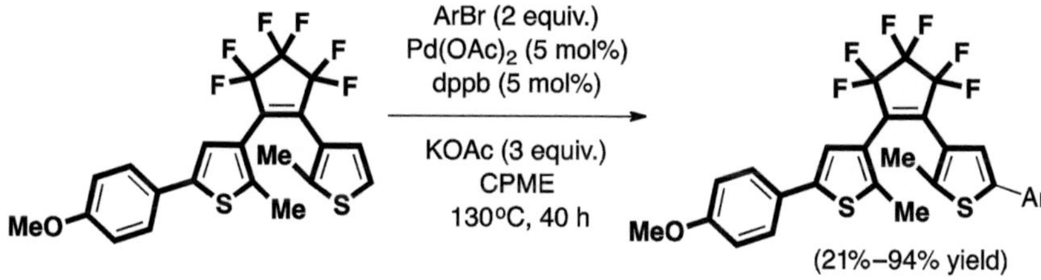

Scheme 3.117 Pd(OAc)$_2$/dppb-catalyzed direct arylation of DTE derivative with (hetero)aryl bromides.[142]

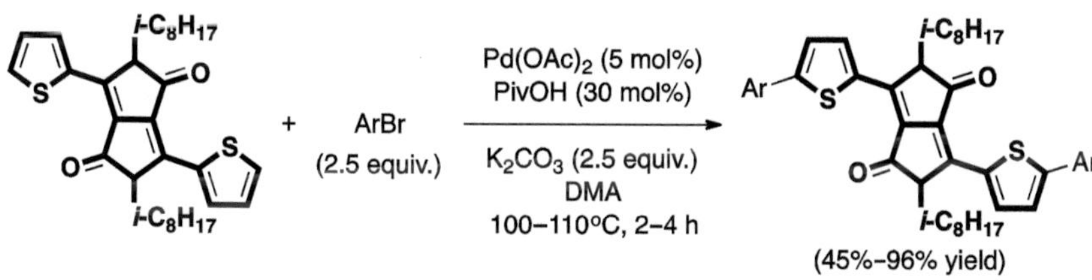

Scheme 3.118 Pd(OAc)$_2$-catalyzed direct arylation of DPP with aryl bromides.[148]

Guerchais and Doucet also developed a protocol for the arylation of DTE derivatives, using aryl bromides as the coupling partners.[142] By employing a Pd(OAc)$_2$/dppb system in cyclopentyl methyl ether (CPME), the DTE derivative could be coupled with various (hetero)aryl bromides giving moderate-to-good yields (Scheme 3.117).

Diketopyrrolopyrrole (DPP) derivatives are widely explored for optoelectronic applications, because of their unique π-conjugated systems as well as their exceptional stability,[143] such as OPVCs,[144] field-effect transistors,[145] dye-sensitized solar cells,[146] and chemical sensors.[147] Shi, Jen, and Chen synthesized a series of DPP-based π-conjugated molecules bearing diverse aryl substituents on the thiophene-DPPs via palladium-catalyzed direct arylation of C—H bonds.[148] The Pd(OAc)$_2$ catalytic system allowed the arylation of DPP with various aryl bromides to obtain the arylated DPP derivatives with excellent yields in short reaction time (Scheme 3.118). In addition, bithiophene-functionalized DPP compounds could be afforded in moderate yields by the coupling of thiophenes with bromo-DPP (Scheme 3.119). Gratifyingly, the arylation of pentafluorobenzene by bromo-DPP was also successfully achieved.

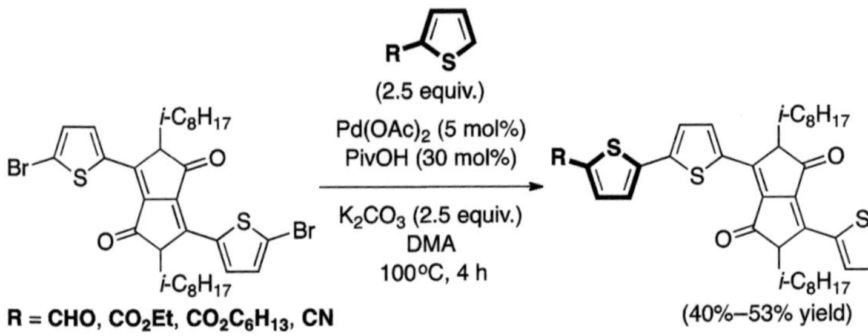

Scheme 3.119 $Pd(OAc)_2$-catalyzed direct arylation of thiophenes with bromo-DPP.[148]

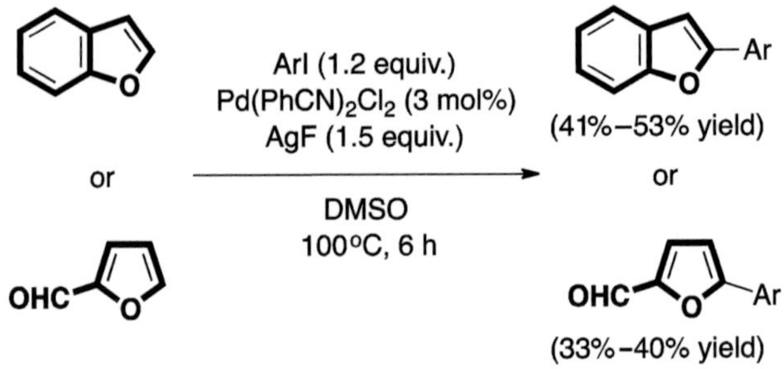

Scheme 3.120 $Pd(PhCN)_2Cl_2$-catalyzed direct α-arylation of benzofuran or 2-formyl furan with aryl iodides.[150]

3.2.2.5 Furans and Benzofurans

Easy access to a variety of arylated (benzo)furans is an important field for research in both organic and green chemistry because of the biological properties of these derivatives.[149]

Mori developed a simple protocol for the direct arylation of benzofuran with several aryl iodides via $Pd(PhCN)_2Cl_2$-catalyzed C—H functionalization (Scheme 3.120).[150] The $Pd(PhCN)_2Cl_2$ catalytic system, consisting of AgF as the additive in DMSO, allowed α-arylation of benzofuran or 2-formyl furan to occur to form the corresponding products in moderate yields. Notably, this protocol also allowed the direct C—Si functionalization of trimethylsilyl benzofuran with aryl iodides.

Doucet subsequently reported an effective $Pd(OAc)_2$/KOAc-catalyzed direct C-3-arylation of 2-substituted benzofurans to prepare 3-arylbenzofuran derivatives.[151] This protocol was carried out using a broad variety of (hetero) aryl bromides as the coupling partners with low catalyst loadings of 0.1—0.2 mol% to furnish the desired products in moderate-to-good yields (Scheme 3.121).

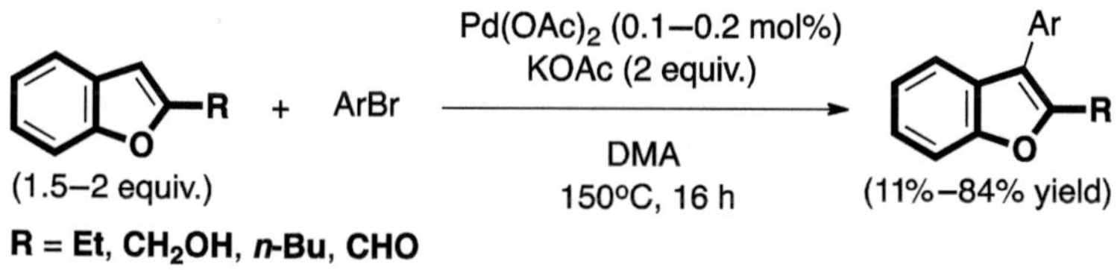

Scheme 3.121 $Pd(OAc)_2$/KOAc-catalyzed direct C-3-arylation of 2-substituted benzofurans with (hetero)aryl bromides.[151]

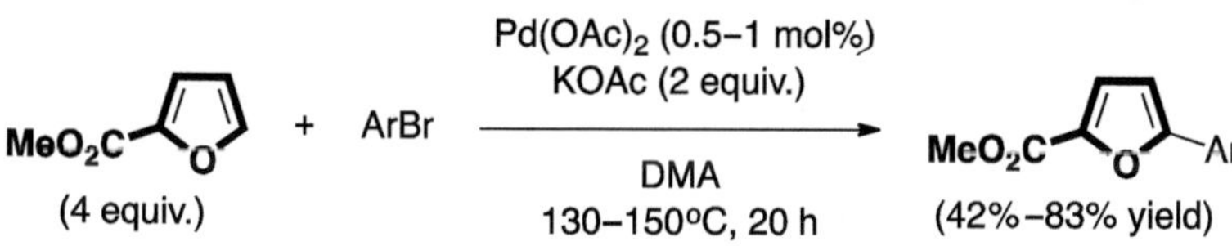

Scheme 3.122 $Pd(OAc)_2$/KOAc-catalyzed direct C-5-arylation of methyl 2-furoate with (hetero)aryl bromides.[152]

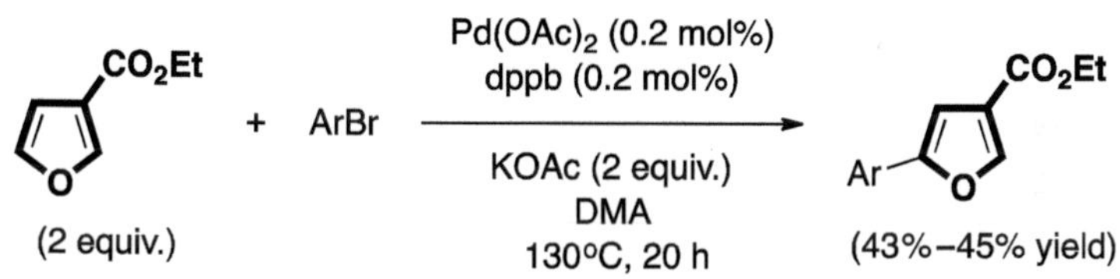

Scheme 3.123 $Pd(OAc)_2$/dppb-catalyzed direct C-5-arylation of ethyl 3-furoate with aryl bromides.[124]

Doucet also reported the direct C-5-arylation of methyl 2-furoate with (hetero)aryl bromides using a $Pd(OAc)_2$/KOAc catalytic system (Scheme 3.122).[152] This protocol allowed the arylation to take place regioselectively at the C-5-position without decarboxylation. It should be noted that the authors also examined the functionalization of the C-2-position by decarboxylative coupling with the addition of KOH and increasing the reaction temperature to 150°C.

As discussed previously, Doucet carried out a study of the influence of substituents on the regioselectivity of the arylation of thiophene via a Pd $(OAc)_2$/dppb system (see Scheme 3.106).[124] The authors also studied the regioselectivity of the arylation of ethyl 3-furoate, the major C-5-arylated products were obtained in moderate yields (Scheme 3.123).

Doucet and Hierso reported the synthesis of novel ferrocenyl diphosphines, **28–31** (Fig. 3.7), that were not sensitive to air and moisture.[153] These phosphines, as ligands, were employed for the direct arylation of *n*-butyl furan with aryl chlorides. In the presence of $Pd(OAc)_2$ as the

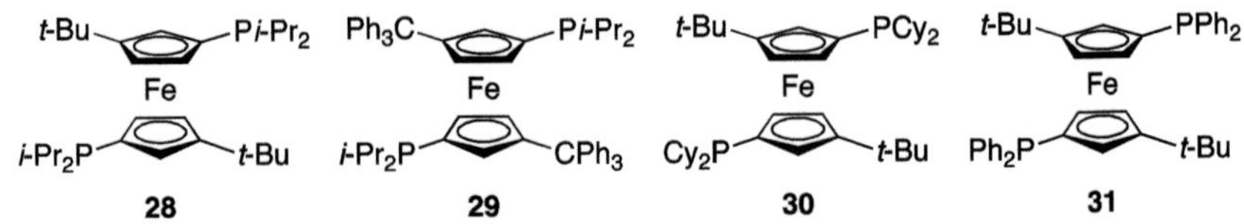

Figure 3.7 Ferrocenyl diphosphines **28–31**.[153]

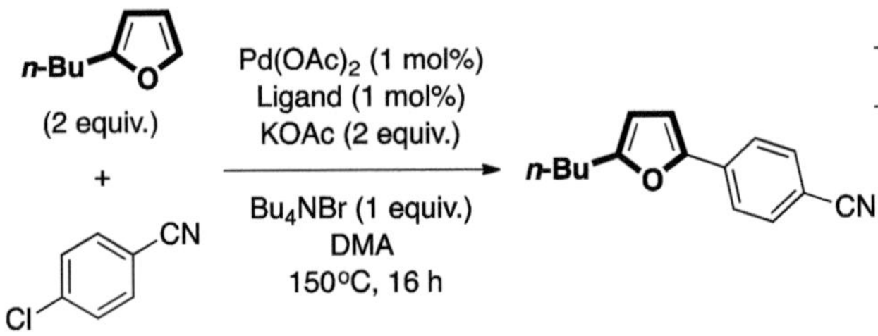

Ligand	Yield (%)
29	88
30	87
31	90
32	79
dppf	59

Scheme 3.124 $Pd(OAc)_2$-catalyzed direct C-5-arylation of *n*-butyl furan with 4-chlorobenzonitrile.[153]

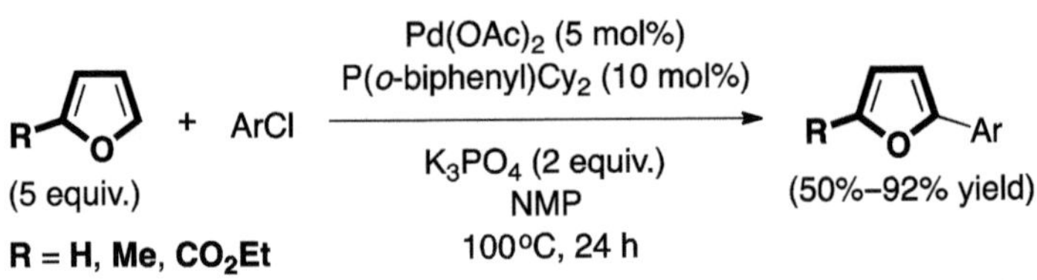

Scheme 3.125 $Pd(OAc)_2$/P(*o*-biphenyl)Cy_2-catalyzed direct arylation of furans with (hetero)aryl chlorides.[66]

catalyst, the arylation took place regioselectively at the C-5-position of *n*-butyl furan to afford the desired product in good yields (Scheme 3.124). In addition, a comparison with the $Pd(OAc)_2$/dppf system was conducted, and a better result for the diphosphines was obtained.

As discussed in the previous section, Daugulis demonstrated the use of a $Pd(OAc)_2$/P(*o*-biphenyl)Cy_2 catalytic system to catalyze the direct arylation of indoles (see Scheme 3.46), pyrroles (see Scheme 3.47), or furans with aryl chlorides.[66] Furan, furan-2-carboxylic acid ethyl ester, and 2-methylfuran could be arylated with various (hetero)aryl chlorides to give the corresponding α-arylated products in moderate-to-good yields (Scheme 3.125). Furthermore, dichloroarenes can be reacted with furan to afford diarylated product in high yields.

Bertounesque and co-workers described an effective synthetic route for 3-arylbenzofurans via $Pd(OAc)_2$/P*t*-$Bu_2Me \cdot HBF_4$-catalyzed direct C-3-arylation of 2-substituted benzofurans with (hetero)aryl bromides.[154]

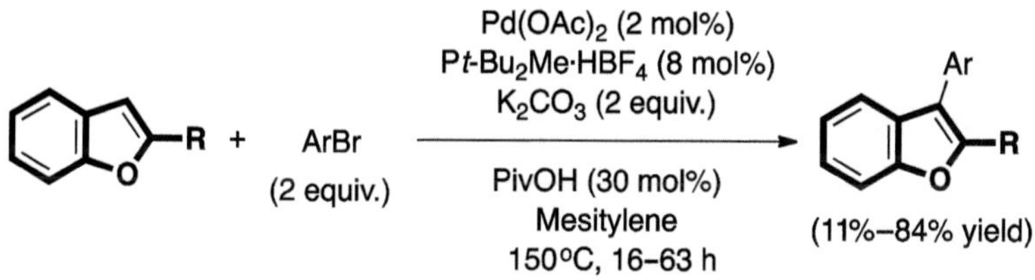

R = C(O)Ph, NO_2, CN, Ac, C(O)t-Bu, CO_2Me, CH_2OH, CHMeOH, CHPhOH

Scheme 3.126 $Pd(OAc)_2$/P*t*-$Bu_2Me \cdot HBF_4$-catalyzed direct C-3-arylation of 2-substituted benzofurans with (hetero)aryl bromides.[154]

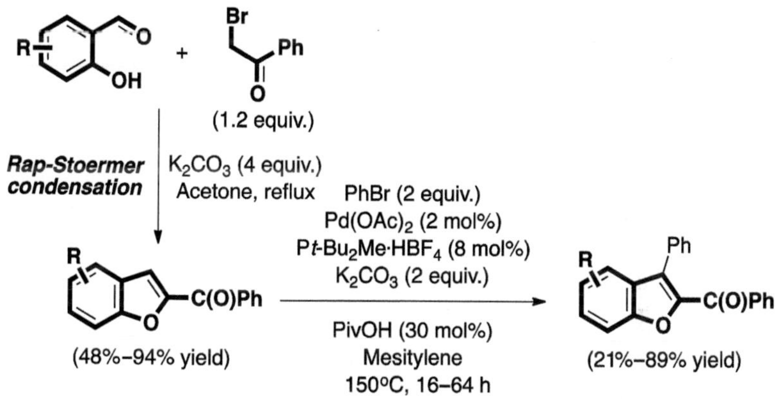

R = 5,6-$(OMe)_2$, 5,6,7-$(OMe)_3$, 4,5,6-$(OMe)_2$, 4-OMe, 4-OH, 4,6-Me_2, 7-CO_2Me

Scheme 3.127 Rap – Stoermer condensation followed by $Pd(OAc)_2$/P*t*-$Bu_2Me \cdot HBF_4$-catalyzed direct C-3-arylation of polysubstituted 2-benzoylbenzofurans with bromobenzene.[154]

This protocol could tolerate a variety of functional groups on the aryl bromides and afford the corresponding products in moderate-to-good yields (Scheme 3.126). In addition, this protocol had also been successfully extended to polysubstituted benzofurans that were synthesized by Rap–Stoermer condensation of substituted salicylaldehyde and bromoacetophenone. Subsequent arylation of the polysubstituted 2-benzoylbenzofurans was carried out to obtain the corresponding 3-arylbenzofurans with moderate-to-good yields (Scheme 3.127).

Since the C-3-position of benzofuran is known to be reactive for direct arylation with aryl halides using palladium catalysts with acetate bases in polar solvents,[151] Doucet described the $PdCl(C_3H_5)(dppb)$/KOAc-catalyzed C-3-arylation of 2-arylbenzofurans with (hetero)aryl bromides.[155] Selective C-3-arylation of 2-arylbenzofurans were observed smoothly to afford desired products in good yields (Scheme 3.128).

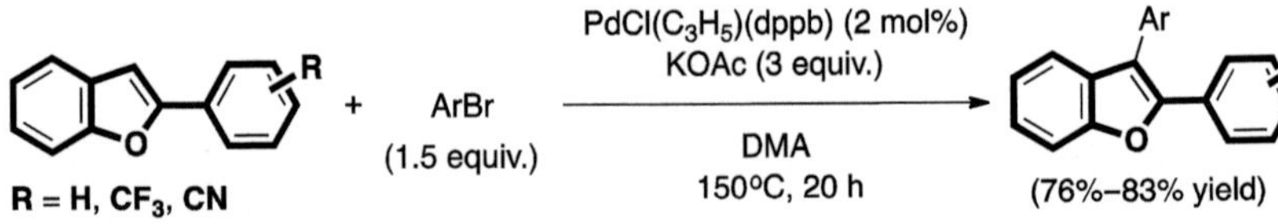

Scheme 3.128 PdCl(C_3H_5)(dppb)/KOAc-catalyzed direct C-3-arylation of 2-arylbenzofurans with (hetero)aryl bromides.[155]

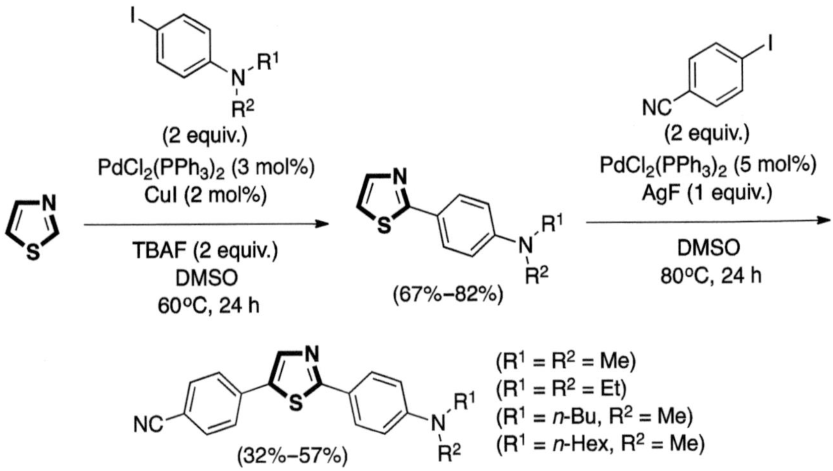

Scheme 3.129 Synthesis of donor-acceptor-type 2,5-diarylthiazoles via $PdCl_2(PPh_3)_3$-catalyzed direct arylation of thiazole with aryl iodides.[157]

3.2.2.6 Thiazoles and Benzothiazoles

(Benzo)thiazoles attached with (hetero)aryl represent privileged structural motifs that are frequently utilized in bioactive compounds and pharmaceuticals.[156]

Mori and co-workers reported a protocol to synthesize donor-acceptor-type 2,5-diarylthiazoles bearing electron-donating *N,N*-dialkylamine and an electron-withdrawing cyano group at the C-2- and C-5-position respectively via palladium-catalyzed direct C—H arylation reactions.[157] The 2,5-diarylthiazoles were first synthesized via the C—H arylation of thiazole at the C-2-position with a $PdCl_2(PPh_3)_3$/CuI catalyst in the presence of tetrabutylammonium fluoride (TBAF) as the activator (Scheme 3.129). The C-5-arylation of the 2-arylated thiazole was then carried out in the presence of AgF to afford the donor-acceptor-type 2,5-diarylthiazoles with *N,N*-dialkylamine groups of different chain lengths. The UV/Vis absorption, photoluminescence, and electrochemical behavior of the donor-acceptor-type 2,5-diarylthiazoles were also studied.

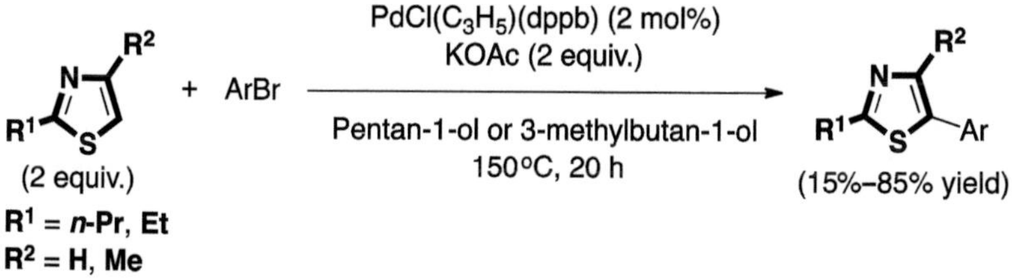

Scheme 3.130 $PdCl(C_3H_5)(dppb)/KOAc$-catalyzed direct arylation of thiazole derivatives with (hetero)aryl bromides in pentan-1-ol or 3-methylbutan-1-ol.[158]

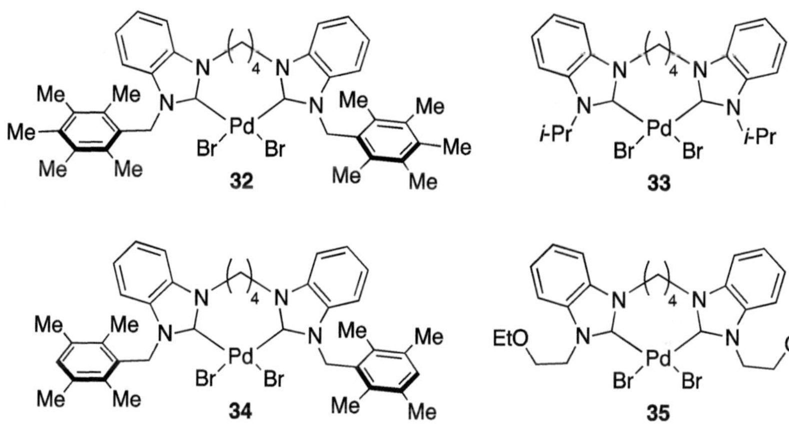

Figure 3.8 Pd–NHC complexes **32–35.**[159]

In developing the eco-compatible reaction conditions for the direct arylation reactions, Doucet employed less toxic alcohol, for the first time, to the $PdCl(C_3H_5)(dppb)/KOAc$-catalyzed direct arylation of thiazoles with (hetero)aryl bromides.[158] Pentan-1-ol or 3-methylbutan-1-ol was chosen as the solvent for the direct arylation reactions, which is moderately soluble in water and miscible with common organic solvents, and is readily biodegradable (Scheme 3.130).

Özdemir synthesized a new set of bis(benzimidazolium) carbene-palladium catalysts (Pd-NHC), **32–35** (Fig. 3.8), which were employed for the direct arylation of benzothiazole by using aryl bromides as the coupling partners.[159] In the presence of K_3PO_4 as the base in NMP, this set of Pd–NHC catalysts could promote the arylation of benzothiazole with good yields (Scheme 3.131).

Itami accomplished a programmed synthesis of privileged arylthiazoles via sequential palladium-catalyzed C–H couplings.[160] This versatile protocol could supply all possible arylthiazole substitution patterns, such as

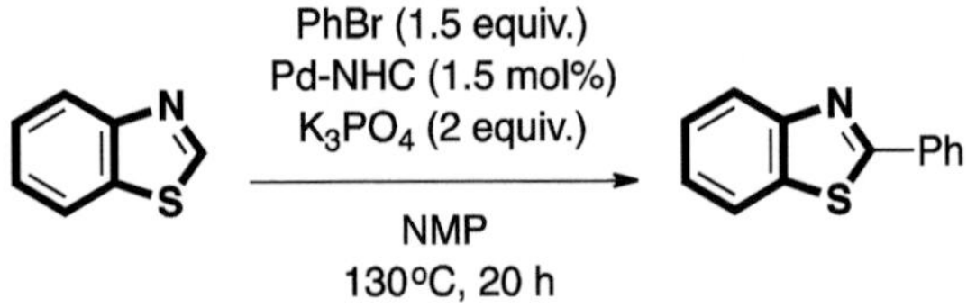

Pd-NHC	Yield (%)
32	89
33	75
34	61
35	65

Scheme 3.131 Pd-NHC-catalyzed direct arylation of benzothiazole with bromobenzene.[159]

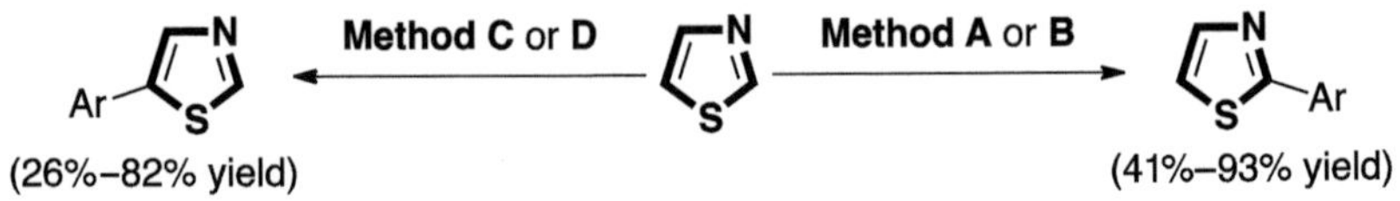

Method A: Thiazole (2 equiv.), ArI (1 equiv.), $Pd(OAc)_2$ (5 mol%), CuI (2 equiv.), DMF, 140°C, 16–40 h
Method B: Thiazole (1.5 equiv.), ArBr (1 equiv.) $Pd(P\textit{t}\text{-}Bu_3)_2$ (2 mol%), LiO*t*-Bu (1.5 equiv.), Dioxane, 100°C, 12 h
Method C: Thiazole (1.5 equiv.), ArI (1 equiv.), $Pd(OAc)_2$ (5 mol%) $P\textit{t}\text{-}Bu_2Me{\cdot}HBF_4$ (10 mo%), Cs_2CO_3 (1.5 equiv.), *t*-amylOH, 80°C, 36 h
Method D: Thiazole (2 equiv.), ArBr (1 equiv.) $Pd(OAc)_2$ (1 mol%), KOAc (2 equiv.), DMA, 130°C, 20 h

Scheme 3.132 Pd-catalyzed direct arylation of thiazole with (hetero)aryl iodides/bromides.[160]

2-aryl, 4-aryl, 5-aryl, 2,4-diaryl, 2,5-diaryl, 4,5-diaryl, and 2,4,5-triaryl from an unfunctionalized thiazole. An unfunctionalized thiazole, which could undergo C-2- or C-5-arylation with various (hetero)aryl halides by employing different catalytic systems to afford the desired products in moderate-to-good yields was used as the starting point (Scheme 3.132). The C-4-arylation could also be accomplished by introducing arylboronic acids as the coupling partners. By combining the developed catalytic system in a sequential manner, triarylthiazoles could be afforded easily with all substitution patterns (Scheme 3.133). It should be highlighted that the authors have generated 150 arylthiazoles by using these methodologies that could be applied to synthesize different biologically active moieties.[161]

3.2.2.7 Oxazoles and Benzoxazoles

Consecutive linked oxazole sequences are found in a variety of biological natural products,[162] such as bisoxazole hennoxazole A[163] (antiherpes simplex virus activity), and the antifungal trisoxazole ulapualide A.[164]

Greaney reported an efficient protocol to synthesize 2,5-disubstituted oxazoles via Negishi cross-coupling[165] followed by palladium-catalyzed

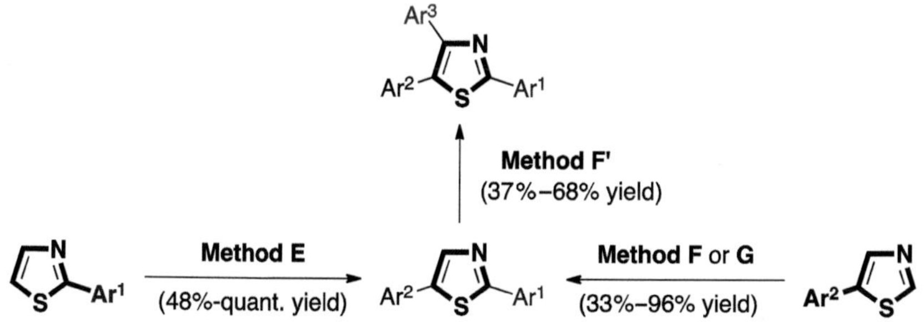

Method E: Thiazole (1.5 equiv.), Ar^2I (1 equiv.), $PdCl_2$(bipy) (5 mol%) Ag_2CO_3 (1 equiv.), Dioxane, 120–150°C, 22–36 h
Method F: Thiazole (1 equiv.), Ar^1I (1.1 equiv.), $Pd(phen)_2(PF_6)_2$ (5 mol%) Cs_2CO_3 (1.1 equiv.), DMA, 130–150°C, 18 h
Method G: Thiazole (1 equiv.), Ar^1I (1.2 equiv.), $PdCl_2(dppf)\cdot CH_2Cl_2$ (5 mol%) PPh_3 (10 mol%), Ag_2CO_3 (2 equiv.), H_2O, 60°C, 24 h
Method F': Thiazole (1 equiv.), Ar^3I (1.1 equiv.), $Pd(phen)_2(PF_6)_2$ (5 mol%) Cs_2CO_3 (1.1 equiv.), DMA, 130°C, 40 h

Scheme 3.133 Synthesis of triarylthiazoles via Pd-catalyzed direct arylation arylthiazoles with (hetero)aryl iodides.[160]

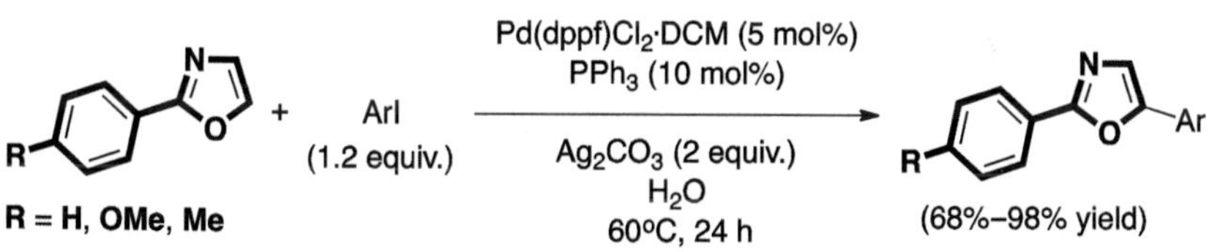

Scheme 3.134 $Pd(dppf)Cl_2\cdot DCM/PPh_3$-catalyzed direct arylation of 2-substituted oxazoles with aryl iodides.[166]

direct arylation.[166] Particularly noteworthy was that using the protocol consisting of $Pd(dppf)Cl_2\cdot DCM/PPh_3$ as the catalytic system in water, a set of 2-aryloxazoles could couple with a wide range of aryl iodides in good-to-excellent yields (Scheme 3.134). Having established this methodology for synthesis of 2,5-diaryloxazoles, the authors constructed two oxazole natural products, balsoxin and texaline, which have antimycobacterial activity against *Mycobacterium tuberculosis*, *M. avium*, and *M. kansasii*.[167]

Greaney then used the same catalytic protocol with $Pd(dppf)Cl_2$ as the catalyst for the direct C-2-arylation of 5-substituted oxazoles with (hetero)aryl iodides (Scheme 3.135).[168] In addition, synthesis of bis- and trisoxazoles could be achieved when 2-triisopropylsilyl-4-iodooxazole was used as the electrophile in the arylation and the Hermann–Beller palladacycle[169] was employed as the catalyst.

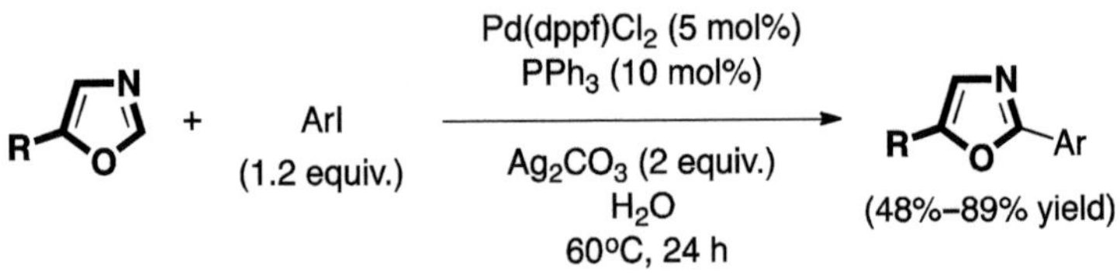

R = Ph, 4-(OMe)C_6H_4, 2-ClC_6H_4, CO_2Et

Scheme 3.135 Pd(dppf)Cl_2/PPh_3-catalyzed direct C-2-arylation of 5-substituted oxazoles with (hetero)aryl iodides.[168]

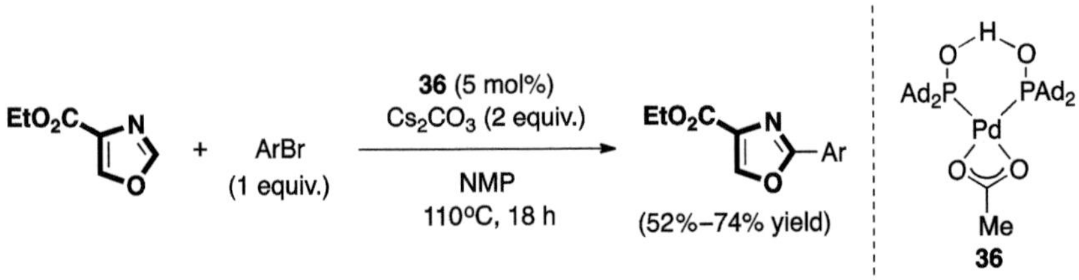

Scheme 3.136 Pd-SPO-catalyzed direct C-2-arylation of ethyl 4-oxazolecarboxylate with (hetero)aryl bromides.[170]

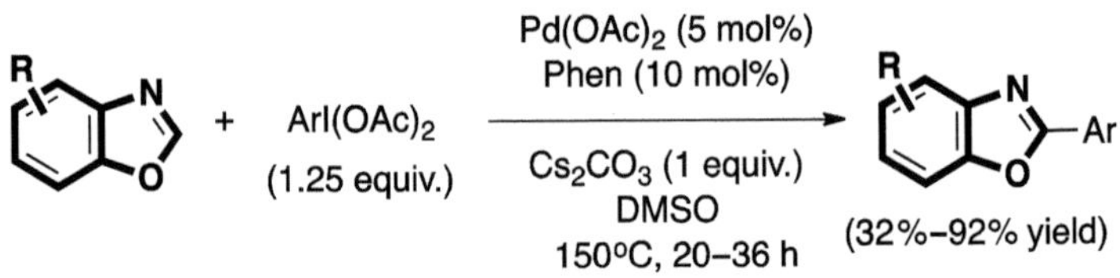

R = 5-Me, 6-Me, 5-*t*-Bu, 5-Cl, 5-NO_2

Scheme 3.137 Pd$(OAc)_2$/phen-catalyzed direct arylation of benzoxazole with iodobenzene diacetates.[171]

Ackermann et al. developed an efficient protocol for the direct C-2-arylation of oxazole with aryl bromides using a palladium complex derived from an air-stable SPO **36**.[170] Ethyl 4-oxazolecarboxylate could be coupled with various (hetero)aryl bromides to afford C-2-arylated products in good yields (Scheme 3.136). Furthermore, this protocol was also suitable for the direct benzylation of oxazole and arylation of oxazoline.

Chen and Cheng developed the Pd$(OAc)_2$/phen-catalyzed direct arylation of benzoxazole with iodobenzene diacetates as the coupling partners, proceeding in moderate-to-good yields (Scheme 3.137).[171] A further study by Fairlamb on the role of the iodobenzene diacetates in the arylation process was performed, showing the degradation to iodobenzene to give the arylated products.[172]

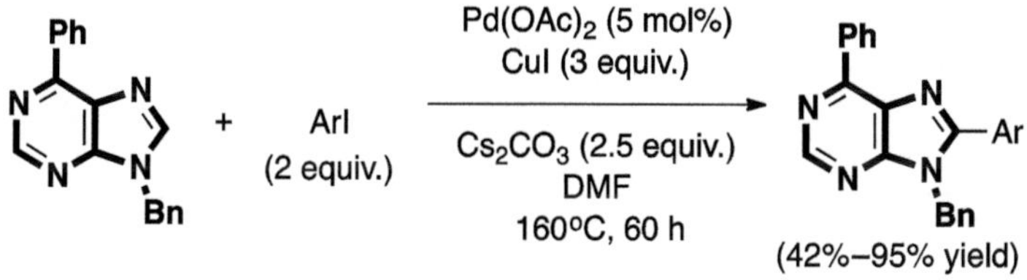

Scheme 3.138 Pd(OAc)$_2$-catalyzed direct C-8-arylation of purine with (hetero)aryl iodides.[177]

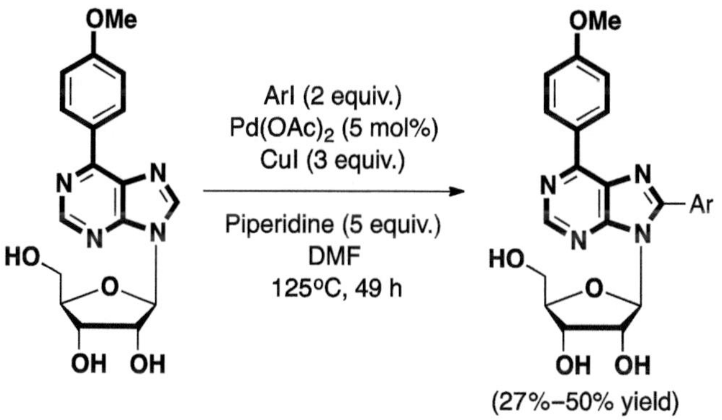

Scheme 3.139 Pd(OAc)$_2$-catalyzed direct C-8-arylation of 6-(4-methoxyphenyl)purine ribonucleoside with aryl iodides.[178]

3.2.2.8 *Fused Heteroarenes*

In recent years, fused ring systems have attracted much attention in the field of drug discovery, since such systems are found in compounds showing a wide range of biological activities.[173] Therefore, many research groups wish to introduce additional functionalities to the fused ring systems, in which direct arylation is one of the most efficient methods for this purpose.

Substituted-purines display a wide range of biological activities, such as inhibition of kinases,[174] tubulin polymerization,[175] and antagonist effects to receptors.[176] Hocek disclosed Pd(OAc)$_2$-catalyzed direct C-8-arylation of purine with (hetero)aryl iodides in the presence of CuI and Cs_2CO_3 (Scheme 3.138).[177] Moreover, the direct arylation of purine nucleosides, for the first time, was also achieved with a similar catalytic system.[178] The use of piperidine as a base made it possible to decrease the reaction temperature and apply this system for arylation of nucleosides. Unprotected purine nucleosides, such as 6-(4-methoxyphenyl)purine ribonucleoside, could be easily arylated by diverse aryl iodides at the C-8-position (Scheme 3.139).

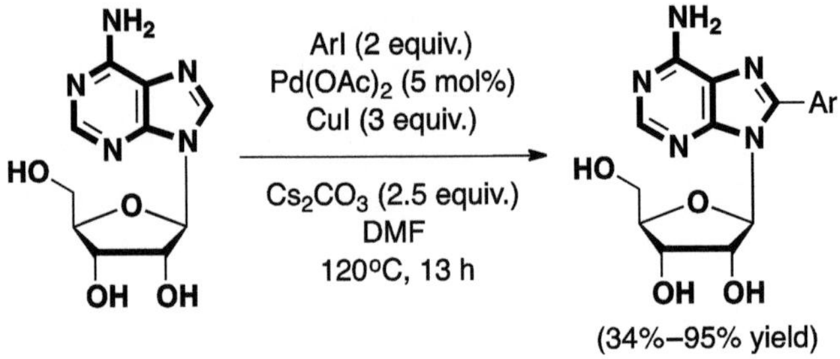

Scheme 3.140 Pd(OAc)$_2$-catalyzed direct C-8-arylation of adenosine with aryl iodides.[179]

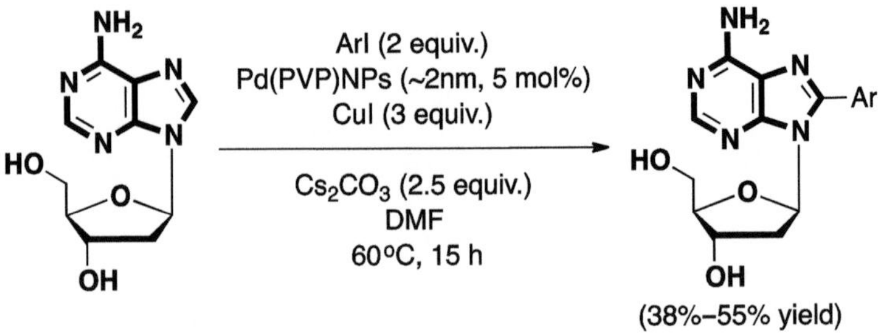

Scheme 3.141 Pd(PVP)NPs-catalyzed direct C-8-arylation of 2′deoxyadenosine with aryl iodides.[180]

Fairlamb also used the same catalytic system for the direct arylation of adenosine with various aryl iodides (Scheme 3.140).[179] The unprotected adenosine could be smoothly coupled to afford the C-8-arylated products in moderate-to-good yields without the reactions with the free $-NH_2$ substituent.

Fairlamb then employed the palladium nanoparticles with polyvinylpyrrolidinone Pd(PVP)NPs for the direct arylation of 2′deoxyadenosine with aryl iodides (Scheme 3.141).[180] The catalytic activity of Pd(PVP) NPs was found to be commensurate with Pd(OAc)$_2$.

Azaindoles, which possess the same [4.3]-bicyclic indene architecture as indoles, but with a second nitrogen in the azine ring, are important subunits in medicinal chemistry.[181] Fagnou and co-workers showed site-selective azaindole arylation at azine and azole rings via *N*-oxide activation.[182] Azaindoles have both an electron-rich azole and electron-deficient azine ring. The authors developed a protocol to selectively functionalize both rings by employing *N*-oxide activation. The arylation of azole ring is herein discussed first, Fagnous found that, by modifying the Larrosa's protocol (see Scheme 3.38)[56] slightly (changing solvent to DMF

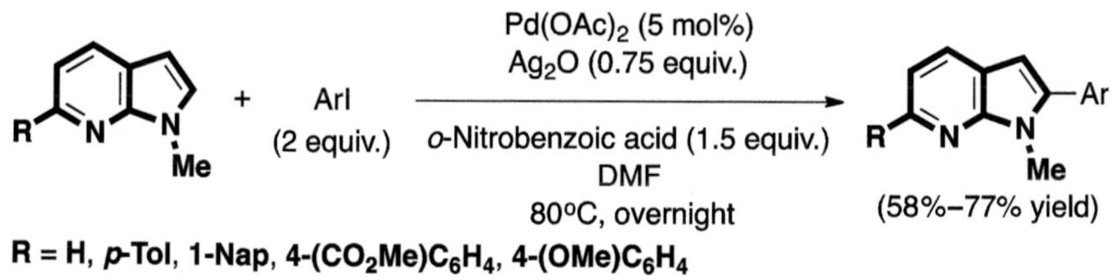

Scheme 3.142 $Pd(OAc)_2$-catalyzed direct C-2-arylation of *N*-methyl-7-azaindole with aryl iodides.[182]

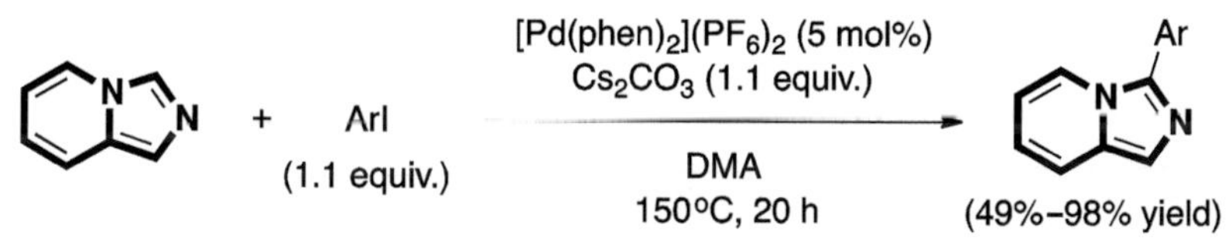

Scheme 3.143 $[Pd(phen)_2](PF_6)_2$-catalyzed direct C-3-arylation of imidazo[1,5-*a*]pyridine with (hetero)aryl iodides.[186]

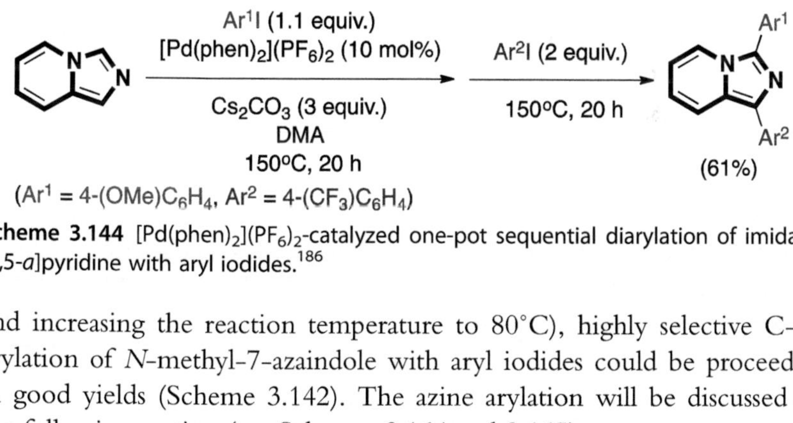

Scheme 3.144 $[Pd(phen)_2](PF_6)_2$-catalyzed one-pot sequential diarylation of imidazo [1,5-*a*]pyridine with aryl iodides.[186]

and increasing the reaction temperature to 80°C), highly selective C-2-arylation of *N*-methyl-7-azaindole with aryl iodides could be proceeded in good yields (Scheme 3.142). The azine arylation will be discussed in the following section (see Schemes 3.164 and 3.165).

Synthesis of imidazo[1,5-*a*]pyridine derivatives has recently attracted increasing attention, due to their bioactive properties. Imidazo[1,5-*a*]pyridine derivatives have been used in pharmaceuticals, such as HIV-protease inhibitors,[183] cardiotonic agents,[184] and aromatase inhibitors in estrogen-dependent diseases.[185] Shibahara and Murai reported the direct C-3-arylation of imidazo[1,5-*a*]pyridine with (hetero)aryl iodides by using a $[Pd(phen)_2](PF_6)_2/Cs_2CO_3$ catalytic system to afford the C-3-arylated products in moderate-to-excellent yields (Scheme 3.143).[186] In addition, the direct one-pot sequential diarylation at the C-3- and C-1-position of unsubstituted imidazo[1,5-*a*]pyridines with different aryl groups was successfully achieved (Scheme 3.144).

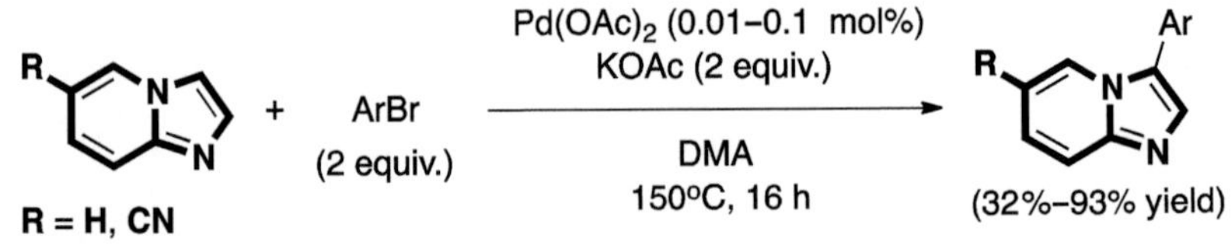

Scheme 3.145 $Pd(OAc)_2/KOAc$-catalyzed direct C-3-arylation of imidazo[1,2-*a*]pyridine with (hetero)aryl bromides.[189]

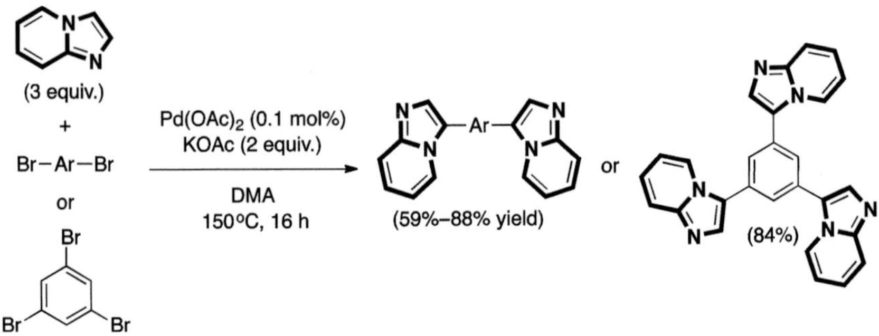

Scheme 3.146 $Pd(OAc)_2/KOAc$-catalyzed direct di- triarylations of imidazo[1,2-*a*]pyridine with di- or tribromobenzene derivatives.[189]

Multisubstituted imidazo[1,2-*a*]pyridines are common structural motifs in natural products and pharmaceuticals.[187] Due to their wide range of biological activities,[188] Doucet developed the direct arylation of imidazo[1,2-*a*]pyridine at the C-3-position with (hetero)aryl bromides by employing a $Pd(OAc)_2/KOAc$ catalytic system (Scheme 3.145).[189] It should be noted that this catalytic system could also be performed effectively at low catalyst loading down to 0.01 mol%. Particularly noteworthy, the direct polyheteroarylation of polybromobenzene, such as di- or tribromobenzene, 1,4-dibromonaphthalene, and 9,10-dibromoanthracene, with imidazo[1,2-*a*]pyridine was also successfully achieved to give the corresponding products in good yields (Scheme 3.146). The polyheteroarylation results were matched with the previous report from Doucet[190] (see Scheme 3.184).

Cao et al. reported an efficient $Pd(OAc)_2/P(n\text{-Bu})Ad_2$-catalyzed C-3-arylation of imidazo[1,2-*a*]pyridines with rarely reported aryl chlorides (Scheme 3.147).[191] Substituted imidazo[1,2-*a*]pyridines, such as methyl, phenyl, trifluoromethyl, chloro, and ester, were applied and coupled with various aryl chlorides to afford the corresponding C-3-arylated products in good yields.

Kim and Lee employed the bimetallic Pd-Fe_3O_4 heterodimer nanocrystals, which are magnetic nanoparticles, for the direct arylation of

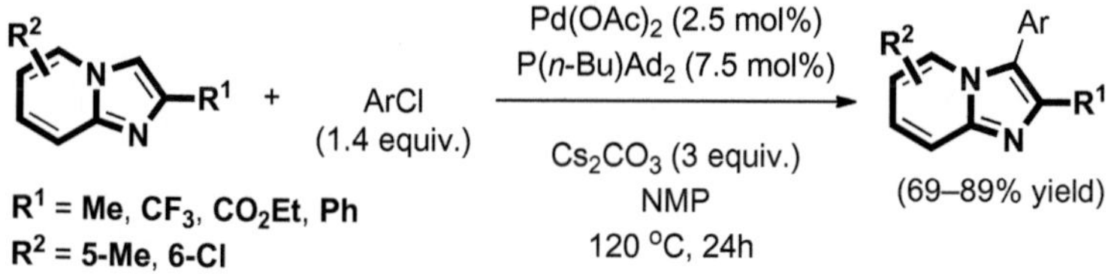

Scheme 3.147 $Pd(OAc)_2/P(n\text{-}Bu)Ad_2$-catalyzed direct C-3-arylation of substituted imidazo[1,2-*a*]pyridines with aryl chlorides.[191]

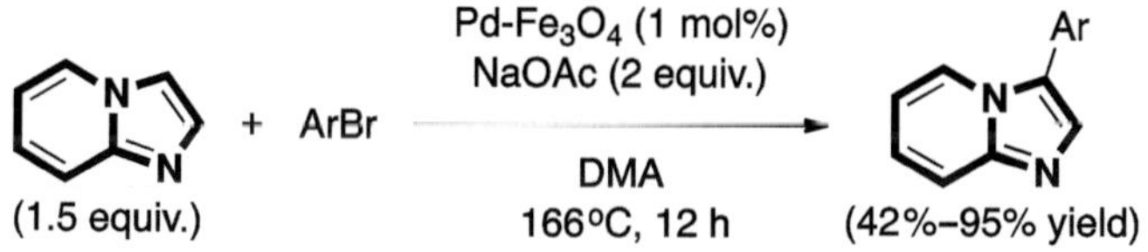

Scheme 3.148 $Pd\text{-}Fe_3O_4$-catalyzed direct C-3-arylation of imidazo[1,2-*a*]pyridine with aryl bromides.[192]

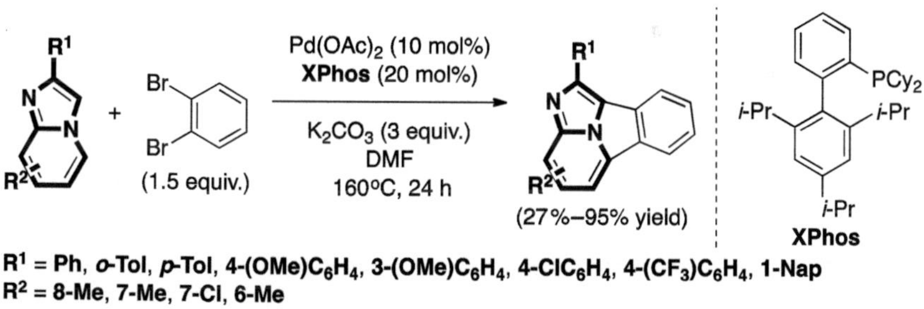

Scheme 3.149 $Pd(OAc)_2$/**XPhos**-catalyzed double direct arylation of 2-arylimidazo [1,2-*a*]pyridines with *o*-dihaloarenes.[194]

imidazo[1,2-*a*]pyridine (Scheme 3.148).[192] The arylation of imidazo[1,2-*a*] pyridine with various aryl bromides gave the corresponding arylated products in good yields with exclusively C-3-regioselectivity. It should be highlighted that the $Pd\text{-}Fe_3O_4$ nanocrystals could be recycled by simple magnetic separation and without loss of catalytic activity even after being recycled 10 times.

Benzo[*a*]imidazo[5,1,2-*cd*]indolizines are novel type of fluorescent dyes.[193] Peng and co-workers reported a highly regiocontrolled $Pd(OAc)_2$/**XPhos**-catalyzed dual arylation to synthesize benzo[*a*]imidazo[5,1,2-*cd*] indolizines from imidazo-[1,2-*a*]pyridines and *o*-dihaloarenes.[194] The process comprised intermolecular C-3-arylation of 2-arylimidazo[1,2-*a*]pyridines followed by an intramolecular C-5-arylation in a highly regioselective fashion to afford benzo[*a*]imidazo[5,1,2-*cd*]indolizine derivatives in moderate-to-good yields (Scheme 3.149).

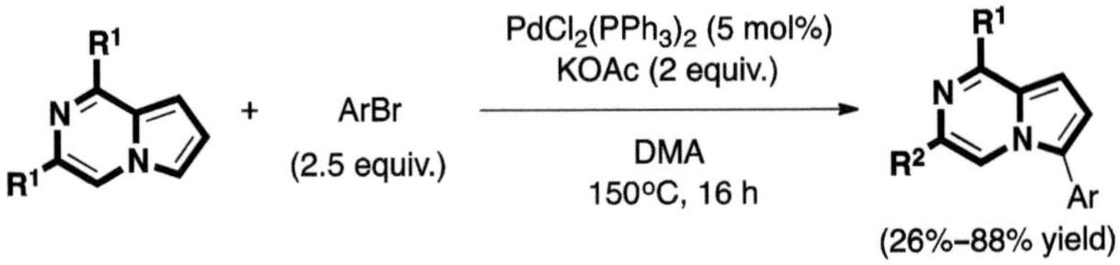

R^1 = Ph, 4-(OMe)C_6H_4, 3-(OMe)C_6H_4, 4-(OBn)C_6H_4, 4-FC_6H_4, 4-ClC_6H_4, Me
R^2 = H, Me

Scheme 3.150 $PdCl_2(PPh_3)_2$-catalyzed direct C-6-arylation of pyrrolo[1,2-*a*]pyrazines with (hetero)aryl bromides.[196]

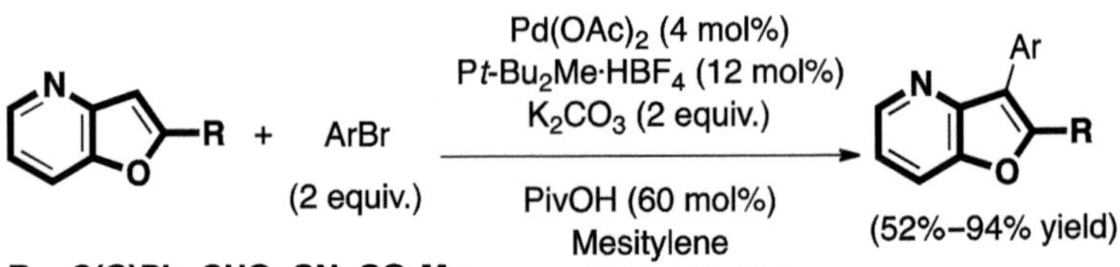

R = C(O)Ph, CHO, CN, CO_2Me

Scheme 3.151 $Pd(OAc)_2$/P*t*-Bu_2Me · HBF_4-catalyzed direct C-3-arylation of 2-substituted furo[3,2-*b*]pyridines with (hetero)aryl bromides.[198]

Pyrrolo[1,2-*a*]pyrazine derivatives have been underevaluated as a drug pharmacophore.[195] Kim described the synthesis of a chemical library based on a pyrrolo[1,2-*a*]pyrazine core.[196] In this approach, the 3-substituted or 1,3-disubstitued pyrrolo[1,2-*a*]pyrazines and (hetero)aryl bromides were used as the coupling partners for direct arylation reactions using $PdCl_2(PPh_3)_2$ as the catalyst (Scheme 3.150). The arylation reactions took place regioselectively at the C-6-position to afford the desired products in moderate-to-good yields.

A furopyridine ring system consists of a furan ring and a pyridine ring, which represent an important pharmacophore in drug discovery.[197] Carrër and Bertounesque reported the first selective $Pd(OAc)_2$/P*t*-Bu_2Me · HBF_4-catalyzed direct arylation of furo[3,2-*b*]pyridine.[198] The arylation would be expected to take place at the C-3-position, yet pyridines are problematic substrates for C—H functionalization due to their electron-deficient character and high Lewis basicity. The direct C-3-arylation of 2-substituted furo[3,2-*b*]pyridines with (hetero)aryl bromides was firstly examined to afford the C-3-arylated products in moderate-to-good yields (Scheme 3.151). Then, the regioselective C-7-arylation of furo [3,2-*b*]pyridines was developed to proceed in moderate-to-good yields (Scheme 3.152). Indeed, the C-7-arylation of furo[3,2-*b*]pyridines was

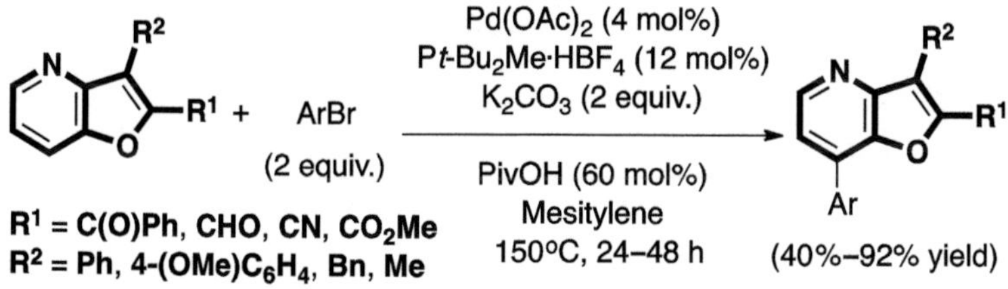

Scheme 3.152 $Pd(OAc)_2/Pt\text{-}Bu_2Me \cdot HBF_4$-catalyzed direct C-7-arylation of 2,3-disubstituted furo[3,2-*b*]pyridines with aryl bromides.[198]

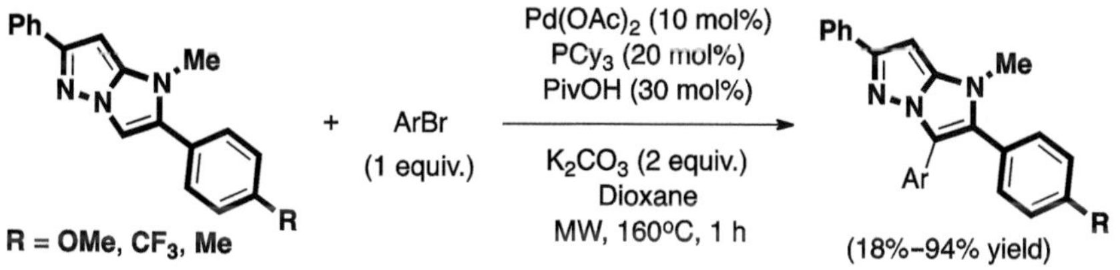

Scheme 3.153 $Pd(OAc)_2/PCy_3$-catalyzed direct C-3-arylation of imidazo[1,2-*b*]pyrazole derivatives with (hetero)aryl bromides.[201]

unexpected since no such reactivity of analogues had ever been described without directing groups. Both C-3 and C-7-arylation tolerated many functional groups and generated a wide range of novel furo[3,2-*b*]pyridine derivatives.

Imidazo[1,2-*b*]pyrazoles have been designed, prepared, and studied as anticancer[199] and antifungal agents.[200] Current strategies for preparing substituted imidazo[1,2-*b*]pyrazole derivatives have attracted much attention. Guillaumet and co-workers reported a rapid approach, for the first time, for synthesizing C-3-arylated imidazo[1,2-*b*]pyrazole derivatives using the $Pd(OAc)_2/PCy_3$ catalytic system with microwave assistance.[201] Substituted imidazo[1,2-*b*]pyrazole derivatives were successfully coupled with various (hetero)aryl bromides to afford regioselective C-3-arylated products in moderate-to-good yields (Scheme 3.153).

Guillaumet then reported the first direct C-7-arylation of imidazo [1,2-*b*]pyrazoles with (hetero)aryl bromides or chlorides (Scheme 3.154).[202] In the presence of a $Pd(OAc)_2/PCy_3 \cdot HBF_4$ system with microwave assistance, a series of (hetero)aryl groups were easily introduced in moderate-to excellent yields. It should be highlighted that double direct C-3- and C-7-diarylation was also achieved by increasing the amount of aryl bromide and base (Scheme 3.155). The corresponding diarylated products were furnished in good yields.

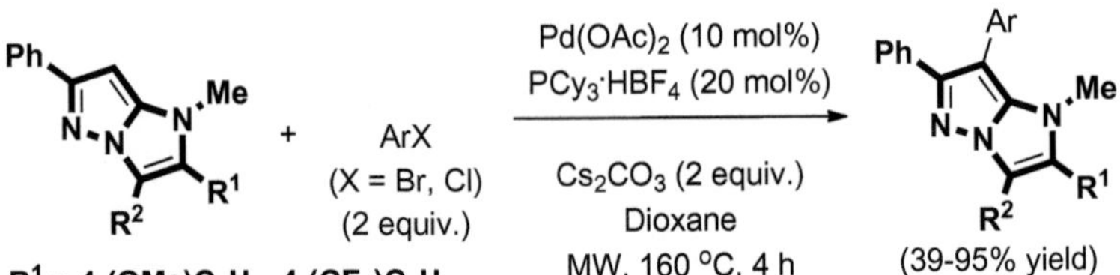

R^1 = 4-(OMe)C_6H_4, 4-(CF_3)C_6H_4
R^2 = *p*-Tol, 4-(OMe)C_6H_4, 4-(CF_3)C_6H_4, 4-Py

Scheme 3.154 $Pd(OAc)_2/PCy_3 \cdot HBF_4$-catalyzed direct C-7-arylation of imidazo[1,2-*b*]pyrazole derivatives with (hetero)aryl bromides/chlorides.[202]

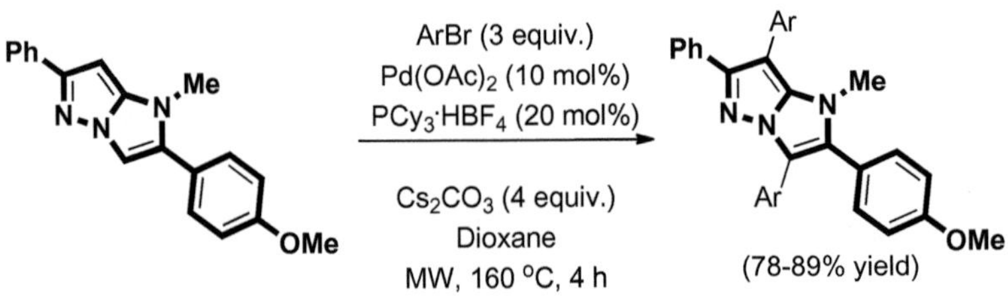

Scheme 3.155 $Pd(OAc)_2/PCy_3 \cdot HBF_4$-catalyzed direct C-3- and C-7-diarylation of imidazo[1,2-*b*]pyrazole derivatives with aryl bromides.[202]

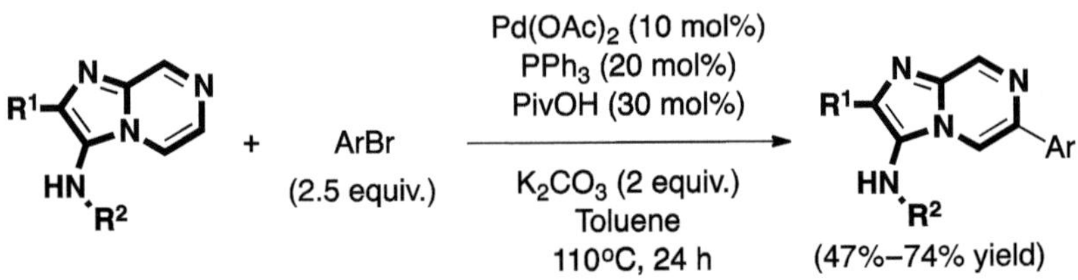

R^1 = Ph, *n*-Pr, 2-Nap, 4-ClC_6H_4, 4-(OMe)C_6H_4
R^2 = *t*-Bu, Cy, 3,4,5-$(OMe)_3C_6H_2$, 4-(OMe)C_6H_4, *n*-Bu

Scheme 3.156 $Pd(OAc)_2/PPh_3$-catalyzed direct C-6-arylation of 3-aminoimidazo[1,2-*a*]pyrazine with aryl bromides.[204]

Imidazo[1,2-*a*]pyrazines and their derivatives have been found to possess a wide range of bioactivities, analogues of this scaffold represent various marketed drugs such as zolpidem and zolimidine.[199b,203] Guchhait et al. reported the use of a $Pd(OAc)_2/PPh_3$ catalytic system for the direct C-6-arylation of 3-aminoimidazo[1,2-*a*]pyrazine with aryl bromides (Scheme 3.156).[204] Indeed, the arylation of this heteroarene suffered from competing C-5 and C-2′-arylation reactions, while this protocol eliminated these competing arylations. Guchhait revealed that the use of K_2CO_3 as the base had a superior effect on providing high regioselectivity C-6-arylated products in good yields.

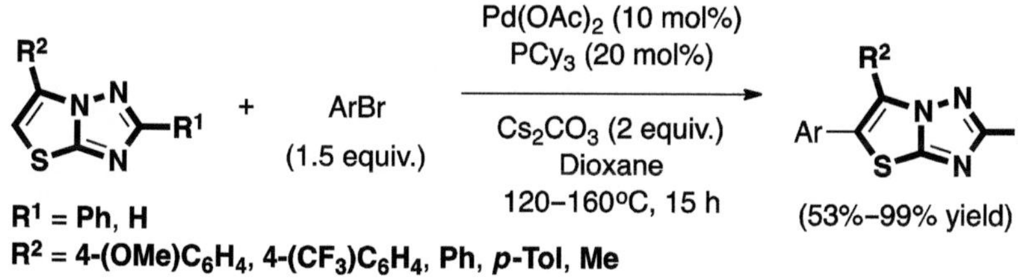

Scheme 3.157 $Pd(OAc)_2/PCy_3$-catalyzed direct C-5-arylation of thiazolo[3,2-*b*]-1,2,4-triazoles with (hetero)aryl bromides.[209]

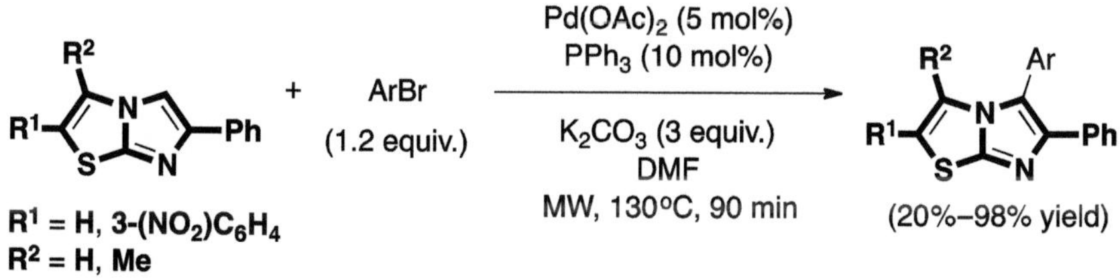

Scheme 3.158 $Pd(OAc)_2/PPh_3$-catalyzed direct C-5-arylation of imidazo[2,1-*b*]thiazoles with aryl bromides.[213]

The thiazolo[3,2-*b*]-1,2,4-triazole core is found as a pharmaceutically important molecule possessing antimicrobial,[205] antibacterial,[206] G-quadruplex stabilizing,[207] and anti-inflammatory[208] activities. Routier and Guillaumet developed, for the first time, an efficient and convenient method for the $Pd(OAc)_2/PCy_3$-catalyzed direct C-5-arylation of mono- or disubstituted thiazolo[3,2-*b*]-1,2,4-triazoles.[209] The polysubstituted thiazolo[3,2-*b*]-1,2,4-triazoles could be afforded in moderate-to-high yields with (hetero)aryl bromides as the coupling partners (Scheme 3.157).

Imidazo[2,1-*b*]thiazoles represent an important class of heteroaromatic compounds that are key structural motifs of numerous bioactive molecules and functional materials,[210] such as tetramisole[211] and pifithrin-β.[212] Jing reported the $Pd(OAc)_2/PPh_3$-catalyzed direct arylation of substituted imidazo[2,1-*b*]thiazoles of aryl bromides under microwave irradiation.[213] Imidazo[2,1-*b*]thiazoles could be regioselectively C-5-arylated to furnish the desired products in moderate-to-excellent yields (Scheme 3.158).

Pyrrolo[2,3-*d*]pyrimidine derivatives display a wide range of biological activities, such as inhibition of protein kinases,[214] thymidylate synthase,[215] and dihydrofolate reductase.[216] Tumkevicius developed an efficient

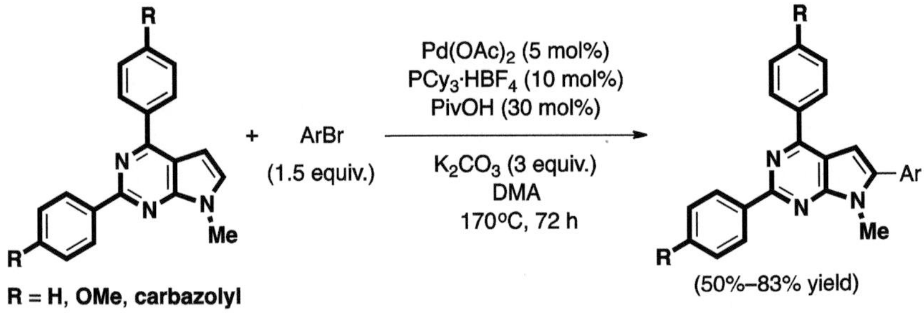

Scheme 3.159 $Pd(OAc)_2/PCy_3 \cdot HBF_4$-catalyzed direct C-6-arylation of 2,4-diarylpyrrolo[2,3-*d*]pyrimidines with aryl bromides.[217]

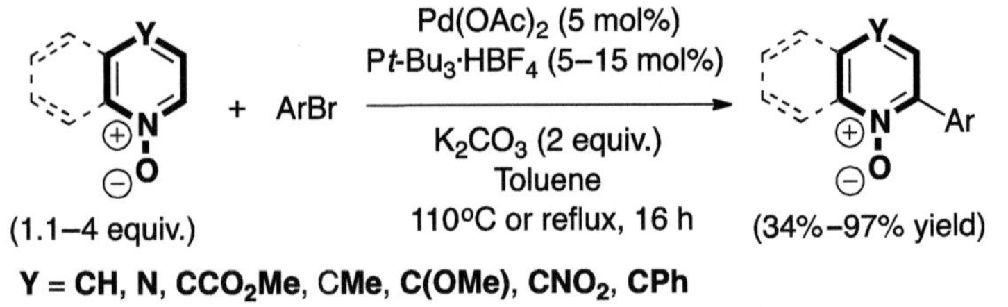

Scheme 3.160 $Pd(OAc)_2/Pt\text{-}Bu_3 \cdot HBF_4$-catalyzed direct arylation of azine *N*-oxides with aryl bromides.[218,219]

method of direct arylation of pyrrolo[2,3-*d*]pyrimidines with aryl bromides.[217] By employing a $Pd(OAc)_2/PCy_3 \cdot HBF_4$ catalytic system, the arylation reactions took place regioselectively at the C-6-position of 2,4-diarylpyrrolo[2,3-*d*]pyrimidines to provide a wide range of 6-arylpyrrolo [2,3-*d*]pyrimidines in moderate-to-good yields (Scheme 3.159).

3.2.2.9 N-Oxides

Fagnou described palladium-catalyzed direct arylation with a broad range of azine and azole *N*-oxides.[218] By employing $Pd(OAc)_2/Pt\text{-}Bu_3 \cdot HBF_4$ as the catalytic system, azine *N*-oxides (e.g., quinoline, pyridine, and pyrazine) underwent arylation with aryl bromides giving moderate-to-excellent yields (Scheme 3.160).[219] Particularly noteworthy was that the authors demonstrated that the *N*-oxide not only imparted a dramatic increase in reactivity in direct arylation at all positions, it also changed the weak thiazole bias for C-5 > C-2-arylation to a reliable C-2 > C-5 > C-4 reactivity profile (Scheme 3.161).[220] This protocol permited regioselective and room temperature arylation of thiazole *N*-oxide at the C-2-position followed by C-5-arylation, and the first examples of C-4-arylation, providing a unique

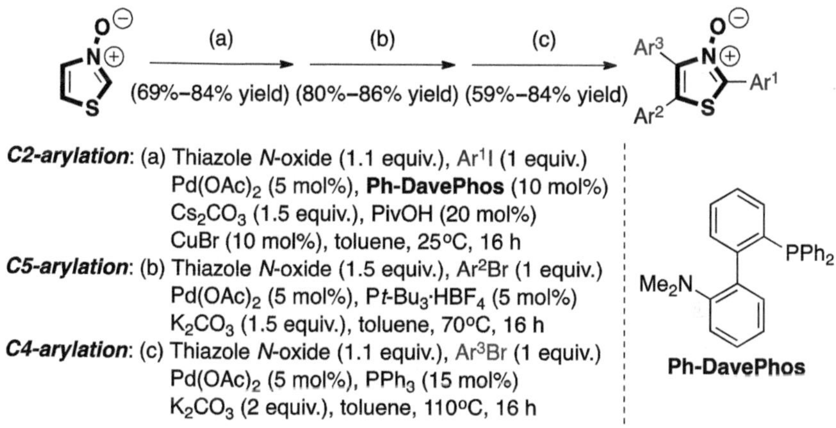

Scheme 3.161 Stepwise $Pd(OAc)_2$-catalyzed direct C-2-C-5-C-4-arylation of thiazole *N*-oxide with aryl halides.[220]

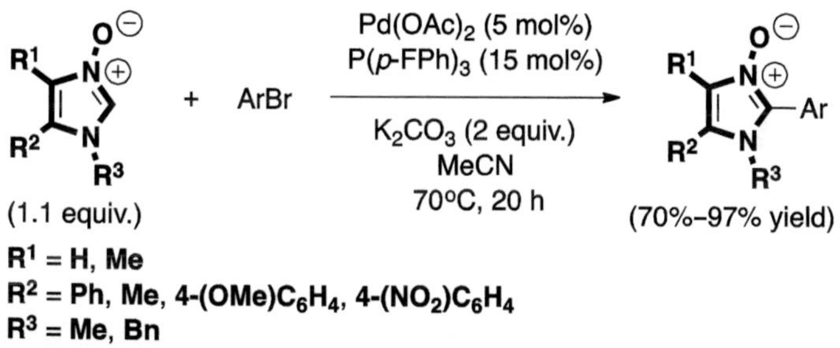

Scheme 3.162 $Pd(OAc)_2/P(p\text{-}FPh)_3$-catalyzed direct C-2-arylation of imidazole *N*-oxides with aryl bromides.[218]

opportunity for functionalization of the thiazole core. Furthermore, Fagnou and co-workers extended these reactions to imidazole *N*-oxides that displayed excellent substrates for direct arylation at the C-2- (Scheme 3.162) and C-4-position (Scheme 3.163).

Fagnou also demonstrated the direct regioselective arylation of *N*-methyl-6- or 7-azaindole *N*-oxides with (hetero)aryl bromides using the $Pd(OAc)_2$/**DavePhos** catalytic system.[182] As discussed previously (see Scheme 3.142), the arylation of *N*-methyl-7-azaindole at the azole ring was successfully achieved. Both 6- and 7-azaindole *N*-oxides could also undergo regioselective direct arylation at the C-7- (Scheme 3.164) and C-6-position (Scheme 3.165) on the azine ring, respectively. Both transformations were compatible with both electron-donating and electron-withdrawing substituents on the aryl bromides to afford the desired products in moderate-to-good yields.

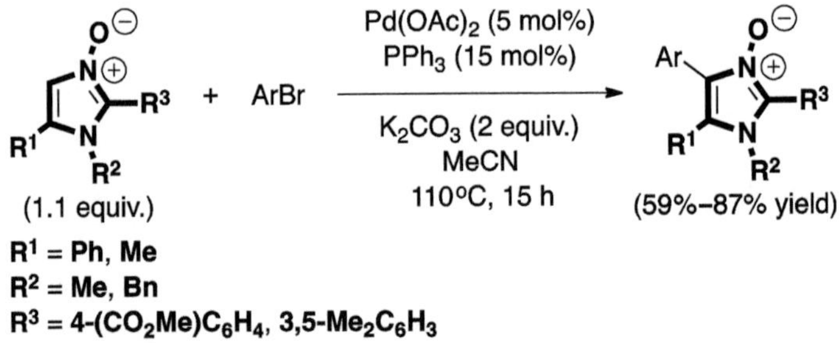

Scheme 3.163 $Pd(OAc)_2/PPh_3$-catalyzed direct C-4-arylation of imidazole *N*-oxides with aryl bromides.[218]

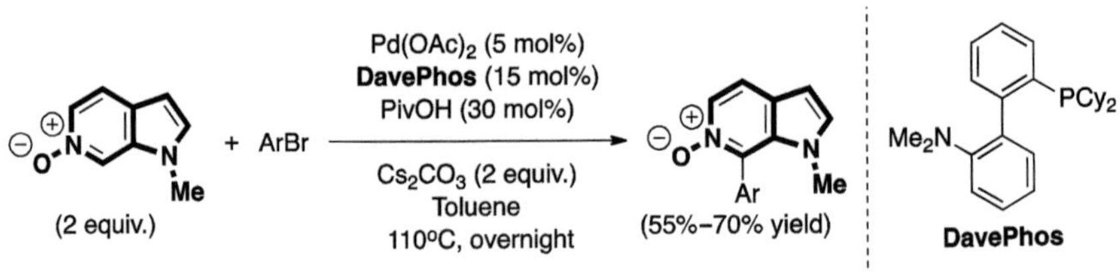

Scheme 3.164 $Pd(OAc)_2$/**DavePhos**-catalyzed direct C-7-arylation of *N*-methyl-6-azaindole *N*-oxides with (hetero)aryl bromides.[182]

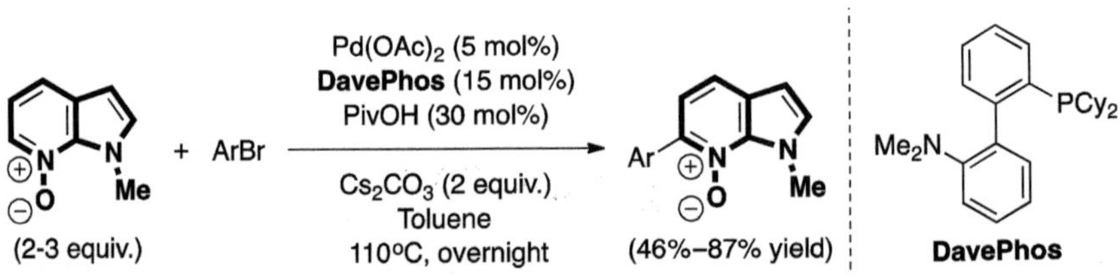

Scheme 3.165 $Pd(OAc)_2$/**DavePhos**-catalyzed direct C-6-arylation of *N*-methyl-7-azaindole *N*-oxides with (hetero)aryl bromides.[182]

Bipyridines constitute an important class of heterocyclics with numerous applications. Bipyridines have found widespread use as chelating ligands in catalysis, and building blocks in natural products with antibiotics.[221] Tzschucke reported a convenient protocol for synthesis of unsymmetrically substituted bipyridines via $Pd(OAc)_2/Pt\text{-}Bu_3$-catalyzed direct arylation of pyridine *N*-oxides.[222] Pyridine *N*-oxides with electron-withdrawing substituents were successfully coupled with 2-, 3-, or 4-bromopyridine to afford the corresponding bipyridine *N*-oxides in

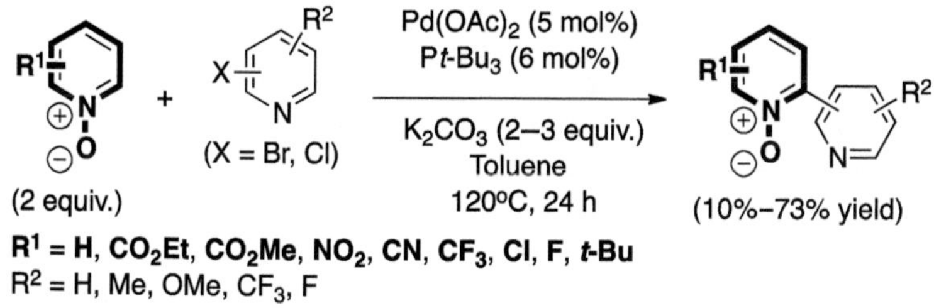

Scheme 3.166 $Pd(OAc)_2/Pt\text{-}Bu_3$-catalyzed direct arylation of pyridine *N*-oxides with halopyridines.[222]

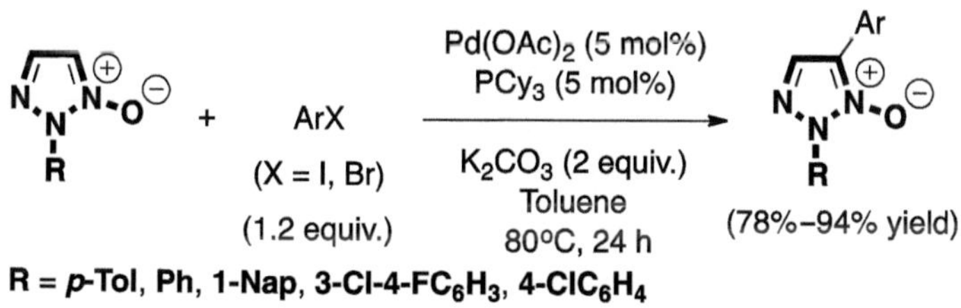

Scheme 3.167 $Pd(OAc)_2/PCy_3$-catalyzed direct C-5-arylation of 1,2,3-triazole *N*-oxides with aryl iodides/bromides.[223]

low-to-good yields (Scheme 3.166). Notably, 4-chloropyridines were compliant with this protocol. The final deoxygenation of bipyridine *N*-oxides was also carried out to produce the desired bipyridines in high yields using hydrogen or sodium borohydride ($NaBH_4$) with palladium on charcoal as a catalyst, or using phosphorus trichloride as a reducing agent.

Kuang demonstrated an efficient and simple methodology for the synthesis of 2,4-diarylated 1,2,3-triazole *N*-oxides via $Pd(OAc)_2/PCy_3$-catalyzed direct C-5-arylation of 2-aryl-1,2,3-triazole *N*-oxides with aryl iodides, bromides, as well as arylboronic acid.[223] Diverse desired C-5-arylated 2-aryl-1,2,3-triazole *N*-oxides were formed in excellent yields (Scheme 3.167). Notably, (hetero)aryl bromides, such as 2-bromopyridine and 2-bromothiophene, were also compatible in this methodology. Furthermore, the arylated products could also undergo deoxygenation using ammonium formate in methanol with palladium on charcoal as a catalyst to afford the desired 1,2,3-triazoles efficiently.

Helicenes possess inherently helical chiralities and their chemistry has been considerably enriched owing to interest in their extraordinary optical and electronic properties.[224] Kamikawa reported the $Pd(OAc)_2$/**SPhos**-catalyzed direct arylation of azahelicene *N*-oxide with (hetero)aryl bromides in good yields (Scheme 3.168).[225]

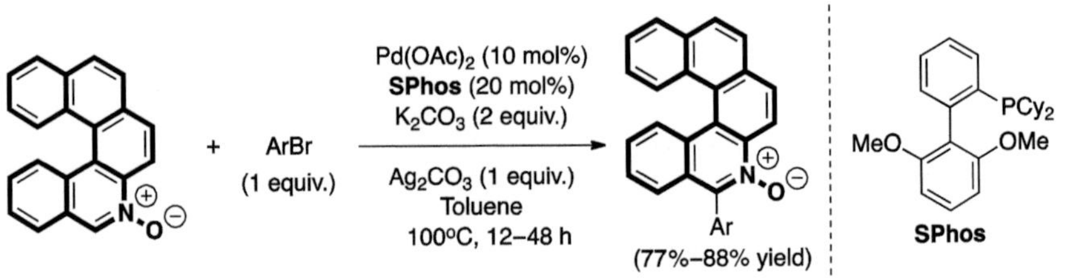

Scheme 3.168 $Pd(OAc)_2$/**SPhos**-catalyzed direct arylation of azahelicene *N*-oxide with (hetero)aryl bromides.[225]

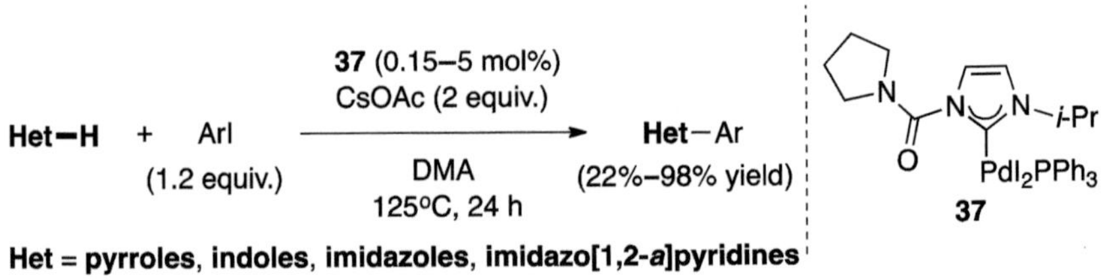

Scheme 3.169 Pd – NHC-catalyzed direct arylation of azoles with (hetero)aryl iodides.[226]

3.2.2.10 Azoles and Other Heteroarenes

Reactions Using Palladium Complexes

Sames reported the synthesis and catalytic investigation of palladium complexes containing imidazolyl carbene ligand (Pd—NHC) of varying steric and electronic properties.[226] Pd—NHC **37** provided direct access to a wide range of SEM-protected azoles with (hetero)aryl iodides to afford the desired products in moderate-to-good yields (Scheme 3.169). It should be noted that *N*-methyl indole and free (NH)-indole were also compatible in this system.

Lee discussed the direct arylation of heteroarenes with (hetero)aryl chlorides using a palladium complex **38** bearing a functionalized NHC and PCy_3.[227] Heteroarenes were coupled with various (hetero)aryl chlorides smoothly to afford the α-arylated products in moderate-to-excellent yields (Scheme 3.170). It is noteworthy that this catalytic system was also highly effective for the coupling of sterically hindered aryl chlorides.

Grisi and co-workers synthesized a new stable palladium(II) complex **39** bearing an unsymmetrical substituted NHC ligand with a pendant hydroxy-functionalized group, and found that **39** with KOAc in DMF was suitable for the direct arylation of furans or thiophenes (Scheme 3.171).[228] Aryl bromides and chlorides were used as the

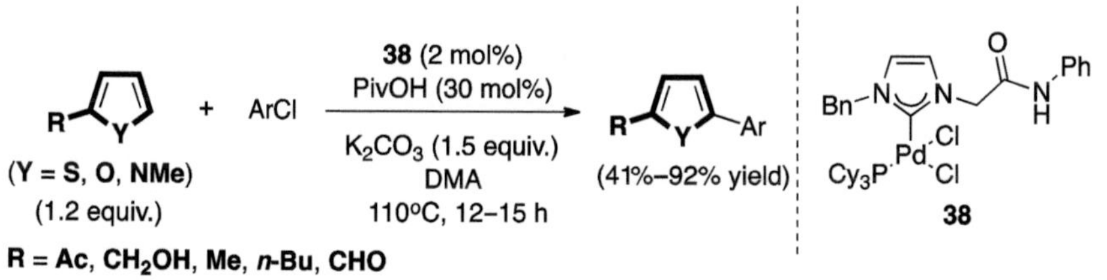

Scheme 3.170 Pd – NHC-catalyzed direct arylation of heteroarenes with (hetero)aryl chlorides.[227]

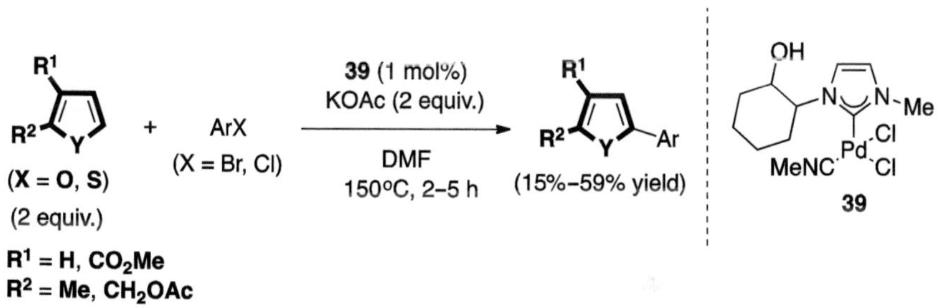

Scheme 3.171 Pd – NHC-catalyzed direct arylation of furans or thiophenes with aryl bromides/chlorides.[228]

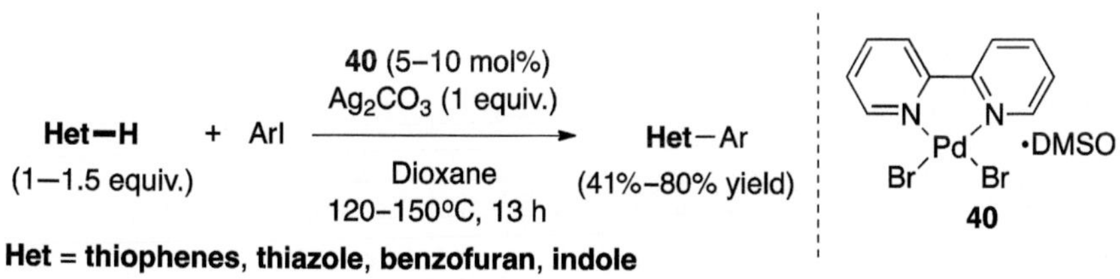

Scheme 3.172 $PdBr_2$(bipy) · DMSO-catalyzed direct arylation of heteroarenes with aryl iodides.[229]

coupling partners to afford the arylated products in acceptable yields, but the limitation was that only electron-withdrawing substituent on aryl chlorides favored the reactions.

Itami described the direct arylation of heteroarenes with aryl iodides using complex $PdBr_2$(bipy) · DMSO **40** as a catalyst and Ag_2CO_3 as a base in dioxane.[229] This protocol was applicable to a range of heteroarenes, such as thiophene, thiazole, benzofuran, and indole (Scheme 3.172).

Punji synthesized an efficient POCN-pincer palladium catalyst **41** for the direct C—H bond arylation of azoles with aryl iodides.[230] The pincer palladium complex **41** efficiently catalyzed the arylation of

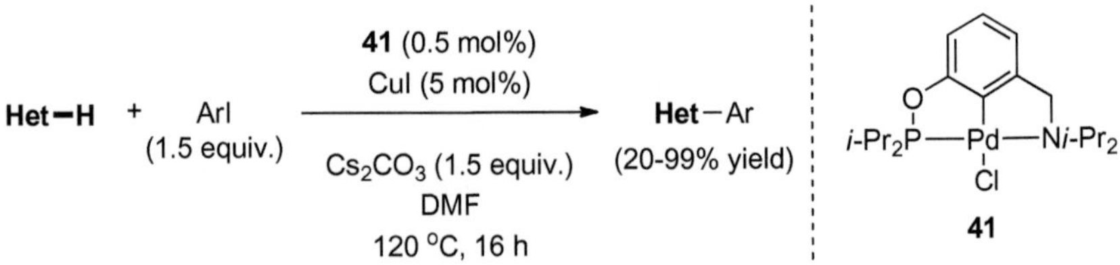

Het = benzothiazole, benzoxazoles, 5-aryloxazoles

Scheme 3.173 (POCN)PdCl-catalyzed direct arylation of azoles with (hetero)aryl iodides.[230]

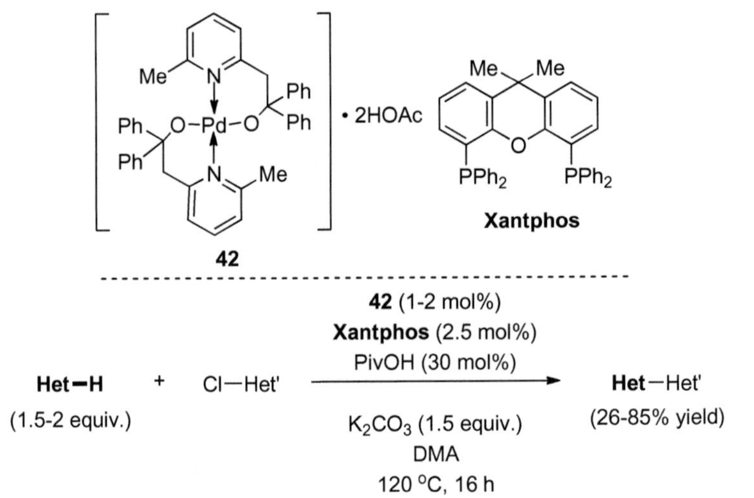

Het = (benzo)thiophenes, (benzo)furans, (benzo)thiazole, imidazole, indoles
Het' = pyridines, quinoxaline, thiophenes, quinoline, isoquinoline

Scheme 3.174 Pd/**Xantphos**-catalyzed direct arylation of heteroarenes with heteroaryl chlorides.[231]

benzothiazoles, benzoxazoles, and 5-aryloxazoles with diverse (hetero)aryl iodides (Scheme 3.173). Notably, this represented the first example of pincer palladium complex being applied for the direct arylation of azoles with low catalyst loading at 0.5 mol%.

Wu and co-workers developed an efficient bis(alkoxo)palladium complex **42** to catalyze the direct arylation of heteroarenes with heteroaryl chlorides.[231] By combining **42** with **Xantphos**, the direct arylation of a series of heteroarenes with various heteroaryl chlorides proceeded smoothly affording desired products in moderate-to-good yields (Scheme 3.174). In addition, this catalytic system allowed two-fold C–H bond direct arylation in one-pot to synthesize 2,5-diheteroarylated thiophene derivatives in good yields (Scheme 3.175).

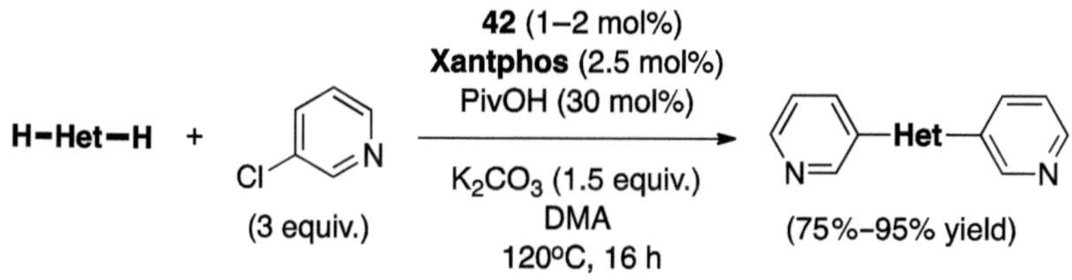

Het = thiophenes, thieno[3,2-*b*]thiophene, 2,2'-bithiophene

Scheme 3.175 Two-fold C—H bond arylation via Pd/**Xantphos**-catalyzed direct arylation of heteroarenes with 3-chloropyridine.[231]

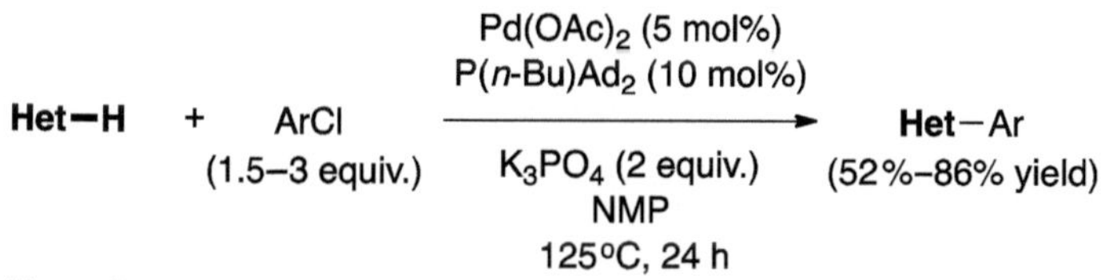

Het = (benzo)thiophene, isoxazole, benzofuran, benzoxazole (benzo)thiazole, imidazole, triazole, caffeine

Scheme 3.176 $Pd(OAc)_2$/P(*n*-Bu)Ad_2-catalyzed direct arylation of heteroarenes with (hetero)aryl chlorides.[232]

Reactions Involving Addition of Phosphine Ligands

Daugulis developed the first efficient and general protocol for the direct arylation of heteroarenes using cheap aryl chlorides as the coupling partners.[232] In the presence of $Pd(OAc)_2$ as the catalyst and P(*n*-Bu)Ad_2 as the ligand, a broad range of heteroarenes could be coupled with (hetero) aryl chlorides in good yields (Scheme 3.176).

In order to develop a protocol that can exemplify mildness, user-friendliness and ease-of-use, Greaney reported the direct arylation of thiazoles, featuring the first direct arylation system that worked "on water."[233] In the presence of Pd(dppf)Cl_2·DCM associated with PPh_3 ligand, an array of 2-arylthiazoles, as well as benzimidazole, benzothiazole, (benz)oxazole, and thiophene, could be coupled with a wide range of (hetero)aryl iodides in good yields (Scheme 3.177). In addition, the reaction was carried out under substantially milder reaction conditions, at 60°C under water, when comparing to those commonly used in heteroarene arylation.

Fagnou reported a broadly applicable protocol for the direct arylation of heteroarenes with aryl bromides.[72a] The $Pd(OAc)_2/PCy_3 \cdot HBF_4$ catalytic system allowed a very broad range of heteroarenes to be coupled with various (hetero)aryl bromides to furnish the desired products in moderate-to-good yields (Scheme 3.178). It should be highlighted that

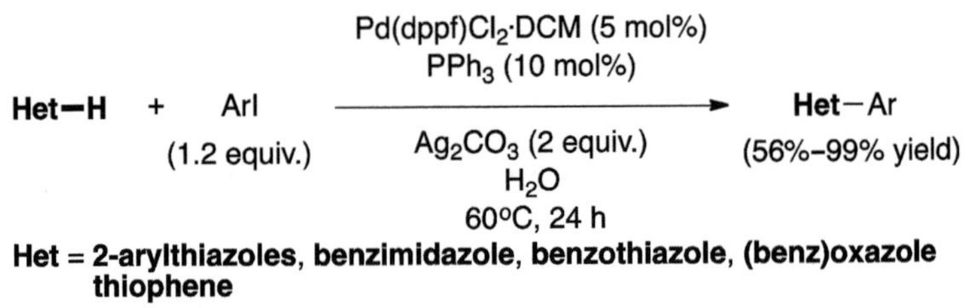

Scheme 3.177 $Pd(dppf)Cl_2 \cdot DCM/PPh_3$-catalyzed direct arylation of heteroarenes with (hetero)aryl iodides.[233]

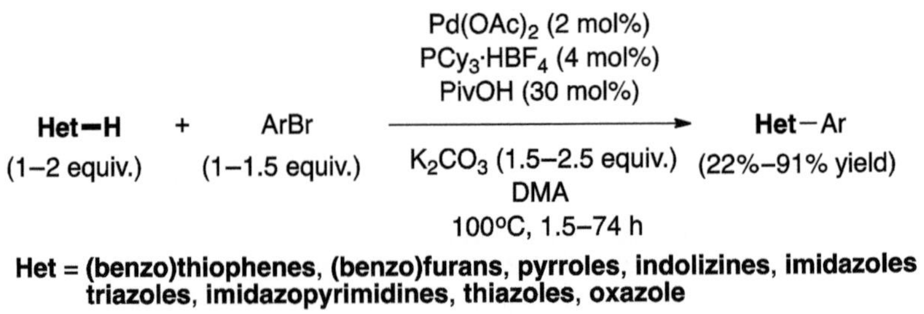

Scheme 3.178 $Pd(OAc)_2/PCy_3 \cdot HBF_4$-catalyzed direct arylation of heteroarenes with (hetero)aryl bromides.[72a]

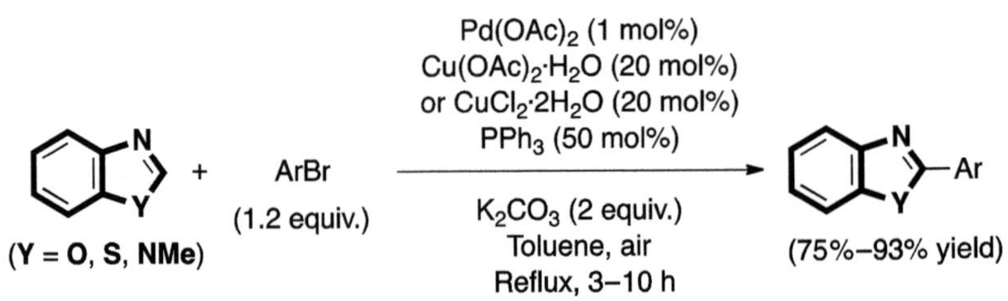

Scheme 3.179 $Pd(OAc)_2/Cu(II)/PPh_3$-catalyzed direct arylation of benzoazoles with (hetero)aryl bromides.[234]

the authors revealed the use of substoichiometric quantities of PivOH was to accelerate and improve the reaction outcomes across a broad range of heteroarenes.

An efficient $Pd(OAc)_2/Cu(II)/PPh_3$ co-catalytic system has been developed by Huang.[234] Both $Cu(OAc)_2 \cdot H_2O$ or $CuCl_2 \cdot 2H_2O$ allowed the direct arylation of benzoxazole, benzothiazole, or benzimidazole with various (hetero)aryl bromides and performed smoothly under air, affording the desired 2-arylated products in good yields (Scheme 3.179).

Traditionally, most of the research has been focused on the study of electron-rich ligands, yet René and Fagnou reported $Pd(OAc)_2$-catalyzed

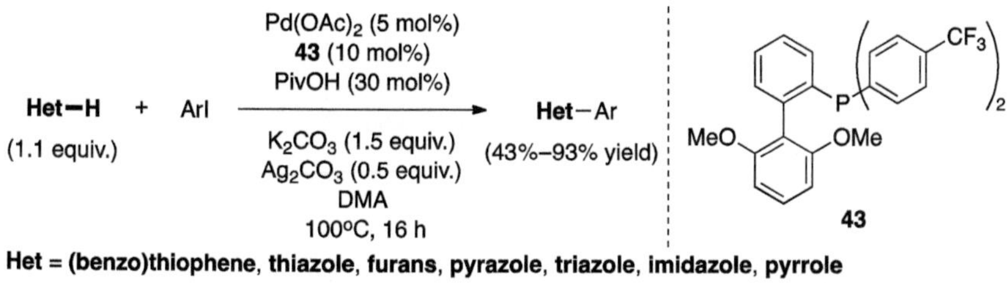

Scheme 3.180 $Pd(OAc)_2$/**43**-catalyzed direct arylation of heteroarenes with aryl iodides using electron-deficient phosphine ligand.[235]

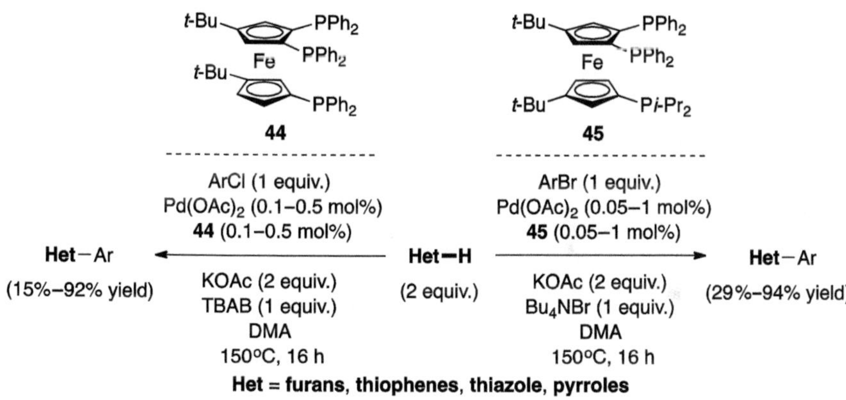

Scheme 3.181 $Pd(OAc)_2$/**44** or **45**-catalyzed direct arylation of heteroarenes with aryl chlorides/bromides.[236,237]

direct arylation of heteroarenes with aryl iodides using a new electron-deficient biarylphosphine ligand **43**[235] showing that a more electron-deficient metal center accelerated the direct arylation reaction. Moreover, a broad range of heteroarenes were proved to be efficient coupling partners to afford the desired products in moderate-to-excellent yields (Scheme 3.180).

Meanwhile, Doucet and Hierso reported the use of ferrocenyl triphosphine ligand **44**[236] and **45**[237] associated with $Pd(OAc)_2$ in the direct arylation of heteroarenes. The coupling of heteroarenes with various aryl chlorides could occur efficiently under low catalyst loadings ranging between 0.1 and 0.5 mol% with the $Pd(OAc)_2$/**44** system, whereas 0.05–1 mol% the $Pd(OAc)_2$/**45** system was used for coupling with aryl bromides (Scheme 3.181).

The palladium-catalyzed direct arylation at the C-3- or C-4-position of heteroarenes are known to be more challenging than at the C-2- or C-5-positions. Hierso and Doucet reported that by combining only 0.5 mol% of $Pd(OAc)_2$ with the new ferrocenyl diphosphine **Sylphos**,

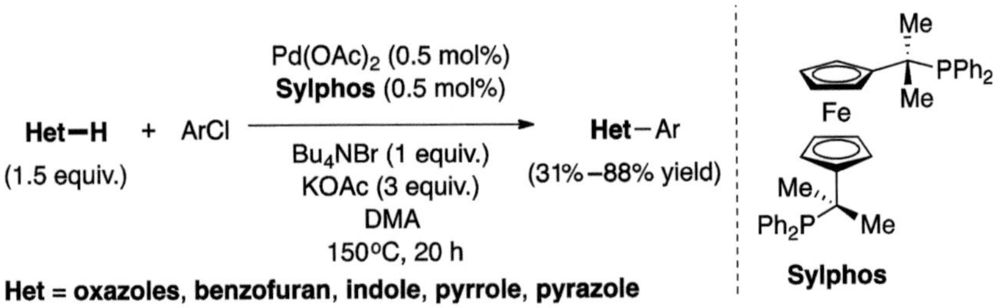

Scheme 3.182 $Pd(OAc)_2$/**Sylphos**-catalyzed direct arylation of heteroarenes with (hetero)aryl chlorides.[238]

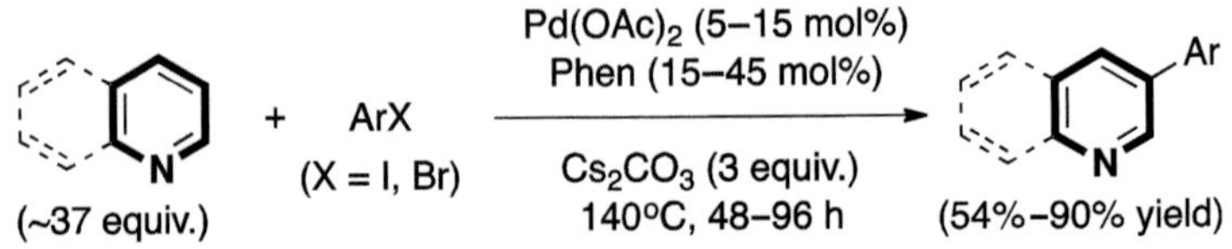

Scheme 3.183 $Pd(OAc)_2$/phen-catalyzed direct C-3-arylation of pyridines with (hetero)aryl iodides/bromides.[240]

the direct C-3- or C-4-arylation of oxazoles, benzofuran, indole, pyrrole, and pyrazole was found to proceed with moderate-to-high yields using various electron-deficient (hetero)aryl chlorides (Scheme 3.182).[238] It should be noted that this was the first example of direct arylation of pyrazole with aryl chlorides.

Arylpyridines and arylpiperidines are important classes of bioactive heterocycles that are difficult to access.[239] Yu reported the first example of $Pd(OAc)_2$/phen-catalyzed C-3-arylation of unprotected pyridines with (hetero)aryl iodides or bromides (Scheme 3.183).[240] This protocol proceeded smoothly to afford C-3-arylated pyridines in moderate-to-good yields. In addition, this protocol provided an expeditious route to an important class of 3-arylpyridines and 3-arylpiperidines frequently found in bioactive compounds.[239] It should be noted that this protocol was also employed in synthesizing the drug molecule (±)-preclamol[241] at a gram scale.

Doucet employed a $Pd(OAc)_2$/dppb catalytic system for the direct polyheteroarylation of dibromobenzene derivatives, such as benzene, biphenyl, fluorene, naphthalene, and anthracene, while $PdCl(C_3H_5)$(dppb) was used for tribromobenzene.[190] It should be noted that the diheteroarylation (Scheme 3.184) and triheteroarylation (Scheme 3.185) were not limit to furans, thiophenes, pyrrole, imidazole, thiazole, or isoxazole to furnish the desired products in both good yields and

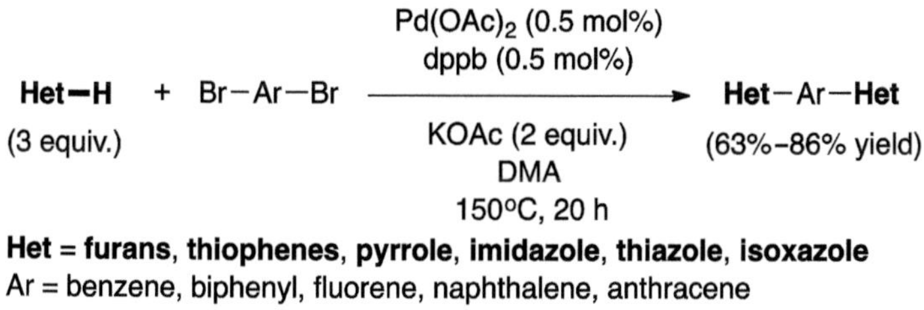

Scheme 3.184 $Pd(OAc)_2$/dppb-catalyzed direct diarylation of heteroarenes with dibromobenzene derivatives.[190]

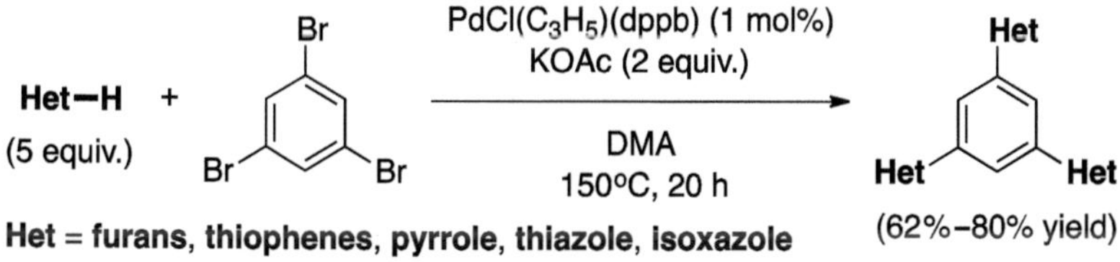

Scheme 3.185 $PdCl(C_3H_5)$(dppb)-catalyzed direct triarylation of heteroarenes with tribromobenzene.[190]

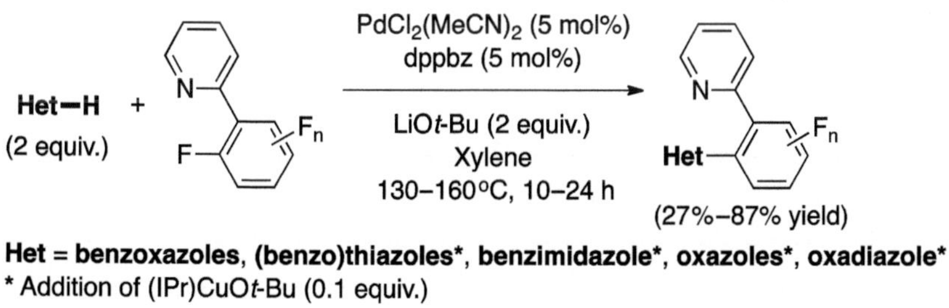

Scheme 3.186 $PdCl_2(MeCN)_2$/dppbz-catalyzed direct arylation of heteroarenes with 2-pyridyl-polyfluoroarenes.[242]

regioselectivities. This simple protocol allowed the synthesis of a wide variety of functionalized conjugated polyheteroarenes in only one step, and provided a powerful tool to material chemists allowing the physical properties of materials to be tuned easily.

Lu and Shen reported the first C–H/C–F coupling between heteroarenes with 2-pyridyl-polyfluoroarenes, in which the selective *ortho*-C–F functionalization was directed by the pyridyl group.[242] In the presence of a $PdCl_2(MeCN)_2$/dppbz catalytic system, reactions of a variety polyfluoroarenes with benzoxazoles occurred smoothly to give the desired products in moderate yields (Scheme 3.186). For other heteroarenes, such as

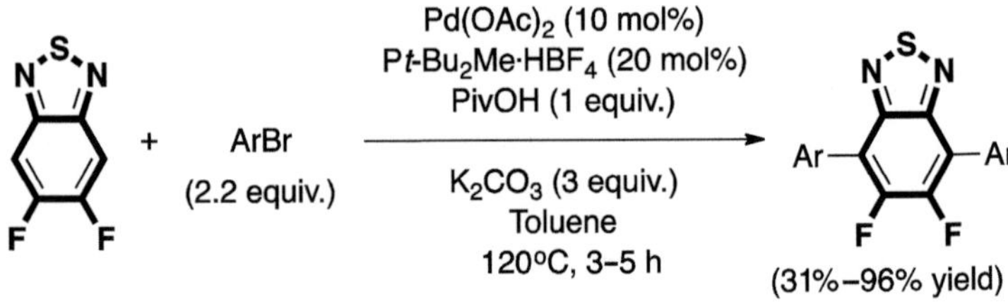

Scheme 3.187 $Pd(OAc)_2$/Pt-$Bu_2Me \cdot HBF_4$-catalyzed direct diarylation of DFBT with (hetero)aryl bromides.[243]

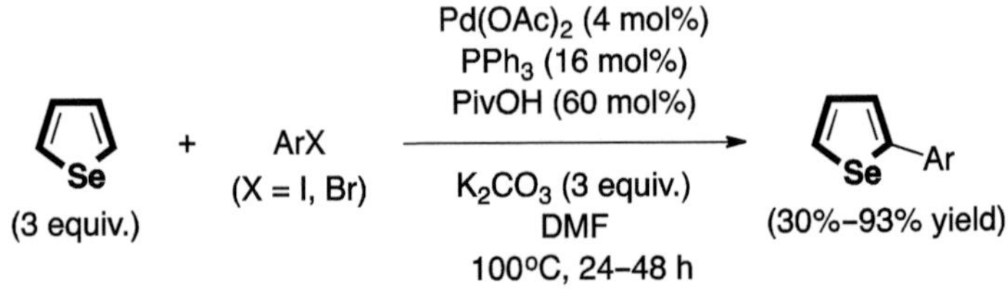

Scheme 3.188 $Pd(OAc)_2$/PPh_3-catalyzed direct C-2-arylation of selenophene with (hetero)aryl iodides/bromides.[245]

thiazoles, benzothiazole, benzimidazole, oxazoles, and oxadiazole, an additive of (IPr)CuO*t*-Bu should be applied in order to have a smooth reaction to furnish the corresponding products in moderate-to-good yields.

Marder reported a controllable protocol for synthesizing 4,7-diaryl-5,6-difluoro-2,1,3-benzothiadiazole via the direct C-4- and C-7-arylation of 5,6-difluoro-2,1,3-benzothiadiazole (DFBT).[243] By employing Fagnou's $Pd(OAc)_2$/P*t*-$Bu_2Me \cdot HBF_4$ catalytic system,[24] diarylation of DFBT with various (hetero)aryl bromides proceeded in moderate-to-good yields to afford the symmetrical diarylated products (Scheme 3.187). Importantly, the diarylated DFBT could be applied to a variety of organic photonic and electronic materials.[244]

Schneider developed an efficient protocol for synthesizing 2-aryl- or 2,5-diarylselenophenes via $Pd(OAc)_2$/PPh_3-catalyzed direct arylation reactions.[245] This system allowed the arylation to be taken place at the C-2-position of selenophene with (hetero)aryl iodides or bromides as the coupling partners to furnish the 2-arylselenophene in moderate-to-good yields (Scheme 3.188). By increasing the amount of aryl iodides and catalyst/ligand loading, diarylation of selenophene at the C-2- and C-5-position could also occur smoothly to afford the symmetrical 2,5-diarylselenophenes.

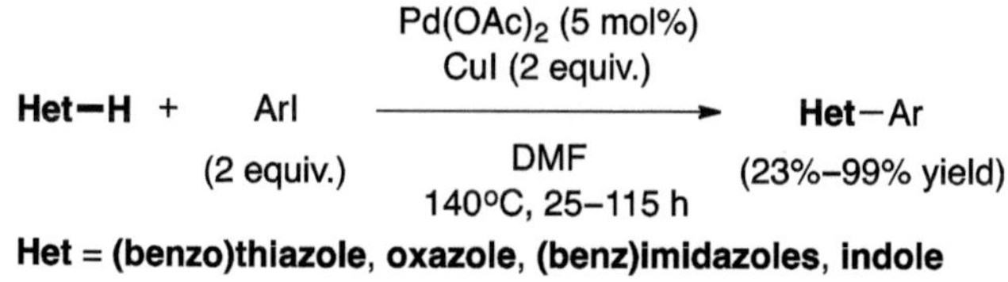

Scheme 3.189 $Pd(OAc)_2$/CuI-catalyzed direct C-2-arylation of heteroarenes with aryl iodides.[246]

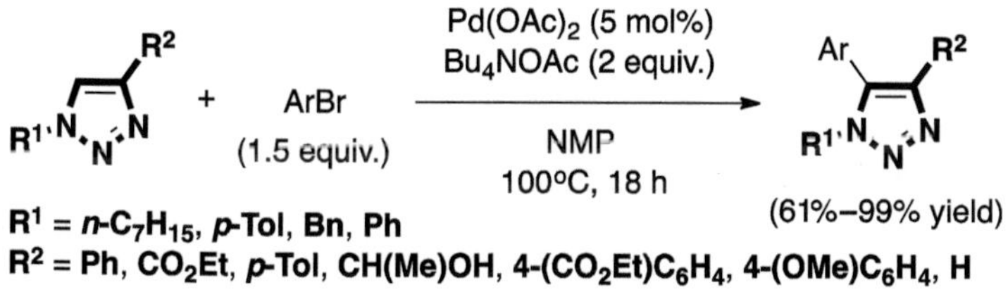

Scheme 3.190 $Pd(OAc)_2$-catalyzed direct C-5-arylation of 1,2,3-triazoles with (hetero) aryl bromides.[247]

Reactions Under Phosphine-Free Conditions or Without Additional Phosphine Ligands

Bellina and Rossi described the first $Pd(OAc)_2$/CuI-mediated C-2-arylation of (benzo)thiazole, oxazole, and imidazoles, as well as free (NH)-(benz)imidazole, and -indole, with aryl iodides to furnish the C-2-arylated products in moderate-to-excellent yields (Scheme 3.189).[246]

Gevorgyan and co-workers developed an efficient $Pd(OAc)_2$-catalyzed direct C-5-arylation of 1,2,3-triazoles with (hetero)aryl bromides.[247] This protocol proceeded smoothly to afford C-5-arylated triazoles in good-to-excellent yields (Scheme 3.190).

You and co-workers reported a new $Pd(OAc)_2$-catalyzed protocol that allowed, for the first time, the direct arylation of azoles with a broad range of (hetero)aryl bromides.[28] This system allowed the direct arylation of azoles, quinoline and pyridine *N*-oxides (Scheme 3.191), as well as pentafluorobenzene (see Scheme 3.13).

Mori showed a facile $Pd(P\textit{t}\text{-}Bu_3)_2$-catalyzed arylation of heteroarenes with (hetero)aryl bromides or chlorides.[248] The reaction proceeded smoothly to afford the corresponding coupling products in moderate-to-excellent yields (Scheme 3.192). Remarkably, the solvent effect was successfully applied to the regioselective arylation of thiazole. It was found that the reaction of 4-bromotoluene and thiazole took place at the C-2-position selectively in dioxane. In addition, Mori described that the use

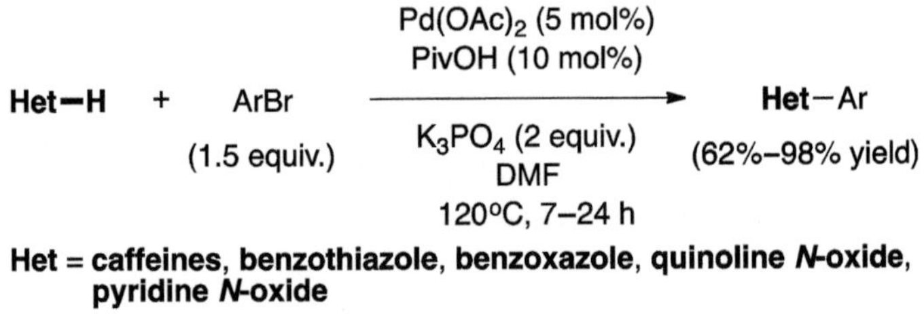

Scheme 3.191 $Pd(OAc)_2$-catalyzed direct arylation of azoles with (hetero)aryl bromides.[28]

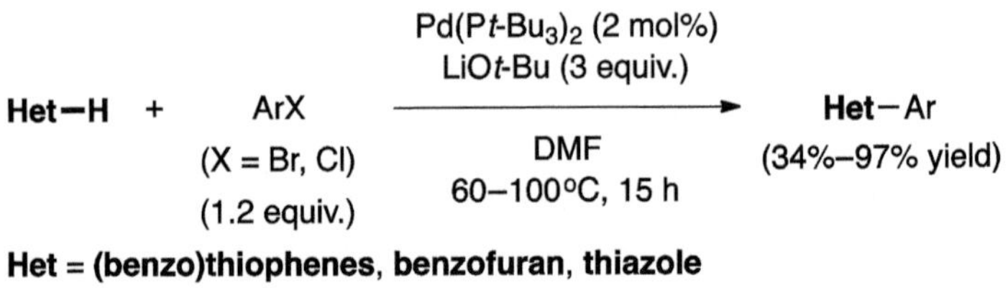

Scheme 3.192 $Pd(Pt\text{-}Bu_3)_2$-catalyzed arylation of heteroarenes with (hetero)aryl bromides/chlorides.[248]

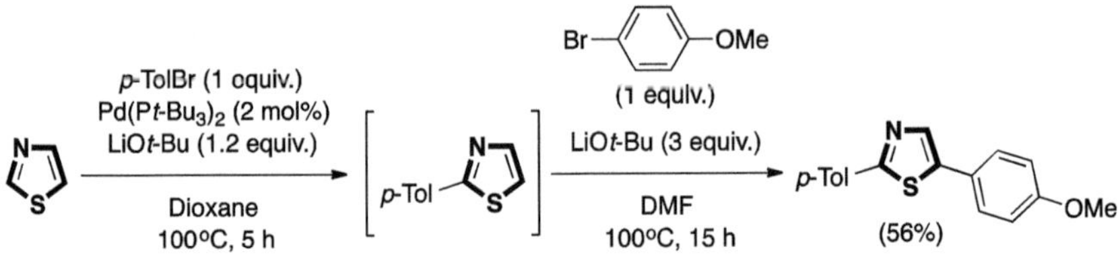

Scheme 3.193 One-pot synthesis of differently substituted 2,5-diarylthiazole via $Pd(Pt\text{-}Bu_3)_2$-catalyzed arylation of thiazole with aryl bromides by switching the solvent.[248]

of a less polar solvent was ineffective for the reaction of an electron-enriched C—H bond. Thus, the reaction allowed one-pot synthesis of differently substituted 2,5-diarylthiazole with the same catalytic system by switching the solvent and the amount of base (Scheme 3.193).

Polyarylated azoles are important motifs in pharmaceuticals and functional materials.[249] Shibahara and Murai reported the direct multiple C—H arylation of heteroarenes using $Pd(phen)_2(PF_6)_2$ as the catalyst.[250] This catalytic system was found to catalyze direct reaction of heteroarenes, such as benzoxazole, benzothiazole, caffeine, and indole, with (hetero)aryl iodides to give the corresponding products in good yields (Scheme 3.194). Moreover, this protocol allowed the synthesis of triarylated azoles consisting of three different aryl groups via one-pot sequential multiple C—H bond

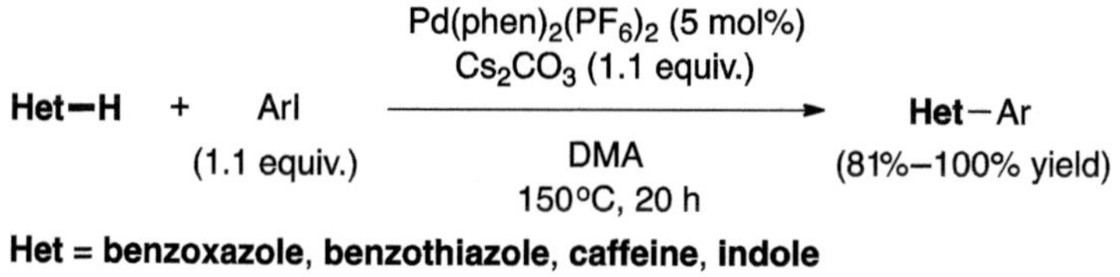

Scheme 3.194 $Pd(phen)_2(PF_6)_2$-catalyzed direct arylation of heteroarenes with (hetero)aryl iodides.[250]

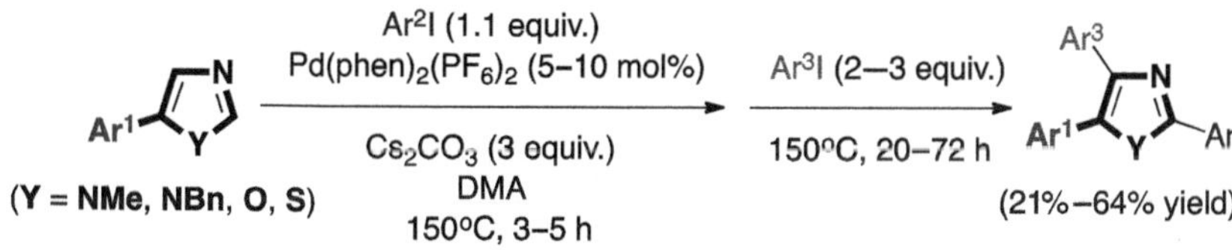

Scheme 3.195 $Pd(phen)_2(PF_6)_2$-catalyzed one-pot sequential diarylation of C-5-monoarylated azoles with (hetero)aryl iodides.[84,251]

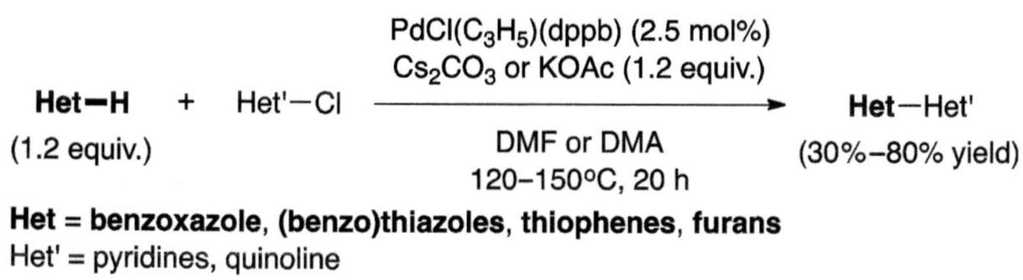

Scheme 3.196 $PdCl(C_3H_5)(dppb)$-catalyzed direct arylation of heteroarenes with chloropyridines or chloroquinoline.[252]

functionalization.[84,251] The one-pot sequential diarylation of C-5-monoarylated azoles was achieved by the simple sequential addition of two different aryl iodides at the C-2- and C-4-position (Scheme 3.195).

The direct coupling of heteroarenes with pyridyl chlorides had been developed by Doucet and co-workers.[252] It was found that with the use of $PdCl(C_3H_5)(dppb)$ as the catalyst, a range of heteroarenes underwent C–H functionalization with 2-, 3-, or 4-chloropyridine or quinoline to afford heteroarylpyridines in moderate-to-good yields (Scheme 3.196).

Both benzylamine and heteroaromatic derivatives present useful biological properties. Abed and Doucet discovered the preparation of a large variety of heteroarylated benzylamine derivatives via a simple PdCl $(C_3H_5)(dppb)$/KOAc-catalyzed direct heteroarylation of bromobenzylacetamide derivatives.[253] A set of heteroarenes could be coupled with 2-, 3-, or 4-bromobenzylacetamide derivatives in moderate-to-good yields (Scheme 3.197).

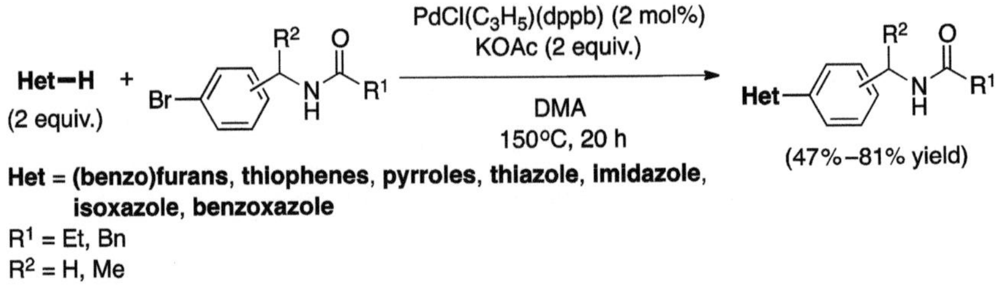

Scheme 3.197 $PdCl(C_3H_5)(dppb)/KOAc$-catalyzed direct arylation of heteroarenes with bromobenzylacetamide derivatives.[253]

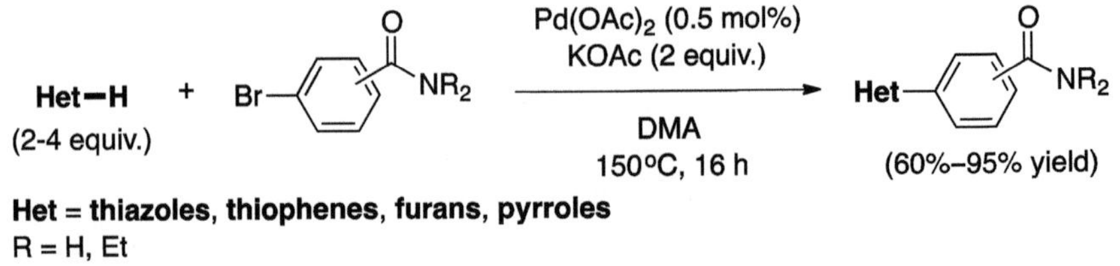

Scheme 3.198 $Pd(OAc)_2/KOAc$-catalyzed direct arylation of heteroarenes with bromobenzamides.[254]

The direct arylation of heteroarenes with 2- or 4-bromobenzamides had been achieved.[254] By employing low loading of a $Pd(OAc)_2/KOAc$ system (0.5 mol%), both primary and tertiary 4-bromobenzamides could be coupled with various heteroarenes to afford C-5-arylated products in good yields (Scheme 3.198). However, higher catalyst loading (2 mol%) of $PdCl(C_3H_5)(dppb)$ should be applied for 2-bromobenzamides and 4-bromo-*N,N*-diisobutylbenzamide. It is noteworthy that no C–N bond formation between two bromobenzamides was observed.

Furthermore, the direct arylation of heteroarenes using bromobenzenes, bearing SO_2R substituents at the C-2- or C-4-position, as the coupling partners was found to proceed in the presence of a $Pd(OAc)_2/KOAc$ catalytic system.[255] With only 0.1–0.5 mol% of $Pd(OAc)_2$ catalyst, various heteroarenes gave good yields and regioselectivities in the direct arylation reactions (Scheme 3.199).

Abed and Doucet employed the $PdCl(C_3H_5)(dppb)/KOAc$ catalytic system for the first direct arylation of heteroarenes bearing *N*-monoalkylcarboxamide CONHR substituents with (hetero)aryl bromides (Scheme 3.200).[256] This protocol allowed a range of heteroarenes to be coupled with a wide variety of (hetero)aryl bromides regioselectively. Moreover, no decarbamoylation by C–N fission was observed.

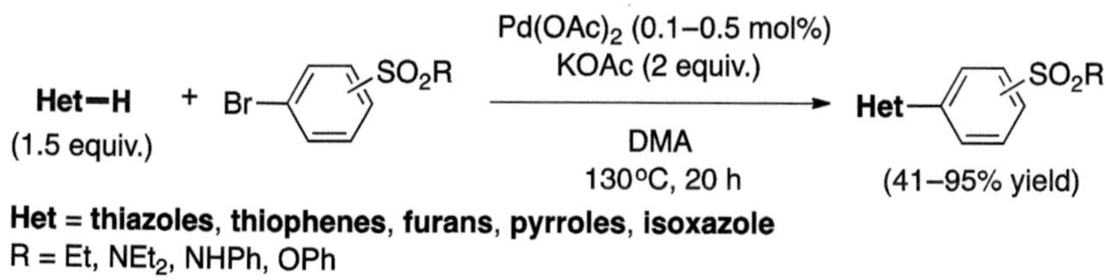

Scheme 3.199 $Pd(OAc)_2$/KOAc-catalyzed direct arylation of heteroarenes with bromobenzenes bearing SO_2R substituents.[255]

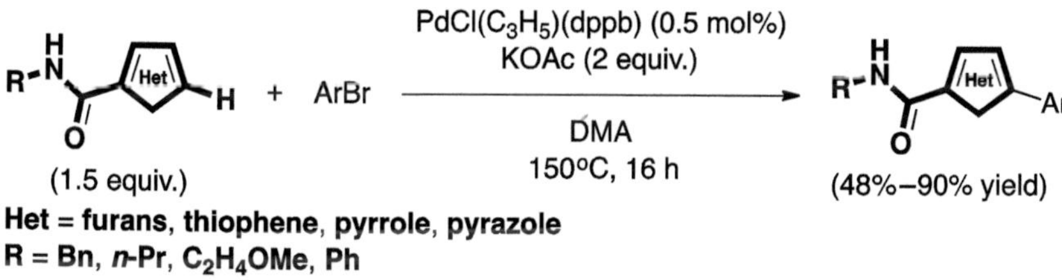

Scheme 3.200 $PdCl(C_3H_5)(dppb)$/KOAc-catalyzed direct arylation of heteroarenes bearing CONHR substituents with (hetero)aryl bromides.[256]

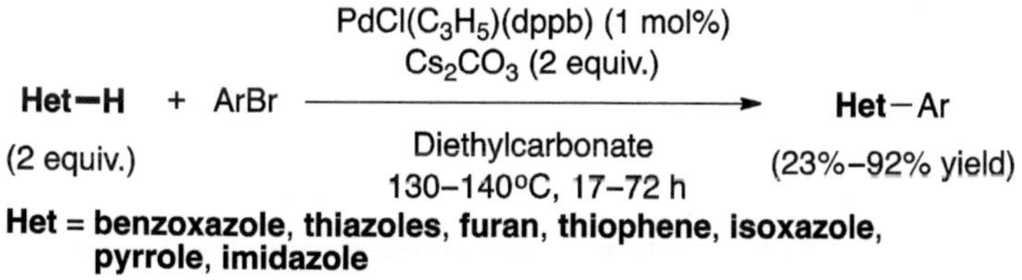

Scheme 3.201 $PdCl(C_3H_5)(dppb)$/KOAc-catalyzed direct arylation of heteroarenes with (hetero)aryl bromides using eco-friendly diethylcarbonate.[258]

Carbonates are polar, aprotic, nontoxic, and biodegradable solvents.[257] Based on these properties, carbonates offer an eco-friendly alternative to standard polar solvents. Hoarau and Doucet concurrently reported the direct arylation of heteroarenes with (hetero)aryl bromides by using $PdCl(C_3H_5)(dppb)$ and the eco-friendly solvent diethylcarbonate (Scheme 3.201).[258] The arylation of benzoxazole and thiazole derivatives gave the best results, while the arylation of furan, thiophene, isoxazole, pyrrole, or imidazole was found to require a higher reaction temperature.

In continuously developing the eco-compatible conditions for direct arylation reactions, Doucet employed "greener"[259] CPME or di-*n*-butyl ether as the solvent, for the first time, for the direct arylation of heteroarenes with (hetero)aryl bromides (Scheme 3.202).[260] CPME presents several advantageous features, such as high hydrophobicity, and its limited

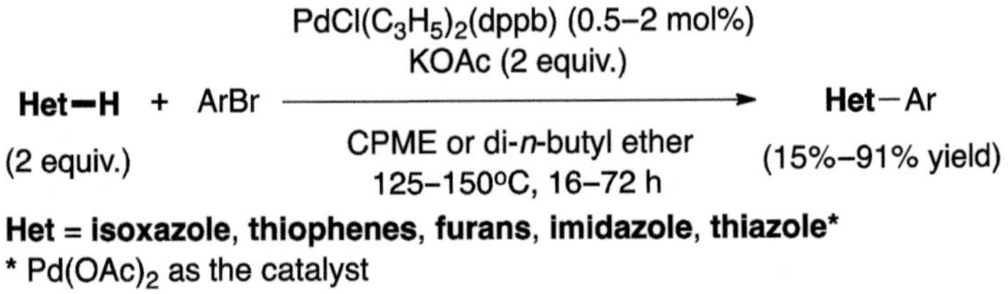

Scheme 3.202 $PdCl(C_3H_5)(dppb)/KOAc$-catalyzed direct arylation of heteroarenes with (hetero)aryl bromides using CPME or di-*n*-butyl ether as the solvent.[260]

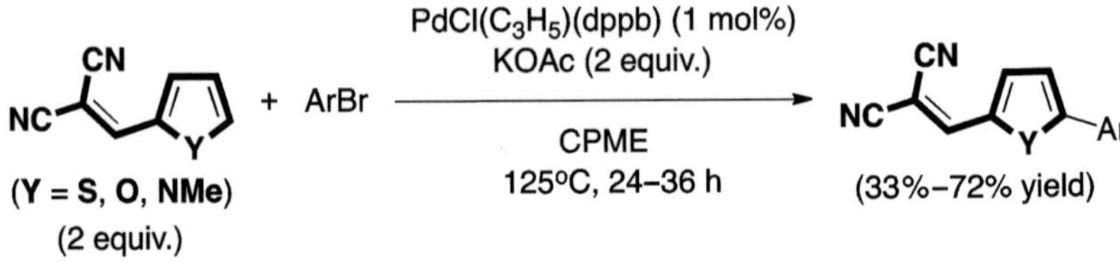

Scheme 3.203 $PdCl(C_3H_5)(dppb)/KOAc$-catalyzed direct C-5-arylation of heteroarenes bearing dicyanovinyls with aryl bromides.[261]

miscibility in water allows easy separation and recovery from water. In the presence of $PdCl(C_3H_5)(dppb)$, the direct arylation of isoxazole, thiophenes, furans, or imidazole proceeded in moderate-to-high yields, whereas thiazole substrates worked better for the arylation reactions when using $Pd(OAc)_2$ as the catalyst.

By using the same catalytic system in CPME, Doucet also demonstrated the direct C-5-arylation of heteroarenes, such as thiophene, furan, and pyrrole, bearing dicyanovinyls at the C-2-position with aryl bromides, allowed the formation of new conjugated systems in moderate-to-good yields in only one step (Scheme 3.203).[261] It should be noted that the use of CPME was crucial to perform such transformations, since a complete decomposition of dicyanovinyls would be observed when using DMA as solvent.

Solvent is one of the major sources of waste in the course of metal-catalyzed direct arylation reactions. Doucet reported that the direct arylation of heteroarenes could be advantageously performed without solvent.[262] The direct C-5-arylation of thiazoles, thiophenes, furans, or pyrroles, and C-4-arylation of oxazoles with various (hetero)aryl bromides as the coupling partners proceeded in moderate-to-high yields (Scheme 3.204). However, only electron-deficient aryl bromides were reactive under the solvent-free conditions. Notably, the influence of the solvent and of the reaction concentration for the direct arylation of heteroarenes were also studied.[263]

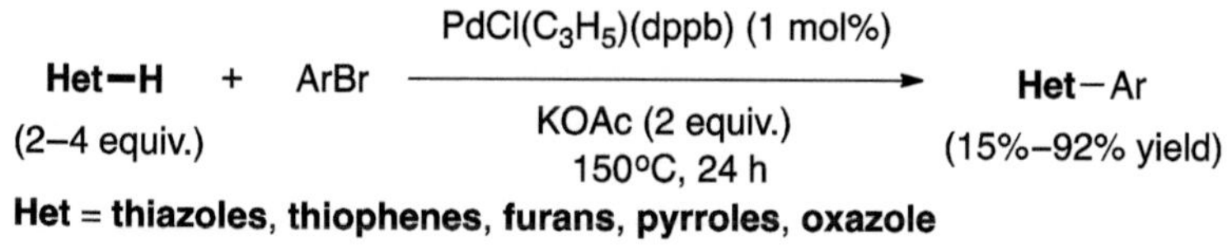

Scheme 3.204 $PdCl(C_3H_5)(dppb)/KOAc$-catalyzed direct arylation of heteroarenes with (hetero)aryl bromides under solvent-free conditions.[262]

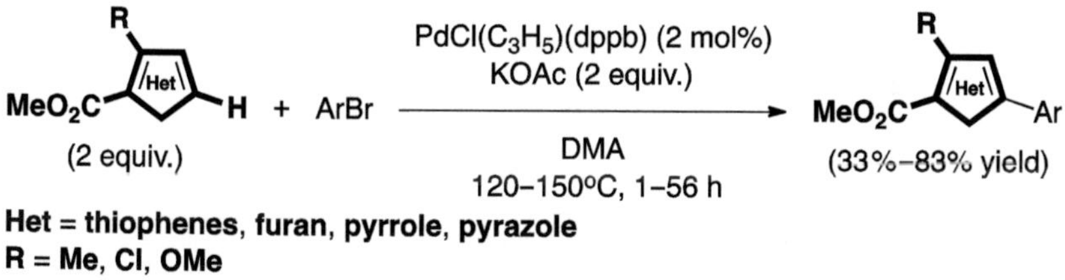

Scheme 3.205 $PdCl(C_3H_5)(dppb)/KOAc$-catalyzed direct arylation of heteroarenes with (hetero)aryl bromides.[264]

Doucet then developed using esters as blocking groups at the C-2-position on 3-substituted heteroarenes, such as thiophenes, furan, pyrrole, and pyrazole, to control the regioselectivity for the direct arylation.[264] By using a $PdCl(C_3H_5)(dppb)/KOAc$ catalytic system, a wide range of (hetero)aryl bromides could be coupled with heteroarenes in moderate-to-good yields with high regioselectivities (Scheme 3.205). This protocol could be followed by easy decarboxylation by simple heating in DMA to produce the expected decarboxylated compound. Thus, a sequential transformation allowed the C-5-arylation, decarboxylation, and C-2-arylation leading to an unsymmetrically 2,5-diarylated-3-substituted product. This method allowed the introduction of aryl groups at C-2 or C-5 in any order.

Doucet extended the use of esters as blocking groups at the C-2-position of bromofuran and bromothiophene derivatives for the synthesis of biheteroarenes via $Pd(OAc)_2$-catalyzed direct C–H functionalization.[265] Methyl 5-bromo-2-furoate and ethyl 5-bromothiophene-2-carboxylate were found to be useful for the direct arylation of heteroarenes. The formation of dimers or oligomers was prohibited since the C-5-position was blocked by ester groups, and therefore allowed the formation of biheteroarenes in good yields (Scheme 3.206). A wide range of heteroarenes could be coupled with these two reagents. Moreover, sequential catalytic C-5-arylation, decarboxylation, and catalytic C-2-arylation reaction was also successfully achieved by this protocol.

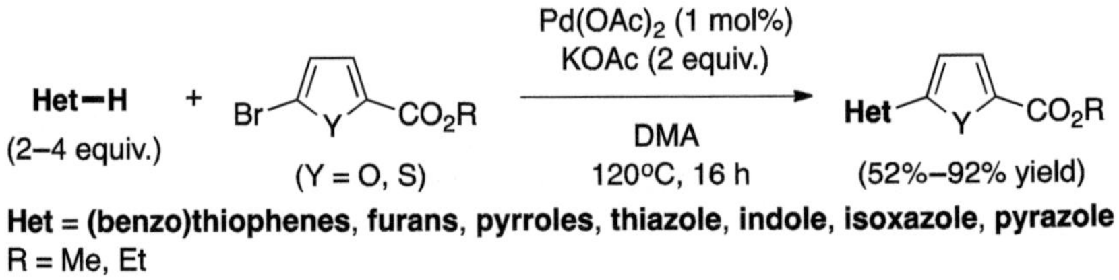

Scheme 3.206 $Pd(OAc)_2$/KOAc-catalyzed direct arylation of heteroarenes with heteroaryl bromides.[265]

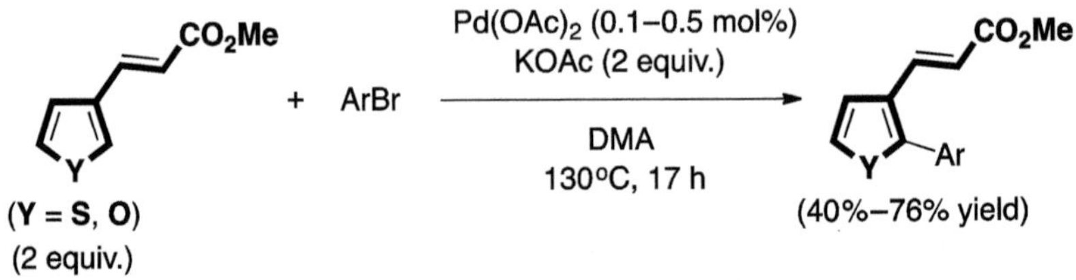

Scheme 3.207 $Pd(OAc)_2$/KOAc-catalyzed direct C-2-arylation of thiophene or furan derivatives bearing acrylate with aryl bromides.[266]

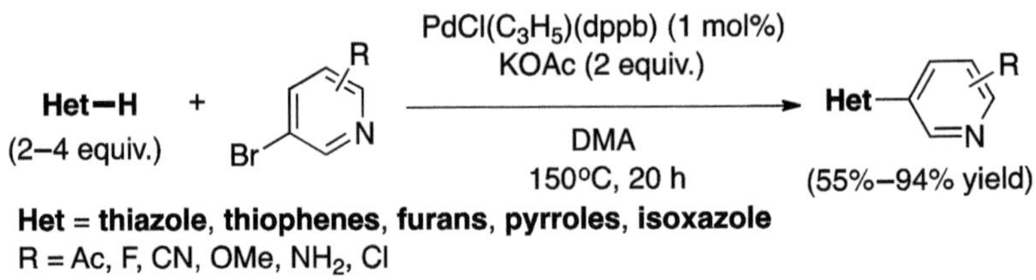

Scheme 3.208 $PdCl(C_3H_5)$(dppb)/KOAc-catalyzed direct arylation of heteroarenes with functionalized bromopyridines.[267]

The $Pd(OAc)_2$/KOAc catalytic system was also found to be active toward the direct arylation of thiophene or furan derivatives bearing an acrylate at the C-3-position (Scheme 3.207).[266] Doucet revealed that the selectivity of the reaction strongly depends on the nature of the base, in which Na_2CO_3 and Li_2CO_3 favored the Heck-type reaction, whereas KOAc promoted regioselective C-2-arylation of the thiophene or furan, and inhibited the Heck-type reaction. In addition, the limitation of these reactions was that only electron-withdrawing substituent on aryl bromides favored the reactions, yet electron-donating one did not.

Doucet developed the direct arylation of a variety of heteroarenes with functionalized pyridyl bromides using a $PdCl(C_3H_5)$(dppb)/KOAc system to give easy access to functionalized heteroarylated pyridines in moderate-to-good yields (Scheme 3.208).[267] Substituents, such as fluoro,

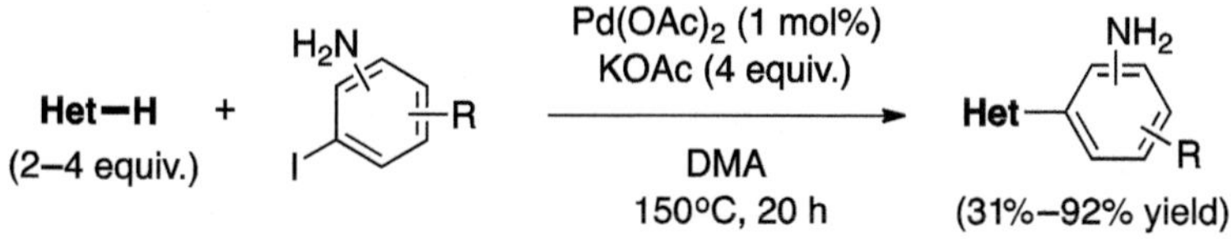

Het = thiazoles, thiophenes, furans, pyrroles, isoxazole
R = H, Cl, F, CO_2Et, CF_3

Scheme 3.209 $Pd(OAc)_2$/KOAc-catalyzed direct arylation of heteroarenes with iodoanilines.[268]

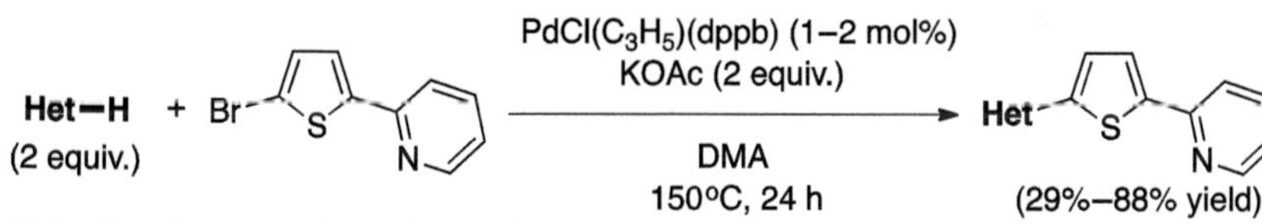

Het = thiophenes, imidazole, pyrrole, furan, isoxazole, benzoxazole

Scheme 3.210 $PdCl(C_3H_5)$(dppb)/KOAc-catalyzed direct arylation of heteroarenes with 2-(5-bromothiophen-2-yl)-pyridine.[269]

acetyl, nitrile, methoxy, amino, and chloro on the bromopyridines were also tolerated. It should be noted that the nature of the substituents influences the yields, in which electron-withdrawing substituents favored the reaction while electron-donating substituents were unfavorable. Furthermore, no significant cleavage of the C–Cl bond was observed for the coupling of chloro-substituted bromopyridines. Therefore, a following Suzuki–Miyaura coupling could take place at the chloro-substituted on the pyridyl ring to access the diarylated pyridines rapidly.

Ammar and Doucet also studied the feasibility of using unprotected iodoanilines as the coupling partners for the direct arylation of heteroarenes.[268] No amination reaction was observed while the $Pd(OAc)_2$/KOAc catalytic system afford the desired products in moderate-to-high yields (Scheme 3.209).

By employing the $PdCl(C_3H_5)$(dppb)/KOAc catalytic system, the direct arylation of heteroarenes with 2-(5-bromothiophen-2-yl)-pyridine proceeded in moderate-to-high yields (Scheme 3.210).[269] It should be highlighted that no functionalization of the C–H bond at the C-3-position of 2-(5-bromothiophen-2-yl)-pyridine was observed. On the other hand, the direct arylation of heteroarenes with 8-bromoquinoline was also achieved using the $Pd(OAc)_2$/CsOAc system (Scheme 3.211).[269] The use of base was found to be crucial for these transformations, in which

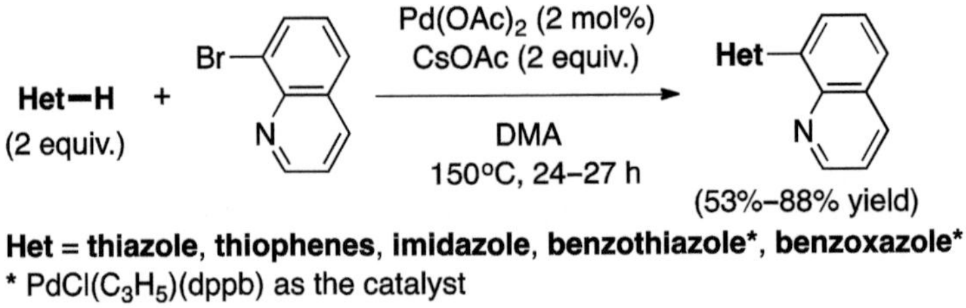

Scheme 3.211 $Pd(OAc)_2$/CsOAc-catalyzed direct arylation of heteroarenes with 8-bromoquinoline.[269]

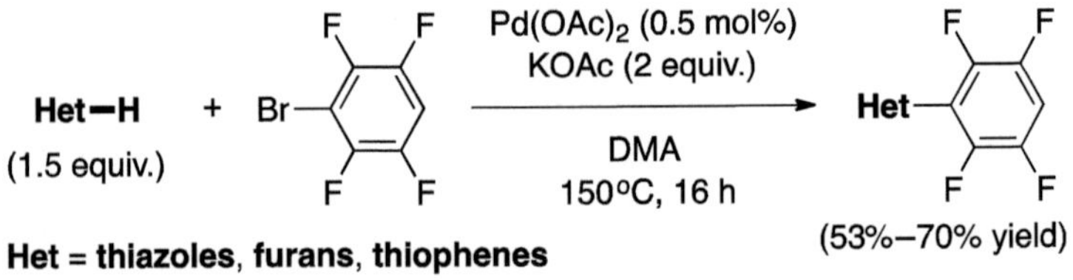

Scheme 3.212 $Pd(OAc)_2$/KOAc-catalyzed direct arylation of heteroarenes with bromopolyfluoroarenes.[270]

CsOAc improved the reaction yields. Notably, these protocols allowed access to polyheteroarenes that are using in the building blocks as chelate ligands.

Furthermore, Doucet also employed bromopolyfluoroarenes as the coupling partners for direct arylation of heteroarenes.[270] The direct arylation allowed selective synthesis of the polyfluoroaryl-heteroarenes in moderate-to-high yields (Scheme 3.212). It should be highlighted that the C–H functionalization took place highly regioselectively at the heteroarene ring instead of the polyfluoroarene ring due to the higher reactivity of C–H bond of heteroarenes as compared to that of bromopolyfluoroarenes. Then, the presence of the C–H bonds on the polyfluoroarene ring of the products could be further functionalized by sequential direct arylation.

Ammar and Doucet continued to extend the scope of the direct arylation of heteroarenes. Direct arylation using bromofluorenes and heteroaromatics as the coupling partners using the $PdCl(C_3H_5)(dppb)$/KOAc catalytic system was also achieved.[271] A wide variety of heteroarenens was successfully employed with 2-bromofluorene derivatives to proceed in moderate-to-good yields (Scheme 3.213). Moreover, the diarylation with 2,7-dibromofluorene or 2,7-dibromofluorenone went smoothly to afford the triaryl derivatives in high yields (Scheme 3.214).

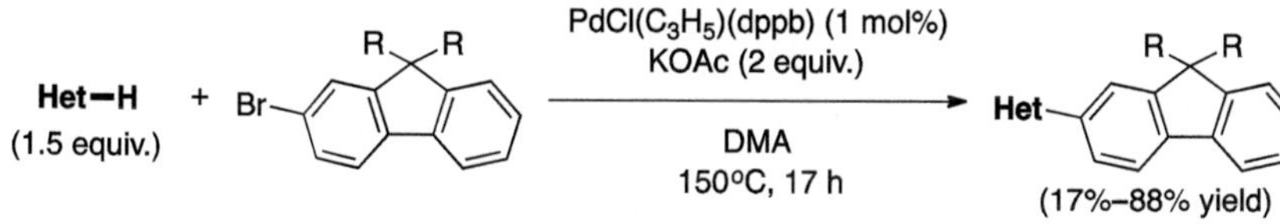

Het = thiazoles, (benzo)thiophenes, (benzo)furans, pyrroles, indole, imidazole, pyrazole, isoxazole, benzoxazole
R = H, $(CH_2)_7CH_3$

Scheme 3.213 $PdCl(C_3H_5)(dppb)/KOAc$-catalyzed direct arylation of heteroarenes with 2-bromofluorene derivatives.[271]

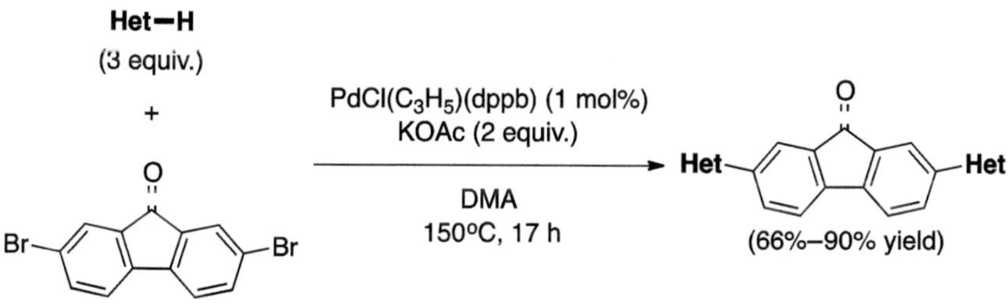

Het = thiazole, thiophenes

Scheme 3.214 $PdCl(C_3H_5)(dppb)/KOAc$-catalyzed direct diarylation of heteroarenes with 2,7-dibromofluorene or 2,7-dibromofluorenone.[271]

Benzothiazol-2-yl- or benzoxazol-2-yl-substituted furan, thiophene, and *N*-methylpyrrole derivatives have long been known to be biologically active, such as in anti-inflammatory activity,[272] and in anti-HIV activity.[273] Ammar and Doucet reported an efficient method for the synthesis of C-5-(hetero)arylated furan, thiophene, and *N*-methylpyrrole derivatives bearing benzothiazolyl or benzoxazolyl moieties at C-2.[274] The C—H functionalization proceeded with $Pd(OAc)_2/KOAc$ and (hetero)aryl bromides as the coupling partners, affording C-5-arylated products in moderate-to-good yields (Scheme 3.215).

Bellina reported a general $Pd(OAc)_2/Bu_4NOAc$-catalyzed direct arylation of azoles with aryl bromides.[275] Oxazole and thiazole reacted at 70°C, whereas *N*-methylimidazoles reacted at 110°C to afford C-5-arylated products in moderate-to-high yields (Scheme 3.216). It should be noted that aryl bromide, including the free hydroxyl group, as the coupling partner was well tolerated and reported for the first time in this protocol. In addition, one-pot synthesis of balsoxin and texaline was successfully achieved which can be isolated from the plant species *Amyris* in the Caribbean.[276]

Scheme 3.215 $Pd(OAc)_2$/KOAc-catalyzed direct C-5-arylation benzothiazolyl or benzoxazolyl heteroarenes with (hetero)aryl bromides.[274]

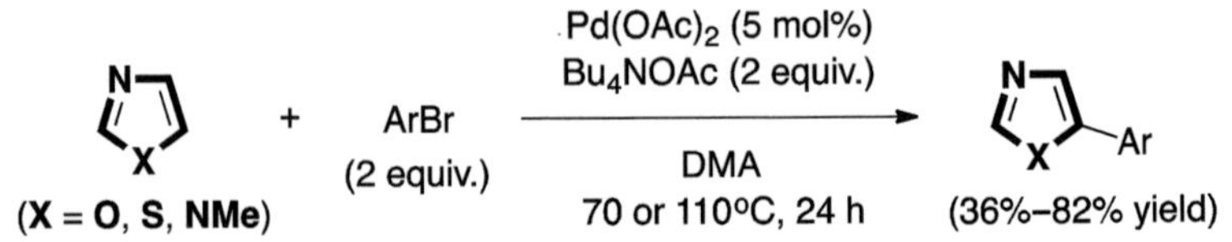

Scheme 3.216 $Pd(OAc)_2/Bu_4NOAc$-catalyzed direct C-5-arylation of azoles with aryl bromides.[275]

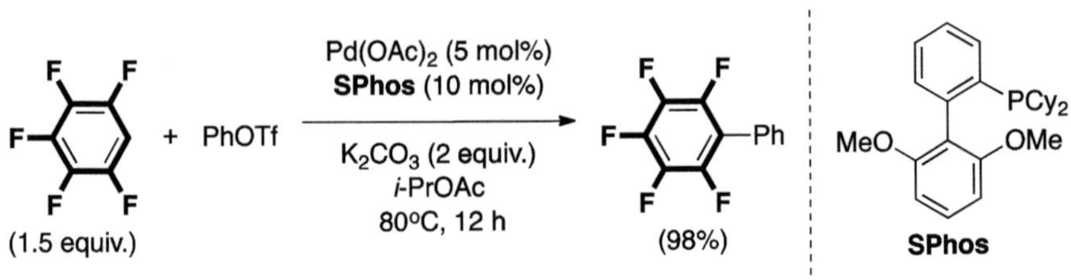

Scheme 3.217 $Pd(OAc)_2$/**SPhos**-catalyzed direct arylation of pentafluorobenzene with phenyl triflate.[25]

3.3 REACTIONS WITH PSEUDOHALIDES

Aryl halides were usually employed as the coupling partners in cross-coupling reactions due to their relatively high reactivity. However, the synthesis of noncommercially available aryl halides may involve multiple steps, and harsh reaction conditions, as well as leading to unwanted waste production. Therefore, the development of pseudohalides as the alternative or complementary electrophile for the coupling reactions began, as well as for C—H functionalization processes.[43]

3.3.1 Palladium-Catalyzed C—H/C—O Cross-Coupling Reactions With Aryl Sulfonates

3.3.1.1 Polyfluoroarenes

Fagnou's group reported an example of $Pd(OAc)_2$/**SPhos**-catalyzed direct arylation of pentafluorobenzene with phenyl triflate (Scheme 3.217).[25]

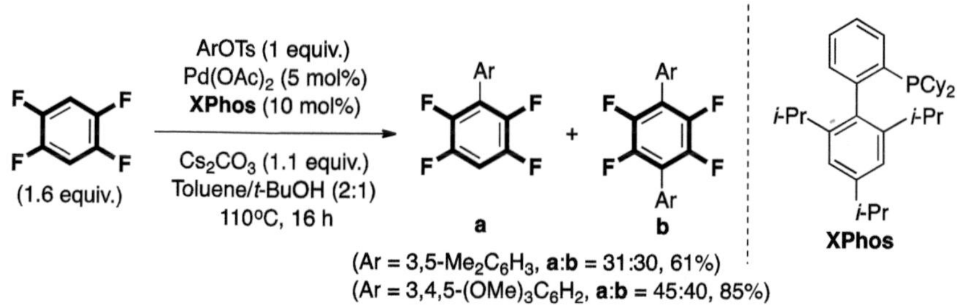

Scheme 3.218 Pd(OAc)$_2$/**XPhos**-catalyzed direct arylation of 1,2,4,5-tetrafluorobenzene with aryl tosylates.[277]

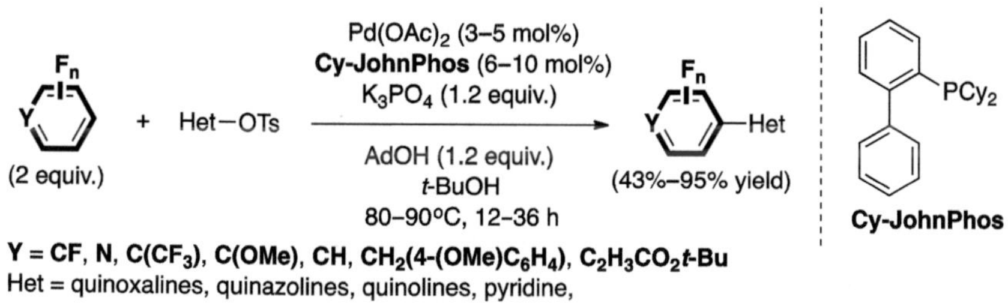

Scheme 3.219 Pd(OAc)$_2$/**Cy-JohnPhos**-catalyzed direct arylation of polyfluorobenzenes with heteroaryl tosylates.[278]

Ackermann et al. developed a Pd(OAc)$_2$/**XPhos** system which was applicable to directly functionalize electron-deficient 1,2,4,5-tetrafluorobenzene with deactivated aryl tosylates (Scheme 3.218).[277] Importantly, this system was also applicable for the direct arylation of heteroaryl *N*-oxides with aryl tosylates/mesylates (see Scheme 3.227).

Zhang described a straightforward Pd(OAc)$_2$/**Cy-JohnPhos**-catalyzed direct arylation of polyfluorobenzenes with heteroaryl tosylates.[278] This system made direct use of various polyfluorobenzenes and inexpensive heteroaryl tosylates to promote corresponding products in moderate-to-high yields (Scheme 3.219).

Using aryl sulfonate as the coupling partner was further improved by Seayad's investigation.[279] Seayad and co-workers employed different catalytic systems to deal with different challenging coupling reactions between polyfluoroarenes and aryl sulfonates. The Pd(OAc)$_2$/**MePhos** efficiently catalyzed the direct arylation of polyfluoroarenes and aryl triflates under nearly ambient conditions (Scheme 3.220). In addition, sterically hindered (hetero)aryl triflates effectively coupled with polyfluoroarenes involving a modified Pd(OAc)$_2$/**RuPhos** system (Scheme 3.221).

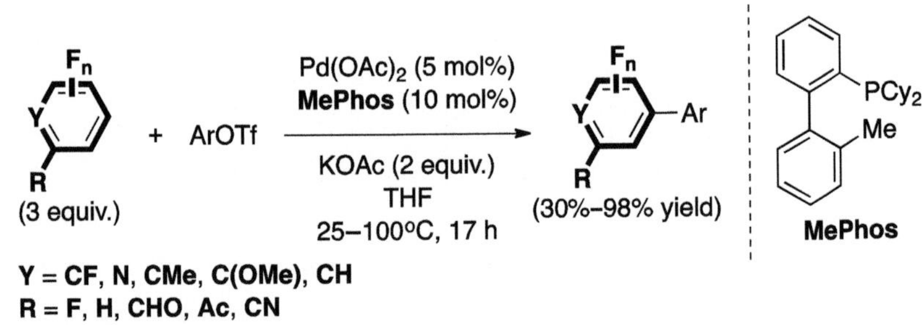

Scheme 3.220 $Pd(OAc)_2$/**MePhos**-catalyzed direct arylation of polyfluoroarenes with aryl triflates.[279]

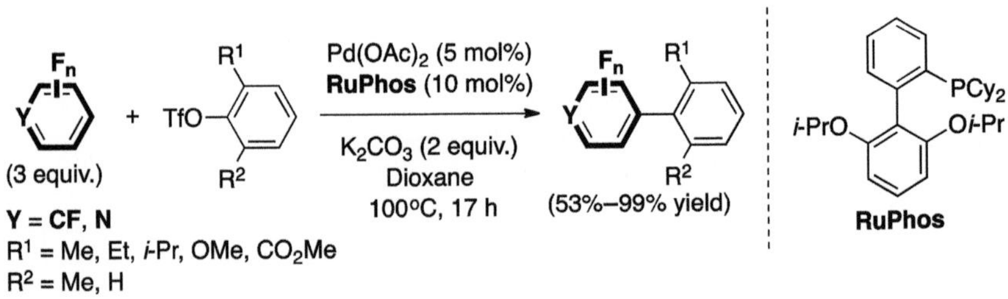

Scheme 3.221 $Pd(OAc)_2$/**RuPhos**-catalyzed direct arylation of polyfluoroareness with sterically hindered (hetero)aryl triflates.[279]

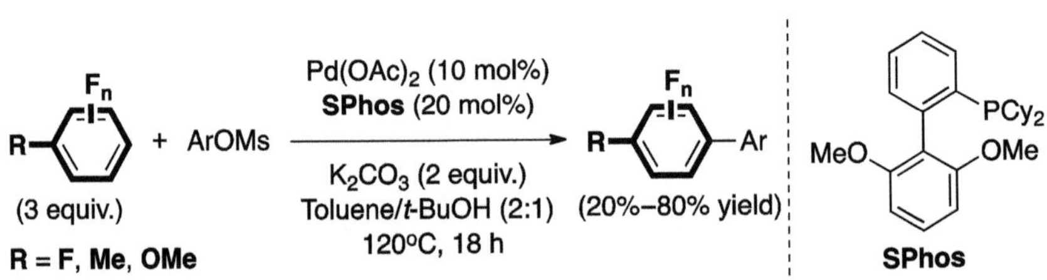

Scheme 3.222 $Pd(OAc)_2$/**SPhos**-catalyzed direct arylation of polyfluoroarenes with aryl mesylates.[279]

Noteworthy, the more attractive with regard to cost, stability and atom economy aryl mesylates could be also used as the coupling partners by employing a higher catalyst loading of $Pd(OAc)_2$ (10 mol%) and ligand **SPhos** (Scheme 3.222).

In the same year, Kwong investigated a general $Pd(OAc)_2$/**CM-phos**-catalyzed direct arylation of polyfluoroarenes with (hetero)aryl tosylates/mesylates.[280] A variety of polyfluoroarenes could be coupled with a wide range of (hetero)aryl tosylates/mesylates in moderate-to-excellent yields under mild reaction conditions (Scheme 3.223).

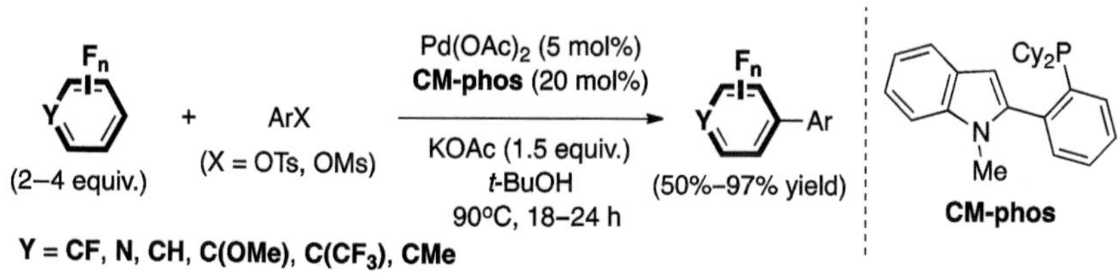

Scheme 3.223 $Pd(OAc)_2$/**CM-phos**-catalyzed direct arylation of polyfluoroarenes with (hetero)aryl tosylates/mesylates.[280]

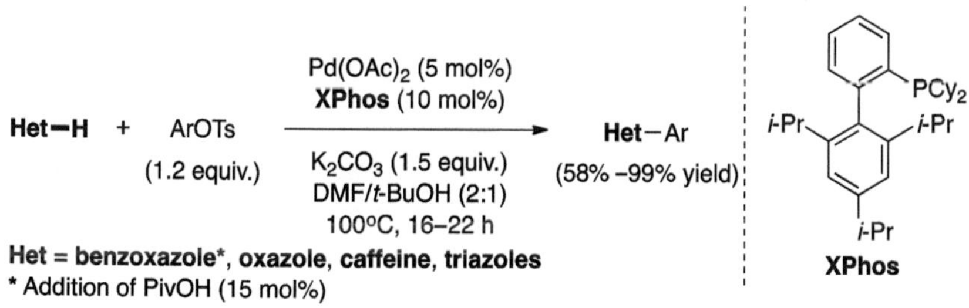

Scheme 3.224 $Pd(OAc)_2$/**XPhos**-catalyzed direct arylation of heteroarenes with aryl tosylates.[281]

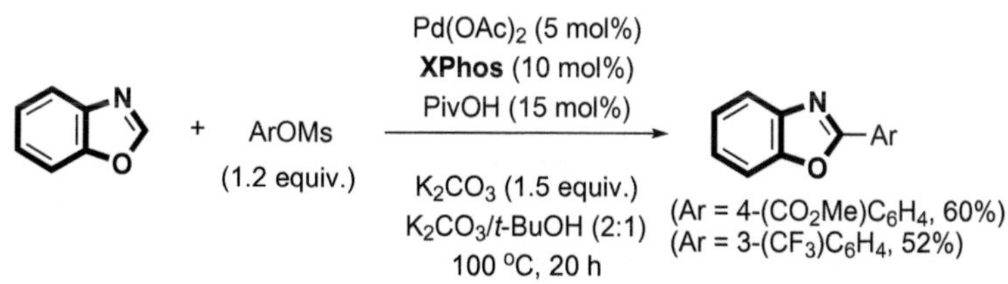

Scheme 3.225 $Pd(OAc)_2$/**XPhos**-catalyzed direct arylation of benzoxazole with aryl mesylates.[281]

3.3.1.2 Heteroarenes

Ackermann et al. disclosed, for the first time, the $Pd(OAc)_2$/**XPhos**-catalyzed direct arylation of heteroarenes using aryl tosylates as the coupling partners, as well as the first atom-economical direct arylation using aryl mesylates as the electrophiles.[281] A variety of aryl tosylates were found to be viable substrates for the direct arylation of heteroarenes, such as benzoxazole, oxazole, caffeine, and triazoles, to afford the corresponding products in moderate-to-excellent yields (Scheme 3.224). Notably, the addition of a catalytic amount of PivOH was beneficial for this reaction. Importantly, direct arylation with aryl mesylates was also accomplished smoothly (Scheme 3.225).

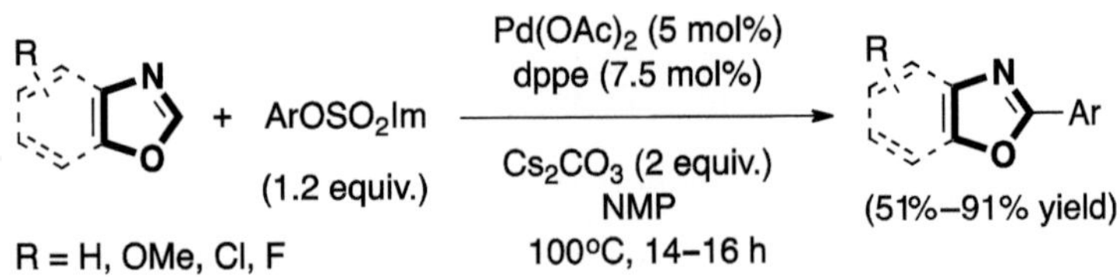

Scheme 3.226 $Pd(OAc)_2$/dppe-catalyzed direct arylation of (benz)oxazoles with imidazolylsulfonates.[282]

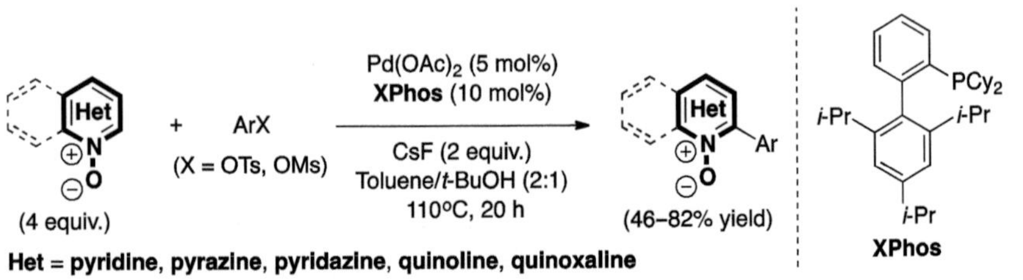

Scheme 3.227 $Pd(OAc)_2$/**XPhos**-catalyzed direct arylation of heteroaryl *N*-oxides with aryl tosylates/mesylates.[277]

Subsequently, Ackermann et al. employed, for the first time, the moisture-stable aryl sulfamates as the electrophiles for the direct arylation processes.[282] A catalytic system comprised of $Pd(OAc)_2$/dppe enabled the direct arylation of (benz)oxazoles with various imidazolylsulfonates in moderate-to-excellent yields (Scheme 3.226).

Later, Ackermann et al. reported the $Pd(OAc)_2$/**XPhos**-catalyzed direct arylation of heteroaryl *N*-oxides with aryl tosylates and mesylates.[277] A variety of heteroaryl *N*-oxides could be coupled with various aryl tosylates/mesylates smoothly (Scheme 3.227).

Kwong disclosed the first general examples of the direct arylation of heteroarenes by using challenging aryl mesylates as the electrophiles.[283] The $Pd(OAc)_2$/**CM-phos** catalytic system enabled the direct arylation of heteroarenes with a range of (hetero)aryl mesylates to afford the corresponding products in moderate-to-excellent yields without the addition of supplementary acid or copper salt additives (Scheme 3.228).

Tan discovered that aryl tosylates could be used to arylate pyridines at the C-3-position using a $Pd(OAc)_2$/phen catalyst system.[284] By employing various pyridines and aryl tosylates as the coupling partners, a variety of C-3-arylated pyridines were successfully synthesized in moderate yields (Scheme 3.229).

Kalyani described the investigation of $Pd(OAc)_2$/dcype (1,2-bis (dicyclohexylphosphino)ethane)-catalyzed direct arylation of benzoxazoles,

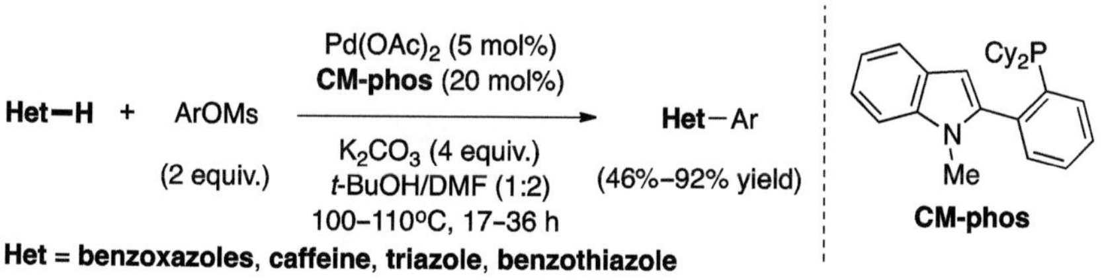

Scheme 3.228 $Pd(OAc)_2$/**CM-phos**-catalyzed direct arylation of heteroarenes with (hetero)aryl mesylates.[283]

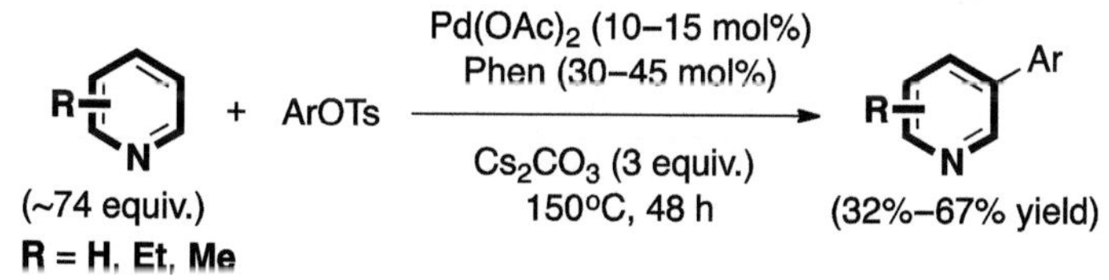

Scheme 3.229 $Pd(OAc)_2$/phen-catalyzed direct C-3-arylation of pyridines with aryl tosylates.[284]

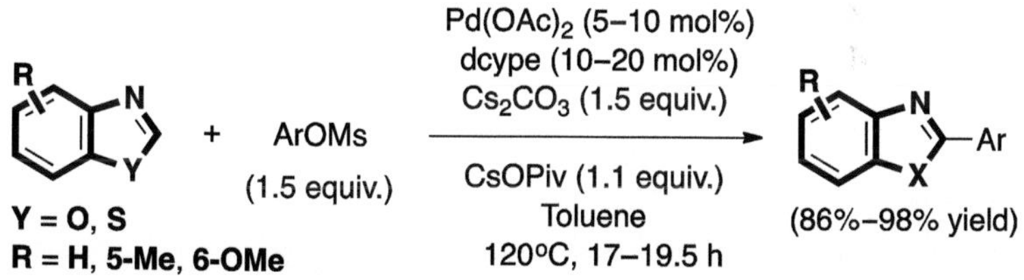

Scheme 3.230 $Pd(OAc)_2$/dcype-catalyzed direct arylation of benzoxazoles or benzothiazole with aryl mesylates.[285]

as well as benzothiazole, using aryl mesylates as the coupling partners.[285] Excellent yields of the arylated products were obtained by using substituted benzoxazoles or benzothiazole, and electron-rich aryl mesylates as the coupling partners (Scheme 3.230). It should be noted that a sequential mesylation/arylation protocol using phenols as substrates was also disclosed, allowing the synthesis of the biaryl products in good yields from phenol derivatives without isolation and purification of the intermediate mesylates.

Very recently, Kwong and Wang reported, for the first time, the direct arylation of imidazo[1,2-*a*]pyridine at the C-3-position with aryl tosylates/mesylates by employing $Pd(OAc)_2$/**SPhos** or **46** (2-(2-(diisopropylphosphino)-phenyl)-1-methyl-1H-indole).[286] The $Pd(OAc)_2$/**SPhos** catalytic system was effective for the coupling using aryl toyslates as the

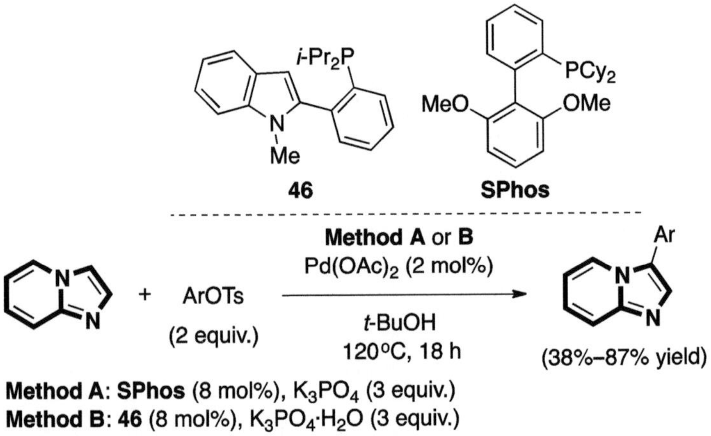

Scheme 3.231 $Pd(OAc)_2$-catalyzed direct arylation of imidazo[1,2-*a*]pyridine with (hetero)aryl tosylates.[286]

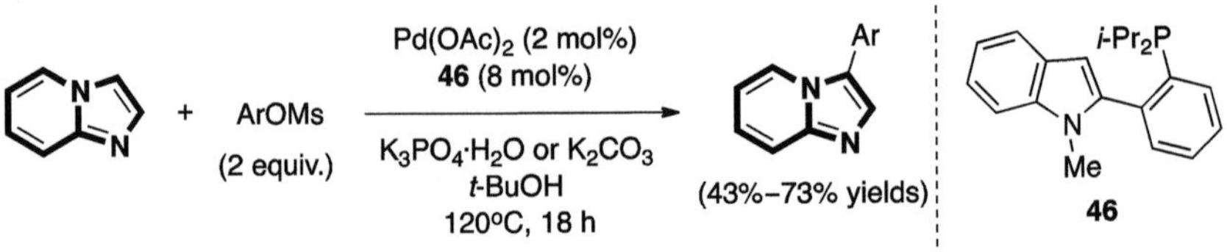

Scheme 3.232 $Pd(OAc)_2$/**46**-catalyzed direct arylation of imidazo[1,2-*a*]pyridine with aryl mesylates.[286]

substrates (Scheme 3.231). Noteworthy, both aryl tosylates and mesylates were also applicable for the arylation of imidazo[1,2-*a*]pyridine by Pd $(OAc)_2$/**46** catalytic system (Schemes 3.231 and 3.232).

3.4 CONCLUSION

The past decade has indicated the rapid development of nondirected palladium-catalyzed C—H functionalization to construct bi(hetero)aryl motifs. Indeed, palladium is the most investigated transition metal in C—H functionalization. It has been successfully utilized to promote all types of reactions discussed in this chapter. A variety of (hetero)arenes and (pseudo)halides can be cross-coupled in both good yields and regioselectivity. From a preparative point of view, the use of simple arenes is highly attractive. However, to achieve the intermolecular arylations with high regioselectivity remains a significant challenge.

The use of the direct C–H functionalization as a route toward the construction of the (hetero)aryl–(hetero)aryl bond is still full of many challenges that need to be resolved in the coming days. Recently, studies have focused on developing milder and greener reaction conditions and more efficient catalytic systems to allow the use of less expensive and industrially attractive substrates bearing no directing group. The solutions to these difficulties will make nondirected C–H functionalization of (hetero)arenes more practical and efficient.

REFERENCES

1. (a) Bringmann, G.; Price Mortimer, A. J.; Keller, P. A.; Gresser, M. J.; Garner, J.; Breuning, M. *Angew. Chem., Int. Ed.* **2005,** *44*, 5384.
   (b) Corbet, J.-P.; Mignani, G. *Chem. Rev.* **2006,** *106*, 2651.
   (c) Beller, M.; Bolm, C. *Transition Metals for Organic Synthesis: Building Blocks and Fine Chemicals*, 2nd rev. and enl. ed.; Wiley-VCH: Weinheim, 2004.
   (d) Balzani, V.; Venturi, M.; Credi, A. *Molecular Devices and Machines: A Journey Into the Nano World;* Wiley-VCH: Weinheim, 2003.
   (e) Ackermann, L. *Modern Arylation Methods;* Wiley-VCH: Weinheim, 2009.
2. (a) Kyoko, T.; Tatsuya, M.; Yoshio, O.; Tamejiro, H. *Tetrahedron Lett.* **1993,** *34*, 8263.
   (b) Denmark, S. E.; Sweis, R. F. *Acc. Chem. Res.* **2002,** *35*, 835.
3. (a) Tamao, K.; Sumitani, K.; Kumada, M. *J. Am. Chem. Soc.* **1972,** *94*, 4374.
   (b) Tamao, K.; Kiso, Y.; Sumitani, K.; Kumada, M. *J. Am. Chem. Soc.* **1972,** *94*, 9268.
4. (a) Baba, S.; Negishi, E. *J. Am. Chem. Soc.* **1976,** *98*, 6729.
   (b) Negishi, E. *Acc. Chem. Res.* **1982,** *15*, 340.
5. (a) Milstein, D.; Stille, J. K. *J. Am. Chem. Soc.* **1978,** *100*, 3636.
   (b) Espinet, P.; Echavarren, A. M. *Angew. Chem., Int. Ed.* **2004,** *43*, 4704.
6. (a) Miyaura, N.; Suzuki, A. *Chem. Rev.* **1995,** *95*, 2457.
   (b) Lloyd-Williams, P.; Giralt, E. *Chem. Soc. Rev.* **2001,** *30*, 145.
7. (a) Ackermann, L.; Vicente, R.; Kapdi, A. R. *Angew. Chem., Int. Ed.* **2009,** *48*, 9792.
   (b) Bellina, F.; Rossi, R. *Tetrahedron* **2009,** *65*, 10269.
8. (a) Alberico, D.; Scott, M. E.; Lautens, M. *Chem. Rev.* **2007,** *107*, 174.
   (b) Gutekunst, W. R.; Baran, P. S. *Chem. Soc. Rev.* **2011,** *40*, 1976.
   (c) Yeung, C. S.; Dong, V. M. *Chem. Rev.* **2011,** *111*, 1215.
9. (a) Seregin, I. V.; Gevorgyan, V. *Chem. Soc. Rev.* **2007,** *36*, 1173.
   (b) Joucla, L.; Djakovitch, L. *Adv. Synth. Catal.* **2009,** *351*, 673.
   (c) Cho, S. H.; Kim, J. Y.; Kwak, J.; Chang, S. *Chem. Soc. Rev.* **2011,** *40*, 5068.
   (d) Hirano, K.; Miura, M. *Synlett* **2011,** 294.
   (e) Zhao, D.; You, J.; Hu, C. *Chem. Eur. J.* **2011,** *17*, 5466.
   (f) Bugaut, X.; Glorius, F. *Angew. Chem., Int. Ed.* **2011,** *50*, 7479.
   (g) Armstrong, A.; Collins, J. C. *Angew. Chem., Int. Ed.* **2010,** *49*, 2282.
10. (a) Zhu, C.; Wang, R.; Falck, J. R. *Chem. Asian J.* **2012,** *7*, 1502.
   (b) Duong, H. A.; Gilligan, R. E.; Cooke, M. L.; Phipps, R. J.; Gaunt, M. J. *Angew. Chem., Int. Ed.* **2011,** *50*, 463.
   (c) Phipps, R. J.; Gaunt, M. J. *Science* **2009,** *323*, 1593.
   (d) Leow, D.; Li, G.; Mei, T.-S.; Yu, J.-Q. *Nature* **2012,** *486*, 518.

11. (a) Rousseau, G.; Breit, B. *Angew. Chem., Int. Ed.* **2011,** *50*, 2450.
    (b) Wang, C.; Chen, H.; Wang, Z.; Chen, J.; Huang, Y. *Angew. Chem., Int. Ed.* **2012,** *51*, 7242.
    (c) Cornella, J.; Righi, M.; Larrosa, I. *Angew. Chem., Int. Ed.* **2011,** *50*, 9429.
    (d) Richter, H.; Beckendorf, S.; Mancheño, O. G. *Adv. Synth. Catal.* **2011,** *353*, 295.
    (e) Yu, M.; Liang, Z.; Wang, Y.; Zhang, Y. *J. Org. Chem.* **2011,** *76*, 4987.
    (f) Chernyak, N.; Dudnik, A. S.; Huang, C.; Gevorgyan, V. *J. Am. Chem. Soc.* **2010,** *132*, 8270.
12. (a) Ashenhurst, J. A. *Chem. Soc. Rev.* **2010,** *39*, 540.
    (b) Lei, A.; Liu, W.; Liu, C.; Chen, M. *Dalton Trans.* **2010,** *39*, 10352.
    (c) Li, B.-J.; Yang, S.-D.; Shi, Z.-J. *Synlett* **2008,** 949.
    (d) Kuhl, N.; Hopkinson, M. N.; Wencel-Delord, J.; Glorius, F. *Angew.Chem., Int. Ed.* **2012,** *51*, 10236.
13. (a) Campeau, L.-C.; Fagnou, K. *Chem. Commun.* **2006,** 1253.
    (b) Daugulis, O.; Do, H.-Q.; Shabashov, D. *Acc. Chem. Res.* **2009,** *42*, 1074.
    (c) McGlacken, G. P.; Bateman, L. M. *Chem. Soc. Rev.* **2009,** *38*, 2447.
14. Lafrance, M.; Fagnou, K. *J. Am. Chem. Soc.* **2006,** *128*, 16496.
15. Qin, C.; Lu, W. *J. Org. Chem.* **2008,** *73*, 7424.
16. Beydoun, K.; Doucet, H. *Catal. Sci. Technol.* **2011,** *1*, 1243.
17. Hong, W.; Qiu, Y.; Yao, Z.; Wang, Z.; Jiang, S. *Tetrahedron Lett.* **2011,** *52*, 4916.
18. Wang, Y.-N.; Guo, X.-Q.; Zhu, X.-H.; Zhong, R.; Cai, L.-H.; Hou, X.-F. *Chem. Commun.* **2012,** *48*, 10437.
19. Zhao, L.; Bruneau, C.; Doucet, H. *Chem. Commun.* **2013,** *49*, 5598.
20. Wu, Z.; Luo, F.; Chen, S.; Li, Z.; Xiang, H.; Zhou, X. *Chem. Commun.* **2013,** *49*, 7653.
21. Mousseau, J. J.; Vallée, F.; Lorion, M. M.; Charette, A. B. *J. Am. Chem. Soc.* **2010,** *132*, 14412.
22. Long, R.; Yan, X.; Wu, Z.; Li, Z.; Xiang, H.; Zhou, X. *Org. Biomol. Chem.* **2015,** *13*, 3571.
23. He, M.; Soulé, J.-F.; Doucet, H. *ChemCatChem* **2014,** *6*, 1824.
24. Lafrance, M.; Rowley, C. N.; Woo, T. K.; Fagnou, K. *J. Am. Chem. Soc.* **2006,** *128*, 8754.
25. Lafrance, M.; Shore, D.; Fagnou, K. *Org. Lett.* **2006,** *8*, 5097.
26. Okamoto, T.; Nakahara, K.; Saeki, A.; Seki, S.; Oh, J. H.; Akkerman, H. B.; Bao, Z.; Matsuo, Y. *Chem. Mater.* **2011,** *23*, 1646.
27. René, O.; Fagnou, K. *Org. Lett.* **2010,** *12*, 2116.
28. Zhao, D.; Wang, W.; Lian, S.; Yang, F.; Lan, J.; You, J. *Chem. Eur. J.* **2009,** *15*, 1337.
29. Chen, F.; Min, Q.-Q.; Zhang, X. *J. Org. Chem.* **2012,** 77, 2992.
30. Zhang, X.; Yu, J.; Yan, G. *Z. Naturforsch. B: Chem. Sci.* **2013,** *68b*, 885.
31. Yuan, D.; Huynh, H. V. *Organometallics* **2012,** *31*, 405.
32. Bernhammer, J. C.; Huynh, H. V. *Organometallics* **2012,** *31*, 5121.
33. Yuan, D.; Huynh, H. V. *Organometallics* **2014,** *33*, 6033.
34. Ma, X.; Liu, Y.; Liu, P.; Xie, J.; Dai, B.; Liu, Z. *Appl. Organomet. Chem.* **2014,** *28*, 180.
35. (a) Miyamura, H.; Kobayashi, S. *Acc. Chem. Res.* **2014,** *47*, 1054.
    (b) Chng, L. L.; Erathodiyil, N.; Ying, J. Y. *Acc. Chem. Res.* **2013,** *46*, 1825.
    (c) Mahmoud, M. A.; Narayanan, R.; El-Sayed, M. A. *Acc. Chem. Res.* **2013,** *46*, 1795.
    (d) Yasukawa, T.; Miyamura, H.; Kobayashi, S. *J. Am. Chem. Soc.* **2012,** *134*, 16963.
    (e) Sankar, M.; Dimitratos, N.; Miedziak, P. J.; Wells, P. P.; Kiely, C. J.; Hutchings, G. J. *Chem. Soc. Rev.* **2012,** *41*, 8099.
36. Cao, M.; Wu, D.; Su, W.; Cao, R. *J. Catal.* **2015,** *321*, 62.
37. Do, H.-Q.; Khan, R. M. K.; Daugulis, O. *J. Am. Chem. Soc.* **2008,** *130*, 15185.
38. Yuen, O. Y.; Charoensak, M.; So, C. M.; Kuhakarn, C.; Kwong, F. Y. *Chem. Asian J.* **2015,** *10*, 857.

39. Fu, Z.; Xiong, Q.; Zhang, W.; Li, Z.; Cai, H. *Tetrahedron Lett.* **2015,** *56*, 123.
40. Abdellaoui, F.; Ammar, H. B.; Soulé, J.-F.; Doucet, H. *Catal. Commun.* **2015,** *71*, 13.
41. Zhao, L.; Yan, T.; Bruneau, C.; Doucet, H. *Catal. Sci. Technol.* **2014,** *4*, 352.
42. (a) García-Cuadrado, D.; de Mendoza, P.; Braga, A. A. C.; Maseras, F.; Echavarren, A. M. *J. Am. Chem. Soc.* **2007,** *129*, 6880.
(b) Gorelsky, S. I.; Lapointe, D.; Fagnou, K. *J. Org. Chem.* **2012,** 77, 658.
(c) Gorelsky, S. I. *Coord. Chem. Rev.* **2013,** *257*, 153.
43. Humphrey, G. R.; Kuethe, J. T. *Chem. Rev.* **2006,** *106*, 2875.
44. (a) Baran, P. S.; Guerrero, C. A.; Ambhaikar, N. B.; Hafensteiner, B. D. *Angew. Chem., Int. Ed.* **2005,** *44*, 606.
(b) Hall, A.; Atkinson, S.; Brown, S. H.; Chessell, I. P.; Chowdhury, A.; Giblin, G. M. P.; Goldsmith, P.; Healy, M. P.; Jandu, K. S.; Johnson, M. R.; Michel, A. D.; Naylor, A.; Sweeting, J. A. *Bioorg. Med. Chem. Lett.* **2007,** *17*, 1200.
45. (a) Larionov, O. V.; de Meijere, A. *Angew. Chem., Int. Ed.* **2005,** *44*, 5664.
(b) Garg, N. K.; Caspi, D. D.; Stoltz, B. M. *J. Am. Chem. Soc.* **2005,** *127*, 5970.
(c) Walsh, C. T.; Garneau-Tsodikova, S.; Howard-Jones, A. R. *Nat. Prod. Rep.* **2006,** *23*, 517.
(d) Bellina, F.; Rossi, R. *Tetrahedron* **2006,** *62*, 7213.
46. Wang, X.; Gribkov, D. V.; Sames, D. *J. Org. Chem.* **2007,** 72, 1476.
47. Roger, J.; Doucet, H. *Adv. Synth. Catal.* **2009,** *351*, 1977.
48. Gryko, D. T.; Vakuliuk, O.; Gryko, D.; Koszarna, B. *J. Org. Chem.* **2009,** *74*, 9517.
49. Jafarpour, F.; Rahiminejadan, S.; Hazrati, H. *J. Org. Chem.* **2010,** *75*, 3109.
50. Bheeter, C. B.; Bera, J. K.; Doucet, H. *Tetrahedron Lett.* **2012,** *53*, 509.
51. Zhao, L.; Bruneau, C.; Doucet, H. *ChemCatChem* **2013,** *5*, 255.
52. Xu, Y.; Zhao, L.; Li, Y.; Doucet, H. *Adv. Synth. Catal.* **2013,** *355*, 1423.
53. Smari, I.; Youssef, C.; Ben Ammar, H.; Ben Hassine, B.; Soulé, J.-F.; Doucet, H. *Tetrahedron* **2015,** *71*, 6586.
54. Özdemir, I.; Gürbüz, N.; Kaloğlu, N.; Doğan, Ö.; Kaloğlu, M.; Bruneau, C.; Doucet, H. *Beilstein J. Org. Chem.* **2013,** *9*, 303.
55. (a) Bernhammer, J. C.; Huynh, H. V. *Organometallics* **2014,** *33*, 1266.
(b) Bernhammer, J. C.; Singh, H.; Huynh, H. V. *Organometallics* **2014,** *33*, 4295.
56. Lebrasseur, N.; Larrosa, I. *J. Am. Chem. Soc.* **2008,** *130*, 2926.
57. Djakovitch, L.; Dufaud, V.; Zaidi, R. *Adv. Synth. Catal.* **2006,** *348*, 715.
58. Djakovitch, L.; Rouge, P.; Zaidi, R. *Catal. Commun.* **2007,** *8*, 1561.
59. Cusati, G.; Djakovitch, L. *Tetrahedron Lett.* **2008,** *49*, 2499.
60. Zhang, Z.; Hu, Z.; Yu, Z.; Lei, P.; Chi, H.; Wang, Y.; He, R. *Tetrahedron Lett.* **2007,** *48*, 2415.
61. Bellina, F.; Benelli, F.; Rossi, R. *J. Org. Chem.* **2008,** *73*, 5529.
62. Ackermann, L. *Synthesis* **2006,** 1557.
63. Ackermann, L.; Barfüßer, S. *Synlett* **2009,** 808.
64. Ruiz-Rodríguez, J.; Albericio, F.; Lavilla, R. *Chem. Eur. J.* **2010,** *16*, 1124.
65. Joucla, L.; Batail, N.; Djakovitch, L. *Adv. Synth. Catal.* **2010,** *352*, 2929.
66. Nadres, E. T.; Lazareva, A.; Daugulis, O. *J. Org. Chem.* **2011,** *76*, 471.
67. Nandi, D.; Jhou, Y.-M.; Lee, J.-Y.; Kuo, B.-C.; Liu, C.-Y.; Huang, P.-W.; Lee, H. M. *J. Org. Chem.* **2012,** 77, 9384.
68. Kawamata, Y.; Tokuji, S.; Yorimitsu, H.; Osuka, A. *Angew. Chem., Int. Ed.* **2011,** *50*, 8867.
69. (a) Senge, M. O.; Fazekas, M.; Notaras, E. G. A.; Blau, W. J.; Zawadzka, M.; Locos, O. B.; Ni Mhuircheartaigh, E. M. *Adv. Mater.* **2007,** *19*, 2737.
(b) Hori, T.; Nakamura, Y.; Aratani, N.; Osuka, A. *J. Organomet. Chem.* **2007,** *692*, 148.
(c) Gust, D.; Moore, T. A.; Moore, A. L. *Acc. Chem. Res.* **2009,** *42*, 1890.
(d) Wasielewski, M. R. *Acc. Chem. Res.* **2009,** *42*, 1910.
(e) Aratani, N.; Kim, D.; Osuka, A. *Acc. Chem. Res.* **2009,** *42*, 1922.
(f) Martinez-Diaz, M. V.; de la Torre, G.; Torres, T. *Chem. Commun.* **2010,** *46*, 7090.

70. (a) Loudet, A.; Burgess, K. *Chem. Rev.* **2007,** *107*, 4891.
(b) Ulrich, G.; Ziessel, R.; Harriman, A. *Angew. Chem., Int. Ed.* **2008,** *47*, 1184.
71. Verbelen, B.; Leen, V.; Wang, L.; Boens, N.; Dehaen, W. *Chem. Commun.* **2012,** *48*, 9129.
72. (a) Liégault, B.; Lapointe, D.; Caron, L.; Vlassova, A.; Fagnou, K. *J. Org. Chem.* **2009,** *74*, 1826.
(b) Liégault, B.; Petrov, I.; Gorelsky, S. I.; Fagnou, K. *J. Org. Chem.* **2010,** *75*, 1047.
73. Fromtling, R. A. *Clin. Microbiol. Rev.* **1988,** *1*, 187.
74. (a) Sharpe, T. R.; Cherkofsky, S. C.; Hewes, W. E.; Smith, D. H.; Gregory, W. A.; Haber, S. B.; Leadbetter, M. R.; Whitney, J. G. *J. Med. Chem.* **1985,** *28*, 1188.
(b) Uçucu, Ü.; Karaburun, N. G.; Işikdağ, İ. *Farmaco* **2001,** *56*, 285.
75. Laufer, S. A.; Zimmermann, W.; Ruff, K. J. *J. Med. Chem.* **2004,** *47*, 6311.
76. (a) Du, H.; He, Y.; Sivappa, R.; Lovely, C. J. *Synlett* **2006,** 965.
(b) Bellina, F.; Cauteruccio, S.; Rossi, R. *Tetrahedron* **2007,** *63*, 4571.
(c) Kamijo, S.; Yamamoto, Y. *Chem. Asian J.* **2007,** *2*, 568.
(d) Bellina, F.; Rossi, R. *Adv. Synth. Catal.* **2010,** *352*, 1223.
77. Bellina, F.; Cauteruccio, S.; Mannina, L.; Rossi, R.; Viel, S. *Eur. J. Org. Chem.* **2006,** 693.
78. Bellina, F.; Cauteruccio, S.; Mannina, L.; Rossi, R.; Viel, S. *J. Org. Chem.* **2005,** *70*, 3997.
79. Bellina, F.; Cauteruccio, S.; Rossi, R. *J. Org. Chem.* **2007,** *72*, 8543.
80. Bellina, F.; Cauteruccio, S.; Di Fiore, A.; Marchetti, C.; Rossi, R. *Tetrahedron* **2008,** *64*, 6060.
81. Bellina, F.; Cauteruccio, S.; Di Fiore, A.; Rossi, R. *Eur. J. Org. Chem.* **2008,** 5436.
82. Joo, J. M.; Touré, B. B.; Sames, D. *J. Org. Chem.* **2010,** *75*, 4911.
83. Kumar, P. V.; Lin, W.-S.; Shen, J.-S.; Nandi, D.; Lee, H. M. *Organometallics* **2011,** *30*, 5160.
84. Shibahara, F.; Yamauchi, T.; Yamaguchi, E.; Murai, T. *J. Org. Chem.* **2012,** 77, 8815.
85. Mahindra, A.; Bagra, N.; Jain, R. *J. Org. Chem.* **2013,** *78*, 10954.
86. Yamauchi, T.; Shibahara, F.; Murai, T. *J. Org. Chem.* **2014,** *79*, 7185.
87. Sueoka, H.; Yasuoka, J.; Nishiyama, A.; Kiuchi, M.; Yamamoto, K.; Sugahara, K.; Ehara, S.; Sakata, K. Patent US 6288061(B1); Welfide Corp., Japan, 2001.
88. Takfaoui, A.; Zhao, L.; Touzani, R.; Soulé, J.-F.; Dixneuf, P. H.; Doucet, H. *Tetrahedron* **2014,** *70*, 8316.
89. Xu, X.; Zhao, L.; Li, Y.; Soulé, J.-F.; Doucet, H. *Adv. Synth. Catal.* **2015,** *357*, 2869.
90. (a) Sliskovic, D. R.; Roth, B. D.; Wilson, M. W.; Hoefle, M. L.; Newton, R. S. *J. Med. Chem.* **1990,** *33*, 31.
(b) Penning, T. D.; Talley, J. J.; Bertenshaw, S. R.; Carter, J. S.; Collins, P. W.; Docter, S.; Graneto, M. J.; Lee, L. F.; Malecha, J. W.; Miyashiro, J. M.; Rogers, R. S.; Rogier, D. J.; Yu, S. S.; Anderson, G. D.; Burton, E. G.; Cogburn, J. N.; Gregory, S. A.; Koboldt, C. M.; Perkins, W. E.; Seibert, K.; Veenhuizen, A. W.; Zhang, Y. Y.; Isakson, P. C. *J. Med. Chem.* **1997,** *40*, 1347.
(c) Moore, K. W.; Bonner, K.; Jones, E. A.; Emms, F.; Leeson, P. D.; Marwood, R.; Patel, S.; Patel, S.; Rowley, M.; Thomas, S.; Carling, R. W. *Bioorg. Med. Chem. Lett.* **1999,** *9*, 1285.(d)S.R. Stauffer, J.A. Katzenellenbogen, *J. Comb. Chem.* **2000**, *2*, 318.
(e) Pozharskiĭ, A. F.; Katritzky, A. R.; Soldatenkov, A. T. *Heterocycles in Life and Society: An Introduction to Heterocyclic Chemistry, Biochemistry, and Applications*, 2nd ed.; Wiley: Chichester, 2011.
91. Goikhman, R.; Jacques, T. L.; Sames, D. *J. Am. Chem. Soc.* **2009,** *131*, 3042.
92. Eicher, T.; Hauptmann, S.; Speicher, A. *The Chemistry of Heterocycles;* Wiley-VCH Verlag GmbH & Co. KGaA, 2004122–184.

93. Fall, Y.; Doucet, H.; Santelli, M. *Synthesis* **2010,** 127.
94. Beladhria, A.; Beydoun, K.; Ammar, H. B.; Salem, R. B.; Doucet, H. *Synthesis* **2011,** 2553.
95. Derridj, F.; Roger, J.; Djebbar, S.; Doucet, H. *Adv. Synth. Catal.* **2012,** *354*, 747.
96. Yan, T.; Chen, L.; Bruneau, C.; Dixneuf, P. H.; Doucet, H. *J. Org. Chem.* **2012,** 77, 7659.
97. Takfaoui, A.; Zhao, L.; Touzani, R.; Dixneuf, P. H.; Doucet, H. *Tetrahedron Lett.* **2014,** *55*, 1697.
98. Mateos, C.; Mendiola, J.; Carpintero, M.; Mínguez, J. M. *Org. Lett.* **2010,** *12*, 4924.
99. Kumpulainen, E. T. T.; Pohjakallio, A. *Adv. Synth. Catal.* **2014,** *356*, 1555.
100. Hattori, K.; Yamaguchi, K.; Yamaguchi, J.; Itami, K. *Tetrahedron* **2012,** *68*, 7605.
101. (a) Chen, H.-S.; Kuo, S.-C.; Teng, C.-M.; Lee, F.-Y.; Wang, J.-P.; Lee, Y.-C.; Kuo, C.-W.; Huang, C.-C.; Wu, C.-C.; Huang, L.-J. *Bioorg. Med. Chem.* **2008,** *16*, 1262.
(b) Wu, C.-C.; Huang, S.-W.; Hwang, T.-L.; Kuo, S.-C.; Lee, F.-Y.; Teng, C.-M. *Br. J. Pharmacol.* **2000,** *130*, 1289.
102. (a) Collot, V.; Dallemagne, P.; Bovy, P. R.; Rault, S. *Tetrahedron* **1999,** *55*, 6917.
(b) Li, S. H.; Shin, D. H.; Chun, Y.-S.; Lee, M. K.; Kim, M.-S.; Park, J.-W. *Mol. Cancer Ther.* **2008,** 7, 3729.
103. Ben-Yahia, A.; Naas, M.; El Kazzouli, S.; Essassi, E. M.; Guillaumet, G. *Eur. J. Org. Chem.* **2012,** 7075.
104. Ye, M.; Edmunds, A. J. F.; Morris, J. A.; Sale, D.; Zhang, Y.; Yu, J.-Q. *Chem. Sci.* **2013,** *4*, 2374.
105. Attaur, R.; Malik, S.; Hasan, S. S.; Choudhary, M. I.; Ni, C.-Z.; Clardy, J. *Tetrahedron Lett.* **1995,** *36*, 1993.
106. Naas, M.; El Kazzouli, Sd; Essassi, E. M.; Bousmina, M.; Guillaumet, G. *J. Org. Chem.* **2014,** *79*, 7286.
107. Gronowits, S.; Hörnfeldt, A.-B. *Thiophenes;* Elsevier: Oxford, 2004.
108. Yanagisawa, S.; Ueda, K.; Sekizawa, H.; Itami, K. *J. Am. Chem. Soc.* **2009,** *131*, 14622.
109. Ueda, K.; Yanagisawa, S.; Yamaguchi, J.; Itami, K. *Angew. Chem., Int. Ed.* **2010,** *49*, 8946.
110. Monguchi, Y.; Kume, A.; Hattori, K.; Maegawa, T.; Sajiki, H. *Tetrahedron* **2006,** *62*, 7926.
111. Gorelsky, S. I.; Lapointe, D.; Fagnou, K. *J. Am. Chem. Soc.* **2008,** *130*, 10848.
112. (a) Quirion, R.; Bowen, W. D.; Itzhak, Y.; Junien, J. L.; Musacchio, J.; Rothman, R. B.; Tsung-Ping, S.; Tam, S. W.; Taylor, D. P. *Trends Pharmacol. Sci.* **1992,** *13*, 85.
(b) Hayashi, T.; Su, T.-P. *CNS Drugs* **2004,** *18*, 269.
(c) Maurice, T.; Su, T.-P. *Pharmacol. Ther.* **2009,** *124*, 195.
(d) Ishikawa, M.; Hashimoto, K. *J. Recept. Ligand Channel Res.* **2010,** *3*, 25.
113. Meyer, C.; Neue, B.; Schepmann, D.; Yanagisawa, S.; Yamaguchi, J.; Wurthwein, E.-U.; Itami, K.; Wunsch, B. *Org. Biomol. Chem.* **2011,** *9*, 8016.
114. (a) Derridj, F.; Roger, J.; Djebbar, S.; Doucet, H. *Org. Lett.* **2010,** *12*, 4320.
(b) Derridj, F.; Si Larbi, K.; Roger, J.; Djebbar, S.; Doucet, H. *Tetrahedron* **2012,** *68*, 7463.
115. (a) Tian, H. K.; Shi, J. W.; He, B.; Hu, N. H.; Dong, S. Q.; Yan, D. H.; Zhang, J. P.; Geng, Y. H.; Wang, F. S. *Adv. Funct. Mater.* **2007,** *17*, 1940.
(b) Fave, C.; Leroux, Y.; Trippé, G.; Randriamahazaka, H.; Noel, V.; Lacroix, J.-C. T. *J. Am. Chem. Soc.* **2007,** *129*, 1890.
(c) Porzio, W.; Destri, S.; Pasini, M.; Giovanella, U.; Ragazzi, M.; Scavia, G.; Kotowski, D.; Zotti, G.; Vercelli, B. *New J. Chem.* **2010,** *34*, 1961.
116. Si Larbi, K.; Djebbar, S.; Doucet, H. *Eur. J. Inorg. Chem.* **2011,** 3493.

117. Beydoun, K.; Doucet, H. *J. Organomet. Chem.* **2011,** *696*, 1749.
118. (a) Bheeter, C. B.; Bera, J. K.; Doucet, H. *J. Org. Chem.* **2011,** *76*, 6407.
(b) Bheeter, C. B.; Bera, J. K.; Doucet, H. *RSC Adv.* **2012,** *2*, 7197.
119. Beladhria, A.; Gottumukkala, A. L.; Youssef, C.; Bheeter, C. B.; Ben Ammar, H.; Ben Salem, R.; Doucet, H. *Catal. Commun.* **2013,** *41*, 119.
120. Beydoun, K.; Zaarour, M.; Williams, J. A. G.; Roisnel, T.; Dorcet, V.; Planchat, A.; Boucekkine, A.; Jacquemin, D.; Doucet, H.; Guerchais, V. *Inorg. Chem.* **2013,** *52*, 12416.
121. Jin, R.; Bheeter, C. B.; Doucet, H. *Beilstein J. Org. Chem.* **2014,** *10*, 1239.
122. Zhao, L.; Bruneau, C.; Doucet, H. *Tetrahedron* **2013,** *69*, 7082.
123. Smari, I.; Ammar, H. B.; Hassine, B. B.; Soulé, J.-F.; Doucet, H. *Synthesis* **2015,** *47*, 3354.
124. Dong, J. J.; Roy, D.; Roy, R. J.; Ionita, M.; Doucet, H. *Synthesis* **2011,** 3530.
125. Tang, D.-T. D.; Collins, K. D.; Glorius, F. *J. Am. Chem. Soc.* **2013,** *135*, 7450.
126. Wang, S.; Hu, D.; Hua, W.; Gu, J.; Zhang, Q.; Jia, X.; Xi, K. *RSC Adv.* **2015,** *5*, 53935.
127. Schipper, D. J.; Fagnou, K. *Chem. Mater.* **2011,** *23*, 1594.
128. (a) Park, S. W.; Son, K.-I.; Ko, M. J.; Kim, K.; Park, N.-G. *Synth. Met.* **2009,** *159*, 2571.
(b) Lee, K.; Park, S. W.; Ko, M. J.; Kim, K.; Park, N.-G. *Nat. Mater.* **2009,** *8*, 665.
129. Zhu, M.; Wang, Q.; Gu, Y.; Cao, X.; Zhong, C.; Ma, D.; Qin, J.; Yang, C. *J. Mater. Chem.* **2011,** *21*, 6409.
130. (a) Anthony, J. E. *Chem. Rev.* **2006,** *106*, 5028.
(b) Wang, C.; Dong, H.; Hu, W.; Liu, Y.; Zhu, D. *Chem. Rev.* **2012,** *112*, 2208.
(c) Holliday, S.; Donaghey, J. E.; McCulloch, I. *Chem. Mater.* **2014,** *26*, 647.
131. Durso, M.; Zambianchi, M.; Zanelli, A.; LoBello, M. G.; De Angelis, F.; Toffanin, S.; Cavallini, S.; Gentili, D.; Tinti, F.; Cavallini, M.; Camaioni, N.; Melucci, M. *Tetrahedron* **2014,** *70*, 6222.
132. Choi, J. H.; Cho, D. W.; Jin, S.-H.; Yoon, U. C. *Bull. Korean Chem. Soc.* **2007,** *28*, 1175.
133. Liu, C.-Y.; Zhao, H.; Yu, H.-h. *Org. Lett.* **2011,** *13*, 4068.
134. (a) Ando, S.; Nishida, J.-i; Tada, H.; Inoue, Y.; Tokito, S.; Yamashita, Y. *J. Am. Chem. Soc.* **2005,** *127*, 5336.
(b) Ie, Y.; Nitani, M.; Uemura, T.; Tominari, Y.; Takeya, J.; Honsho, Y.; Saeki, A.; Seki, S.; Aso, Y. *J. Phys. Chem. C* **2009,** *113*, 17189.
135. (a) Zou, Y.; Najari, A.; Berrouard, P.; Beaupré, S.; Réda Aïch, B.; Tao, Y.; Leclerc, M. *J. Am. Chem. Soc.* **2010,** *132*, 5330.
(b) Piliego, C.; Holcombe, T. W.; Douglas, J. D.; Woo, C. H.; Beaujuge, P. M.; Fréchet, J. M. J. *J. Am. Chem. Soc.* **2010,** *132*, 7595.
136. Wu, Q.; Wang, M.; Qiao, X.; Xiong, Y.; Huang, Y.; Gao, X.; Li, H. *Macromolecules* **2013,** *46*, 3887.
137. (a) Najari, A.; Beaupré, S.; Berrouard, P.; Zou, Y.; Pouliot, J.-R.; Lepage-Pérusse, C.; Leclerc, M. *Adv. Funct. Mater.* **2011,** *21*, 718.
(b) Wang, D. H.; Pron, A.; Leclerc, M.; Heeger, A. J. *Adv. Funct. Mater.* **2013,** *23*, 1297.
138. Chang, S.-Y.; Lin, P.-H.; Liu, C.-Y. *RSC Adv.* **2014,** *4*, 35868.
139. (a) Raymo, F. M.; Tomasulo, M. *Chem. Soc. Rev.* **2005,** *34*, 327.
(b) Gust, D.; Moore, T. A.; Moore, A. L. *Chem. Commun.* **2006,** 1169.
(c) Irie, M. *Bull. Chem. Soc. Jpn.* **2008,** *81*, 917.
140. (a) Matsuda, K.; Irie, M. *J. Photochem. Photobiol. C: Photochem. Rev.* **2004,** *5*, 169.
(b) Morimoto, M.; Irie, M. *Chem. Commun.* **2005,** 3895.
(c) Tian, H.; Wang, S. *Chem. Commun.* **2007,** 781.

141. Kamiya, H.; Yanagisawa, S.; Hiroto, S.; Itami, K.; Shinokubo, H. *Org. Lett.* **2011,** *13*, 6394.
142. Beydoun, K.; Boixel, J.; Guerchais, V.; Doucet, H. *Catal. Sci. Technol.* **2012,** *2*, 1242.
143. (a) Qu, S.; Tian, H. *Chem. Commun.* **2012,** *48*, 3039.
(b) Biniek, L.; Schroeder, B. C.; Nielsen, C. B.; McCulloch, I. *J. Mater. Chem.* **2012,** *22*, 14803.
144. (a) Beaujuge, P. M.; Fréchet, J. M. J. *J. Am. Chem. Soc.* **2011,** *133*, 2009.
(b) Hu, X.; Shi, M.; Zuo, L.; Nan, Y.; Liu, Y.; Fu, L.; Chen, H. *Polymer* **2011,** *52*, 2559.
(c) Lin, Y.; Cheng, P.; Li, Y.; Zhan, X. *Chem. Commun.* **2012,** *48*, 4773.
(d) Hu, X.; Zuo, L.; Fu, W.; Larsen-Olsen, T. T.; Helgesen, M.; Bundgaard, E.; Hagemann, O.; Shi, M.; Krebs, F. C.; Chen, H. *J. Mater. Chem.* **2012,** *22*, 15710.
145. (a) Chen, H.; Guo, Y.; Yu, G.; Zhao, Y.; Zhang, J.; Gao, D.; Liu, H.; Liu, Y. *Adv. Mater.* **2012,** *24*, 4618.
(b) Li, H.; Mei, J.; Ayzner, A. L.; Toney, M. F.; Tok, J. B. H.; Bao, Z. *Org. Electron.* **2012,** *13*, 2450.
146. Qu, S.; Qin, C.; Islam, A.; Wu, Y.; Zhu, W.; Hua, J.; Tian, H.; Han, L. *Chem. Commun.* **2012,** *48*, 6972.
147. (a) Yamagata, T.; Kuwabara, J.; Kanbara, T. *Tetrahedron Lett.* **2010,** *51*, 1596.
(b) Deng, L.; Wu, W.; Guo, H.; Zhao, J.; Ji, S.; Zhang, X.; Yuan, X.; Zhang, C. *J. Org. Chem.* **2011,** *76*, 9294.
148. Liu, S.-Y.; Shi, M.-M.; Huang, J.-C.; Jin, Z.-N.; Hu, X.-L.; Pan, J.-Y.; Li, H.-Y.; Jen, A. K. Y.; Chen, H.-Z. *J. Mater. Chem. A* **2013,** *1*, 2795.
149. Aebi, J.; Lengsfeld, H.; Dehmlow, H.; Morand, O.; Himber, J.; Schmid, G.; Maerki, H.-P.; Ji, Y.-H. Eur. Pat. Appl. EP 778271, 1997.
150. Matsuda, S.; Takahashi, M.; Monguchi, D.; Mori, A. *Synlett* **2009,** 1941.
151. Ionita, M.; Roger, J.; Doucet, H. *ChemSusChem* **2010,** *3*, 367.
152. Fu, H. Y.; Doucet, H. *Eur. J. Org. Chem.* **2011,** 7163.
153. Mom, S.; Beaupérin, M.; Roy, D.; Royer, S.; Amardeil, R.; Cattey, H.; Doucet, H.; Hierso, J. C. *Inorg. Chem.* **2011,** *50*, 11592.
154. Carrër, A.; Brinet, D.; Florent, J.-C.; Rousselle, P.; Bertounesque, E. *J. Org. Chem.* **2012,** *77*, 1316.
155. Loukotova, L.; Yuan, K.; Doucet, H. *ChemCatChem* **2014,** *6*, 1303.
156. Taubes, G. *Science* **2008,** *321*, 356.
157. Shikuma, J.; Mori, A.; Masui, K.; Matsuura, R.; Sekiguchi, A.; Ikegami, H.; Kawamoto, M.; Ikeda, T. *Chem. Asian J.* **2007,** *2*, 301.
158. Bensaid, S.; Laidaoui, N.; Abed, D. E.; Kacimi, S.; Doucet, H. *Tetrahedron Lett.* **2011,** *52*, 1383.
159. Demir, S.; Özdemir, İ.; Arslan, H.; VanDerveer, D. *J. Organomet. Chem.* **2011,** *696*, 2589.
160. Tani, S.; Uehara, T. N.; Yamaguchi, J.; Itami, K. *Chem. Sci.* **2014,** *5*, 123.
161. Lohrey, L.; Uehara, T. N.; Tani, S.; Yamaguchi, J.; Humpf, H.-U.; Itami, K. *Eur. J. Org. Chem.* **2014,** 3387.
162. Yeh, V. S. C. *Tetrahedron* **2004,** *60*, 11995.
163. Ichiba, T.; Yoshida, W. Y.; Scheuer, P. J.; Higa, T.; Gravalos, D. G. *J. Am. Chem. Soc.* **1991,** *113*, 3173.
164. Roesener, J. A.; Scheuer, P. J. *J. Am. Chem. Soc.* **1986,** *108*, 846.
165. Reeder, M. R.; Gleaves, H. E.; Hoover, S. A.; Imbordino, R. J.; Pangborn, J. J. *Org. Process Res. Dev.* **2003,** 7, 696.
166. Ohnmacht, S. A.; Mamone, P.; Culshaw, A. J.; Greaney, M. F. *Chem. Commun.* **2008,** 1241.
167. Rastogi, N.; Abaul, J.; Goh, K. S.; Devallois, A.; Philogène, E.; Bourgeois, P. *FEMS Immunol. Med. Microbiol.* **1998,** *20*, 267.

168. Flegeau, E. F.; Popkin, M. E.; Greaney, M. F. *Org. Lett.* **2008,** *10*, 2717.
169. Herrmann, W. A.; Brossmer, C.; Öfele, K.; Reisinger, C.-P.; Priermeier, T.; Beller, M.; Fischer, H. *Angew. Chem., Int. Ed. Engl.* **1995,** *34*, 1844.
170. Ackermann, L.; Barfüsser, S.; Kornhaass, C.; Kapdi, A. R. *Org. Lett.* **2011,** *13*, 3082.
171. Yu, P.; Zhang, G.; Chen, F.; Cheng, J. *Tetrahedron Lett.* **2012,** *53*, 4588.
172. Williams, T. J.; Fairlamb, I. J. S. *Tetrahedron Lett.* **2013,** *54*, 2906.
173. (a) Nicolaou, K. C.; Bulger, P. G.; Sarlah, D. *Angew. Chem., Int. Ed.* **2005,** *44*, 4442.
(b) Hughes, R. A.; Moody, C. J. *Angew. Chem., Int. Ed.* **2007,** *46*, 7930.
(c) Kim, O.; Jeong, Y.; Lee, H.; Hong, S.-S.; Hong, S. *J. Med. Chem.* **2011,** *54*, 2455.
(d) Wan, J.; Zheng, C.-J.; Fung, M.-K.; Liu, X.-K.; Lee, C.-S.; Zhang, X.-H. *J. Mater. Chem.* **2012,** *22*, 4502.
174. Gray, N. S.; Wodicka, L.; Thunnissen, A.-M. W. H.; Norman, T. C.; Kwon, S.; Espinoza, F. H.; Morgan, D. O.; Barnes, G.; LeClerc, S.; Meijer, L.; Kim, S.-H.; Lockhart, D. J.; Schultz, P. G. *Science* **1998,** *281*, 533.
175. Chang, Y.-T.; Wignall, S. M.; Rosania, G. R.; Gray, N. S.; Hanson, S. R.; Su, A. I.; Merlie, J.; Moon, H.-S.; Sangankar, S. B.; Perez, O.; Heald, R.; Schultz, P. G. *J. Med. Chem.* **2001,** *44*, 4497.
176. Chang, L. C. W.; Spanjersberg, R. F.; von Frijtag Drabbe Künzel, J. K.; Mulder-Krieger, T.; Brussee, J.; Ijzerman, A. P. *J. Med. Chem.* **2006,** *49*, 2861.
177. Čerňa, I.; Pohl, R.; Klepetářová, B.; Hocek, M. *Org. Lett.* **2006,** *8*, 5389.
178. Čerňa, I.; Pohl, R.; Hocek, M. *Chem. Commun.* **2007,** 4729.
179. Storr, T. E.; Firth, A. G.; Wilson, K.; Darley, K.; Baumann, C. G.; Fairlamb, I. J. S. *Tetrahedron* **2008,** *64*, 6125.
180. Baumann, C. G.; De Ornellas, S.; Reeds, J. P.; Storr, T. E.; Williams, T. J.; Fairlamb, I. J. S. *Tetrahedron* **2014,** *70*, 6174.
181. (a) Fernández, D.; Ahaidar, A.; Danelón, G.; Cironi, P.; Marfil, M.; Pérez, O.; Cuevas, C.; Albericio, F.; Joule, A. J.; Álvarez, M. *Monatsh. Chem.* **2004,** *135*, 615.
(b) Fresneda, P. M.; Delgado, S.; Francesch, A.; Manzanares, I.; Cuevas, C.; Molina, P. *J. Med. Chem.* **2006,** *49*, 1217.
182. Huestis, M. P.; Fagnou, K. *Org. Lett.* **2009,** *11*, 1357.
183. Kim, D.; Wang, L.; Hale, J. J.; Lynch, C. L.; Budhu, R. J.; MacCoss, M.; Mills, S. G.; Malkowitz, L.; Gould, S. L.; DeMartino, J. A.; Springer, M. S.; Hazuda, D.; Miller, M.; Kessler, J.; Hrin, R. C.; Carver, G.; Carella, A.; Henry, K.; Lineberger, J.; Schleif, W. A.; Emini, E. A. *Bioorg. Med. Chem. Lett.* **2005,** *15*, 2129.
184. Davey, D.; Erhardt, P. W.; Lumma, W. C.; Wiggins, J.; Sullivan, M.; Pang, D.; Cantor, E. *J. Med. Chem.* **1987,** *30*, 1337.
185. Browne, L. J.; Gude, C.; Rodriguez, H.; Steele, R. E.; Bhatnager, A. *J. Med. Chem.* **1991,** *34*, 725.
186. Yamaguchi, E.; Shibahara, F.; Murai, T. *Chem. Lett.* **2011,** *40*, 939.
187. (a) Leopoldo, M.; Lacivita, E.; Passafiume, E.; Contino, M.; Colabufo, N. A.; Berardi, F.; Perrone, R. *J. Med. Chem.* **2007,** *50*, 5043.
(b) Puerstinger, G.; Paeshuyse, J.; De Clercq, E.; Neyts, J. *Bioorg. Med. Chem. Lett.* **2007,** *17*, 390.
(c) Ismail, M. A.; Arafa, R. K.; Wenzler, T.; Brun, R.; Tanious, F. A.; Wilson, W. D.; Boykin, D. W. *Bioorg. Med. Chem.* **2008,** *16*, 683.
(d) Chernyak, N.; Gevorgyan, V. *Angew. Chem., Int. Ed.* **2010,** *49*, 2743.
188. (a) Rupert, K. C.; Henry, J. R.; Dodd, J. H.; Wadsworth, S. A.; Cavender, D. E.; Olini, G. C.; Fahmy, B.; Siekierka, J. J. *Bioorg. Med. Chem. Lett.* **2003,** *13*, 347.
(b) Hamdouchi, C.; Zhong, B.; Mendoza, J.; Collins, E.; Jaramillo, C.; De Diego, J. E.; Robertson, D.; Spencer, C. D.; Anderson, B. D.; Watkins, S. A.; Zhang, F.; Brooks, H. B. *Bioorg. Med. Chem. Lett.* **2005,** *15*, 1943.
189. Fu, H. Y.; Chen, L.; Doucet, H. *J. Org. Chem.* **2012,** 77, 4473.

190. Baloch, M.; Roy, R. J.; Roy, D.; Beydoun, K.; Doucet, H. *RSC Adv.* **2011,** *1*, 1527.
191. Cao, H.; Lin, Y.; Zhan, H.; Du, Z.; Lin, X.; Liang, Q.-M.; Zhang, H. *RSC Adv.* **2012,** *2*, 5972.
192. Lee, J.; Chung, J.; Byun, S. M.; Kim, B. M.; Lee, C. *Tetrahedron* **2013,** *69*, 5660.
193. Stasyuk, A. J.; Banasiewicz, M.; Ventura, B.; Cyranski, M. K.; Gryko, D. T. *New J. Chem.* **2014,** *38*, 189.
194. Wang, H.; Chen, C.; Huang, Z.; Yao, L.; Li, B.; Peng, J. *Synthesis* **2015,** *47*, 2457.
195. (a) Micheli, F.; Bertani, B.; Bozzoli, A.; Crippa, L.; Cavanni, P.; Di Fabio, R.; Donati, D.; Marzorati, P.; Merlo, G.; Paio, A.; Perugini, L.; Zarantonello, P. *Bioorg. Med. Chem. Lett.* **2008,** *18*, 1804.
(b) Guillon, J.; Le Borgne, M.; Rimbault, C.; Moreau, S.; Savrimoutou, S.; Pinaud, N.; Baratin, S.; Marchivie, M.; Roche, S.; Bollacke, A.; Pecci, A.; Alvarez, L.; Desplat, V.; Jose, J. *Eur. J. Med. Chem.* **2013,** *65*, 205.
196. Park, S.; Jung, Y.; Kim, I. *Tetrahedron* **2014,** *70*, 7534.
197. (a) Wishka, D. G.; Walker, D. P.; Yates, K. M.; Reitz, S. C.; Jia, S.; Myers, J. K.; Olson, K. L.; Jacobsen, E. J.; Wolfe, M. L.; Groppi, V. E.; Hanchar, A. J.; Thornburgh, B. A.; Cortes-Burgos, L. A.; Wong, E. H. F.; Staton, B. A.; Raub, T. J.; Higdon, N. R.; Wall, T. M.; Hurst, R. S.; Walters, R. R.; Hoffmann, W. E.; Hajos, M.; Franklin, S.; Carey, G.; Gold, L. H.; Cook, K. K.; Sands, S. B.; Zhao, S. X.; Soglia, J. R.; Kalgutkar, A. S.; Arneric, S. P.; Rogers, B. N. *J. Med. Chem.* **2006,** *49*, 4425.
(b) Debenham, J. S.; Madsen-Duggan, C. B.; Toupence, R. B.; Walsh, T. F.; Wang, J.; Tong, X.; Kumar, S.; Lao, J.; Fong, T. M.; Xiao, J. C.; Huang, C. R. R. C.; Shen, C.-P.; Feng, Y.; Marsh, D. J.; Stribling, D. S.; Shearman, L. P.; Strack, A. M.; Goulet, M. T. *Bioorg. Med. Chem. Lett.* **2010,** *20*, 1448.
(c) Ren, L.; Wenglowsky, S.; Miknis, G.; Rast, B.; Buckmelter, A. J.; Ely, R. J.; Schlachter, S.; Laird, E. R.; Randolph, N.; Callejo, M.; Martinson, M.; Galbraith, S.; Brandhuber, B. J.; Vigers, G.; Morales, T.; Voegtli, W. C.; Lyssikatos, J. *Bioorg. Med. Chem. Lett.* **2011,** *21*, 1243.
198. Carrër, A.; Rousselle, P.; Florent, J.-C.; Bertounesque, E. *Adv. Synth. Catal.* **2012,** *354*, 2751.
199. (a) Keskin, O.; Bahar, I.; Jernigan, R. L.; Beutler, J. A.; Shoemaker, R. H.; Sausville, E. A.; Covell, D. G. *Anti-Cancer Drug Des.* **2000,** *15*, 79.
(b) Baviskar, A. T.; Madaan, C.; Preet, R.; Mohapatra, P.; Jain, V.; Agarwal, A.; Guchhait, S. K.; Kundu, C. N.; Banerjee, U. C.; Bharatam, P. V. *J. Med. Chem.* **2011,** *54*, 5013.
200. (a) Guo, M.-J.; Li, D.; Sun, Y.-H.; Cheng, J.; Zhang, Z.-G. *Chin. J. Chem.* **2004,** *22*, 1413.
(b) Li, M.; Yang, H.; Zhao, G.; Wen, L.; Cao, W.; Zhang, S. *J. Heterocycl. Chem.* **2005,** *42*, 209.
(c) Abdelhamid, A. O.; Abdelall, E. K. A.; Zaki, Y. H. *J. Heterocycl. Chem.* **2010,** *47*, 477.
201. Grosse, S.; Pillard, C.; Massip, S.; Léger, J. M.; Jarry, C.; Bourg, S.; Bernard, P.; Guillaumet, G. *Chem. Eur. J.* **2012,** *18*, 14943.
202. Grosse, S.; Pillard, C.; Bernard, P.; Guillaumet, G. *Synlett* **2013,** *24*, 2095.
203. (a) Zimmermann, P. J.; Brehm, C.; Buhr, W.; Palmer, A. M.; Volz, J.; Simon, W.-A. *Bioorg. Med. Chem.* **2008,** *16*, 536.
(b) Mitchell, S. A.; Danca, M. D.; Blomgren, P. A.; Darrow, J. W.; Currie, K. S.; Kropf, J. E.; Lee, S. H.; Gallion, S. L.; Xiong, J.-M.; Pippin, D. A.; DeSimone, R. W.; Brittelli, D. R.; Eustice, D. C.; Bourret, A.; Hill-Drzewi, M.; Maciejewski, P. M.; Elkin, L. L. *Bioorg. Med. Chem. Lett.* **2009,** *19*, 6991.
204. Guchhait, S. K.; Kandekar, S.; Kashyap, M.; Taxak, N.; Bharatam, P. V. *J. Org. Chem.* **2012,** *77*, 8321.

205. Karthikeyan, M. S. *Eur. J. Med. Chem.* **2009,** *44*, 827.
206. Barbuceanu, S.-F.; Almajan, G. L.; Saramet, I.; Draghici, C.; Tarcomnicu, A. I.; Bancescu, G. *Eur. J. Med. Chem.* **2009,** *44*, 4752.
207. El Bakali, J.; Klupsch, F.; Guédin, A.; Brassart, B.; Fontaine, G.; Farce, A.; Roussel, P.; Houssin, R.; Bernier, J.-L.; Chavatte, P.; Mergny, J.-L.; Riou, J.-F.; Hénichart, J.-P. *Bioorg. Med. Chem. Lett.* **2009,** *19*, 3434.
208. Tozkoparan, B.; Aytaç, S. P.; Gürsoy, Ş.; Aktay, G. *Med. Chem. Res.* **2012,** *21*, 192.
209. Le Meur, M.; Bourg, S.; Massip, S.; Marchivie, M.; Jarry, C.; Guillaumet, G.; Routier, S. *Eur. J. Org. Chem.* **2014,** 3704.
210. (a) Scribner, A.; Meitz, S.; Fisher, M.; Wyvratt, M.; Leavitt, P.; Liberator, P.; Gurnett, A.; Brown, C.; Mathew, J.; Thompson, D.; Schmatz, D.; Biftu, T. *Bioorg. Med. Chem. Lett.* **2008,** *18*, 5263.
(b) Andreani, A.; Granaiola, M.; Locatelli, A.; Morigi, R.; Rambaldi, M.; Varoli, L.; Calonghi, N.; Cappadone, C.; Farruggia, G.; Stefanelli, C.; Masotti, L.; Nguyen, T. L.; Hamel, E.; Shoemaker, R. H. *J. Med. Chem.* **2012,** *55*, 2078.
211. Raeymaekers, A. H. M.; Allewijn, F. T. N.; Vandenberk, J.; Demoen, P. J. A.; Van Offenwert, T. T. T.; Janssen, P. A. J. *J. Med. Chem.* **1966,** *9*, 545.
212. Walton, M. I.; Wilson, S. C.; Hardcastle, I. R.; Mirza, A. R.; Workman, P. *Mol. Cancer Ther.* **2005,** *4*, 1369.
213. Zhu, Y.-S.; Shi, B.; Fang, R.; Wang, X.; Jing, H. *Org. Biomol. Chem.* **2014,** *12*, 5773.
214. (a) Lin, Q.; Meloni, D.; Pan, Y.; Xia, M.; Rodgers, J.; Shepard, S.; Li, M.; Galya, L.; Metcalf, B.; Yue, T.-Y.; Liu, P.; Zhou, J. *Org. Lett.* **2009,** *11*, 1999.
(b) Wang, T.; Ledeboer, M. W.; Duffy, J. P.; Pierce, A. C.; Zuccola, H. J.; Block, E.; Shlyakter, D.; Hogan, J. K.; Bennani, Y. L. *Bioorg. Med. Chem. Lett.* **2010,** *20*, 153.
(c) Chen, Z.; Venkatesan, A. M.; Dehnhardt, C. M.; Ayral-Kaloustian, S.; Brooijmans, N.; Mallon, R.; Feldberg, L.; Hollander, I.; Lucas, J.; Yu, K.; Kong, F.; Mansour, T. S. *J. Med. Chem.* **2010,** *53*, 3169.
215. Kumar, V. P.; Frey, K. M.; Wang, Y.; Jain, H. K.; Gangjee, A.; Anderson, K. S. *Bioorg. Med. Chem. Lett.* **2013,** *23*, 5426.
216. Gibson, C. L.; Huggan, J. K.; Kennedy, A.; Kiefer, L.; Lee, J. H.; Suckling, C. J.; Clements, C.; Harvey, A. L.; Hunter, W. N.; Tulloch, L. B. *Org. Biomol. Chem.* **2009,** *7*, 1829.
217. Dodonova, J.; Tumkevicius, S. *RSC Adv.* **2014,** *4*, 35966.
218. Campeau, L.-C.; Stuart, D. R.; Leclerc, J.-P.; Bertrand-Laperle, M.; Villemure, E.; Sun, H.-Y.; Lasserre, S.; Guimond, N.; Lecavallier, M.; Fagnou, K. *J. Am. Chem. Soc.* **2009,** *131*, 3291.
219. Campeau, L.-C.; Schipper, D. J.; Fagnou, K. *J. Am. Chem. Soc.* **2008,** *130*, 3266.
220. Campeau, L.-C.; Bertrand-Laperle, M.; Leclerc, J.-P.; Villemure, E.; Gorelsky, S.; Fagnou, K. *J. Am. Chem. Soc.* **2008,** *130*, 3276.
221. (a) Mongin, F.; Trécourt, F.; Gervais, B.; Mongin, O.; Quéguiner, G. *J. Org. Chem.* **2002,** *67*, 3272.
(b) Bringmann, G.; Reichert, Y.; Kane, V. V. *Tetrahedron* **2004,** *60*, 3539.
(c) Yuan, C.; Chang, C.-T.; Axelrod, A.; Siegel, D. *J. Am. Chem. Soc.* **2010,** *132*, 5924.
222. Duric, S.; Tzschucke, C. C. *Org. Lett.* **2011,** *13*, 2310.
223. Liu, W.; Li, Y.; Wang, Y.; Kuang, C. *Eur. J. Org. Chem.* **2013,** 5272.
224. (a) Shen, Y.; Chen, C.-F. *Chem. Rev.* **2012,** *112*, 1463.
(b) Gingras, M. *Chem. Soc. Rev.* **2013,** *42*, 968.
(c) Gingras, M.; Felix, G.; Peresutti, R. *Chem. Soc. Rev.* **2013,** *42*, 1007.
(d) Gingras, M. *Chem. Soc. Rev.* **2013,** *42*, 1051.
225. Kaneko, E.; Matsumoto, Y.; Kamikawa, K. *Chem. Eur. J.* **2013,** *19*, 11837.
226. Touré, B. B.; Lane, B. S.; Sames, D. *Org. Lett.* **2006,** *8*, 1979.

227. Ghosh, D.; Lee, H. M. *Org. Lett.* **2012,** *14*, 5534.
228. Mariconda, A.; Grisi, F.; Costabile, C.; Falcone, S.; Bertolasi, V.; Longo, P. *New J. Chem.* **2014,** *38*, 762.
229. Yanagisawa, S.; Itami, K. *Tetrahedron* **2011,** *67*, 4425.
230. Khake, S. M.; Soni, V.; Gonnade, R. G.; Punji, B. *Dalton Trans.* **2014,** *43*, 16084.
231. Li, Y.; Wang, J.; Yan, B.; Huang, M.; Zhu, Y.; Wu, Y.; Wu, Y. *Tetrahedron* **2015,** *71*, 2729.
232. Chiong, H. A.; Daugulis, O. *Org. Lett.* **2007,** *9*, 1449.
233. Turner, G. L.; Morris, J. A.; Greaney, M. F. *Angew. Chem., Int. Ed.* **2007,** *46*, 7996.
234. Yan, X.-M.; Mao, X.-R.; Huang, Z.-Z. *Heterocycles* **2011,** *83*, 1371.
235. René, O.; Fagnou, K. *Adv. Synth. Catal.* **2010,** *352*, 2116.
236. Roy, D.; Mom, S.; Beaupérin, M.; Doucet, H.; Hierso, J.-C. *Angew. Chem., Int. Ed.* **2010,** *49*, 6650.
237. Roy, D.; Mom, S.; Lucas, D.; Cattey, H.; Hierso, J.-C.; Doucet, H. *Chem. Eur. J.* **2011,** *17*, 6453.
238. Roy, D.; Mom, S.; Royer, S.; Lucas, D.; Hierso, J.-C.; Doucet, H. *ACS Catal.* **2012,** *2*, 1033.
239. Colpaert, F.; Mangelinckx, S.; De Kimpe, N. *J. Org. Chem.* **2011,** *76*, 234.
240. Ye, M.; Gao, G.-L.; Edmunds, A. J. F.; Worthington, P. A.; Morris, J. A.; Yu, J.-Q. *J. Am. Chem. Soc.* **2011,** *133*, 19090.
241. (a) Amat, M.; Cantó, M.; Llor, N.; Escolano, C.; Molins, E.; Espinosa, E.; Bosch, J. *J. Org. Chem.* **2002,** *67*, 5343.
(b) Macchia, M.; Cervetto, L.; Demontis, G. C.; Longoni, B.; Minutolo, F.; Orlandini, E.; Ortore, G.; Papi, C.; Sbrana, A.; Macchia, B. *J. Med. Chem.* **2003,** *46*, 161.
(c) Chang, M.-Y.; Hsu, R.-T.; Chen, H.-P.; Lin, P.-J. *Heterocycles* **2006,** *68*, 1173.
242. Yu, D.; Lu, L.; Shen, Q. *Org. Lett.* **2013,** *15*, 940.
243. Zhang, J.; Chen, W.; Rojas, A. J.; Jucov, E. V.; Timofeeva, T. V.; Parker, T. C.; Barlow, S.; Marder, S. R. *J. Am. Chem. Soc.* **2013,** *135*, 16376.
244. (a) Pawlicki, M.; Collins, H. A.; Denning, R. G.; Anderson, H. L. *Angew. Chem., Int. Ed.* **2009,** *48*, 3244.
(b) Dalton, L. R.; Sullivan, P. A.; Bale, D. H. *Chem. Rev.* **2010,** *110*, 25.
(c) Hagfeldt, A.; Boschloo, G.; Sun, L.; Kloo, L.; Pettersson, H. *Chem. Rev.* **2010,** *110*, 6595.
(d) Henson, Z. B.; Mullen, K.; Bazan, G. C. *Nat. Chem.* **2012,** *4*, 699.
245. Rampon, D. S.; Wessjohann, L. A.; Schneider, P. H. *J. Org. Chem.* **2014,** *79*, 5987.
246. (a) Bellina, F.; Cauteruccio, S.; Rossi, R. *Eur. J. Org. Chem.* **2006,** 1379.
(b) Bellina, F.; Calandri, C.; Cauteruccio, S.; Rossi, R. *Tetrahedron* **2007,** *63*, 1970.
247. Chuprakov, S.; Chernyak, N.; Dudnik, A. S.; Gevorgyan, V. *Org. Lett.* **2007,** *9*, 2333.
248. Tamba, S.; Okubo, Y.; Tanaka, S.; Monguchi, D.; Mori, A. *J. Org. Chem.* **2010,** *75*, 6998.
249. (a) Sohda, K.-y; Nagai, K.; Yamori, T.; Suzuki, K.-i; Tanaka, A. *J. Antibiot.* **2005,** *58*, 27.
(b) Cummings, C. G.; Ross, N. T.; Katt, W. P.; Hamilton, A. D. *Org. Lett.* **2009,** *11*, 25.
(c) Lefranc, D.; Ciufolini, M. A. *Angew. Chem., Int. Ed.* **2009,** *48*, 4198.
(d) Demirbas, A.; Sahin, D.; Demirbas, N.; Karaoglu, S. A. *Eur. J. Med. Chem.* **2009,** *44*, 2896.
(e) Sarojini, B. K.; Krishna, B. G.; Darshanraj, C. G.; Bharath, B. R.; Manjunatha, H. *Eur. J. Med. Chem.* **2010,** *45*, 3490.
250. Shibahara, F.; Yamaguchi, E.; Murai, T. *Chem. Commun.* **2010,** *46* (11), 2471–2473.
251. Shibahara, F.; Yamaguchi, E.; Murai, T. *J. Org. Chem.* **2011,** *76*, 2680.
252. Derridj, F.; Roger, J.; Geneste, F.; Djebbar, S.; Doucet, H. *J. Organomet. Chem.* **2009,** *694*, 455.
253. Laidaoui, N.; Miloudi, A.; Abed, D. E.; Doucet, H. *Synthesis* **2010,** 2553.

254. Chen, L.; Bruneau, C.; Dixneuf, P. H.; Doucet, H. *ChemCatChem* **2013,** *5*, 1956.
255. Bheeter, C. B.; Jin, R.; Bera, J. K.; Doucet, H. *RSC Adv.* **2013,** *3*, 5987.
256. Laidaoui, N.; Roger, J.; Miloudi, A.; Abed, D. E.; Doucet, H. P. *Eur. J. Org. Chem.* **2011,** 4373.
257. (a) Sakakura, T.; Kohno, K. *Chem. Commun.* **2009,** 1312.
(b) Schäffner, B.; Schäffner, F.; Verevkin, S. P.; Börner, A. *Chem. Rev.* **2010,** *110*, 4554.
258. Dong, J. J.; Roger, J.; Verrier, C.; Martin, T.; Le Goff, R.; Hoarau, C.; Doucet, H. *Green Chem.* **2010,** *12*, 2053.
259. Fischmeister, C.; Doucet, H. *Green Chem.* **2011,** *13*, 741.
260. Beydoun, K.; Doucet, H. *ChemSusChem* **2011,** *4*, 526.
261. Bouazizi, Y.; Beydoun, K.; Romdhane, A.; Ben Jannet, H.; Doucet, H. *Tetrahedron Lett.* **2012,** *53*, 6801.
262. Bensaid, S.; Doucet, H. *ChemSusChem* **2012,** *5*, 1559.
263. Bensaid, S.; Doucet, H. *C. R. Chimie* **2014,** *17*, 1184.
264. Chen, L.; Bruneau, C.; Dixneuf, P. H.; Doucet, H. *Green Chem.* **2012,** *14*, 1111.
265. Fu, H. Y.; Zhao, L.; Bruneau, C.; Doucet, H. *Synlett* **2012,** *23*, 2077.
266. Chen, L.; Bruneau, C.; Dixneuf, P. H.; Doucet, H. *Tetrahedron* **2013,** *69*, 4381.
267. (a) Bensaid, S.; Doucet, H. *Tetrahedron* **2012,** *68*, 7655.
(b) Baloch, M.; Roy, D.; Bensaid, S.; Guerchais, V.; Doucet, H. *Eur. J. Inorg. Chem.* **2012,** 4454.
268. Beladhria, A.; Beydoun, K.; Ammar, H. B.; Salem, R. B.; Doucet, H. *Synthesis* **2012,** *44*, 2264.
269. Laroche, J.; Beydoun, K.; Guerchais, V.; Doucet, H. *Catal. Sci. Technol.* **2013,** *3*, 2072.
270. Yan, T.; Chen, L.; Bruneau, C.; Dixneuf, P. H.; Doucet, H. *J. Org. Chem.* **2013,** *78*, 4177.
271. Smari, I.; Zhao, L.; Yuan, K.; Ammar, H. B.; Doucet, H. *Catal. Sci. Technol.* **2014,** *4*, 3723.
272. Dunwell, D. W.; Evans, D.; Hicks, T. A. *J. Med. Chem.* **1975,** *18*, 1158.
273. Racane, L.; Tralic-Kulenovic, V.; Fiser-Jakic, L.; Boykin, D. W.; Karminski-Zamola, G. *Heterocycles* **2001,** *55*, 2085.
274. Abdellaoui, F.; Youssef, C.; Ben Ammar, H.; Soulé, J.-F.; Doucet, H. *Synthesis* **2014,** *46*, 3341.
275. Bellina, F.; Lessi, M.; Manzini, C. *Eur. J. Org. Chem.* **2013,** 5621.
276. (a) Burke, B.; Parkins, H.; Talbot, A. M. A. *Heterocycles* **1979,** *12*, 349.
(b) Domínguez, X. A.; Fuente, Gdl; González, A. G.; Reina, M.; Timón, I. *Heterocycles* **1988,** *27*, 35.
277. Ackermann, L.; Fenner, S. *Chem. Commun.* **2011,** *47*, 430.
278. Fan, S.; Yang, J.; Zhang, X. *Org. Lett.* **2011,** *13*, 4374.
279. Chang, J. W. W.; Chia, E. Y.; Chai, C. L. L.; Seayad, J. *Org. Biomol. Chem.* **2012,** *10*, 2289.
280. Lee, D. S.; Choy, P. Y.; So, C. M.; Wang, J.; Lau, C. P.; Kwong, F. Y. *RSC Adv.* **2012,** *2*, 9179.
281. Ackermann, L.; Althammer, A.; Fenner, S. *Angew. Chem., Int. Ed.* **2009,** *48*, 201.
282. Ackermann, L.; Barfüsser, S.; Pospech, J. *Org. Lett.* **2010,** *12*, 724.
283. So, C. M.; Lau, C. P.; Kwong, F. Y. *Chem. Eur. J.* **2011,** *17*, 761.
284. Dai, F.; Gui, Q.; Liu, J.; Yang, Z.; Chen, X.; Guo, R.; Tan, Z. *Chem. Commun.* **2013,** *49*, 4634.
285. Ferguson, D. M.; Rudolph, S. R.; Kalyani, D. *ACS Catal.* **2014,** *4*, 2395.
286. Choy, P. Y.; Luk, K. C.; Wu, Y.; So, C. M.; Wang, L.-l; Kwong, F. Y. *J. Org. Chem.* **2015,** *80*, 1457.

CHAPTER 4

Palladium-Catalyzed Directed Arylation of Unactivated $C(sp^3)-H$ Bonds

Bin Liu, Ye-Qiang Han, Fang Hu and Bing-Feng Shi
Zhejiang University, Hangzhou, China

Contents

4.1 INTRODUCTION

Over the past few decades, palladium-catalyzed C—H arylation has emerged as an efficient, eco-friendly, and straightforward alternative to traditional cross-coupling reactions. This strategy obviates the requirement of prefunctionalization of the starting materials and minimizes the generation of extra metallic wastes, just providing an atom and step economical pathway to access various arylated organic molecules.[1] Particularly, the transition metal-catalyzed arylation of $C(sp^2)-H$ bonds has been well investigated. These transformations have also been applied to the total synthesis and late-stage arylation of natural products and biologically important organic molecules.[2,3] In contrast, the arylation of unactivated $C(sp^3)-H$ is more challenging, largely due to the significantly low

Strategies for Palladium-Catalyzed Non-Directed and Directed C—H Bond Functionalization
DOI: http://dx.doi.org/10.1016/B978-0-12-805254-9.00004-9

reactivity of aliphatic C(sp^3)—H bonds and the difficulty of distinguishing one bond among those chemically similar entities.[4] To address the reactivity and selectivity issues, chelation-assisted strategy has been recognized as a universal and powerful method for the chemoselective functionalization of proximal C—H bonds, mostly through a kinetically favored five-membered palladacycle. Pd-catalyzed, site-selective arylation of unactivated C(sp^3)—H bonds have been investigated comprehensively with the assistance of various directing groups (DGs). According to the modes of coordination, the DGs could be divided into two categories, namely monodentate DGs (Fig. 4.1) and bidentate DGs (Fig. 4.2). The arylation reagents that have been used for C(sp^3)—H arylation include aryl halides, diarylhyperiodonium salts, arylmetallic reagents, and simple arenes.

In this chapter, we aim to provide a comprehensive overview on Pd-catalyzed directed arylation of unactivated C(sp^3)—H bonds. For clarity,

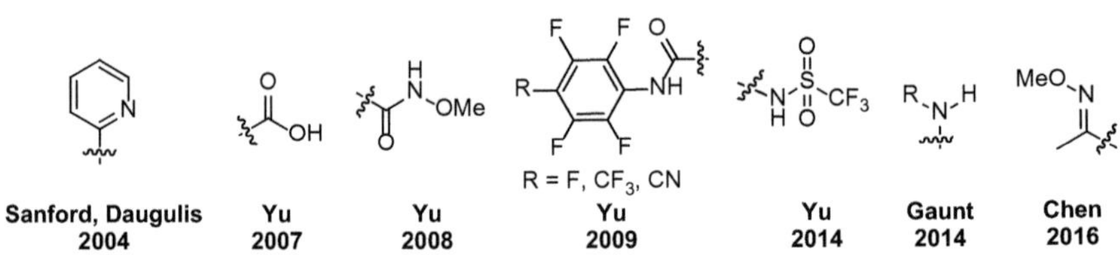

Figure 4.1 Monodentate DGs used in C(sp^3)—H arylation.

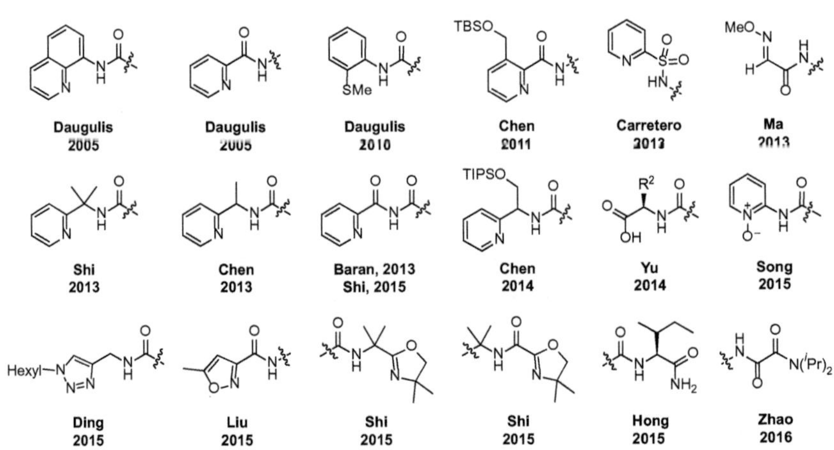

Figure 4.2 Bidentate DGs used in C(sp^3)—H arylation.

this chapter is classified by the nature of DGs, including monodentate DGs, bidentate DGs, and the state-of-the-art strategy using transient DGs. In consideration of the challenge and importance of stereochemistry, Pd-catalyzed asymmetric arylation of C(sp^3)—H bonds is presented in Section 4.5.

4.2 PALLADIUM-CATALYZED C(sp^3)—H ARYLATION DIRECTED BY MONODENTATE DGs

In 2004, Sanford and co-workers reported the Pd-catalyzed pyridyl-directed oxygenation of unactivated C—H bonds. They proposed that the C—H oxygenation reaction proceed via C—O reductive elimination from a putative Pd(IV) intermediate via a Pd(II)/Pd(IV) catalytic cycle.[5] They hypothesized that C—H arylation could be achieved by replacing $PhI(OAc)_2$ with iodine(III) arylating agents via an analogous mechanistic pathway. Based on this hypothesis, they achieved the Pd-catalyzed arylation of a broad range of aromatic C—H bonds with iodine(III) reagent $[Ph_2I]BF_4$. In addition, they also found that Pd-catalyzed arylation of benzylic C(sp^3)—H bonds of 8-methylquinolines could also be achieved. Further mechanistic studies indicated that a bimetallic high oxidation state palladium species, such as a mixed-valent Pd(II)—Pd(IV) species or a Pd(III)—Pd(III) dimer, was involved as a key intermediate (Scheme 4.1).[6]

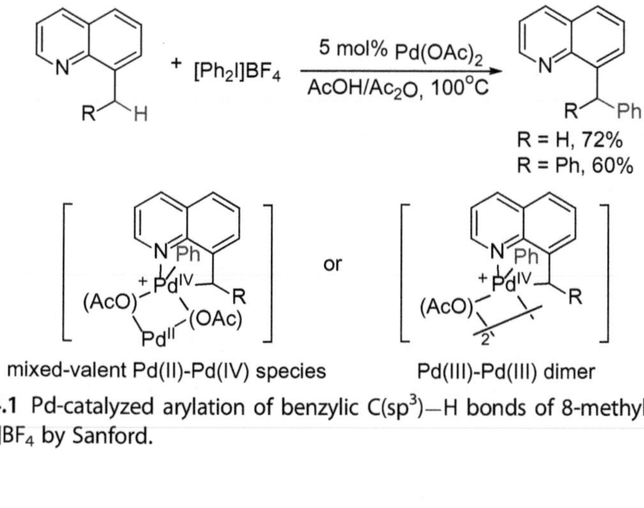

Scheme 4.1 Pd-catalyzed arylation of benzylic C(sp^3)—H bonds of 8-methylquinolines with $[Ph_2I]BF_4$ by Sanford.

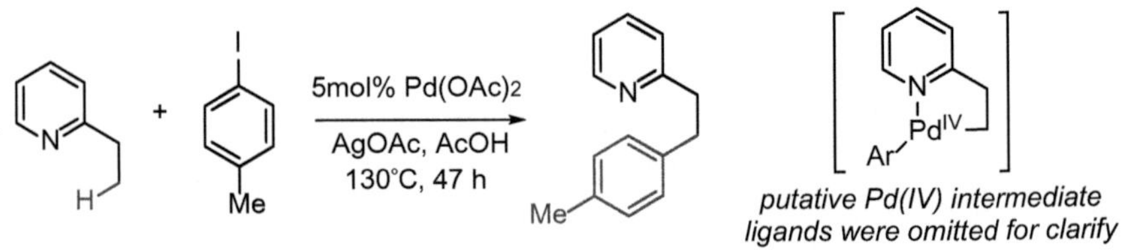

Scheme 4.2 Pd-catalyzed pyridyl-directed arylation of unactivated C(sp^3) – H bonds with ArI reported by Gaugulis.

Compared to diarylhyperiodonium salts, aryl halides are cheaper and more accessible. In 2005, the Daugulis group reported the Pd-catalyzed arylation of substituted anilides with aryl iodides.[7] Inspired by this result, they further extended this protocol to pyridyl-directed C–H arylation and found that C(sp^3)–H arylation of 2-ethylpyridine with aryl iodides could also be achieved.[8] The reaction was proposed to proceed via the reductive elimination from a possible Pd(IV) intermediate involving a Pd(II)/Pd(IV) catalytic cycle (Scheme 4.2).

The above-mentioned arylation reactions occurred via a Pd(II)/Pd(IV) catalytic cycle through oxidative addition to the resulting palladacycle, followed by reductive elimination from the putative aryl–Pd(IV) intermediates. During the mechanistic studies of the role of tri-*o*-tolylphosphane-bound palladium complexes in traditional cross-coupling reactions, Hartwig reported a seminal example of stoichiometric arylation of the Herrmann palladacycle with phenylstannane through transmetalation.[9] However, the first C–H activation/C–C coupling reaction with organometallic reagents using catalytic palladium catalyst was not realized until 2006 by the Yu group.[10] These reactions were proposed to occur via a Pd(II)/Pd(0) catalytic cycle through transmetalation between the resulting palladacyle with organometallic reagents, followed by reductive elimination. In 2007, Yu and co-workers successfully extended this protocol to the arylation of C(sp^3)–H bonds in simple carboxylic acids with phenylboronic esters (Scheme 4.3).[11] The arylation reaction proceeded with excellent mono-selectivity. It was found that the formation of potassium salts of carboxylic acids was crucial for the success of this protocol. They also found that the arylation of aliphatic carboxylic acids could also be achieved using aryl iodides as the coupling partners, albeit with a mixture of mono- and diarylation products.

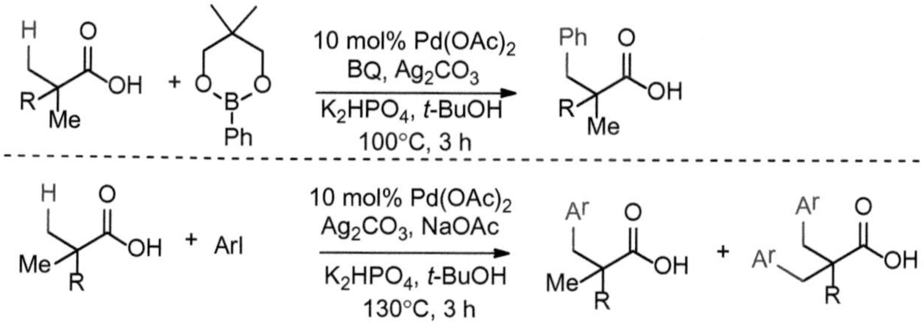

Scheme 4.3 Pd(II)-catalyzed carboxylate-directed arylation of β-C(sp^3)—H bond by Yu.

However, the above-mentioned arylation protocol was limited, with narrow substrate scope and the use of extra silver salts as oxidant. To address these limitations and make this protocol more practical, Yu and co-workers further derivatized carboxylic acids to *O*-methyl hydroxamic acids with stronger binding ability. They found that air could serve as an efficient oxidant to replace the silver salts (Scheme 4.4).[12] Moreover, CONHOMe group can be easily converted to esters and amides or reduced to alkane fragments.

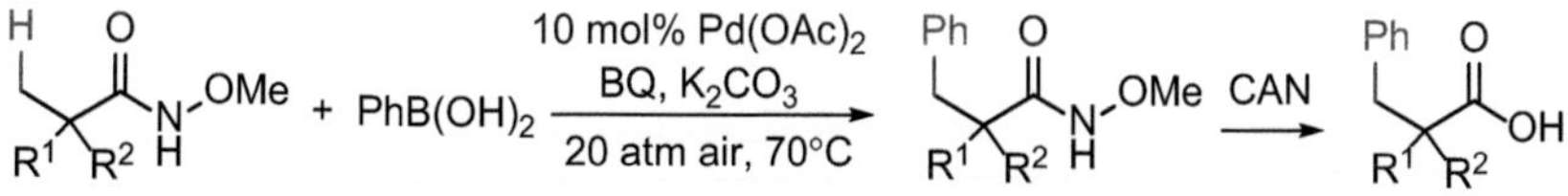

Scheme 4.4 Pd(II)-catalyzed *N*-methoxylamide-directed arylation of β-C(sp^3)—H bonds with phenylboronic acid.

In attempting to achieve the intermolecular arylation of β-C(sp^3)—H bonds of *O*-methyl hydroxamic acids using an Pd(0)/PPh_3/ArI catalytic system, Yu and co-workers found that the N—H bond of the CONHOMe moiety underwent amination with aryl iodide. They rationalized that the undesired amination pathway could be suppressed by reducing the nucleophilicity of the amide N—H bond. To this end, they introduced the CONH-C_6F_5 as an efficient DG to promote the β-C(sp^3)—H arylation with Cy-JohnPhos as ligand and CsF as base (Scheme 4.5).[13] This method provided a straightforward method to obtain various β-arylated aliphatic amides in high yields. The acidity of the N—H bond in the DG was found

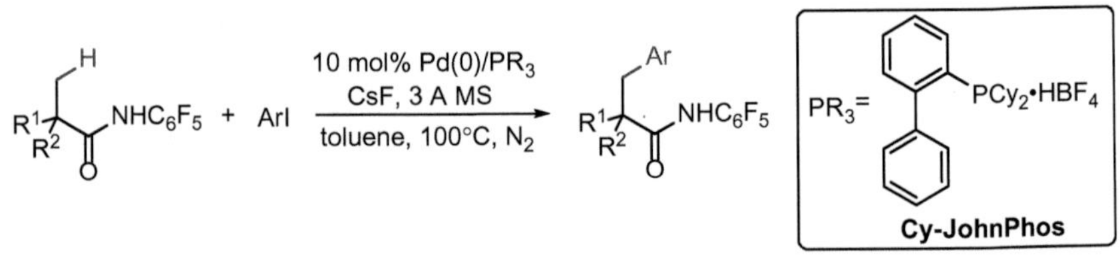

Scheme 4.5 Pd(0)-catalyzed CONH-C_6F_5-directed arylation of β-C(sp^3)—H bonds with ArI.

to be crucial for reactivity. It was proposed that the reaction occurred via a Pd(0)/Pd(II) catalytic cycle.

Although the weakly coordinating *N*-arylamide ($CONHAr_F$) auxiliary developed by the Yu group served as an efficient DG in Pd-catalyzed functionalization of primary C(sp^3)—H bonds of aliphatic acids, the activation of more inert secondary C—H bonds using weakly coordinating DGs remained undeveloped. Based on their previous achievement of a "mutually repulsive" pyridine ligand that promotes the activation of electron-deficient arenes,[14] the Yu group reported the first example of *N*-arylamide-directed Pd(II)-catalyzed arylation of β-methylene C(sp^3)—H bonds enabled by a mutually repulsive and electron-rich quinoline ligand (Scheme 4.6).[15]

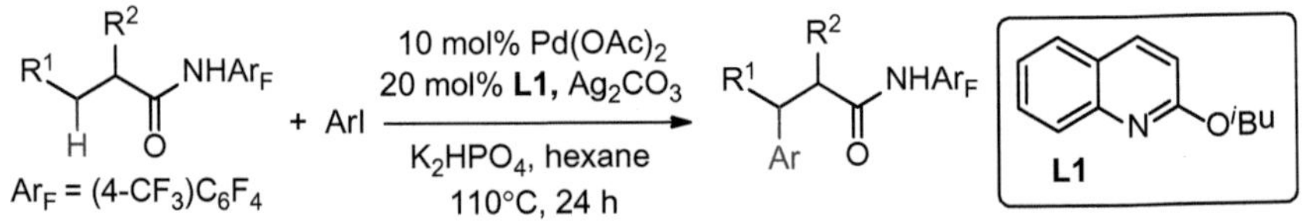

Scheme 4.6 *N*-arylamide-directed Pd(II)-catalyzed arylation of methylene C(sp^3)—H bonds enabled by ligand.

Later, the Yu group reported the ligand-controlled C(sp^3)—H arylation of α-amino acid derivatives, which relied on the choice of appropriate ligands to control the selectivity of primary or secondary C(sp^3)—H using the weakly coordinating *N*-arylamide DG (Scheme 4.7A).[16] This method achieved the monoarylation of primary C(sp^3)—H bonds with pyridine-based ligands and the arylation of more steric hindrance secondary C(sp^3)—H bonds with quinoline-based ligands. The sequential arylation of alanine derivative with two different aryl iodides in one-pot enabled the straightforward synthesis of various chiral *β*-Ar-*β*-Ar′-amino acids (Scheme 4.7B).

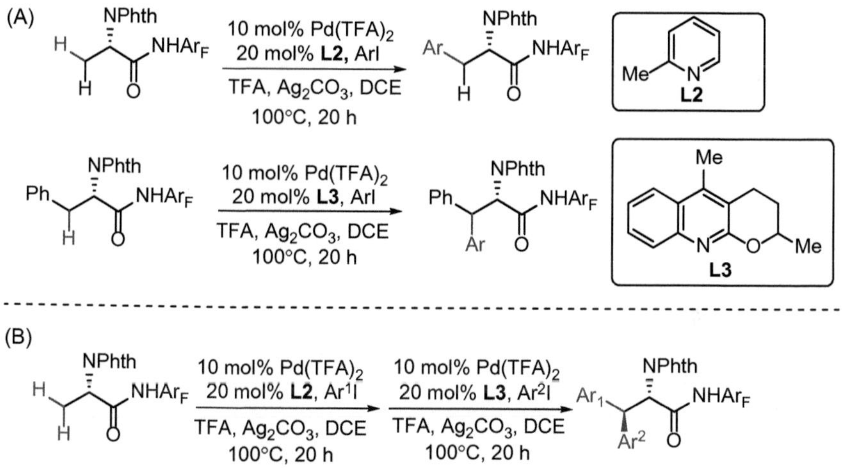

Scheme 4.7 *N*-Arylamide-directed ligand-controled C(sp^3)—H arylation of α-amino acid derivatives.

Very recently, the Yu group reported a Pd-catalyzed sequential triple C—H activation enabled by a weakly coordinating pyrazole and an *N*-arylamide group.[17] The cascade reaction was proved to procced via pyrazole-directed C(sp^3)—H arylation, followed by the *N*-arylamide-directed *ortho*-C(sp^2)—H arylation. Pd(OTf)$_2$(MeCN)$_4$ was found to be the most effective catalyst (Scheme 4.8). This method can tolerate a large

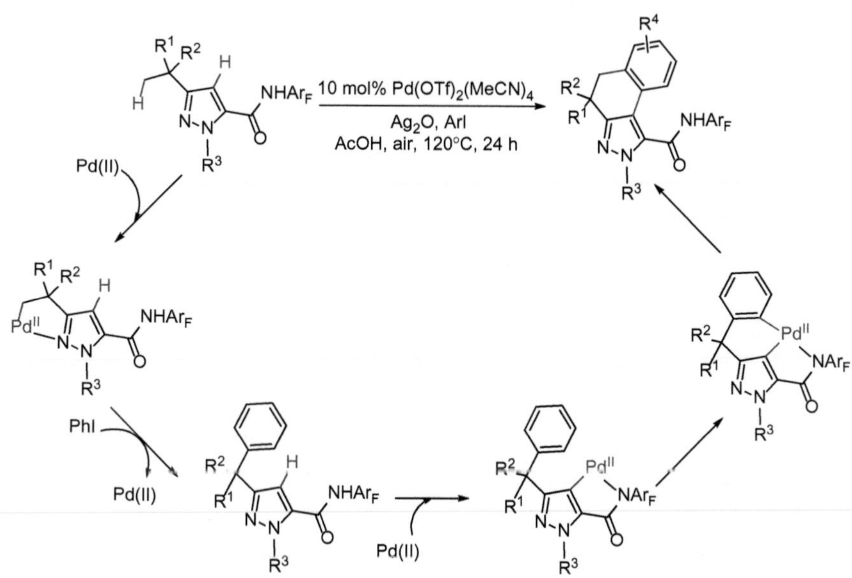

Scheme 4.8 Orchestrated triple C—H arylation reactions with a pyrazole and an amide DG.

scope of functional groups with good regioselectivity and yields, providing an efficient strategy for the synthesis of complex benzo[*e*]indazole derivatives.

The Yu group has widely used mono-*N*-protected amino acids (MPAAs) to promote a wide range of C–H functionalization reactions directed by weak coordinating DGs, such as –COOH, –OH, –CN, and –OMe.[18] Inspired by these precedents, Yu and co-workers demonstrated the first example of MPAA-enabled, Pd-catalyzed γ-C(sp^3)–H arylation of triflyl-protected aliphatic amines with arylboron reagents (Scheme 4.9).[19] No arylation product was observed in the absence of MPAA ligand, indicating that MPAA ligand was crucial for the success of this reaction. The reaction was proposed to occur via a Pd(II)/Pd(0) catalytic cycle. Notably, they found that arylboron reagents could be replaced by aryl iodides, which implied that the MPAA ligand could also promote the arylation of C(sp^3)–H bonds via a Pd(II)/Pd(IV) pathway.

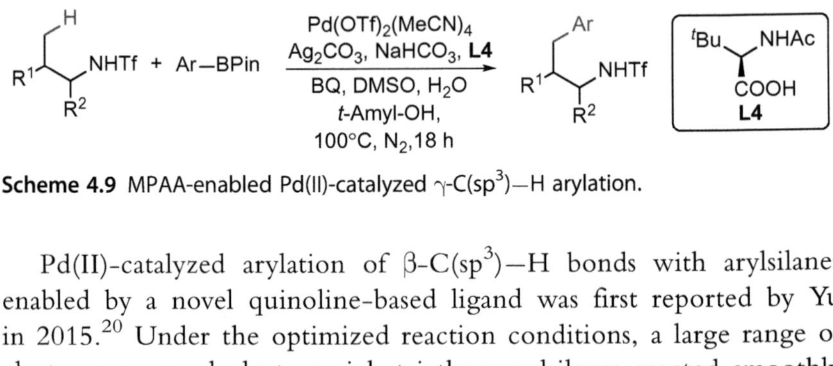

Scheme 4.9 MPAA-enabled Pd(II)-catalyzed γ-C(sp^3)–H arylation.

Pd(II)-catalyzed arylation of β-C(sp^3)–H bonds with arylsilanes enabled by a novel quinoline-based ligand was first reported by Yu in 2015.[20] Under the optimized reaction conditions, a large range of electron-poor and electron-rich triethoxyarylsilanes reacted smoothly with alanine-derived amide in high yields. Other aliphatic acids were also compatible with this protocol by simply tuning the ligand and base (Scheme 4.10A). Interestingly, when ibuprofen-derived amide was used as the substrate, γ-C(sp^2)–H arylation occurred preferentially under this Pd(II)/Pd(0) catalytic system. In contrast, β-C(sp^3)–H arylation could be successfully achieved using phenyl iodide as the coupling partner and 2-methylpyridine as the ligand via a Pd(II)/Pd(IV) catalytic cycle, as reported before by the same group (Scheme 4.10B).

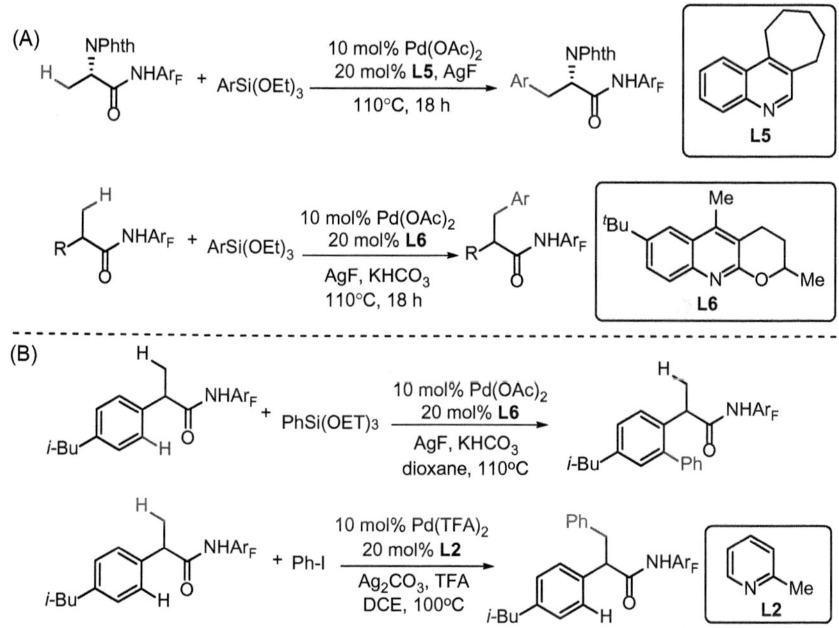

Scheme 4.10 Ligand-promoted C(sp^3)—H arylation with arylsianes.

The Yu group have developed a simple *N*-methoxyamide DG to direct C(sp^3)—H arylation of aliphatic acids; however, this arylation protocol was not compatible with those containing α-hydrogen atoms (Scheme 4.4).[12] Because *N*-methoxyamide DG was cheaper and easier to install and remove than other types of amide DGs, Yu and co-workers have successfully achieved the selective mono- and di-arylation of a wide range of carboxylic acids using the simple *N*-methoxyamide auxiliary as DG. The use of pyridine-type ligands was crucial for the success of this reaction.[21] They tested various pyridine-based ligands and found that 2-picoline was the best for monoarylation of alanine derivative. Aryl iodides bearing both electron-withdrawing and electron-donating groups in *ortho-*, *meta-*, and *para-*positions were compatible with the reaction. Moreover, 2,6-lutidine was found to be the best ligand to enable the arylation of secondary C(sp^3)—H bonds (Scheme 4.11).

One-pot synthesis of 4-aryl-2-quinolines via palladium-catalyzed ligand-promoted triple C—H functionalization with the assist of *N*-arylamide was reported by Yu and co-workers (Scheme 4.12).[22] Preliminary mechanistic

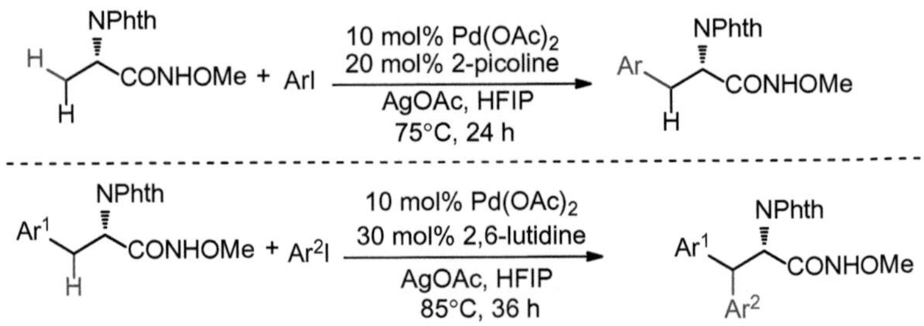

Scheme 4.11 Ligand-enabled *N*-methoxyamide-directed C(sp^3)—H arylation of α-amino acid derivatives.

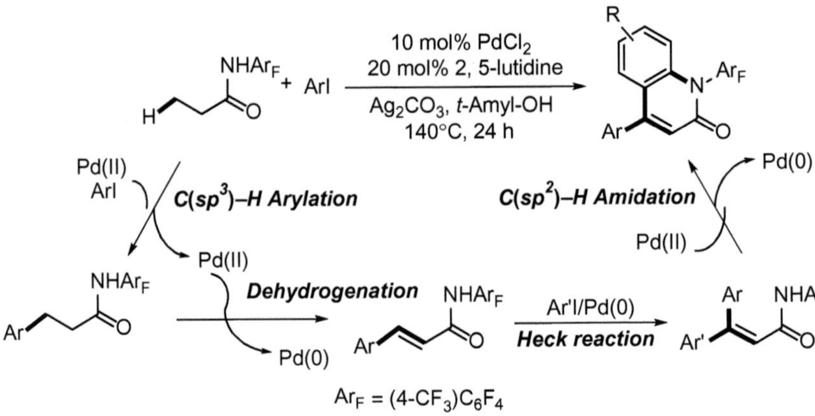

Scheme 4.12 Ligand-promoted CONH—p-$CF_3C_6F_4$-directed triple C—H activation.

studies suggested that the reaction might involve sequential β-C(sp^3)—H arylation followed by dehydrogenation, Heck coupling with a second aryl iodide, and finally intramolecular C—H amidation.

Although β-C(sp^3)—H arylation of aliphatic acid derivatives via a five-membered palladacycle has been extensively investigated, γ-C(sp^3)—H arylation via a six-membered palladacycle remained a great challenge. Recently, Yu and co-workers reported a Pd(II)-catalyzed ligand-promoted arylation of γ-methyl-C(sp^3)—H bonds directed by *N*-arylamide DG.[23] The γ-arylation of methyl C(sp^3)—H bonds attached to quaternary carbon centers of aliphatic amides occurred smoothly with quinoline-based ligand **L3** (Scheme 4.13A). The protocol could also be applied to γ-arylation of valine, isoleucine, and *tert*-leucine derivatives, when another

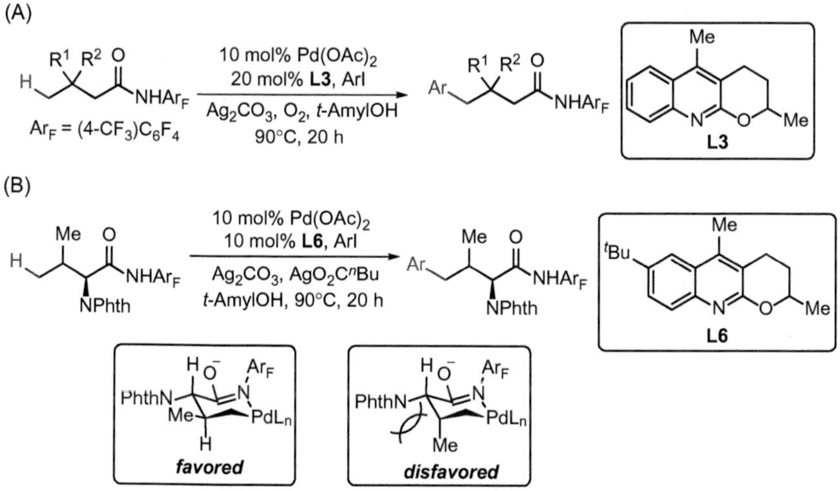

Scheme 4.13 Ligand-promoted γ-C(sp^3)—H arylation of aliphatic amides.

quinoline-based ligand **L6** was used under modified conditions. It is worth noting that the arylation of α-amino acids proceeded with high diastereoselectivity (dr>20:1), owing to the favored *trans* orientation of the methyl and *N*-Phth group (Scheme 4.13B).

The introduction of the adamantyl group can dramatically increase the lipophilicity and metabolic stability of drugs, therefore, the direct modification of adamantine derivatives has received tremendous attention. Inspired by Yu's work on *N*-arylamide-directed functionalization of C(sp^3)—H bonds,[13,24] Wang reported a Pd(0)-catalyzed arylation of methylene C(sp^3)—H bonds of adamantyl scaffold with aryl iodides (Scheme 4.14).[25]

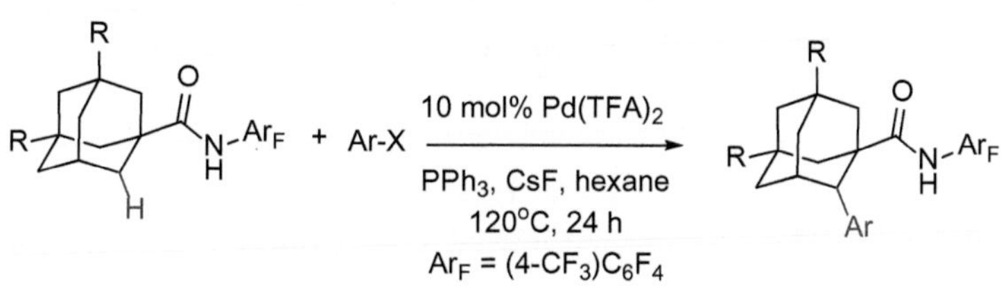

Scheme 4.14 Pd(0)-catalyzed C(sp^3) – H arylation of adamantine derivatives.

Recently, the Gaunt group found that the steric hindered secondary aliphatic amines can undergo C—H activation via an unusual four-membered palladacycle.[26] They achieved the Pd-catalyzed C—H carbonylation and amination to strained nitrogen hetercycles and C—H acetoxylation to protected amino alcohols. Based on these results, they further extended this novel C—H activation mode to Pd(II)-catalyzed C(sp^3)—H arylation of secondary aliphatic amines with arylboronic esters promoted by a MPAA ligand.[27] A range of piperidine derivatives and arylboronic esters were well tolerated, giving the arylated products in moderate-to-good yields (Scheme 4.15A).

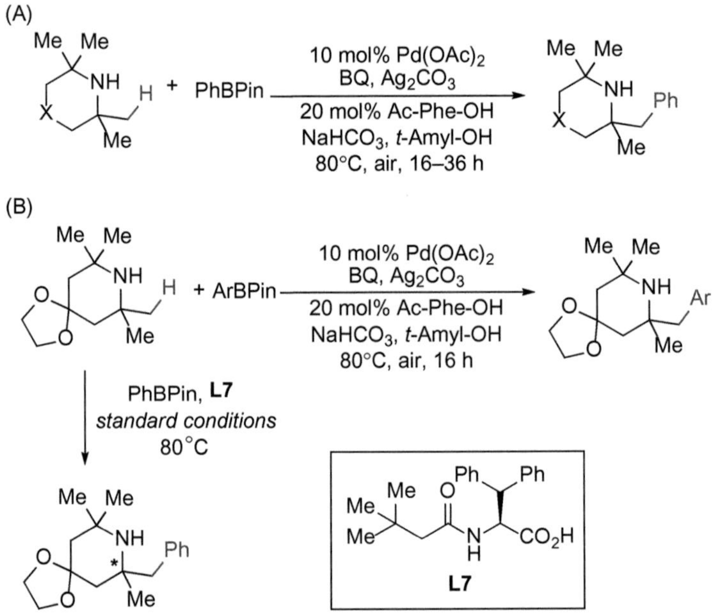

Scheme 4.15 MPAA-enabled palladium-catalyzed β-C(sp^3)—H arylation of aliphatic amines.

Inspired by Yu's work on the Pd-catalyzed enantioselective C—H activation using MPAA ligands,[28] they further examined the potential for amine-directed enantioselective C(sp^3)—H arylation. A number of MPAA ligands were assessed under otherwise identical conditions at 60°C and a β-phenyl-phenylalanine protected with a bulky acyl side chain (**L7**) was found to give the best enantioselectivity (Scheme 4.15B, 60% ee). Further optimizations were needed with the help of the best understanding of the mechanistic studies.

Gaunt and co-workers further applied this sterically promoted Pd-catalyzed C—H activation to primary amino alcohols, simply by

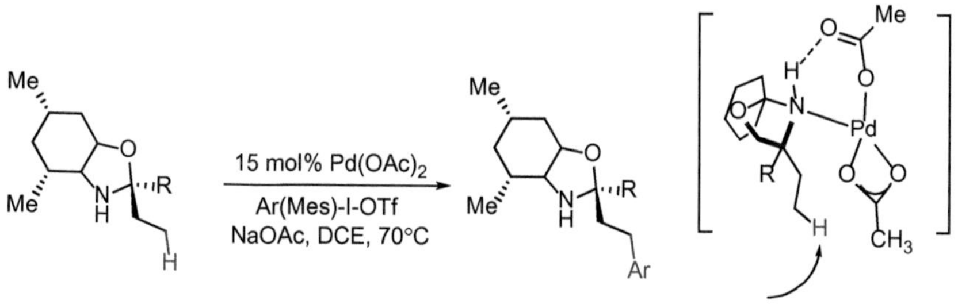

Scheme 4.16 Pd(II)-catalyzed γ-C(sp^3)—H arylation of amino alcohol derivatives.

converting the incompatible primary amino alcohols into the corresponding *N,O*-ketals with cyclohexanone to temporarily generate sterically hindered secondary amines. They found that γ-methyl C(sp^3)—H arylation of a range of *N,O*-ketals occurred smoothly with various diaryliodonium triflates (Scheme 4.16).[29] Calculations revealed that a hydrogen bond between amine and acetate ligands orients the γ-C—H bond ideally to interact with the palladium catalyst.

Up to now, C(sp^3)—H functionalization of saturated cyclic amines was limited to the functionalization of the activated C—H bonds α to nitrogen, or of C—H bonds on an exocyclic alkyl group. Recently, Sanford and co-workers reported a palladium-catalyzed transannular C(sp^3)—H arylation of various alicyclic amines at remote positions.[30] They rationalized that the boat conformation of the cyclic amines could oriented the remote C—H bonds ideally for C—H cleavage (Scheme 4.17A). Initially, they selected 3-azabicyclo[3.1.0]hexane as a model substrate, since its bicyclic core would prearrange it in a boat-like conformation (Scheme 4.17B). They have also successfully extended this reaction to piperidine and various alicyclic amine derivatives with slightly modified reaction conditions. The transannular C—H arylation of several biologically important molecules, such as pharmaceutical candidate amitifadine, the drug varenicline, and the natural product cytisine, has been achieved (Scheme 4.17C).

More recently, Chen reported a Pd(II)-catalyzed oxime-directed arylation of β-C(sp^3)—H bonds with diaryliodonium salts, providing practical access to β-arylated oximes from/to ketones and aldehydes (Scheme 4.18).[31] They have also applied this arylation protocol to the late-stage functionalization of natural products.

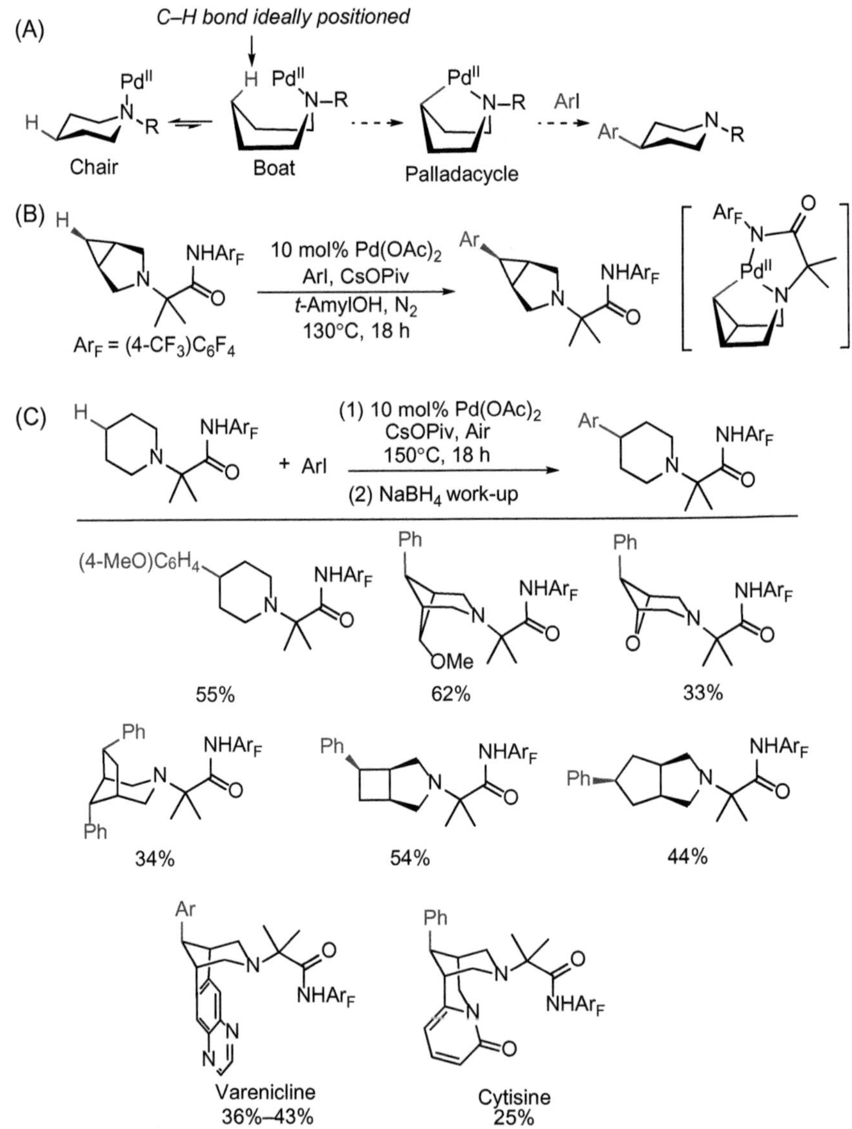

Scheme 4.17 Pd(II)-catalyzed transannular $C(sp^3)$—H arylation of alicyclic amines.

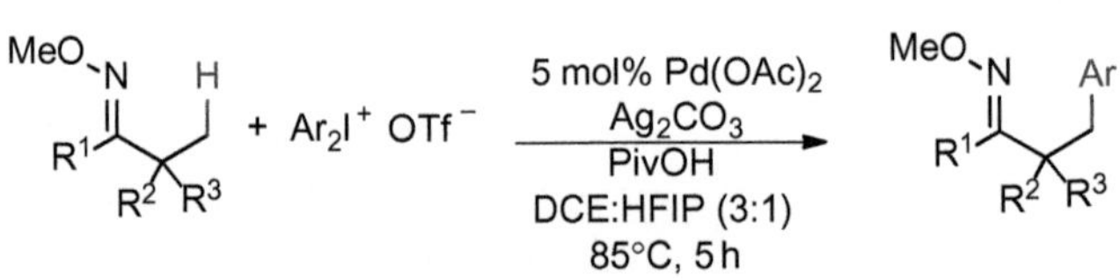

Scheme 4.18 Pd(II)-catalyzed oxime-directed β-arylation with diaryliodonium salts.

4.3 PALLADIUM-CATALYZED C(sp^3)—H ARYLATION DIRECTED BY BIDENTATE DGs

Palladium-catalyzed methyl C(sp^3)—H bond arylation of 2-ethylpridine with aryl iodides was reported by Daugulis (Scheme 4.2);[8] however, the difficulty of removing or modifying the pyridyl DG has limited its synthetic application. They envisioned that if the pyridyl group could be replaced by an amide-tethered pyridyl-based DG, the practicability and generality of this protocol will be greatly improved. To this end, the removable, bidentate 8-aminoquinoline (AQ) and picolinic acid (PA) were first introduced by Daugulis and co-workers.[32] Pd-catalyzed arylation of primary C(sp^3)—H bonds, and even the more challenging secondary C(sp^3)—H bonds, were elegantly demonstrated by the use of AQ or PA as bidentate DGs (Scheme 4.19). It was proposed that the bidentate, monoanionic auxiliaries could facilitate both the C—H activation and the oxidative addition step by stabilizing the putative Pd(IV) intermediates. However, a mixture of mono- and diarylated products was obtained when the arylation of primary C(sp^3)—H bonds using AQ auxiliary was carried out.

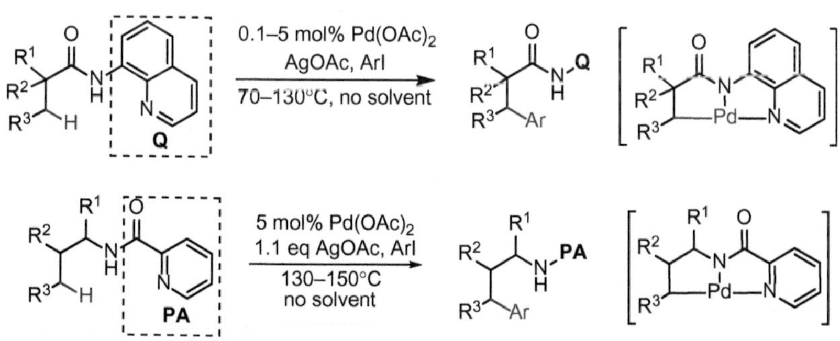

Scheme 4.19 Pd(II)-catalyzed C(sp^3)—H arylation directed by AQ and PA bidentate DGs.

To address this chemoselectivity issue, the same group has reported the selective monoarylation of primary C(sp^3)—H bonds by using 2-methylthioaniline auxiliary (MT) in 2010. They have also demonstrated the use of inorganic bases instead of silver salts as halide scavengers (Scheme 4.20).[33] Reaction of 8-aminoquinoline pivalamide with stoichiometric palladium acetate in nitrile gave the corresponding bicyclic palladacycle, the structure of which was unambitiously verified by single-crystal X-ray diffraction analysis.

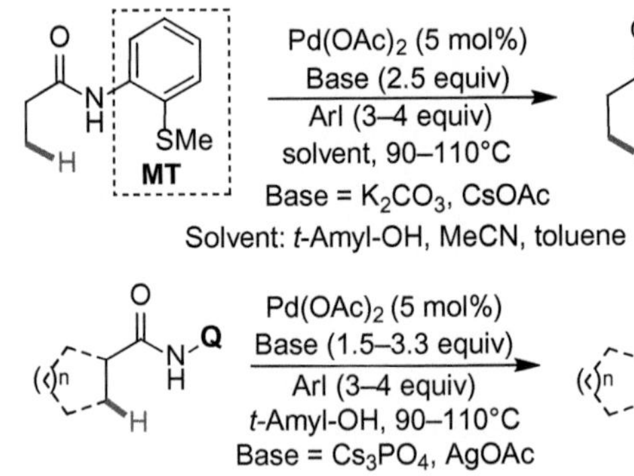

Scheme 4.20 Pd(II)-catalyzed β-C(sp^3)—H arylation of aliphatic acids directed by MT and AQ auxiliaries.

In 2006, Corey and co-workers reported the Pd-catalyzed β-C(sp^3)—H arylation of α-amino acid derivatives employing AQ auxiliary (Scheme 4.21).[34] An unexpected γ-C(sp^3)—H bond arylation was observed when isoleucine was used as a substrate, which implied that the γ-C(sp^3)—H bond could be activated if it was difficult to activate the *β*-C—H bond. Biarylation of primary C(sp^3)—H bonds was observed when the alanine 8-aminoquinoline amide was treated with palladium acetate.

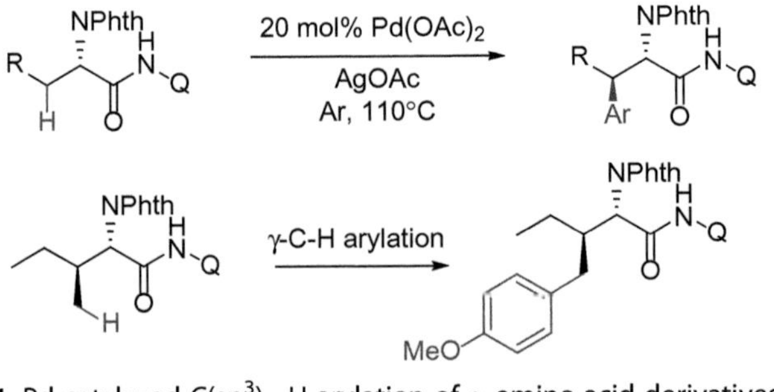

Scheme 4.21 Pd-catalyzed C(sp^3)—H arylation of α-amino acid derivatives employing AQ auxiliary by Corey.

Although the pioneering work of Corey showcases that AQ auxiliary can be efficiently used in the direct arylation of α-amino acid derivatives, it is difficult to control the mono- and diarylation when alanine derivative was used as substrate. In 2012, the monoarylation of primary C(sp^3)—H bonds of *N*-pathaloyalanine derivatives directed by MT auxiliary was reported by Daugulis (Scheme 4.22).[35] A broad range of substituted

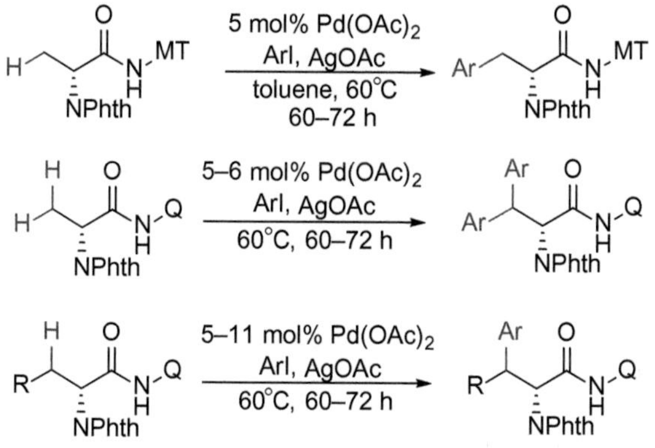

Scheme 4.22 Pd(II)-catalyzed chemoselective arylation of α-amino acid derivatives by Daugulis.

phenylalanines bearing synthetically useful functional groups can be prepared using this protocol. Moreover, they found that the AQ auxiliary was very efficient for the diarylation of methyl and monoarylation of methylene C(sp^3)—H bonds. Finally, both MT and AQ auxiliarys could be readily removed by treatment with $BF_3 \cdot Et_2O$ in MeOH at 100°C, leading to the corresponding methyl esters. However, a slight loss of enantiopurity of the arylation products was observed in this protocol.

Inspired by early studies of Daugulis[32] and Corey,[34] Chen and co-workers have achieved the total synthesis of celogenin by the use of AQ-directed arylation as the key step.[36] As part of their continuing interest in Pd-catalyzed C(sp^3)—H functionalization, they reported the synthesis of benzo-rings by pallidum-catalyzed intramolecular coupling of an aryl iodide and β-methylene C—H bond under mild conditions.[37] They found that *ortho*-phenyl benzoic acid served as a critical ligand for the efficiency of the catalytic process, and even allowed the reaction proceed at room temperature (Scheme 4.23).

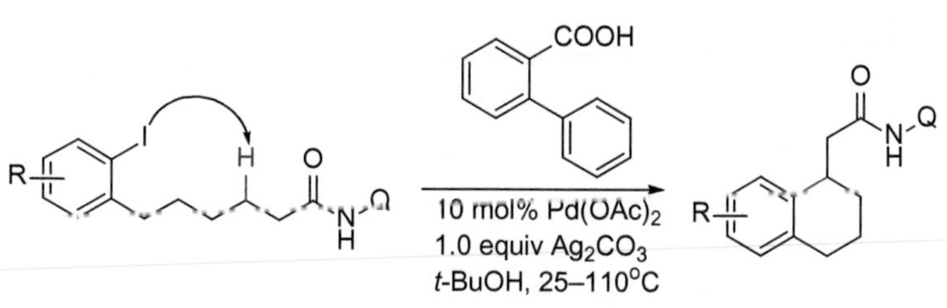

Scheme 4.23 Intramolecular arylation of methylene β-C(sp³)—H bonds.

Inspired by the pioneering work of Daugulis,[32] a number of bidentate, monoanionic DGs have been developed and widely employed in the direct functionalization of unactivated $C(sp^3)$—H bonds (Fig. 4.2).[4a,4c–e,4p] In order to achieve the challenging methylene $C(sp^3)$—H alkoxylation and avoid the undesired alkoxylation of the quinolone $C(sp^2)$—H bond on the AQ auxiliary, Shi and co-workers have developed a removable bidentate DG derived from 2(pyridine-2-yl)isopropylamine (PIP-amine), which exhibited superior reactivity in Pd-catalyzed functionalization of unactivated methylene $C(sp^3)$—H bonds.[4p,38] They successfully achieved the β-monoarylation of alanine derivative using PIP auxiliary (Scheme 4.24).[38a] Various aryl iodides containing different electron-poor and electron-rich groups were compatible with this protocol, affording the desired arylalanines in good yields, high selectivity, and complete retention of chirality. It is noteworthy that the PIP auxiliary can be removed through a mild *N*-nitrosylation/hydrolysis sequence in high yield without reducing enantiopurity.

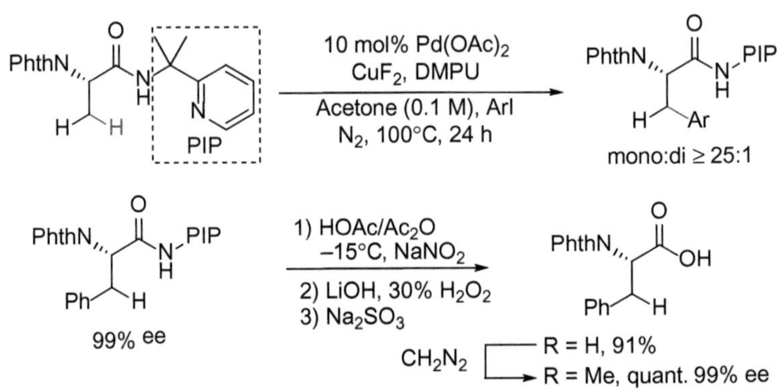

Scheme 4.24 PIP-directed arylation of β-$C(sp^3)$—H bonds of α-amino acids.

In 2013, the Shi group reported Pd(II)-catalyzed direct arylation of primary and secondary $C(sp^3)$—H bonds with diarylhyperiodonium directed by AQ auxiliary (Scheme 4.25).[39] Control experiments revealed

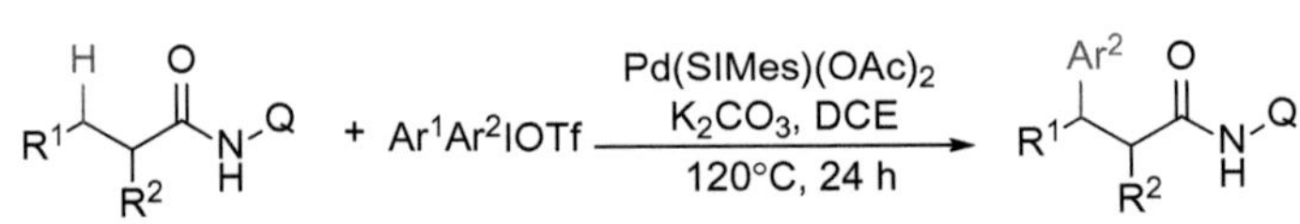

Scheme 4.25 Pd-catalyzed $C(sp^3)$—H arylation with diarylhyperiodonium directed by AQ auxiliary.

that aryl iodides were less reactive than diarylhyperiodonium triflates under the same conditions. The solubility of the counteranion of diaryhyperiodonium salts was found to play an important role in the efficiency of this reaction.

However, the above-mentioned C(sp^3)—H arylation reactions mainly rely on the use of aryl iodides or diarylhyperiodonium salts, the use of readily available and cost-effective aryl bromides remains a great challenge. Compared to the widely used AQ auxiliary, the nitrogen atom on the PIP auxiliary is more electron-rich and sterically bulky. Thus, the Shi group hypothesized that this may promote the oxidative addition of the less reactive aryl bromides to the resulting Pd(II)-intermediate after C—H activation. Based on this hypothesis, they achieved the arylation of methylene C(sp^3)—H bonds with aryl bromides assisted by PIP DG.[38c] Potassium carbonate was used as the halide scavenger without the need for any silver salts. A broad range of aryl bromides were well tolerated under the optimized reaction condition: 10 mol% $Pd(OAc)_2$, 2.5 equivalent K_2CO_3 and 20 mol% PivOH in *t*-BuOH at 120°C for 24 h (Scheme 4.26). In the same year, Zeng and co-workers also reported a Pd-catalyzed arylation of unactivated C(sp^3)—H bonds with aryl bromides promoted by AQ auxiliary.[40]

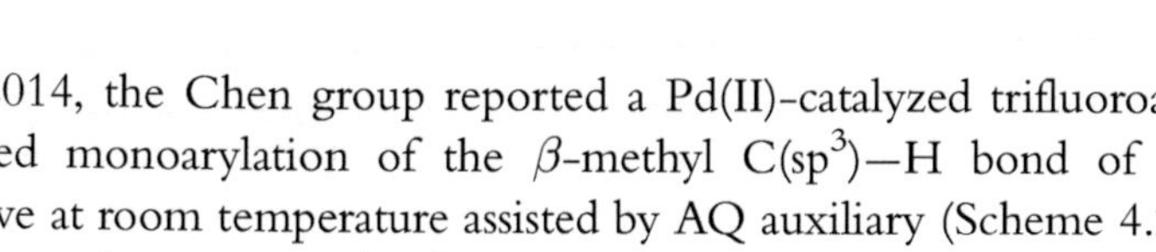

Scheme 4.26 Pd-catalyzed methylene C(sp^3)—H arylation with aryl bromides directed by PIP auxiliary.

In 2014, the Chen group reported a Pd(II)-catalyzed trifluoroacetate-promoted monoarylation of the β-methyl C(sp^3)—H bond of alanine derivative at room temperature assisted by AQ auxiliary (Scheme 4.27A).[41] This protocol represents the first example of Pd-catalyzed intermolecular arylation of inert C(sp^3)—H bonds at room temperature, although the reaction proceeded with prolonged reaction time. Almost at the same time, Shi and co-workers demonstrated the Pd-catalyzed monoarylation of alanine derivative using AQ auxiliary under completely different reaction conditions.[42] The optimized reaction conditions were found to be 10 mol% $Pd(OAc)_2$, 1.5 equiv $AgBF_4$ in ^{t}BuOH at 75°C. The protocol features excellent functional group tolerance, gram scale synthesis,

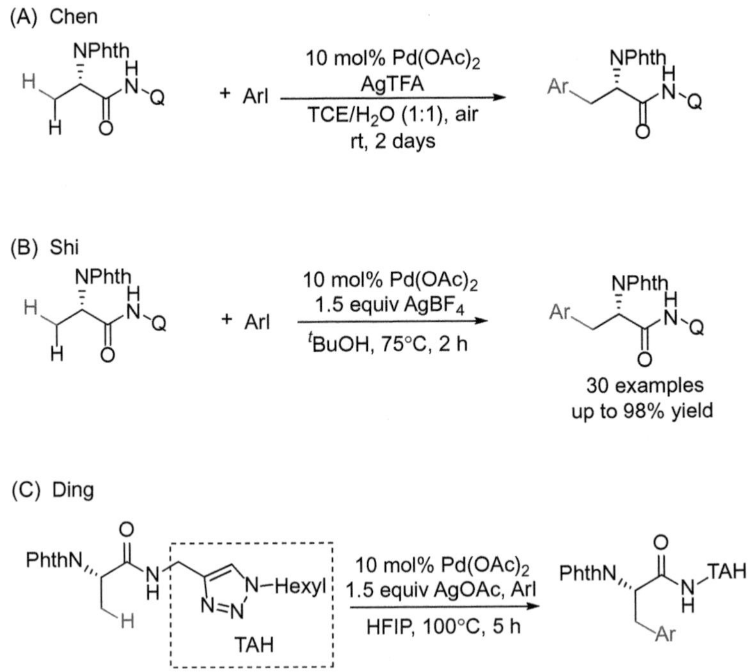

Scheme 4.27 Pd-catalyzed monoarylation of alanine derivative directed by bidentate DGs.

and complete retention of configuration (Scheme 4.27B). In 2015, the Ding group reported the Pd-catalyzed monoarylation of alanine derivative using 1,2,3-triazole-type auxiliary (TAH) as a removable DG (Scheme 4.27C).[43]

In 2014, Chen and co-workers achieved the total synthesis of hibispeptins A by the use of Pd-catalyzed γ-methyl C(sp^3)—H arylation as the key step. In their initial attempts, they found that γ methyl C(sp^3)—H arylation of isoleucin with sterically hindered aryl iodides bearing *ortho*-substituent using PA auxiliary was not efficient. After screening a number of bidentate DGs, they discovered that the introduction of a novel pyridylmethylamine-based DG (PR) could enable the use of sterically hindered *ortho*-substituted aryl iodides as coupling partners (Scheme 4.28A).[44] To further investigate the efficiency of pyridylmethylamine-type DGs in C(sp^3)—H functionalization, they discovered that the 2-(2-pyridyl)ethylamine (PE) could enable the Pd-catalyzed β-C(sp^3)—H arylation of aliphatic carboxamides with sterically hindered aryl iodides (Scheme 4.28B).[45]

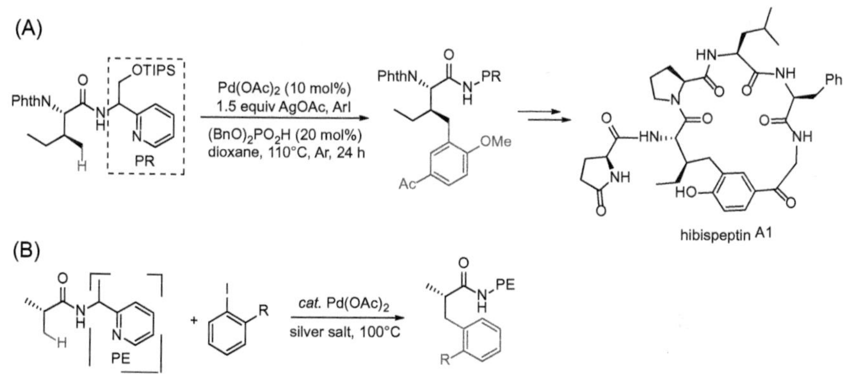

Scheme 4.28 Pd-catalyzed C(sp^3)—H arylation with sterically hindered aryl iodides directed by pyridylmethylamine-type DGs.

Despite β-functionalization of aliphatic acid derivatives using various amides as DGs being well investigated, the amide linkers are relatively stable and difficult to remove. In 2015, the Shi group discovered that readily available and cost-effective 2-picolinamide was an efficient DG for Pd-catalyzed β-C(sp^3)—H arylation with an imide linker (Scheme 4.29).[46] As expected, the imide linker was readily hydrolyzed with LiOH/H_2O_2 at room temperature, affording the desired aliphatic acids in high yields.

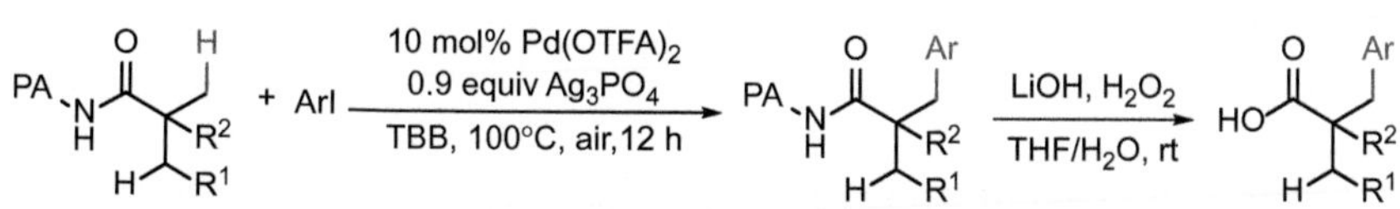

Scheme 4.29 Pd-catalyzed C(sp^3)—H arylation of imides enabled by PA auxiliary.

Previously, Yu and co-workers discovered that MPAA ligands could coordinate to palladium complexes in a bidentate manner to drastically accelerate C—H activation reactions.[18a,18d,28a,28c] They reasoned that native amino acids embedded in the peptide backbone could also coordinate to palladium catalyst to activate proximate C(sp^3)—H bonds. As expected, Pd-catalyzed C(sp^3)—H arylation at the *N*-terminus of various di-, tri-, and tetrapeptides with aryl iodides was achieved without the introduction of any DG (Scheme 4.30).[47]

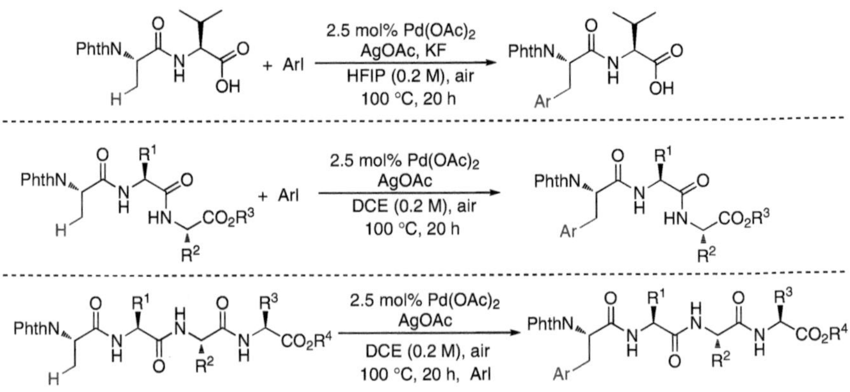

Scheme 4.30 Amino acid-directed C(sp^3)—H arylation of di-, tri-, and tetrapeptides.

In 2015, the Yu group further extended this protocol to the arylation of *O*-protected α-hydroxy aliphatic acid derivatives with amino acid as the DG. They found that the use of weakly acidic hexafluoroisopropanol (HFIP) as solvent is crucial for the success of this reaction. A broad range of β-aryl-α-hydroxy carboxylic acids were obtained in moderate-to-good yields with the retention of configuration (Scheme 4.31).[48] Notably, β-arylation of *O*-protected lactic acids using PIP or AQ as DGs has been reported by Baudoin and Shi, respectively.[49]

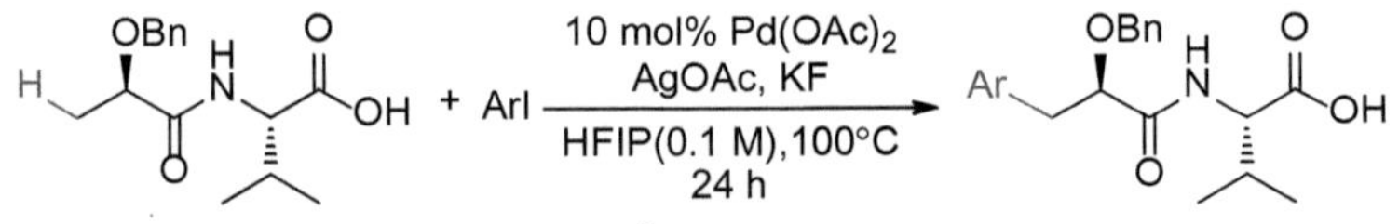

Scheme 4.31 Amino acid-directed C(sp^3)—H arylation of lactic acid.

Song and co-workers reported a Pd-catalyzed β-C(sp^3)—H arylation of aliphatic amides using 2-aminopyridine-1-oxide (PyO) as bidentate auxiliary. The arylation reactions occurred in the presence of K_2HPO_4·H_2O in the DMSO without conventional silver salts (Scheme 4.32).[50]

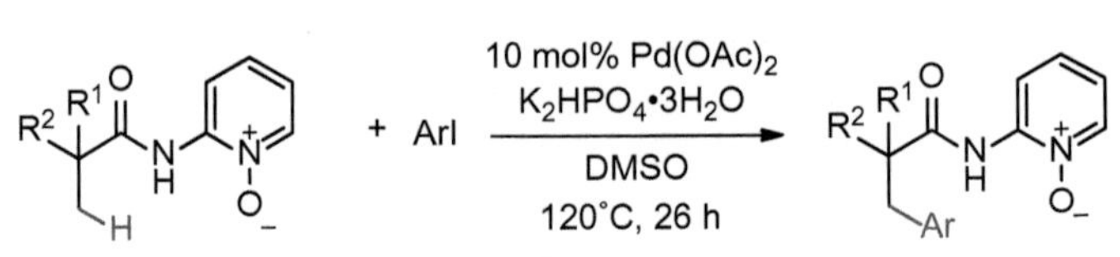

Scheme 4.32 Palladium-catalyzed β-C(sp^3)—H arylation of aliphatic amides directed by PyO auxiliary.

Zeng and Lu also reported the C(sp^3)–H arylation of aliphatic amides directed by PyO auxiliary, albeit in the presence of silver acetate.[51]

On the basis of Daugulis' early report,[32] the Chen group systematically investigated the palladium-catalyzed γ-C(sp^3)–H arylation of aliphatic amines with the assistance of PA auxiliary in 2011.[52] They successfully achieved the regio- and stereoselective arylation under mild conditions with a variety of aryl iodides and heteroaryl iodides (Scheme 4.33). However, the PA auxiliary was difficult to remove. Thus, they prepared a modified picolinic acid (PAr) with a methylene hydroxyl group installed at the *ortho* position of picolinamide to accelerate the cleavage under mild conditions. A formal synthesis of (+)-obafluorin was achieved using this arylation protocol.

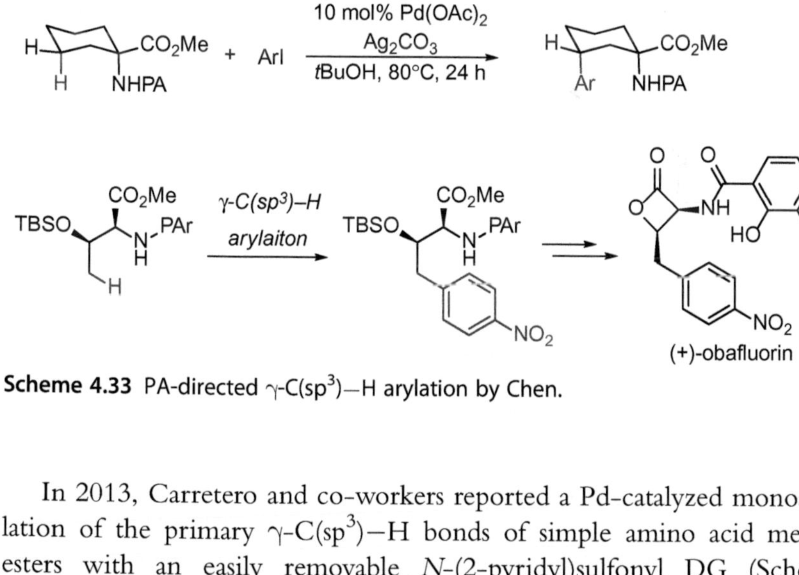

Scheme 4.33 PA-directed γ-C(sp^3)–H arylation by Chen.

In 2013, Carretero and co-workers reported a Pd-catalyzed monoarylation of the primary γ-C(sp^3)–H bonds of simple amino acid methyl esters with an easily removable *N*-(2-pyridyl)sulfonyl DG (Scheme 4.34).[52] Various amino acid derivatives, including β-amino acid and α-quaternary amino acid derivatives were compatible with this arylation protocol, providing γ-arylated products in moderate-to-good yields. Moreover, an unprecedented C(sp^3)–H arylation of dipeptides was also demonstrated, illustrating the synthetic potential of the protocol in the functionalization of peptide derivatives (Scheme 4.34).[53]

In the same year, the Ma group reported the Pd-catalyzed γ-C(sp^3)–H arylation of substituted 2-aminobutanates employing a

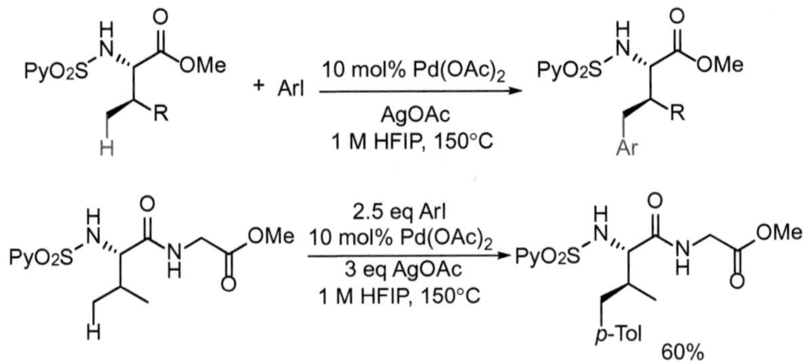

Scheme 4.34 Pd(II)-catalyzed γ-arylation of methyl C(sp^3)—H bonds directed by *N*-(2-pyridyl)-sulfonyl auxiliary.

novel 2-methoxyiminoacetyl (MIA) auxiliary.[54] Both electron-rich and electron-poor aryl iodides were well tolerated, giving the desired products in good yields with the retention of configuration. Notably, the MIA auxiliary could be easily removed by treatment with 1 N KOH in dioxane at room temperature. It could also be readily transformed into a glycine moiety, which could be used in the rapid synthesis of peptides bearing a glycine residue. (Scheme 4.35).

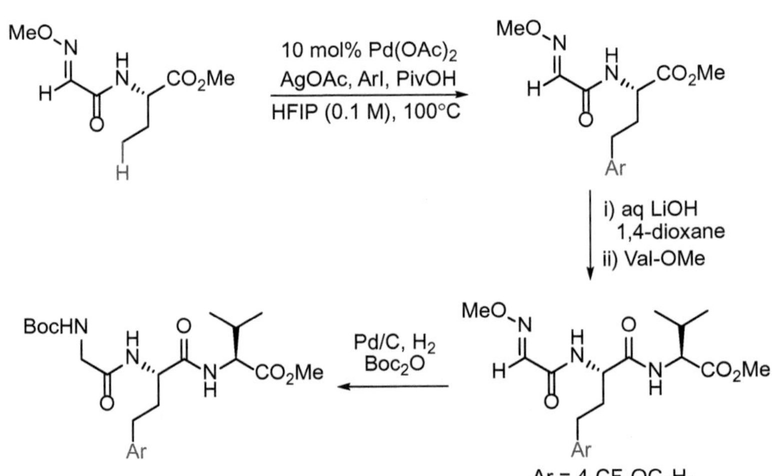

Scheme 4.35 Pd(II)-catalyzed γ-C(sp^3)—H arylation directed by MIA auxiliary.

The Babu and Shuto groups also reported Pd(II)-catalyzed arylation of secondary $C(sp^3)$—H bonds of cyclopropane derivatives with the assistance of AQ or MT auxiliaries, respectively.[55] Shortly after, the Bull and Zhang groups independently reported the Pd-catalyzed arylation of proline with aryl iodides assisted by AQ auxiliary.[56] In the same year, Charette and co-workers reported the Pd-catalyzed arylation of cyclopropanes directed by PA auxiliary.[57] The Wang group reported a PA-enabled, Pd-catalyzed remote $C(sp^3)$—H arylation of 3-pinanamine.[58] Notably, this represents a rare example of the direct functionalization of an aliphatic amine at the remote δ position via a six-membered palladacycle.[59] More recently, Pd-catalyzed γ-$C(sp^3)$—H arylation of rimantadine derivatives with various aryl iodides using PA as the bidentate auxiliary and Boc-L-Ileu-OH as the ligand was reported by the Zhang group.[60]

Despite these elegant precedents in the γ-$C(sp^3)$ − H arylation of aliphatic amine derivatives, the direct arylation of unactivated γ-methylene $C(sp^3)$ − H bonds of simple amine derivatives remains undeveloped. To this end, the Shi group developed a removable oxazoline-carboxylate bidentate auxiliary, which has been used in the arylation of unactivated γ-methylene $C(sp^3)$ − H bonds of linear aliphatic amines.[61] A wide range of aryl iodides with various substituents were compatible with this protocol (Scheme 4.36).

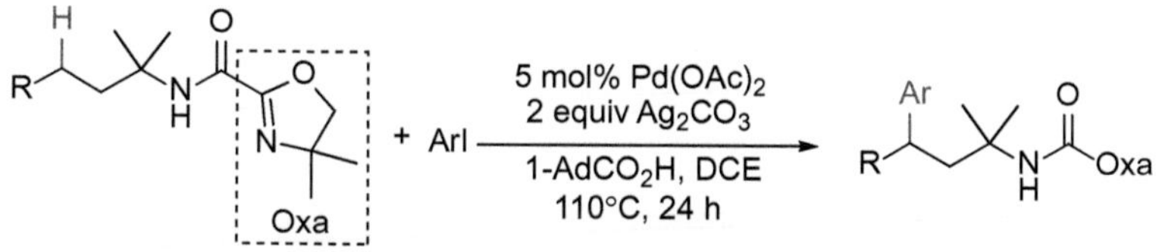

Scheme 4.36 Pd-catalyzed arylation of unactivated γ-methylene $C(sp^3)$—H bonds directed by oxazoline-carboxylate.

In 2015, the Liu group reported the γ-$C(sp^3)$—H arylation of aliphatic amine using a new bidentate DG derived from the isoxazole-3-carboxamide (MICA), which could be readily removed and recovered under very mild conditions (Scheme 4.37A).[62] Very recently, Zhao and co-workers reported a Pd-catalyzed γ-$C(sp^3)$—H arylation of aliphatic amine directed by an oxalyl amide (OA) group (Scheme 4.37B).[63]

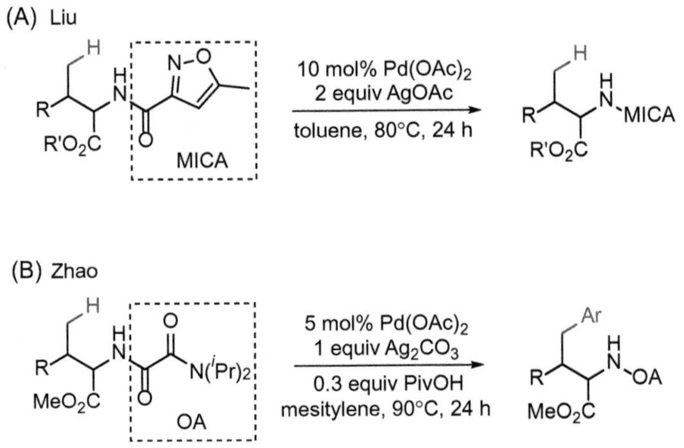

Scheme 4.37 Pd(II)-catalyzed γ-C(sp^3)–H arylation directed by MICA or OA auxiliary.

4.4 PALLADIUM-CATALYZED C(sp^3)–H ARYLATION USING TRANSIENT DGs

Though coordination-assisted C(sp^3)–H activation shows great efficiency and site-selectivity as described above, a major shortcoming of this strategy is the requirement for extra steps to install and remove the stoichiometric amount of the DGs. Therefore, it is highly desirable to develop an auxiliary that can be reversibly linked to the substrate and can serve as an efficient DG. Based on previous works on the development of MPAA ligands c[18a,18d,28a,28c] and Pd-catalyzed C(sp^3)–H bond functionalization of peptides using amino acids as bidentate DG,[47] Yu and co-workers reasoned that the amino acid could be reversibly tethered to an aldehyde or ketone substrate via an imine linkage under appropriate conditions. Notably, the imine moiety and the carboxylate could act as temporary bidentate DGs coordinating with palladium to facilitate subsequent C–H bond functionalization. After extensive investigations, they found that the arylation occurred smoothly when 40 mol% glycine was used as the transient DG. Arylation of a broad range of aldehydes and ketones were achieved in moderate-to-high yields. Control experiments revealed that no arylation product was observed in the absence of glycine, or when *N*-protected glycine was used, indicating that the temporarily formed imine-tether was crucial for promoting C–H functionalization. Moreover, Pd-catalyzed enantioselective arylation of methylene C(sp^3)–H bonds with 20 mol% L-*tert*-leucine as the transient DG was also demonstrated (Scheme 4.38).[64]

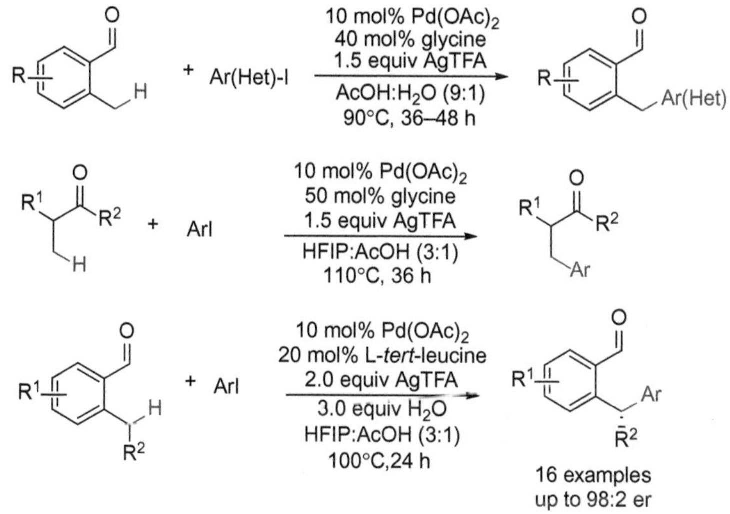

Scheme 4.38 Pd-catalyzed C(sp^3)—H arylation using α-amino acid as the transient DG.

Inspired by this pioneering study, the Hu group reported the arylation of substituted 2-methylbenzaldehyde using acetohydrazine as a transient DG.[65]

4.5 PALLADIUM-CATALYZED ASYMMETRIC C(sp^3)—H ARYLATION

4.5.1 Palladium-Catalyzed Enantioselective C(sp^3)—H Arylation With Chiral Ligands

In 2011, the Yu group reported the first enantioselective C(sp^3)—H arylation of cyclopropanes with arylboron reagents using MPAA as chiral ligand.[66] After extensive screenings, CONH-*p*-CNC_6F_4 auxiliary was found to be the best weakly coordinating DG, and the sterically bulky MPAA **L8** was the optimal chiral ligand (Scheme 4.39). Although high

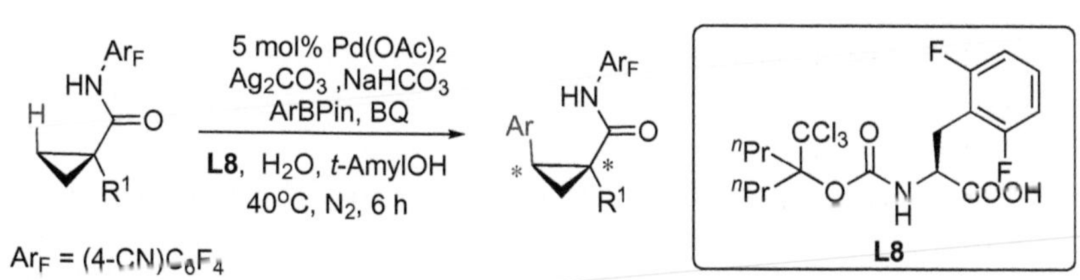

Scheme 4.39 Pd(II)-catalyzed enantioselective arylation of cyclopropanes with arylboron reagents.

enantioselectivity has been achieved, the reaction was limited to the relatively acidic cyclopropyl C—H bonds.

To overcome this limitation, Yu and co-workers reported the enantioselective arylation of cyclobutanecarboxylic acid derivatives with arylboron reagents using MPAA **L9** as chiral ligand (Scheme 4.40).[67] These examples showcase that MPAAs are powerful chiral ligands for Pd-catalyzed enantioselective $C(sp^3)$—H functionalization.

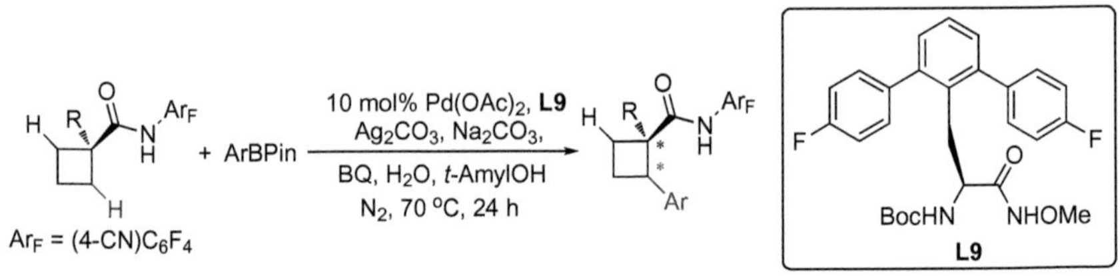

Scheme 4.40 Pd(II)-catalyzed enantioselective arylation of cyclobutanecarboxamides with arylboron reagents.

The first enantioselective $C(sp^3)$—H arylation using MPAA ligand via a Pd(II)/Pd(IV) catalytic cycle was disclosed by Yu and co-workers in 2015 (Scheme 4.41).[68] Arylation of cyclopropylmethylamines with various aryl iodides was achieve in high enantioselectivity (up to 99.5% *ee*), providing an efficient and straightforward access to chiral *cis*-aryl-cyclopropymethylamines. This method could also be scaled-up to 5 mmol without erosion of yield and enantioselectivity (89% yield and 98.6% *ee*). It is worth noting that MPAA ligand can override substrate-controlled diastereoselectivity in substrates containing chiral center.

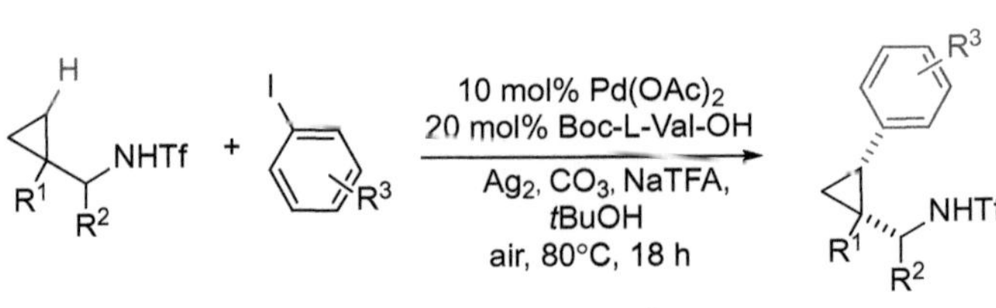

Scheme 4.41 Pd(II)-catalyzed enantioselective $C(sp^3)$—H arylation via a Pd(II)/Pd(IV) catalytic cycle.

So far, Pd-catalyzed enantioselective C—H functionalization has relied on the use of MPAA ligands. In 2015, the Duan group reported the first example of Pd-catalyzed enantioselective arylation of methylene $C(sp^3)$—H bonds of aliphatic amides directed by AQ auxiliary using

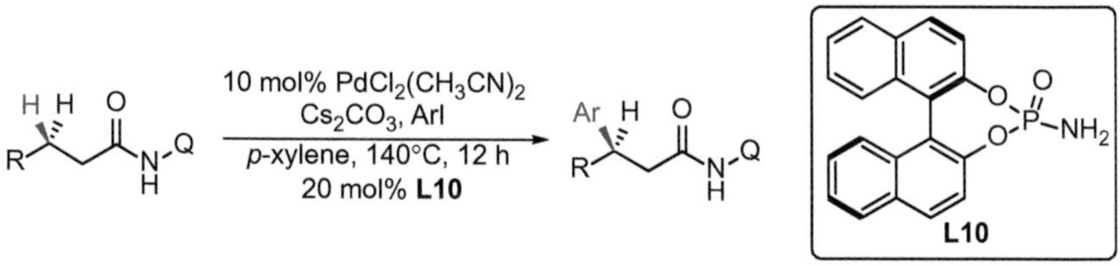

Scheme 4.42 Pd(II)-catalyzed enantioselective C(sp^3) – H arylation using phosphoric amide as chiral ligand.

phosphoric amides/acids as chiral ligands (Scheme 4.42).[69] Control experiments revealed that these ligands not only had a prominent effect on the control of enantioselectivity, but they also accelerated the reaction rate.

4.5.2 Palladium-Catalyzed Asymmetric C(sp^3)—H Arylation With Chiral Auxiliaries

Although the diastereoselective arylation of unactivated C(sp^3)—H bonds in chiral substrates, such as α-amino acids[4b,4c] and lactic acid,[48,49b] has been extensively investigated, asymmetric C(sp^3)—H arylation using a chiral auxiliary remained undeveloped.[70] The Hong group reported the first asymmetric C(sp^3)—H arylation of cyclopropanes using isoleucine amide as chiral bidentate auxiliary (Scheme 4.43).[71] They assumed that the diastereoselectivity was established at the oxidative addition step under the Curtin – Hammett rule, since the C—H cleavage of both diastereotopic hydrogens is relatively facile.

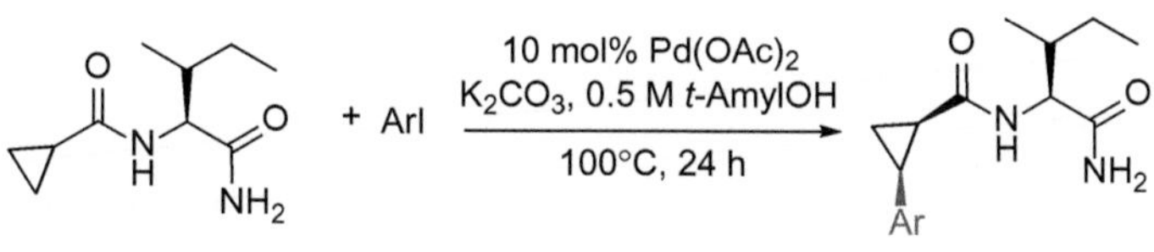

Scheme 4.43 Pd-catalyzed asymmetric C(sp^3) – H arylation using isoleucine amide as chiral auxiliary.

In 2015, the Shi group reported the diastereoselective arylation of methylene C(sp^3)—H bonds using a novel α-amino oxazolinyl bidentate auxiliary. They found that the choice of counterion could drastically affect the diastereoselectivity, and good diastereoselectivity was obtained with the addition of NaOPiv (Scheme 4.44, dr up to 9:1).[72]

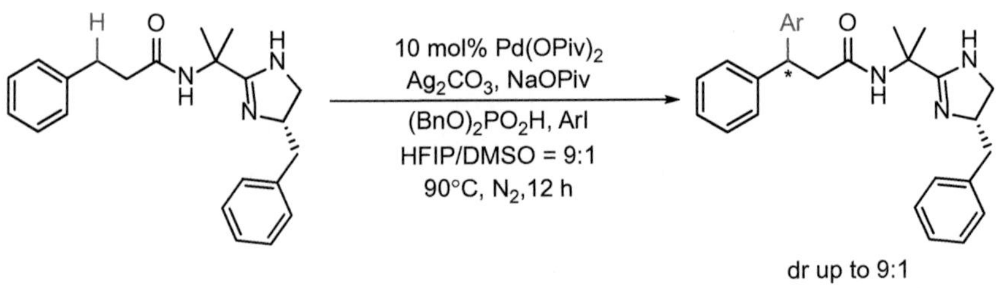

Scheme 4.44 Pd-catalyzed diastereoselective arylation of methylene $C(sp^3)$–H bonds using chiral oxazolinyl auxiliary.

4.6 CONCLUSIONS AND OUTLOOK

In conclusion, palladium-catalyzed arylation of unactivated $C(sp^3)$–H bonds directed by monodentate or bidentate DGs has been widely investigated over the past decades. These elegant methods offer powerful and straightforward strategies to synthesis of highly functionalized building blocks. However, several fundamental challenges, such as the use of high catalyst loading (5–10 mol%), elevated temperature, complicated catalytic system with stoichiometric additives or metal oxidants, and extra steps to install and remove the stoichiometric DGs, still need to be addressed.

REFERENCES

1. (a) Alberico, D.; Scott, M. E.; Lautens, M. Aryl–Aryl bond formation by transition-metal-catalyzed direct arylation. *Chem. Rev.* **2007,** *107* (1), 174–238.
(b) Seregin, I. V.; Gevorgyan, V. Direct transition metal-catalyzed functionalization of heteroaromatic compounds. *Chem. Soc. Rev.* **2007,** *36* (7), 1173–1193.
(c) Li, B.-J.; Yang, S.-D.; Shi, Z.-J. Recent advances in direct arylation via palladium-catalyzed aromatic C–H activation. *Synlett* **2008,** *2008* (07), 949–957.
(d) Daugulis, O.; Do, H.-Q.; Shabashov, D. Palladium- and copper-catalyzed arylation of carbon–hydrogen bonds. *Acc. Chem. Res.* **2009,** *42* (8), 1074–1086.
(e) Chen, X.; Engle, K. M.; Wang, D.-H., Yu, J.-Q. Palladium(II)-catalyzed C–H activation/C–C cross-coupling reactions: versatility and practicality. *Angew. Chem. Int. Ed.* **2009,** *48* (28), 5094–5115.
(f) Zhang, M.; Zhang, Y.; Jie, X.; Zhao, H.; Li, G.; Su, W. Recent advances in directed C–H functionalizations using monodentate nitrogen-based directing groups. *Org. Chem. Front.* **2014,** *1* (7), 843–895.
(g) Zhao, D.; You, J.; Hu, C. Recent progress in coupling of two heteroarenes. *Chem. Eur. J.* **2011,** *17* (20), 5466–5492.
(h) Ackermann, L. Carboxylate-assisted transition-metal-catalyzed C–H bond functionalizations: mechanism and scope. *Chem. Rev.* **2011,** *111* (3), 1315–1345.
(i) Lyons, T. W.; Sanford, M. S. Palladium-catalyzed ligand-directed C–H functionalization reactions. *Chem. Rev.* **2010,** *110* (2), 1147–1169.
(j) Colby, D. A.; Bergman, R. G.; Ellman, J. A. Rhodium-catalyzed C–C bond formation via heteroatom-directed C–H bond activation. *Chem. Rev.* **2010,** *110* (2), 624–655.

(k) Hirano, K.; Miura, M. Direct carbon—hydrogen bond functionalization of heterocyclic compounds. *Synlett* **2011,** *2011* (03), 294—307.
(l) Ackermann, L.; Vicente, R.; Kapdi, A. R. Transition-metal-catalyzed direct arylation of (hetero)arenes by C—H bond cleavage. *Angew. Chem. Int. Ed.* **2009,** *48* (52), 9792—9826.

2. (a) Yamaguchi, J.; Yamaguchi, A. D.; Itami, K. C—H bond functionalization: emerging synthetic tools for natural products and pharmaceuticals. *Angew. Chem. Int. Ed.* **2012,** *51* (36), 8960—9009.
(b) McMurray, L.; O'Hara, F.; Gaunt, M. J. Recent developments in natural product synthesis using metal-catalysed C-H bond functionalisation. *Chem. Soc. Rev.* **2011,** *40* (4), 1885—1898.
(c) Gutekunst, W. R.; Baran, P. S. C-H functionalization logic in total synthesis. *Chem. Soc. Rev.* **2011,** *40* (4), 1976—1991.

3. (a) Amaike, K.; Itami, K.; Yamaguchi, J. Synthesis of triarylpyridines in thiopeptide antibiotics by using a C—H arylation/ring-transformation strategy. *Chem. Eur. J.* **2016,** *22* (13), 4384—4388.
(b) Ueda, K.; Amaike, K.; Maceiczyk, R. M.; Itami, K.; Yamaguchi, J. β-selective C—H arylation of pyrroles leading to concise syntheses of lamellarins C and I. *J. Am. Chem. Soc.* **2014,** *136* (38), 13226—13232.

4. (a) Rouquet, G.; Chatani, N. Catalytic functionalization of C(sp^2)—H and C(sp^3)—H bonds by using bidentate directing groups. *Angew. Chem. Int. Ed.* **2013,** *52* (45), 11726—11743.
(b) Lu, X.; Xiao, B.; Shang, R.; Liu, L. Synthesis of unnatural amino acids through palladium-catalyzed C(sp^3)—H functionalization. *Chin. Chem. Lett.* **2016,** *27* (3), 305—311.
(c) He, G.; Wang, B.; Nack, W. A.; Chen, G. Syntheses and transformations of α-amino acids via palladium-catalyzed auxiliary-directed sp^3 C—H functionalization. *Acc. Chem. Res.* **2016,** *49* (4), 635—645.
(d) Rit, R. K.; Yadav, M. R.; Ghosh, K.; Sahoo, A. K. Reusable directing groups [8-aminoquinoline, picolinamide, sulfoximine] in C(sp^3)—H bond activation: present and future. *Tetrahedron* **2015,** *71* (26—27), 4450—4459.
(e) Daugulis, O.; Roane, J.; Tran, L. D. Bidentate, monoanionic auxiliary-directed functionalization of carbon-hydrogen bonds. *Acc. Chem. Res.* **2015,** *48* (4), 1053—1064.
(f) Corbet, M.; De Campo, F. 8-Aminoquinoline: a powerful directing group in metal-catalyzed direct functionalization of C-H bonds. *Angew. Chem. Int. Ed.* **2013,** *52* (38), 9896—9898.
(g) Zhang, B.; Guan, H.; Liu, B.; Shi, B.-F. Transition-metal-catalyzed arylation of unactivated C(sp^3)—H bonds assisted by bidentate directing groups. *Chin. J. Org. Chem.* **2014,** *34* (8), 1487—1498.
(h) Schnürch, M.; Dastbaravardeh, N.; Christakakou, M.; Haider, M. Recent advances in palladium-catalyzed C(sp^3)—H activation for the formation of Carbon—Carbon and Carbon—Heteroatom bonds. *Synthesis* **2014,** *46* (11), 1421—1439.
(i) He, J.; Zhao, Z. Shanghai synchrotron radiation facility. *Natl. Sci. Rev.* **2014,** *1* (2), 171—172.
(j) Fernández-Ibáñez, M. Á. Ligands control reactivity and selectivity in palladium-catalyzed functionalization of unactivated C(sp^3)—H bonds. *ChemCatChem* **2014,** *6* (8), 2188—2190.
(k) Girard, S. A.; Knauber, T.; Li, C. J. The cross-dehydrogenative coupling of C(sp^3)—H bonds: a versatile strategy for C—C bond formations. *Angew. Chem. Int. Ed.* **2014,** *53* (1), 74—100.

(l) Baudoin, O. Transition metal-catalyzed arylation of unactivated C(sp^3)—H bonds. *Chem. Soc. Rev.* **2011,** *40* (10), 4902—4911.
(m) Li, H.; Li, B.-J.; Shi, Z.-J. Challenge and progress: palladium-catalyzed sp^3 C—H activation. *Catal. Sci. Technol.* **2011,** *1* (2), 191—206.
(n) Wasa, M.; Engle, K. M.; Yu, J.-Q. Cross-coupling of C(sp^3)—H bonds with organometallic reagents via Pd(II)/Pd(0) catalysis. *Isr. J. Chem.* **2010,** *50* (5—6), 605—616.
(o) Jazzar, R.; Hitce, J.; Renaudat, A.; Sofack-Kreutzer, J.; Baudoin, O. Functionalization of organic molecules by transition-metal-catalyzed C(sp^3)—H activation. *Chem. Eur. J.* **2010,** *16* (9), 2654—2672.
(p) Shi, B.-F.; Yin, X.-S. α,α-dimethyl-2-pyridinemethanamine. *Encyclopedia of reagents for organic synthesis;* John Wiley & Sons, Ltd., 2016.

5. (a) Dick, A. R.; Hull, K. L.; Sanford, M. S. A highly selective catalytic method for the oxidative functionalization of C—H bonds. *J. Am. Chem. Soc.* **2004,** *126* (8), 2300—2301.
(b) Desai, L. V.; Hull, K. L.; Sanford, M. S. Palladium-catalyzed oxygenation of unactivated sp^3 C—H bonds. *J. Am. Chem. Soc.* **2004,** *126* (31), 9542—9543.
6. (a) Deprez, N. R.; Sanford, M. S. Synthetic and mechanistic studies of Pd-catalyzed C—H arylation with diaryliodonium salts: evidence for a bimetallic high oxidation state Pd intermediate. *J. Am. Chem. Soc.* **2009,** *131* (31), 11234—11241.
(b) Powers, D. C.; Xiao, D. Y.; Geibel, M. A. L.; Ritter, T. On the mechanism of palladium-catalyzed aromatic C—H oxidation. *J. Am. Chem. Soc.* **2010,** *132* (41), 14530—14536.
(c) Powers, D. C.; Geibel, M. A. L.; Klein, J.; Ritter, T. Bimetallic palladium catalysis: direct observation of Pd(III)-Pd(III) intermediates. *J. Am. Chem. Soc.* **2009,** *131* (47), 17050—17051.
(d) Powers, D. C.; Benitez, D.; Tkatchouk, E.; Goddard, W. A.; Ritter, T. Bimetallic reductive elimination from dinuclear Pd(III) complexes. *J. Am. Chem. Soc.* **2010,** *132* (40), 14092—14103.
7. Daugulis, O.; Zaitsev, V. G. Anilide *ortho*-arylation by using C—H activation methodology. *Angew. Chem. Int. Ed.* **2005,** *44* (26), 4046—4048.
8. Shabashov, D.; Daugulis, O. Catalytic coupling of C—H and C—I bonds using pyridine as a directing group. *Org. Lett.* **2005,** *7* (17), 3657—3659.
9. (a) Louie, J.; Hartwig, J. F. A Route to Pd^o from Pd^{II} metallacycles in animation and cross-coupling chemistry. *Angew. Chem. Int. Ed. Engl.* **1996,** *35* (20), 2359—2361.
(b) Herrmann, W. A.; Brossmer, C.; Öfele, K.; Reisinger, C.-P.; Priermeier, T.; Beller, M.; Fischer, H. Palladacycles as structurally defined catalysts for the heck olefination of chloro- and bromoarenes. *Angew. Chem. Int. Ed. Engl.* **1995,** *34* (17), 1844—1848.
(c) Beller, M.; Fischer, H.; Herrmann, W. A.; Öfele, K.; Brossmer, C. Palladacycles as Efficient Catalysts for Aryl Coupling Reactions. *Angew. Chem. Int. Ed. Engl.* **1995,** *34* (17), 1848—1849.
10. (a) Chen, X.; Goodhue, C. E.; Yu, J.-Q. Palladium-catalyzed alkylation of sp^2 and sp^3 C—H bonds with methylboroxine and alkylboronic acids: two distinct C—H activation pathways. *J. Am. Chem. Soc.* **2006,** *128* (39), 12634—12635.
(b) Chen, X.; Li, J.-J.; Hao, X.-S.; Goodhue, C. E.; Yu, J.-Q. Palladium-catalyzed alkylation of aryl C—H bonds with sp^3 organotin reagents using benzoquinone as a crucial promoter. *J. Am. Chem. Soc.* **2006,** *128* (1), 78—79.
11. Giri, R.; Maugel, N.; Li, J. J.; Wang, D. H.; Breazzano, S. P.; Saunders, L. B.; Yu, J.-Q. Palladium-catalyzed methylation and arylation of sp^2 and sp^3 C—H bonds in simple carboxylic acids. *J. Am. Chem. Soc.* **2007,** *129* (12), 3510—3511.

12. Wang, D. H.; Wasa, M.; Giri, R.; Yu, J.-Q. Pd(II)-catalyzed cross-coupling of sp^3 C—H Bonds with sp^2 and sp^3 boronic acids using air as the oxidant. *J. Am. Chem. Soc.* **2008,** *130* (23), 7190—7191.
13. Wasa, M.; Engle, K. M.; Yu, J.-Q. Pd(0)/PR_3-catalyzed intermolecular arylation of sp^3 C—H bonds. *J. Am. Chem. Soc.* **2009,** *131* (29), 9886—9887.
14. Zhang, Y.-H.; Shi, B.-F.; Yu, J.-Q. Pd(II)-catalyzed olefination of electron-deficient arenes using 2,6-dialkylpyridine ligands. *J. Am. Chem. Soc.* **2009,** *131* (14), 5072—5074.
15. Wasa, M.; Chan, K. S.; Zhang, X. G.; He, J.; Miura, M.; Yu, J.-Q. Ligand-enabled methylene C(sp^3)—H bond activation with a Pd(II) catalyst. *J. Am. Chem. Soc.* **2012,** *134* (45), 18570—18572.
16. He, J.; Li, S.; Deng, Y.; Fu, H.; Laforteza, B. N.; Spangler, J. E.; Homs, A.; Yu, J.-Q. Ligand-controlled C(sp^3)—H arylation and olefination in synthesis of unnatural chiral α-amino acids. *Science* **2014,** *343* (6176), 1216—1220.
17. Yang, W.; Ye, S.; Fanning, D.; Coon, T.; Schmidt, Y.; Krenitsky, P.; Stamos, D.; Yu, J.-Q. Orchestrated triple C—H activation reactions using two directing groups: rapid assembly of complex pyrazoles. *Angew. Chem. Int. Ed.* **2015,** *54* (8), 2501—2504.
18. (a) Wang, D.-H.; Engle, K. M.; Shi, B.-F.; Yu, J.-Q. Ligand-enabled reactivity and selectivity in a synthetically versatile aryl C—H olefination. *Science* **2010,** *327* (5963), 315—319.
    (b) Baxter, R. D.; Sale, D.; Engle, K. M.; Yu, J.-Q.; Blackmond, D. G. Mechanistic rationalization of unusual kinetics in Pd-Catalyzed C—H olefination. *J. Am. Chem. Soc.* **2012,** *134* (10), 4600—4606.
    (c) Engle, K. M.; Thuy-Boun, P. S.; Dang, M.; Yu, J.-Q. Ligand-accelerated cross-coupling of C(sp^2)—H bonds with arylboron reagents. *J. Am. Chem. Soc.* **2011,** *133* (45), 18183—18193.
    (d) Engle, K. M.; Wang, D.-H.; Yu, J.-Q. Ligand-accelerated C—H activation reactions: evidence for a switch of mechanism. *J. Am. Chem. Soc.* **2010,** *132* (40), 14137—14151.
19. Chan, K. S.; Wasa, M.; Chu, L.; Laforteza, B. N.; Miura, M.; Yu, J.-Q. Ligand-enabled cross-coupling of C(sp^3)-H bonds with arylboron reagents via Pd(II)/Pd(0) catalysis. *Nat. Chem.* **2014,** *6* (2), 146—150.
20. He, J.; Takise, R.; Fu, H.; Yu, J.-Q. Ligand-enabled cross-coupling of C(sp^3)—H bonds with arylsilanes. *J. Am. Chem. Soc.* **2015,** *137* (14), 4618—4621.
21. Chen, G.; Shigenari, T.; Jain, P.; Zhang, Z.; Jin, Z.; He, J.; Li, S.; Mapelli, C.; Miller, M. M.; Poss, M. A.; Scola, P. M.; Yeung, K. S.; Yu, J.-Q. Ligand-enabled β-C—H arylation of α-amino acids using a simple and practical auxiliary. *J. Am. Chem. Soc.* **2015,** *137* (9), 3338—3351.
22. Deng, Y.; Gong, W.; He, J.; Yu, J.-Q. Ligand-enabled triple C—H activation reactions: one-pot synthesis of diverse 4-Aryl-2-quinolinones from propionamides. *Angew. Chem. Int. Ed.* **2014,** *53* (26), 6692—6695.
23. Li, S.; Zhu, R.-Y.; Xiao, K.-J.; Yu, J.-Q. Ligand-enabled arylation of γ-C—H bonds. *Angew. Chem. Int. Ed.* **2016,** *55* (13), 4317—4321.
24. Engle, K. M.; Mei, T.-S.; Wasa, M.; Yu, J.-Q. Weak coordination as a powerful means for developing broadly useful C—H functionalization reactions. *Acc. Chem. Res.* **2011,** *45* (6), 788—802.
25. Lao, Y.-X.; Wu, J.-Q.; Chen, Y.; Zhang, S.-S.; Li, Q.; Wang, H. Palladium-catalyzed methylene C(sp^3)—H arylation of the adamantyl scaffold. *Org. Chem. Front.* **2015,** *2* (10), 1374—1378.
26. McNally, A.; Haffemayer, B.; Collins, B. S. L.; Gaunt, M. J. Palladium-catalysed C—H activation of aliphatic amines to give strained nitrogen heterocycles. *Nature* **2014,** *510* (7503), 129—133.

27. He, C.; Gaunt, M. J. Ligand-enabled catalytic C—H arylation of aliphatic amines by a four-membered-ring cyclopalladation pathway. *Angew. Chem. Int. Ed.* **2015,** *54* (52), 15840—15844.
28. (a) Shi, B.-F.; Maugel, N.; Zhang, Y.-H.; Yu, J.-Q. Pd^{II}-catalyzed enantioselective activation of $C(sp^2)$—H and $C(sp^3)$—H bonds using monoprotected amino acids as chiral ligands. *Angew. Chem. Int. Ed.* **2008,** *47* (26), 4882—4886.
    (b) Engle, K. M.; Yu, J.-Q. Developing ligands for palladium(II)-catalyzed C—H functionalization: intimate dialogue between ligand and substrate. *J. Org. Chem.* **2013,** *78* (18), 8927—8955.
    (c) Shi, B.-F.; Zhang, Y.-H.; Lam, J. K.; Wang, D.-H.; Yu, J.-Q. Pd(II)-catalyzed enantioselective C—H olefination of diphenylacetic acids. *J. Am. Chem. Soc.* **2010,** *132* (2), 460—461.
29. Calleja, J.; Pla, D.; Gorman, T. W.; Domingo, V.; Haffemayer, B.; Gaunt, M. J. A steric tethering approach enables palladium-catalysed C—H activation of primary amino alcohols. *Nat. Chem.* **2015,** 7 (12), 1009—1016.
30. Topczewski, J. J.; Cabrera, P. J.; Saper, N. I.; Sanford, M. S. Palladium-catalysed transannular C—H functionalization of alicyclic amines. *Nature* **2016,** *531* (7593), 220—224.
31. Peng, J.; Chen, C.; Xi, C. β-Arylation of oxime ethers using diaryliodonium salts through activation of inert $C(sp^3)$—H bonds using a palladium catalyst. *Chem. Sci.* **2016,** 7 (2), 1383—1387.
32. Zaitsev, V. G.; Shabashov, D.; Daugulis, O. Highly regioselective arylation of sp^3 C—H bonds catalyzed by palladium acetate. *J. Am. Chem. Soc.* **2005,** *127* (38), 13154—13155.
33. (a) Shabashov, D.; Daugulis, O. Auxiliary-assisted palladium-catalyzed arylation and alkylation of sp^2 and sp^3 carbon—hydrogen bonds. *J. Am. Chem. Soc.* **2010,** *132* (11), 3965—3972.
    (b) Nadres, E. T.; Santos, G. I. F.; Shabashov, D.; Daugulis, O. Scope and limitations of auxiliary-assisted, palladium-catalyzed arylation and alkylation of sp^2 and sp^3 C—H bonds. *J. Org. Chem.* **2013,** *78* (19), 9689—9714.
34. Reddy, B. V.; Reddy, L. R.; Corey, E. J. Novel acetoxylation and C—C coupling reactions at unactivated positions in α-amino acid derivatives. *Org. Lett.* **2006,** *8* (15), 3391—3394.
35. Tran, L. D.; Daugulis, O. Nonnatural amino acid synthesis by using carbon-hydrogen bond functionalization methodology. *Angew. Chem. Int. Ed.* **2012,** *51* (21), 5188—5191.
36. Feng, Y.; Chen, G. Total synthesis of celogentin C by stereoselective C—H activation. *Angew. Chem. Int. Ed.* **2010,** *49* (5), 958—961.
37. Feng, Y.; Wang, Y., Landgraf, B.; Liu, S.; Chen, G. Facile benzo-ring construction via palladium-catalyzed functionalization of unactivated sp^3 C—H bonds under mild reaction conditions. *Org. Lett.* **2010,** *12* (15), 3414—3417.
38. (a) Zhang, Q.; Chen, K.; Rao, W.; Zhang, Y.; Chen, F. J.; Shi, B.-F. Stereoselective synthesis of chiral α-amino-β-lactams through palladium(II)-catalyzed sequential monoarylation/amidation of $C(sp^3)$—H bonds. *Angew. Chem. Int. Ed.* **2013,** *52* (51), 13588—13592.
    (b) Chen, F.-J.; Zhao, S.; Hu, F.; Chen, K.; Zhang, Q.; Zhang, S.-Q.; Shi, B.-F. Pd (II)-catalyzed alkoxylation of unactivated $C(sp^3)$—H and $C(sp^2)$—H bonds using a removable directing group: efficient synthesis of alkyl ethers. *Chem. Sci.* **2013,** *4* (11), 4187—4192.
    (c) Zhang, Q.; Yin, X. S.; Zhao, S.; Fang, S. L.; Shi, B.-F. Pd(II)-Catalyzed arylation of unactivated methylene $C(sp^3)$—H bonds with aryl halides using a removable auxiliary. *Chem. Commun.* **2014,** *50* (61), 8353—8355.

(d) Zhang, Q.; Yin, X.-S.; Chen, K.; Zhang, S.-Q.; Shi, B.-F. Stereoselective synthesis of chiral β-fluoro α-amino acids via Pd(II)-catalyzed fluorination of unactivated methylene C(sp^3)—H bonds: scope and mechanistic studies. *J. Am. Chem. Soc.* **2015,** *137* (25), 8219—8226.

39. Pan, F.; Shen, P. X.; Zhang, L. S.; Wang, X.; Shi, Z.-J. Direct arylation of primary and secondary sp^3 C—H bonds with diarylhyperiodonium salts via Pd catalysis. *Org. Lett.* **2013,** *15* (18), 4758—4761.
40. Wei, Y.; Tang, H.; Cong, X.; Rao, B.; Wu, C.; Zeng, X. Pd(II)-catalyzed intermolecular arylation of unactivated C(sp^3)—H bonds with aryl bromides enabled by 8-aminoquinoline auxiliary. *Org. Lett.* **2014,** *16* (8), 2248—2251.
41. Wang, B.; Nack, W. A.; He, G.; Zhang, S.-Y.; Chen, G. Palladium-catalyzed trifluoroacetate-promoted mono-arylation of the β-methyl group of alanine at room temperature: synthesis of β-arylated α-amino acids through sequential C—H functionalization. *Chem. Sci.* **2014,** *5* (10), 3952—3957.
42. Chen, K.; Zhang, S.-Q.; Xu, J.-W.; Hu, F.; Shi, B.-F. A general and practical palladium-catalyzed monoarylation of β-methyl C(sp^3)—H of alanine. *Chem. Commun.* **2014,** *50* (90), 13924—13927.
43. Zhang, G.; Xie, X.; Zhu, J.; Li, S.; Ding, C.; Ding, P. Pd(II)-catalyzed C(sp^3)—H arylation of amino acid derivatives with click-triazoles as removable directing groups. *Org. Biomol. Chem.* **2015,** *13* (19), 5444—5449.
44. He, G.; Zhang, S.-Y.; Nack, W. A.; Pearson, R.; Rabb-Lynch, J.; Chen, G. Total synthesis of hibispeptin A via Pd-Catalyzed C(sp^3)—H arylation with sterically hindered aryl iodides. *Org. Lett.* **2014,** *16* (24), 6488—6491.
45. Zhang, X.; He, G.; Chen, G. Palladium-catalyzed β-C(sp^3)—H arylation of phthaloyl alanine with hindered aryl iodides: synthesis of complex β-aryl α-amino acids. *Org. Biomol. Chem.* **2016,** *14* (24), 5511—5515.
46. (a) Zhang, Y. F.; Zhao, H. W.; Wang, H.; Wei, J. B.; Shi, Z.-J. Readily removable directing group assisted chemo- and regioselective C(sp^3)—H activation by palladium catalysis. *Angew. Chem. Int. Ed.* **2015,** *54* (46), 13686—13690.
    (b) Gutekunst, W. R.; Baran, P. S. Applications of C—H functionalization logic to cyclobutane synthesis. *J. Org. Chem.* **2014,** *79* (6), 2430—2452.
47. Gong, W.; Zhang, G.; Liu, T.; Giri, R.; Yu, J.-Q. Site-selective C(sp^3)—H functionalization of di-, tri-, and tetrapeptides at the N-terminus. *J. Am. Chem. Soc.* **2014,** *136* (48), 16940—16946.
48. Toba, T.; Hu, Y.; Tran, A. T.; Yu, J.-Q. β-C(sp^3)—H arylation of alpha-hydroxy acid derivatives utilizing amino acid as a directing group. *Org. Lett.* **2015,** *17* (24), 5966—5969.
49. (a) Dailler, D.; Danoun, G.; Baudoin, O. A general and scalable synthesis of aeruginosin marine natural products based on two strategic C(sp^3)—H activation reactions. *Angew. Chem. Int. Ed.* **2015,** *54* (16), 4919—4922.
    (b) Chen, K.; Li, X.; Zhang, S.-Q.; Shi, B.-F. Palladium-catalyzed C(sp^3)—H arylation of lactic acid: efficient synthesis of chiral β-aryl-α-hydroxy acids. *Org. Chem. Front.* **2016,** *3* (2), 204—208.
50. Zhang, S.-K.; Yang, X.-Y.; Zhao, X.-M.; Li, P.-X.; Niu, J.-L.; Song, M.-P. Palladium-catalyzed selective β-arylation of aliphatic amides using a removable *N,O*-bidentate auxiliary. *Organometallics* **2015,** *34* (17), 4331—4339.
51. Liu, J.; Xie, Y.; Zeng, W.; Lin, D.; Deng, Y.; Lu, X. Pd(II)-catalyzed pyridine *N*-oxides directed arylation of unactivated C(sp^3)—H bonds. *J. Org. Chem.* **2015,** *80* (9), 4618—4626.
52. He, G.; Chen, G. A practical strategy for the structural diversification of aliphatic scaffolds through the palladium-catalyzed picolinamide-directed remote functionalization of unactivated C(sp^3)—H bonds. *Angew. Chem. Int. Ed.* **2011,** *50* (22), 5192—5196.

53. Rodríguez, N.; Romero-Revilla, J. A.; Fernández-Ibáñez, M. Á.; Carretero, J. C. Palladium-catalyzed N-(2-pyridyl)sulfonyl-directed $C(sp^3)-H$ γ-arylation of amino acid derivatives. *Chem. Sci.* **2013,** *4* (1), 175−179.
54. Fan, M.; Ma, D. Palladium-catalyzed direct functionalization of 2-aminobutanoic acid derivatives: application of a convenient and versatile auxiliary. *Angew. Chem. Int. Ed.* **2013,** *52* (46), 12152−12155.
55. (a) Parella, R.; Gopalakrishnan, B.; Babu, S. A. Auxiliary-enabled Pd-catalyzed direct arylation of methylene $C(sp^3)-H$ bond of cyclopropanes: highly diastereoselective assembling of di- and trisubstituted cyclopropanecarboxamides. *Org. Lett.* **2013,** *15* (13), 3238−3241.
(b) Hoshiya, N.; Kobayashi, T.; Arisawa, M.; Shuto, S. Palladium-catalyzed arylation of cyclopropanes via directing group-mediated $C(sp^3)-H$ bond activation to construct quaternary carbon centers: synthesis of *cis*- and *trans*-1,1,2-trisubstituted chiral cyclopropanes. *Org. Lett.* **2013,** *15* (24), 6202−6205.
56. (a) Affron, D. P.; Davis, O. A.; Bull, J. A. Regio- and stereospecific synthesis of C-3 functionalized proline derivatives by palladium catalyzed directed $C(sp^3)-H$ arylation. *Org. Lett.* **2014,** *16* (18), 4956−4959.
(b) Feng, R.; Wang, B.; Liu, Y.; Liu, Z.; Zhang, Y. Efficient synthesis of cis-3-substituted prolines by bidentate-assisted palladium catalysis. *Eur. J. Org. Chem.* **2015,** *2015* (1), 142−151.
57. Roman, D. S.; Charette, A. B. C−H functionalization of cyclopropanes: a practical approach employing a picolinamide auxiliary. *Org. Lett.* **2013,** *15* (17), 4394−4397.
58. Cui, W.; Chen, S.; Wu, J.-Q.; Zhao, X.; Hu, W.; Wang, H. Palladium-catalyzed remote $C(sp^3)-H$ arylation of 3-pinanamine. *Org. Lett.* **2014,** *16* (16), 4288−4291.
59. Xu, J.-W.; Zhang, Z.-Z.; Rao, W.-H.; Shi, B.-F. Site-selective alkenylation of δ-$C(sp^3)-H$ bonds with alkynes via a six-membered palladacycle. *J. Am. Chem. Soc.* **2016,** *138* (34), 10750−10753.
60. Fan, Z.; Shu, S.; Ni, J.; Yao, Q.; Zhang, A. Ligand-promoted Pd(II)-catalyzed functionalization of unactivated $C(sp^3)-H$ bond: regio- and stereoselective synthesis of arylated rimantadine derivatives. *ACS Catal.* **2016,** *6* (2), 769−774.
61. Ling, P. X.; Fang, S. L.; Yin, X. S.; Chen, K.; Sun, B. Z.; Shi, B.-F. Palladium-catalyzed arylation of unactivated γ-methylene $C(sp^3)-H$ and δ-C−H bonds with an oxazoline-carboxylate auxiliary. *Chem. Eur. J.* **2015,** *21* (48), 17503−17507.
62. Pasunooti, K. K.; Banerjee, B.; Yap, T.; Jiang, Y.; Liu, C.-F. Auxiliary-directed Pd-catalyzed γ-$C(sp^3)-H$ bond activation of α-aminobutanoic acid derivatives. *Org. Lett.* **2015,** *17* (24), 6094−6097.
63. Han, J.; Zheng, Y.; Wang, C.; Zhu, Y.; Huang, Z.-B.; Shi, D.-Q.; Zeng, R.; Zhao, Y. Pd-catalyzed coupling of γ-$C(sp^3)-H$ bonds of oxalyl amide-protected amino acids with heteroaryl and aryl iodides. *J. Org. Chem.* **2016,** *81* (13), 5681−5689.
64. Zhang, F. L.; Hong, K.; Li, T. J.; Park, H.; Yu, J.-Q. Functionalization of $C(sp^3)-H$ bonds using a transient directing group. *Science* **2016,** *351* (6270), 252−256.
65. Ma, F.; Lei, M.; Hu, L. Acetohydrazone: a transient directing group for arylation of unactivated $C(sp^3)-H$ bonds. *Org. Lett.* **2016,** *18* (11), 2708−2711.
66. Wasa, M.; Engle, K. M.; Lin, D. W.; Yoo, E. J.; Yu, J.-Q. Pd(II)-catalyzed enantioselective C−H activation of cyclopropanes. *J. Am. Chem. Soc.* **2011,** *133* (49), 19598−19601.
67. Xiao, K. J.; Lin, D. W.; Miura, M.; Zhu, R. Y.; Gong, W.; Wasa, M.; Yu, J.-Q. Palladium(II)-catalyzed enantioselective $C(sp^3)-H$ activation using a chiral hydroxamic acid ligand. *J. Am. Chem. Soc.* **2014,** *136* (22), 8138−8142.
68. Chan, K. S.; Fu, H. Y.; Yu, J.-Q. Palladium(II)-catalyzed highly enantioselective C−H arylation of cyclopropylmethylamines. *J. Am. Chem. Soc.* **2015,** *137* (5), 2042−2046.

69. Yan, S. B.; Zhang, S.; Duan, W. L. Palladium-catalyzed asymmetric arylation of $C(sp^3)$—H bonds of aliphatic amides: controlling enantioselectivity using chiral phosphoric amides/acids. *Org. Lett.* **2015,** *17* (10), 2458–2461.
70. Giri, R.; Chen, X.; Yu, J.-Q. Palladium-catalyzed asymmetric iodination of unactivated C—H bonds under mild conditions. *Angew. Chem. Int. Ed.* **2005,** *44* (14), 2112–2115.
71. Kim, J.; Sim, M.; Kim, N.; Hong, S. Asymmetric C—H functionalization of cyclopropanes using an isoleucine-NH_2 bidentate directing group. *Chem. Sci.* **2015,** *6* (6), 3611–3616.
72. Chen, K.; Li, Z. W.; Shen, P. X.; Zhao, H. W.; Shi, Z.-J. Development of modifiable bidentate amino oxazoline directing group for Pd-catalyzed arylation of secondary C—H bonds. *Chem. Eur. J.* **2015,** *21* (20), 7389–7393.

CHAPTER 5

C—H Bond Functionalization at the Benzene Core of Indoles and Indolines (Excluding C-2 and C-3)

Lin-Yu Jiao[1] and Martin Oestreich[2]
[1]Northwest University, Xi'an, Shaanxi, PR China
[2]Technical University of Berlin, Berlin, Germany

Contents

In the area of C—C or C—Heteroatom bond-forming reactions that proceed by transition metal-catalyzed C—H bond activation, palladium catalysts stand out for their functional-group tolerance and, hence, broad synthetic utility.[1–10] Owing to the prevalence of heterocyclic compounds in synthetic chemistry, the development of ways to make, as well as modify, heterocycles continues to be an active area of research.

As a heterocycle frequently found in natural and man-made biologically active compounds, indole-nuclei-containing structures have garnered tremendous attention during the past decades.[1,11] Thus, efficient functionalization of indole derivatives continues to attract attention from both academia and industry, especially with regard to C—H bond activation based on transition metal catalysis. Consequently, significant efforts have been devoted toward not only developing new strategies for their facile synthesis, but also modifying these structures.[12,13] Over the course of the

Strategies for Palladium-Catalyzed Non-Directed and Directed C—H Bond Functionalization
DOI: http://dx.doi.org/10.1016/B978-0-12-805254-9.00005-0

past few years, several methods for directed C—H bond functionalization at the benzene core of indoles and indolines have been realized.[11,14]

This chapter describes the recent advances in this rapidly developing field under palladium catalysis, yet it does not aim to be comprehensive. As palladium-catalyzed C—H bond functionalization at the more reactive C-2 or C-3 position has been competently summarized recently,[15–17] it is not covered in this chapter. We nevertheless hope that this chapter will serve as a handy reference for chemists interested in developing novel approaches for C—C bond construction at existing indole-nuclei.

5.1 PALLADIUM-CATALYZED C—H BOND FUNCTIONALIZATION OF INDOLE

Prior to the popularization of transition metal-catalyzed C—H bond activation, approaches to build new C—C bond and C—Heteroatom bonds at the benzene ring of indole involved prefunctionalization in the form of iodides, bromides, and triflates as the electrophilic coupling partners.[18] Several established transformations were applied, e.g., the Suzuki—Miyaura reaction using arylboronic acids,[19] the Negishi coupling using organozinc compounds,[20] as well as Buchwald—Hartwig aminations by the use of free amines.[21–23]

5.1.1 C—H Bond Alkenylation of Indole at the C-4 Position

An often-used traditional method for the construction of conjugated structures is the Mizoroki—Heck reaction, that is the cross-coupling between aryl (or vinyl) (pseudo-)halides and, usually, terminal alkenes to afford 1,2-disubstituted alkenes with exclusive *trans* selectivity under non-oxidative conditions.[24–27] Alternatively, the same net transformation is achieved by the use of unactivated aromatic C—H bonds under oxidative conditions, known as the Fujiwara—Moritani reaction or oxidative Heck reaction.

4-Substituted indoles, 4-alkenylated indoles in particular, are important intermediates for the preparation of ergot alkaloids and related heterocyclic compounds.[28–30] In 2008, the Yu group disclosed a palladium (II)-catalyzed one-pot intramolecular amination of arylethylamines into indolines and tetrahydroisoquinolines.[31] In this report, a twofold Fujiwara—Moritani reaction at both the C-2 and C-4 positions of indole was included as an isolated example. Compared to the C-4 position, the C—H bond at the C-2 position was more reactive. With $Pd(OAc)_2$ as the catalyst, AgOAc as the oxidant was effective for promoting this oxidative dialkenylation in DCE/DMF (20:1) as the solvent combination in moderate yield (Fig. 5.1).

Based on the aforementioned observation, a great deal of effort has been devoted to developing the selective C—H bond functionalization at the C-4 position of indole substrates.

Inspired by Yu's work, Jia and co-workers realized a selective alkenylation of tryptophan under palladium catalysis.[32] Regioselective C-4 oxidative alkenylation of *N*-TIPS protected tryptophan was achieved through an NHTf directing-group strategy,[33–39] which led to full conversions in acceptable yields. The protecting group, solvent, reaction temperature, and catalyst loading were crucial for success. The bulky TIPS group at the indole nitrogen atom was superior to other protecting groups such as Piv and Boc. In addition, toluene improved the yield dramatically after degassing. This method exhibits broad scope and tolerates several functional groups (Fig. 5.2). The utility of this selective alkenylation reaction

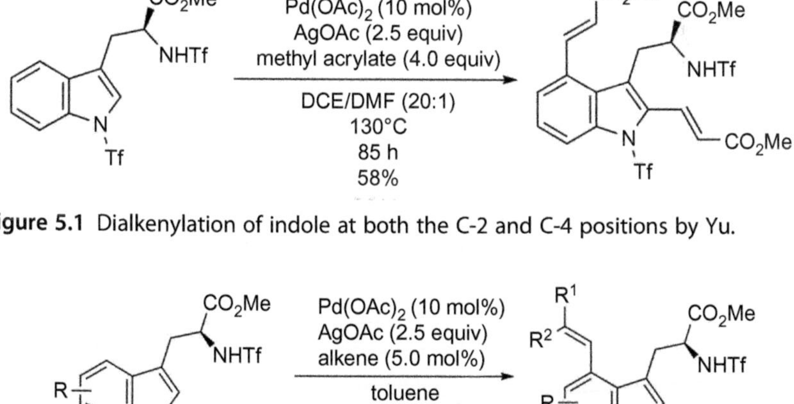

Figure 5.1 Dialkenylation of indole at both the C-2 and C-4 positions by Yu.

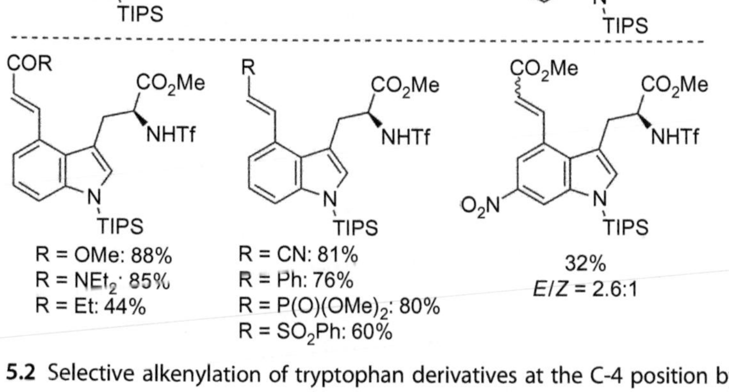

Figure 5.2 Selective alkenylation of tryptophan derivatives at the C-4 position by Jia.

was demonstrated in the total synthesis of (+)-*cis*- and (−)-*trans*-Clavicipitic acid[40–43] hemiterpene indole alkaloids derived from tryptophan (not shown).

5.1.2 C—H Bond Arylation and Alkenylation of Indole at the C-7 Position

As mentioned before, indole contains other more reactive sites than the C—H bond at C-7. Therefore, addressing this position represents a great challenge. As a consequence, to exclude the possibility of C—H bond activation at C-2 and C-3, indolines were used as precursors that could subsequently be converted into the corresponding indoles by facile oxidation. However, considerable effort was not only devoted to indolines, but also to new directing groups at the nitrogen atom to make indoles susceptible to site-selective C—H functionalization at C-7.

Notably, Shi and co-workers recently reported a highly selective, directed arylation of indoles at the C-7 position under palladium(II) catalysis;[44] arylboronic acids were chosen as coupling partners. An easily removable, bulky phosphinoyl directing group, which had been shown to be effective in the C—H bond alkenylation of anilines by the Oestreich group shortly before,[45] provided the desired C-7-arylated indole with good-to-excellent yield and regioselectivity (Fig. 5.3). A screening of directing groups was critical, as it seemed that steric effects and electronic biases favoring the C—H bonds at C-2, C-3, and C-7 needed to be well balanced. Other directing groups, however, exhibited moderate or even no directing effect toward the C-7 position. The pyridine-type ligand was also found to be crucial.

Additionally, with the aid of the same directing group, the alkenylation was also realized regioselectively by the same group.[44] An unsubstituted indole underwent alkenylation in moderate yield and with decent regioselectivity (C-7:C-3 = 86:14). After the introduction of a substituent at the C-3 position, C-7 alkenylation occurred exclusively, further highlighting the utility of the novel directing group (Fig. 5.4).

5.2 PALLADIUM-CATALYZED C—H BOND FUNCTIONALIZATION OF INDOLINES

As part of this promising research area the modification of indolines, especially at the C-4, C-6, and C-7 positions, is synthetically attractive but had been less intensively studied previously. Parallel to the rapid

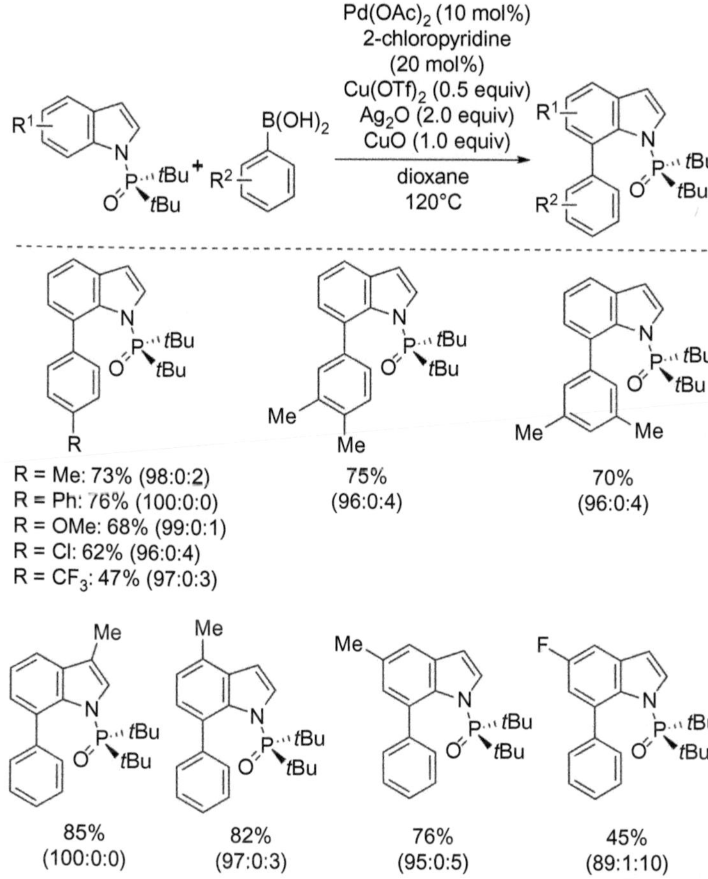

Figure 5.3 Selective C—H bond arylation of indole at the C-7 position with the aid of a bulky phosphinoyl directing group by Shi. (Ratio in the parentheses refers to C-7:C-2:C-3 arylation.)

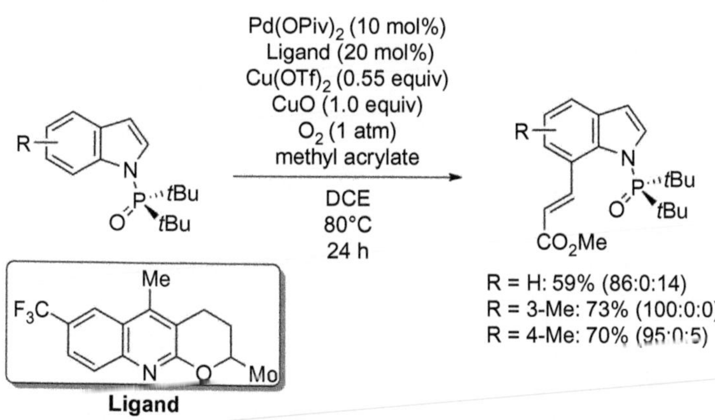

Figure 5.4 Selective C—H bond alkenylation of indole at the C-7 position under palladium catalysis by Shi. (Ratio in the parentheses refers to C-7:C-2:C-3 alkenylation.)

development of organometallic chemistry in the past years, hundreds of catalyst/directing group/oxidant combinations have been discovered and applied, also enabling transition metal-catalyzed indoline C—H bond activation at C-7.[9,13] Accordingly, various protocols were elaborated, and approaches related to palladium catalysis are discussed here.

5.2.1 C—H Bond Functionalization at the C-7 Position of Indoline

As the most basic transformation, the palladium-catalyzed C—C bond coupling of indoline at the C-7 position comprises several kinds of reactions such as $C(sp^2)$—$C(sp^2)$ bond formation (arylation/alkenylation as well as acylation/carbonylation), and $C(sp^2)$—$C(sp^3)$ bond formation (alkylation). These are individually present in the following sections.

5.2.1.1 C—H Bond Arylation of Indoline at the C-7 Position

The conventional procedures for the arylation at the C-7 position of indoline mainly involve commonly used organometallic reagents. More than two decades ago, Flippin and co-workers established a protocol for the construction of biaryl compounds at room temperature under Ullmann reaction conditions.[46] In 2009, Leonori and Coldham reported a similar arylation through a Negishi-type cross-coupling reaction.[47] Despite the attractiveness of the transmetalation pathway, the intramolecular C-7 arylation of indoline was also achieved by homolytic aromatic substitution (HAS) promoted by KO*t*Bu in a transition metal-free fashion, as disclosed by Bisai and co-workers in 2012.[48] It was observed that KO*t*Bu was solely responsible for the biaryl coupling mediated by aryl radicals involving radical anions.

About a decade ago, the Sanford group accomplished a single example of a direct oxidative Suzuki—Miyaura coupling between an unfunctionalized indoline and a hypervalent iodine(III) reagent under acidic conditions at 100°C (Fig. 5.5).[49] This example was an advance, but the yield was moderate. The authors employed the widely used acetyl group as the directing group to control the site-selectivity in favor of the C-7 position. Based on experimental observations, a palladium(II)/palladium(IV) catalytic cycle was proposed, in which the hypervalent iodine(III) reagent plays a dual role, both as coupling partner and terminal oxidant to generate the palladium(IV) intermediate.

Two years later, Shi and co-workers reported bond formation between *N*-acetyl indoline and phenyl boronic acid catalyzed by $Pd(OAc)_2$ at 120°C.

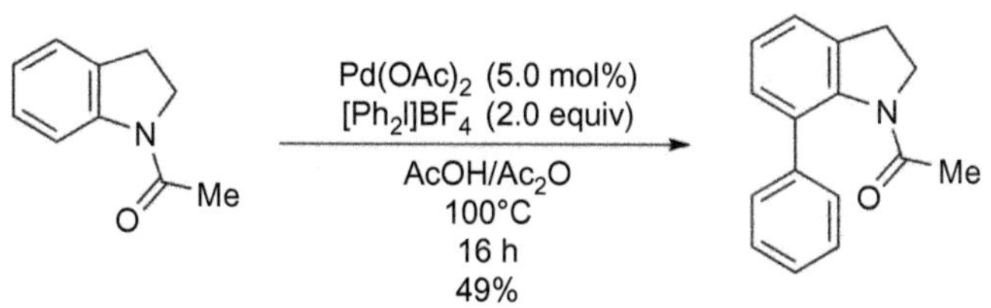

Figure 5.5 Oxidative Suzuki—Miyaura arylation between an indoline and a hypervalent iodine(III) reagent by Sanford.

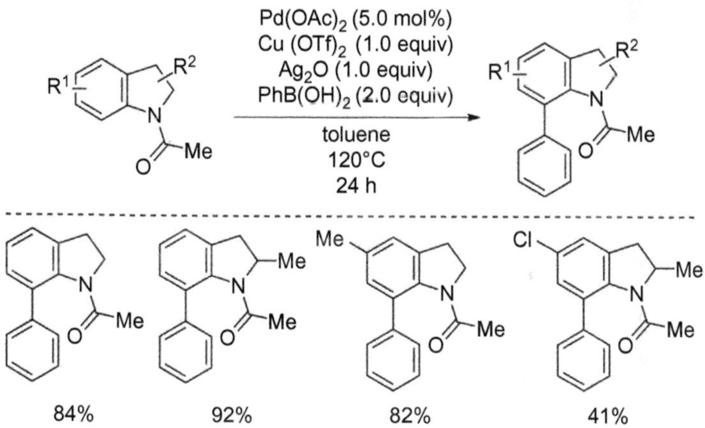

Figure 5.6 Oxidative Suzuki—Miyaura arylation between indolines with an acetyl directing group and phenyl boronic acids by Shi.

$Cu(OAc)_2$ and Ag_2O served as oxidant and co-oxidant, respectively, to promote this transformation efficiently in 84% yield.[50] The more hindered 2-methyl indoline reacted in excellent yield, but indolines bearing electron-withdrawing groups on the benzene core gave moderate yields (Fig. 5.6). Preliminary investigations into the reaction mechanism were also part of this work.

Similar but more exciting progress was made by Lipshutz and coworkers in 2010. In this case, the use of a carbamoyl instead of an acetyl group as a directing group made the Suzuki—Miyaura arylation proceed under exceptionally mild conditions (room temperature) without loss of reactivity (Fig. 5.7).[51] The urea directing group had been introduced before by the Lloyd—Jones and Booker—Milburn groups in the carboamination of dienes involving C—H bond activation,[52] and the C—H bond methoxycarbonylation of aniline derivatives.[53] Later, this directing group was independently utilized by other groups in C—H bond

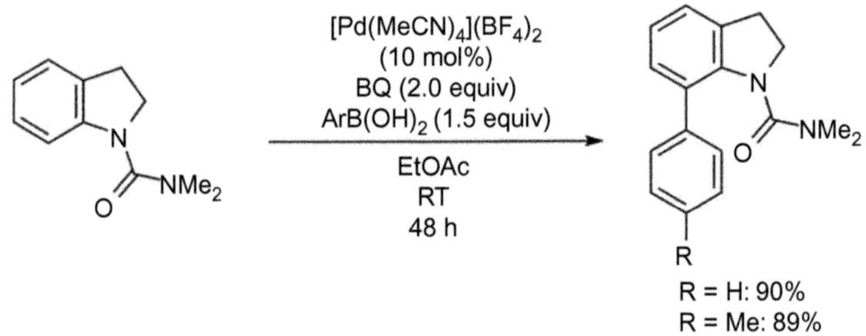

Figure 5.7 Oxidative Suzuki–Miyaura arylation between indolines containing a urea group and phenyl boronic acids catalyzed by a cationic palladium(II) salt by Lipshutz.

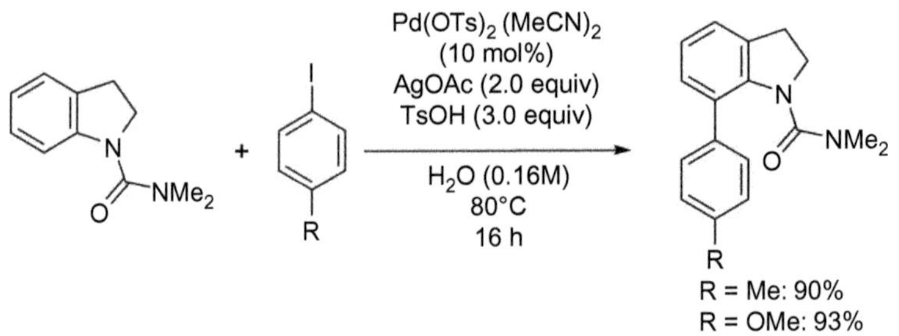

Figure 5.8 Oxidative C-7 arylation between parent indoline and aryl iodides in aqueous medium by Fan.

functionalization.[54–58] It is noteworthy that palladium salts other than cationic $[Pd(MeCN)_4](BF_4)_2$ were ineffective. 1,4-Benzoquinone (BQ) was used as the terminal oxidant and, at the same time, to promote the reductive elimination step. Compared to the acetyl directing group, installation of this urea group and subsequent removal were straightforward, and this enhanced its synthetic practicality.

Aside from Suzuki–Miyaura-type reactions, the high-yielding reaction between parent indoline and aryl iodides was also disclosed recently by the Fan group (Fig. 5.8).[59] The catalysis proceeded smoothly in aqueous medium and was found to be compatible with different functional groups at the arene coupling partner. However, no reaction was observed when aryl bromides were used instead. This methodology extended the coupling partner scope of C-7 arylation of indoline.

In these above-mentioned examples, at least one coupling partner required prefunctionalization.[60] Examples of cross-dehydrogenative couplings[61] (CDCs) had been elusive at that time. In 2013, Oestreich and co-workers accomplished such arylation of indolines at the C-7 position with a CDC protocol.[62] $Na_2S_2O_8$ was chosen as the optimal oxidant, and the combination of $Pd(OAc)_2$ as catalyst with TFA as additive favored the formation of the target compounds. Several di- and trisubstituted indolines were subjected to the optimized conditions in moderate to excellent yields (Fig. 5.9). However, the procedure required harsh reaction conditions, under which less substituted indolines would undergo oxidation to the corresponding indoles.

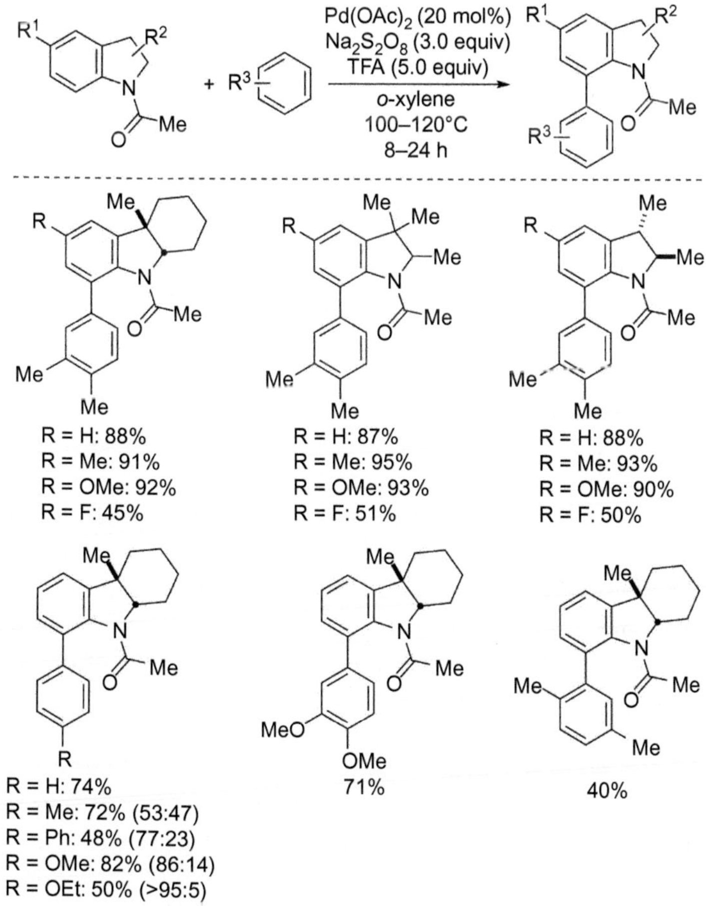

Figure 5.9 Palladium(II)-catalyzed cross-dehydrogenative arylation of indolines at the C-7 position mediated by $Na_2S_2O_8$ as terminal oxidant by Oestreich. (Ratio in the parentheses refers to *para:meta* arylation.)

Figure 5.10 Amide rotamer equilibrium controlled by repulsion and attraction.

Interestingly, the acetyl directing group alone is not sufficient to allow for this transformation. NMR experiments revealed that the substituent at the C-2 position was a critical factor for reactivity and regiocontrol. Indoline and 3-methyl indoline as substrates were reluctant to yield the desired coupling. Instead, indoline-to-indole oxidation was observed under oxidative reaction conditions. Accordingly, these authors proposed an attraction (Fig. 5.10, left) and repulsion (Fig. 5.10, right) equilibrium. The repulsion in one rotamer between the lone electron pair of the amide oxygen atom and the C-2 substituent pushes the amide isomer toward the other rotamer, in which the same oxygen atom approaches the C—H bond at the C-7 position of the indoline to engage in hydrogen bonding. This proposal was supported by assignment of both rotamers through chemical exchange in 2D EXSY and 2D NOESY NMR measurements.

The limitations mentioned above were overcome by the same group shortly thereafter by using a urea directing group instead of the acetyl group, which enabled the same transformation under much milder conditions (50°C vs 100°C or higher).[63] As a result, the use of either Cu $(OAc)_2$ in an open flask or dioxygen (balloon) tolerated simple indolines without substitution at the C-2 or C-3 position, thereby substantially expanding the substrate scope to more general motifs. This methodology was then successfully scaled-up to gram scale with slightly diminished conversion and yield (Fig. 5.11). It is worthy of mention that other common directing groups, even a urea with a free NH group at the terminus, failed to facilitate the desired C—H bond arylation.

5.2.1.2 C—H Bond Alkenylation of Indoline at the C-7 Position

Compared with the aforementioned cross-dehydrogenative arylation of indolines at the C-7 position, the direct C—H bond alkenylation was even rarer previously, but it has been well studied in the past 3 years.

Early examples were disclosed by Somei and co-workers in the 1980s.[64,65] To achieve this goal, the authors designed an interesting strategy with the use of a suitable "ligand" at the nitrogen atom of the

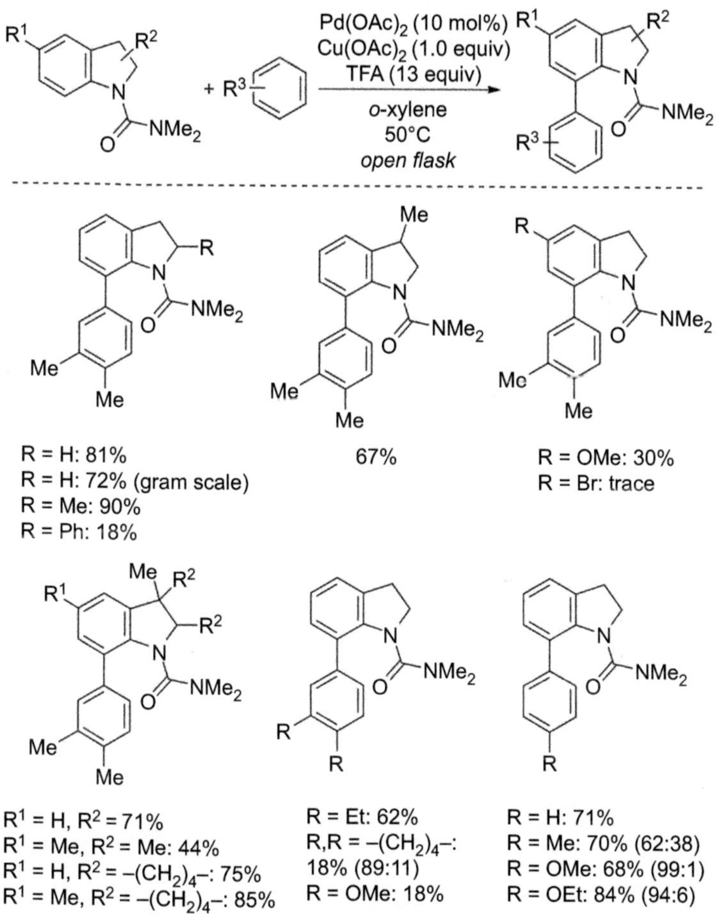

Figure 5.11 Palladium(II)-catalyzed cross-dehydrogenative arylation of indolines at the C-7 position mediated by $Cu(OAc)_2$ or dioxygen as terminal oxidant by Oestreich. (Ratio in the parentheses refers to *para:meta* arylation.)

indoline substrate. Its purpose was to coordinate to a metal and guide the metal close to the C—H bond at the C-7 position. Based on this proposal, the acetyl group was found to be suitable. The indoline underwent thallation with $Tl(TFA)_3$ followed by the addition of a catalytic amount of $Pd(OAc)_2$ and an electron-deficient alkene as coupling partner. The yield was, however, only 4%.

In early 2013, Arrayás and Carretero reported an oxidative alkenylation of carbazoles.[66] In this report, one example of an indoline was included. Their protocol relied on a 2-pyridylsulfonyl group connected to the nitrogen atom as a directing group. $Pd(OAc)_2$ (10 mol%) and *N*-fluoro-2,4,6-trimethylpyridinium triflate (2.0 equiv) were used as catalyst and terminal oxidant, respectively, in DCE at 110°C (Fig. 5.12).

By the use of the previously employed urea directing group, a protocol for the direct alkenylation of indolines at the C-7 position was disclosed by the Oestreich group shortly afterwards.[67] This catalytic system proved to be broadly applicable, mild, and efficient. Both α,β-unsaturated acceptors and styrenes participated smoothly in this direct C—H bond functionalization. Interestingly, substituted α,β-unsaturated acceptors also reacted, providing good-to-excellent regioselectivities, albeit yields were somewhat lower (Fig. 5.13).

Notably, the reactive urea with a free NH group at its terminus also reacted under the optimized reaction conditions. However, instead of forming the desired alkenylated indoline, a six-membered ring compound was isolated (Fig. 5.14). It was assumed that the starting material underwent a domino procedure involving an intermolecular C—H bond alkenylation followed by an intramolecular C—N bond formation by conjuagte addition. The connectivities were assigned by nOe measurements. This observation was in accordance with a few former reports.[68–70]

In 2015, a similar work by Wang and co-workers appeared using the same strategy, in which molecular oxygen was employed as the terminal oxidant.[71] A 2-pyridylsulfonyl group was used as the directing group, exclusively furnishing the C-7-alkenylated indolines. The catalytic system

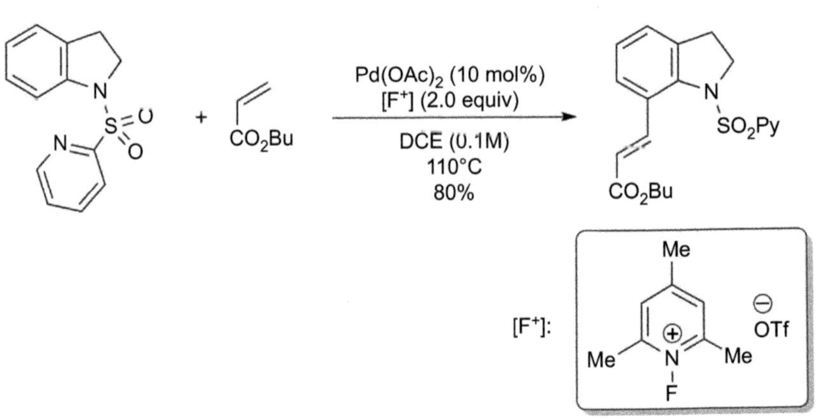

Figure 5.12 C-7 alkenylation of indoline directed by a 2-pyridylsulfonyl group by Arrayás and Carretero.

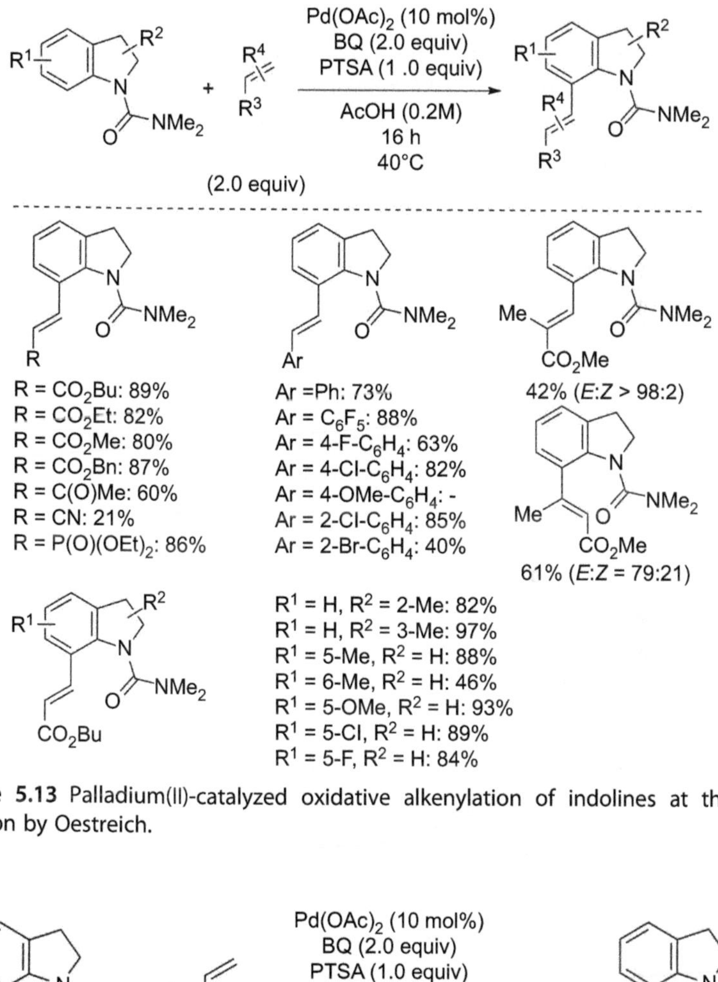

Figure 5.13 Palladium(II)-catalyzed oxidative alkenylation of indolines at the C-7 position by Oestreich.

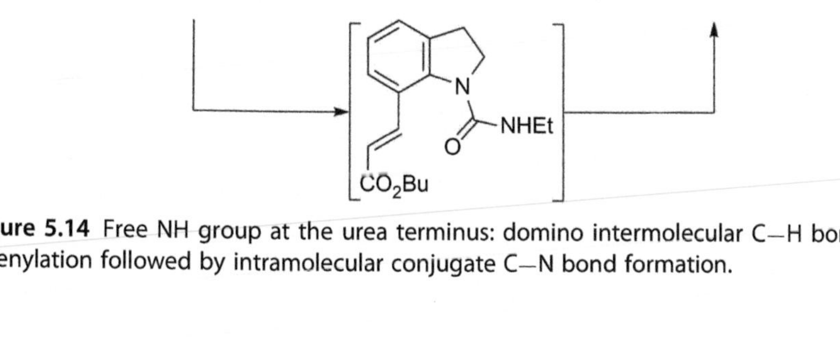

Figure 5.14 Free NH group at the urea terminus: domino intermolecular C–H bond alkenylation followed by intramolecular conjugate C–N bond formation.

tolerated a variety of alkenes, including both mono- and disubstituted unsaturated compounds, in spite of steric effects (Fig. 5.15). It is interesting to note that with the aid of TFA as an additive, catalytically active Pd $(O_2CCF_3)^+$ was easily generated in situ, believed to undergo cyclopalladation through coordination of palladium(II) to the nitrogen atom of the directing group.

Additional examples with other transition metals such as rhodium(III) (the Loh group[72] and the Antonchick group[73]) and iridium(III) (the Shibata group[74]) appeared in the literature. Directing groups such as acetyl, carbamoyl, pyridyl, and 2-pyridylsulfonyl are all well tolerated.

5.2.1.3 C—H Bond Acylation of Indoline at the C-7 Position

Acylated indole nuclei are other common functional and structural motifs existing in numerous naturally occurring and synthetic molecules.[75–78] For C-3-acylated indoles, the conventional preparation methods typically involve Friedel—Crafts-type reactions,[79,80] Vilsmeier—Haack reactions,[81] or electrophilic metalation. However, the C-2-acylated indole is usually obtained through transition metal-catalyzed C—H bond activation by using

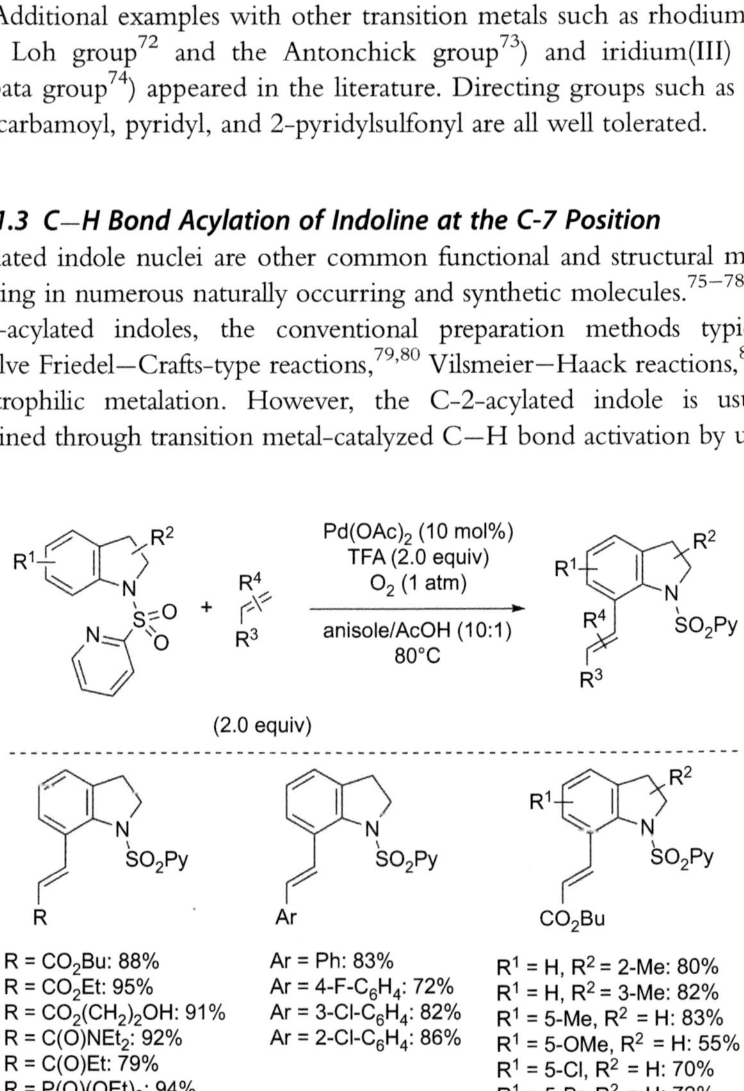

Figure 5.15 Palladium-catalyzed direct C-7 alkenylation of indolines using O_2 as oxidant by Wang.

aldehydes[82] or α-keto acids[83] as the acylation reagents. Interestingly, the C—H bond acylation at the C-7 position of indole is again challenging, and it is difficult to override the inherent C-3 as well as C-2 selectivity.

In turn, this challenging transformation is again within reach through C-7-selective acylation of indolines followed by facile indoline-to-indole oxidation. Recently, a decarboxylative acylation of indolines for the synthesis of C-7-carbonylated indoles was developed by Kim and co-workers.[84] These authors employed indolines with an easily removable benzoyl group as substrates and α-keto acids as acylating reagents. The combination of palladium catalyst/strong oxidant [$Pd(TFA)_2$/$(NH_4)_2S_2O_8$] was confirmed to be effective in this transformation. It is noteworthy that halogen groups on either the indoline substrate or the α-keto acid reagent were well tolerated under the oxidative setup, providing opportunities for further functional-group manipulations (Fig. 5.16). In addition, the versatility and practicality of this reaction was demonstrated by a scale-up experiment to form the desired C-7 acylation indole in one-pot.

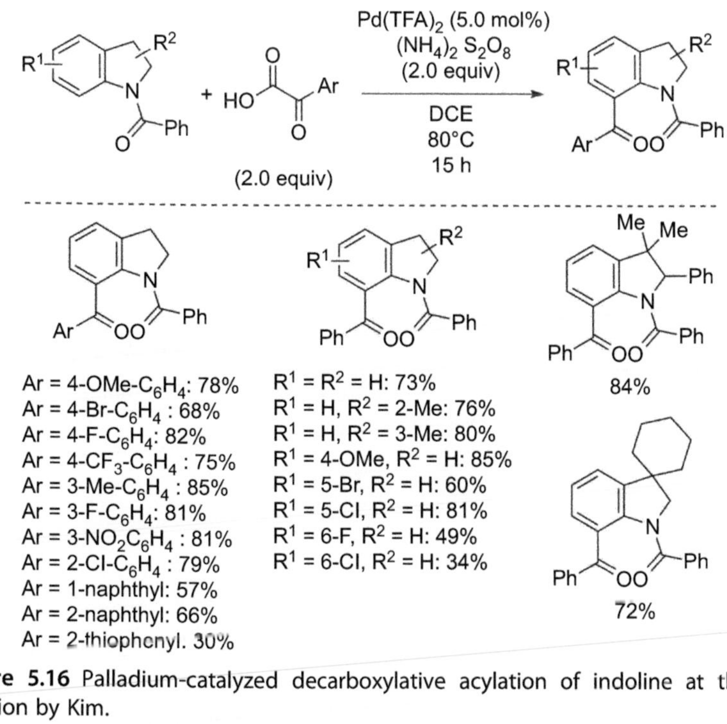

Figure 5.16 Palladium-catalyzed decarboxylative acylation of indoline at the C-7 position by Kim.

An oversimplified catalytic cycle was proposed (Fig. 5.17). Initially, with the aid of the benzoyl directing group, coordination of starting indoline to the palladium(II) catalyst followed by cyclopalladation at C-7 results in a six-membered palladacycle. This intermediate then reacts with $(NH_4)_2S_2O_8$ and the α-keto acid to generate a dimeric palladium(III) or a monomeric palladium(IV) complex concomitant with decarboxylation. The C-7 acylated indoline is released through reductive elimination.

Aldehydes also serve as acylating reagents. In 2015, the same group reported a palladium(II)-mediated C – H-bond-activation protocol for the direct C-7 acylation of indolines with aldehydes.[85] Unlike the previous report, *tert*-butoxycarbonyl was employed as an effective directing group. The substrate scope was broad and included not only aromatic and aliphatic aldehydes, but also aliphatic alcohols albeit with low reactivity (Fig. 5.18).

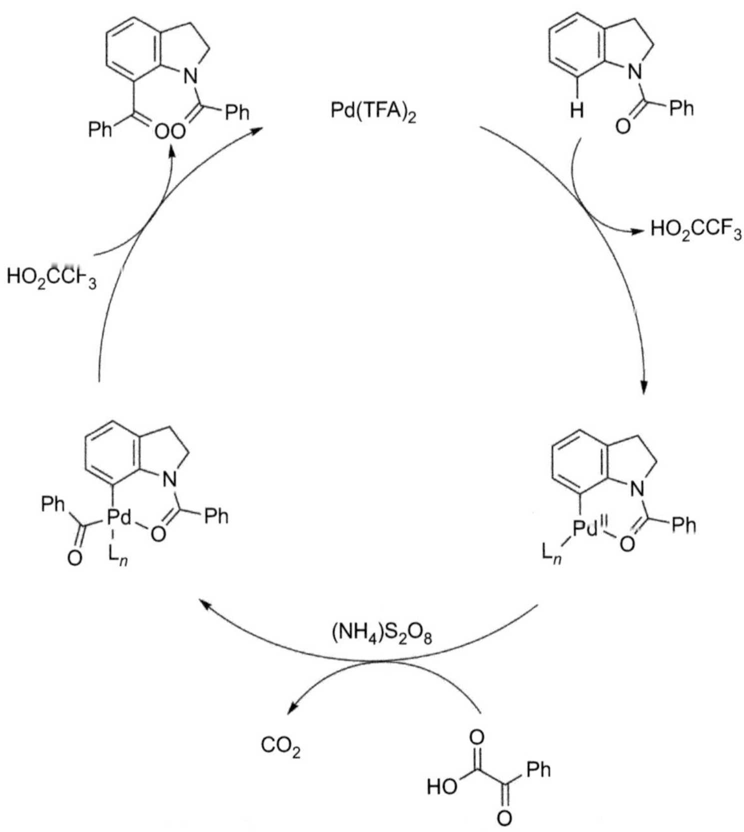

Figure 5.17 Oversimplified catalytic cycle of decarboxylative acylation.

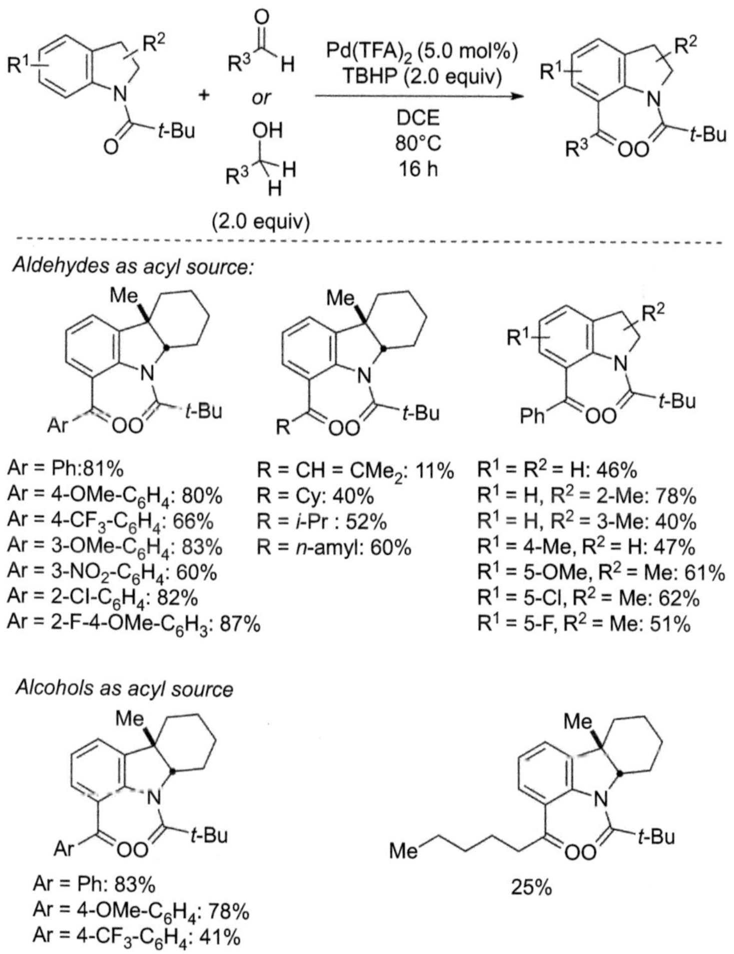

Figure 5.18 Palladium-catalyzed direct acylation of indoline using aldehydes and alcohols as acyl sources at the C-7 position by Kim.

Although detailed data on the mechanism of the catalytic system were not disclosed, the kinetic isotope effect (KIE, $k_H/k_D = 2.79$) indicated that the C—H bond cleavage at the C-7 position might be involved in the rate-determining step (RDS) (Fig. 5.19).

5.2.1.4 C—H Bond Carbonylation of Indoline at the C-7 Position

Pyrroloquinazolinediones are a class of structures having great potential as biologically active compounds.[86] Basically, all known methods for their

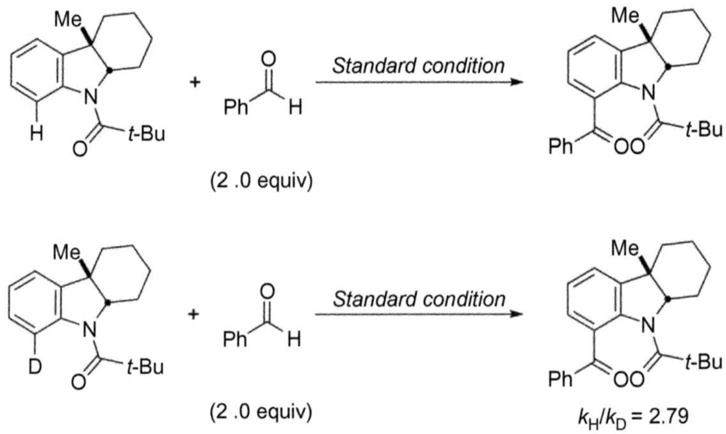

Figure 5.19 Kinetic isotope experiments.

preparation involve indoline-7-carboxylic acids or carboxylates. Just recently, an interesting sequential C-7 selective C—H carbonylation/cyclization was disclosed by Li and Wang under palladium(II) catalysis.[87] Various pyrroloquinazolinediones were prepared by the directed coupling of indolines and carbon monoxide (CO). Ureas with a free NH terminus seemed an ideal directing group. Even indoline bearing an electron-withdrawing carboxyl group underwent the desired coupling in high yield. The steric effect was highlighted by using 6-substituted indolines, resulting in decreased yields (Fig. 5.20). This method proved complementary to the previously reports in pyrroloquinazolinediones synthesis.

5.2.1.5 C—H Bond Alkylation of Indoline at the C-7 Position

Regarding the alkylation of indole derivatives, the C-2 and C-3 alkylation of indole is achieved by various methods such as Friedel—Crafts alkylation, allylic alkylation, conjugate addition, and C—H bond functionalization.[88,89] However, there have been relatively few reports on the C-7 alkylation of indole.

In 2013, the Sanford group disclosed a palladium-catalyzed oxidative C—H bond alkylation between arenes and trifluoroborate reagents.[90] In this report, an example of indoline was included. With the aid of an acetyl directing group attached at the nitrogen atom of indoline, a methyl group was selectively installed at the C-7 position in the presence of MnF_3 as terminal oxidant. A combination of TFE/H_2O/AcOH (8:1:1) was employed as solvent mixture (Fig. 5.21).

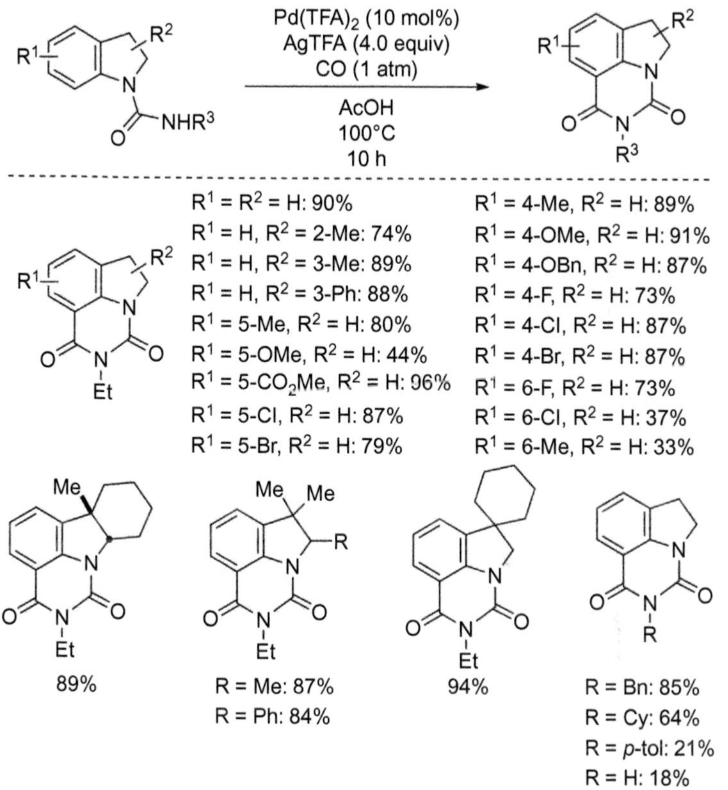

Figure 5.20 Palladium-catalyzed sequential carbonylation/cyclization at the C-7 position of indoline by Li and Wang.

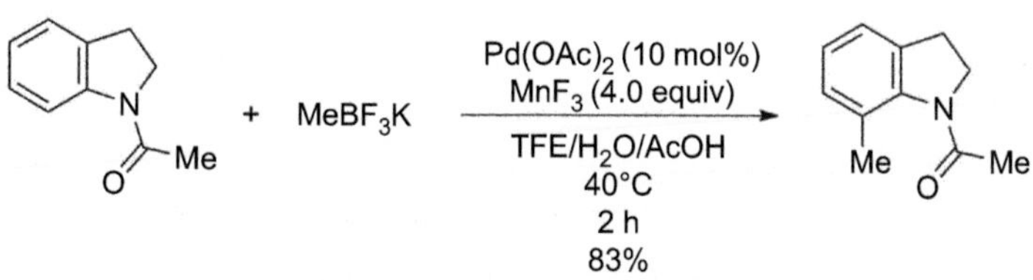

Figure 5.21 Palladium-catalyzed direct indoline C-7 methylation using a trifluoroborate as methyl source by Sanford.

As a very strong oxidant, MnF_3 is believed to promote oxidative coupling through one of the following two pathways; high valent palladium intermediates are generated in both cases (Fig. 5.22). The first pathway involves the formation of a highly reactive alkyl radical (R•) through the oxidation of RBF_3K, followed by the reaction between this radical and

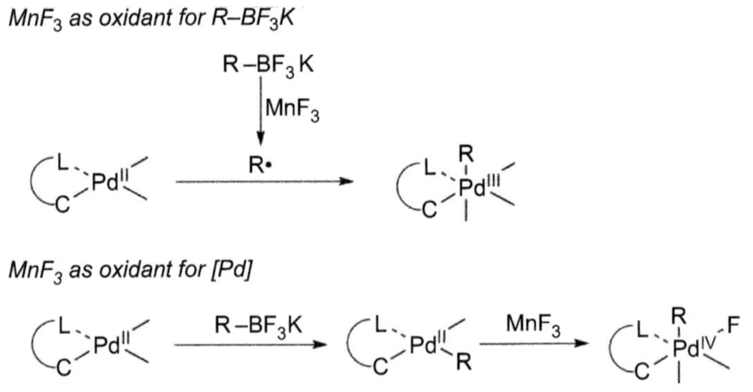

Figure 5.22 Two conceivable pathways involving MnF_3 as terminal oxidant.

the palladacycle to generate a palladium(III) intermediate (Fig. 5.22, top). Alternatively, it could also involve oxidation of palladium(II) to palladium (IV) by MnF_3 directly. Precisely, "F^+", generated in situ from MnF_3, promotes this oxidation. MnF_3 is an important F^+ source in fluorination reactions of organic compounds.[91] In this case, the palladium(II)/(IV) mechanism becomes predominant (Fig. 5.22, bottom).[92]

One year later, Jain and co-workers presented a palladium(II)-catalyzed formation of $C(sp^2)$—$C(sp^3)$ bonds through decarboxylative cross-coupling of indolines at the C-7 position.[93] As the terminal oxidant, PhI $(OAc)_2$ was found to be optimal to promote the transformation under mild, solvent-free conditions. Full conversion was observed at 40°C, and 80°C, as well as 100°C, but lower reaction temperatures led to significantly improved yields. Also, indoline-to-indole oxidation at elevated temperatures was eventually suppressed. A number of secondary or tertiary α-substituted, cyclic, and acylic aliphatic carboxylic acids were introduced as the alkyl sources in the absence of any solvent (Fig. 5.23). The generality of this method was further demostrated by using 2-phenylpyridines and aromatic azo compounds, leading to moderate-to-good yield and highly *ortho*-selectivity (not shown). To collect evidence for or against a radical mechanism, the addition of radical scavengers such as ascorbic acid was found to reduce the yield dramatically (not shown). Thus, it can be concluded that a radical mechanism is involved in this palladium(II)-mediated transformation.

5.2.2 C—C Bond Formation at the C-6 Position of Indoline

Compared to C—H bond functionalization at the *ortho* position of an existing directing group, examples related to selective *meta*- or remote

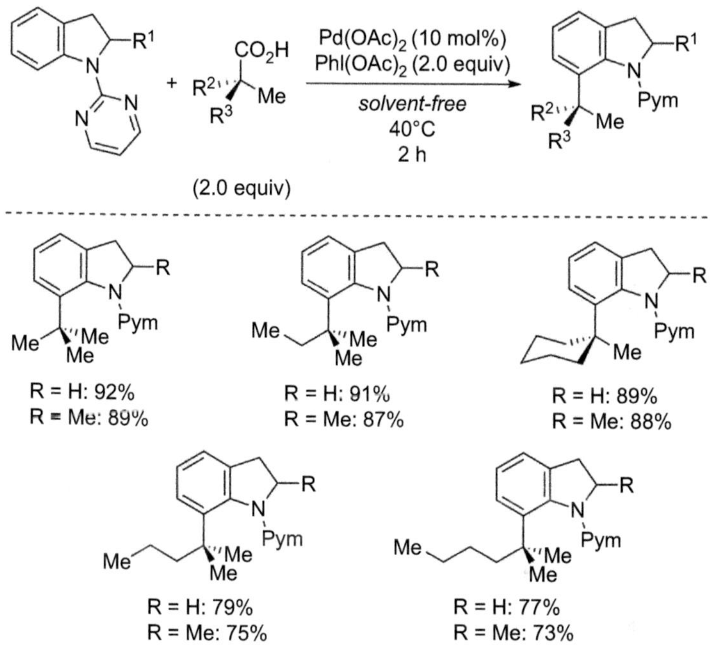

Figure 5.23 Palladium-catalyzed decarboxylative alkylation of indolines at the C-7 position by Jain.

C—H bond activation remain scarce due to the inability of functional groups to strongly direct the activation to a single remote site. Just recently, several different nitrile-containing templates were designed and synthesized to achieve the selective C—H bond functionalization of a wide range of arenes and heteroarenes.[94]

In 2014, Movassaghi and Yu accomplished unprecedented alkenylation, arylation, and acetoxylation at the C-6 position of indolines with a U-shaped nitrile template (Fig. 5.24).[95] With $Pd(OAc)_2$ as the catalyst precursor, 2.5 equiv of alkene, 3.0 equiv of AgOAc, and 20 mol% of Ac-Gly-OH were applied as the alkenylation reagent, oxidant, and effective ligand, respectively. With this optimized setup in hand, a series of indoline substrates bearing a variety of substituents at the C-4 or C-5 position were evaluated. Unlike C-4-substituted indolines, the alkenylation of C-5-substituted substrates always required an elevated reaction temperature (80°C) due to their low reactivity. Generally, the C-6 alkenylated products were obtained in good-to-excellent yield and with high regioselectivity (Fig. 5.25).

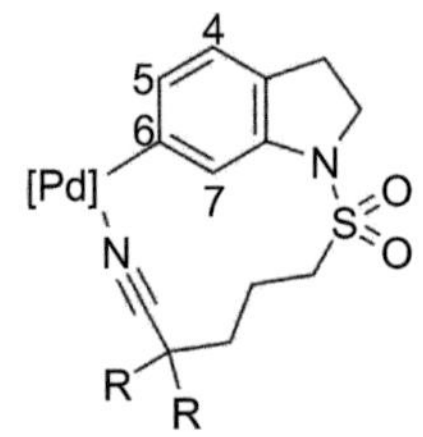

Figure 5.24 The design of a directing template for indolines.

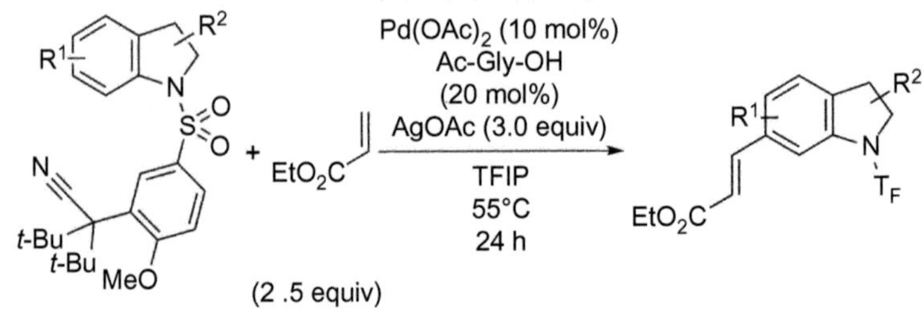

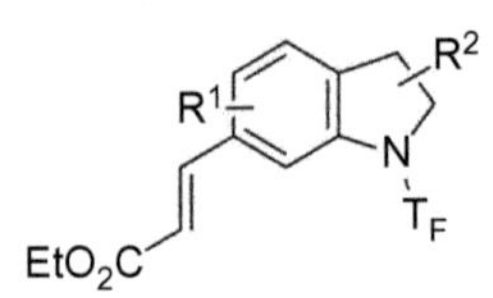

$R^1 = R^2$ = H: 75% C6:others > 20:1
R^1 = H, R^2 = 2-Me: 66% C6:C5 = 16:1
R^1 = H, R^2 = 2-cyclopentyl: 83% C6:C5 = 14:1
R^1 = H, R^2 = 3-Me: 71% C6:others > 20:1
R^1 = 4-Me, R^2 = H: 70% C6:others > 20:1
R^1 = 4-F, R^2 = H: 80% C6:C5 = 7.4:1
R^1 = 4-Cl, R^2 = H: 67% C6:others > 20:1
R^1 = 5-Me, R^2 = H: 57% C6:others > 20:1
R^1 = 5-F, R^2 = H: 58% C6:others > 20:1
R^1 = 5-Cl, R^2 = H: 68% C6:others > 20:1
R^1 = 5-CO_2Me, R^2 = H: 76% C6:others > 20:1

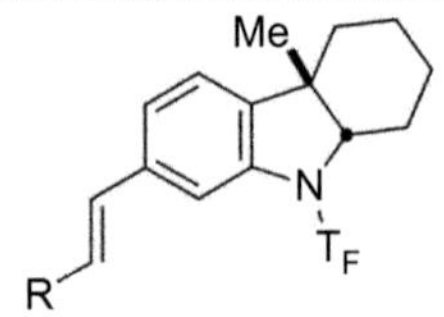

R = CO_2Bu: 74%
C6:others > 20:1
R = CO_2Bn: 78%
C6:C5 = 20:1
R =COMe: 75%
C6:others > 20:1
R = $CONMe_2$: 81%
C6:C5 = 20:1
R = $P(O)(OEt)_2$: 64%
C6:C5 = 17:1

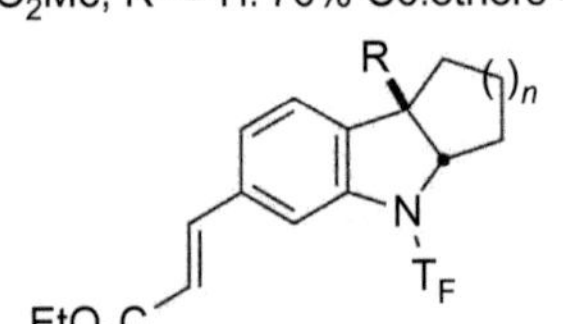

R = H, n = 1: 60% C6:others > 20:1
R = H, n = 2: 75% C6:C5 = 18:1
R = Me, n = 2: 80% C6:others > 20:1
R = $(CH_2)_2CO_2$Me, n = 2: 90% C6:others > 20:1

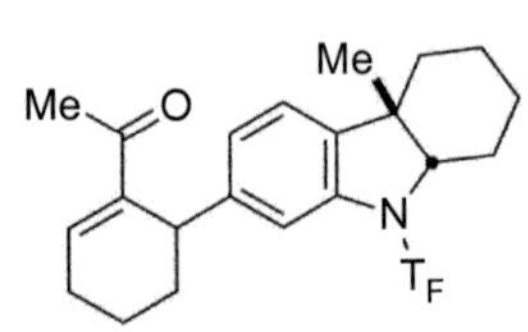

65% C6:others > 20:1

Figure 5.25 Palladium-catalyzed C-6 alkenylation of indolines directed by a U-shaped template by Movassaghi and Yu.

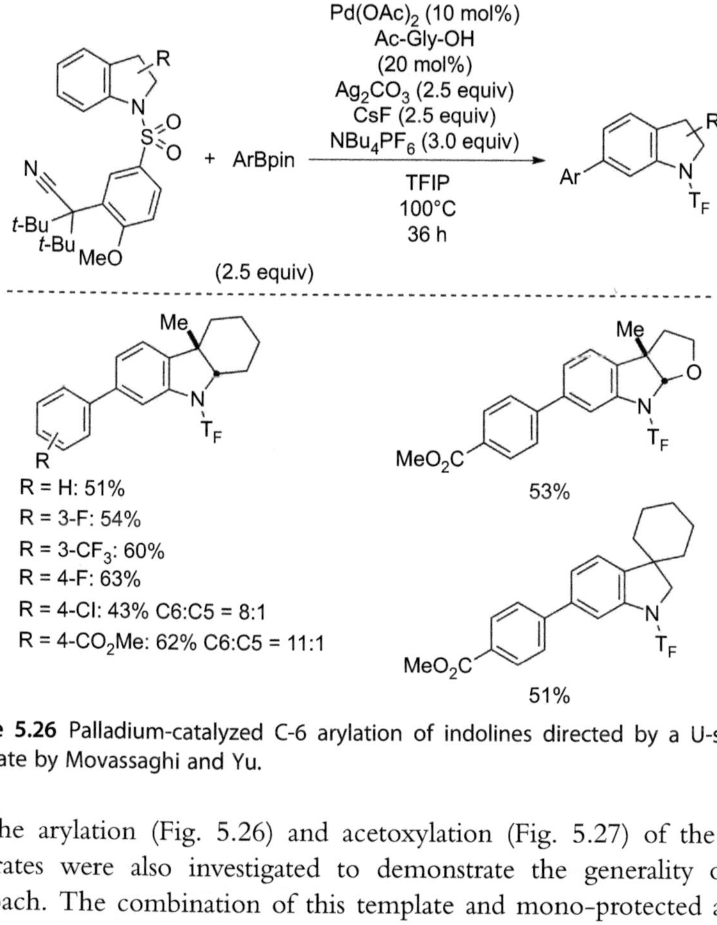

Figure 5.26 Palladium-catalyzed C-6 arylation of indolines directed by a U-shaped template by Movassaghi and Yu.

The arylation (Fig. 5.26) and acetoxylation (Fig. 5.27) of the same substrates were also investigated to demonstrate the generality of the approach. The combination of this template and mono-protected amino acid ligands were utilized to promote these transformations with arylboronic acid pinacol esters and iodobenzene diacetate as coupling partners, respectively. These achievements clearly offer opportunities for a wide range of chemical manipulations to meet synthetic needs.

5.3 CONCLUSIONS

In summary, as a timely and fast-developing area, the palladium-catalyzed C—H bond functionalization of indoles and indolines have been extensively studied during the past years. Transformations that had stretched the limits of known methodologies have now become feasible. This chapter provides a brief summary for these recent advancements in the C—H

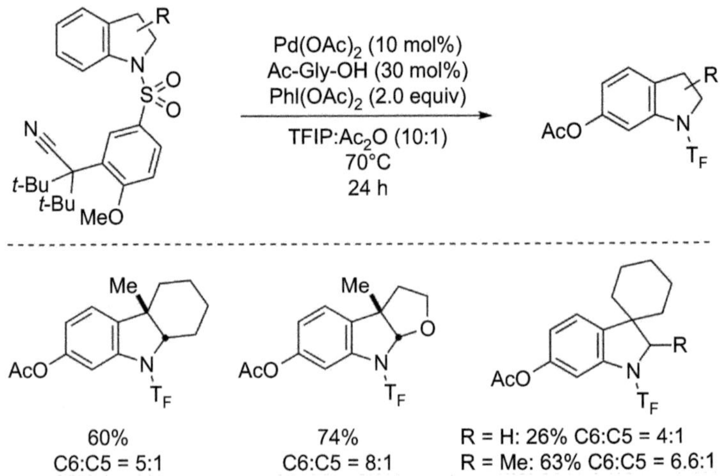

Figure 5.27 Palladium-catalyzed C-6 acetoxylation of indolines directed by a U-shaped template by Movassaghi and Yu.

bond functionalization of indole-nuclei-containing compounds. Newly developed strategies are focused on searching for new types of directing groups to assist targeted C—H bond activation in a regioselective fashion. It is obvious that palladium catalysis still dominates the C—C bond formation reactions. However, it cannot be neglected that other transition metals, such as rhodium, ruthenium, and iridium, as well as nickel, are currently attracting considerable attention. For C-7-selective direct alkenylation of indoline, several examples have already been mentioned above.[72–74] Moreover, other transition metals enabled the C-7-selective anulation of indoline,[70,96,97] alkylation of indoline[98,99] and indole,[100] alkynylation of indoline,[72,101,102] allylation of indoline,[103,104] cyanation of indoline,[105] amidation of indoline,[106–108] thiolation/selenation of indoline,[109] as well as borylation of indole,[110–114] providing opportunities for further functionalizations of the indole-nuclei-containing compounds.

LIST OF ABBREVIATIONS

Ac Acetyl
BQ 1,4-Benzoquinone
CDC Cross-Dehydrogenative Coupling
DCE Dichloroethane
DG Directing Group
DMF *N,N*-Dimethylformamide
EXSY EXchange SpectroscopY

HAS Homolytic Aromatic Substitution
KIE Kinetic Isotope Effects
nOe Nuclear Overhauser Effect
NOESY Nuclear Overhauser Enhancement SpectroscopY
OTf Triflate
Ph Phenyl
Py Pyridyl
RDS Rate-Determining Step
TBHP *tert*-Butyl Hydroperoxide
TIPS Triisopropylsilane

REFERENCES

1. Seregin, I. V.; Gevorgyan, V. *Chem. Soc. Rev.* **2007,** *36*, 1173.
2. Li, C.-J. *Acc. Chem. Res.* **2009,** *42*, 335.
3. McGlacken, G. P.; Bateman, L. M. *Chem. Soc. Rev.* **2009,** *38*, 2447.
4. Cho, S. H.; Kim, J. Y.; Kwak, J.; Chang, S. *Chem. Soc. Rev.* **2011,** *40*, 5068.
5. McMurray, L.; O'Hara, F.; Gaunt, M. J. *Chem. Soc. Rev.* **2011,** *40*, 1885.
6. Shi, W.; Liu, C.; Lei, A. *Chem. Soc. Rev.* **2011,** *40*, 2761.
7. Yeung, C. S.; Dong, V. M. *Chem. Rev.* **2011,** *111*, 1215.
8. Mousseau, J. J.; Charette, A. B. *Acc. Chem. Res.* **2013,** *46*, 412.
9. Hussain, I.; Singh, T. *Adv. Synth. Catal.* **2014,** *356*, 1661.
10. Yang, J. *Org. Biomol. Chem.* **2015,** *13*, 1930.
11. Walker, S. R.; Carter, E. J.; Huff, B. C.; Morris, J. C. *Chem. Rev.* **2009,** *109*, 3080.
12. Torborg, C.; Beller, M. *Adv. Synth. Catal.* **2009,** *351*, 3027.
13. Gutekunst, W. R.; Baran, P. S. *Chem. Soc. Rev.* **2011,** *40*, 1976.
14. Bandini, M.; Eichholzer, A. *Angew. Chem. Int. Ed.* **2009,** *48*, 9608.
15. Joucla, L.; Djakovitch, L. *Adv. Synth. Catal.* **2009,** *351*, 673.
16. Cacchi, S.; Fabrizi, G. *Chem. Rev.* **2011,** *111*, PR215.
17. Sandtorv, A. H. *Adv. Synth. Catal.* **2015,** *357*, 2403.
18. Han, F.-S. *Chem. Soc. Rev.* **2013,** *42*, 5270.
19. Miyaura, N.; Suzuki, A. *Chem. Rev.* **1995,** *95*, 2457.
20. Negishi, E. *Acc. Chem. Res.* **1982,** *15*, 340.
21. Hartwig, J. F. *Acc. Chem. Res.* **1998,** *31*, 852.
22. Muci, A. R.; Buchwald, S. L. *Top. Curr. Chem.* **2002,** *219*, 131.
23. Surry, D. S.; Buchwald, S. L. *Chem. Sci.* **2011,** *2*, 27.
24. Mc Cartney, D.; Guiry, P. J. *Chem. Soc. Rev.* **2011,** *40*, 5122.
25. Beccalli, E. M.; Broggini, G.; Martinelli, M.; Sottocornola, S. *Chem. Rev.* **2007,** *107*, 5318.
26. Chen, X.; Engle, K. M.; Wang, D.-H.; Yu, J.-Q. *Angew. Chem. Int. Ed.* **2009,** *48*, 5094.
27. Beletskaya, I. P.; Cheprakov, A. V. *Chem. Rev.* **2000,** *100*, 3009.
28. Oppolzer, W.; Grayson, J. I.; Wegmann, H.; Urrea, M. *Tetrahedron* **1983,** *39*, 3695.
29. Hatanaka, N.; Ozaki, O.; Matsumoto, M. *Tetrahedron Lett.* **1986,** *27*, 3169.
30. Wallwey, C.; Li, S.-M. *Nat. Prod. Rep.* **2011,** *28*, 496.
31. Li, J.-J.; Mei, T.-S.; Yu, J.-Q. *Angew. Chem. Int. Ed.* **2008,** *47*, 6452.
32. Liu, Q.; Li, Q.; Ma, Y.; Jia, Y. *Org. Lett.* **2013,** *15*, 4528.
33. Engle, K. M.; Mei, T.-S.; Wasa, M.; Yu, J.-Q. *Acc. Chem. Res.* **2012,** *45*, 788.
34. Neufeldt, S. R.; Sanford, M. S. *Acc. Chem. Res.* **2012,** *45*, 936.
35. Rouquet, G.; Chatani, N. *Angew. Chem. Int. Ed.* **2013,** *52*, 11726.

36. Zhang, F.; Spring, D. R. *Chem. Soc. Rev.* **2014,** *43*, 6906.
37. Zhang, M.; Zhang, Y.; Jie, X.; Zhao, H.; Li, G.; Su, W. *Org. Chem. Front.* **2014,** *1*, 843.
38. Chen, Z.; Wang, B.; Zhang, J.; Yu, W.; Liu, Z.; Zhang, Y. *Org. Chem. Front.* **2015,** *2*, 1107.
39. Huang, Z.; Lim, H. N.; Mo, F.; Young, M. C.; Dong, G. *Chem. Soc. Rev.* **2015,** *44*, 7764.
40. Harrington, P. J.; Hegedus, L. S.; McDaniel, K. F. *J. Am. Chem. Soc.* **1987,** *109*, 4335.
41. Yokoyama, H.; Otaya, K.; Kobayashi, H.; Miyazawa, M.; Yamaguchi, S.; Hirai, Y. *Org. Lett.* **2000,** *2*, 2427.
42. Makabe, H.; Kong, L. K.; Hirota, M. *Org. Lett.* **2003,** *5*, 27.
43. Ku, J.-M.; Jeong, B.-S.; Jew, S.-S.; Park, H.-G. *J. Org. Chem.* **2007,** *72*, 8115.
44. Yang, Y.; Qiu, X.; Zhao, Y.; Mu, Y.; Shi, Z. *J. Am. Chem. Soc.* **2016,** *138*, 495.
45. Jiao, L.-Y.; Ferreira, A. V.; Oestreich, M. *Chem. Asian J.* **2016,** *11*, 367.
46. Parnes, J. S.; Carter, D. S.; Kurz, L. J.; Flippin, L. A. *J. Org. Chem.* **1994,** *59*, 3497.
47. Leonori, D.; Coldham, I. *Adv. Synth. Catal.* **2009,** *351*, 2619.
48. De, S.; Ghosh, S.; Bhunia, S.; Sheikh, J. A.; Bisai, A. *Org. Lett.* **2012,** *14*, 4466.
49. Deprez, N. R.; Sanford, M. S. *J. Am. Chem. Soc.* **2009,** *131*, 11234.
50. Shi, Z.; Li, B.; Wan, X.; Cheng, J.; Fang, Z.; Cao, B.; Qin, C.; Wang, Y. *Angew. Chem. Int. Ed.* **2007,** *46*, 5554.
51. Nishikata, T.; Abela, A. R.; Huang, S.; Lipshutz, B. H. *J. Am. Chem. Soc.* **2010,** *132*, 4978.
52. Houlden, C. E.; Bailey, C. D.; Ford, J. G.; Gagne, M. R.; Lloyd-Jones, G. C.; Booker-Milburn, K. I. *J. Am. Chem. Soc.* **2008,** *130*, 10066.
53. Houlden, C. E.; Hutchby, M.; Bailey, C. D.; Ford, J. G.; Tyler, S. N.; Gagne, M. R.; Lloyd-Jones, G. C.; Booker-Milburn, K. I. *Angew. Chem. Int. Ed.* **2009,** *48*, 1830.
54. Rauf, W.; Thompson, A. L.; Brown, J. M. *Chem. Commun.* **2009,** 3874.
55. Nishikata, T.; Abela, A. R.; Lipshutz, B. H. *Angew. Chem. Int. Ed.* **2010,** *49*, 781.
56. Schipper, D. J.; Hutchinson, M.; Fagnou, K. *J. Am. Chem. Soc.* **2010,** *132*, 6910.
57. Wang, L.; Liu, S.; Li, Z.; Yu, Y. *Org. Lett.* **2011,** *13*, 6137.
58. Willwacher, J.; Rakshit, S.; Glorius, F. *Org. Biomol. Chem.* **2011,** *9*, 4736.
59. Jiang, Z.; Zhang, L.; Dong, C.; Su, X.; Li, H.; Tang, W.; Xu, L.; Fan, Q. *RSC Adv.* **2013,** *3*, 1025.
60. Littke, A. F.; Fu, G. C. *Angew. Chem. Int. Ed.* **2002,** *41*, 4176.
61. Wu, Y.; Wang, J.; Mao, F.; Kwong, F. Y. *Chem. Asian J.* **2014,** *9*, 26.
62. Jiao, L.-Y.; Oestreich, M. *Chem. Eur. J.* **2013,** *19*, 10845.
63. Jiao, L.-Y.; Smirnov, P.; Oestreich, M. *Org. Lett.* **2014,** *16*, 6020.
64. Somei, M.; Saida, Y.; Funamoto, T.; Ohta, T. *Chem. Pharm. Bull.* **1987,** *35*, 3146.
65. Somei, M.; Kawasaki, T. *Chem. Pharm. Bull.* **1989,** *37*, 3426.
66. Urones, B.; Arrayás, R. G.; Carretero, J. C. *Org. Lett.* **2013,** *15*, 1120.
67. Jiao, L.-Y.; Oestreich, M. *Org. Lett.* **2013,** *15*, 5374.
68. Miura, M.; Tsuda, T.; Satoh, T.; Pivsa-Art, S.; Nomura, M. *J. Org. Chem.* **1998,** *63*, 5211.
69. Kim, B. S.; Lee, S. Y.; Youn, S. W. *Chem. Asian J.* **2011,** *6*, 1952.
70. Wang, X.; Tang, H.; Feng, H.; Li, Y.; Yang, Y.; Zhou, B. *J. Org. Chem.* **2015,** *80*, 6238.
71. Yang, D.; Mao, S.; Gao, Y.-R.; Guo, D.-D.; Guo, S.-H.; Li, B.; Wang, Y.-Q. *RSC Adv.* **2015,** *5*, 23727.
72. Yang, X. F.; Hu, X. H.; Feng, C.; Loh, T. P. *Chem. Commun.* **2015,** *51*, 2532.
73. Song, Z.; Samanta, R.; Antonchick, A. P. *Org. Lett.* **2013,** *15*, 5662.
74. Pan, S.; Wakaki, T.; Ryu, N.; Shibata, T. *Chem. Asian J.* **2014,** *9*, 1257.

75. Saxton, J. E. *Nat. Prod. Rep.* **1992,** *9*, 393.
76. Sturino, C. F.; Lachance, N.; Boyd, M.; Berthelette, C.; Labelle, M.; Li, L.; Roy, B.; Scheigetz, J.; Tsou, N.; Brideau, C.; Cauchon, E.; Carriere, M. C.; Denis, D.; Greig, G.; Kargman, S.; Lamontagne, S.; Mathieu, M. C.; Sawyer, N.; Slipetz, D.; O'Neill, G.; Wang, Z.; Zamboni, R.; Metters, K. M.; Young, R. N. *Bioorg. Med. Chem. Lett.* **2006,** *16*, 3043.
77. Ondeyka, J. G.; Helms, G. L.; Hensens, O. D.; Goetz, M. A.; Zink, D. L.; Tsipouras, A.; Shoop, W. L.; Slayton, L.; Dombrowski, A. W.; Polishook, J. D.; Ostlind, D. A.; Tsou, N. N.; Ball, R. G.; Singh, S. B. *J. Am. Chem. Soc.* **1997,** *119*, 8809.
78. Miller, D. D.; Bamborough, P.; Christopher, J. A.; Baldwin, I. R.; Champigny, A. C.; Cutler, G. J.; Kerns, J. K.; Longstaff, T.; Mellor, G. W.; Morey, J. V.; Morse, M. A.; Nie, H.; Rumsey, W. L.; Taggart, J. J. *Bioorg. Med. Chem. Lett.* **2011,** *21*, 2255.
79. Ottoni, O.; Neder, A. D. V. F.; Dias, A. K. B.; Cruz, R. P. A.; Aquino, L. B. *Org. Lett.* **2001,** *3*, 1005.
80. Guchhait, S. K.; Kashyap, M.; Kamble, H. *J. Org. Chem.* **2011,** *76*, 4753.
81. Bergman, J.; Venemalm, L. *Tetrahedron Lett.* **1987,** *28*, 3741.
82. Zhou, B.; Yang, Y.; Li, Y. *Chem. Commun.* **2012,** *48*, 5163.
83. Pan, C.; Jin, H.; Liu, X.; Cheng, Y.; Zhu, C. *Chem. Commun.* **2013,** *49*, 2933.
84. Kim, M.; Mishra, N. K.; Park, J.; Han, S.; Shin, Y.; Sharma, S.; Lee, Y.; Lee, E. K.; Kwak, J. H.; Kim, I. S. *Chem. Commun.* **2014,** *50*, 14249.
85. Shin, Y.; Sharma, S.; Mishra, N. K.; Han, S.; Park, J.; Oh, H.; Ha, J.; Yoo, H.; Jung, Y. H.; Kim, I. S. *Adv. Synth. Catal.* **2015,** *357*, 594.
86. (a) Masumoto, S.; Kitano, M.; Ohashi, N. *PCT Int. Appl.* **2001**. WO 2001/079206A1.
    (b) Ellsworth, E. L.; Showalter, H. D. H.; Powell, S. A.; Sanchez, J. P.; Kerschen, J. A.; Stier, M. A.; Tran, T. P. *PCT Int. Appl.* **2002**. WO 2002/102793A2.
    (c) Barbara, B.; Wolfgang, G.; Georg, J.; Antonio, R.; Daniel, R.; Eric, V. *PCT Int. Appl.* **2015**. WO 2015/044075A1.
87. Wang, P.-L.; Li, Y.; Ma, L.; Luo, C.-G.; Wang, Z.-Y.; Lan, Q.; Wang, X.-S. *Adv. Synth. Catal.* **2016,** *358*, 1048.
88. Cacchi, S.; Fabrizi, G. *Chem. Rev.* **2005,** *105*, 2873.
89. Tsogoeva, S. B. *Eur. J. Org. Chem.* **2007,** 1701.
90. Neufeldt, S. R.; Seigerman, C. K.; Sanford, M. S. *Org. Lett.* **2013,** *15*, 2302.
91. Boltalina, O. V.; Borschevskii, A. Y.; Sidorov, L. N.; Street, J. M.; Taylor, R. *Chem. Commun.* **1996,** 529.
92. Vigalok, A. *Organometallics* **2011,** *30*, 4802.
93. Premi, C.; Dixit, A.; Jain, N. *Org. Lett.* **2015,** *17*, 2598.
94. Sharma, R.; Thakur, K.; Kumar, R.; Kumar, I.; Sharma, U. *Catal. Rev.* **2015,** *57*, 345.
95. Yang, G.; Lindovska, P.; Zhu, D.; Kim, J.; Wang, P.; Tang, R. Y.; Movassaghi, M.; Yu, J.-Q. *J. Am. Chem. Soc.* **2014,** *136*, 10807.
96. Manoharan, R.; Jeganmohan, M. *Org. Biomol. Chem.* **2015,** *13*, 9276.
97. Yang, X.-F.; Hu, X.-H.; Loh, T.-P. *Org. Lett.* **2015,** *17*, 1481.
98. Ai, W.; Yang, X.; Wu, Y.; Wang, X.; Li, Y.; Yang, Y.; Zhou, B. *Chem. Eur. J.* **2014,** *20*, 17653.
99. Pan, S.; Ryu, N.; Shibata, T. *Adv. Synth. Catal.* **2014,** *356*, 929.
100. Wang, L.; Li, Z.; Qu, X.; Peng, W.-M.; Hu, S.-Q.; Wang, H.-B. *Tetrahedron Lett.* **2015,** *56*, 6214.
101. Jin, N.; Pan, C.; Zhang, H.; Xu, P.; Cheng, Y.; Zhu, C. *Adv. Synth. Catal.* **2015,** *357*, 1149.
102. Wu, Y.; Yang, Y.; Zhou, B.; Li, Y. *J. Org. Chem.* **2015,** *80*, 1946.
103. Park, J.; Mishra, N. K.; Sharma, S.; Han, S.; Shin, Y.; Jeong, T.; Oh, J. S.; Kwak, J. H.; Jung, Y. H.; Kim, I. S. *J. Org. Chem.* **2015,** *80*, 1818.

104. Sharma, S.; Shin, Y.; Mishra, N. K.; Park, J.; Han, S.; Jeong, T.; Oh, Y.; Lee, Y.; Choi, M.; Kim, I. S. *Tetrahedron* **2015,** *71*, 2435.
105. Mishra, N. K.; Jeong, T.; Sharma, S.; Shin, Y.; Han, S.; Park, J.; Oh, J. S.; Kwak, J. H.; Jung, Y. H.; Kim, I. S. *Adv. Synth. Catal.* **2015,** *357*, 1293.
106. Pan, C.; Abdukader, A.; Han, J.; Cheng, Y.; Zhu, C. *Chem. Eur. J.* **2014,** *20*, 3606.
107. Shin, K.; Chang, S. *J. Org. Chem.* **2014,** *79*, 12197.
108. Hou, W.; Yang, Y.; Ai, W.; Wu, Y.; Wang, X.; Zhou, B.; Li, Y. *Eur. J. Org. Chem.* **2015,** 395.
109. Xie, W.; Li, B.; Wang, B. *J. Org. Chem.* **2016,** *81*, 396.
110. Paul, S.; Chotana, G. A.; Holmes, D.; Reichle, R. C.; Maleczka, R. E., Jr.; Smith, M. R., III *J. Am. Chem. Soc.* **2006,** *128*, 15552.
111. Lo, W. F.; Kaiser, H. M.; Spannenberg, A.; Beller, M.; Tse, M. K. *Tetrahedron Lett.* **2007,** *48*, 371.
112. Robbins, D. W.; Boebel, T. A.; Hartwig, J. F. *J. Am. Chem. Soc.* **2010,** *132*, 4068.
113. Wang, C.; Sperry, J. *J. Org. Chem.* **2012,** 77, 2584.
114. Loach, R. P.; Fenton, O. S.; Amaike, K.; Siegel, D. S.; Ozkal, E.; Movassaghi, M. *J. Org. Chem.* **2014,** *79*, 11254.

CHAPTER 6

Palladium-Catalyzed Carbonylative and Carboxylative C—H Functionalization Reactions: Importance and Role of Regioselectivity

Prashant Gautam and Bhalchandra M. Bhanage
Institute of Chemical Technology, Mumbai, Maharashtra, India

Contents

Strategies for Palladium-Catalyzed Non-Directed and Directed C—H Bond Functionalization
DOI: http://dx.doi.org/10.1016/B978-0-12-805254-9.00006-2

6.1 INTRODUCTION

6.1.1 Introduction to Carbonylation Chemistry

Carbonylation generally refers to a group of closely connected reactions in which a molecule of CO is incorporated into a substrate. This either takes place by the insertion of CO into an existing bond [C—X (X=Cl, Br, I, and H)], or by the addition and insertion of CO into unsaturated compounds such as alkynes or alkenes in the presence of different nucleophiles.[1] The latter reaction is more closely related to the hydroformylation reaction,[2] in which a formyl group gets attached to an olefinic double bond. Even though hydroformylation comes under the class of reactions which are termed carbonylation, it is however, treated separately due to its immense industrial significance. Hence, carbonylation reactions are of great interest to synthetic chemists as they allow the construction of new bonds, along with the simultaneous introduction of a carbonyl group. In the late 1930s and early 1940s, W. Reppe at BASF worked immensely in this area and also coined the term "carbonylation." Over the last five decades, carbonylation reactions have gained tremendous importance in different branches of chemistry. In the present era, hydroformylation and alcohol carbonylation represent core industrial technology and are the most significant branch of carbonylation chemistry.

6.1.2 Classification of Types of Carbonylation Reactions

Carbonylation reactions are generally classified according to the class of substrates employed and the reaction parameters (catalytic system, additive, and/or oxidant) effecting the carbonylative transformation. In the 1960s and 1970s, Heck and co-workers discovered the reaction of aryl and vinyl halides with CO.[3] This class of carbonylation has wide application in the industry and laboratory as it can introduce the carbonyl group with high efficiency and excellent selectivity, thus resulting in the synthesis of carboxylic acid and its derivatives. Later the reactivity of aryl/vinyl halides was extended to include pseudohalides (organic triflates, tosylates, mesylates, and fluorosulfonates) as the electrophilic coupling partners. The nucleophiles range from organometallic reagents, amines, alcohols, water, hydrides, to C—H bonds (alkynes, alkenes, aromatic C—H bonds). Other than intermolecular carbonylation reactions, the electrophile and the nucleophile can be present on the same molecule, which allows for intramolecular carbonylation reactions. This allows the synthesis of lactones, lactams, and other heterocycles. When amines are employed as nucleophiles, double carbonylation may also take place, in addition to mono carbonylation. All these reactions generally proceed in the presence of palladium catalysts, and are termed "classical carbonylation" reactions (Scheme 6.1).

When two nucleophiles (similar or different) are coupling partners, they can be carbonylatively coupled with the assistance of an oxidant. These reactions typically proceed at milder reaction conditions *vis-a-vis* classical carbonylation reactions. This is primarily due to the absence of the oxidative addition step of aryl halide (electrophile) to the metal, which is usually inhibited by the large concentration of CO. These class of reactions are termed "oxidative carbonylation" reactions (Scheme 6.2).

In classical and oxidative carbonylation reactions, if the nucleophile happens to be a C—H bond, the reaction is termed as a "carbonylative C—H functionalization". The C—H bond can be a part of an alkane, alkene, alkyne, or a (hetero)aromatic system. Such a transformation is extremely advantageous due to the ubiquitous presence of C—H bonds. More importantly, the transformation can be brought about without functionalizing the substrate molecule. For example, in classical carbonylation reactions, the aryl halide is prepared from the corresponding arene, so it would be advantageous to use the arene directly for carbonylation, as it brings down the overall cost of the process. This increases the step economy of the protocol and makes it environmentally benign by reducing the amount of waste generated (Scheme 6.3).

In 1980, Fujiwara and co-workers first documented the carbonylation of arenes promoted by $Pd(OAc)_2$ resulting in the synthesis of aromatic carboxylic acids[4] in low yields (26%—48%) as compared to $Pd(OAc)_2$. In this protocol, the arene substrates were used as solvents, and the reaction was not catalytic in nature. Later the same group reported that the

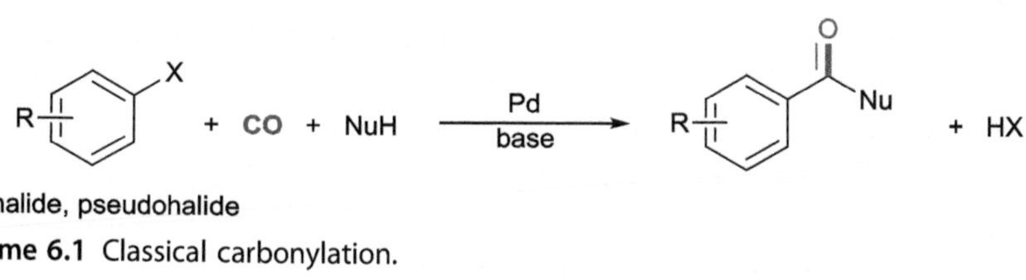

Scheme 6.1 Classical carbonylation.

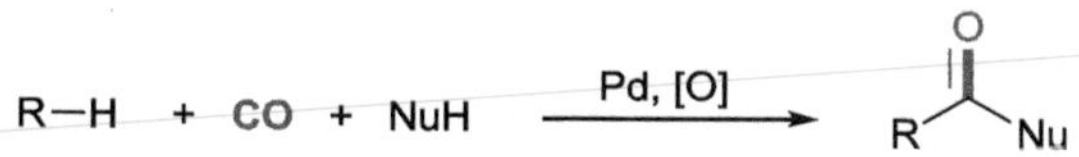

Scheme 6.2 Oxidative carbonylation.

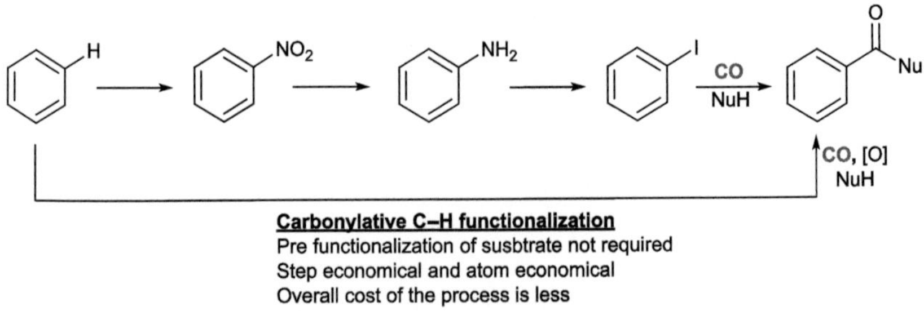

Scheme 6.3 Comparison between carbonylative C—H functionalization and conventional carbonylation.

reaction could be carried out in trifluoroacetic acid and benzene could be transformed into benzoic acid under atmospheric pressure of CO in presence of catalytic amounts of $Pd(OAc)_2$ and $K_2S_2O_8$ as the oxidant.[5]

This paved the way for the carbonylation of C—H bonds in the coming decades. However, in spite of the inherent advantages of such a transformation, it has its own share of drawbacks and challenges which shall be discussed in the next section.

6.1.3 Difference Between Carbonylative C—H Functionalization and Its Noncarbonylative Counterpart

The presence of CO in a carbonylative C—H functionalization makes its chemistry very different from the noncarbonylative version. Considering the reactions to be catalyzed by palladium, there are several disadvantages of the C—H carbonylation which are discussed: (1) The strong binding ability of CO results in the inhibition of the electrophilic attack of Pd(II) species on the aromatic C—H bonds by occupying the active sites. (2) Since CO is a π-acidic ligand, the unique π-back bonding decreases the electron density on the Pd-center. (3) The reducing nature of CO might facilitate the reduction of Pd(II) species to Pd(0), which results in the inhibition of the electrophilic attack. (4) Since C—H bonds are more prevalent in a molecule, there are issues with regards to selective functionalization of a C—H bond. Thus, in the case of substituted arenes, a mixture of products is obtained. These problems are generally overcome by incorporating a directing group in the substrate.

The present review will focus on the regioselective aspects of the carbonylative C(sp^2)—H functionalization in aromatic systems under the assistance of directing groups. The carbonylation of C—H bonds in unactivated heteroaromatic systems takes place regioselectively in the absence of directing groups and will also be touched upon. There, palladium-catalyzed insertion of CO_2 in C—H bonds is significant, and extremely challenging, and will be covered in the final section. The carbonylative C—H functionalization of alkanes, alkenes, and alkynes using gaseous CO and CO surrogates have been documented in seminal reviews,[6] and will not be covered.

6.2 CARBONYLATIVE C—H FUNCTIONALIZATION OF ARENES IN THE PRESENCE OF DIRECTING GROUP

6.2.1 Introduction to Directing Groups

Over the last two decades, palladium-catalyzed carbonylative C(sp^2)—H functionalization has become prevalent. Although it involves the use of the C—H bond present in arenes as the starting point, the difficulty in the cleavage of the strong C—H bond, followed by CO insertion and controlling the regioselectivity, pose challenges. However, these can be circumvented through the incorporation of a directing group in the substrate. The coordination (chelation) of a heteroatom present in the directing group with the palladium center and the close proximity of the C—H bond to it, ensure the regioselective cleavage of the C—H bond. Hence, the coordination of the palladium atom with the directing group is an important step involved in this class of C—H functionalization reactions (Scheme 6.4).

Various functional groups, containing heteroatoms, such as aliphatic amines, amides, amidines, carboxylic acids, hydroxyl, and phosphoryl groups, and nitrogen containing heterocyclic moieties have been reported to be used as directing groups for carbonylative C(sp^2)—H functionalization (Scheme 6.5). The next section will discuss the C(sp^2)—H functionalization in arenes in detail, and has been classified according to the directing group employed.

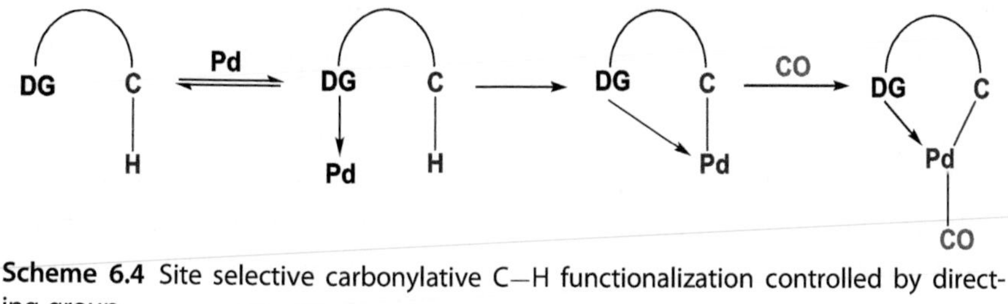

Scheme 6.4 Site selective carbonylative C—H functionalization controlled by directing group.

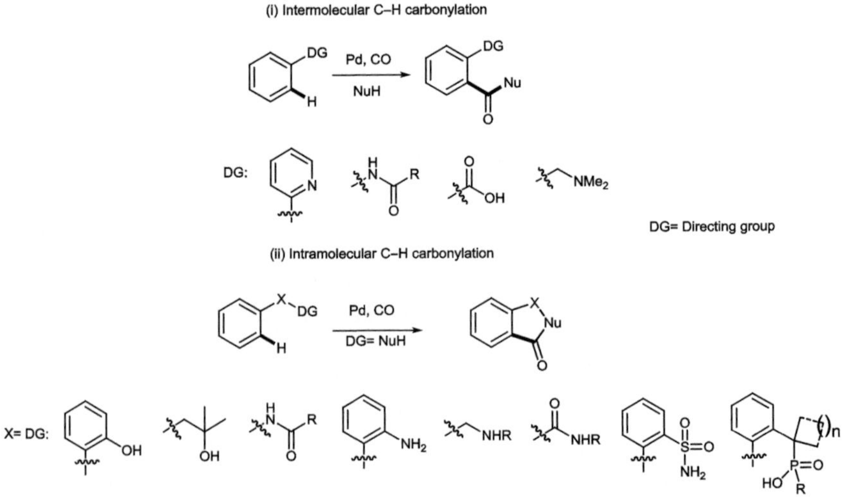

Scheme 6.5 Representative examples of directing groups in intermolecular and intramolecular carbonylative C(sp^2)—H functionalization.

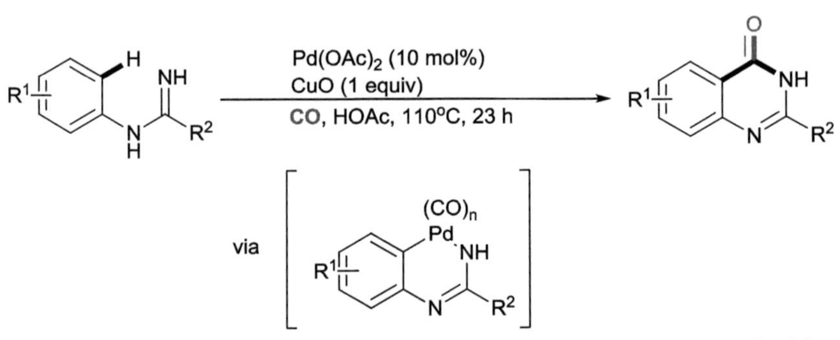

Scheme 6.6 Synthesis of 2-aryl(alkyl) quinazolin-4(3*H*)-ones from *N*-arylamidines through intramolecular carbonylative C(sp^2)—H functionalization.

6.2.2 Regioselective Functionalization of Arenes in the Presence of Directing Groups

6.2.2.1 Amidine and Oxime as Directing Groups

The synthesis of 2-aryl(alkyl) quinazolin-4(3*H*)-ones from *N*-arylamidines through intramolecular carbonylative C(sp^2)—H functionalization was reported by Zhu and co-workers.[7] The reaction could be carried out under an atmospheric pressure of CO and in the presence of CuO as an oxidant. Since *N*-arylamidines can be synthesized from anilines and nitriles, this protocol provides access to the quinazolin-4(3*H*)-one moiety in an atom-economical and step-economical fashion (Scheme 6.6).

Mechanistically, the initial chelation of the amidine nitrogen with Pd(II) forms intermediate **2**. This is followed by reversible cyclopalladation to generate **3**. CO inserts into the C—Pd bond in **3**, generating a seven-membered palladacycle **4**. Subsequently reductive elimination leads to the product **5** and releases Pd(0), which is then reoxidized by CuO with the aid of HOAc (Scheme 6.7).

Jiang and co-workers reported the synthesis of benzo[*d*][1,2]oxazin-1-ones and 3-methyleneisoindolin-1-ones by intramolecular carbonylation of the C(sp^2)—H bond with aromatic oximes as directing groups.[8] Interestingly, the formation of benzo[*d*][1,2]oxazin-1-one involves the hydroxyl-directed *ortho*-C(sp^2)—H carbonylation pathway in which AgOAc is required as an oxidant, whereas the formation of 3-methyleneisoindolin-1-one involves the activation of the vinyl C(sp^2)—H bond/*ortho*-Csp^2—H carbonylation pathway and the N—OH group of the oxime could be used as an internal oxidant (Scheme 6.8).

As far as the mechanism is concerned, the formation of benzo[*d*][1,2]oxazin-1-one involves an oxime hydroxyl directing *ortho*-C(sp^2)—H bond cleavage by Pd(II) to form a six-membered palladacycle intermediate **9**. CO coordination followed by migratory insertion generates a seven-membered cyclic intermediate **10**. Reductive elimination from **10** leads to the desired product **6** and concurrent formation of a Pd(0) species, which is oxidized to Pd(II) by AgOAc and air to complete the catalytic

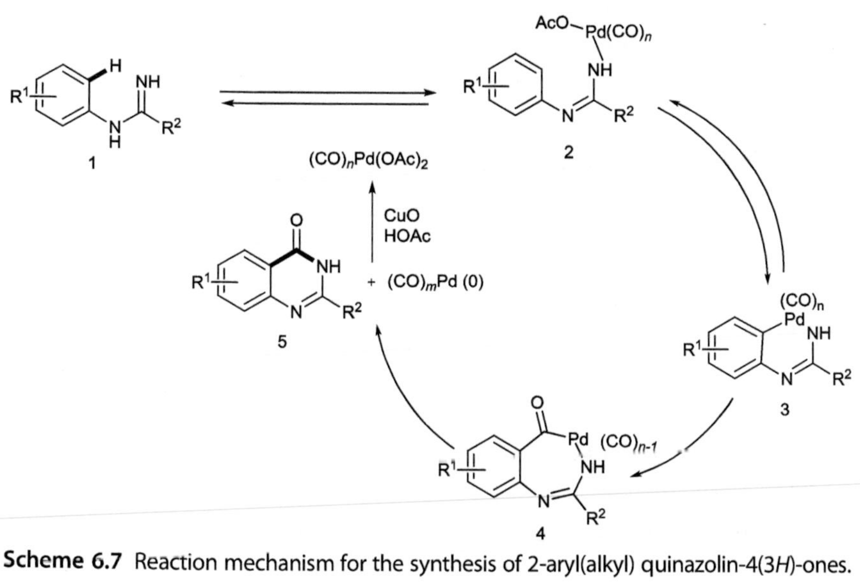

Scheme 6.7 Reaction mechanism for the synthesis of 2-aryl(alkyl) quinazolin-4(3*H*)-ones.

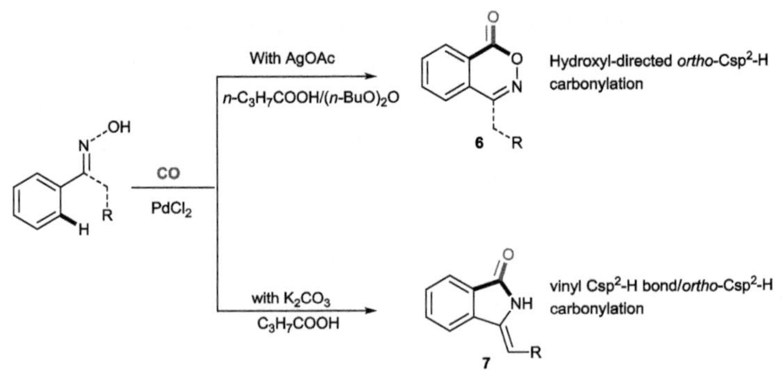

Scheme 6.8 Palladium-catalyzed C(sp^2)—H carbonylation of aromatic oximes.

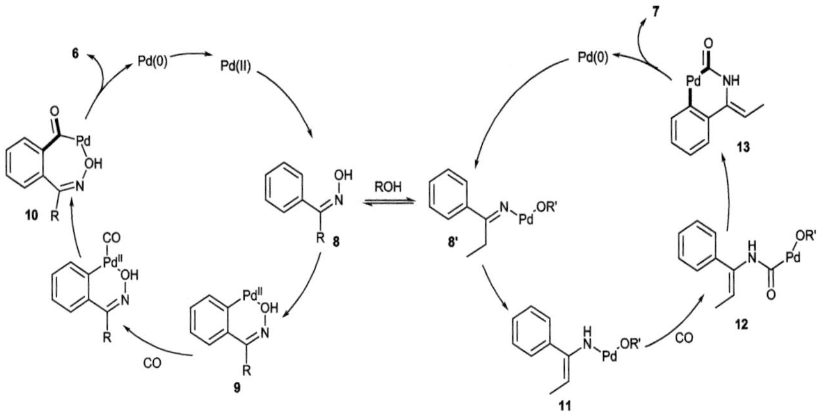

Scheme 6.9 Mechanism for the formation of benzo[*d*][1,2]oxazin-1-ones and 3-methyleneisoindolin-1-ones.

cycle. On the other hand, the oxidative addition of **8** results in intermediate **8′** which undergoes tautomerization to generate the enamine-derived amido-Pd(II) species **11**. CO coordination followed by insertion leads to the formation of **12**. This is followed by C—H bond cleavage which results in the formation of palladacyclic complex **13**. Subsequent reductive elimination affords the 3-methyleneisoindolin-1-one product **7** and the active Pd(0) catalyst is regenerated (Scheme 6.9).

6.2.2.2 Amide and Sulfonamide as Directing Groups

In 2009, Houlden et al. reported the synthesis of cyclic imidates and anthranilates from *N*-aryl urea derivatives as directing groups through carbonylative *ortho* C(sp^2)—H functionalization.[9] After the initial optimization, it was found that in the presence of $[Pd(OTs)_2(MeCN)_2]$ as the palladium catalyst and CH_2Cl_2 as the solvent, the aryl urea derivatives were readily converted into cyclic imidates within 5 h and at 18°C. However, when CH_2Cl_2 was replaced with MeOH:THF (1:1), the corresponding anthranilates were obtained in good yields. However, control experiments showed that the formation of anthranilates takes place through the formation of cyclic imidates. The cyclic imidates, being labile, undergo a solvolytic ring-opening sequence leading to the synthesis of anthranilates (Scheme 6.10).

Yu and co-workers documented the ortho C—H carboxylation of anilide derivatives leading to the synthesis of *N*-acyl anthranilic acids.[10] The reaction proceeded efficiently in the presence of $Pd(OAc)_2$ as the catalyst, benzoquinone as the oxidant, acetic acid/dioxane (2:1) as the solvent system, and under an atmosphere of CO. The addition of TsOH. H_2O as an additive was crucial for the synthesis of anthranilic acids in high yields. Notably, different acetanilide and benzanilide derivatives, and *N*-phenyl pyrrolidinone, could be carboxylated. The benzanilide

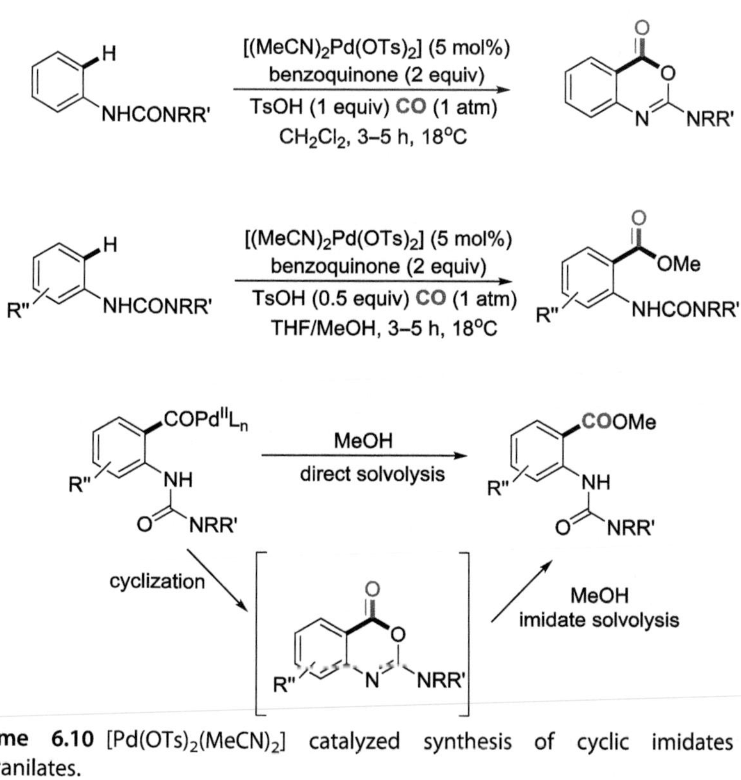

Scheme 6.10 $[Pd(OTs)_2(MeCN)_2]$ catalyzed synthesis of cyclic imidates and anthranilates.

derivatives could be selectively carboxylated at the ortho C—H bond of the aniline fragment, even in the presence of *ortho* C—H bonds of the benzoyl ring. The *N*-benzoyl anthranilic acids could be cyclized and transformed into benzoxazinones and quinazolinones by treatment with Ac_2O and PCl_3 in presence of aniline respectively (Scheme 6.11).

The proposed mechanism involves the attachment of the OTs anion to Pd(II) in the form of $L_2Pd(OTs)X$ on the C—H activation step, thereby increasing the reactivity. Subsequently, the C—H insertion intermediate **15** inserts CO to generate palladacycle **16**, which then undergoes reductive elimination with *p*-toluenesulfonate to give mixed anhydride species **17** (Scheme 6.12, path a). On the other hand, intermediate **18** could release a molecule of *p*-TsOH via enolization of the acidic N—H bond to generate a second intermediate **19**, which subsequently undergoes reductive elimination to generate benzoxazinone **20** (Scheme 6.12, path b). The final carboxylic acid product **14** thus arises from the subsequent hydrolysis of these anhydride and benzoxazinone intermediate products by water present in the reaction system.

In 2011, Wrigglesworth et al. documented the synthesis of *N*-alkoxyphthalimides from *N*-alkoxybenzamides through ortho carbonylative

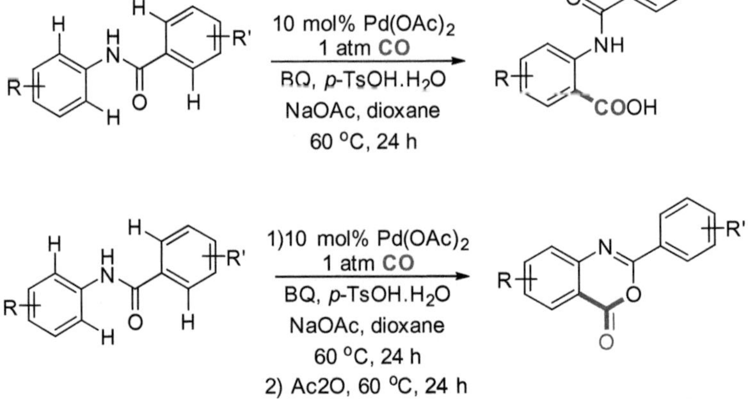

Scheme 6.11 Palladium-catalyzed ortho C—H carboxylation of acetanilide and benzanilide derivatives.

C(sp^2)—H activation.[11] The phthalimides were isolated in excellent yields and formation of exclusive products was observed, except in the case of *m*-methyl and *m*-methoxy-*N*-methoxybenzamide where formation of regioisomers were observed (Scheme 6.13).

The sulfonamide functional group is an important structural motif present in many drug scaffolds, especially sulfa drugs. It also acts as a directing group for C—H functionalization reactions, and hence can prove to be useful for the synthesis of potential drug candidates. In 2011, Yu and co-workers thus developed a protocol for the synthesis of a novel analogue of celecoxib, as a potential cyclooxygenase-II (COX-2)-specific inhibitor, through the palladium-catalyzed ortho C(sp^2)—H functionalization[12] (Scheme 6.14).

6.2.2.3 Amine as a Directing Group

The amine functionality is known to undergo facile cyclometalation with palladium, and hence finds use as an advantageous directing group in the carbonylative C(sp^2)—H functionalization reactions for the synthesis of a diverse set of molecules. In 2004, Orito and co-workers reported the

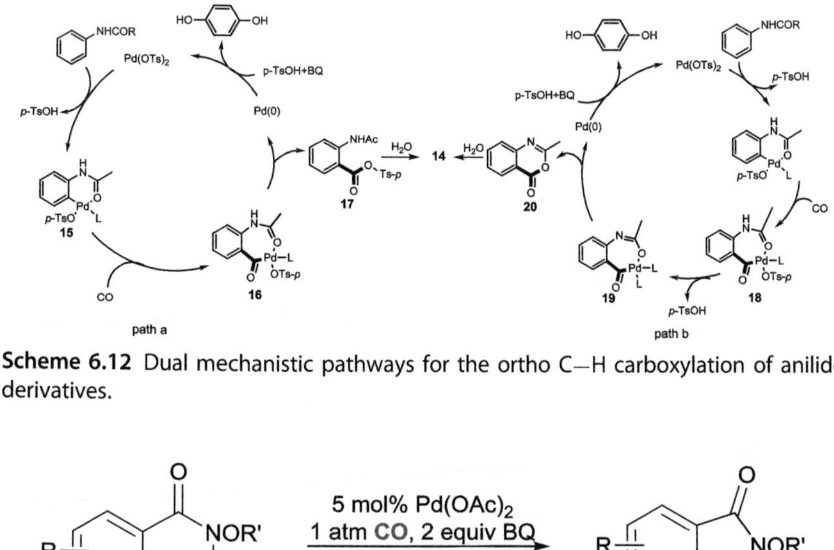

Scheme 6.12 Dual mechanistic pathways for the ortho C—H carboxylation of anilide derivatives.

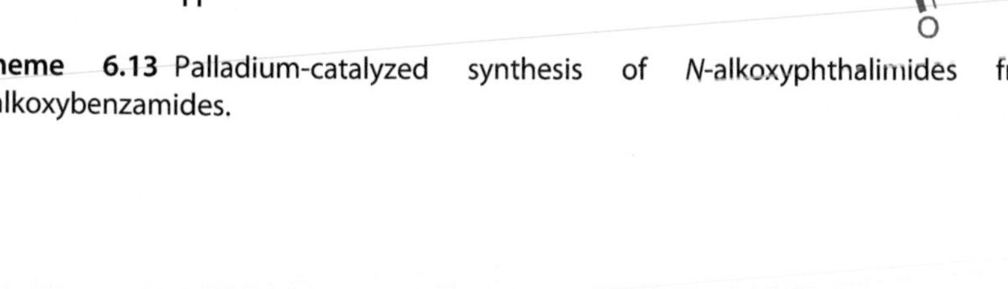

Scheme 6.13 Palladium-catalyzed synthesis of *N*-alkoxyphthalimides from *N*-alkoxybenzamides.

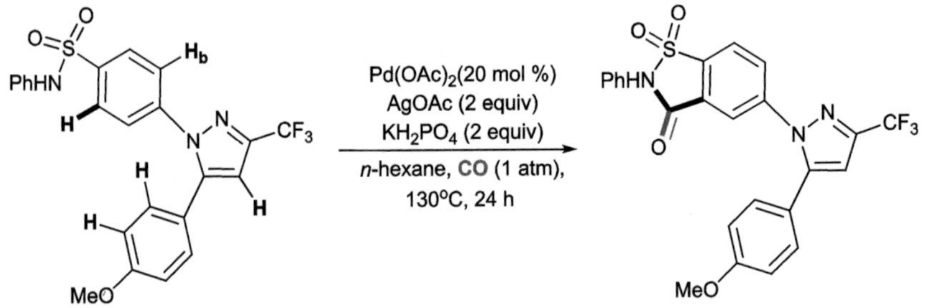

Scheme 6.14 Sulphonamide-directed synthesis of celecoxib analogue.

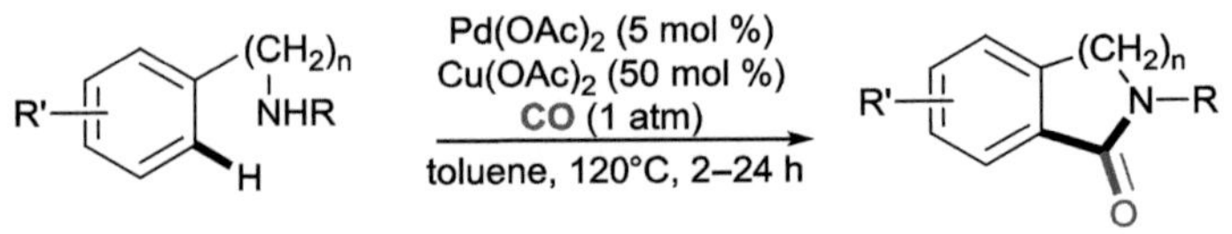

Scheme 6.15 Palladium-catalyzed synthesis of benzolactams from *N*-alkyl-ω-arylalkylamines.

synthesis of five- and six-membered benzolactams from *N*-alkyl-ω-arylalkylamines.[13] The reaction was catalyzed by $Pd(OAc)_2$ in the presence of Cu $(OAc)_2$, and under an atmosphere of CO gas containing 0.5 molar equivalent of O_2. It was observed that benzylic amines underwent carbonylation at a much faster rate as compared to the corresponding phenethylamines. Moreover, the rate of the five-membered ring formation was 11 times greater than that of the six-membered ring formation. The presence of a methylenedioxy group at the 3,4-positions increased the reaction rate by 10-fold in the six-membered ring closure (Scheme 6.15).

In 2010, Shi and co-workers reported the *ortho*-alkoxycarbonylation of *N,N*-dimethylbenzylamines.[14] The optimization studies revealed that the combination of $PdCl_2$ as the palladium precursor and $Cu(OAc)_2$ as the oxidant in 2,2,2-trifluoroethanol (TFEol) and acetic acid in a 4:1 ratio as the solvent system gave a 50% yield of the product. However, the addition of LiCl as an additive had a profound effect, and the product yield increased to 70%. The protocol was extended for the synthesis of various esters using different alcohols as nucleophiles, as well as differently substituted *N,N*-dimethylbenzylamines. Further, the benzylamine functionality could be converted into a methyl group through catalytic hydrogenation using Pd/C. Thus, the synthesis of ortho-methyl esters from *N,N*-dimethylbenzylamines through the carbonylation—hydrogenation sequence could be achieved in a single pot in 41% yield (Scheme 6.16).

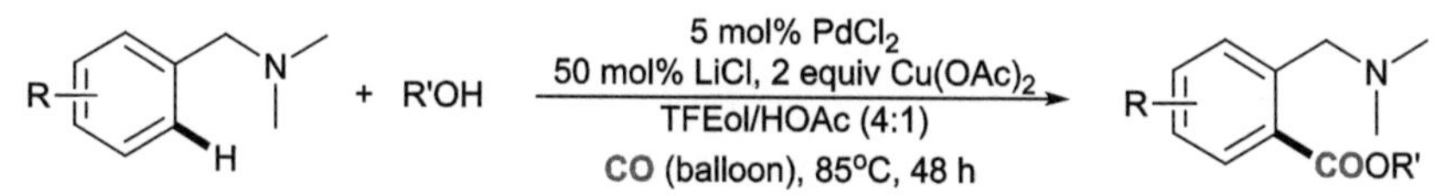

Scheme 6.16 Palladium-catalyzed ortho-alkoxycarbonylation of *N,N*-dimethylbenzylamines.

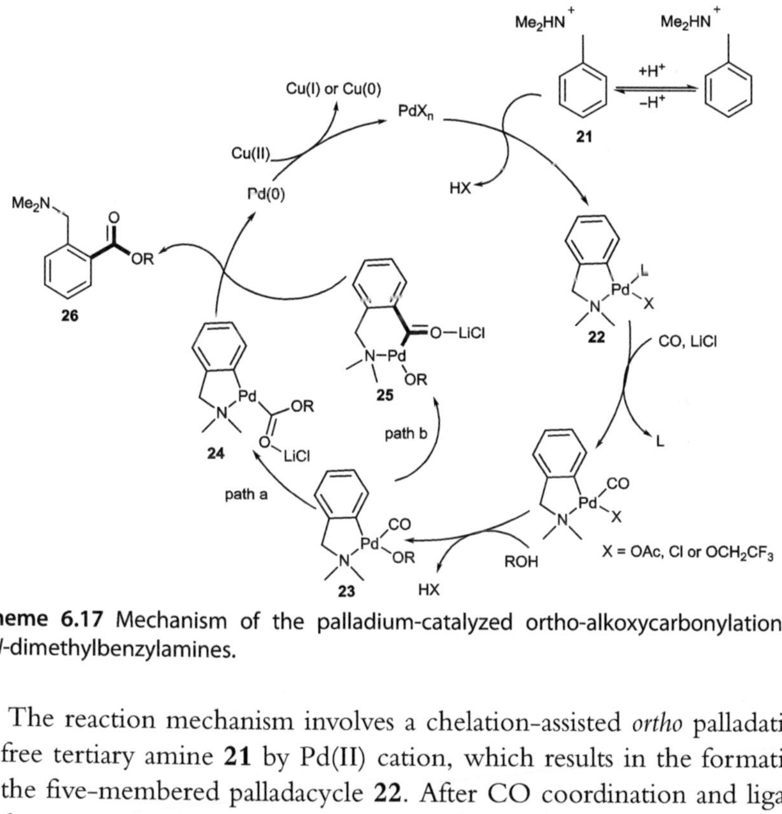

Scheme 6.17 Mechanism of the palladium-catalyzed ortho-alkoxycarbonylation of *N,N*-dimethylbenzylamines.

The reaction mechanism involves a chelation-assisted *ortho* palladation of free tertiary amine **21** by Pd(II) cation, which results in the formation of the five-membered palladacycle **22**. After CO coordination and ligand exchange, intermediate **23** transforms into **24** through path a, or **25** through path b, under the assistance of LiCl. The reductive elimination of **24** or **25** gives the desired product **26** and Pd(0), which is oxidized to Pd(II) by Cu(II) to regenerate the catalyst (Scheme 6.17).

Gaunt and co-workers reported Pd(II)-catalyzed carbonylative C—H functionalization in the β-arylethylamine motif, thus leading to the synthesis of dihydro-2-quinolone derivatives.[15] $Pd(OAc)_2$ in conjunction with benzoquinone and O_2 as oxidants, sodium acetate as an additive, and acetic acid as the solvent under an atmosphere of CO at room temperature, formed part of the optimized reaction conditions. Moreover, the use of acetic acid suppressed the CO-mediated reduction of Pd(II) to inactive palladium black. A range of alkyl—*N*-aryl substrates could be carbonylated, and the protocol was only effective for *N*-aryl compounds (Scheme 6.18)

In 2011, Urriolabeitia and co-workers reported the synthesis of orthopalladated methyl arylglycinate derivatives.[16] These orthopalladated complexes were further subjected to carbonylation, thus resulting in the synthesis of substituted (1*H*)-isoindolin-1-one-3-carboxylates (Scheme 6.19).

Guan and co-workers developed a Pd-catalyzed regioselective C—H bond carbonylation of *N*-alkyl anilines resulting in the synthesis of isatoic anhydrides.[17] The reaction proceeded in the presence of $Cu(OAc)_2$ as an oxidant and KI as an additive, and tolerated a wide range of functional groups (Scheme 6.20).

The mechanism involves the electrophilic palladation of the C—H bond of *N*-methyl aniline **27** by $Pd(OAc)_2$ under a CO atmosphere to

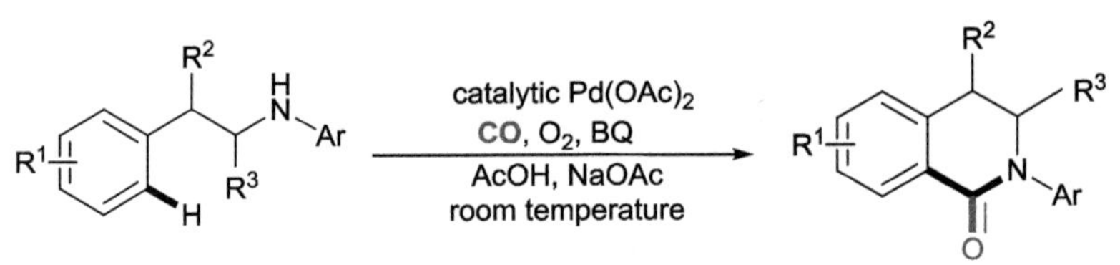

Scheme 6.18 Palladium-catalyzed C—H functionalization of β-arylethylamine derivatives.

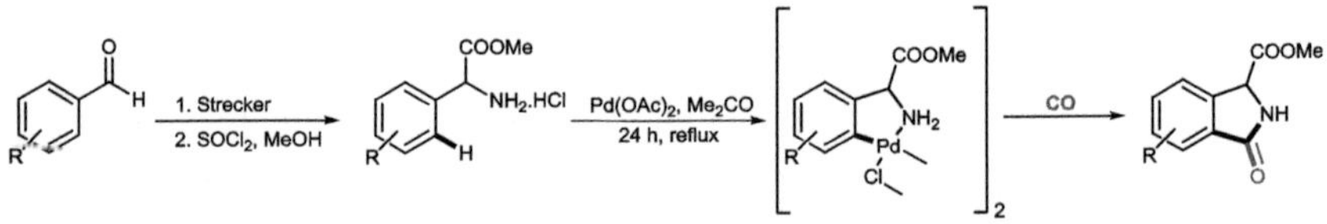

Scheme 6.19 Synthesis of (1*H*)-isoindolin-1-one derivatives by carbonylation of orthopalladated complexes.

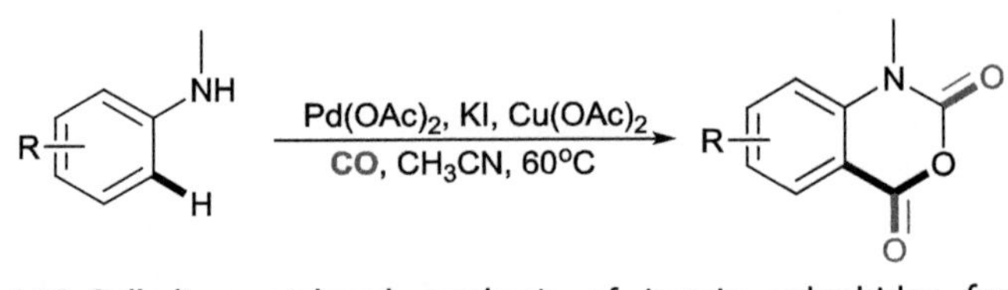

Scheme 6.20 Palladium-catalyzed synthesis of isatoic anhydrides from *N*-alkyl anilines by carbonylative C—H functionalization.

form a dimeric palladium intermediate **28**. This was a key intermediate in the reaction, and was characterized by 1H NMR, mass spectrometry, and X-ray crystallography. Insertion of CO to intermediate **28**, followed by reductive elimination, gives *N*-methylanthranilic acid **29** and Pd(0). Pd(0) was subsequently assumed to be oxidized by $Cu(OAc)_2$ to generate the $Pd(OAc)_2$ catalyst. The metathesis of *N*-methylanthranilic acid **29** and $Pd(OAc)_2$ generates the intermediate **30**. Coordination and migratory insertion of CO to **30** generates **31**. The nucleophilic attack of the amino group on the acylpalladium moiety thus gives the isatoic anhydride **32** and Pd(0). Finally, Pd(0) is oxidized by $Cu(OAc)_2$ to regenerate the $Pd(OAc)_2$ catalyst (Scheme 6.21).

In 2013, Zhu and co-workers reported a Pd-catalyzed $C(sp^2)$—H aminocarbonylation of unprotected *o*-arylanilines under an atmospheric pressure of CO resulting in the synthesis of (NH)-phenanthridinones.[18] The protocol could also be extended to the synthesis of phenanthridinones from ortho heteroarene substituted anilines, as well as *N*-alkyl protected *o*-arylanilines. The optimization revealed that the combination

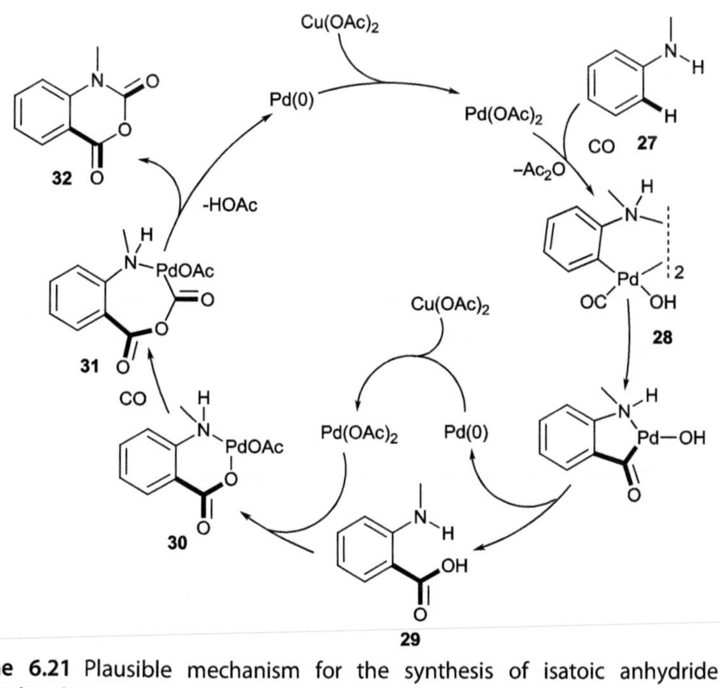

Scheme 6.21 Plausible mechanism for the synthesis of isatoic anhydride from *N*-methyl aniline.

of $Pd(MeCN)_2Cl_2$, $Cu(TFA)_2$, and 1 equivalent of trifluoroacetic acid was crucial for getting high yields (Scheme 6.22).

Initial chelation of the amino group with the CO ligated palladium(II) complex forms intermediate **33**. Subsequent electrophilic cyclopalladation on the nonaniline ring results in intermediate **34**. This is followed by migratory insertion of coordinated CO into the aryl—Pd bond. Reductive elimination from the seven-membered palladacycle **35** leads to the amino-carbonylation product **36** and simultaneous formation of Pd(0) species, which is reoxidized to the active palladium(II) complex by $Cu(TFA)_2$ in the presence of trifluoroacetic acid (TFA) and CO (Scheme 6.23).

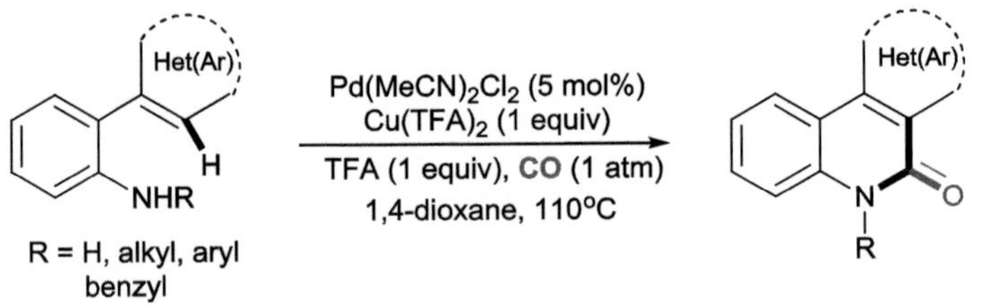

Scheme 6.22 Palladium-catalyzed C—H aminocarbonylation of *o*-arylanilines.

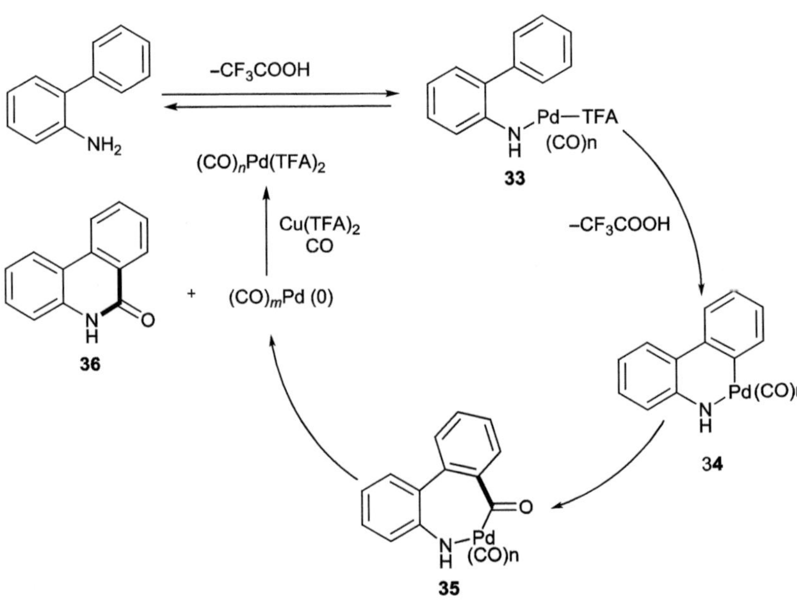

Scheme 6.23 Plausible mechanism of the palladium-catalyzed C—H aminocarbonylation of *o*-arylanilines.

Chuang and co-workers reported the palladium-catalyzed oxidative carbonylation of *N*-sulfonyl-2-aminobiaryls through carbonylative C—H functionalization resulting in the synthesis of phenanthridinone derivatives.[19] Notably, the reaction proceeded under trifluoroacetic acid free conditions, and did not also require addition of acid additives (Scheme 6.24).

Albert et al. developed an NH_2-directed Pd(II)-catalytic carbonylation of quaternary aromatic a-amino esters resulting in the synthesis of benzolactams.[20] The steric hindrance around the amino group is important for the success of the protocol. The stoichiometric cyclometalation of a variety of amino acid esters was studied to evaluate the influence of different variables (size of the metallacycle, aromatic ring substituents, and steric bulk) in the process. Additionally, a complete kinetico-mechanistic study of the cyclopalladation process was also carried out. The experimental results indicate that the full substitution of the carbon in the α position of the amino acid esters plays an important role in the cyclopalladation reaction. The steric hindrance around the amino group seems to be crucial due to which the six-membered lactams were favored over five-membered lactams (Scheme 6.25).

Lei and co-workers reported the palladium-catalyzed oxidative double carbonylation of C—H bonds of anilines thus leading to the synthesis of isatins (indoline-2,3-diones).[21] Interestingly, in the presence of $PdCl_2(PPh3)_2$ as the catalyst and $Cu(OPiv)_2$ as the oxidant, the reaction proceeds under atmospheric pressure of CO (Scheme 6.26).

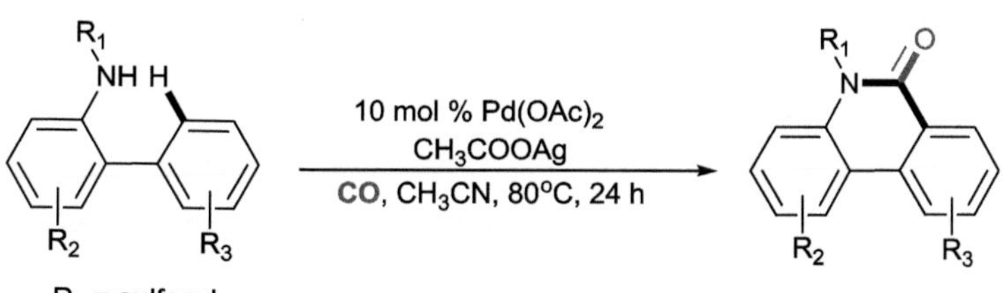

Scheme 6.24 Palladium-catalyzed oxidative aminocarbonylation of *N*-sulfonyl-2-aminobiaryls.

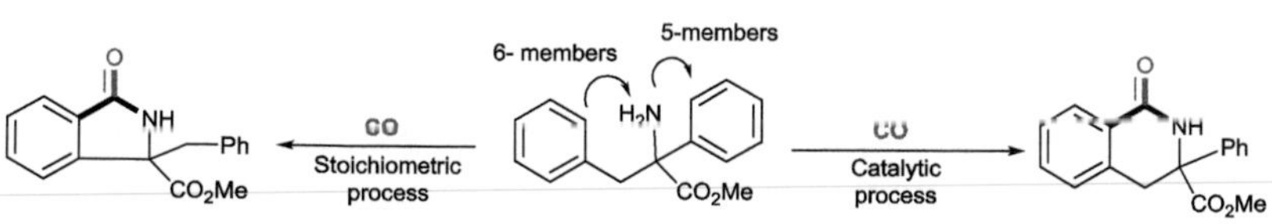

Scheme 6.25 Palladium-catalyzed NH_2-directed carbonylation of quaternary aromatic α-amino esters.

A possible mechanism for this reaction has been outlined in Scheme 6.27. First, aryl N—H activation by the palladium complex forms the intermediate **38** from **37**. Then, coordination and insertion of CO into **38** affords the carbamoyl intermediate **39**. Then **39** undergoes another CO insertion to give the intermediate **40**. The C—H activation of **40** generates a six-membered cyclic carbamoyl intermediate **41**, which is followed by a reductive elimination reaction to give the product **42**. Finally, the Pd(0) species is reoxidized to the Pd(II) catalyst in the presence of $Cu(OPiv)_2$. Density functional theory investigations were carried out which indicate that the mechanism depicted in Scheme 6.27 is favorable.

In 2015, Guan and co-workers reported the ortho C—H carbonylation of *N*-alkylanilines leading to the synthesis of *o*-aminobenzoates.[22] In the protocol, various aliphatic alcohols and phenol were tolerated,

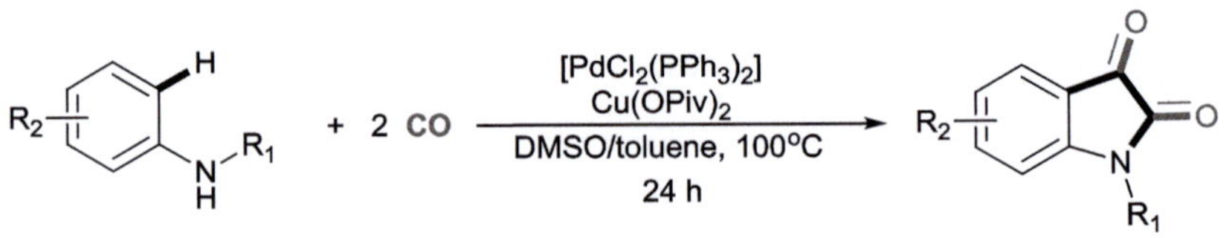

Scheme 6.26 Palladium-catalyzed oxidative double carbonylation of aniline.

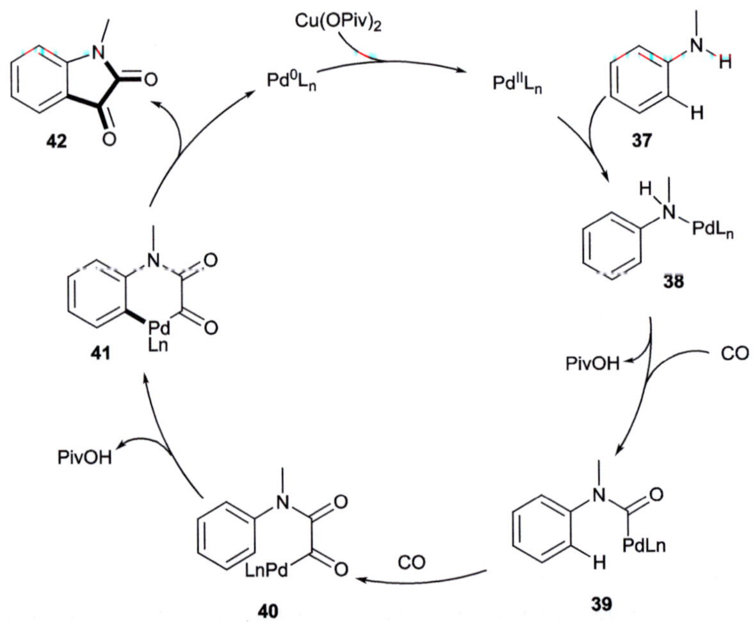

Scheme 6.27 Proposed mechanism for the formation of isatin from aniline.

affording the corresponding *o*-aminobenzoates under balloon pressure of CO (Scheme 6.28).

Almost at the same time, Lei's group documented the ortho C—H carbonylation of N-substituted anilines with alcohols resulting in the synthesis of *o*-aminobenzoates.[23] Lei's protocol, however, used molecular oxygen as the terminal oxidant in combination with palladium and copper (Scheme 6.29).

6.2.2.4 Carboxylic Acid as a Directing Group

In 2008, Giri and Yu reported the first carbonylative C—H functionalization using carboxylic acid as a directing group.[24] The methodology involved the regioselective *ortho* carboxylation of benzoic acid and phenylacetic acid derivatives resulting in the synthesis of dicarboxylic acids. The cyclopalladated intermediate involved in the reaction was isolated and characterized by X-ray crystallography (Scheme 6.30).

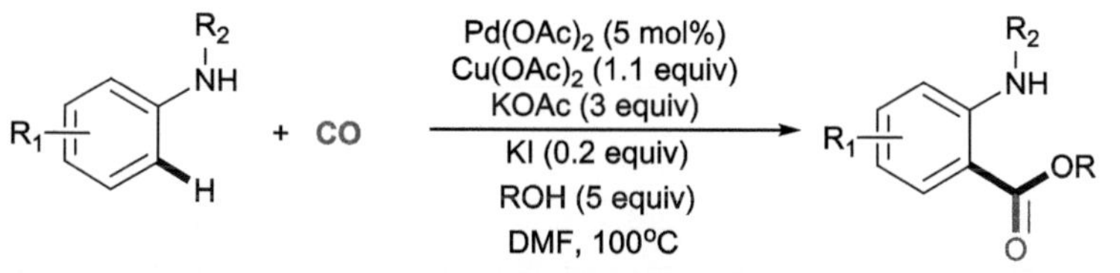

Scheme 6.28 Palladium-catalyzed ortho C—H carbonylation of *N*-alkylanilines.

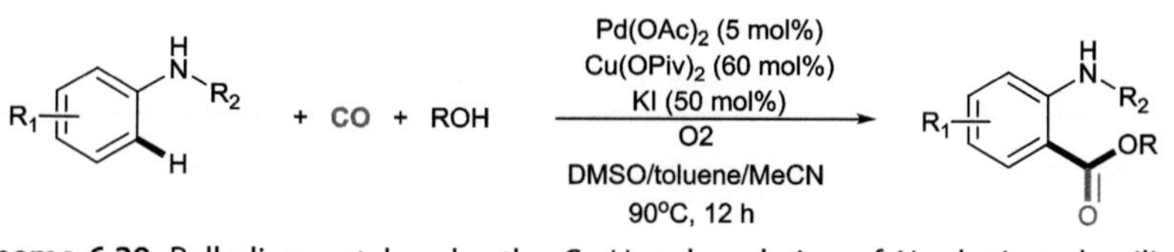

Scheme 6.29 Palladium-catalyzed ortho C—H carbonylation of *N*-substituted anilines using molecular oxygen as the terminal oxidant.

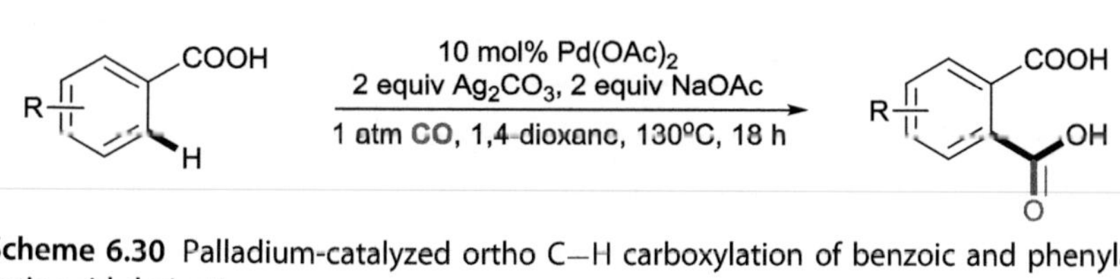

Scheme 6.30 Palladium-catalyzed ortho C—H carboxylation of benzoic and phenylacetic acid derivatives.

6.2.2.5 Nitrogen Containing Heterocycle as a Directing Group

Nitrogen containing heterocycles are useful directing groups and have been reported extensively in C—H functionalization reactions. As far as carbonylative C—H functionalization is concerned, they have been mainly used in conjunction with rhodium and ruthenium. Palladium-catalyzed carbonylative C—H functionalizations using a heterocycle moiety as a directing group are relatively few and shall be discussed in this section. In 2013, Liu and Shi reported the alkoxycarbonylation of arenes and heteroarenes with the assistance of the pyridyl functionality as the directing group.[25] The reaction proceeded in the presence of $CuBr_2$ as the oxidant, NaOAc as the additive, and under an atmosphere of CO (Scheme 6.31).

The plausible mechanism involves coordination of **43** to palladium acetate which is followed by C—H activation to form the dimeric palladacycle **45**. Subsequently, CO and alcohol coordinate to Pd via ligand exchange and release of AcOH. At this point two possible pathways exist for the formation of carboxylic ester **44** from intermediate **46**. (1) migratory insertion of CO into the Pd—C bond to form intermediate **47** which is followed by C—O bond reductive elimination. (2) Migratory insertion of CO into the Pd—OR bond to form intermediate **48** followed by C—C bond reductive elimination to give **44** (Scheme 6.32).

In 2014, the same group documented the alkoxycarbonylation of 2-phenoxypyridines under an atmosphere of CO and O_2 in the presence of AgTFA as an oxidant.[26] The protocol thus allowed the C—H functionalization of phenol derivatives with the assistance of the pyridyl functionality as a directing group. The protocol was also extended for the late stage modification of complex molecules by the derivatization of estrone (Scheme 6.33).

Zhu and co-workers documented the synthesis of 11*H*-pyrido[2,1-*b*]quinazolin-11-ones from *N*-aryl-2-aminopyridines.[27] Interestingly, this was the first example where the pyridyl functionality acted as a directing

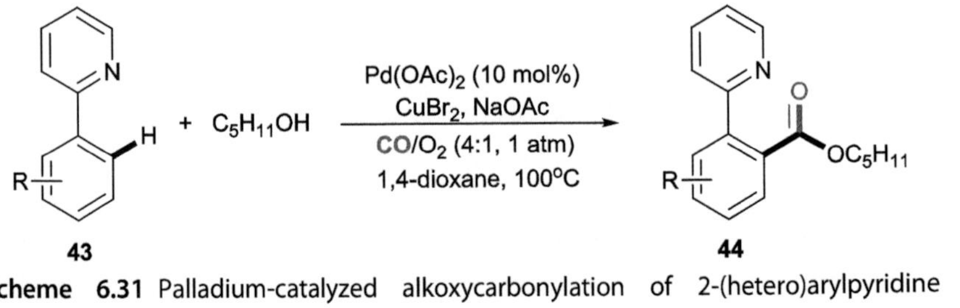

43 44

Scheme 6.31 Palladium-catalyzed alkoxycarbonylation of 2-(hetero)arylpyridine derivatives.

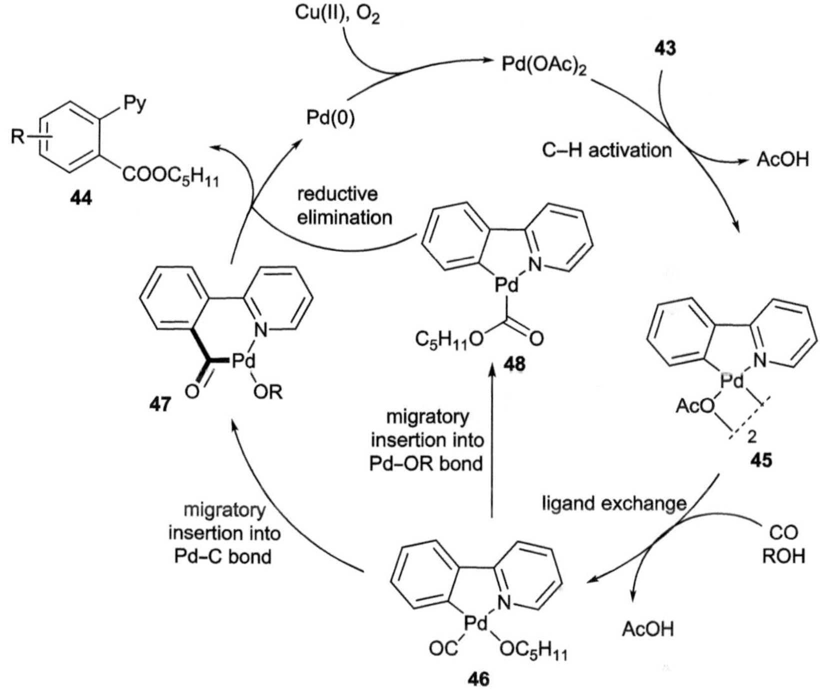

Scheme 6.32 Mechanism of the palladium-catalyzed alkoxycarbonylation of 2-(hetero)arylpyridine derivatives.

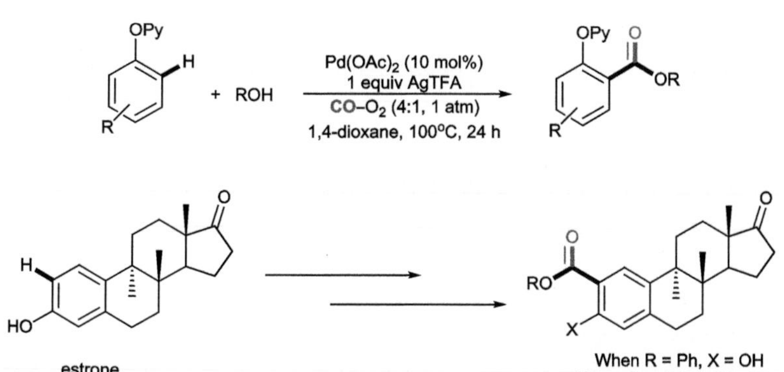

Scheme 6.33 Palladium-catalyzed alkoxycarbonylation of 2-phenoxypyridines.

group and an intramolecular nucleophile in a palladium-catalyzed $C(sp^2)$—H carbonylation. The C—H functionalization proceeded under an atmospheric pressure of CO in the presence of $Pd(OAc)_2$ and $K_2S_2O_8$ as an oxidant in TFA (Scheme 6.34).

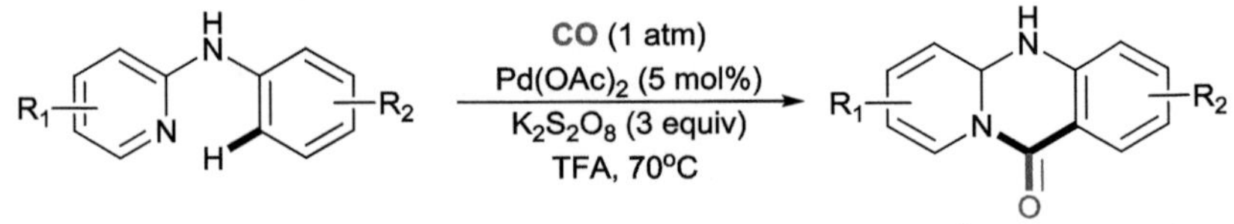

Scheme 6.34 Palladium-catalyzed pyridine directed intramolecular C—H carbonylation.

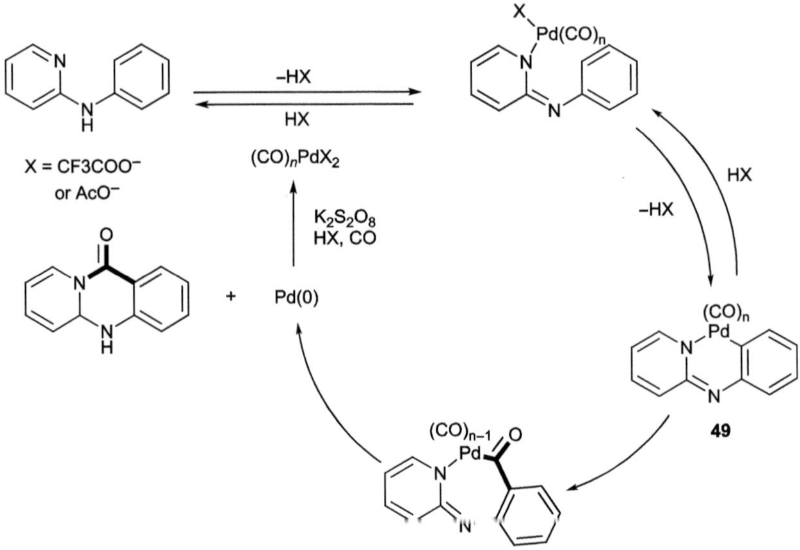

Scheme 6.35 Mechanism of the pyridine directed intramolecular C—H carbonylation.

The proposed mechanism involved initial chelation of the pyridine nitrogen with the CO ligated Pd(II) complex. This is followed by electrophilic cyclopalladation on the phenyl ring to generate intermediate **49**, which is probably a rate-limiting step. Migratory insertion of a coordinated CO into the aryl—Pd bond produces a seven-numbered palladacycle **50**. Reductive elimination leads to the corresponding pyridocarbonylation product with the simultaneous formation of Pd(0), which is reoxidized to Pd(II) by $K_2S_2O_8$ in the presence of CO and TFA (Scheme 6.35).

6.2.2.6 Hydroxyl Functionality as a Directing Group

In 2011, Yu and co-workers reported the Pd(II)-catalyzed *ortho*-C(sp^2)—H carbonylation of phenethyl alcohols promoted by amino acid ligands.[28] The transformation resulted in the synthesis of the 1-isochromanone functionality which is part of many natural products and biologically active compounds. More importantly, this protocol represents a case in which the directing group and the coupling partner are fully incorporated into the target molecule without any further synthetic operation (Scheme 6.36).

In the proposed mechanism, the hydroxyl group directs *ortho*-C—H bond cleavage by Pd(II). Subsequently, a molecule of CO binds to the [Pd(II)—Ar] species. This is followed by a 1,1-migratory insertion of CO into the Pd—C bond to forge the new C—C bond, generating a seven-membered cyclic intermediate. Nucleophilic attack of the hydroxyl group on the carbonyl moiety gives the final product (Scheme 6.37).

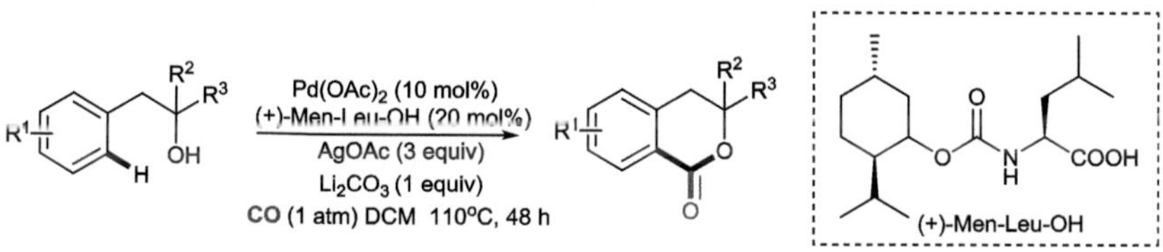

Scheme 6.36 Palladium-catalyzed C—H carbonylation of phenethyl alcohols.

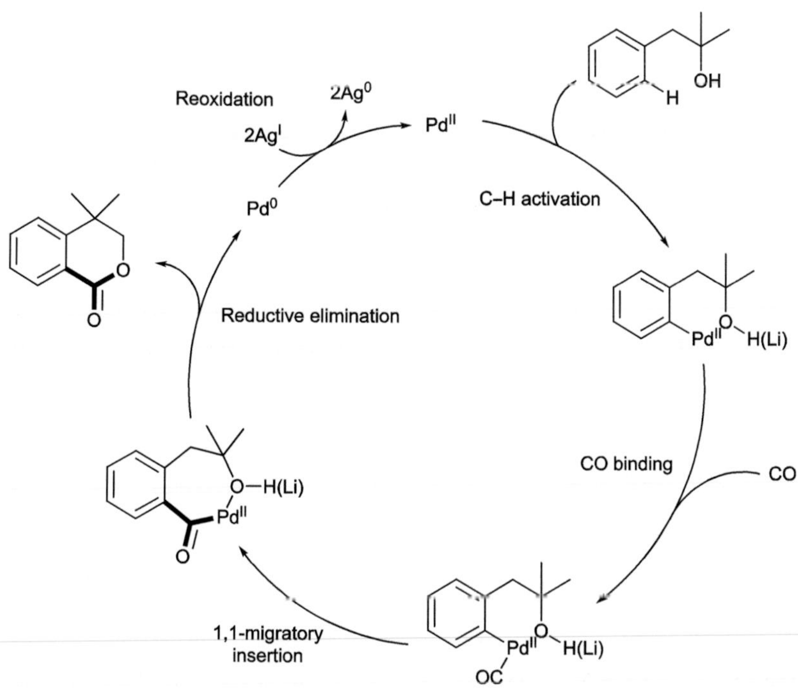

Scheme 6.37 Proposed mechanism of the C—H carbonylation of phenethyl alcohols.

In 2013, Shi and co-workers reported the Pd-catalyzed C—H carbonylation of 2-arylphenol derivatives resulting in the synthesis of various dibenzopyranone derivatives.[29] The transformation could be effected in the presence of $Pd(OAc)_2$ as a catalyst and $Cu(OAc)_2$ as a catalytic oxidant under an atmospheric pressure of CO and O_2 (Scheme 6.38).

The proposed mechanism involves generation of intermediate **52** by the chelation of the hydroxy group of **51** with the CO-ligated palladium(II) complex. Subsequent C—H activation occurs through electrophilic cyclopalladation, resulting in the six-membered palladacycle **53**. Then the migratory insertion of coordinated CO into the aryl—Pd bond forms the seven-membered palladacycle **54**. Reductive elimination from intermediate **54** leads to the formation of lactone **55** and concurrent formation of a Pd(0), which is reoxidized to palladium(II) by $Cu(OAc)_2$ in the presence of O_2 (Scheme 6.39).

Immediately, Lee et al. reported the synthesis of benzopyranones by a similar carbonylative C—H functionalization of 2-arylphenols,[30] as reported by Shi's group. However, as compared to Shi's protocol, this transformation could be effected without the addition of acid, base, or ligand, and could be carried out at lower temperature (Scheme 6.40).

Lee and co-workers reported the palladium-catalyzed carbonylative $C(sp^2)$—H functionalization of phosphonic and phosphinic acids leading to the synthesis of oxaphosphorinanone oxides, which are novel heterocyclic moieties.[31] The hydroxyl functionality in the phosphonic/phosphinic acids provided the required directing group assistance. The reaction was catalyzed by $Pd(OAc)_2$ in combination with $PhI(OAc)_2$ and AgOAc as oxidants (Scheme 6.41).

The plausible reaction mechanism is initiated by coordination of the hydroxyl group in the phosphonic/phosphinic acids to Pd(II) to provide palladium(II) phosphonate/phosphinate respectively. This is followed by *ortho*-metalation to provide palladacycle intermediate **56**. CO

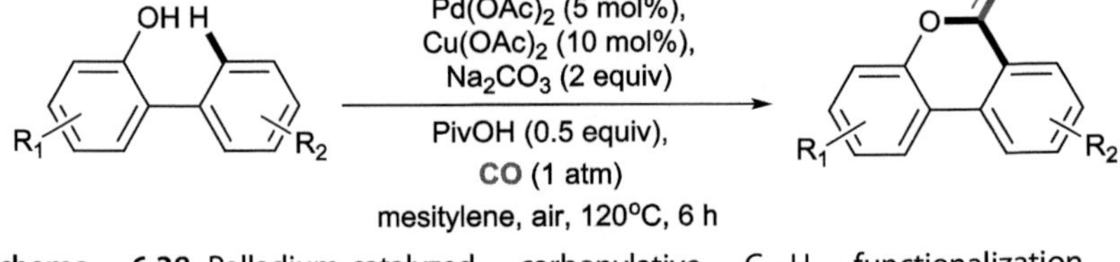

Scheme 6.38 Palladium-catalyzed carbonylative C—H functionalization of 2-phenylphenols.

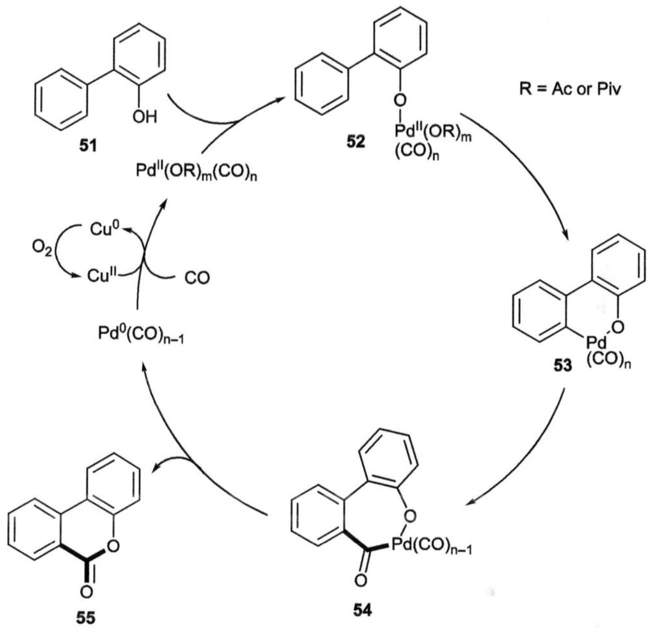

Scheme 6.39 Mechanism for the palladium-catalyzed carbonylative C—H functionalization of 2-phenylphenols.

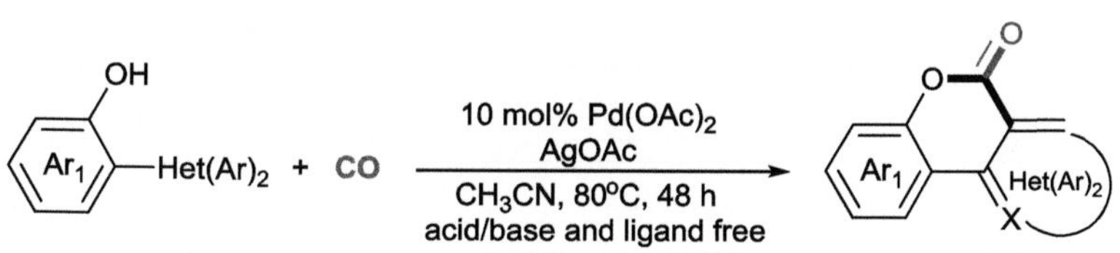

Scheme 6.40 Palladium-catalyzed synthesis of benzopyranones through C—H carbonylation of 2-arylphenols under acid, base, and ligand free conditions.

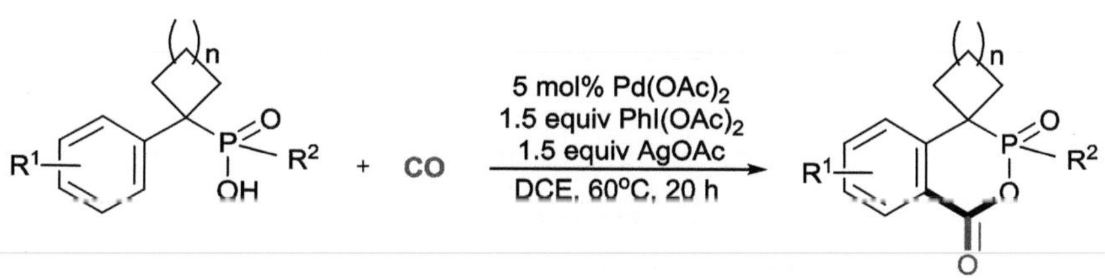

Scheme 6.41 Palladium-catalyzed C—H carbonylation of phosphonic and phosphinic acids.

coordination, migratory insertion, and reductive elimination from **57** results in the carbonylated products, oxaphosphorinanone oxides **58** (Scheme 6.42).

In 2015, Wang and Gevorgyan showed that salicylic acid derivatives could be synthesized by the palladium-catalyzed, silanol directed C—H carbonylation of phenols.[32] The methodology involves transformation of the phenol to phenoxy silanols. This then undergoes a carbonylative $C(sp^2)$—H functionalization under the directing group assistance of the hydroxyl group of the silanol to generate the silacycle **59**. Desilylation of the silacycle thus affords the salicyclic acid derivatives. The conditions for the synthesis of the silacycle from phenoxy silanol were optimized and proceeded efficiently in the presence of $Pd(OAc)_2$, Boc-Leu-OH as a mono-protected amino acid ligand, and AgOAc as the oxidant in 1,2-dichloroethane as the solvent. The desilylation could be effected using tetrabutylammonium fluoride. The ^{18}O labeling studies showed that the lactone oxygen in the silacycle originates from the hydroxyl group in the silanol. This protocol was also applied for the synthesis of complex phenol molecules (Scheme 6.43).

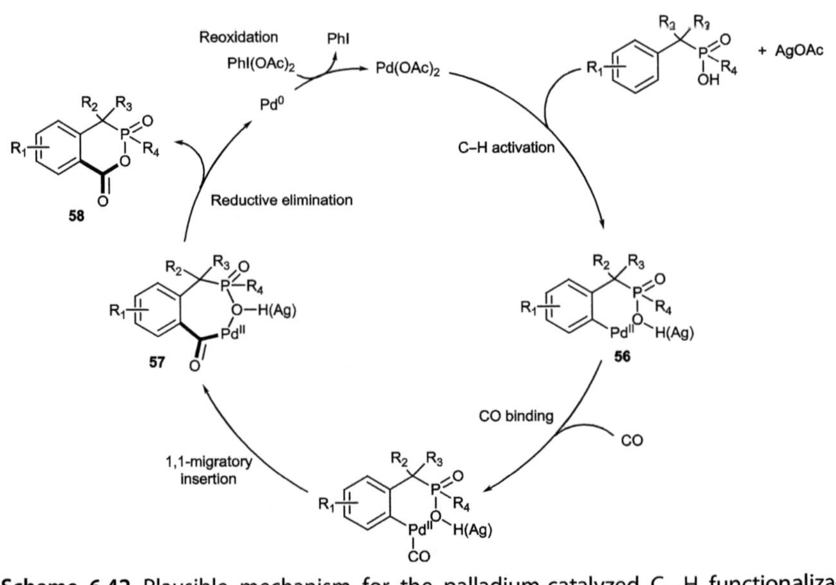

Scheme 6.42 Plausible mechanism for the palladium-catalyzed C—H functionalization of phosphonic and phosphinic acids.

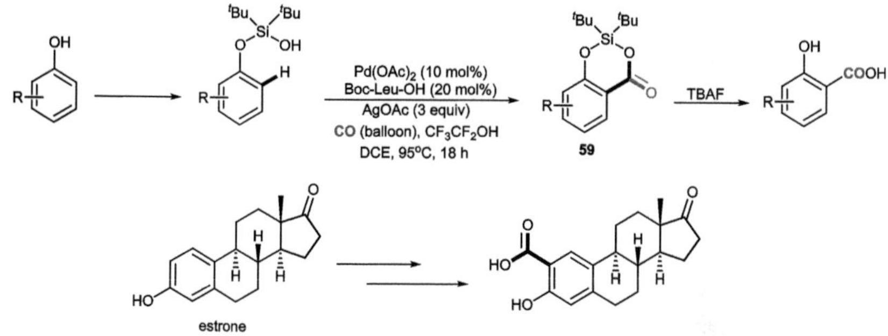

Scheme 6.43 Palladium-catalyzed synthesis of salicylic acid derivatives from phenols through the silanol directed C—H carbonylation-desilylation sequence.

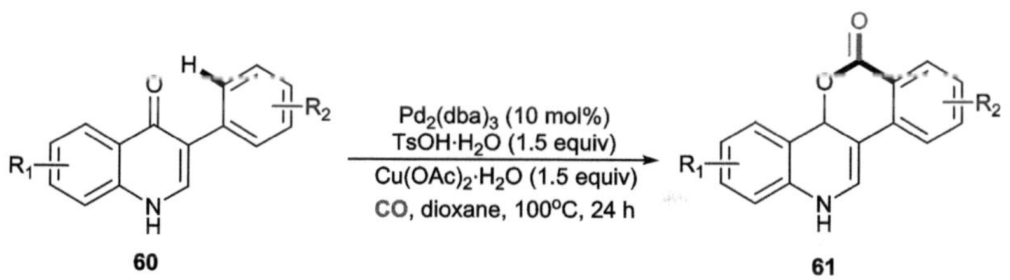

Scheme 6.44 Palladium-catalyzed synthesis of 6*H*-isochromeno[4,3-*c*]-quinolin-6-ones.

6.2.2.7 *Ketone as a Directing Group*

The use of the ketone functionality for directing group assistance in carbonylative C—H functionalization is less documented. In 2014, Jiang and co-workers reported the synthesis of 6*H*-isochromeno[4,3-*c*]-quinolin-6-ones from 3-phenylquinolin-4(1*H*)-ones through the carbonylative C—H functionalization strategy.[33] Thus, this methodology allowed for the increase in molecular complexity through the synthesis of polycyclic aromatic hydrocarbons. The reaction was catalyzed by $Pd_2(dba)_3$ in the presence of acid additive $TsOH \cdot H_2O$ and oxidant $Cu(OAc)_2$ (Scheme 6.44).

The proposed mechanism involves in situ formation of $Pd(OTs)_2$ by the reaction of $Pd_2(dba)_3$ with $Cu(OAc)_2$ and $TsOH \cdot H_2O$. Subsequently, aryl C—H activation by $Pd(OTs)_2$ affords the arylpalladium species **62** which further reacts with CO to form the intermediate **63**. Then, **63** is transformed into **64** assisted by expulsion of *p*-TsOH. Finally, reductive elimination from **64** gives the desired carbonylation product **61**, along with formation of Pd(0). Pd(0) is then oxidized to Pd(II) in the presence of $Cu(OAc)_2$ and $TsOH \cdot H_2O$ (Scheme 6.45).

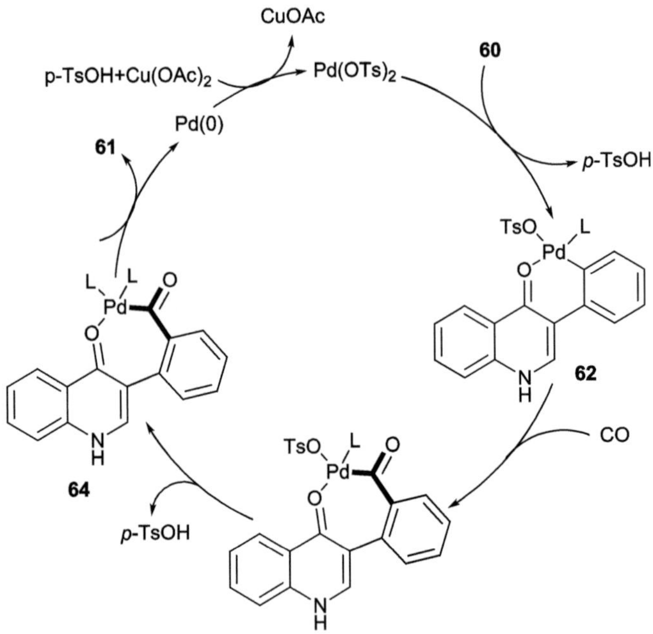

Scheme 6.45 Mechanism for the formation of 6*H*-isochromeno[4,3-*c*]-quinolin-6-ones by the carbonylative C—H functionalization.

6.3 CARBONYLATIVE C(sp^2)—H FUNCTIONALIZATION OF HETEROARENES

6.3.1 Introduction

In heterocyclic aromatic systems, one or more carbon atoms of the aromatic system are replaced by a heteroatom, typically a nitrogen, oxygen, or a sulfur atom. This greatly alters the electronic properties of the ring. Moreover, the presence of heteroatoms could lead to the coordination of palladium to these atoms, thus sometimes resulting in catalytic poisoning. Hence, the carbonylative C(sp^2)—H functionalization of heteroaromatic systems is relatively challenging as compared to the aromatic counterpart. However, heterocyclic rings are central building blocks present in drugs, natural products, agrochemicals, and other fine and specialty chemicals. Therefore, the C—H functionalization of heteroarenes is an advantageous protocol in synthetic chemistry. Many strategies have been developed for the efficient and regioselective palladium-catalyzed carbonylative C—H functionalization of heteroaromatic systems and shall be discussed in the next section. The reactions have been classified according to the type of heteroarene being carbonylated.

6.3.2 Carbonylative C—H Functionalization of Heteroarenes Containing a Single Heteroatom

In 1980, Fujiwara and co-workers documented the first example of the palladium-promoted carbonylation of heteroarenes.[4] Furan and thiophene were transformed to their corresponding carboxylic acids in one step, with excellent regioselectivity at the second position. Interestingly, the reaction was not catalytic and proceeded in the absence of oxygen. The formation of a benzoyl PdOAc σ-complex as a key intermediate was proposed (Scheme 6.46).

Later, Itahara carried out the palladium-promoted regioselective carbonylation of thiophenes and indoles.[34] The reactions were carried out in acetic acid reflux in the presence of $Na_2S_2O_8$ as an oxidant and bubbling CO (Scheme 6.47).

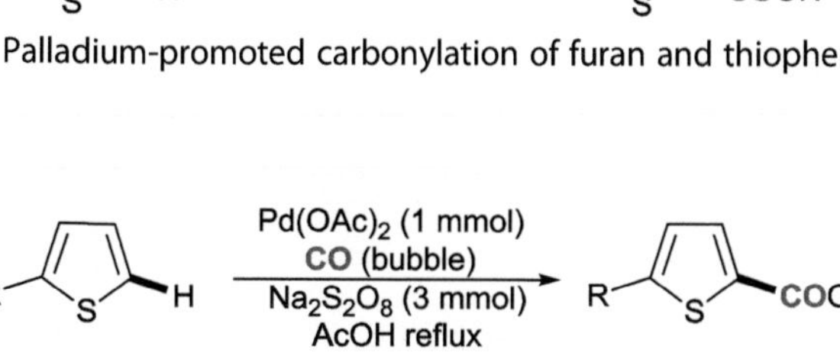

Scheme 6.46 Palladium-promoted carbonylation of furan and thiophene.

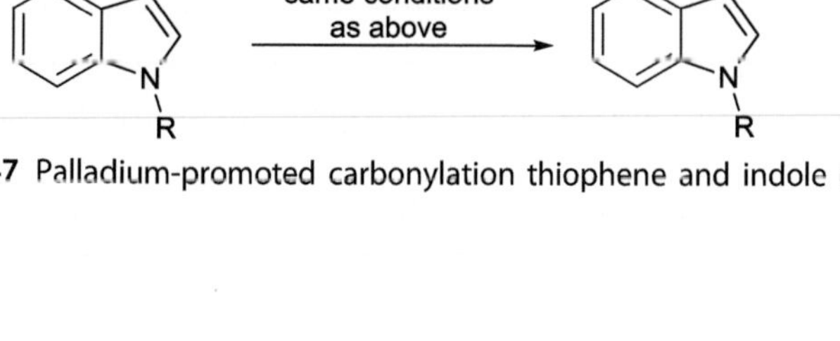

Scheme 6.47 Palladium-promoted carbonylation thiophene and indole under acetic acid reflux.

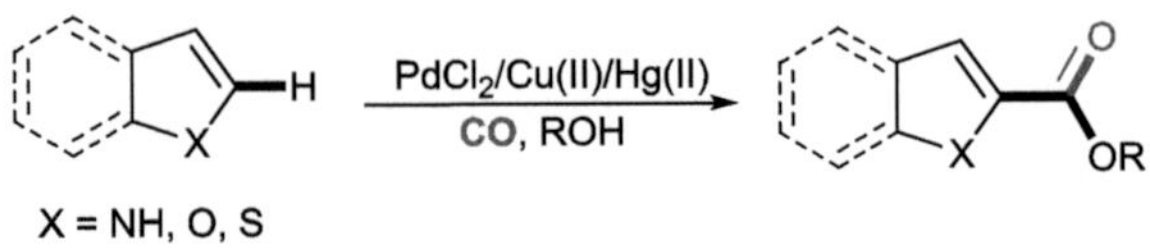

Scheme 6.48 Palladium-catalyzed regioselective alkoxycarbonylation of C—H bonds in heteroarenes.

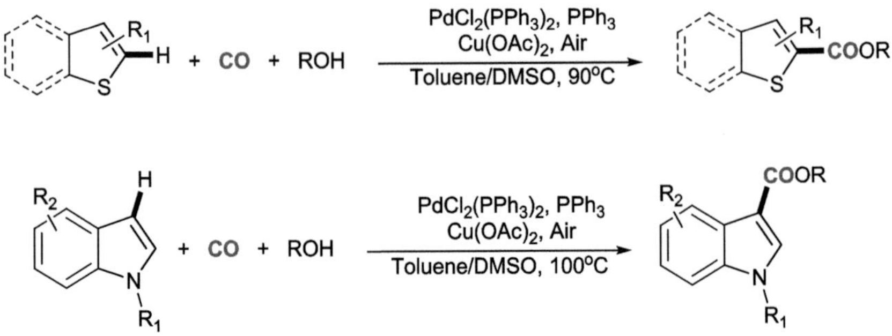

Scheme 6.49 Palladium-catalyzed oxidative C—H carbonylation of heteroarenes with alcohols.

In 1986, Dixneuf and co-workers reported the palladium-catalyzed alkoxycarbonylation of C—H bonds of thiophene, furan, benzofuran, and pyrrole leading to the synthesis of different esters using a $PdCl_2$-mercury salt-copper(II) system at room temperature.[35] The reaction proceeded with excellent regioselectivity as the C-2 position in the heteroaromatic systems could be carbonylated. The effect of different mercury and copper salts was studied, and it was proposed that the in situ formation of mercury derivatives of the heteroarenes was a key step in the C—H activation (Scheme 6.48).

In 2011, Lei's group reported the regioselective oxidative C—H carbonylation of heteroarenes with alcohols as the coupling partners leading to the synthesis of a range of heterocyclic esters under balloon pressure of CO.[36] Thiophenes and benzothiophenes could be carbonylated at the C-2 position. On the other hand, when *N*-alkyl/aryl indoles were employed as substrates, the carbonylation took place at the C-3 position (Scheme 6.49).

Since the electron-rich carbon atom on each aromatic ring is regioselectively carbonylated, the reaction probably takes place through an electrophilic palladation process. The electrophilic palladation of indole **66** at the C-3 position with the Pd(II) species **65** afforded the intermediate **67**. This further reacted with CO and ROH to produce the final oxidative

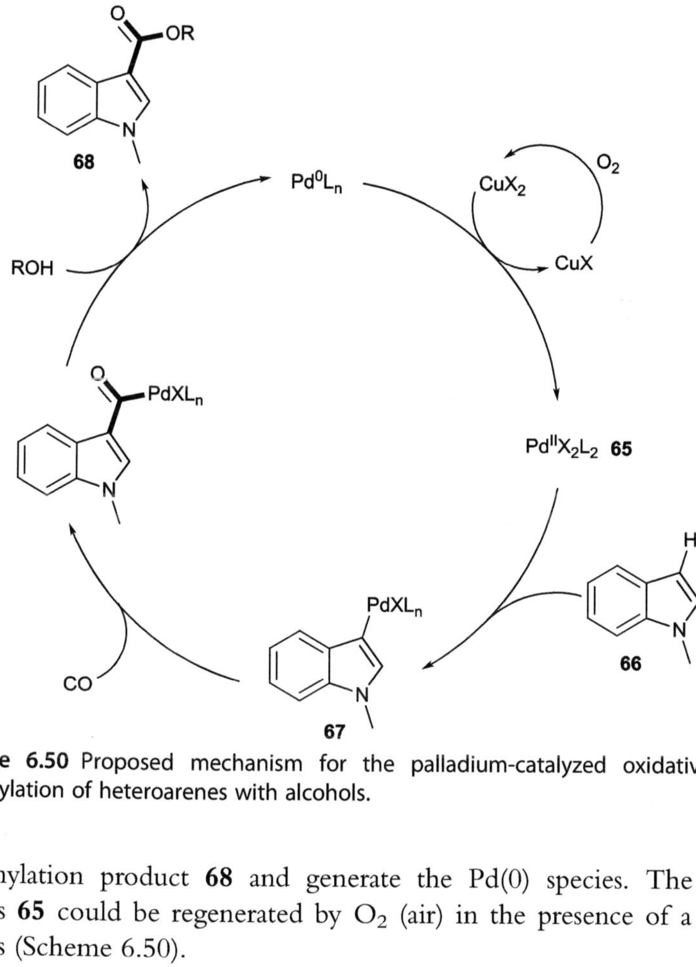

Scheme 6.50 Proposed mechanism for the palladium-catalyzed oxidative C—H carbonylation of heteroarenes with alcohols.

carbonylation product **68** and generate the Pd(0) species. The Pd(II) species **65** could be regenerated by O_2 (air) in the presence of a Cu(II) species (Scheme 6.50).

Later, Li and co-workers reported a novel strategy for the C—H carbonylation of indoles to synthesize indole-3-carboxylates.[37] As compared to Lei's protocol, this strategy involved an oxidative iodination followed by carbonylation in the presence of $Pd(OAc)_2$, base, and I_2 under an atmospheric pressure of CO. Interestingly, even *N*-free indoles could be regioselectively carbonylated at the C-3 position, and the protocol could be used for the synthesis of alcohol and phenol esters. The current strategy was also employed for the synthesis of tropisetron, a 5 hydroxytryptamine receptor (5-HT3) antagonist (Scheme 6.51).

The mechanism involved the oxidation of 1*H*-indole **69** to 3-iodo-1*H*-indole **70**. A reactive Pd(0) species generated from CO reduction of Pd(II) preferentially initiates oxidative addition affording intermediate **71**

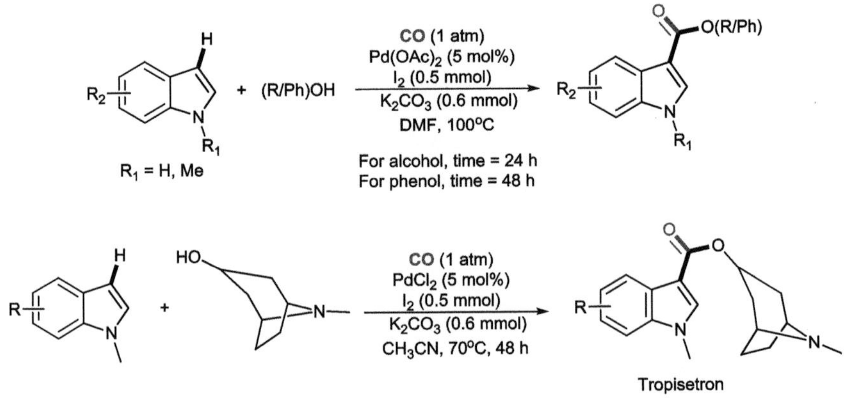

Scheme 6.51 Palladium-catalyzed C—H carbonylation of indoles with alcohols and phenols.

rather than coordinating to the N-H position of **70**, which usually leads to a carbamate or urea product. This is followed by insertion of CO and alcohol to generate the Pd(II) acyl intermediate **72** that undergoes reductive elimination to release the ester **73** and simultaneously regenerate the Pd(0) species (Scheme 6.52).

The same protocol was employed by Li's group for the C—H aminocarbonylation of indole derivatives.[38] After carrying out an elaborate optimization study, it was clear that indole-3-α-ketoamides and indole-3-amides can be synthesized by slightly modifying the reaction conditions. Double carbonylation could be effected under high pressure of CO (4 MPa), while mono carbonylation proceeded under atmospheric pressure of CO and high temperature (100°C). This protocol was also used for the synthesis of an HIV-1 inhibitor (Scheme 6.53).

The synthesis of indole-3-alkynones was also reported by the same group applying the Pd/Cu/I_2/base system.[39] The alkynone products could be easily transformed to 1,2,3-triazole derivatives (Scheme 6.54).

Guan and co-workers reported the C − H carbonylation of indoles with arylboronic acids for the synthesis of indol-3-yl aryl ketones.[40] After the reaction optimization, the combination of $Pd(OAc)_2$, I_2, KOH, along with pyridine and cesium fluoride as additives under balloon pressure of CO, were found to afford the ketones in high yield (Scheme 6.55).

Mechanistically it was proposed that iodination of indole in the presence of base resulted in the formation of 3-iodoindole **74**. The sequential Pd(0)-catalyzed carbonylation of 3-iodoindole **74** and phenyboronic acid

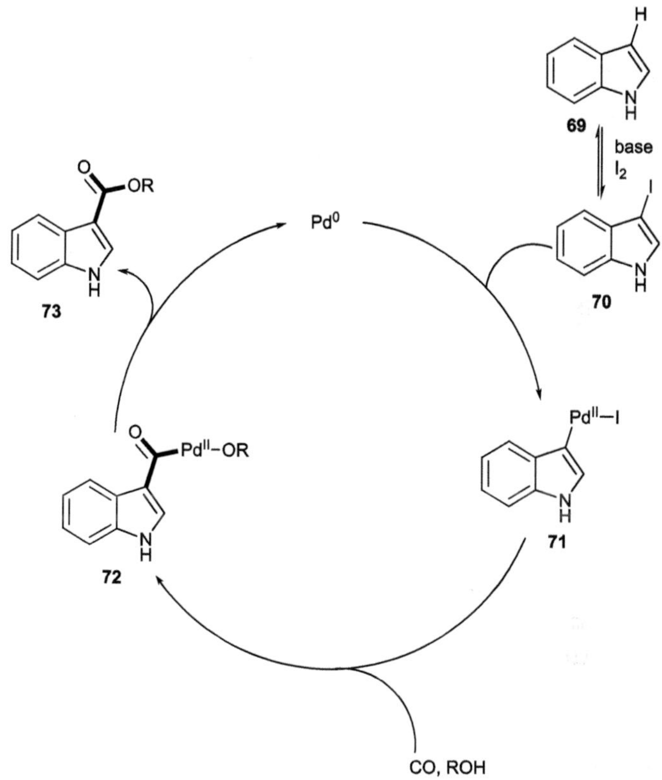

Scheme 6.52 Proposed mechanism for the palladium-catalyzed C—H carbonylation of indoles with alcohols and phenols.

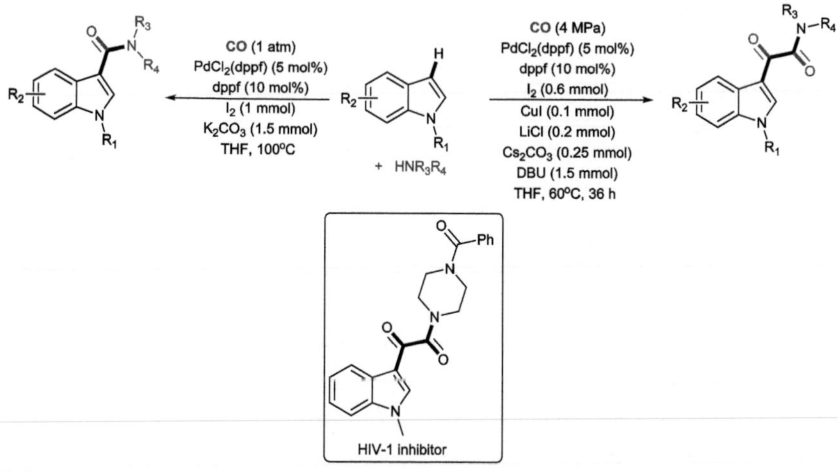

Scheme 6.53 Palladium-catalyzed aminocarbonylation of indole derivatives.

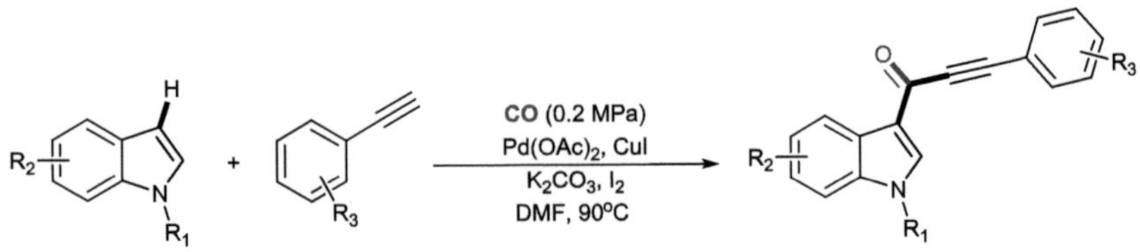

Scheme 6.54 Palladium-catalyzed C—H carbonylation of indoles for the synthesis of indole-3-alkynones.

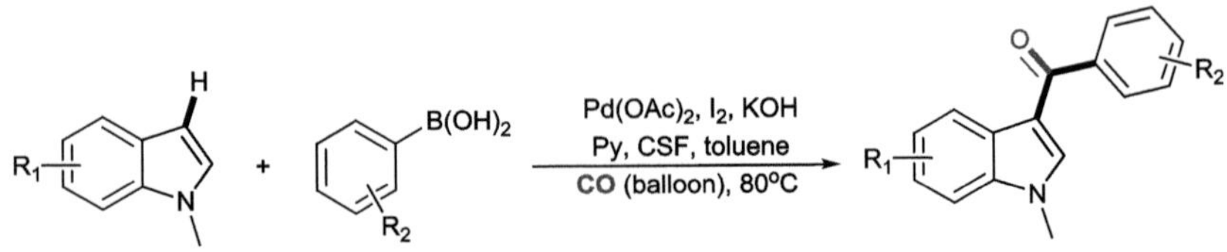

Scheme 6.55 Palladium-catalyzed C—H carbonylation of indoles with arylboronic acids for the synthesis of indol-3-yl aryl ketones.

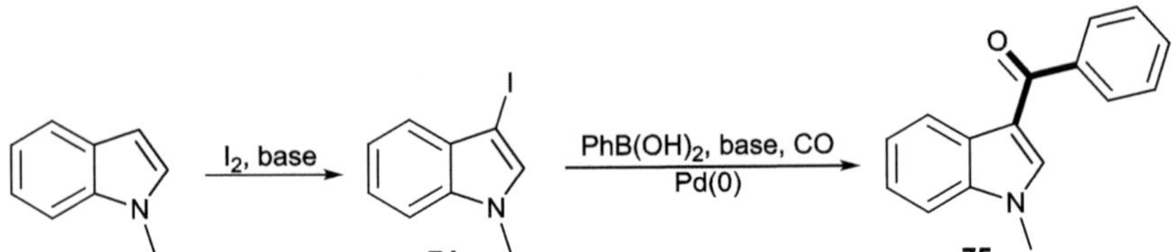

74 75

Scheme 6.56 Mechanism of the palladium-catalyzed C—H carbonylation of indoles with arylboronic acids.

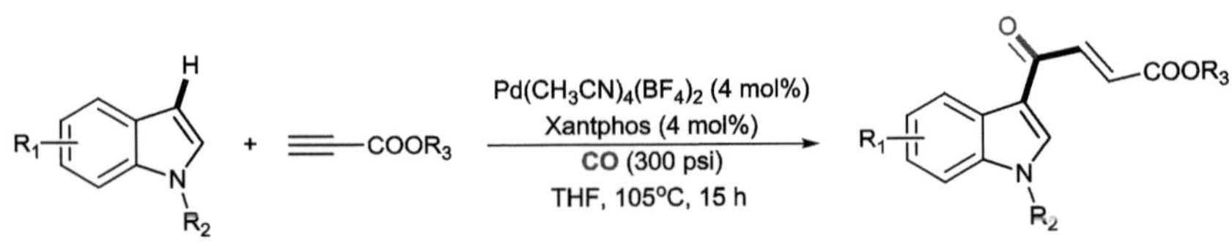

Scheme 6.57 Palladium-catalyzed synthesis of linear unsaturated ketones through C—H carbonylation of indoles with alkynes.

in the presence of CO gives the final indol-3-yl phenyl ketone **75**. This is similar to the mechanism proposed by Li's group for the C—H carbonylation of indole derivatives (Scheme 6.56).

Zeng and Alper reported the synthesis of linear α,β-unsaturated ketones through the carbonylative $C(sp^2)$—H functionalization of indoles with alkynes.[41] The catalytic system comprising of $Pd(CH_3CN)_4(BF_4)_2$ and Xantphos results in affording a linear regioselectivity at the C-3 position (Scheme 6.57).

6.3.3 Carbonylative C—H Functionalization of Heteroarenes Containing Multiple Heteroatoms

The first example for the palladium-catalyzed C—H carbonylation of heteroarene bearing two heteroatoms with aryl halide was documented in 2010 by Beller's group.[42] In the presence of a palladium—copper system, oxazole, thiazole, benzoxazole, benzthiazole, and benzimidazole could be regioselectively carbonylated at the second position leading to the synthesis of di-(hetero)aryl ketones (Scheme 6.58).

The proposed mechanism involves formation of the arylpalladium(II) complex **76**, which produces the key arylpalladium(II) complex **77** after CO insertion. Next, complex **79** is generated by transmetalation with the preformed copper/arene species **78** resulting from the heteroarene. Reductive elimination of the ketone from complex **79** completes the catalytic cycle and regenerates the active palladium species (Scheme 6.59).

Recently, Tjutrins and Arndtsen reported a copper free, palladium-catalyzed carbonylative $C(sp^2)$—H functionalization of a diverse range of heterocycles with aryl iodides.[43] Notably, this protocol avoided the need to employ premetalated heterocycles. Pyrrole, indole, imidazole, furan, benzoxazole, and benzimidazole derivatives could be regioselectively carbonylated in the presence of $Pd(P^tBu_3)_2$, which happens to be the most active catalyst for this transformation. This is primarily because of the large cone angle of P^tBu_3 (Scheme 6.60).

As far as the mechanism is concerned, the possibility of an aroyl iodide intermediate was proposed. The aroyl iodide (ArCO-I) itself can undergo reductive elimination (Scheme 6.62, path A). However, aroyl iodides are much more reactive and electrophilic products, and have not been previously observed in Pd-catalyzed carbonylation chemistry. There is also a possibility that the palladium-aroyl intermediates of carbonylation could directly react with the heterocycle, such as via a Friedel—Crafts reaction with the aroyl ligand (path B), or the electrophilic palladation of the heterocycle (path C), analogous to known noncarbonylative reactions (Scheme 6.61).

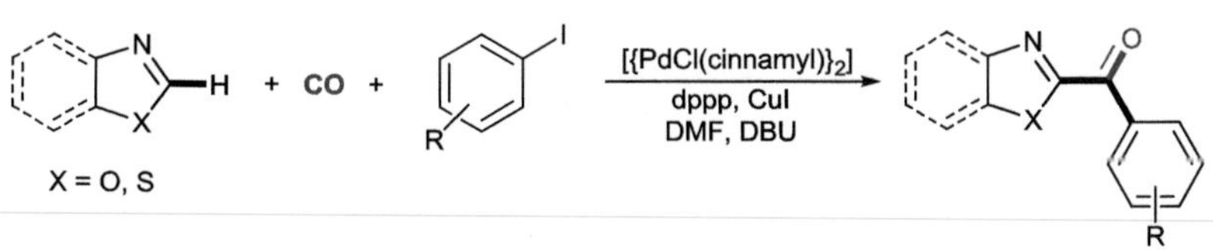

Scheme 6.58 Palladium-catalyzed C—H carbonylation of heteroarenes with aryl iodide.

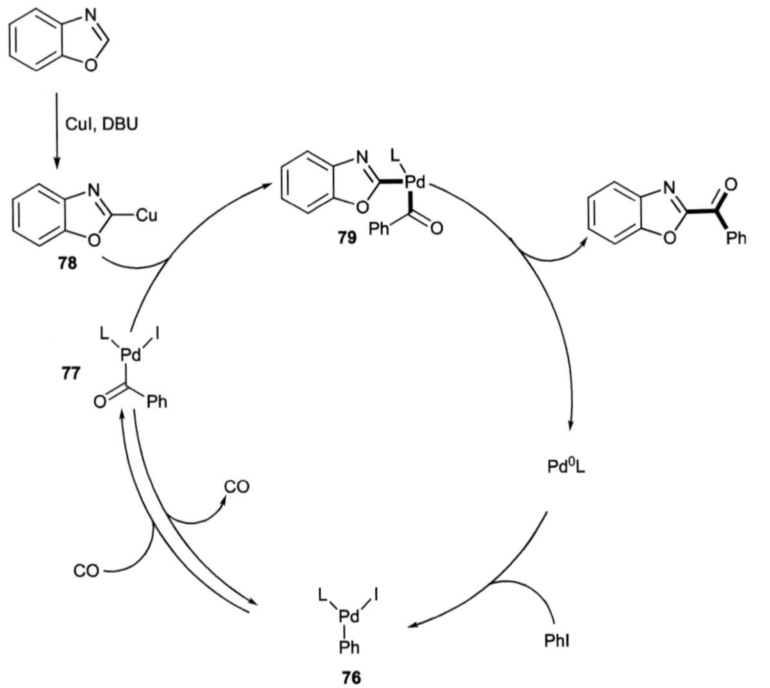

Scheme 6.59 Proposed mechanism for the palladium-catalyzed C—H carbonylation of heteroarenes with aryl iodide.

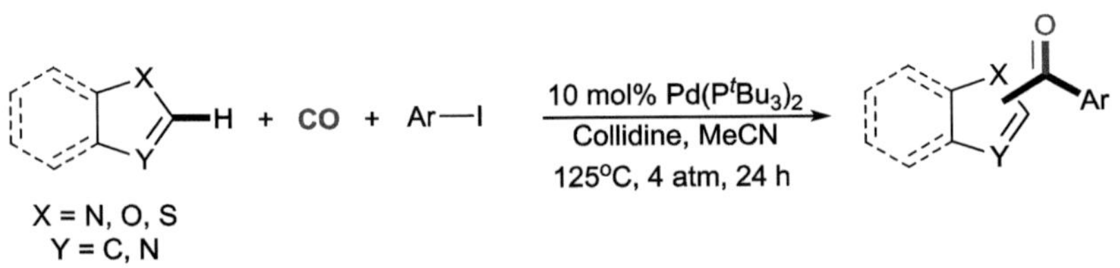

Scheme 6.60 Pd(P^{t}Bu$_3$)$_2$-catalyzed C—H carbonylation of heteroaromatic systems with aryl iodide.

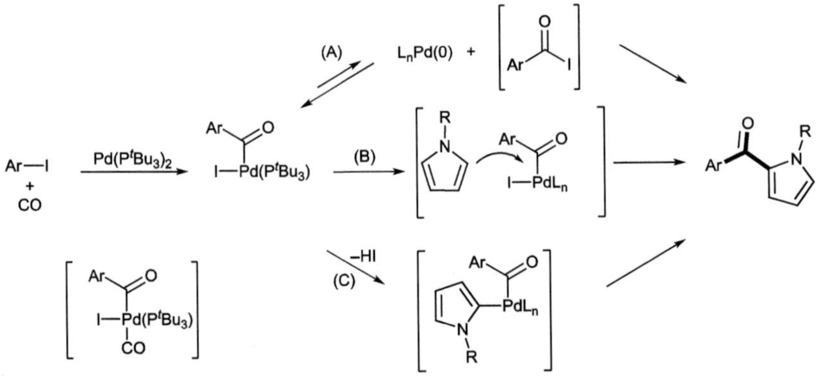

Scheme 6.61 Mechanistic possibility for the carbonylative C—H functionalization of heterocycles with aryl iodide.

6.4 CARBOXYLATIVE C—H BOND FUNCTIONALIZATION

6.4.1 Introduction

The industrialization age has ensured that the amount of CO_2 released into the atmosphere has steadily increased. CO_2 is one of the principal components of greenhouse gases, which in turn are mainly responsible for the rise in atmospheric temperature and the unusual changes in global climate. The increase in the combustion of fossil fuels in order to meet the world's energy demand is one of the primary factors responsible for the increase in the atmospheric CO_2 concentration. Hence, there is an urgent need to reduce CO_2 emissions and develop efficient systems for its capture from the atmosphere. Devising efficient methods for the utilization of carbon dioxide for production of chemicals cannot drastically bring down the level of atmospheric carbon dioxide. Nevertheless, CO_2 is a valuable C1 building block in organic synthesis, primarily due to its abundance, availability, nontoxicity, and recyclability. However, very few industrial processes have been developed which make use of CO_2 as a raw material. This is mainly attributed to its kinetic and thermodynamic stability, and the fact that carbon exists in its highest oxidative state. However, this can be overcome by making use of high energy starting materials or energy rich substrates such as strong nucleophilic organometallic reagents (organolithium and Grignard reagents), epoxides, and aziridines.[44] In such a case, cooperative activation of carbon dioxide with such substrates in the presence of a suitable catalytic system can be realized, which in turn makes the synthetic protocol less energy intensive. The use of transition metal catalytic systems in combination with energy rich substrates is a convenient method for the activation of CO_2. The transition metal-catalyzed activation of CO_2 mainly relies on the coordination of CO_2 to the metal. The chelating mode between the transition metal center and CO_2 has been investigated,[45] and is shown in Scheme 6.62.

Complexes D and E are stable metal complexes in which the oxygen atoms of CO_2 chelate with the metal, thus rendering the transition metal catalyst inactive. Hence, methods have been developed in which metal coordinates with CO_2 through the modes shown in A, B, and C.

Given the independent advantages of C—H functionalization and CO_2 fixation, the utilization of CO_2 for the functionalization of C—H bonds is of tremendous significance, and would represent an ideal transformation for the synthesis of carboxylic acid and its derivatives.

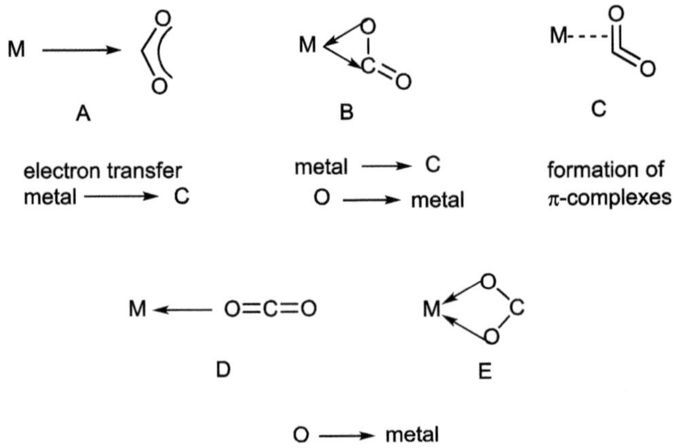

Scheme 6.62 Different coordination modes between CO_2 and transition metal.

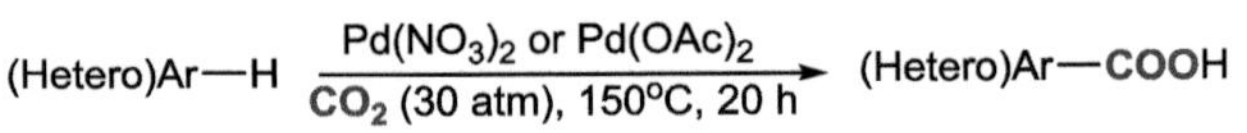

Scheme 6.63 Palladium-catalyzed $C(sp^2)$–H functionalization of aromatic compounds through CO_2 activation.

However, given the inherent challenges which this transformation possesses, there are only a few reports in the literature. The carboxylation of sp, sp^2, and sp^3 C–H bonds have been reported and are catalyzed mainly by copper, gold, and silver.[44c] To the best of our knowledge, there are only two reports where palladium has been used for C–H functionalization through CO_2 activation which shall be discussed in the next section

6.4.2 Palladium-catalyzed Carboxylative C–H Functionalization

In 1984, Fujiwara and co-workers documented the first palladium-catalyzed carboxylation of aromatic $C(sp^2)$–H bonds of benzene, anisole, thiophene, and furan to give the corresponding aromatic acids.[46] The reaction was carried out using $Pd(NO_3)_2$ or $Pd(OAc)_2$ as the catalyst at 150°C for 20 h under 30 atm of CO_2. Interestingly, the addition of *t*-BuOOH as an oxidant increases the yield of the product (Scheme 6.63).

In 2013, Iwasawa's group reported the first palladium-catalyzed carboxylation of alkenyl bonds.[47] In the protocol, 2-hydroxystyrenes could be transformed to the corresponding coumarin derivatives using $Pd(OAc)_2$ as the catalyst, Cs_2CO_3 as the base, and under an atmosphere of CO_2 (Scheme 6.64).

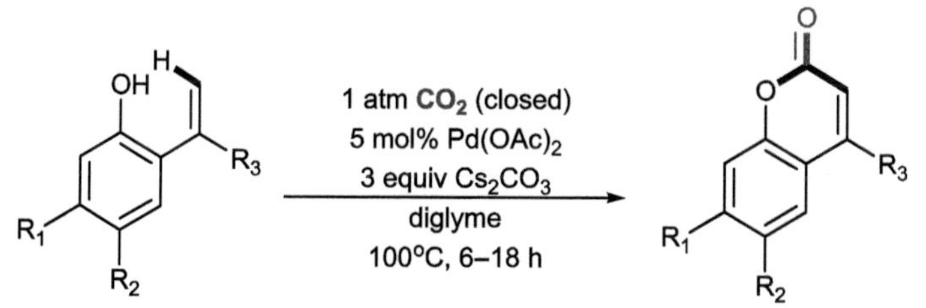

Scheme 6.64 Palladium-catalyzed carboxylation of alkenyl bond in 2-hydroxystyrene.

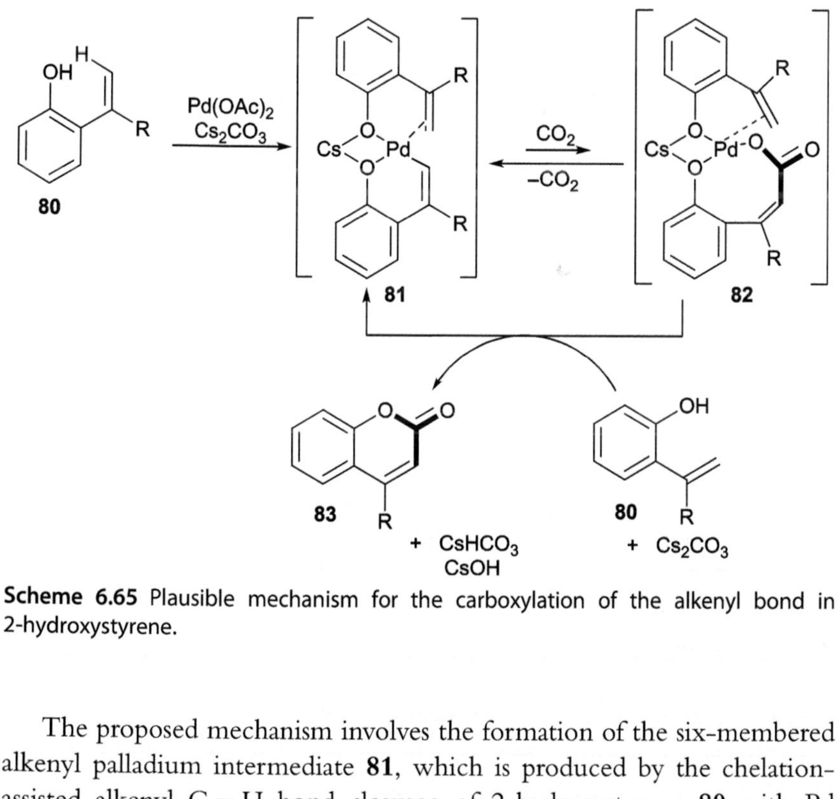

Scheme 6.65 Plausible mechanism for the carboxylation of the alkenyl bond in 2-hydroxystyrene.

The proposed mechanism involves the formation of the six-membered alkenyl palladium intermediate **81**, which is produced by the chelation-assisted alkenyl C – H bond cleavage of 2-hydroxystyrene **80** with Pd $(OAc)_2$ along with coordination of the second molecule of 2-hydroxystyrene **80** as its cesium salt. The intermediate **81** was isolated and characterized by X-ray crystallography. Subsequently, alkenyl palladium(II) **81** undergoes reversible nucleophilic carboxylation to generate palladium carboxylate intermediate **82**. This in turn reacts with another molecule of 2-hydroxystyrene **80** and base to give coumarin **83**, with regeneration of the cyclometalated intermediate **81** (Scheme 6.65).

6.5 CONCLUSION AND FUTURE OUTLOOK

This chapter highlights the progress made in the area of palladium-catalyzed carbonylative C(sp^2)—H functionalization of arenes and heteroarenes. It also touches upon the extremely challenging insertion of CO_2 into C—H bonds under palladium catalysis. The importance of directing group assistance in the regioselective carbonylation of arenes is presented. The regioselective carbonylation of unactivated heteroarenes in the absence of directing groups is also documented. The role of the active palladium species, oxidants, additives, and the formation of important intermediates in the reaction mechanism have been described in detail.

The number of examples reported in the case of carbonylative C(sp^2)—H functionalization is relatively less as compared to the noncarbonylative C(sp^2)—H functionalization. This implies that this area of research is challenging, and there is still much scope left. The role of the directing group in regioselective carbonylation is extremely important. However, in intermolecular C—H carbonylation, the directing group invariably ends up in the product. Most of the examples under directing group assistance involve regioselective *ortho* C—H carbonylation. The development of a suitable directing group, catalytic system, and a deeper understanding of the reaction mechanism would help in accomplishing the more challenging, remote *meta* C—H carbonylation. Moreover, devising directing groups which can be easily removed after the carbonylation step would make the addition of molecular complexity at an advanced stage in the synthesis of a molecule convenient. The use of the cheaper aryl bromides and pseudohalides in those C—H carbonylation reactions where one of the coupling partner is an aryl halide needs to be encouraged. This would greatly help in bringing down the cost of the process and making it more sustainable. Selective C—H carbonylation in complex heterocyclic moieties would help in the synthesis of important drug candidates and natural products in a convenient, step-economical, and atom-economical fashion. The palladium-catalyzed C—H functionalization with CO_2 is still a niche area, and efforts need to be targeted on finding newer protocols along with catalyst and ligand design.

REFERENCES

1. (a) Barnard, C. F. *J. Organomet.* **2008,** *27*, 5402—5422.
   (b) Brennfuhrer, A.; Neumann, H.; Beller, M. *Angew. Chem. Int. Ed.* **2009,** *48*, 4114—4133.
   (c) Liu, Q.; Zhang, H.; Lei, A. *Angew. Chem. Int. Ed.* **2011,** *50*, 10788—10799.

(d) Wu, X.-F.; Neumann, H.; Beller, M. *Chem. Soc. Rev.* **2011,** *40*, 4986–5009.
(e) Wu, X.-F.; Neumann, H.; Beller, M. *Chem. Rev.* **2013,** *113*, 1–35.
(f) Wu, X.-F.; Neumann, H.; Beller, M. *ChemSusChem* **2013,** *6*, 229–241.
2. Franke, R.; Selent, D.; Börner, A. *Chem. Rev.* **2012,** *112*, 5675–5732.
3. (a) Heck, R. F. *J. Am. Chem. Soc.* **1967,** 5546–5548.
(b) Schoenberg, A.; Bartoletti, I.; Heck, R. F. *J. Org. Chem.* **1974,** *39*, 3318–3326.
4. Fujiwara, Y.; Kawauchi, T.; Taniguchi, H. *J. Chem. Soc. Chem. Commun.* **1980,** 220–221.
5. (a) Yaniguchi, Y.; Yamaoka, Y.; Nakata, K.; Takaki, K.; Fujiwara, Y. *Chem. Lett.* **1995,** 345–346.
(b) Lu, W.; Yamaoka, Y.; Taniguchi, Y.; Kitamura, T.; Takaki, K.; Fujiwara, Y. *J. Organomet. Chem.* **1999,** *580*, 290–294.
6. (a) Wu, L.; Fang, X.; Liu, Q.; Jackstell, R.; Beller, M.; Wu, X.-F. *ACS Catal* **2014,** *4*, 2977–2989.
(b) Gautam, P.; Bhanage, B. M. *Catal. Sci. Technol.* **2015,** *5*, 4663–4702.
(c) Wu, L.; Liu, Q.; Jackstell, R.; Beller, M. *Angew. Chem. Int. Ed.* **2014,** *53*, 6310–6320.
7. Ma, B.; Wang, Y.; Peng, J.; Zhu, Q. *J. Org. Chem.* **2011,** *76*, 6362–6366.
8. Xu, Y.; Hu, W.; Tang, X.; Zhao, J.; Wu, W.; Jiang, H. *Chem. Commun.* **2015,** *51*, 6843–6846.
9. Houlden, C. E.; Hutchby, M.; Bailey, C. D.; Gair Ford, J.; Tyler, S. N. G.; Gagne, M. R.; Lloyd-Jones, G. C.; Booker-Milburn, K. I. *Angew. Chem. Int. Ed.* **2009,** *48*, 1830–1833.
10. Giri, R.; Lam, J. K.; Yu, J.-Q. *J. Am. Chem. Soc.* **2010,** *132*, 686–693.
11. Wrigglesworth, J. W.; Cox, B.; Lloyd-Jones, G. C.; Booker-Milburn, K. I. *Org. Lett.* **2011,** *13*, 5326–5329.
12. Dai, H.-X.; Stepan, A. F.; Plummer, M. S.; Zhang, Y.-H.; Yu, J.-Q. *J. Am. Chem. Soc.* **2011,** *133*, 7222–7228.
13. Orito, K.; Horibata, A.; Nakamura, T.; Ushito, H.; Nagasaki, H.; Yuguchi, M.; Yamashita, S.; Tokuda, M. *J. Am. Chem. Soc.* **2004,** *126*, 14342–14343.
14. Li, H.; Cai, G.-X.; Shi, Z.-J. *Dalton Trans.* **2010,** *39*, 10442–10446.
15. Haffemayer, B.; Gulias, M.; Gaunt, M. J. *Chem. Sci.* **2011,** *2*, 312–315.
16. Nieto, S.; Sayago, F.; Laborda, P.; Soler, T.; Cativiela, C.; Urriolabeitia, E. *Tetrahedron* **2011,** *67*, 4185–4191.
17. Guan, Z.-H.; Chen, M.; Ren, Z.-H. *J. Am. Chem. Soc.* **2012,** *134*, 17490–17493.
18. Liang, D.; Hu, Z.; Peng, J.; Huang, J.; Zhu, Q. *Chem. Commun.* **2013,** *49*, 173–175.
19. Rajeshkumar, V.; Lee, T.-H.; Chuang, S.-C. *Org. Lett.* **2013,** *15*, 1468–1471.
20. Albert, J.; Ariza, X.; Calvet, T.; Font-Bardia, M.; Garcia, J.; Granell, J.; Lamela, A.; López, B.; Martinez, M.; Ortega, L., et al. *Organometallics* **2013,** *32*, 649–659.
21. Li, W.; Duan, Z.; Zhang, X.; Zhang, H.; Wang, M.; Jiang, R.; Zeng, H.; Liu, C.; Lei, A. *Angew. Chem. Int. Ed.* **2014,** *54*, 1893–1896.
22. Chen, M.; Ren, Z.-H.; Wang, Y.-Y.; Guan, Z.-H. *J. Org. Chem.* **2015,** *80*, 1258–1263.
23. Li, W.; Duan, Z.; Jiang, R.; Lei, A. *Org. Lett.* **2015,** *17*, 1397–1400.
24. Giri, R.; Yu, J.-Q. *J. Am. Chem. Soc.* **2008,** *130*, 14082–14083.
25. Shi, B.; Liu, B.-F. *Synlett* **2013,** *24*, 2274–2278.
26. Liu, B.; Jiang, H.-Z.; Shi, B.-F. *Org. Biomol. Chem.* **2014,** *12*, 2538.
27. Liang, D.; He, Y.; Zhu, Q. *Org. Lett.* **2014,** *16*, 2748–2751.
28. Lu, Y.; Leow, D.; Wang, X.; Engle, K. M.; Yu, J.-Q. *Chem. Sci.* **2011,** *2*, 967.
29. Luo, S.; Luo, F.-X.; Zhang, X.-S.; Shi, Z.-J. *Angew. Chem. Int. Ed.* **2013,** *52*, 10598–10601.
30. Lee, T. H.; Jayakumar, J.; Cheng, C.-H.; Chuang, S.-C. *Chem. Commun.* **2013,** *49*, 11797–11799.

31. Shin, S.; Jeong, Y.; Jeon, W. H.; Lee, P. H. *Org. Lett.* **2014,** *16*, 2930—2933.
32. Wang, Y.; Gevorgyan, V. *Angew. Chem. Int. Ed.* **2015,** *54*, 2255—2259.
33. Ji, F.; Li, X.; Wu, W.; Jiang, H. *J. Org. Chem.* **2014,** *79*, 11246—11253.
34. (a) Itahara, T. *Chem. Lett.* **1982,** *11*, 1151.
    (b) Itahara, T. *Chem. Lett.* **1983,** *12*, 127.
35. Jaouhari, R.; Dixneuf, P.; Lécolier, S. *Tetrahedron Lett.* **1986,** *27*, 6315—6318.
36. Zhang, H.; Liu, D.; Chen, C.; Liu, C.; Lei, A. *Chem. Eur. J* **2011,** *17*, 9581—9585.
37. Lang, R.; Shi, L.; Li, D.; Xia, C.; Li, F. *Org. Lett.* **2012,** *14*, 4130—4133.
38. Xing, Q.; Shi, L.; Lang, R.; Xia, C.; Li, F. *Chem. Commun.* **2012,** *48*, 11023—11025.
39. Li, D.; Shan, S.; Shi, L.; Lang, R.; Xia, C.; Li, F. *Chin. J. Chem.* **2013,** *34*, 185—192.
40. Zhao, M.-N.; Ran, L.; Chen, M.; Ren, Z.-H.; Wang, Y.-Y.; Guan, Z.-H. *ACS Catal* **2015,** *5*, 1210—1213.
41. Zeng, F.; Alper, H. *Org. Lett.* **2013,** *15*, 2034—2037.
42. Wu, X.-F.; Anbarasan, P.; Neumann, H.; Beller, M. *Angew. Chem. Int. Ed.* **2010,** *49*, 7316—7319.
43. Tjutrins, J.; Arndtsen, B. A. *J. Am. Chem. Soc.* **2015,** *137*, 12050—12054.
44. (a) Huang, K.; Sun, C.-L.; Shi, Z.-J. *Chem. Soc. Rev.* **2011,** *40*, 2435—2452.
    (b) Liu, Q.; Wu, L.; Jackstell, R.; Beller, M. *Nat. Commun.* **2015,** *6*, 5933.
    (c) Yu, D.; Teong, S. P.; Zhang, Y. *Coord. Chem. Rev.* **2015,** *293-294*, 279—291.
45. Behr, A. *Angew. Chem. Int. Ed.* **1988,** *27*, 661—678.
46. Sugimoto, H.; Kawata, I.; Taniguchi, H.; Fujiwara, Y. *J. Organomet. Chem.* **1984,** *266*, C44—C46.
47. Sasano, K.; Takaya, J.; Iwasawa, N. *J. Am. Chem. Soc.* **2013,** *135*, 10954—10957.

CHAPTER 7

Flow Chemistry Perspective for C—H Bond Functionalization

Gabriele Laudadio and Timothy Noël
Eindhoven University of Technology, Eindhoven, The Netherlands

Contents

7.1 INTRODUCTION

Cross-coupling chemistry has attracted a lot of attention in the last three decades to facilitate carbon – carbon and carbon – heteroatom bond formation which enables the rapid synthesis of pharmaceuticals, agrochemicals, and materials.[1] More recently, C—H activation has emerged as a new avenue in cross-coupling chemistry and allows the use of unfunctionalized starting materials.[2] However, despite the apparent utility of this transformation, C—H activation reactions are still in their infancy, and the state of the art is still far from ideal. The applicability of C—H activation chemistry has been limited mainly by the inert nature of the carbon – hydrogen bond (the bond dissociation energies of aromatic C—H are around 110 kcal/mol, and of aliphatic C—H around 105 kcal/mol).[2a] In order to cleave the C—H bond, harsh reaction conditions, long reaction times, and high catalyst loadings are typically required.

Strategies for Palladium-Catalyzed Non-Directed and Directed C—H Bond Functionalization
DOI: http://dx.doi.org/10.1016/B978-0-12-805254-9.00007-4

In the last decade, continuous-flow microreactors have been used as an interesting new tool to boost chemical reactions by using harsh reaction conditions in a safe and reliable fashion.[3] Advantages, such as excellent heat- and mass-transfer, safety of operation, and ease of scale-up, are associated with such reactors, and continuous-flow technology allows performance of reactions under conditions that cannot be easily achieved in batch.[4] The ability to use elevated temperatures in a safe and reliable way may assist the deprotonation step, which is required to enable C–H functionalization. In addition, excellent interfacial contact between gaseous reactants (e.g., oxygen) and the liquid phase can be expected in micro flow. This feature of microreactor technology allows for an enhancement of the mass-transfer from the gas to the liquid phase, which is advantageous for reoxidation processes (e.g., Pd(0) to Pd(II) oxidation).[3e,5]

7.2 ADVANTAGES OF USING CONTINUOUS-FLOW MICROREACTOR TECHNOLOGY FOR C–H ACTIVATION

Microreactor technology has gained growing interest during this last decade from the chemical and processing engineer community. This is due to the important advantages that can be achieved by continuous processing in such miniaturized devices (Table 7.1, advantages).[6] In fact, the use of continuous-flow microreactors allows for precise control over reaction and residence time and other important process parameters (e.g., temperature, pressure, gas – liquid interfacial area).[4] Such advantages lead ultimately to a higher reproducibility compared to the more classical batch techniques. The small dimensions of these systems (generally with an internal diameter of 100–1000 μm and a few hundred microliters in volume) provide a

Table 7.1 Advantages and disadvantages of continuous-flow microreactors

Advantages	Disadvantages
High mass transfer	Solid-forming issues
High heat transfer	Engineering knowledge
Safe processes	Initial investment
Novel Process Windows	Multidisciplinary
Precise control over reaction parameters	
Efficient scalability	
Inline spectroscopic tools	
Integrated multistep synthesis	

means to carry out hazardous and explosive reactions with a high degree of safety (i.e., the power of an explosion is correlated to the total amount of hazardous material in the reactor). In the same way, toxic and/or highly reactive compounds can be produced in situ and consumed immediately in a follow-up reaction, as the storage of large amounts of these hazardous species is avoided. Furthermore, the high area-to-volume ratio provide good heat and mass transfer, drastically reducing reaction times and allowing fast and high-controlled exothermic reactions. Therefore, microreactors can be safely used to enable harsh process conditions (e.g., high temperatures and high pressures), also known as *Novel Process Windows*.[3a,7] The possibility to include *inline* analytical spectrophotometric detectors is a powerful way to monitor and optimize the reaction rapidly. Moreover, new strategies which allow scalability of flow chemistry without any further optimization of the reaction parameters have been developed; this includes using prolonged operation times, and adopting a numbering-up strategy or a smart scale-out strategy.[8]

Despite the apparent advantages of microflow processes, the field has met with skepticism in the past (Table 7.1, disadvantages). This can be partly attributed to the lack of engineering skills of many chemists. A fundamental understanding of the basic engineering principles, such as transport phenomena and process control, are key to design a successful process. This different point of view can cost the researcher a lot in terms of time and scientific attitude change. In addition, the introduction of new parameters to control (e.g., mixing, heat dissipation, multiphase characteristics, solid formation) has to be considered. In addition, the initial investment to equip a laboratory with proper flow equipment like pumps, microreactors, etc., can be quite high. Nevertheless, start-up companies noticed the increasing interest in continuous-flow chemistry and numerous types of microcapillaries, with their specific microfluidic fittings, in a wide variety of materials and dimensions are currently on the market. This creates a cheaper alternative to the expensive silicon or stainless-steel microreactors.

One of the most important threats for flow chemistry is the presence of solid particles which can compromise the reaction itself by irreversible clogging of the microchannel. Usually, flow chemists try to apply reaction conditions in which the homogeneity of the reaction mixture is guaranteed. However, strategies have been developed to overcome clogging problems in microchannels, such as ultrasonication and encapsulation of the solids forming reaction in segmented flow.[9]

7.3 CROSS-DEHYDROGENATIVE COUPLING IN CONTINUOUS FLOW

One of the most important C—H activation protocols developed so far is catalytic cross-dehydrogenative coupling (CDC). This strategy allows the connection of two different C—H bonds under oxidative conditions, bypassing any prefunctionalization of the coupling partners and obtaining, theoretically, only water as a by-product.[10]

In 2014, Noël et al. developed a very robust continuous-flow protocol to establish a selective C3-olefination of indoles.[11] Molecular oxygen was used to reoxidize Pd(II) to Pd(0). In microflow, a segmented flow regime is obtained upon merging the liquid and gas stream providing a very intense contact between the liquid and the gaseous phase, and thus high gas – liquid mass-transfer coefficients are obtained under such flow regimes. Due to the excellent access to molecular oxygen in flow,[12] premature catalyst deactivation to palladium black can be avoided and high reaction rates are observed, leading to a substantial reduction of the reaction times (from 4 h to 10 min in most cases). Furthermore, continuous-flow microreactor technology provides accurate control of the temperature and oxygen pressure, and allows the preparation of a small library of C3-olefinated indoles using acrylates, acrylamides, styrene, and simple alkenes as coupling partners (Fig. 7.1).[11]

In a very recent paper, Noël et al. established the palladium-catalyzed homocoupling of the bulk chemical *o*-xylene, in order to prepare 3,4,3′,4′-tetramethylbiphenyl, a precursor of the Upilex-S polyamide polymer and a potential intermediate in the synthesis of metal organic frameworks.[13] Starting from the results obtained by Stahl et al., who designed this reaction achieving good selectivity but very low yields (i.e., 7%) over a long reaction time,[14] Noël et al. transferred this synthesis in flow to obtain a scalable protocol with reduced reaction times (Fig. 7.2). The flow method uses catalytic amounts of $Pd(OAc)_2$, trifluoroacetic acid (TFA), trifluoromethyl sulfonic acid (TfOH), 2-fluoropyridine (2-F-py) and $Cu(OAc)_2$, and acetic acid as the solvent. Again, molecular oxygen was used to establish a segmented flow regime and to enable efficient reoxidation of the catalyst. Elevated temperatures (110°C) and pressures (40 bar O_2 pressure) were used to accelerate the reaction (from 17 h in batch to 40 min in flow) leading to 41% of the target compound (31% isolated yield) with an overall selectivity of 60%.[13]

The CDC can also be used to establish carbon – heteroatom bonds.[15] There is only a single example of this transformation using a

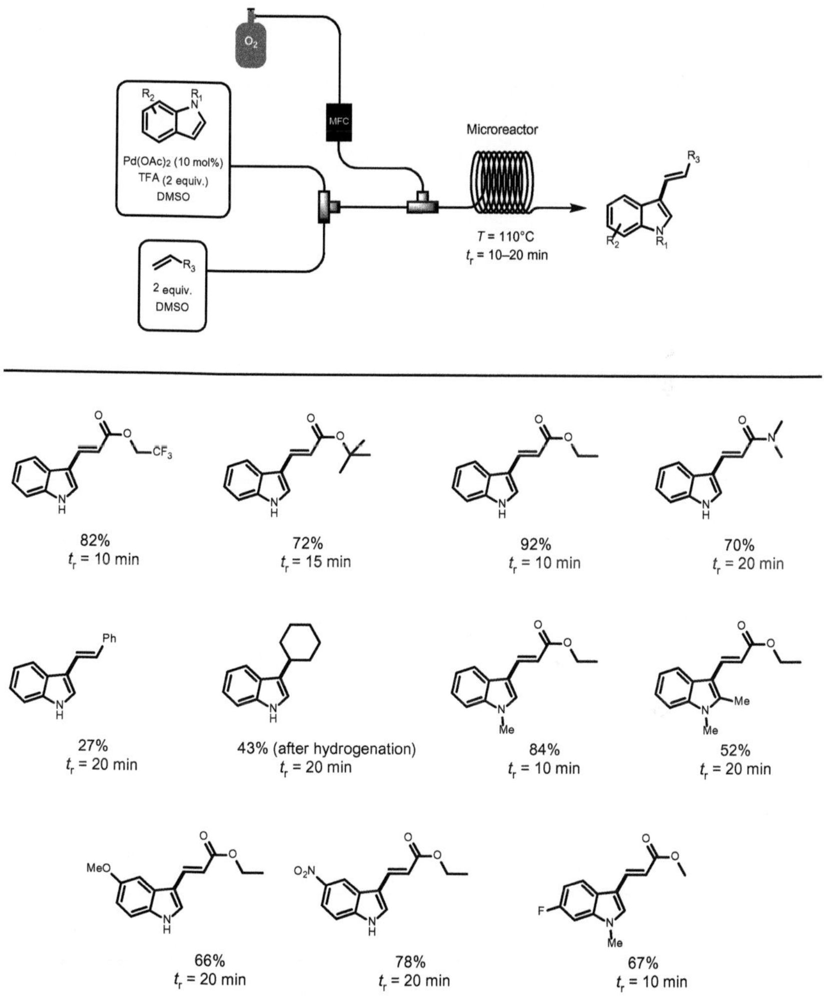

Figure 7.1 C3-olefination of indoles through continuous-flow cross-dehydrogenative coupling.

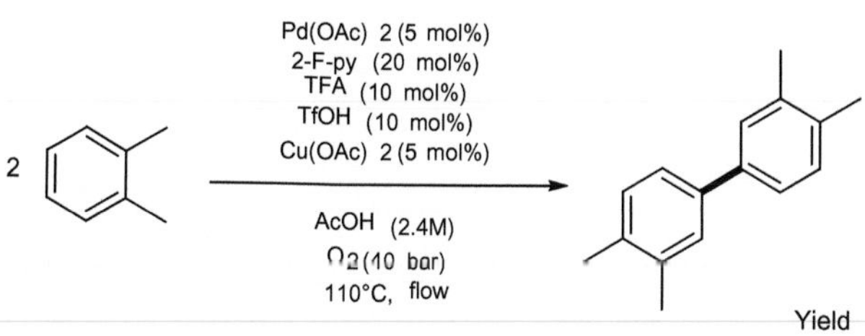

	Batch ($Pd(OAc)_2$ 0.1 mol%)	Flow
Yield	7%	**31%**
Reaction time	17 h	**40 min**

Figure 7.2 Homocoupling of *o*-xylene through continuous-flow cross-dehydrogenative coupling.

continuous-flow reactor. Kappe et al. designed a copper-catalyzed C—O bond formation directly activating α-C—H bond of ethers in the presence of *tert*-butyl hydroperoxide (TBHP) as the oxidant.[16] At first, the reaction was optimized in a microwave batch reactor. These experiments demonstrated that an increase in reaction temperatures and pressures was advantageous for this transformation. Unfortunately, this technique is not scalable because of the low penetration capacity of microwaves. However, optimized reaction conditions in a microwave reactor can be efficiently scaled employing continuous-flow microreactors, as such reactors provide excellent heat transfer properties ("microwave-to-flow" paradigm).[17] Consequently, similar results were obtained using stainless steel capillaries as compared to the microwave batch experiments. In the flow setup, the reagents are merged with the oxidant in a micromixer and the combined liquid stream is subsequently introduced in a 20 mL microreactor at a temperature of 130°C (Fig. 7.3). The lower yields in flow can be rationalized by the single addition of the peroxide, while in batch this was introduced in portions over time.[16]

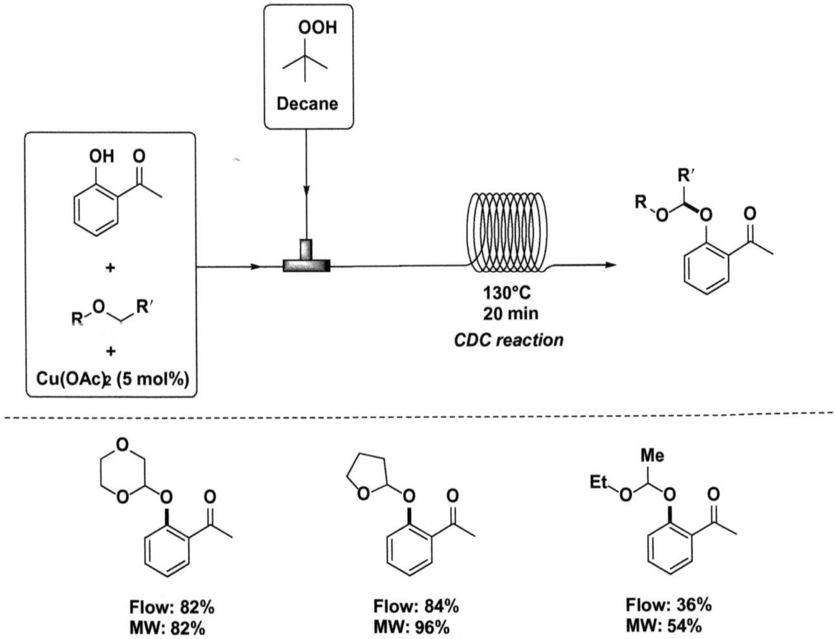

Figure 7.3 Continuous-flow cross-dehydrogenative coupling via a direct α-C—H bond activation of ethers.

7.4 RH-CATALYZED C—H ACTIVATION REACTIONS IN CONTINUOUS FLOW

Davies et al. developed a polymer/oxide hollow fiber made of cellulose acetate (a common polymer used in commercial hollow fibers) and a commercially available porous amorphous silica.[18] On the silica membrane an analogue of the well-known [Rh_2(*S*-DOSP)$_4$] catalyst was attached via a radical coupling between a styryl-functionalized silica and a styryl-bearing catalyst (Fig. 7.4C). With these fibers, the researchers obtained a hollow fiber reactor capped at the end. The reactants were fed through the central capillary and the solvent was directed perpendicularly toward the reactant supply (Fig. 7.4B). This supported catalyst was tested in cyclopropanation reactions and a C—H functionalization reaction (Fig. 7.4A). The results obtained in flow are quite good in terms of yield and diastereoselectivity, but fail to reach those results obtained with a homogeneous catalyst.[19]

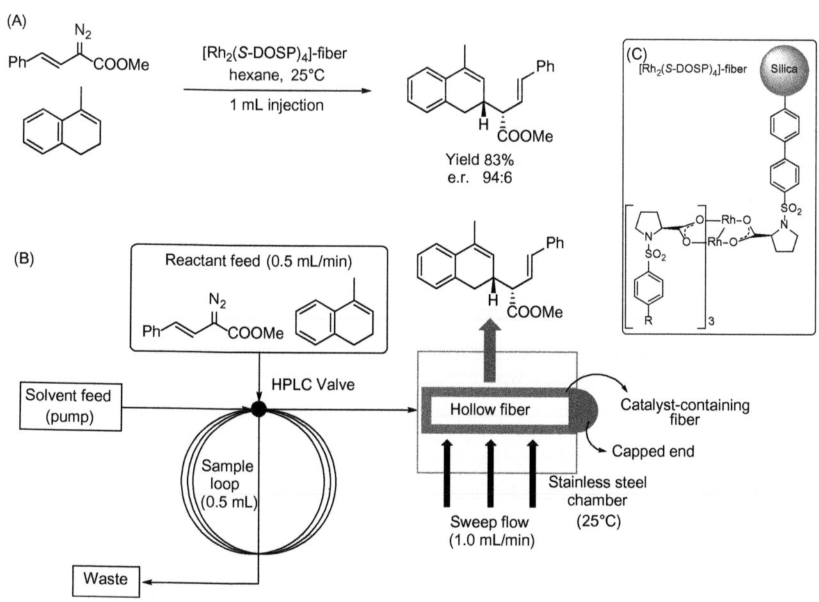

Figure 7.4 (A) C—H functionalization reaction, followed by a retro-Cope rearrangement; (B) Schematic representation of the hollow-fiber reactor setup for the C—H functionalization reaction/retro-Cope rearrangement reaction; (C) Supported catalyst used in the hollow-fiber reactor.

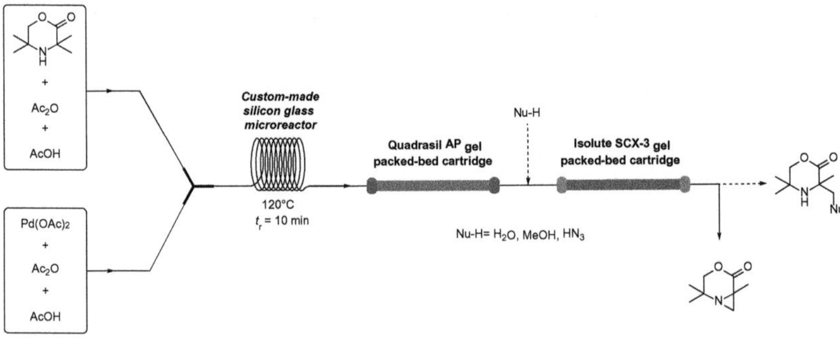

Figure 7.5 Palladium-catalyzed C(sp^3)–H activation to prepare aziridines in flow.

7.5 PALLADIUM-CATALYZED C(SP^3)–H ACTIVATION REACTIONS IN CONTINUOUS FLOW

Gaunt and Lapkin et al. designed a very efficient flow setup for the synthesis of aziridine derivatives of 3,3,5,5-tetramethylmorpholin-2-one.[20] Starting from a deep mechanistic understanding of this reaction (via kinetic analysis and DFT calculations),[21] the authors optimized the reaction conditions in semi-batch reactors. notably, they could obtain full conversion in custom-made silicon glass microfluidic reactors in only 10 min at a temperature of 120°C, with a very low amount of palladium catalyst (0.5 mol% of $Pd(OAc)_2$, instead of 1.0 mol% for the corresponding batch conditions). Moreover, the importance of acetic acid in this reaction was underscored as it allows the formation of the unreactive bis(amine) palladium complex by the protonation of the starting material to be avoided. After reaction, the palladium species was successfully scavenged by a QuadraSil AP cartridge, and the product was subsequently purified using a "catch and release" technique, employing a Isolute SCX-3 gel cartridge.[20]

With this setup, it was possible to obtain the desired aziridine of 3,3,5,5-tetramethylmorpholin-2-one in very high purity. Also, a productivity of 0.77 g/h (STY=046 kg/L*h) was obtained. Alternatively, a subsequent nucleophilic attack with water, methanol, or hydrazoic acid can be achieved in flow (Fig. 7.5).[20]

7.6 RU-CATALYZED C–H ACTIVATION REACTIONS IN CONTINUOUS FLOW

Schnürch et al. explored the possibility of using 2-phenylpyridine as a starting material in C–H activation reactions in flow.[22] An important optimization of this reaction was required in order to obtain flow-compatible conditions. In particular, it was necessary to screen a broad range of bases to achieve a homogeneous reaction mixture and thus avoid clogging issues. Using dichloro-(*p*-cymene)ruthenium(II) as the catalyst, DBU as base, PPh_3 as additional ligand, and NMP as solvent at a temperature of 160°C, the authors realized the C—H *ortho*-arylation of 2-phenylpyridine with bromobenzene derivatives. Interestingly, the reaction displays a different selectivity in *mono*- or *di*-arylation depending on the coupling partner (Fig. 7.6A). For example, electron-donating groups on the aryl bromide make the reaction generally mono-selective, while electron withdrawing groups give access to mainly disubstitution. Another important aspect of this reaction is that, when using a monoarylated compound as starting material, the reaction becomes particularly slow which is probably caused by steric effects during the precoordination of the ruthenium catalyst with the pyridine nitrogen (Fig. 7.6B).[22]

(A)

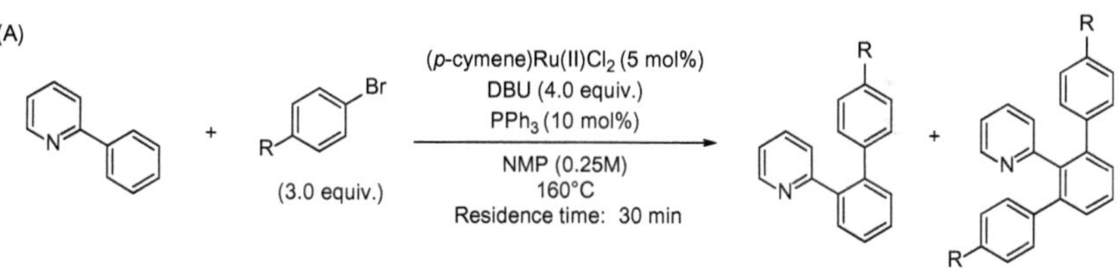

R	Mono	Di
H	26	69
OMe	59	0
CF_3	3	71

(B)

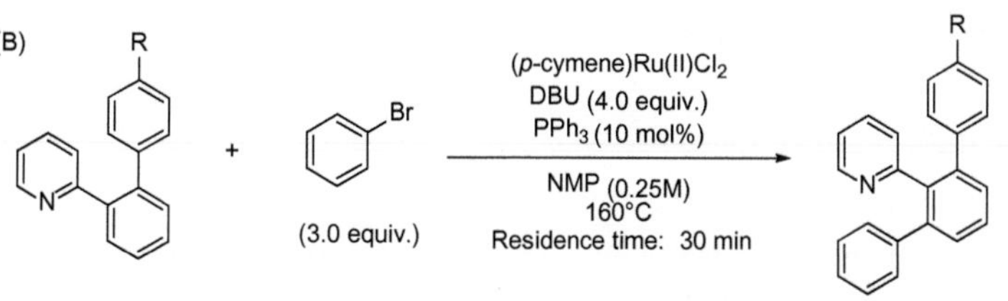

R = H (44% 7.5 mol% Ru),
OMe (24%, 7.5 mol% Ru)

Figure 7.6 (A) Continuous-flow C—H *ortho*-arylation of 2-phenylpyridine; (B) Continuous-flow C—H *ortho*-arylation of a *mono*-arylated 2-phenylpyridine.

7.7 INTRAMOLECULAR C—H ACTIVATION REACTIONS IN CONTINUOUS FLOW

Li et al. successfully transferred an intramolecular palladium-catalyzed direct arylation of 2-halobenzyl phenyl ethers to prepare biaryl scaffolds from batch to continuous-flow microreactors, using a very simple and accessible flow design composed of two syringe pumps, a T-mixer, and a capillary coil (Fig. 7.7A).[23] In the optimization of the reaction conditions, a solution of *N,N*-dimethylacetamide (DMA)/water (10:1) provided the highest conversion. Furthermore, during the base screening, potassium acetate was chosen providing the most selective reaction conditions. In order to prevent the degradation of the catalyst, a small amount of tetrabutylammonium bromide (TBAB) was also added. Nevertheless, it was noticed that the mixture was not completely homogeneous, causing clogging of the microreactor. This problem was efficiently overcome by the use of ultrasonication. The reaction was tested with aryl iodides, bromides, and chlorides, and several biaryl and heterocyclic compounds were prepared in flow (Fig. 7.7B). However, to achieve complete conversions long reaction times were still necessary (3—6 h).[23]

7.8 ORTHO-DIRECTED HYDROGEN/DEUTERIUM ISOTOPE EXCHANGE VIA IRIDIUM-CATALYZED C—H ACTIVATION

C—H activation can be used to prepare stable isotopically-labeled internal standards (STILs) which are used for the quantification of analytes with LC-MS/MS.[24] Noël and Vliegen et al. developed an iridium-catalyzed *ortho*-directed hydrogen/deuterium isotope exchange reaction to deuterate *N*-(4-methoxyphenyl)-*N*-methylbenzamide employing continuous-flow reactors.[25] A solid phase immobilized iridium catalyst was prepared using polystyrene-based triphenylphosphine and Crabtree's catalyst (Fig. 7.8A). This catalyst was subsequently used in continuous stirred tank reactors (CSTR) and in micro packed-bed reactors (PBR) (Fig. 7.8B). With these continuous flow reactors, deuterium was efficiently incorporated, obtaining a conversion of 54% with a maximum deuterium labeling of M+7, while in tank reactors the conversion reached 16%. Finally, the commercially available H-Cube Pro was used which allowed the use of higher deuterium pressures. In this way, deuterium was incorporated with a similar conversion as compared to the micro PBR, and the maximum deuterium labeling was M+3 (Fig. 7.8C).[25]

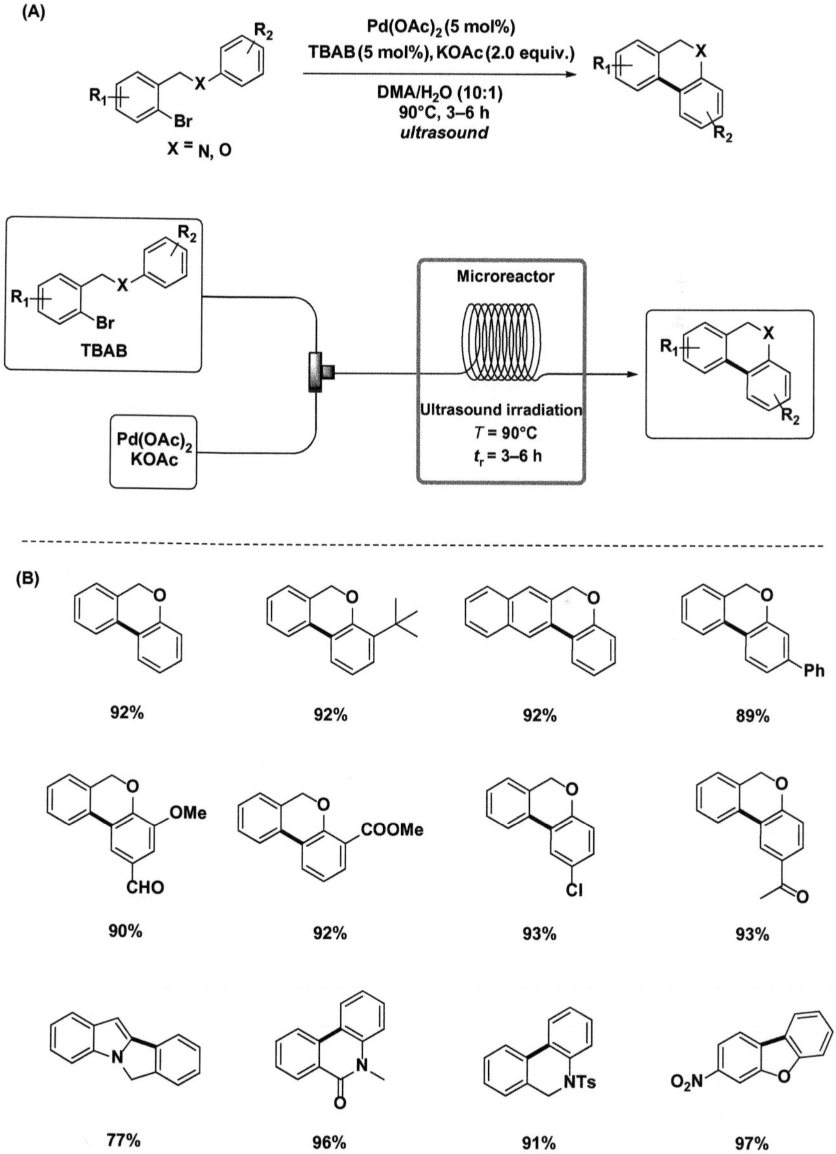

Figure 7.7 (A) Continuous-flow direct arylation to form heterocyclic biaryls, with a schematic representation of the flow setup; (B) Compounds prepared with this technique.

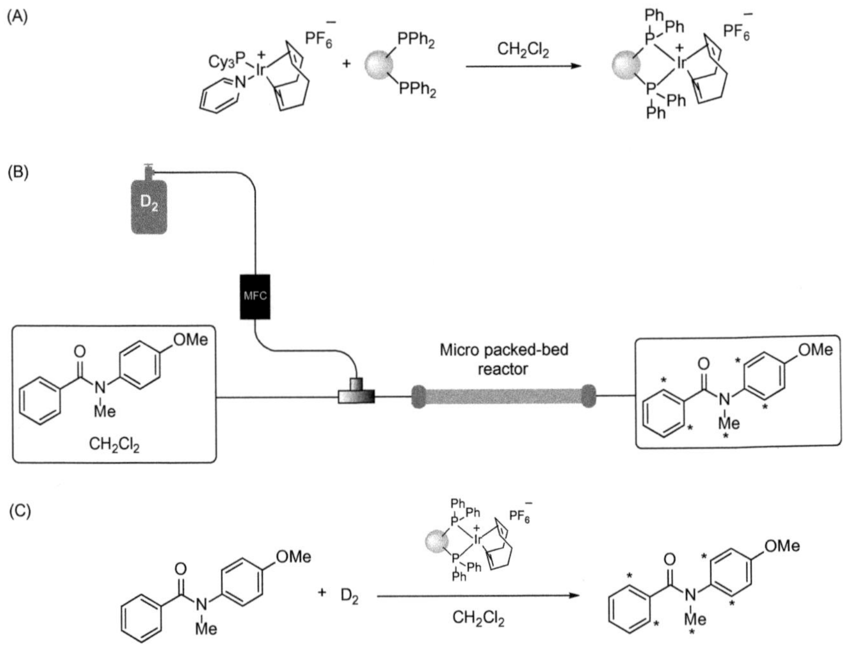

Reactor	Conversion	Max. deuterium label
CSTR	16	M+2
PBR	54	M+7
H-Cube Pro	64	M+3

Figure 7.8 (A) Preparation of the immobilized Crabtree catalyst; (B) Schematic representation of the micro packed/bed reactor setup; (C) Iridium-catalyzed *ortho*-directed hydrogen isotope exchange reaction in the three different reactors.

7.9 CONCLUSION

In contrast to classical cross-coupling reactions, the use of continuous-flow reactors for C–H activation reactions has been only marginally explored so far. However, as demonstrated in this chapter, flow reactors provide many advantages to boost C–H activation chemistry, such as increased gas – liquid mass transfer, high temperature, high pressure, and scalability. Consequently, we anticipate that many more flow examples will appear in the near future.

REFERENCES

1. (a) Negishi, E. *Handbook of Organopalladium Chemistry Handbook of Organopalladium Chemistry;* Wiley, 2002.
   (b) Oestreich, M. *The Mizoroki–Heck Reaction;* Wiley, 2008605.

(c) de Meijere, A.; Brase, S.; Oestreich, M. *Metal-Catalyzed Cross-Coupling Reactions and More;* Wiley, 2014.
2. (a) Jia, C.; Kitamura, T.; Fujiwara, Y. Catalytic Functionalization of Arenes and Alkanes via C—H Bond Activation. *Acc. Chem. Res.* **2001,** *34*, 633—639.
(b) Davies, H. M. L.; Manning, J. R. Catalytic C—H Functionalization by Metal Carbenoid and Nitrenoid Insertion. *Nature* **2008,** *451*, 417—424.
(c) Chen, X.; Engle, K. M.; Wang, D. H.; Yu, J.-Q. Palladium(II)-Catalyzed C—H Activation/C—C Cross-Coupling Reactions: Versatility and Practicality. *Angew. Chem. Int. Ed. Engl.* **2009,** *48*, 5094—5115.
(d) Engle, K. M.; Mei, T.-S.; Wasa, M.; Yu, J.-Q. Developing Broadly Useful C—H Functionalization Reactions. *Acc. Chem. Res.* **2012,** *45*, 788—802.
(e) Hartwig, J. F. Evolution of C—H Bond Functionalization from Methane to Methodology. *J. Am. Chem. Soc.* **2016,** *138*, 2—24.
3. (a) Hessel, V.; Kralisch, D.; Kockmann, N.; Noël, T.; Wang, Q. Novel Process Windows for Enabling, Accelerating, and Uplifting Flow Chemistry. *ChemSusChem* **2013,** *6*, 746—789.
(b) Noël, T.; Buchwald, S. L. Cross-Coupling in Flow. *Chem. Soc. Rev.* **2011,** *40*, 5010.
(c) Jas, G.; Kirschning, A. Continuous Flow Techniques in Organic Synthesis. *Chem. Eur. J.* **2003,** *9*, 5708—5723.
(d) Wegner, J.; Ceylan, S.; Kirschning, A. Ten Key Issues in Modern Flow Chemistry. *Chem. Commun.* **2011,** *47*, 4583—4592.
(e) Noël, T.; Hessel, V. *Cross-Coupling Chemistry in Continuous Flow;* In New Trends in Cross-Coupling: Theory and Applications; The Royal Society of Chemistry, 2015610—644.
4. Noël, T.; Su, Y.; Hessel, V. Beyond Organometallic Flow Chemistry: The Principles Behind the Use of Continuous-Flow Reactors for Synthesis. *Top. Organomet. Chem.* **2016,** *57*, 1—41.
5. Gemoets, H. P. L.; Su, Y.; Shang, M.; Hessel, V.; Luque, R.; Noël, T. Liquid Phase Oxidation Chemistry in Continuous-Flow Microreactors. *Chem. Soc. Rev.* **2016,** *45*, 83—117.
6. (a) Noël, T.; Wang, X.; Hessel, V. Accelerating Photoredox Catalysis in Continuous Microflow. *Chim. Oggi* **2013,** *31*, 10—14.
(b) Denčić, I.; Noël, T.; Meuldijk, J.; de Croon, M.; Hessel, V. Micro Reaction Technology for Valorization of Biomolecules Using Enzymes and Metal Catalysts. *Eng. Life Sci.* **2013,** *13*, 326—343.
(c) Malet-Sanz, L.; Susanne, F. Continuous Flow Synthesis. A Pharma Perspective. *J. Med. Chem.* **2012,** *55*, 4062—4098.
(d) Hartman, R. L.; McMullen, J. P.; Jensen, K. F. Deciding Whether to go with the Flow: Evaluating the Merits of Flow Reactors for Synthesis. *Angew. Chem. Int. Ed. Engl.* **2011,** *50*, 7502—7519.
7. (a) Hessel, V.; Gursel, I. V.; Wang, Q.; Noël, T.; Lang, J. R. Potenzialanalyse von Milli- und Mikroprozesstechniken für die Verkürzung von Prozessentwicklungszeiten - Chemie und Prozessdesign als Intensivierungsfelder. *Chem. Ing. Tech.* **2012,** *84*, 660—684.
(b) Hessel, V.; Gürsel, I.; Wang, Q.; Noël, T.; Lang, J. Potential Analysis of Smart Flow Processing and Micro Process Technology for Fastening Process Development: Use of Chemistry and Process Design as Intensification Fields. *Chem. Eng. Technol.* **2012,** *35*, 1184—1204.
8. (a) Mcmullen, J. P.; Jensen, K. F. Integrated Microreactors for Reaction Automation: New Approaches to Reaction Development. *Annu. Rev. Anal. Chem.* **2010,** *3*, 19—42.
(b) Caygill, G.; Zanfir, M.; Gavriilidis, A. Scalable Reactor Design for Pharmaceuticals and Fine Chemicals Production. 1: Potential Scale-Up Obstacles. *Org. Process Res. Dev.* **2006,** *10*, 539—552.
(c) Kockmann, N.; Gottsponer, M.; Roberge, D. M. Scale-Up Concept of Single-Channel Microreactors from Process Development to Industrial Production. *Chem. Eng. J.* **2011,** *167*, 718—726.

9. (a) Flowers, B. S.; Hartman, R. L. Particle Handling Techniques in Microchemical Processes. *Challenges* **2012,** *3*, 194—211.
   (b) Jensen, K. F.; Reizman, B. J.; Newman, S. G. Tools for Chemical Synthesis in Microsystems. *Lab Chip* **2014,** *14*, 3206.
10. Le Bras, J.; Muzart, J. Intermolecular Dehydrogenative Heck Reactions. *Chem. Rev.* **2011,** *111*, 1170—1214.
11. Gemoets, H. P. L.; Hessel, V.; Noël, T. Aerobic C—H Olefination of Indoles via a Cross-Dehydrogenative Coupling in Continuous Flow. *Org. Lett.* **2014,** *16*, 5800—5803.
12. Hessel, V.; Angeli, P.; Gavriilidis, A.; Löwe, H. Gas − Liquid and Gas − Liquid − Solid Microstructured Reactors: Contacting Principles and Applications. *Ind. Eng. Chem. Res.* **2005,** *44*, 9750—9769.
13. Erdmann, N.; Su, Y.; Bosmans, B.; Hessel, V.; Noël, T. Palladium-Catalyzed Aerobic Oxidative Coupling of o-Xylene in Flow: A Safe and Scalable Protocol for Cross-Dehydrogenative Coupling. *Org. Process Res. Dev.* **2016,** *20*, 831—835.
14. Wang, D.; Izawa, Y.; Stahl, S. S. Pd-Catalyzed Aerobic Oxidative Coupling of Arenes: Evidence for Transmetalation Between Two Pd(II)-Aryl Intermediates. *J. Am. Chem. Soc.* **2014,** *136*, 9914—9917.
15. (a) Wang, X.; Lu, Y.; Dai, H.-X.; Yu, J.-Q. Pd(II)-Catalyzed Hydroxyl-Directed C—H Activation/C—O Cyclization: Expedient Construction of Dihydrobenzofurans. *J. Am. Chem. Soc.* **2010,** *132*, 12203—12205.
   (b) Davies, H. M. L.; Bois, J. D.; Yu, J.-Q.; Zhang, S.-Y.; Zhang, F.-M.; Tu, Y.-Q. Direct Sp^3 α-C—H Activation and Functionalization of Alcohol and Etherw. *Chem. Soc. Rev.* **2011,** *40*, 1937—1949.
16. Kumar, G. S.; Pieber, B.; Reddy, K. R.; Kappe, C. O. Copper-Catalyzed Formation of C-O Bonds by Direct α-C—H Bond Activation of Ethers Using Stoichiometric Amounts of Peroxide in Batch and Continuous-Flow Formats. *Chem. Eur. J.* **2012,** *18*, 6124—6128.
17. Glasnov, T. N.; Kappe, C. O. The Microwave-to-Flow Paradigm: Translating High-Temperature Batch Microwave Chemistry to Scalable Continuous-Flow Processes. *Chem. Eur. J.* **2011,** *17*, 11956—11968.
18. Moschetta, E. G.; Negretti, S.; Chepiga, K. M.; Brunelli, N. A.; Labreche, Y.; Feng, Y.; Rezaei, F.; Lively, R. P.; Koros, W. J.; Davies, H. M. L.; Jones, C. W. Composite Polymer/Oxide Hollow Fiber Contactors: Versatile and Scalable Flow Reactors for Heterogeneous Catalytic Reactions in Organic Synthesis. *Angew. Chem. Int. Ed. Engl.* **2015,** *54*, 6470—6474.
19. Chepiga, K. M.; Feng, Y.; Brunelli, N. A.; Jones, C. W.; Davies, H. M. L. Silica-Immobilized Chiral Dirhodium(II) Catalyst for Enantioselective Carbenoid Reactions. *Org. Lett.* **2013,** *15*, 6136—6139.
20. Zakrzewski, J.; Smalley, A. P.; Kabeshov, M. A.; Gaunt, M. J.; Lapkın, A. A. Continuous-Flow Synthesis and Derivatization of Aziridines Through Palladium-Catalyzed C(sp^3) − H Activation. *Angew. Chem. Int. Ed. Engl.* **2016,** *55*, 8878—8883.
21. Smalley, A. P.; Gaunt, M. J. Mechanistic Insights into the Palladium-Catalyzed Aziridination of Aliphatic Amines by C-H Activation. *J. Am. Chem. Soc.* **2015,** *137*, 10632—10641.
22. Christakakou, M.; Schön, M.; Schnürch, M.; Mihovilovic, M. Arylation of Pyridines via Suzuki-Miyaura Cross-Coupling and Pyridine—Directed C-H Activation Using a Continuous-Flow Approach. *Synlett* **2013,** *24*, 2411—2418.
23. Zhang, L.; Geng, M.; Teng, P.; Zhao, D.; Lu, X.; Li, J.-X. Ultrasound-Promoted Intramolecular Direct Arylation in a Capillary Flow Microreactor. *Ultrason. Sonochem.* **2012,** *19*, 250—256.
24. Vogeser, M.; Seger, C. Pitfalls Associated with the Use of Liquid Chromatography-Tandem Mass Spectrometry in the Clinical Laboratory. *Clin. Chem.* **2010,** *56*, 1234—1244.
25. Habraken, E.; Haspeslagh, P.; Vliegen, M.; Noël, T. Iridium(I)-Catalyzed Ortho-Directed Hydrogen Isotope Exchange in Continuous-Flow Reactors. *J. Flow Chem.* **2015,** *5*, 2—5.

CHAPTER 8

Directed *Meta*-Selective C—H Bond Functionalizations

Gang Li
Fujian Institute of Research on the Structure of Matter, Chinese Academy of Sciences, Fuzhou, Fujian, China

Contents

8.1 INTRODUCTION

Site-selectivity control of C—H functionalization reactions is of paramount inportance, as well as being an outstanding challenge in developing synthetically useful direct C—H transformation methodologies.[1–7] For aromatic compounds, the study of *ortho-*, *meta-*, or *para*-selective C—H functionalization would be very attractive for developing step- and atom-economical organic syntheses (Scheme 8.1).[8–13] To date, proximity-induced *ortho*-C—H functionalizations of (hetero)arenes have been intensely studied with transition metal catalysts such as Pd(II), Rh(III), Ru(II), and Ir(III) with the assistance of directing group (DG) (Scheme 8.1A), or by directed *ortho* metalation.[14–27] In contrast, so far, only a small number of approaches are reported for addressing the challenges of *meta*-C—H functionalizations of (hetero)arenes with transition metal catalysts (Scheme 8.1B).[1–7] One representative approach is substrate-controlled *meta*-C—H functionalizations by taking advantage of a substrate's inherent steric and/or electronic properties.[28–51] Other approaches include chelating group promoted

Strategies for Palladium-Catalyzed Non-Directed and Directed C—H Bond Functionalization
DOI: http://dx.doi.org/10.1016/B978-0-12-805254-9.00008-6

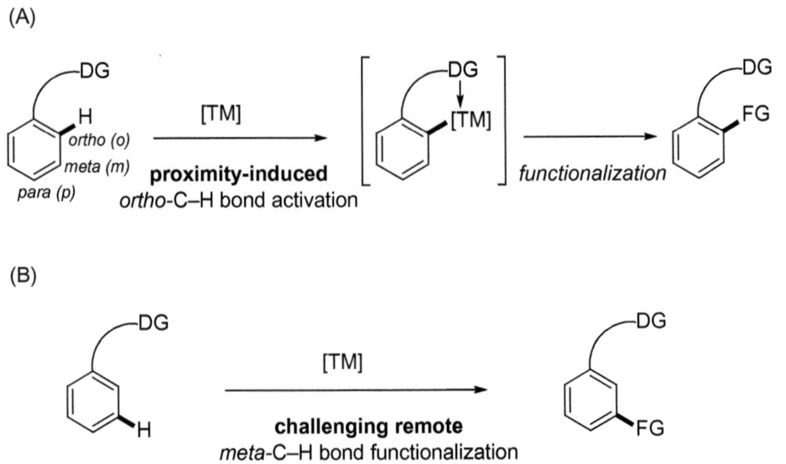

Scheme 8.1 Directed site-selective C—H bond functionalization. (A) Directed *ortho*-C—H bond functionalization. (B) Directed *meta*-C—H bond functionalization.

Cu(II)-catalyzed *meta*-C—H arylation,[52–53] ruthenium(II) complex facilitated *meta*-C—H functionalizations via *ortho*-C—H metalation,[54–60] hydrogen bonding between the substrate and the catalyst assisted *meta*-C—H borylation,[61] and in situ generated imines directed formal Ir-catalyzed *meta*-C—H borylation of aromatic aldehydes.[62]

Herein, we summarize three other types of *meta*-C—H functionalizations assisted by DGs that proved to be quite versatile using palladium catalysts. The first approach is *meta*-C—H functionalization of (hetero)arenes assisted by DGs that consist of predominantly nitrile-based ones.[63–79] The second approach is the very recently developed norbornene-mediated *meta*-C—H alkylation and arylation with an *ortho*-directing group.[80–83] Lastly, formal *meta*-C—H functionalization using a traceless DG will be presented.[84–87] Reports disclosed until May 2016 will be discussed in this chapter.

8.2 DIRECTING GROUP ASSISTED *META*-C—H FUNCTIONALIZATION

8.2.1 Nitrile-based Directing Group Assisted *Meta*-C—H Functionalization

In directed C—H activation, σ-chelating DGs have proved to be powerful in promoting diverse *ortho*-C—H functionalizations.[14–27] However, it has been very challenging to extend the proximity-induced reactivity to

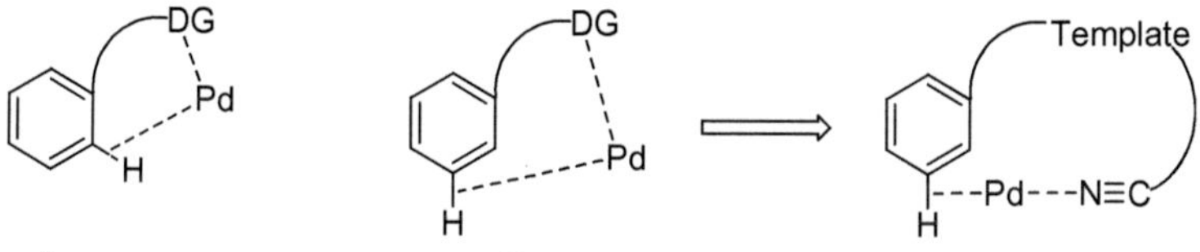

Scheme 8.2 Directed *meta*-selective C—H bond functionalization. (A) Well-studied directed *ortho*-C—H activation. (B) Directed *meta*-C—H activation. (C) Remote *meta*-C—H activation with CN-based DG.

meta-C—H functionalization, since such transformation may proceed via strained cyclophane-like pre-transition state (Scheme 8.2 structure **B**). In 2012, Yu and co-workers reported two types of rationally designed U-shaped nitrile-based DGs (or called templates) that successfully led to remote *meta*-C—H functionalization of arenes (Scheme 8.2 structure **C**).[63] The linear nitrile functionality was presumed to weakly coordinate to the palladium center in an end-on fashion, making the cyclophane-like pre-transition state of *meta*-C—H activation step the less strained. The weakly coordinating nitrile group may also "catch and release" the Pd(II) catalyst to deliver it close to the target *meta*-C—H bond, leading to highly effective concentration of the Pd(II) catalyst. The latter relay scenario would avoid the need to proceed via a cyclophane-like pre-transition state altogether. This seminal work inspired the discovery of a series of cleavable nitrile-based *meta*-directing groups which assisted *meta*-C—H olefination, arylation, and oxygenation of (hetero)arenes.[63–76]

8.2.1.1 Olefination

In their seminal report, Yu and co-workers first engineered an effective nitrile-based DG which was attached to the toluene derivatives through a removable benzyl (Bn) ether linkage (Scheme 8.3).[63] A key factor in the DG design is to improve the reactivity via the Thorpe—Ingold effect by installing two isobutyl groups at the α-position adjacent to the nitrile group. With $Pd(OPiv)_2$ as the catalyst and AgOPiv as the oxidant, this DG enabled *meta*-C—H Fujiwara—Moritani-type olefination of a broad range of substrates,[88–94] overriding the intrinsic electronic and steric biases of the substrates. Notably, this approach was applicable to di- and tri-substituted olefin coupling partners, which are often not reactive in

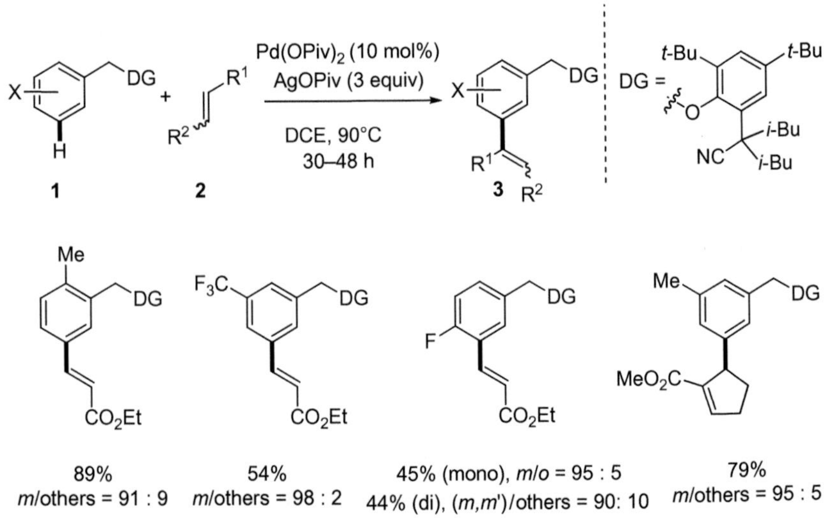

Scheme 8.3 Directing group assisted *meta*-C—H olefination of toluene derivatives.

ortho-C—H olefination reactions. Importantly, removal of the DG was facile via hydrogenolysis with a Pd/C catalyst. It is worth noting that the cyclophane-like 12-membered pre-transition state was first encountered in directed C—H functionalization reactions.

In the same report, a readily cleavable and recyclable 2,2′-azanediyldibenzonitrile DG was designed to attach to hydrocinnamic acid via the amide linkage, enabling *meta*-C—H olefination of hydrocinnamic acid derivatives, core structures of some drug molecules (Scheme 8.4). It was found the above reaction conditions were not suitable for this new type of substrate. After extensive reaction condition screening, it was discovered that the simple mono-*N*-protected amino acid (MPAA) *N*-acetyl glycine (Ac-Gly-OH) ligand significantly accelerated the reaction, and improved the selectivity as well with the optimal HFIP (Hexafluoroisopropanol) solvent which was crucial for the full conversion of the substrate.[95,96] This new DG not only overrode the intrinsic electronic biases of the substrate (**7**), but also was able to bypass the *ortho*-methyl groups (**8**). Intriguingly, unprecedented site-selectivity was observed with biphenyl substrate in the *meta*-selective C—H olefination of the remote aryl (Ar) ring of this synthetically useful substrate (**9**). Moreover, drug molecule Baclofen was also diversified with this chemistry (**10**). Importantly, the benzonitrile DG could be easily cleaved by hydrolysis using LiOH as the base at room temperature, affording the *meta*-olefinated hydrocinnamic acid and the recycled DG (Scheme 8.4B).

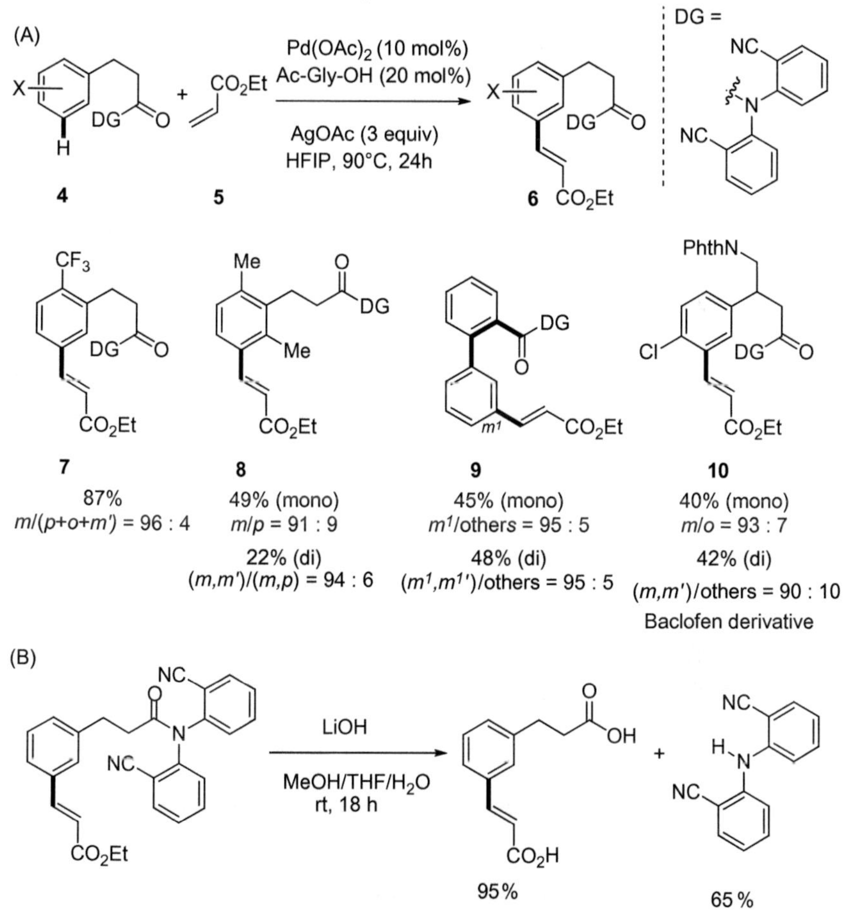

Scheme 8.4 *Meta*-C—H olefination of hydrocinnamic acid derivatives.

In a follow-up mechanistic study, Houk, Yu, and Wu revealed a mechanism which explains the dual roles of the amino acid ligand in boosting reactivity and selectivity.[96] Such dual roles were fulfilled by stabilizing monomeric Pd complexes and serving as the internal base for proton abstraction through a concerted metalation deprotonation (CMD) pathway of the rate-determining C—H activation step.[95–97]

Further study demonstrated the versatility of the above-mentioned benzonitrile DG for enabling diverse *meta* C H functionalizations.[64,65,68] Interestingly, the benzonitrile was able to accommodate a smaller ring size of macropalladacycle by olefinating the *meta*-C—H bond of phenylacetic acids smoothly (Scheme 8.5).[68] In this reaction, Yu and co-workers

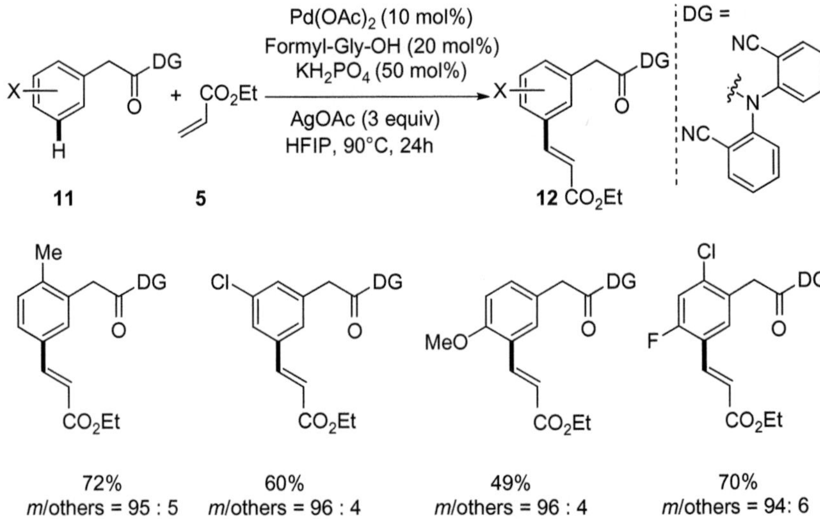

Scheme 8.5 *Meta*-C—H olefination of phenylacetic acid derivatives.

identified the N-formyl-protected glycine (Formyl-Gly-OH) as a better ligand than Ac-Gly-OH.

Remarkably, unprecedented *meta*-C—H functionalization of electron-rich phenol derivatives was also made possible with the same benzonitrile DG (Scheme 8.6) by Yu and co-workers, providing a synthetically useful route for *meta*-functionalizing α-phenoxycarboxylic acids, the parent

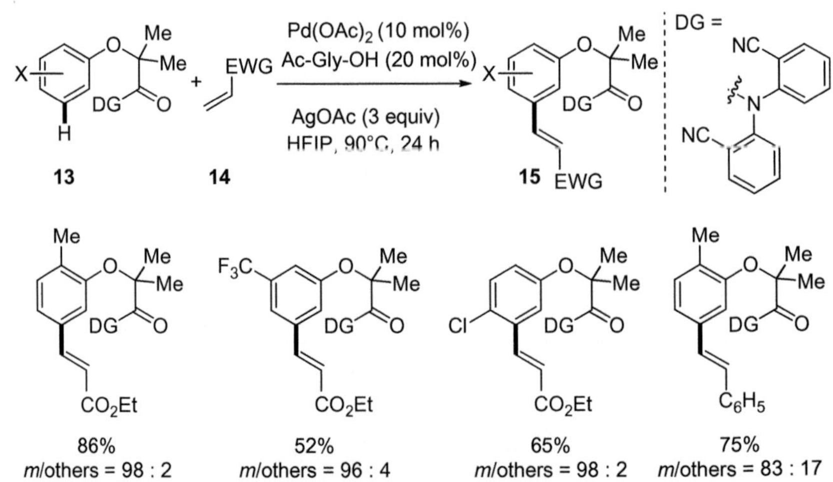

Scheme 8.6 *Meta*-C—H olefination of phenol derivatives.

structure of a fibrate class of drug molecules.[64] Notably, styrene derivatives containing electron-withdrawing groups (EWGs) were viable olefin coupling partners with this class of substrate.

A catalytic cycle was proposed to explain the observed *meta*-selectivity with the above-mentioned end-on nitrile-based DGs (Scheme 8.7).[64,68] The linear coordinating nitrile group delivers the Pd(II) catalyst to the vicinity of the *meta*-C—H bond of the substrate **A**, generating macropalladacycle **B** via insertion of the Pd(II) species into the target C—H bond. Following coordination of **B** with olefin **C**, the resulting complex **D** undergoes 1,2-migratory insertion to give intermediate **E**. It should be noted that when the reaction involves the use of mono-protected amino acid ligand, the nitrile group is likely displaced by the olefin **C** prior to the migratory insertion due to the bisdentate coordination mode of the amino acid ligand. Product **F** is then formed by β-hydride elimination of **E**. Reductive elimination of **G** affords Pd(0), which is reoxidized to Pd(II) by a silver salt to re-enter the catalytic cycle. A significant isotope

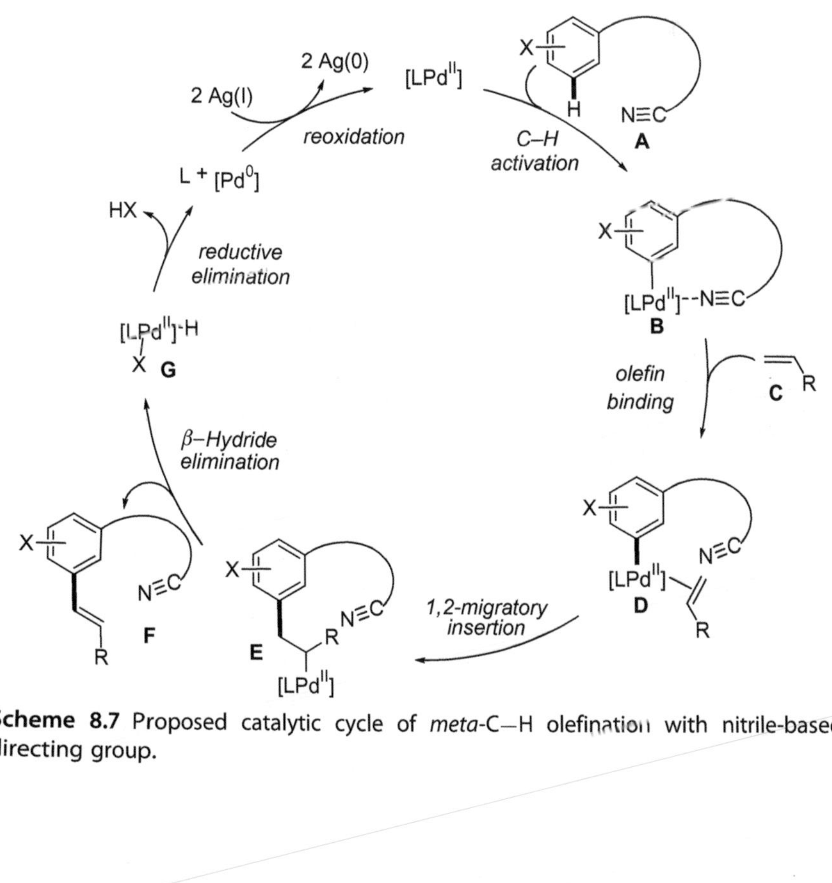

Scheme 8.7 Proposed catalytic cycle of *meta*-C—H olefination with nitrile-based directing group.

effect ($k_H/k_D = 3.8$) suggested the cleavage of the C—H bond may be involved in the rate-determining step.[64]

The power of nitrile-based DG assisted *meta*-C—H activation chemistry was further demonstrated with highly strained bicyclic tetrahydroquinolines as well as aniline derivatives,[66] which contain the amine substituents that are well-known strong *ortho/para* directors in electrophilic aromatic substitution reactions and electrophilic palladation (Scheme 8.8). In search of the optimal *meta*-directing group, Yu and co-workers found a fluorine substituent in the auxiliary scaffold led to a significant change in the conformation, resulting in the best *meta*-selectivity with tetrahydroquinoline under similar reaction conditions for the above-mentioned arenes (Scheme 8.8A).

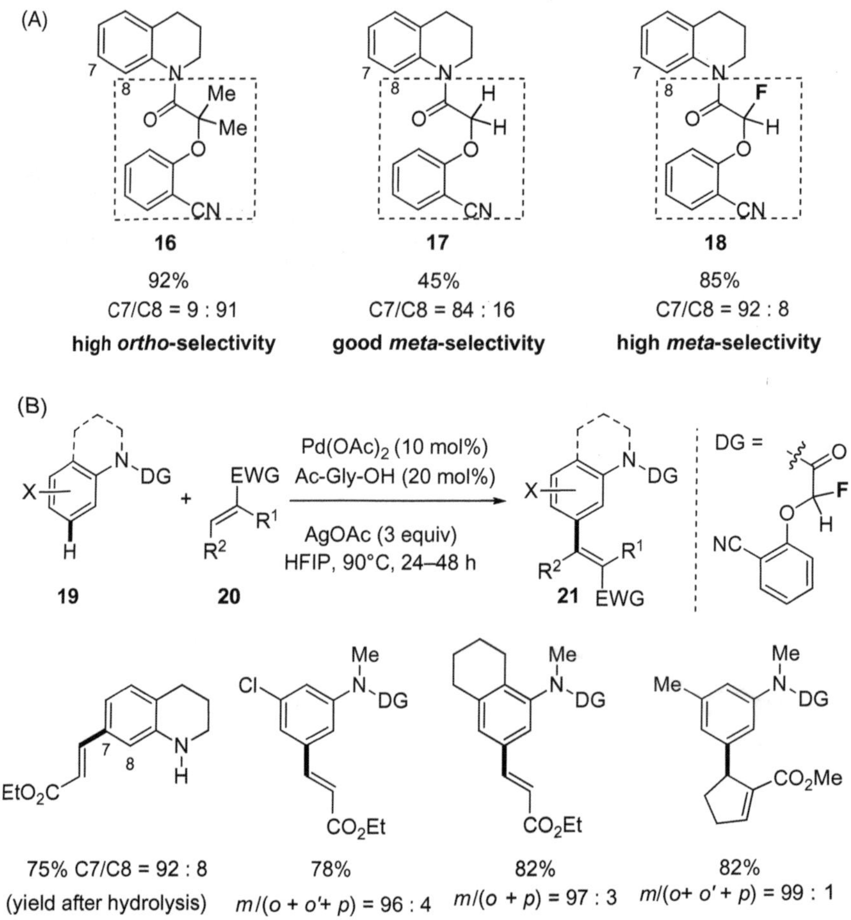

Scheme 8.8 *Meta*-C—H olefination of aniline and tetrahydroquinoline derivatives.

In contrast, when the substituents in the auxiliary scaffold were methyl groups, high *ortho*-C—H activation was observed. Again, the MPAA ligand Ac-Gly-OH was shown to enhance reactivity and site-selectivity. Thus, the new fluorine-containing nitrile-based DG smoothly promoted the *meta*-C—H olefination of tetrahydroquinolines and anilines in good yields, overriding the intrinsic electronic and steric biases of these substrates (Scheme 8.8B).

Subsequently, to accommodate indoline, Yu and co-workers engineered a new removable nitrile-based sulfonamide DG, while the previous DG for tetrahydroquinolines and anilines was not feasible presumably due to the stronger electron-donating ability of the nitrogen atom and the new skeleton (Scheme 8.9A).[67] Such linkage is crucial for the

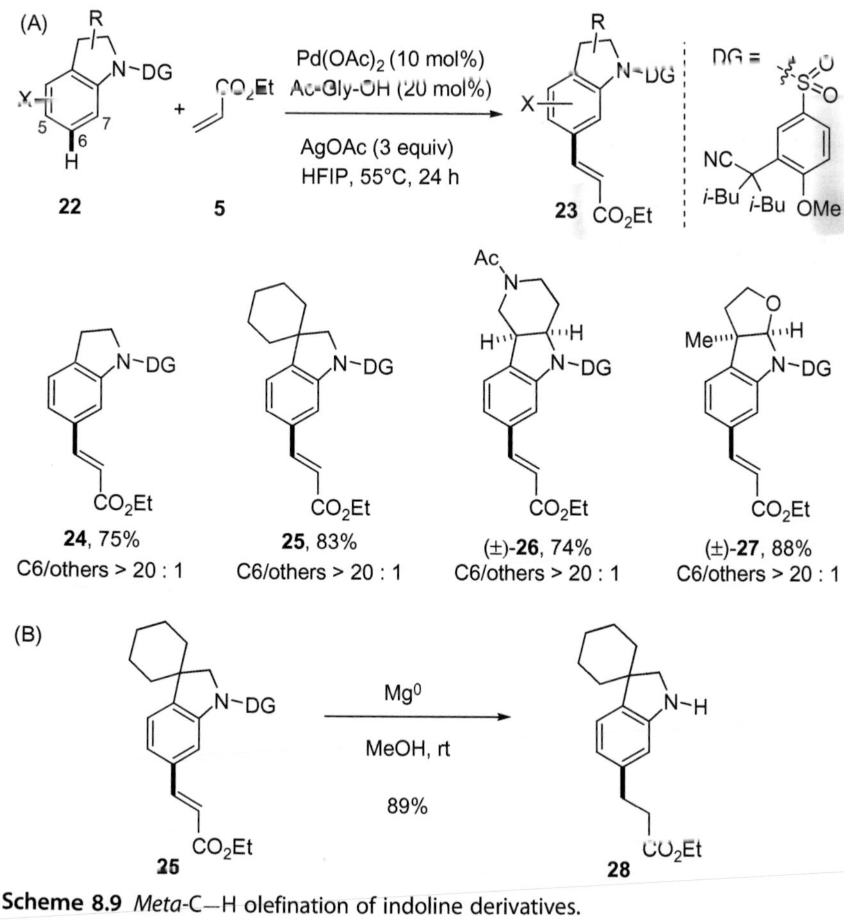

Scheme 8.9 *Meta*-C—H olefination of indoline derivatives.

meta-selective C—H functionalization of electron-rich indolines, since the electronically withdrawing sulfonyl group can reduce the high electronic bias of the substrates. Thus, under similar reaction conditions for *meta*-C—H olefination of anilines, a broad range of synthetically useful and advanced indoline analogues were efficiently *meta*-olefinated. Importantly, the sulfonamide DG was readily removed at room temperature with magnesium turnings in methanol, which simultaneously reduced the newly installed olefins to afford *meta*-alkylated indoline derivatives (Scheme 8.9B).

Inspired by the pioneering work by Yu, Tan and co-workers reported an effective bulky di-isopropyl silyl ether tethered nitrile-based *meta*-directing group,[69] which was synthetically practical since it could be easily introduced into the benzyl alcohol substrates and cleaved in situ with TBAF under mild conditions (Scheme 8.10). Under similar reaction conditions employed in the above-mentioned Yu's *meta*-C—H olefinations, a range of *meta*-olefinated benzyl alcohols were smoothly generated with in situ cleavage of the directing auxiliary.

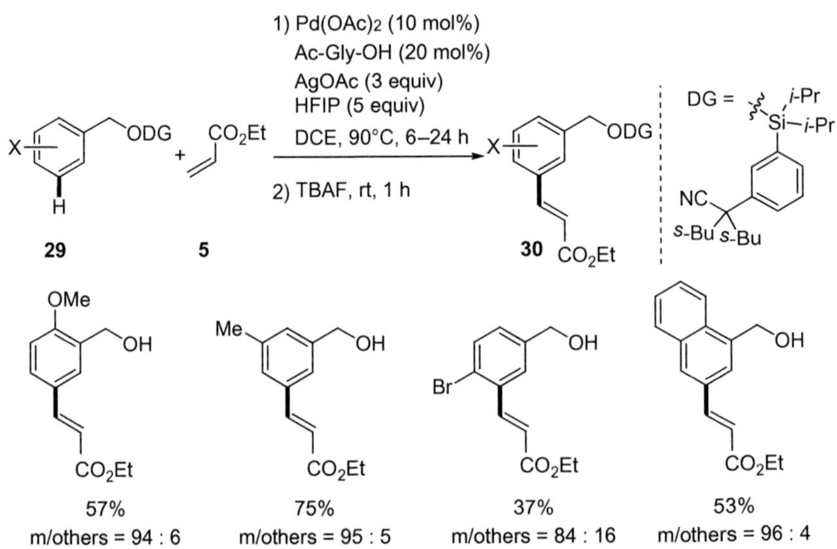

Scheme 8.10 *Meta*-C—H olefination of benzyl alcohol derivatives.

In 2014, Maiti and co-workers disclosed a *meta*-C—H olefination of useful phenylacetic acid derivatives of medicinal interest with a simple 2-hydroxybenzonitrile DG.[70] Notably, the DG effectively suppressed the

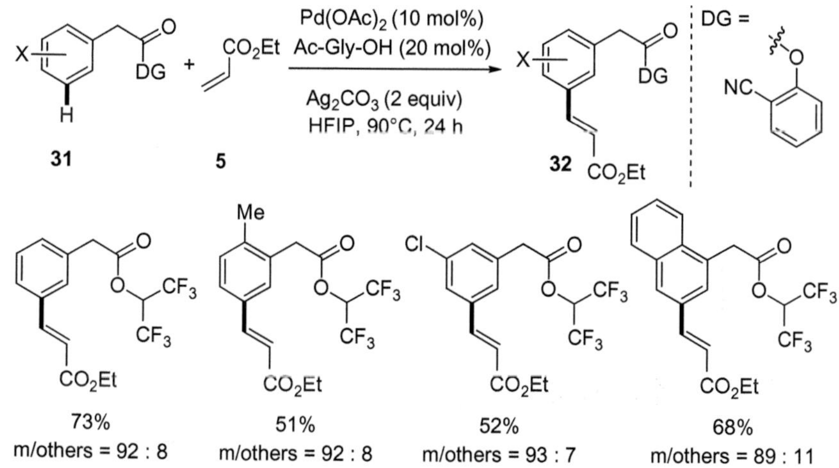

Scheme 8.11 *Meta*-C—H olefination of phenylacetic acid derivatives with 2-hydroxybenzonitrile directing group.

di-*meta*-olefination, leading to good yields of *meta*-olefinated products whose DG was removed in situ via trans-esterification with the solvent HFIP (Scheme 8.11).

The increasing applications of direct C—H transformations in organic synthesis demand accessibility to adjustable site selectivities with a common directing functionality. In 2015, Li and co-workers reported a *remote*-selective regiodivergent *ortho*- and *meta*-C—H functionalizations protocol with phenylethylamines, which was enabled by using a novel and simple 2-cyanobenzoyl group as the common original directing functionality (Scheme 8.12).[71] It was found the *N*-methyl-phenylethylamides **33** produced with commercially available 2-cyanobenzoic acid underwent efficient *meta*-C—H olefination in the presence of $Pd(OAc)_2$ and Ac-Gly-OH under nitrogen atmosphere (Scheme 8.12A). Interestingly, *remote*-selective *ortho*-olefination occurred with secondary amide **35** leaving the proximal aromatic *ortho*-C—H bond intact (Scheme 8.12B). This was proved to proceed with an imidamide intermediate that was formed through cyclization of the cyanobenzoyl motif. The *remote*-selective regiodivergent C—H activation was then exemplified via sequential C—H olefinations. Thus, the desired *meta*-directing nitrile group was reconstructed simultaneously with methylation by using LiHMDS to afford **38** after hydrogenation. Under standard reaction conditions, tetra-substituted phenylethylamide **40** was generated with tri-substituted

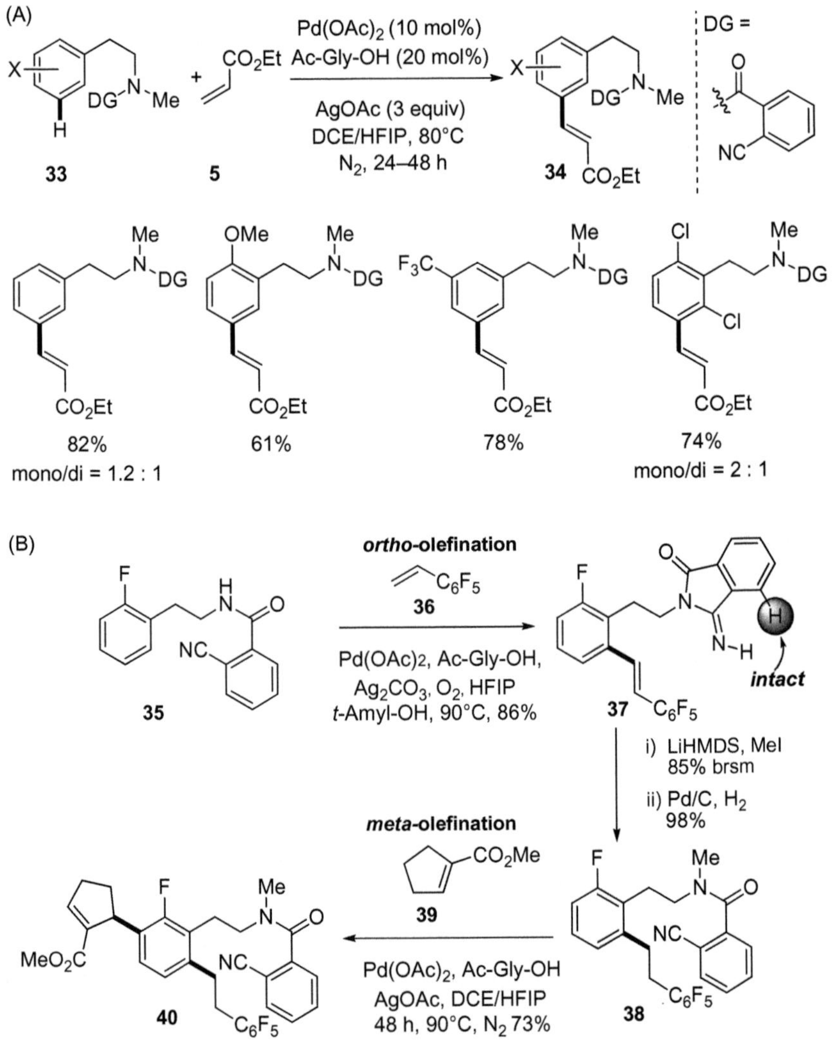

Scheme 8.12 *Meta*-C—H olefination of phenylethylamines with 2-cyanobenzoyl directing group and *remote*-selective regiodivergent C—H olefinations.

olefin **39**, enabling the building of complexity in a concise manner. Further development of this novel remote regiodivergent C—H functionalization is believed to empower C—H functionalization as a more versatile synthetic tool.

Sequential double *meta*-C—H olefination was made possible for the first time by Maiti and co-workers (Scheme 8.13) in 2015.[72] With the

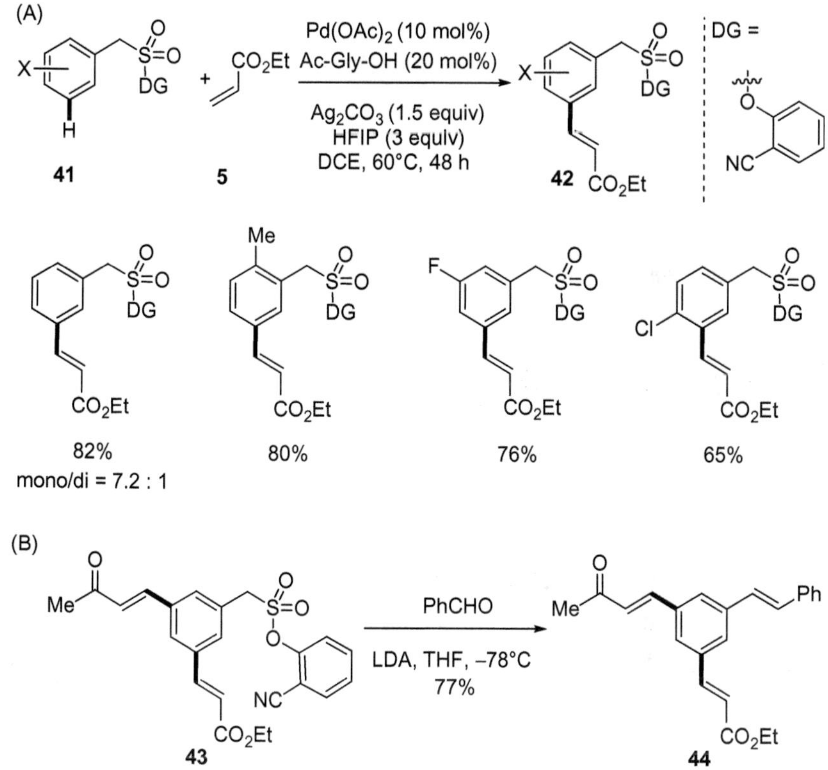

Scheme 8.13 *Meta*-C—H olefination of benzylsulfonyl ester derivatives.

above-mentioned commercially available 2-hydroxybenzonitrile DG, a broad range of benzylsulfonyl ester derivatives were *meta*-selectively olefinated with various types of olefin coupling partners (Scheme 8.13A). This protocol also enabled *meta*-selective homo-diolefination and sequential hetero-diolefination of benzylsulfonyl ester derivatives, product of which could be converted to 1,3,5-trialkenylated compounds using known modified Julia olefination conditions (Scheme 8.13B).

Subsequently, the sequential double *meta*-C—H olefination strategy was applied to synthetically versatile benzyl silanes by the Maiti group under similar reaction conditions (Scheme 8.14A).[74] The silyl tether was advantageous due to its easy installation and easy removal under standard TBAF conditions. Importantly, late-stage modification of the C—Si bond under various conditions could produce benzaldehyde and benzyl alcohol derivatives (Scheme 8.14B), showcasing the synthetic potential of this method.

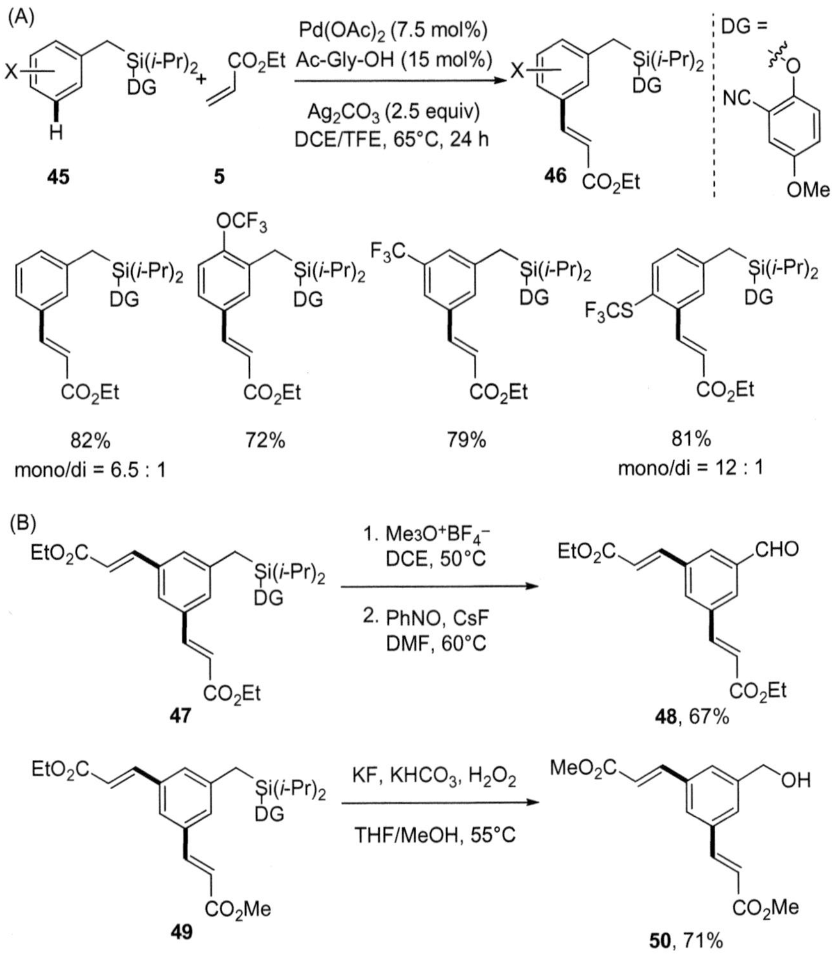

Scheme 8.14 *Meta*-C—H olefination of benzyl silane derivatives.

Furthermore, room-temperature *meta*-C—H olefination of benzylic phosphonate ester was also reported by Maiti and co-workers recently, using the powerful 2-hydroxybenzonitrile DG (Scheme 8.15A).[76] Moreover, the *meta*-selective homo-diolefination and sequential hetero-diolefination was also viable, which required elevated temperatures to proceed. Notably, the phosphonate linkage could be readily converted to alkenes by the modified Horner—Wadsworth—Emmons reaction to afford 1,3,5-trialkenylated arenes (Scheme 8.15B).

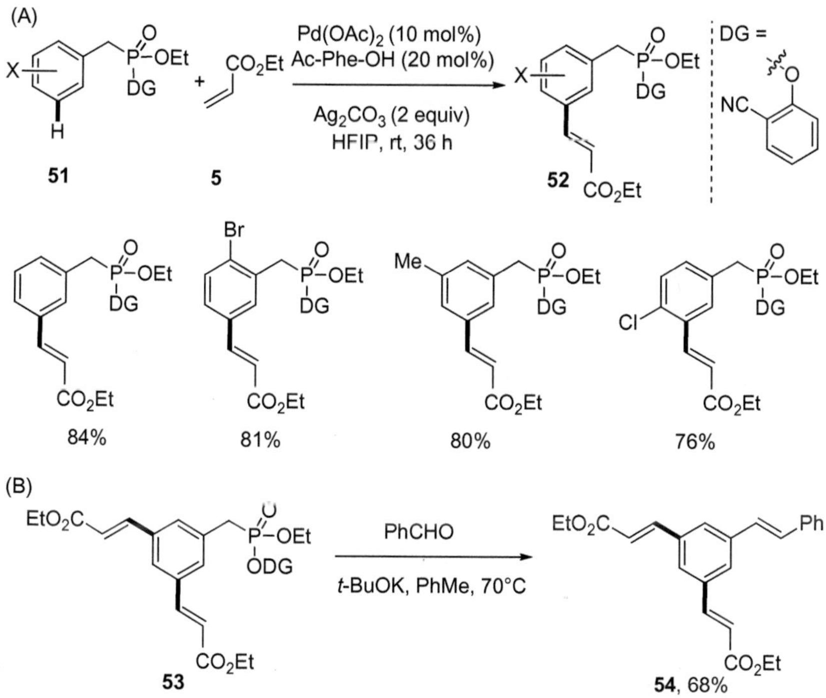

Scheme 8.15 *Meta*-C—H olefination of phosphonate ester derivatives.

Benzoic acids are highly important structural motifs and precursors in drug discovery and material sciences. Traditionally, benzoic acids were *meta*-functionalized by electrophilic aromatic substitution under harsh conditions, as they are generally deactivated towards this reaction. Though nitrile-based *meta*-directing groups were successfully applied to all the above-mentioned electron-rich/neutral arenes, this strategy had remained ineffective for enabling *meta*-C—H functionalization of benzoic acids, possibly due to the low reactivity of electron-deficient arenes towards palladation in this type of reaction.

In early 2016, Li and co-workers disclosed the first effective nosyl protected 2-cyano-phenylethylamine DG that promoted *meta*-C—H olefination as well as acetoxylation (see Section 8.2.1.3), with a broad range of electron-poor benzoic acid derivatives regardless of their substitution patterns (Scheme 8.16).[73] Notably, the new protocol featured using environmentally benign molecular oxygen as the terminal oxidant, while costly silver salt oxidants were required in all previous chelation-assisted

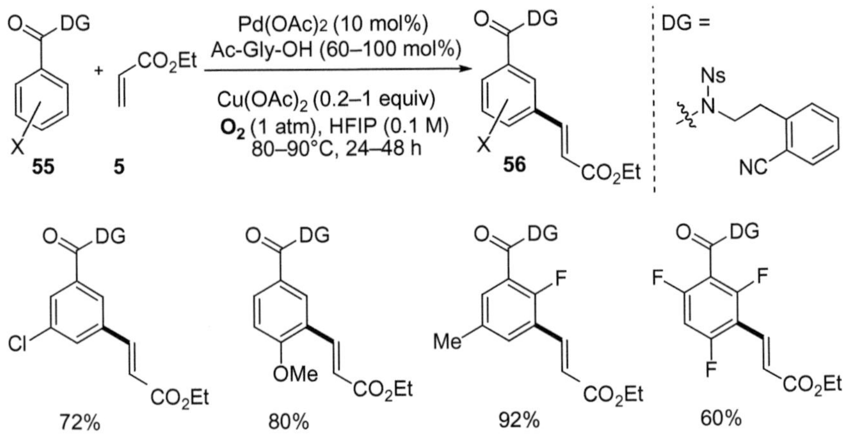

Scheme 8.16 *Meta*-C—H olefination of benzoic acid derivatives.

meta-C—H olefinations. Remarkably, this protocol tolerated the challenging tri-substituted substrates, which was not viable in previous transition metal-catalyzed *meta*-C—H functionalizations of (hetero)arenes through known approaches. Interestingly, ligand could be used to tune the mono- versus di-olefination selectivity. Thus, a good ratio of mono- over di-olefination could be obtained by using Formyl-Gly-OH instead of Ac-Gly-OH in the presence of inorganic bases like KH_2PO_4 or K_2HPO_4. Importantly, the sulfonamide directing auxiliary could be prepared on a large scale, and readily cleaved and recycled under very mild conditions (Scheme 8.17).

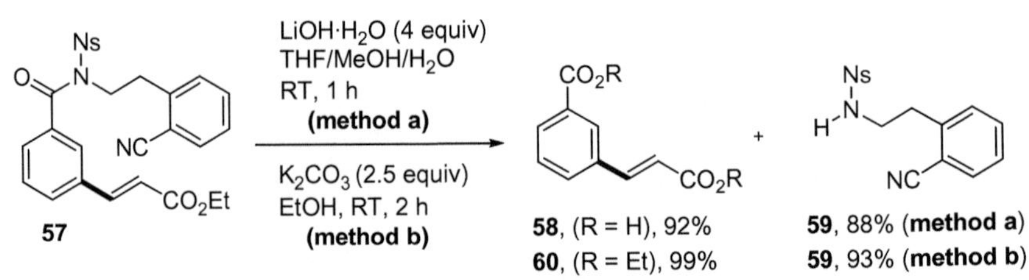

Scheme 8.17 Removal and recovery of the directing group for benzoic acids.

8.2.1.2 Arylation

Biaryl compounds are frequently found in numerous pharmaceuticals and agrochemicals, as well as functional molecules.[98] Thus, a direct and selective construction method for biaryl compounds is highly desired. In 2013, Yu and co-workers reported the first example of Pd-catalyzed cross-coupling of *meta*-C—H bonds with arylboronic esters with phenylpropanoic acid (Scheme 8.18) and phenolic derivatives (Scheme 8.19).[65] The combination of the nitrile-based DGs and MPAA ligand (Ac-Gly-OH) previously used for *meta*-C—H olefination of hydrocinnamic acid was

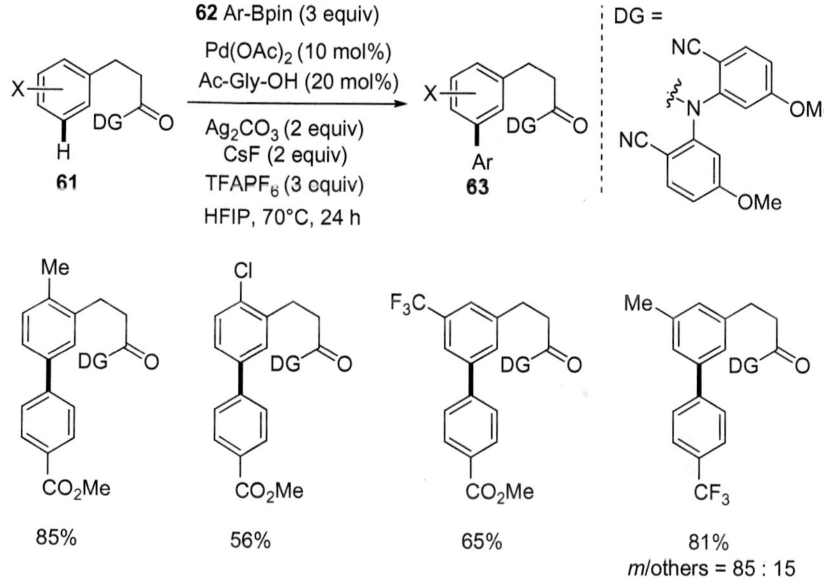

Scheme 8.18 *Meta*-C—H arylation of phenylpropanoic acid derivatives.

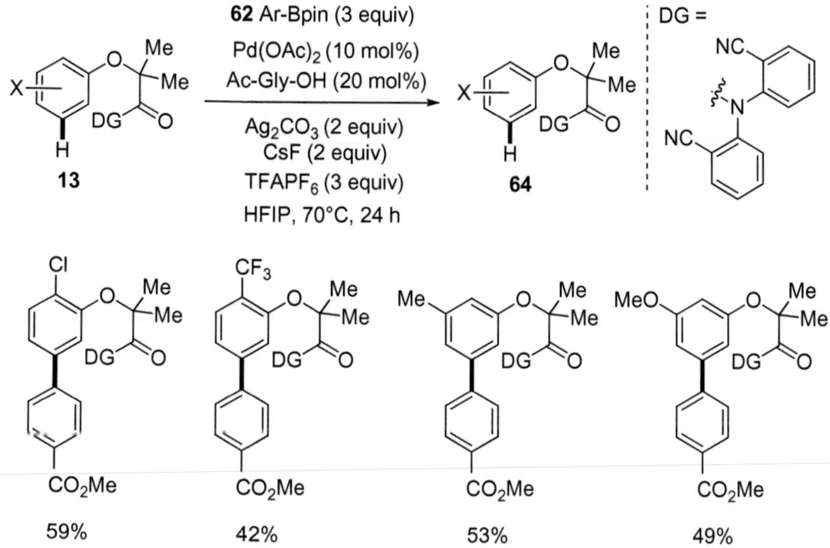

Scheme 8.19 *Meta*-C—H arylation of phenolic derivatives.

crucial for the reaction to proceed. It was also shown that tetrabutylammonium (TBA) salt $TBAPF_6$ dramatically increased the reaction yield, which was believed to result from the ability of TBA surfactants to prevent undesired agglomeration of Pd(0) species to form unreactive palladium black.

Subsequently, the above protocol was applied to indoline derivatives of medicinal interest with previously reported nitrile-based sulfonamide DG by the Yu group (Scheme 8.20).[67]

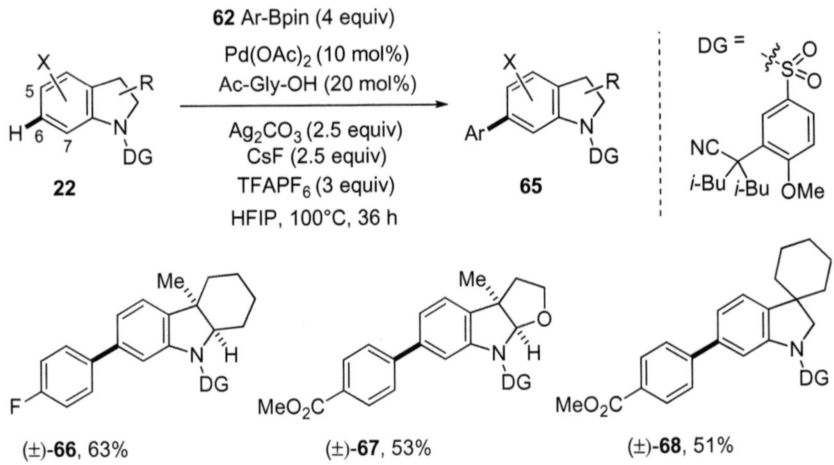

Scheme 8.20 *Meta*-C—H arylation of phenolic derivatives.

8.2.1.3 Oxygenation

The potential of nitrile-based *meta*-directing group to induce C—heteroatom bond formation via a different catalytic cycle was demonstrated by *meta*-C—H acetoxylation and hydroxylation, which proceed through a Pd(II)/Pd(IV) redox cycle rather than the Pd(II)/Pd(0) catalytic cycle in previous *meta*-C—H olefination and arylation. In this context, *meta*-C—H acetoxylation was first reported with aniline (Scheme 8.21) and benzylamine (Scheme 8.22) derivatives by Yu and co-workers in 2014.[66]

Though this type of reaction was generally not as efficient as *meta*-C—H olefination, it proceeded with various substituted anilines at excellent levels of *meta*-selectivity (90%—98%) using $PhI(OAc)_2$ as the oxidant and acetic anhydride as the additive (Scheme 8.21). The protocol

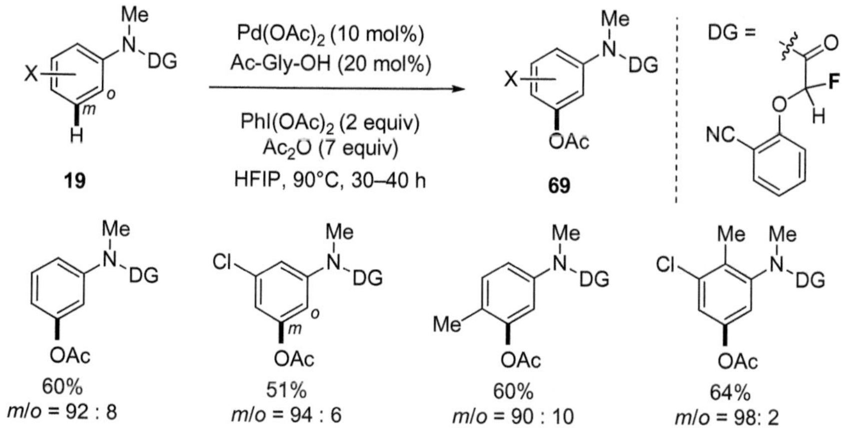

Scheme 8.21 *Meta*-C—H acetoxylation of aniline derivatives.

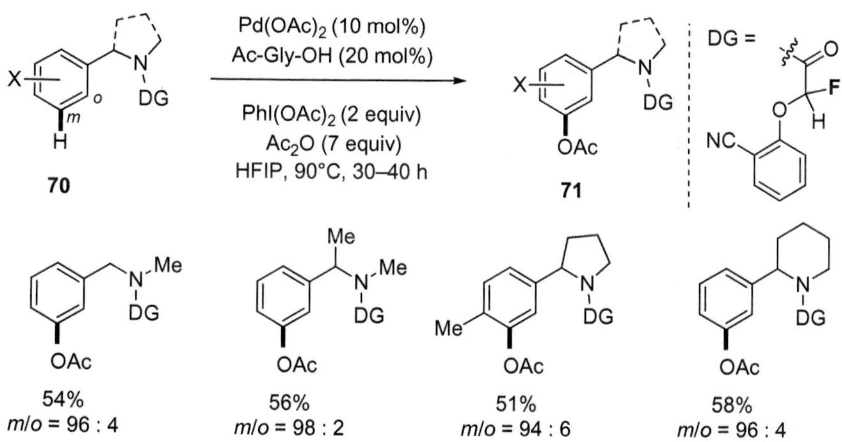

Scheme 8.22 *Meta*-C—H acetoxylation of benzylamine derivatives.

was further extended to benzylamine derivatives, which are medicinally important heterocycles (Scheme 8.22).

Subsequently, Yu and co-workers attempted the *meta*-C—H acetoxylation with a few indolines, products of which possess important biological activities (Scheme 8.23).[67] Using the above established reaction conditions, although substantial amounts of undesired *para*-acetoxylated indolines were produced due to the electrophilic palladation at the electron-rich C-5 position, *meta*-acetoxylated indolines were obtained as the major products using a sulfonamide DG.

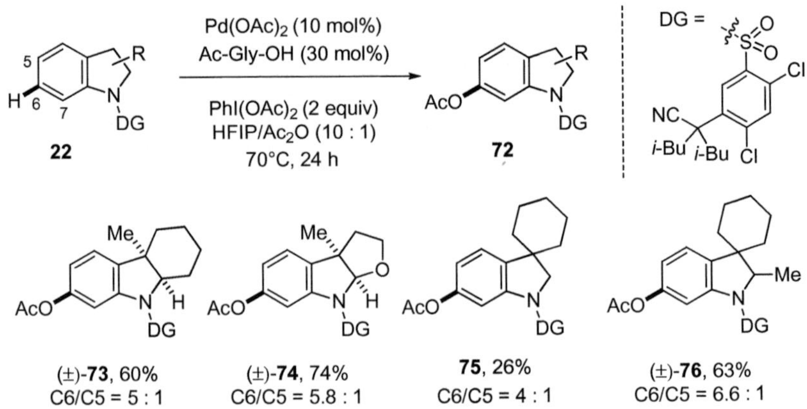

Scheme 8.23 *Meta*-C—H acetoxylation of indoline derivatives.

In 2016, using the nosyl protected 2-cyano-phenylethylamine as the directing auxiliary, electron-deficient benzoic acid derivatives with different substitution patterns were *meta*-acetoxylated successfully by Li and co-workers (Scheme 8.24).[73] Remarkably, the DG could be readily cleaved under mild basic conditions, with concomitant methanolysis of the acetoxy group, product of which could be triflated to enable access to five synthetically useful *meta*-functionalized benzoic acid derivatives (Scheme 8.25).

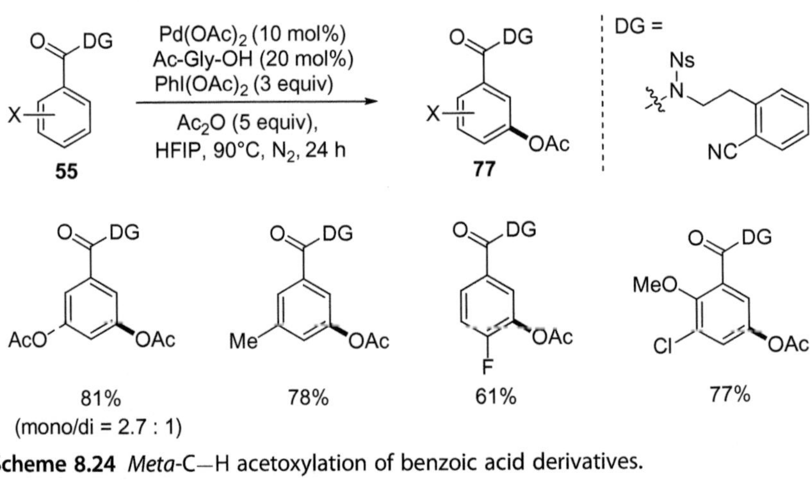

Scheme 8.24 *Meta*-C—H acetoxylation of benzoic acid derivatives.

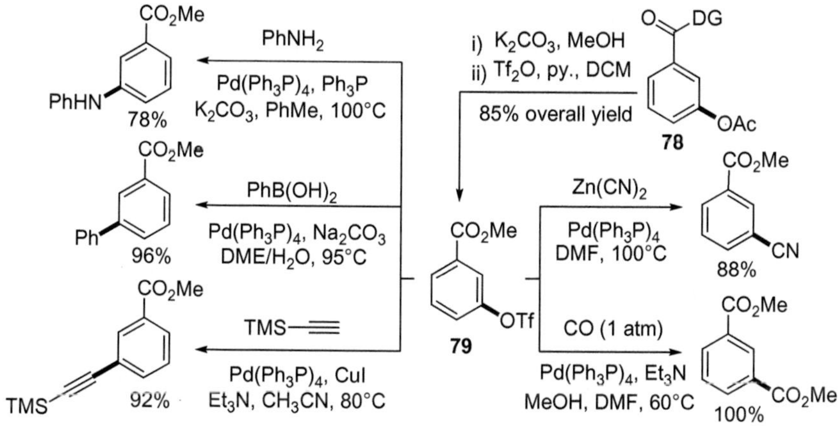

Scheme 8.25 Synthetic elaboration of *meta*-C—H acetoxylation product of benzoic acids.

Recently, Maiti and co-workers reported a *meta*-C—H acetoxylation of benzylsulfonyl ester derivatives under similar conditions, but with the *N*-*tert*-butyloxycarbonyl-alanine (Boc-Ala-OH) instead of Ac-Gly-OH as the MPPA ligand (Scheme 8.26).[75] Moreover, a change of $PhI(OAc)_2$ to $PhI(TFA)_2$ led to *meta*-C—H hydroxylation instead of acetoxylation with *N*-formyl glycine as the ligand, which is probably due to the easier hydrolysis of the trifluoroacetate than the acetate intermediate under the reaction conditions (Scheme 8.27).

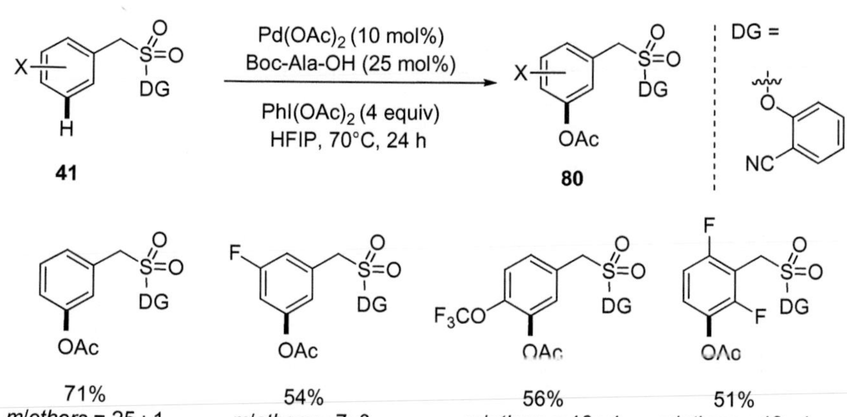

Scheme 8.26 *Meta*-C—H acetoxylation of benzylsulfonyl ester derivatives.

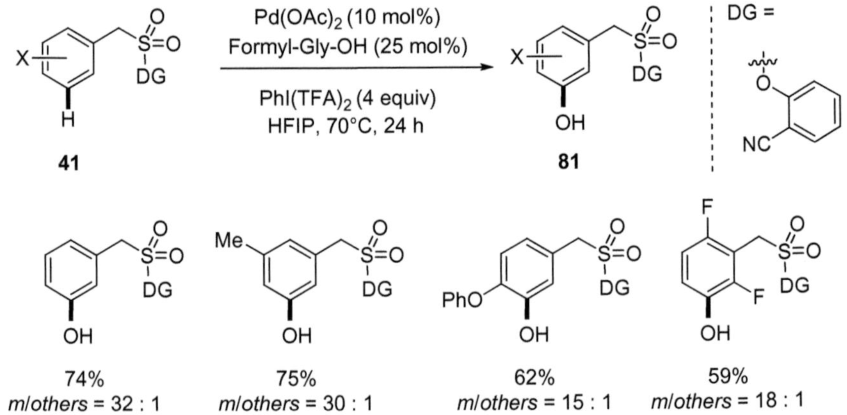

Scheme 8.27 *Meta*-C—H hydroxylation of benzylsulfonyl ester derivatives.

Using the 2-hydroxybenzonitrile DG, Maiti and co-workers also reported an efficient *meta*-C—H acetoxylation of benzylic phosphonate esters (Scheme 8.28).[76] Similarly to the benzylsulfonyl ester derivatives, switching the oxidant from $PhI(OAc)_2$ to $PhI(TFA)_2$ afforded the *meta*-C—H hydroxylation of benzylic phosphonate esters with Boc-Ala-OH as the ligand (Scheme 8.29).

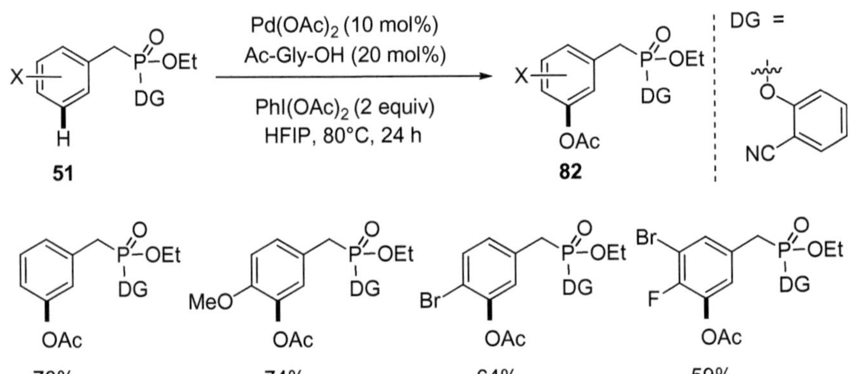

Scheme 8.28 *Meta*-C—H acetoxylation of phosphonate ester derivatives.

8.2.2 Nonnitrile-based Directing Group Assisted *Meta*-C—H Functionalization

Inspired by using the U-shaped nitrile-based DGs to achieve remote *meta*-C—H activation via the recognition of distance and geometry,

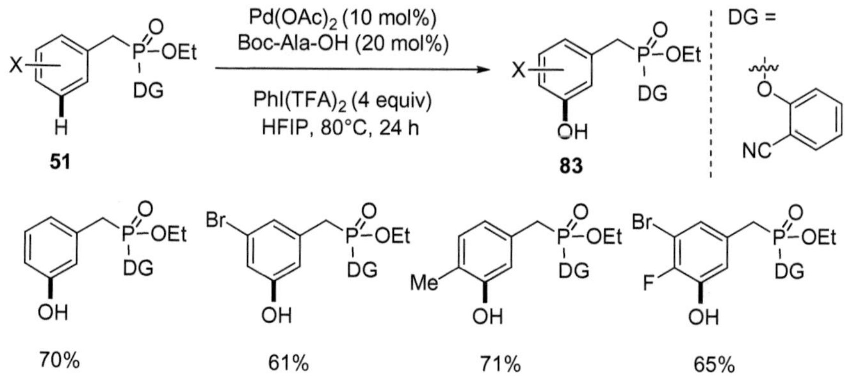

Scheme 8.29 *Meta*-C—H hydroxylation of phosphonate ester derivatives.

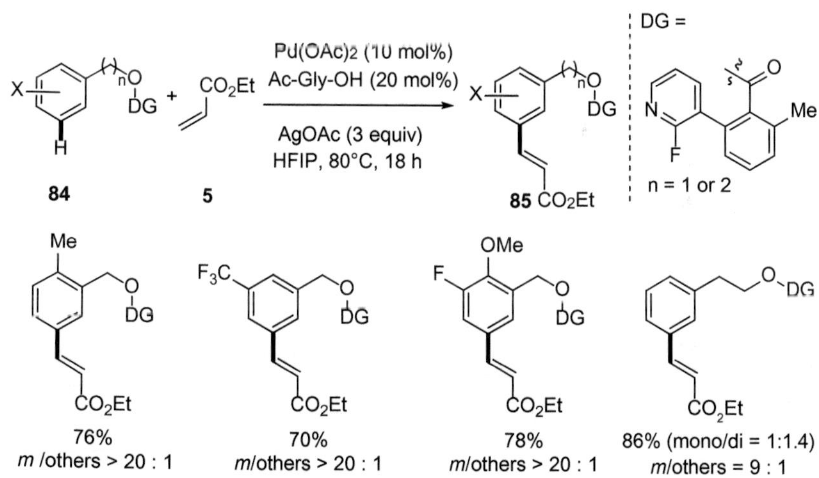

Scheme 8.30 *Meta*-C—H olefination of benzyl and phenyl ethyl alcohol derivatives.

Yu and co-workers engineered a pyridine-based DG for *meta*-C—H olefination (Scheme 8.30) and iodination (Scheme 8.31) of benzyl and phenyl ethyl alcohols.[77] This remarkable breakthrough demonstrated that with appropriate distance and geometry not only weakly coordinating nitrile group could lead to *meta*-C—H activation, but also the conventional strongly coordinating group such as the extensively employed *ortho*-directing pyridyl group could induce similar or even new *meta*-C—H transformations, which paved a new way for developing various unexplored *meta*-C—H transformations.[7] In this regard, a new *meta*-C—H iodination reaction that was not compatible with previous nitrile-based

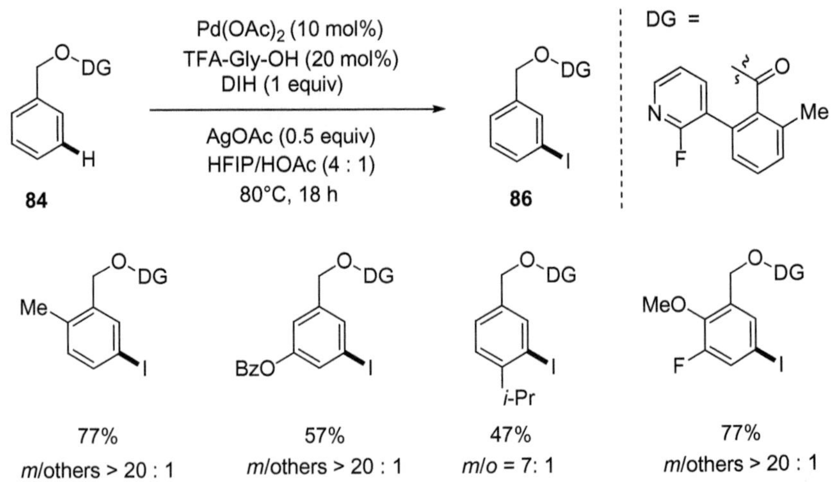

Scheme 8.31 *Meta*-C—H iodination of benzyl alcohol derivatives.

DGs was realized with the 2-fluoropyridine containing DG using DIH (1,3-diiodo-5,5-dimethylhydantoin) as the iodination reagent, affording aryl iodides that are precious intermediates for a wide range of carbon—carbon and carbon—heteroatom bond formation reactions such as Heck and Suzuki reactions (Scheme 8.31).

In 2015, Wang, Zhang, and co-workers reported a method for *meta*-selective arylation of *O*-β-naphthyl carbamates using $Pd(OAc)_2$ as the metal catalyst and aryl boronic acid as the coupling partner (Scheme 8.32).[78]

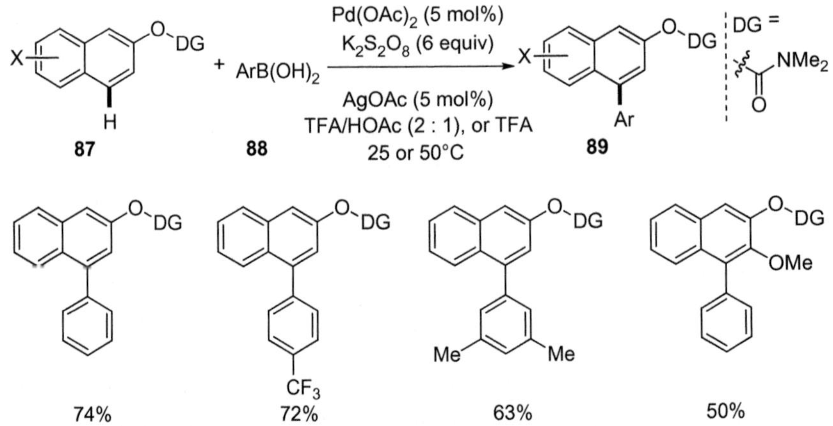

Scheme 8.32 *Meta*-C—H arylation of *O*-β-naphthyl carbamates.

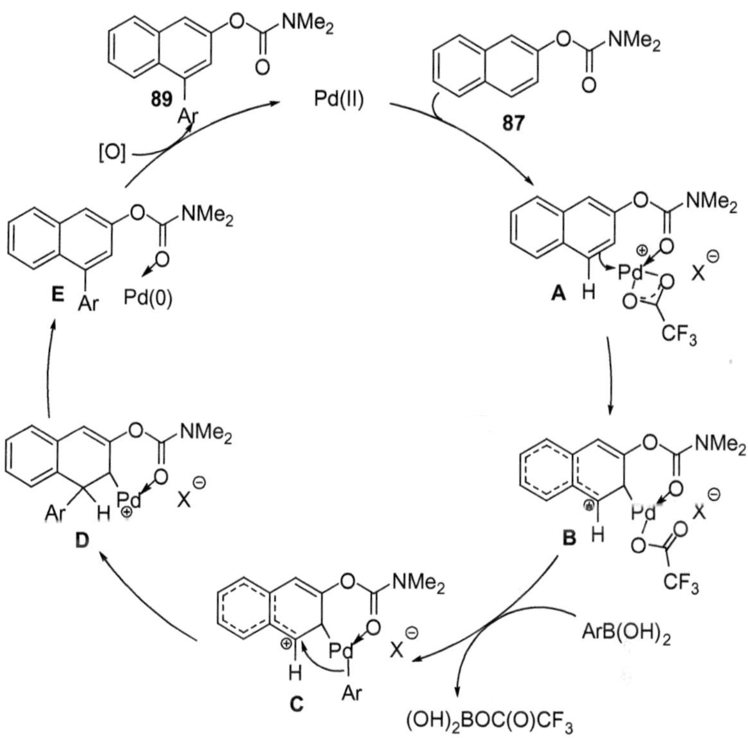

Scheme 8.33 Proposed catalytic cycle of *meta*-C—H arylation of *O*-β-naphthyl carbamates.

Mechanistic study revealed that the carbamate group-assisted *ortho*-carbopalladation occurred first, followed by *meta*-direct arylation, representing a new strategy for accomplishing the *meta*-C—H arylation of arenes (Scheme 8.33). It was also found that the cleavage of *ortho*-C—H bond was not necessary to realize the *meta*-C—H arylation, since *O*-3-methoxy-2-naphthyl carbamate also led to the desired *meta*-arylation product.

Subsequently, *meta*-arylation of 2-naphthyl ureas was also disclosed by Wang and co-workers under similar reaction conditions, but with $Cu(OAc)_2$ and $PhCO_3Bu$ as the new oxidant combination (Scheme 8.34).[79] However, *ortho*-arylated products were obtained with phenyl ureas, indicating that the regioselectivity is aromaticity-dependent on the corresponding aryl substituents.

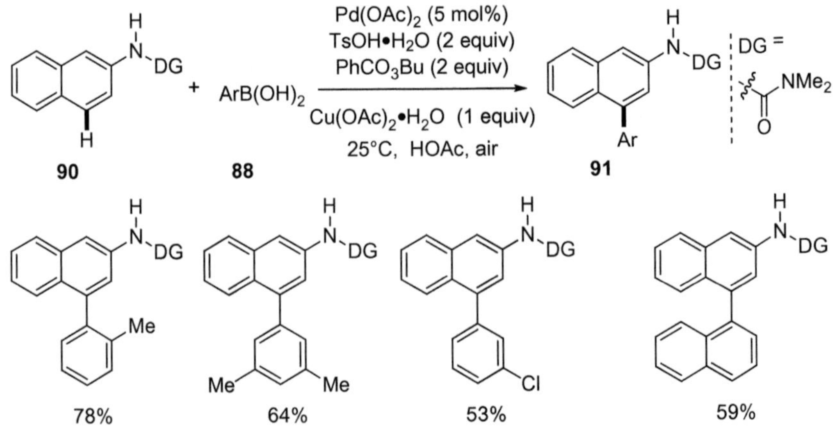

Scheme 8.34 *Meta*-C—H arylation of 2-naphthyl ureas.

8.3 *ORTHO*-DIRECTING GROUP-ASSISTED AND NORBORNENE-MEDIATED *META*-C—H FUNCTIONALIZATION

The catalytic cross-coupling cascade Catellani reaction using norbornene as the transient mediator has inspired various synthetic useful direct C—H functionalizations of aryl iodides.[99–103] For example, Catellani and co-workers disclosed a formal *meta*-C—H arylation to produce unsymmetrical homobiaryl products via the Catellani-type reaction with a delayed hydrogenolysis using benzyl alcohol as hydrogen transfer reagent in 2005 (Scheme 8.35).[101]

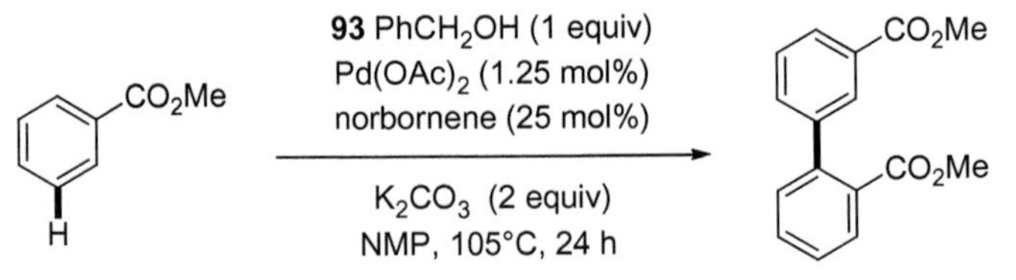

Scheme 8.35 Synthesis of *meta*-substituted arenes via the Catellani reaction.

In 2015, the distinct reactivity of norbornene in site-selective C—H functionalization of arenes was further explored in a remarkable DG-assisted *meta*-C—H activation reaction by Yu and co-workers.[80] Promoted with pyridine- or quinoline-based ligands and mediated by norbornene, phenylacetic acid derivatives attached with a common *ortho*-directing group were *meta*-C—H alkylated or arylated with electrophiles including methyl iodide, ethyliodoacetate, benzyl bromide, and aryl iodides (Scheme 8.36). In the proposed catalytic cycle (Scheme 8.37), the formation of the key side product **99** was found to be substantially reduced by using electron-rich pyridine-based ligand **98** which also facilitated the

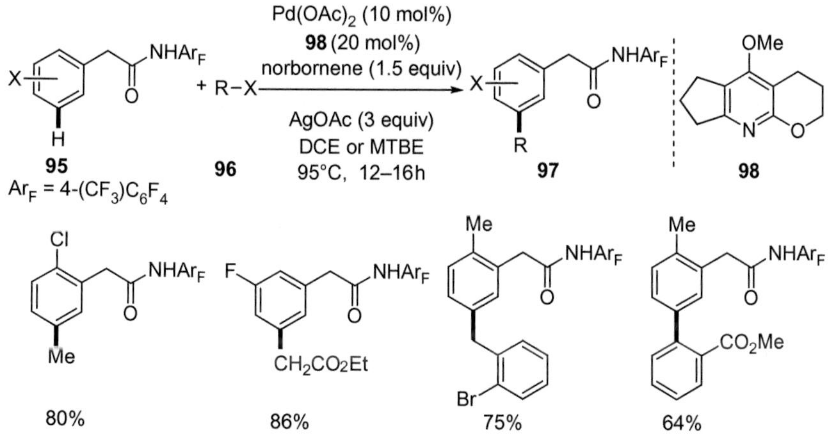

Scheme 8.36 Norbornene-mediated *meta*-C—H functionalization of phenylacetic acid derivatives.

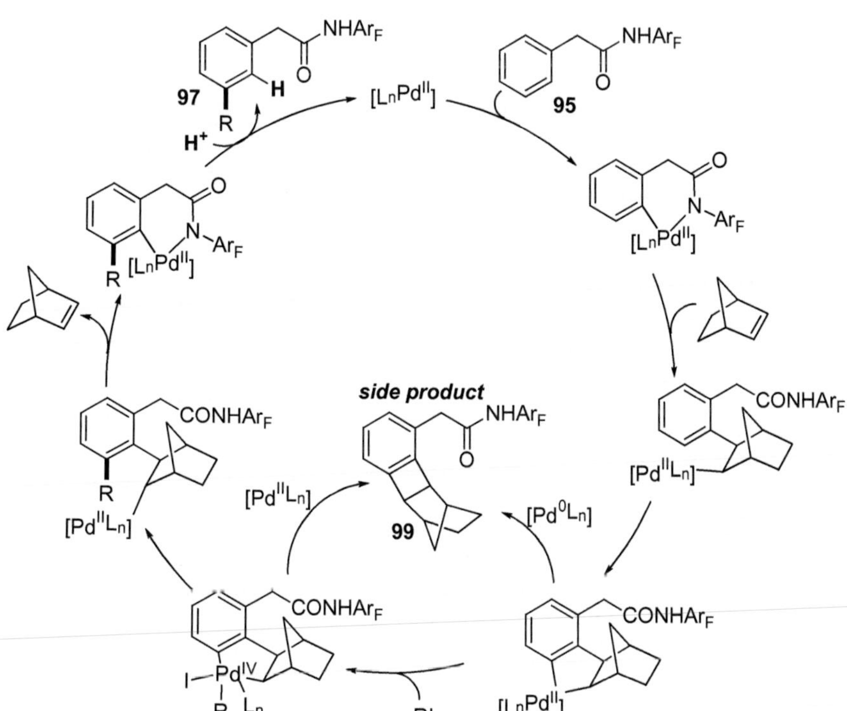

Scheme 8.37 Proposed catalytic cycle for norbornene-mediated *meta*-C—H functionalization.

first key steps including *ortho*-C—H activation, 1,2-migratory insertion, and *meta*-C—H activation in the reaction.

Almost simultaneously, Dong and co-workers reported a norbornene-mediated *meta*-C—H arylation of tertiary amine **100** with $Pd(OAc)_2$ as the metal catalyst and commercially available $AsPh_3$ as the ligand under the promotion of an "acetate cocktail" (Scheme 8.38).[81] The interesting "acetate cocktail," a combination of LiOAc hydrate, CsOAc hydrate, and $Cu(OAc)_2$ hydrate in acetic acid, was proved to improve the reaction rate. Notably, the easily installed tertiary amine group could be converted to other synthetically useful functional groups (FGs) readily (Scheme 8.39).

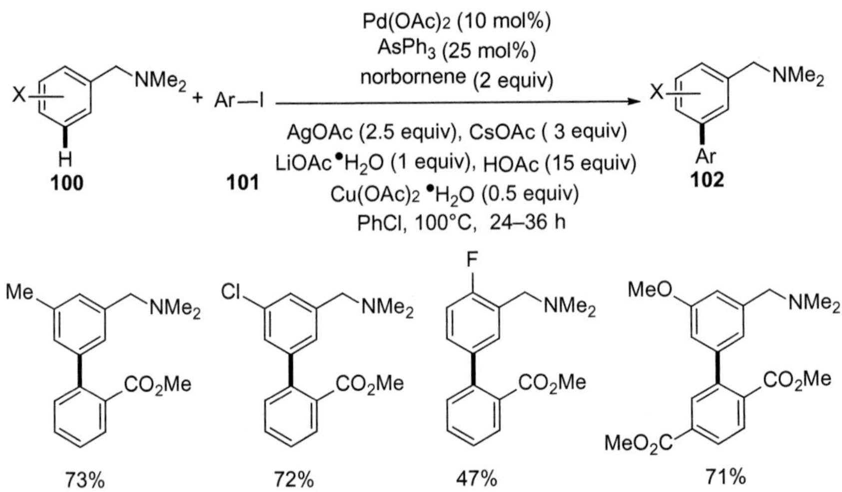

Scheme 8.38 Norbornene-mediated *meta*-C—H functionalization of benzylamines.

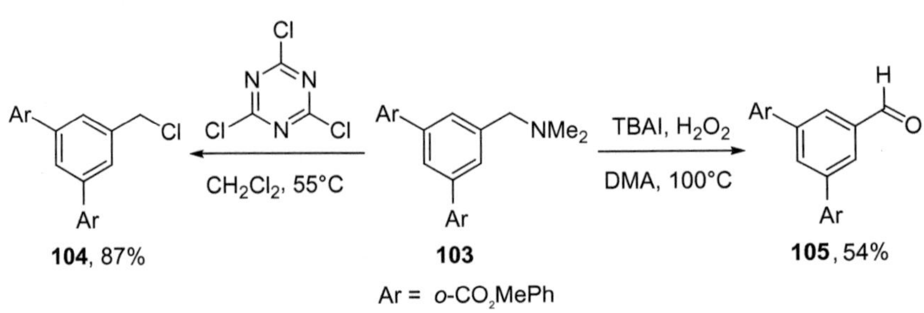

Scheme 8.39 Derivatization of the *meta*-arylation product of benzylamine.

In the above-mentioned norbornene-mediated *meta*-C—H functionalizations, the reactions were limited to aryl iodide coupling partners bearing *ortho*-substituents, and alkylations with β-hydrogen-containing alkyl iodides only led to low yields of desired products. In a follow-up work,[82] Yu and co-workers identified 2-carbomethoxynorbornene as a more effective transient mediator that promoted unprecedented *meta*-C—H alkylation of phenyl acetamides **95** with a wide range of alkyl iodides (Scheme 8.40), as well as arylation (Scheme 8.41) with previously incompatible aryl iodides using a quinoline-based ligand **107**.

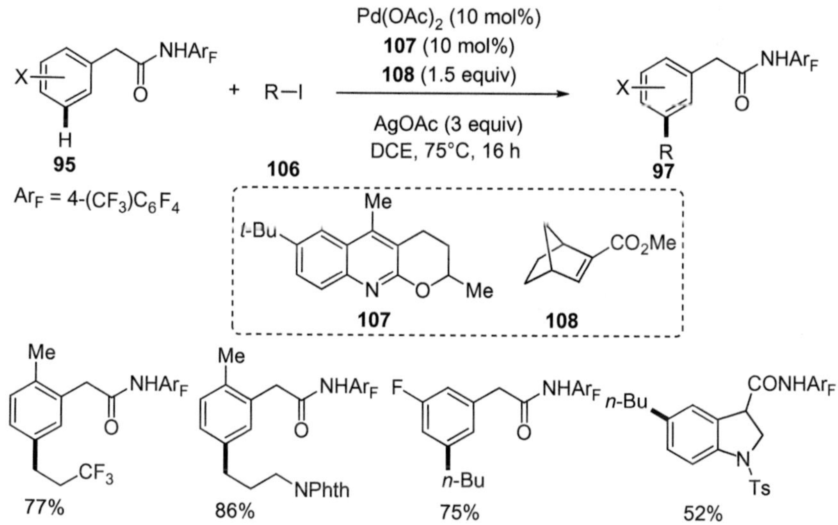

Scheme 8.40 Norbornene-mediated *meta*-C—H alkylation.

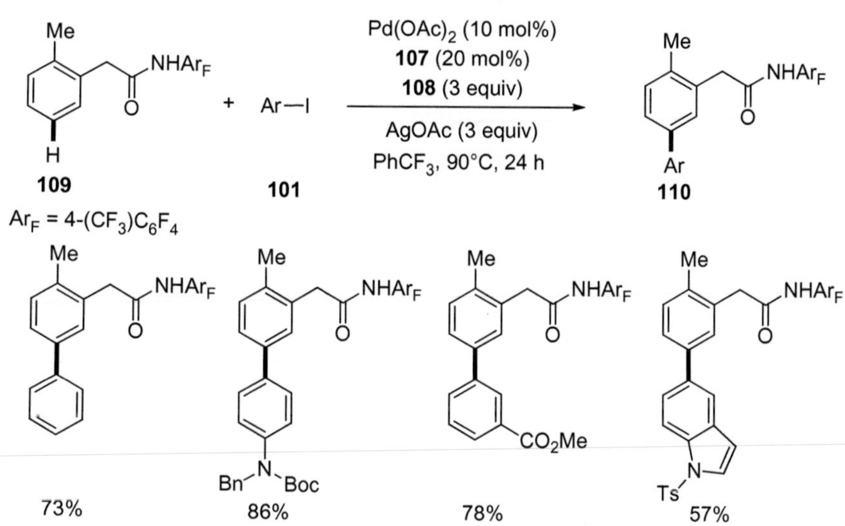

Scheme 8.41 Norbornene-mediated *meta*-C—H arylation.

Most recently, Zhao, Shi, and co-workers disclosed an oxalyl amide DG-assisted highly selective *meta*-arylation of *β*-arylethylamine derivatives mediated by norbornene with a broad range of aryl iodides (Scheme 8.42).[83] Moreover, thiophene derivatives were also viable with this reaction. Notably, this is the first report on bidentate DG-assisted norbornene-mediated *meta*-C—H functionalization.

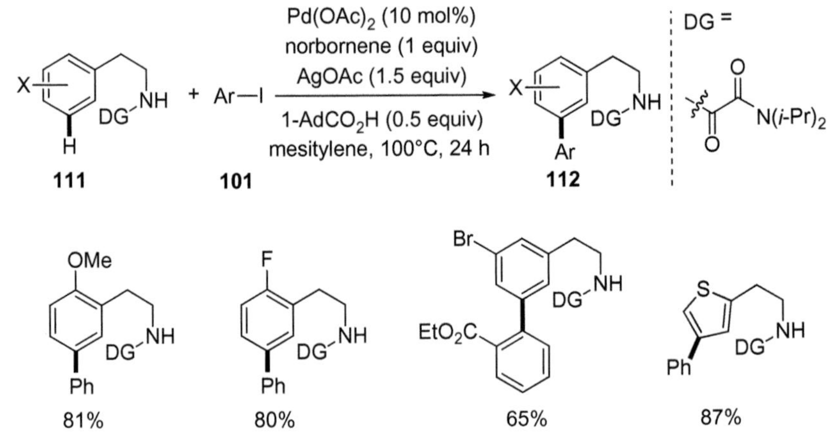

Scheme 8.42 Norbornene-mediated *meta*-C—H arylation of *β*—arylethyamides.

8.4 FORMAL *META*-C—H FUNCTIONALIZATION USING A TRACELESS DIRECTING GROUP

In 2011, Larrosa and co-workers disclosed a method for formal *meta*-C—H arylation using carboxylic acids as a traceless *ortho*-directing group (Scheme 8.43) with aryl iodides as coupling partners, which was compatible with a wide range of *meta*-substituents (Scheme 8.44).[84] Importantly, this tandem directed C—H arylation/protodecarboxylation process overrode electronic bias of the benzoic acids, representing an efficient alternative route to generate *meta*-substituted biaryls. Notably, this approach is also applicable using other transition metals (TMs).[104–109] Remarkably, as depicted by the proposed working mode (Scheme 8.43), this method successfully bypassed the undesired protodecarboxylation of the starting *ortho*-substituted benzoic acid and decarboxylative *ipso*-arylation process.

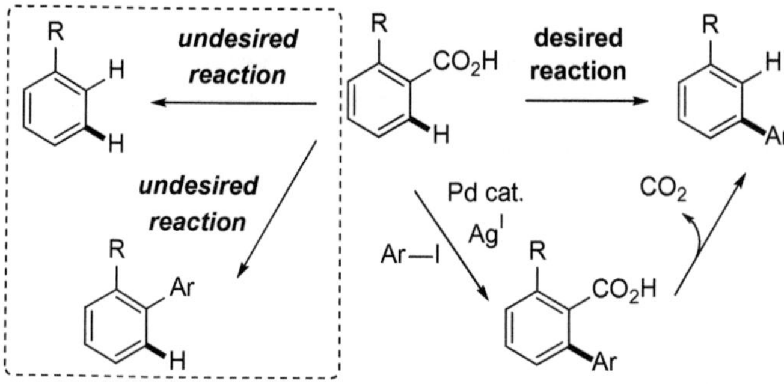

Scheme 8.43 Tandem *ortho*-selective arylation/protodecarboxylation process for formal *meta*-selective C—H arylation.

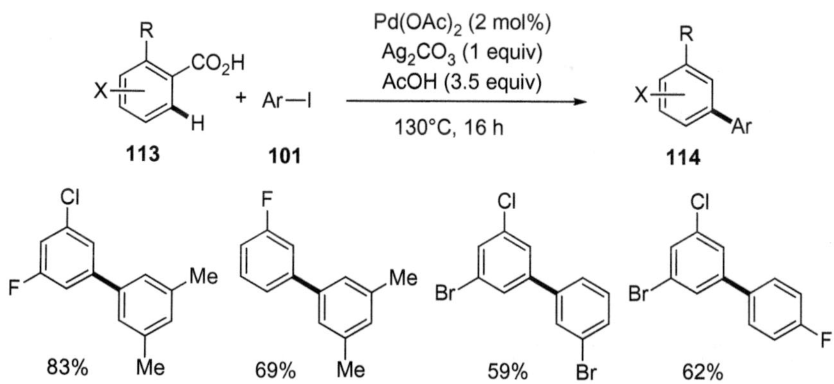

Scheme 8.44 Formal *meta*-C—H arylation using carboxylic acid as a traceless directing group.

Subsequently, a one-pot direct *meta*-arylation of phenols using carbon dioxide as a transient DG was reported by Larrosa and co-workers, which was compatible with a variety of FGs both in the phenol and in the iodoarene coupling partner (Scheme 8.45).[85] After installation of the carboxylic acid motif via the Kolbe—Schmitt reaction, a tandem directed C—H arylation/protodecarboxylation process occurred with PEPPSI-IPr as the catalyst. This unique strategy represented an ingenious method for accessing *meta*-arylated phenols of substantial synthetic utility. However, the protocol was not compatible with *para*-substituted phenols or *ortho*-substituted iodoarene coupling partners.

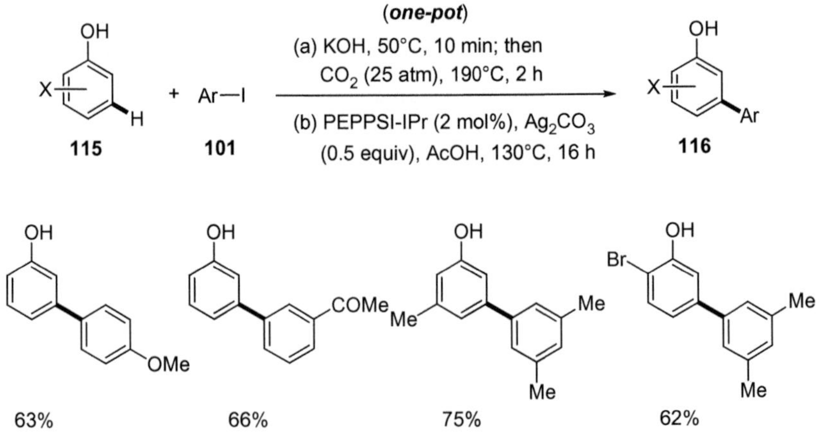

Scheme 8.45 Formal *meta*-C–H arylation of phenols.

In a follow-up work, since the Kolbe–Schmitt reaction used for generating a transient carboxylic DG suffered from harsh reaction conditions, the readily available salicylic acids were used directly as the substrate for producing *meta*-arylphenols by the Larrosa group (Scheme 8.46).[86] This method was further highlighted by efficient conversion of the *meta*-arylphenols to a variety of *meta*-functionalized biaryls.

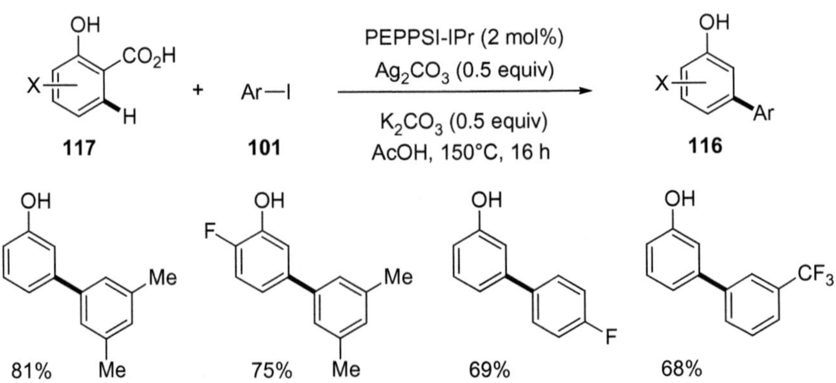

Scheme 8.46 Access to *meta*-arylphenols from salicylic acids.

Recently, a new method for accessing *meta*-arylated phenol derivatives was reported by the same research group via a cascade sequence of oxidation/arylation/protodecarboxylation of salicylaldehydes which could be generated easily from phenol derivatives under mild conditions by an *ortho*-selective formulation (Scheme 8.47).[87]

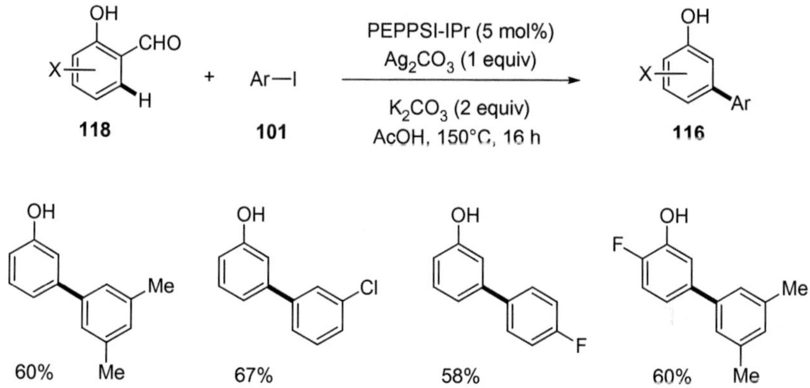

Scheme 8.47 Access to *meta*-arylphenols from salicylaldehydes.

8.5 CONCLUSION

Selectivity is one of the most important central themes in organic synthesis. Due to the ubiquity of C—H bonds in organic molecules, site selectivity is one of the most significant challenges that need to be effectively addressed before the widespread application of this atom- and step-economical synthetic methodology both in academia and industry. Over the past decades, tremendous advances in the *ortho*-selective C—H functionalizations have been achieved through chelation-assisted and proximity-induced C—H activation of arenes. Strategies of chelation-assisted *meta*-selective C—H activation have only been developed in very recent years. Although study of such fascinating strategies still remains in its infancy, it holds great opportunities for discovering a series of synthetically useful transformations, as well as uncovering new informative mechanistic insights into C—H activation. For example, although the *meta*-directing group-assisted C—H functionalizations have been compatible with various arenes, they are limited to *meta*-C—H olefination, arylation, oxygenation, and a single example of iodination. Thus, new transformations need to be developed with this powerful strategy under new catalytic conditions, or with new carefully designed DGs. In addition, attempts should be made to study whether these *meta*-directing groups could be used in a catalytic amount by attaching them to the substrates in situ or via a secondary interaction.[61,62] Moreover, only halide coupling partners could be utilized in the *meta*-C—H transformations via the Catellani-type reaction or with a traceless DG to afford *meta*-C—H alkylation and arylation. Hence, these two innovative strategies bear great potential for developments of new types of transformation, as well as with new classes of substrates. Furthermore, the chelation-assisted *meta*-C—H activation strategies have been most successfully exploited with palladium catalysts, leaving great opportunities for employing other metal catalysts with such strategies.[104–109] Overall, significant and exciting future progress could be anticipated in the research field of *meta*-C—H functionalization in the coming years.

ABBREVIATIONS

Ac	Acetyl
Ala	L-alanine
Ar	Aryl
Boc	*tert*-Butyloxycarbonyl
Bn	Benzyl
Bu	Butyl
cat	Catalytic
CMD	Concerted metalation deprotonation
DCE	1,2-Dichloroethane
DG	Directing group
DIH	1,3-diiodo-5,5-dimethylhydantoin
DMF	*N,N*-Dimethylformamide
DMPU	1,3-Dimethyl-3,4,5,6-tetrahydro-2(1H)-pyrimidinone
equiv	Equivalent
EDG	Electron-donating group
EWG	Electron-withdrawing group
FG	Functional group
Gly	Glycine
HFIP	Hexafluoroisopropanol
KIE	Kinetic isotope effect
L	Ligand
LDA	Lithium diisopropylamide
m	meta
Me	Methyl
MPAA	Mono-*N*-protected amino acid
NMP	*N*-Methylpyrrolidinone
o	ortho
p	para
PEPPSI	[1,3-Bis(2,6-Diisopropylphenyl)imidazol-2-ylidene](3-chloropyridyl) palladium(II) dichloride
Ph	Phenyl
Phe	L-phenylalanine
Pin	Pinacol
Piv	Pivaloyl
Pr	Propyl

TBA Tetrabutylammonium
TFA Trifluoroacetic acid
TFAPF$_6$ Tetrabutylammonium hexafluorophosphate
TFE Trifluoroethanol
THF Tetrahydrofuran
TM Transition metal

REFERENCES

1. Li, J.; Sarkar, S. D.; Ackermann, L. *Top. Organomet. Chem.* **2016,** *55*, 217.
2. Yang, J. *Org. Biomol. Chem.* **2015,** *13*, 1930.
3. Dey, A.; Agasti, S.; Maiti, D. *Org. Biomol. Chem.* **2016,** *14*, 5440.
4. Truong, T.; Daugulis, O. *Angew. Chem., Int. Ed.* **2012,** *51*, 11677.
5. Julia-Hernandez, F.; Simonetti, M.; Larrosa, I. *Angew. Chem. Int. Ed.* **2013,** *52*, 11458.
6. Martinez, C.; Muniz, K. *ChemCatChem* **2013,** *5*, 3502.
7. Frost, C. G.; Paterson, A. J. *ACS Cent. Sci.* **2015,** *1*, 418.
8. Davies, H. M.; Manning, J. R. *Nature* **2008,** *451*, 417.
9. McMurray, L.; O'Hara, F.; Gaunt, M. J. *Chem. Soc. Rev.* **2011,** *40*, 1885.
10. Gutekunst, W. R.; Baran, P. S. *Chem. Soc. Rev.* **2011,** *40*, 1976.
11. Chen, D. Y. K.; Youn, S. W. *Chem. Eur. J.* **2012,** *18*, 9452.
12. Yamaguchi, J.; Yamaguchi, A. D.; Itami, K. *Angew. Chem., Int. Ed.* **2012,** *51*, 8960.
13. Wencel-Delord, J.; Glorius, F. *Nat. Chem.* **2013,** *5*, 369.
14. Lyons, T. W.; Sanford, M. S. *Chem. Rev.* **2010,** *110*, 1147.
15. Satoh, T.; Miura, M. *Synthesis (Mass).* **2010,** 3395.
16. Colby, D. A.; Bergman, R. G.; Ellman, J. A. *Chem. Rev.* **2010,** *110*, 624.
17. Yeung, C. S.; Dong, V. M. *Chem. Rev.* **2011,** *111*, 1215.
18. Hartwig, J. F. *Acc. Chem. Res.* **2012,** *45*, 864.
19. Engle, K. M.; Mei, T.-S.; Wasa, M.; Yu, J.-Q. *Acc. Chem. Res.* **2012,** *45*, 788.
20. Patureau, F. W.; Wencel-Delord, J.; Glorius, F. *Aldrichimica Acta* **2012,** *45*, 31.
21. Song, G.; Wang, F.; Li, X. *Chem. Soc. Rev.* **2012,** *41*, 3651.
22. Rouquet, G.; Chatani, N. *Angew. Chem., Int. Ed.* **2013,** *52*, 11726.
23. Ackermann, L. *Acc. Chem. Res.* **2014,** *47*, 281.
24. Zhang, X.-S.; Chen, K.; Shi, Z.-J. *Chem. Sci.* **2014,** *5*, 2146.
25. Zhang, M.; Zhang, Y.; Jie, X.; Zhao, H.; Li, G.; Su, W. *Org. Chem. Front.* **2014,** *1*, 843.
26. Daugulis, O.; Roane, J.; Tran, L. D. *Acc. Chem. Res.* **2015,** *48*, 1053.
27. Chen, Z.; Wang, B.; Zhang, J.; Yu, W.; Liu, Z.; Zhang, Y. *Org. Chem. Front.* **2015,** *2*, 1107.
28. Kuhl, N.; Hopkinson, M. N.; Wencel-Delord, J.; Glorius, F. *Angew. Chem., Int. Ed.* **2012,** *51*, 10236.
29. Cho, J.-Y.; Tse, M. K.; Holmes, D.; Maleczka, R. E.; Smith, M. R. *Science* **2002,** *295*, 305.
30. Ishiyama, T.; Takagi, J.; Ishida, K.; Miyaura, N.; Anastasi, N. R.; Hartwig, J. F. *J. Am. Chem. Soc.* **2002,** *124*, 390.
31. Ishiyama, T.; Takagi, J.; Hartwig, J. F.; Miyaura, N. *Angew. Chem., Int. Ed.* **2002,** *41*, 3056.
32. Hull, K. L.; Sanford, M. S. *J. Am. Chem. Soc.* **2007,** *129*, 11904.
33. Zhang, Y.-H.; Shi, B.-F.; Yu, J.-Q. *J. Am. Chem. Soc.* **2009,** *131*, 5072.
34. Zhao, X.; Yeung, C. S.; Dong, V. M. *J. Am. Chem. Soc.* **2010,** *132*, 5837.
35. Emmert, M. H.; Cook, A. K.; Xie, Y. J.; Sanford, M. S. *Angew. Chem., Int. Ed.* **2011,** *50*, 9409.
36. Lyons, T. W.; Hull, K. L.; Sanford, M. S. *J. Am. Chem. Soc.* **2011,** *133*, 4455.

37. Patureau, F. W.; Nimphius, C.; Glorius, F. *Org. Lett.* **2011,** *13*, 6346.
38. Kubota, A.; Emmert, M. H.; Sanford, M. S. *Org. Lett.* **2012,** *14*, 1760.
39. Ball, L. T.; Lloyd-Jones, G. C.; Russell, C. A. *Science* **2012,** *337*, 1644.
40. Wencel-Delord, J.; Nimphius, C.; Wang, H.; Glorius, F. *Angew. Chem., Int. Ed.* **2012,** *51*, 13001.
41. Wencel-Delord, J.; Nimphius, C.; Patureau, F. W.; Glorius, F. *Angew. Chem., Int. Ed.* **2012,** *51*, 2247.
42. Zhou, L.; Lu, W. *Organometallics.* **2012,** *31*, 2124.
43. Wang, Y.-N.; Guo, X.-Q.; Zhu, X.-H.; Zhong, R.; Cai, L.-H.; Hou, X.-F. *Chem. Commun.* **2012,** *48*, 10437.
44. Wagner, A. M.; Hickman, A. J.; Sanford, M. S. *J. Am. Chem. Soc.* **2013,** *135*, 15710.
45. Cook, A. K.; Emmert, M. H.; Sanford, M. S. *Org. Lett.* **2013,** *15*, 5428.
46. Shrestha, R.; Mukherjee, P.; Tan, Y.; Litman, Z. C.; Hartwig, J. F. *J. Am. Chem. Soc.* **2013,** *135*, 8480.
47. Boursalian, G. B.; Ngai, M.-Y.; Hojczyk, K. N.; Ritter, T. *J. Am. Chem. Soc.* **2013,** *135*, 13278.
48. Cong, X.; Tang, H.; Wu, C.; Zeng, X. *Organometallics* **2013,** *32*, 6565.
49. Cheng, C.; Hartwig, J. F. *Science* **2014,** *343*, 853.
50. Lei, Y.; Qiu, R.; Zhang, L.; Xu, C.; Pan, Y.; Qin, X.; Li, H.; Xu, L.; Deng, Y. *ChemCatChem* **2015,** 7, 1275.
51. Cheng, C.; Hartwig, J. F. *J. Am. Chem. Soc.* **2015,** *137*, 592.
52. Phipps, R. J.; Gaunt, M. *J. Sci.* **2009,** *323*, 1593.
53. Duong, H. A.; Gilligan, R. E.; Cooke, M. L.; Phipps, R. J.; Gaunt, M. J. *Angew. Chem., Int. Ed.* **2011,** *50*, 463.
54. Saidi, O.; Marafie, J.; Ledger, A. E. W.; Liu, P. M.; Mahon, M. F.; Kociok-Köhn, G.; Whittlesey, M. K.; Frost, C. G. *J. Am. Chem. Soc.* **2011,** *133*, 19298.
55. Ackermann, L.; Hofmann, N.; Vicente, R. *Org. Lett.* **2011,** *13*, 1875.
56. Hofmann, N.; Ackermann, L. *J. Am. Chem. Soc.* **2013,** *135*, 5877.
57. Frost, C.; Reynolds, W.; Liu, P.; Kociok-Köhn, G. *Synlett* **2013,** *24*, 2687.
58. Paterson, A. J.; St John-Campbell, S.; Mahon, M. F.; Press, N. J.; Frost, C. G. *Chem. Commun.* **2015,** *51*, 12807.
59. Li, J.; Warratz, S.; Zell, D.; De Sarkar, S.; Ishikawa, E. E.; Ackermann, L. *J. Am. Chem. Soc.* **2015,** *137*, 13894.
60. Teskey, C. J.; Lui, A. Y. W.; Greaney, M. F. *Angew. Chem., Int. Ed.* **2015,** *54*, 11677.
61. Kuninobu, Y.; Ida, H.; Nishi, M.; Kanai, M. *Nat. Chem.* **2015,** 7, 712.
62. Bisht, R.; Chattopadhyay, B. *J. Am. Chem. Soc.* **2016,** *138*, 84.
63. Leow, D.; Li, G.; Mei, T.-S.; Yu, J.-Q. *Nature* **2012,** *486*, 518.
64. Dai, H.-X.; Li, G.; Zhang, X.-G.; Stepan, A. F.; Yu, J.-Q. *J. Am. Chem. Soc.* **2013,** *135*, 7567.
65. Wan, L.; Dastbaravardeh, N.; Li, G.; Yu, J.-Q. *J. Am. Chem. Soc.* **2013,** *135*, 18056.
66. Tang, R.-Y.; Li, G.; Yu, J.-Q. *Nature* **2014,** *507*, 215.
67. Yang, G.; Lindovska, P.; Zhu, D.; Kim, J.; Wang, P.; Tang, R.-Y.; Movassaghi, M.; Yu, J.-Q. *J. Am. Chem. Soc.* **2014,** *136*, 10807.
68. Deng, Y.; Yu, J.-Q. *Angew. Chem., Int. Ed.* **2015,** *54*, 888.
69. Lee, S.; Lee, H.; Tan, K. L. *J. Am. Chem. Soc.* **2013,** *135*, 18778.
70. Bera, M.; Modak, A.; Patra, T.; Maji, A.; Maiti, D. *Org. Lett.* **2014,** *16*, 5760.
71. Li, S.; Ji, H.; Cai, L.; Li, G. *Chem. Sci.* **2015,** *6*, 5595.
72. Bera, M.; Maji, A.; Sahoo, S. K.; Maiti, D. *Angew. Chem., Int. Ed.* **2015,** *54*, 8515.
73. Li, S.; Cai, L.; Ji, H.; Yang, L.; Li, G. *Nat. Commun.* **2016,** 7, 10443.
74. Patra, T.; Watile, R.; Agasti, S.; Naveen, T.; Maiti, D. *Chem. Commun.* **2016,** *52*, 2027.
75. Maji, A.; Bhaskararao, B.; Singha, S.; Sunoj, R. B.; Maiti, D. *Chem. Sci.* **2016,** 7, 3147.

76. Bera, M.; Sahoo, S. K.; Maiti, D. *ACS Catal.* **2016,** *6*, 3575.
77. Chu, L.; Shang, M.; Tanaka, K.; Chen, Q.; Pissarnitski, N.; Streckfuss, E.; Yu, J.-Q. *ACS Cent. Sci.* **2015,** *1*, 463.
78. Zhang, J.; Liu, Q.; Liu, X.; Zhang, S.; Jiang, P.; Wang, Y.; Luo, S.; Li, Y.; Wang, Q. *Chem. Commun.* **2015,** *51*, 1297.
79. Jiang, P.; Li, F.; Xu, Y.; Liu, Q.; Wang, J.; Ding, H.; Yu, R.; Wang, Q. *Org. Lett.* **2015,** *17*, 5918.
80. Wang, X.-C.; Gong, W.; Fang, L.-Z.; Zhu, R.-Y.; Li, S.; Engle, K. M.; Yu, J.-Q. *Nature* **2015,** *519*, 334.
81. Dong, Z.; Wang, J.; Dong, G. *J. Am. Chem. Soc.* **2015,** *137*, 5887.
82. Shen, P. X.; Wang, X. C.; Wang, P.; Zhu, R. Y.; Yu, J. Q. *J. Am. Chem. Soc.* **2015,** *137*, 11574.
83. Han, J.; Zhang, L.; Zhu, Y.; Zheng, Y.; Chen, X.; Huang, Z. B.; Shi, D. Q.; Zhao, Y. *Chem. Commun.* **2016,** *52*, 6903.
84. Cornella, J.; Righi, M.; Larrosa, I. *Angew. Chem., Int. Ed.* **2011,** *50*, 9429.
85. Luo, J.; Preciado, S.; Larrosa, I. *J. Am. Chem. Soc.* **2014,** *136*, 4109.
86. Luo, J.; Preciado, S.; Larrosa, I. *Chem. Commun.* **2015,** *51*, 3127.
87. Luo, J.; Preciado, S.; Araromi, S. O.; Larrosa, I. *Chem. Asian. J.* **2016,** *11*, 347.
88. Moritani, I.; Fujiwara, Y. *Tetrahedron. Lett.* **1967,** *8*, 1119.
89. Jia, C.; Kitamura, T.; Fujiwara, Y. *Acc. Chem. Res.* **2001,** *34*, 633.
90. Satoh, T.; Miura, M. *Chem. Eur. J.* **2010,** *16*, 11212.
91. Le Bras, J.; Muzart, J. *Chem. Rev.* **2011,** *111*, 1170.
92. Yeung, C. S.; Dong, V. M. *Chem. Rev.* **2011,** *111*, 1215.
93. Zhou, L.; Lu, W. *Chem. Eur. J.* **2014,** *20*, 634.
94. Kozhushkov, S. I.; Ackermann, L. *Chem. Sci.* **2013,** *4*, 886.
95. Engle, K. M.; Wang, D.-H.; Yu, J.-Q. *J. Am. Chem. Soc.* **2010,** *132*, 14137.
96. Cheng, G. J.; Yang, Y. F.; Liu, P.; Chen, P.; Sun, T. Y.; Li, G.; Zhang, X. H.; Houk, K. N.; Yu, J. Q.; Wu, Y. D. *J. Am. Chem. Soc.* **2014,** *136*, 894.
97. Lapointe, D.; Fagnou, K. *Chem. Lett.* **2010,** *39*, 1118.
98. Hassan, J.; Sévignon, M.; Gozzi, C.; Schulz, E.; Lemaire, M. *Chem. Rev.* **2002,** *102*, 1359.
99. Catellani, M.; Motti, E.; Della Ca', N. *Acc. Chem. Res.* **2008,** *41*, 1512.
100. Della, C. N.; Fontana, M.; Motti, E.; Catellani, M. *Acc. Chem. Res.* **2016,** *49*, 1389.
101. Deledda, S.; Motti, E.; Catellani, M. *Can. J. Chem.* **2005,** *83*, 741.
102. Wilhelm, T.; Lautens, M. *Org. Lett.* **2005,** 7, 4053.
103. Dong, Z.; Dong, G. *J. Am. Chem. Soc.* **2013,** *135*, 18350.
104. Mochida, S.; Hirano, K.; Satoh, T.; Miura, M. *Org. Lett.* **2010,** *12*, 5776.
105. Bhadra, S.; Dzik, W. I.; Gooßen, L. J. *Angew. Chem. Int. Ed.* **2013,** *52*, 2959.
106. Zhang, Y.; Zhao, H.; Zhang, M.; Su, W. *Angew. Chem. Int. Ed.* **2015,** *54*, 3817.
107. Lee, D.; Chang, S. *Chem. Eur. J.* **2015,** *21*, 5364.
108. Shi, X.-Y.; Liu, K.-Y.; Fan, J.; Dong, X.-F.; Wei, J.-F.; Li, C.-J. *Chem. Eur. J.* **2015,** *21*, 1900.
109. Kumar, N. Y.; Bechtoldt, A.; Raghuvanshi, K.; Ackermann, L. *Angew. Chem. Int. Ed.* **2016,** *55*, 6929.

CHAPTER 9

Recent Advances in Distal Aliphatic sp^3 C—H Functionalization

Aniruddha Dey, Uttam Dhawa and Debabrata Maiti
Indian Institute of Technology Bombay, Mumbai, Maharashtra, India

Contents

9.1 INTRODUCTION

Emergence of C—H functionalization chemistry as a ubiquitous tool for simplifying contemporary organic synthesis has been realized in recent years. The associated novelty and innate comprehensive features of this chemical tributary have unleashed a course of diversification in synthetic transformations. This has been reciprocated in making use of simple and easier routes for production of value-added molecules with complex frameworks in areas of pharmaceutical, material science, polymer chemistry and drug development. The last decade bears testimony to a growing interest for streamlining C—H functionalization in the modification of sp^3/sp^2 carbon centers.[1] Transition metal-based catalysis has paved the way for transforming the unactivated carbon cores in both aliphatic and aromatic molecules into ideal functionalized centers. Above all, functionalization of

Strategies for Palladium-Catalyzed Non-Directed and Directed C—H Bond Functionalization
DOI: http://dx.doi.org/10.1016/B978-0-12-805254-9.00009-8

the sp^3 C—H bond turns out to be quite challenging. This is attributed to the complete absence of any polarizable π-electron cloud in such bonds and any low lying σ^* orbital. Hence, an incoming polarizing electrophile would not find a ready swap for the hydrogen as an easily achievable target. Employment of organometallic catalysts has helped in providing solution to such a challenge, thereby allowing an assortment of functional groups to be successfully incorporated into sp^3 C—H bonds. A major criteria to ascertain the potential of the employed functionalization strategies involved the reaction proceeding in a highly regioselective manner, coupled with a good yield of the final products. In order to meet this, chemists unveiled a collection of approaches for carrying out the C—H functionalization. Eminent research groups, like that of Daugulis, Chen and others, have contributed significantly in devising these celebrated strategies that can perform unchallenged sp^3 C—H functionalization with high regio-, chemo-, and enantioselectivity in a high-yielding manner.[2] Undoubtedly, these approaches are superior to the classical organic synthetic transformations which relied mostly upon electronic and steric bias.

Effective role of the nature of the metal catalysts and the associated oxidants for selective C—H functionalization phenomenon have been presented through a large number of scientific reports. Importantly, the role of ligands in determining the regioselective and stereoselective outcome of the functionalization were strongly realized. Moreover, their structural and electronic aspects were found to be influential in these reactions. Relative to all, a complementary combination of a palladium catalyst and a corresponding suitable ligand has been found to be the most thoroughly explored reaction system. In terms of aliphatic distal functionalization, incorporation of functional groups at locations that are β- with respect to the pre-existing functionality within the molecules have been widely conducted.[3] A reach beyond the β- position required further scientific investigations, since functionalization at unactivated distal locations within acyclic, aliphatic molecules are severely challenging. Adoption of a diverse number of strategies has helped in successful installation of functional moieties at the remote γ-, δ-, and even at the ε- position in aliphatic positions (Fig. 9.1).

Beside the conventional cross-coupling reactions, use of directing group-enabled C—H functionalization protocols at remote locations within molecules have become increasingly popular. The latter functions by weak coordination interactions with the metal pre-catalyst followed by selective delivery the electrophilic moiety to the distal target bonds that are otherwise not easily accessible. Use of directing groups in facilitating a chelation

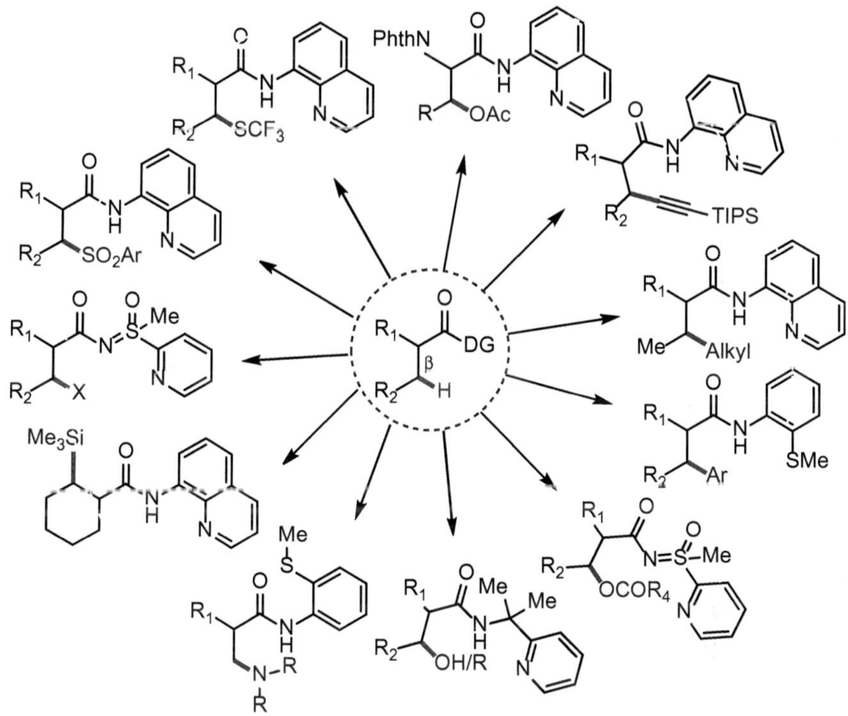

Figure 9.1 An assortment of β-C(sp^3)—H functionalizations in an aliphatic system.

approach-based functionalization at remote aliphatic carbon centers has been investigated. Highly selective transformations involving C—C, C—O, C—N, etc., types of bond formations at the sp^3 carbon cores have been widely reported.[4] In this chapter, we present a brief discussion on distal aliphatic sp^3 C—H functionalization, thereby highlighting their importance in modern-day synthetic chemistry. We hope that this will aid in creating an awareness about the current developments in the areas concerned, also encouraging younger minds to forge their scientific temperament to take a lead in the futuristic advancements of organometallic catalysis and relative areas of interest.

9.2 SCOPE OF FUNCTIONALIZATION

Scientific reports on aliphatic C—H functionalization are abundant, and an abridged discussion involving each one of them is beyond the scope of this chapter. Consequently, the current discussion circumscribes the recent developments on distal C—H functionalization that majorly involve γ-C—H functionalization and beyond (e.g., δ, ε, etc.). This chapter deals with the protocols underlined mainly by palladium-based catalytic systems. A vast scope of functionalization like arylation, alkenylation, amination, borylation, etc., have been duly covered. The various aspects of the transformation and the designing of the corresponding reaction condition have been delineated concomitantly with the gradual evolution of the methodological strategies.

9.2.1 Arylation

Adoption of transition metal complex catalysis in yielding γ-C—H transformation was undertaken by Daugulis and co-workers in 2005.[2a] Reports confirming oxidative carbon – carbon bond formation at unreactive sp^3 C—H sites were exceptionally rare during those years. This was important, as it could allow shorter reaction sequences with distinctive regioselectivities. Daugulis resorted to the use of picolinamide-based directing auxiliary in performing γ-C—H arylation of amine derivatives (Scheme 9.1A). The palladium-catalyzed reaction proceeded through

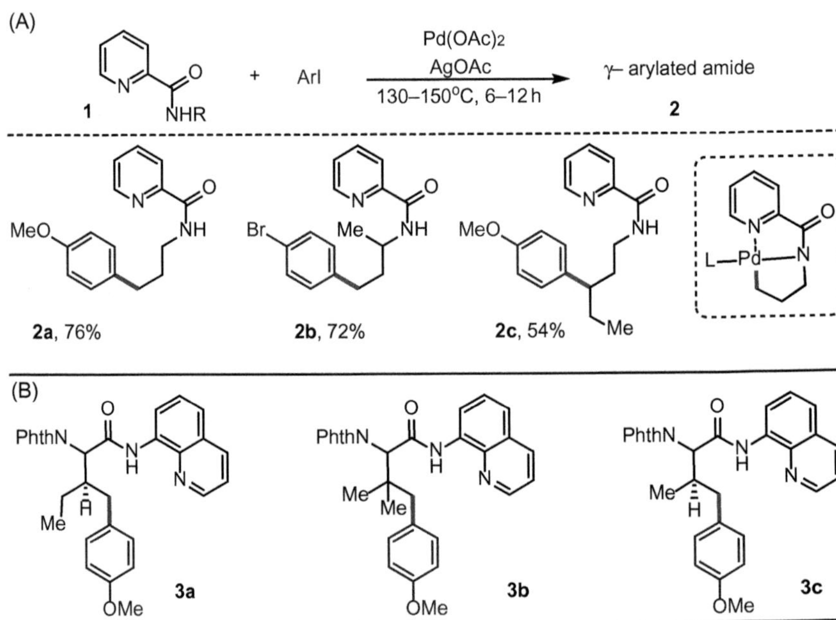

Scheme 9.1 (A). Palladium-catalyzed γ-arylation using Daugulis' picolinamide auxiliary. (B). Corey's γ-arylation *N*-phthaloyl-protected amino acid derivatives.

five-membered palladacycle intermediates with aryl iodides as the efficient coupling partners. This was confirmed through a structural characterization of an isolated palladated crystal.

Selective monoarylation of the free methyl groups was obtained, indicating a relatively faster reactivity of the primary C—H bonds compared to the secondary C—H bonds. Few years later, Daugulis extended the utility of the same protocol in performing a remote δ-C—H activation allowing formation of cyclized products.[5] The auxiliary facilitated strategy was subsequently developed by various research groups for a variety of synthetic purposes. In 2006, Corey reported Pd-catalyzed distal C—C bond formation to produce a broad range of unnatural (S)-α-amino acids (Scheme 9.1B).[6] While demonstrating the work on C—H arylation of the β-C(sp^3)—H bond using 8-aminoquinoline as a directing group, it was revealed that a prolonged reaction time could afford γ-C(sp^3)—H activation of amino acid derivatives such as those of isoleucine, leucine and valine. This could be attributed to the diminished rate of β-C—H activation by the metal catalyst at a tertiary carbon center. This caused a competitive γ-C—H activation phenomenon to become operative. Interestingly, a comparatively easier way of synthesis of such γ-arylated *N*-phthaloyl-protected amino acid derivatives was hardly known at that time.

In 2008, Buchwald and co-workers reported a palladium-catalyzed transformation for γ-arylation of α,β- or β,γ- unsaturated ketones (Scheme 9.2).[7] Starting from dienolates and using this arylation strategy, they performed a two-step, one-pot synthesis of ketoindolines with bromoanilines as coupling partners. Dienolates are less nucleophilic compared to enolates. They are relatively vulnerable towards self-Michael condensation. A combination of such factors, along with the possibility for generation of regioisomeric products during selective arylation of these moieties prohibited execution of selective functionalization on dienolates. Through an effective use of a ligand controlled strategy and a proper choice of a base, Buchwald demonstrated the formation of quaternary carbon centers at the γ-position of β,γ-unsaturated ketones. The nature of the ligand was found to be influential in determining the regioselectivity of the functionalization. Moreover, the relative rate of reductive elimination from α- and γ-positions, as well as steric and electronic environment around the Pd center, affected the selectivity. Consequently, using Cs_2CO_3 as a base and a bidentate ligand like **1,2-bis (diphenylphosphino)ethane** (dppe) with a small bite angle provided the best γ-selectivity with an 84% yield. Greater acidity and lower

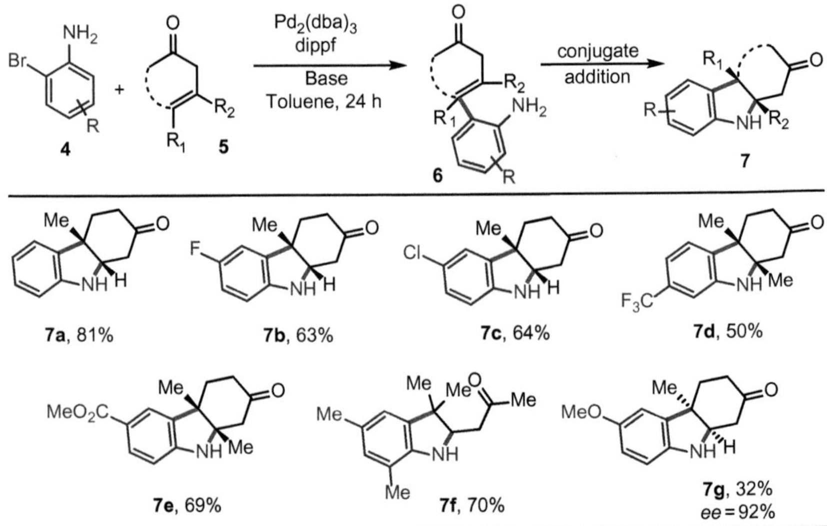

Scheme 9.2 Palladium-catalyzed ketoindoline synthesis via γ-arylation of unsaturated ketones.

tendency to self-condensation for β,γ-unsaturated ketones allowed higher yields of products relative to α,β-unsaturated ketones. Conjugate addition of bromoaniline derivatives to the γ-functionalized ketones furnished ketoindolines with the use of 1,1′-bis(di-iso-propylphosphino)ferrocene (dippf) as the ligand in the presence of a base like Cs_2CO_3 (for electron-rich/neutral anilines) or K_2CO_3 (for electron-poor anilines). Changing the ligand to DTBM-Segphos produced (**L1**) an enantiomeric excess up to 92% for indoline synthesis. The *t*-butyl groups on the P-bound phenyl ring of the ligand were crucial in controlling this facial selectivity.

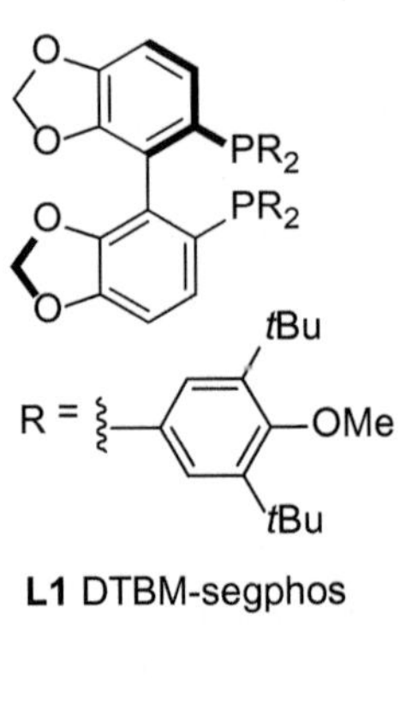

L1 DTBM-segphos

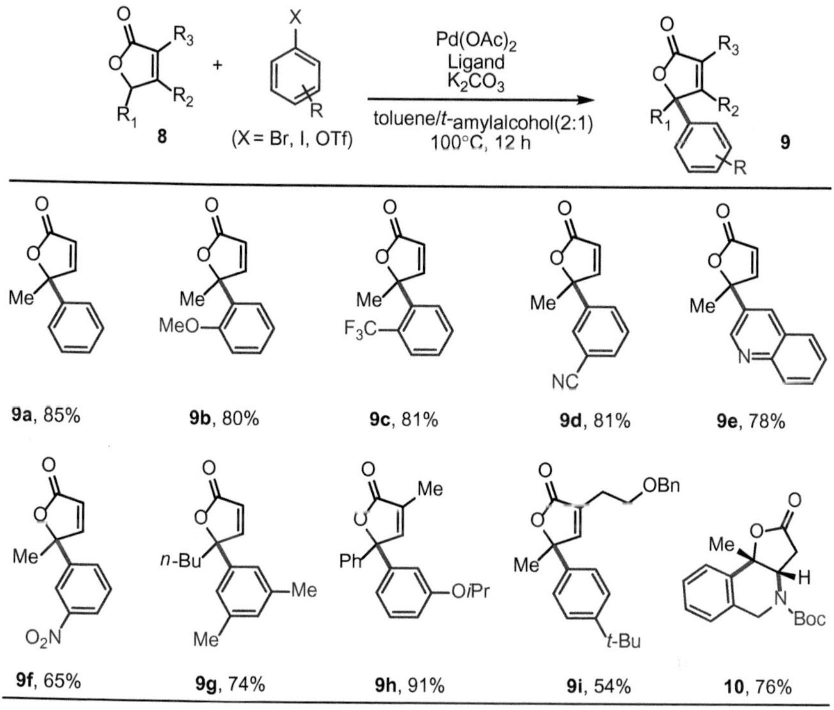

Scheme 9.3 Palladium-catalyzed γ-arylation of substituted butenolides.

Unsaturated γ-butyrolactones, also known as butenolides, are important core structures of several natural products and bio-active molecules. Their dienolates are generally more nucleophilic at the γ-position. Moreover, the β,γ-unsaturated isomer is more stable than the α,β- unsaturated isomer, which allows an efficient arylation to be performed at the γ-position. In 2009, Buchwald performed γ-arylation of butenolides through a palladium-catalyzed cross-coupling process (Scheme 9.3).[8] In the presence of a suitable ligand, base and solvent, γ-arylation was carried out with a variety of γ-substituted butenolides and aryl coupling partners (like aryl bromides, aryl chloride, and aryl triflates) with excellent selectivity ratio (>98:2). While electron-rich and neutral aryl partners required moderately hindered biaryl phosphines like MePhos as the ligand source, electron-deficient analogues required **L2** as the desired ligand. However, butenolides can easily dimerize in the presence of a base through Michael reactions, or can get decomposed. Thus, a mild base like K_2CO_3 was used for the reaction. The solvent combination of toluene/*tert*-amyl alcohol was chosen to prevent decomposition of the dienolate. Efficacy of the transformation was further displayed by synthesis of the tricyclic tetrahydroisoquinolinone **10**. This report on a direct γ-arylation of unsaturated esters was significant at a time of increasing popularity for cross-coupling reactions between ester enolates and aryl/vinyl halides.

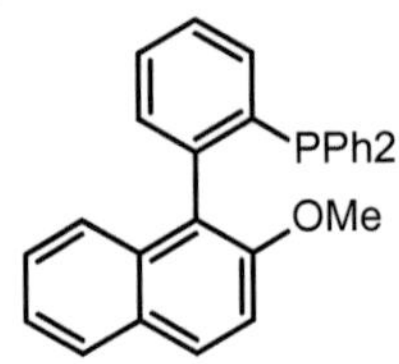

L2

Chen and co-workers had explored milder and more general reaction conditions for Pd-catalyzed remote γ-C(sp^3)—H bond activation using an easily removable picolinamide (PA) as a directing group.[9] A highly mono-selective arylation and alkenylation of γ-C—H bonds of a variety of aliphatic substrates including cyclohexanes were performed. Electron-donating substituents on the pyridine ring of PA tend to increase the yield of arylation, whereas electron-withdrawing substituents decrease the yield. A kinetically favored five-membered palladacycle controlled the γ-regioselectivity for arylation that involves a Pd(II)/(IV) catalytic cycle. Substrate scope showed a range of aryl iodide and aliphatic amines to undergo arylation with moderate-to-excellent yield with tolerance towards various functional groups (Scheme 9.4).

This picolinamide-directed γ-C(sp^3)—H alkenylation was also employed in performing a γ-C—H alkenylation by using vinyl iodides (Scheme 9.5). Various cyclic disubstituted vinyl iodides yielded the corresponding alkenylated product in good-to-moderate yield at elevated temperatures and in the presence of 0.1 equivalent of benzoquinone. Synthetic utility of the protocol was further highlighted by synthesis of the natural product (+)-obafluorin. Removal of the picolinamide linkage required slight modification on the auxiliary structure.

Installation of an *ortho*-methyloxysilyl group allowed a successful cleavage of the modified picolinamide core through intramolecular acyl transfer under acidic conditions.

The year of 2012 saw the development of an enantioselective Heck reaction coupled with a redox-relay strategy for the generation of remote chiral centers in aliphatic molecules. Sigman and co-workers

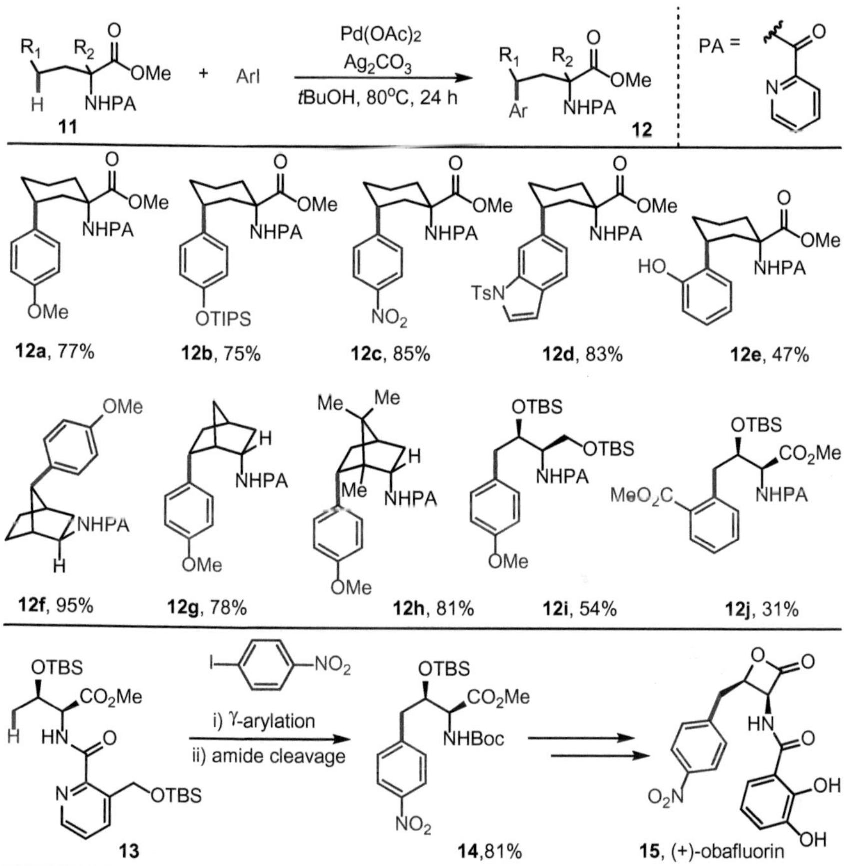

Scheme 9.4 Palladium-catalyzed γ-arylation of cyclohexane derivatives using aryl iodides.

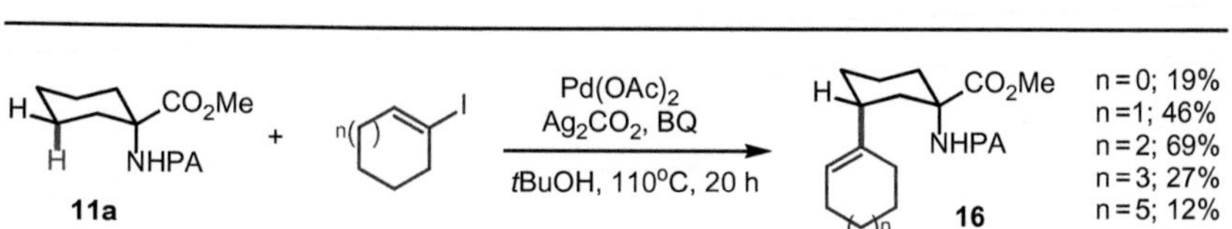

Scheme 9.5 Palladium-catalyzed γ-functionalization of cyclohexane derivatives using vinyl iodides.

presented a palladium-catalyzed asymmetric Heck reaction of acyclic, alkenyl alcohols (Scheme 9.6).[10] Addition of organometallics to the alkene was followed by oxidation of the alcohol by migratory insertion of the metal through the alkyl chain that rendered formation of β-, γ-, or δ-aryl carbonyl products from the alkenyl alcohols with excellent

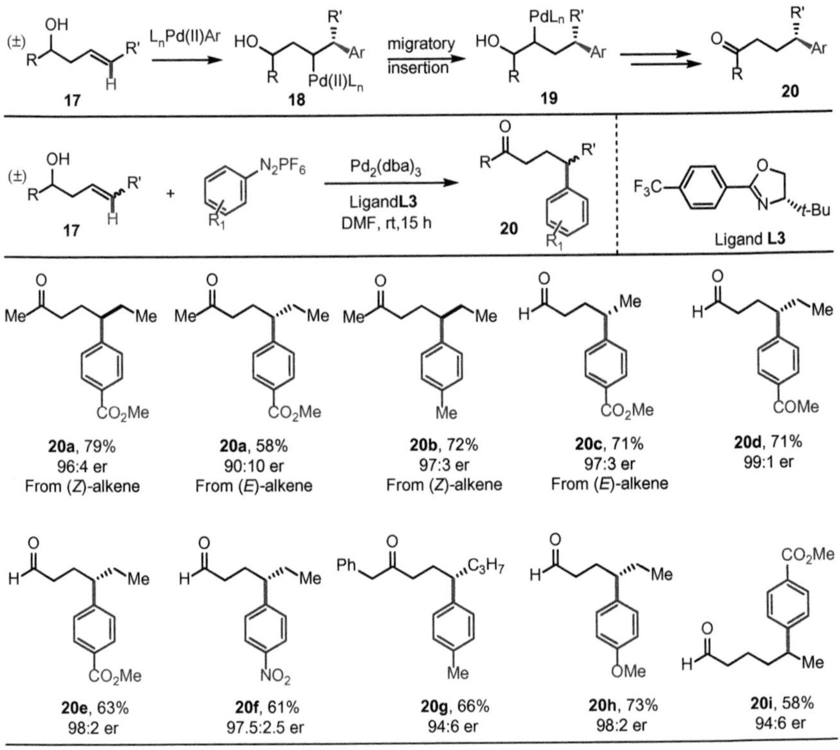

Scheme 9.6 Palladium-catalyzed enantioselective γ-arylation of alkenyl alcohols with aryldiazonium salts.

enantioselectivity(~er = 97:3). An electrophilic catalyst was realized for favoring the relay-based migration of the unsaturation towards the alcohol over dissociation of metal from the alkene system. This necessitated aryldiazonium salts as the arene source and a concomitant use of chiral pyridine/oxazoline ligands, owing to their two-coordination site with distinct electronic properties allowing effective asymmetric catalysis. Screening of the substrate scopes revealed higher enantioselectivity with greater substitution on the saturated side of the allylic alcohol. (*Z*)-alkenes were found to be better than (*E*)-alkenes. Besides a variety of aryldiazonium partners, arylation with nitro- and iodo-containing arenes also proceeded successfully with undiminished regioselectivity. Interestingly, applying the protocol towards bis-homoallylic alcohols containing δ,ε-unsaturation yielded δ-arylated carbonyl compounds with high enantioselectivity.

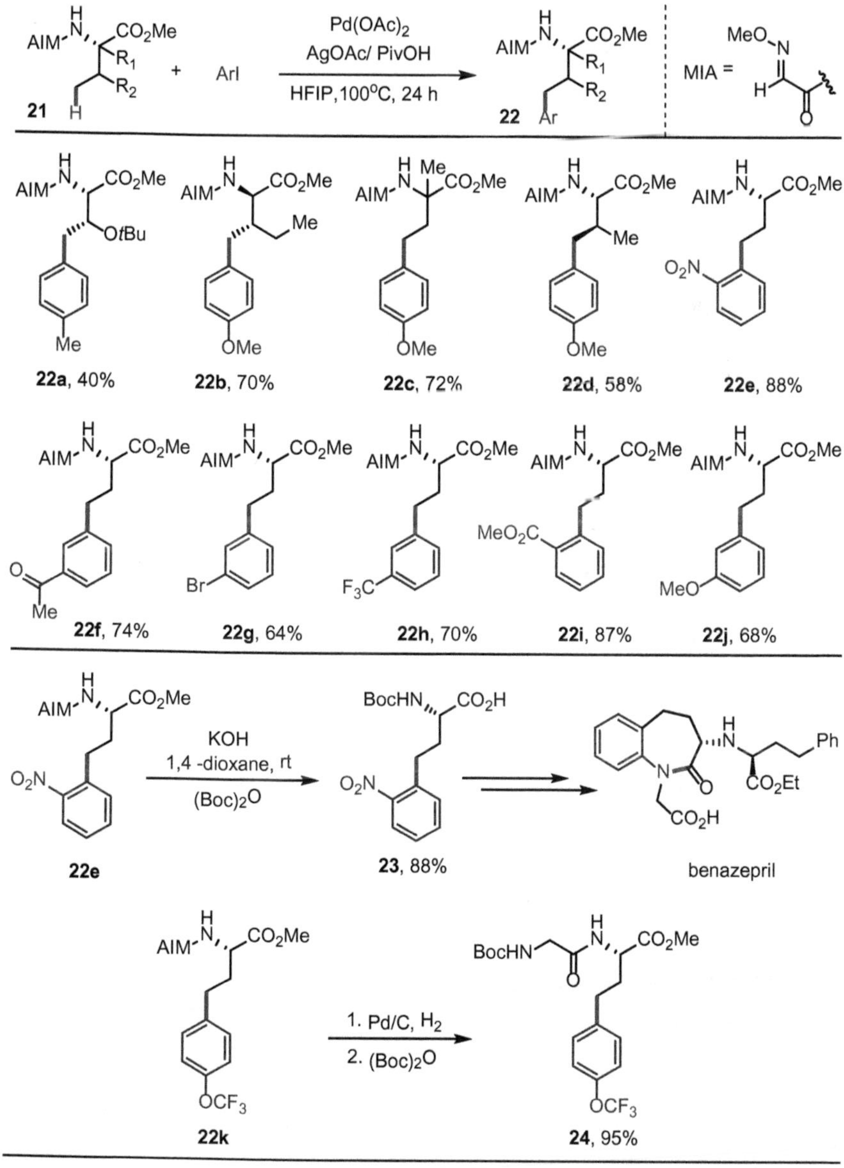

Scheme 9.7 Palladium-catalyzed γ-arylation of amines with 2-methoxyiminoacetyl auxiliary.

A few years later, Ma and co-workers reported the use of 2-methoxyiminoacetyl (MIA) as a powerful auxiliary to facilitate γ-C—H arylation of amines (Scheme 9.7).[11] With aryl iodides as the coupling partner, the corresponding γ-arylated amides could be easily molded into a variety of homophenylalanines and peptides. Synthetic methods for preparation of homophenylamines are worthwhile in order to understand their structure – activity relationship with peptidomimetics. This had significantly helped in the synthesis of important antihypertensive drugs like enalapril and benazepril. Successful γ-C—H arylation of 2-aminobutanates was performed with a variety of electron-rich and electron-poor aryl halides in the presence of palladium salt as the pre-catalyst. The importance of this protocol lies in the fact that the MIA auxiliary is easily removable under mild basic conditions at room temperature. A subsequent protection of the resulting amino acid with $(Boc)_2O$ provided products with high enantiopurity. Moreover, transformation of the MIA group into a suitable glycine moiety could be done that can further effect dipeptide formation via hydrogenation.

In 2013, Carretero and co-workers reported distal γ-C(sp^3)—H bond activation of amino ester derivatives by using a *N*-(2-pyridyl)sulfonyl directing group (Scheme 9.8).[12] The directing group can be easily installed and removed. Under the optimized condition, they tested the scope of the reaction with respect to the substitution at the aryl iodide along with other allo-isoleucine derivatives.

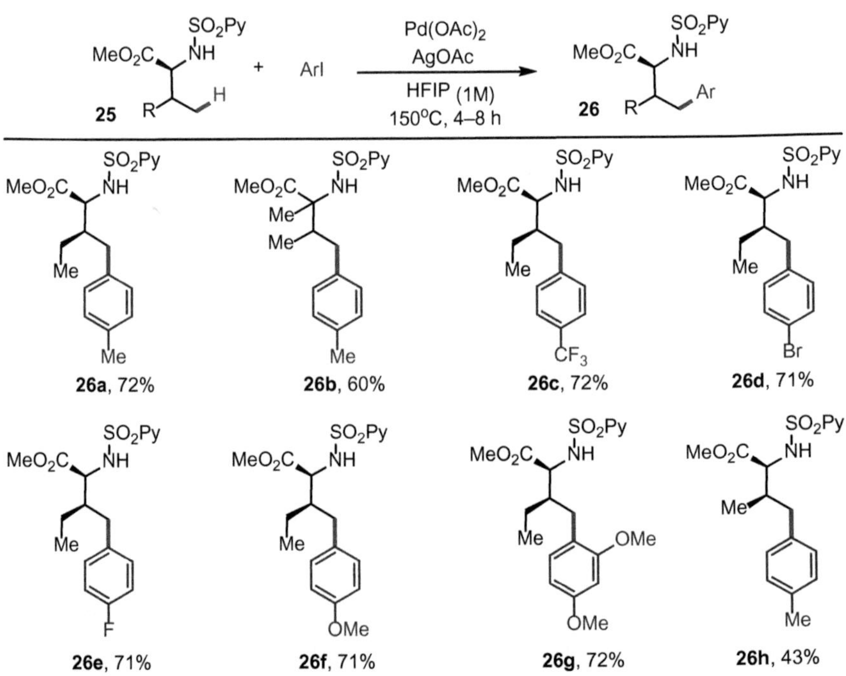

Scheme 9.8 Palladium-catalyzed γ-arylation of aminoesters with *N*-(2-pyridyl)sulfonyl auxiliary.

This showed that electron-rich and electron-poor aryl iodides were both equally tolerated. A bimetallic Pd^{II} γ-metalated complex was isolated and characterized which provided the role of the (2-pyridyl)sulfonyl unit in the arylation process. The 2-pyridylsulfonyl directing group was very easily cleaved by using Zn in THF-NH_4Cl for 16 h at 60°C.

In 2013, Mazet and co-workers independently presented a palladium-catalyzed γ-arylation of unactivated γ-substituted α,β-unsaturated aldehydes, thereby leading to generation of quaternary stereocenters (Scheme 9.9).[13] Previously, Miura and co-workers reported the γ-arylation of α-substituted α,β unsaturated aldehydes.[14] However, a remote γ-arylation with generation of a quaternary carbon center at the vinylogous position awaited revelation. Development of the reaction conditions was emphasized to prevent the possibility of self-aldol condensation for this unsaturated aldehyde. A variety of electron-rich and electron-poor aryl bromides were effectively cross-coupled with low catalyst loadings, besides providing excellent functional group tolerance. Substrate scope was performed with various α,β- unsaturated aldehydes and a high E/Z selectivity of >10:1. Both the palladium source and t-Bu_3P ligand were essential in maintaining the stereoselectivity of the reaction and providing an augmented reactivity. Additionally, γ-arylation

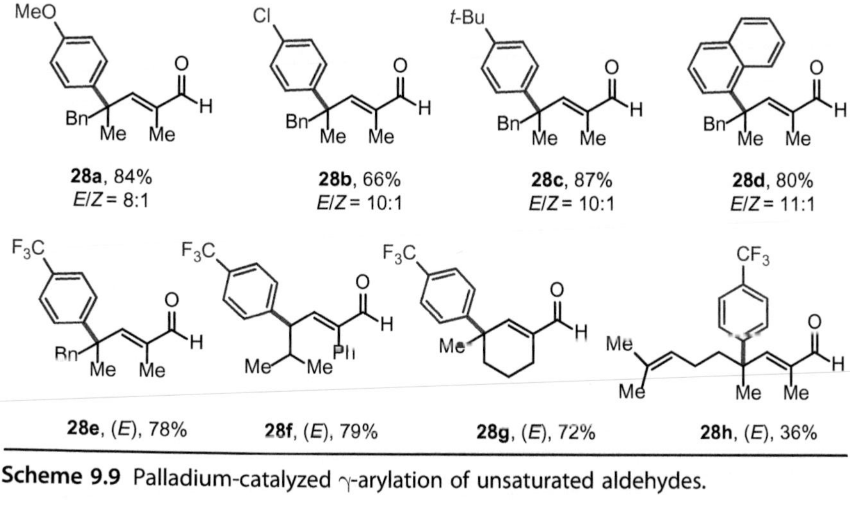

Scheme 9.9 Palladium-catalyzed γ-arylation of unsaturated aldehydes.

was further extended beyond the vinylogous position to perform ε-arylation with excellent regioselectivity.

Following this, in 2014, Sigman and co-workers revisited their arylation protocol with a similar palladium-catalyzed remote C–H arylation of tri-substituted alkenes with aryl boronic acids (Scheme 9.10).[15] This time they applied the enantioselective intermolecular Heck-type reaction for generation of distal quaternary stereocenters in aliphatic chains. Small molecules with quaternary carbon centers form the core structural motifs in various secondary metabolites and other pharmaceutical agents. A stereoselective generation of such centers within molecules is therefore of great significance. In the presence of $Pd(CH_3CN)_2(OTs)$ as the catalyst and **L3** as the ligand, arylation at the γ-position proceeded with >15:1 site-selectivity (γ:β) in the presence of $Cu(OTf)_2$ and 3 Å molecular sieves as additives. A wide range of acyclic, non-conjugated tri-substituted alkenyl alcohols were selectively arylated with enhanced enantioselectivity at either of the β-, γ-, δ-, ε-, or ζ-positions, depending upon the chain length, resulting in the corresponding remotely arylated carbonylated products bearing remote chiral quaternary centers. Electron-rich aryl boronic acids provided higher yields than the electron-poor ones, reflecting the greater nucleophilicity facilitated migratory insertion of the in situ formed

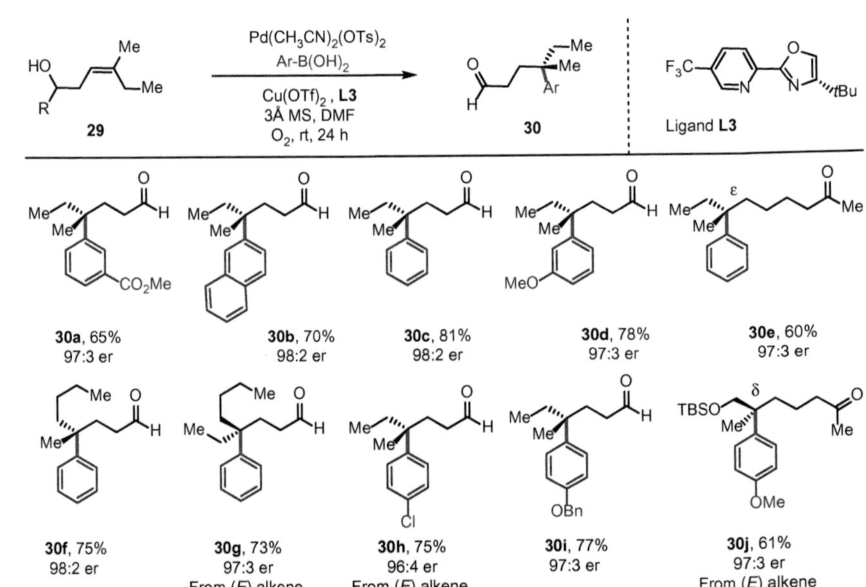

Scheme 9.10 Palladium-catalyzed enantioselective γ-arylation of tri-substituted alkenes.

alkene complex. Enantioselectivity remained independent of the steric and electronic nature of the substrates. Further mechanistic insights supported a "chain-walking" redox relay type Heck reaction to be operative in the transformation concerned.

Later in the same year, Chen and co-workers extended the scope of such a picolinamide auxiliary directed strategy by demonstrating the use of sterically hindered aryl coupling partners in C—H arylation.[16] Chen devised a methodology for synthesizing a crosslink between isoleucine (Ile) and 4-hydroxyphenacylamine (Hpa) residue by a palladium-catalyzed auxiliary directed γ-C—H arylation strategy (Scheme 9.11). This crosslink serves as an important structural feature of hibispeptins that are used in the formation of antipyretics and anthelmintics. Investigation was initiated with N-linked picolinamide auxiliaries for performing γ-C—H arylation on Ile by different aryl iodides. Steric disfavor of methyl γ-C—H arylation was posed by the α-CO_2Me and β-Et groups, leading to some methylene arylation. Arylation was also performed with Carratero's auxiliary as in **31b**,

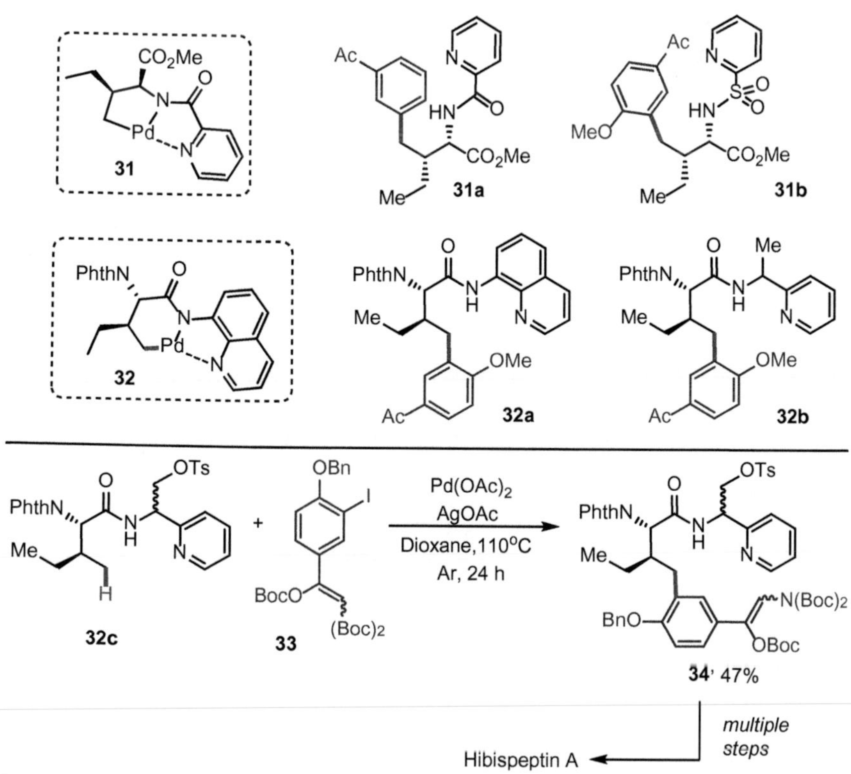

Scheme 9.11 Palladium-catalyzed γ-arylation with sterically hindered aryl iodides.

which underwent arylation but in low yields. Prevention of the unwanted methylene arylation was done by switching to the use of a C-linked auxiliary such as Daugulis' 8-aminoquinoline group. Owing to the lower yield of products obtained with the latter, a further leap was made to utilize pyridylmethylamine-based auxiliary groups. Among them, 2-pyridylethylamine (PE) auxiliary provided γ-C—H arylation of Ile giving a 70% yield. Installation of TIPS-protected hydroxyl on the free methyl of PE allowed an easy intramolecular cleavage strategy to furnish the carboxylic acid. This modified auxiliary was then used for the required γ-C − H arylation, finally paving the way for total synthesis of hibispeptin A.

Recently, in 2016, Yu and co-workers developed a novel directing group-based strategy in performing γ-C—H arylation of aliphatic carboxylic acids (Scheme 9.12).[17] Yu demonstrated the use of tricyclic quinoline-based ligands to be essential in this γ-C—H arylation protocol. Till then, γ-C—H arylation of simple acids was not developed in the absence of sterically bulky phthalimido groups. Yu emphasized the role of ligands in affecting the γ-arylation of amino acid derivatives such as valine, isoleucine, and *tert*-leucine. A thorough screening of the ligands showed electron-rich ligands to be more effective towards the transformation, e.g., **L4**, **L5**. Various aryl iodides including both electron-donating and electron-withdrawing substituents were found to be compatible with the reaction condition. In all cases, stereochemistry of the α-carbon centers were duly maintained, also obtaining a high level of diastereoselectivity (d.r $>$ 20:1). This could be attributed to the favored *trans*-conformation and disfavored *cis*-conformation with respect to the methyl and phthalimide groups. Notably, tetrahydropyran and substituted norbornane structural motif were also consistent with the arylation reaction. Removal of the directing group under $LiOH/H_2O_2$ conditions furnished the free carboxylic acid.

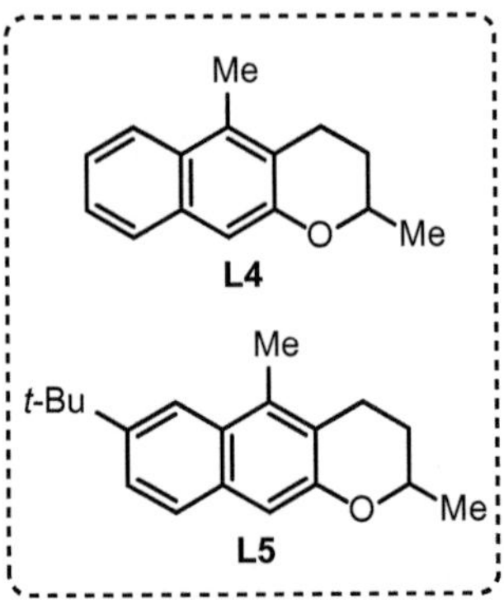

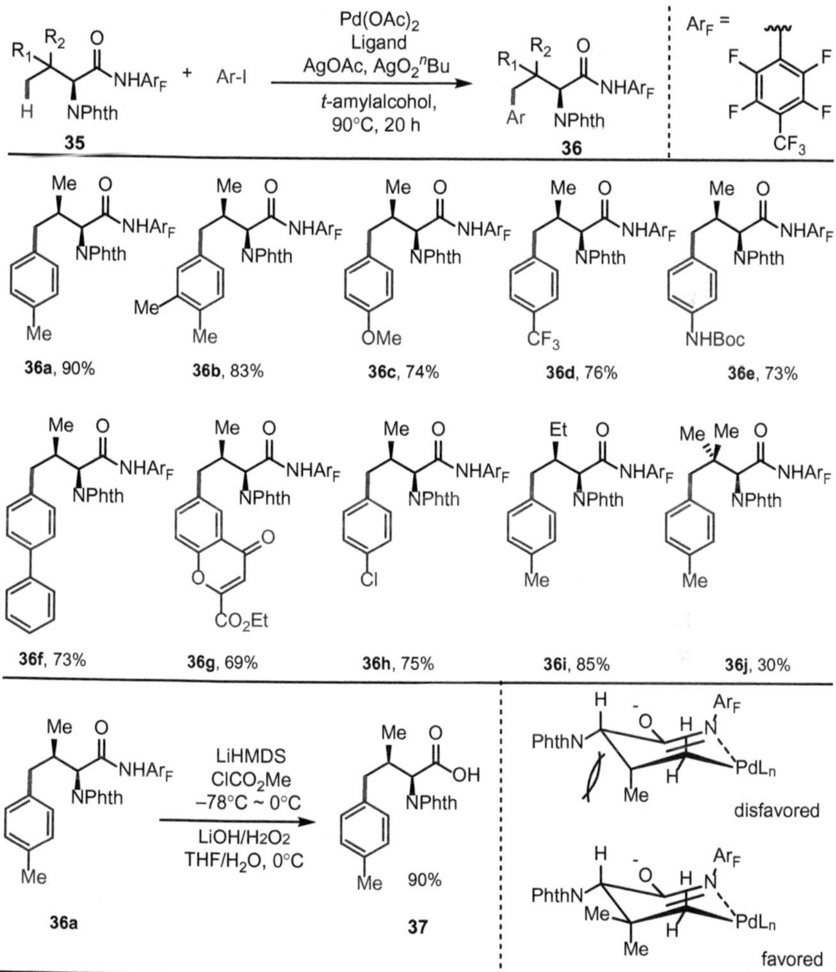

Scheme 9.12 Palladium-catalyzed γ-arylation of *N*-protected amino acids.

9.2.2 Alkylation

In 2013, Chen and co-workers reported palladium-catalyzed alkylation of the γ-C(sp^3)—H bond of aliphatic amine with primary alkyl iodides (Scheme 9.13).[18] Despite the ample reports on sp^2 C—H alkylation, palladium-catalyzed alkylation reactions of distal aliphatic C—H bonds with significant synthetic relevance are rare. A picolinamide-directed remote γ/δ-C(sp^3)—H alkylation was performed with a variety of amine substrates (like threonine, alloisoleucine, norbornene, etc.) and linear alkyl halides, with good yields and excellent selectivity. In all cases,

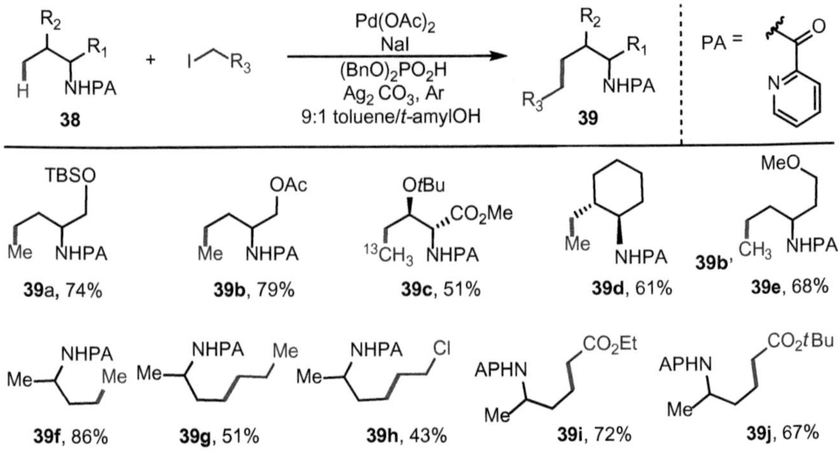

Scheme 9.13 Palladium-catalyzed γ-alkylation of aliphatic amines.

MeI and α-iodoacetic ester served as efficient alkylating agents. NaI aided in alkylation with alkyl iodides in toluene/*t*-amylalcohol, whereas $CuCl_2$ was effective for alkylations using α-iodoacetic ester in *t*-amylalcohol. Addition of Ag_2CO_3 in the reaction was purposive, since Ag^+ helped in scavenging the I^- ions thereby improving the catalytic turnover and increasing the conversion. However, the amount of Ag^+ ions was kept in check by the addition of a low cost organic acid, namely dibenzyl phosphate $(BnO)_2PO_2H$, which would prevent decomposition of the electrophile by acting as a Phase Transfer Catalyst (PTC). Primary γ-C–H bonds were found to be more reactive than secondary γ-C–H bonds for this system.

9.2.3 Olefination

Synthesis of starting materials containing quaternary carbon centers had attracted wide-spread attention, owing to their augmented utility in complex structure formation. With regard to this, construction of α-quaternary centers from pivalic acid which relies on the development of asymmetric β-C–H functionalization reactions has been under investigation. Formation of β-quaternary centers will undoubtedly depend on γ-C–H functionalization that has remained elusive. As a solution to this problem, Yu and co-workers presented their first report on γ-C–H olefination for derivatized β-quaternary amides (Scheme 9.14).[16] The necessity of a quinoline-based ligand system was realized for the transformation concerned. A strategized engineering of the ligand

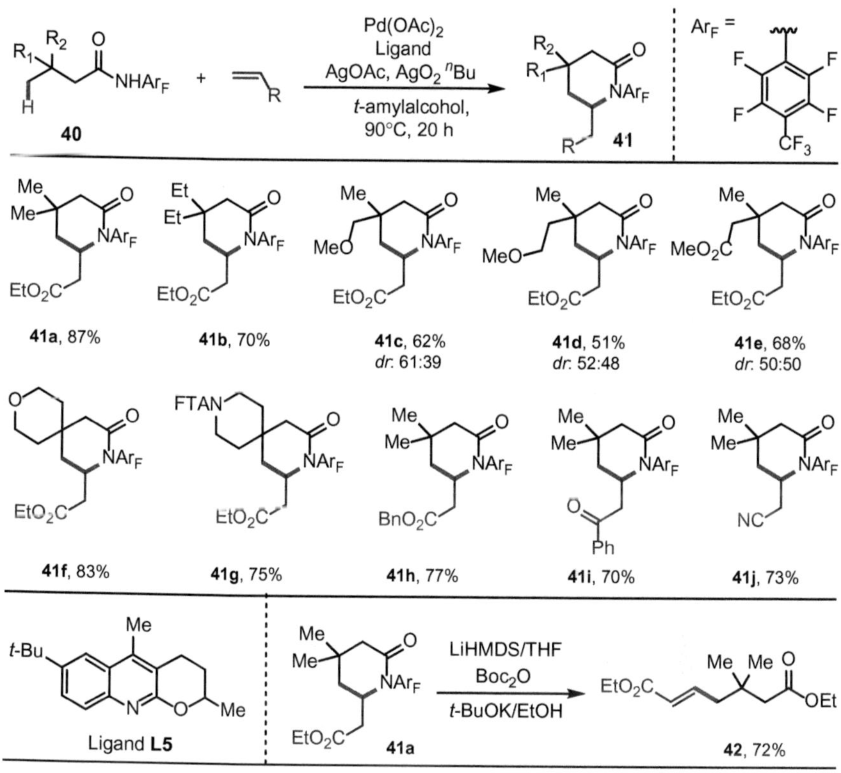

Scheme 9.14 Palladium-catalyzed γ-olefination of β-quaternary amides.

design was made by employing pyran containing fused quinoline ligand systems. By suitable modification, the lone pairs of oxygen were constrained to the plane of the π-system of the aromatic ring. This rendered an electron-rich ligand **L5** that would affect the reactivity of the metal pre-catalyst in the +2 oxidation state. Using TEMPO as a co-oxidant, and in presence of doubled equivalents of $NaHCO_3$, mono-selective γ-C—H olefinated products were obtained with 95% yield on reacting activated acrylates with aliphatic amides under an oxygen atmosphere. Substrate scope was performed by variation of both the amides and the olefins. Amides with γ- and δ- methoxy groups, γ-esters and spirocyclic compounds furnished γ-olefinated products with good selectivity and yields. Robustness of the reaction was further shown by performing gram scale synthesis of the γ-olefinated products. A one-pot deprotection procedure post the olefination reaction rendered the synthesis of γ-olefinated ester.

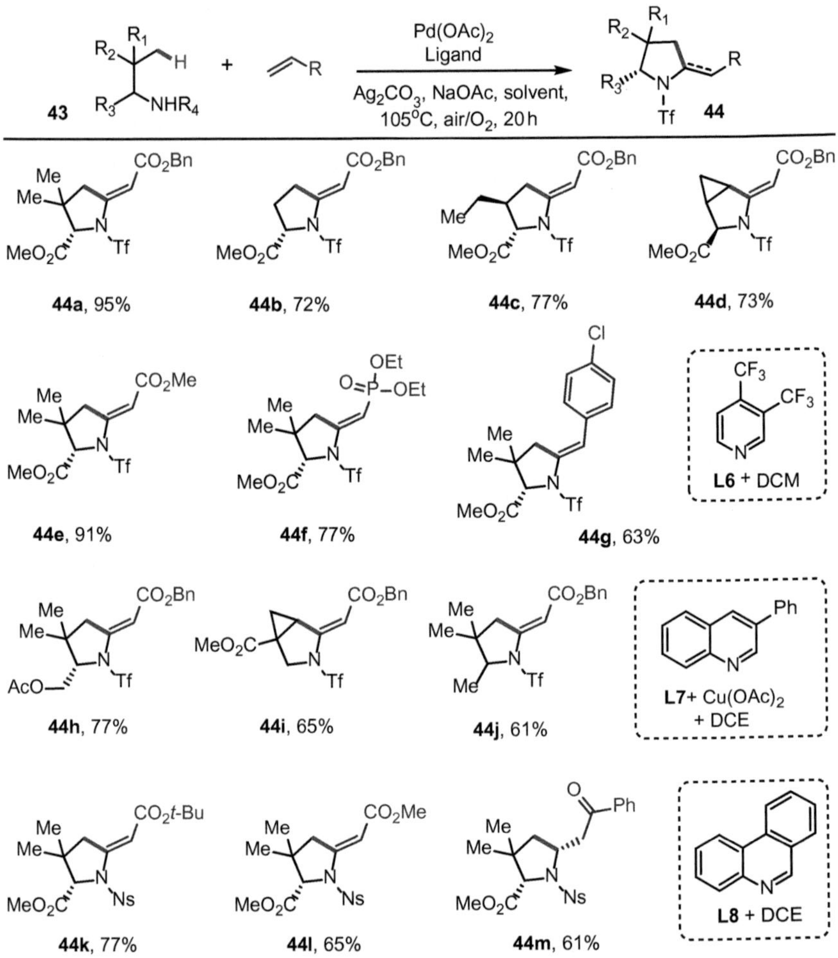

Scheme 9.15 Palladium-catalyzed γ-olefination of alkyl amines.

In 2016, Yu and co-workers revisited their ligand controlled aliphatic γ-C—H functionalization strategy for performing γ-C—H olefination of alkyl amine derivatives in palladium-catalyzed reaction medium (Scheme 9.15).[19] Accelerated by pyridine/quinoline based ligands, this strategy proved to be effective for Tf (triflic)- and Ns(Nosyl)-protected aliphatic amines. Activation of γ-C—H by Tf-protected alkyl amines and subsequent cross-coupling with arylborons has been previously found to proceed smoothly using monoprotected amino acids (MPAA). However, MPAAs did not aid in the reaction this time by any means.

Consequently, Yu resorted to designing pyridine-based ligands that could perform a selective γ-C—H olefination of such amines.

The γ-C—H olefinated product underwent an intramolecular palladium-catalyzed aza – Wacker oxidative cyclization reaction to provide pyrrolidines. Tf-protected α-amino acids like L-valine and L-*tert*-leucine provided the desired products in excellent yields. Olefination of γ-methyl groups were preferred relative to γ-methylene groups. Tf-protected β-amino alcohols and acids underwent the transformation in the presence of quinoline-based ligands. Importantly, an array of α,β unsaturated olefins and styrenes were shown as effective coupling partners. Use of phenanthridine ligands allowed the protocol to be effective for even Ns- protected alkyl amines. Synthetic application was highlighted by the preparation of chiral-pyrrolidine based bio-active molecules and natural products.

9.2.4 Cyclization/amination

A novel way to synthesize azetidine, pyrrolidine, and indoline compounds by using Pd-catalyzed C—H amination reactions at distal γ and δ positions was discovered. Synthetic procedures for aliphatic N-heterocycles date back to the Hofmann – Löffler – Freytag reaction and other "outer-sphere mechanisms."[20] In 2011, Chen and co-workers showed various picolinamide substrates with primary γ-H undergoing intramolecular amination to give azetidines in good-to-moderate yields (Scheme 9.16).[21] $PhI(OAc)_2$ served as an effective oxidant in toluene as a solvent. The cyclization

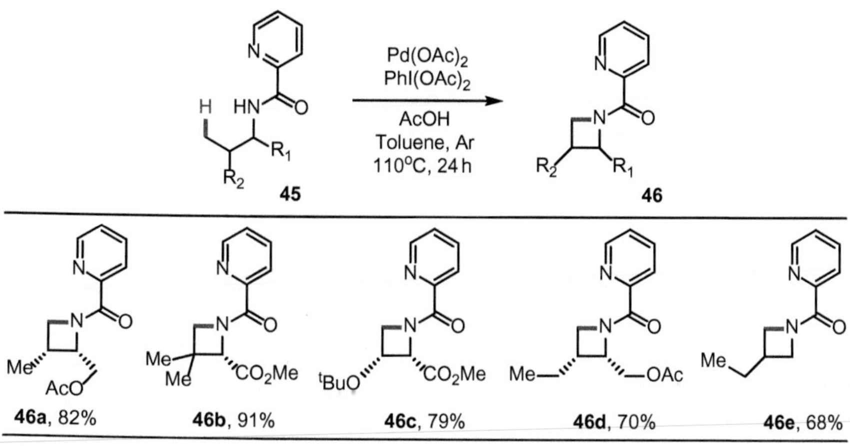

Scheme 9.16 Palladium-catalyzed intramolecular γ-amination in the synthesis of azetidines.

proceeded in a facile manner under an inert Argon atmosphere with catalyst loading as low as 2.5%. Formation of a Pd(IV) intermediate was presupposed by $PhI(OAc)_2$ oxidation of the in situ formed palladacycle intermediate.

An imminent C—O/C—N reductive elimination allowed the formation of the cyclized product. Further, steric demands of the substrate and the associated OAc ligands affected the rate of the reductive elimination step. Torsional strain inflicted by the β-substituents and the γ-C—H bond during the bond reorganization process for the out-of-palladacycle-plane C—O formation kinetically favored the N-cyclization over the ring strain. The same reaction condition was further employed in the synthesis of pyrrolidines from aliphatic substrates with primary δ-C(sp^3)—H bonds (Scheme 9.17). This would mean generation of a five-membered product from a kinetically less favored six-membered palladacycle intermediate. The driving force of the C—N reductive elimination from the Pd^{IV} center was presumed to allow the δ-selectivity over the already obtained γ-functionalization. Cyclization of a leucine derivative proceeded with a diastereotopic ratio of 7:1 with 82% yield. Indoline derivatives were also obtained by cyclization of **47d** and **47e**. Moreover, deuteration experiments performed in the absence of $PhI(OAc)_2$ led to the conclusion for determining the relative reactivities to be in the order: primary γ-C—H> primary δ-C—H> secondary and tertiary γ-C—H bonds. Importantly, within a few days of the publication of Chen's paper,

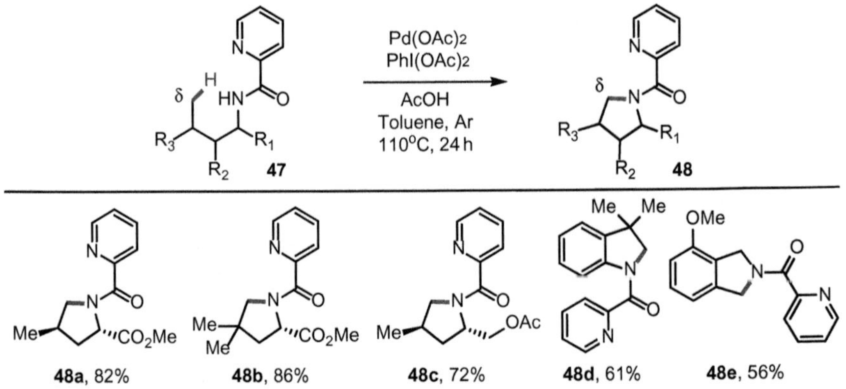

Scheme 9.17 Palladium-catalyzed intramolecular δ-amination in the synthesis of pyrrolidines.

Daugulis presented a similar report on an auxiliary facilitated δ-sp^3 C—H/N—H cyclization.[5]

In 2014, Zhao and co-workers revealed the use of a N,O-bidentate auxiliary in a palladium-catalyzed distal aliphatic C—H bond functionalization (Scheme 9.18).[22] By use of an oxalyl amide directing group, Zhao performed an intramolecular δ-C—H/N—H cyclization of aliphatic amines leading to the formation of pyrrolidines. The weakly coordinating auxiliary allowed the remote C—H activation via formation of a six-membered palladacycle. Substrates bearing two or three primary δ-C(sp^3)—H bonds underwent cyclization smoothly to give the corresponding pyrrolidine products in good-to-excellent yields. Substrates with or without any α-substituent proceeded with the cyclization in moderate yields with uncompromised selectivity. The novelty of the auxiliary lie in its easy installation and removal under mild reaction conditions.

9.2.5 Borylation

Organoboranes are key intermediates in a variety of organic reactions and serve as synthetic precursors for a wide range of functionalized molecules. Although significant developments have been made on metal-catalyzed C—H borylation reactions, formation of regioselective C—B linkages at unactivated aliphatic carbon centers has remained as challenging owing to the high reactivity of C—B bonds in the presence of metal salts, other functional moieties, oxidants, etc. In 2014, Shi and co-workers reported a novel picolinamide-directing group-based strategy in performing a selective borylation at remote unactivated sp^3 C—H centers involving a Pd(0)/Pd(II) catalytic system and using aryl boronic esters as

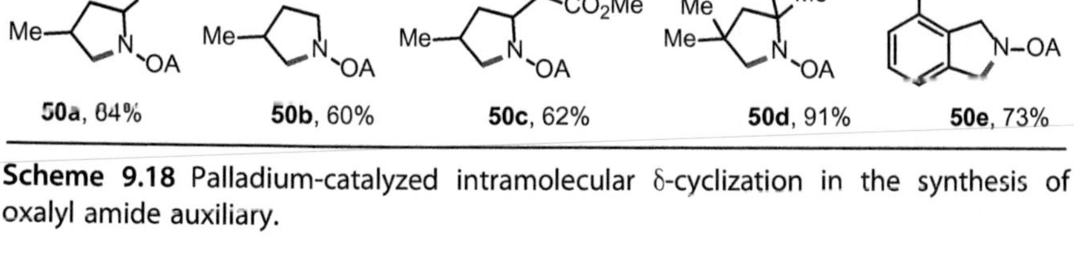

Scheme 9.18 Palladium-catalyzed intramolecular δ-cyclization in the synthesis of oxalyl amide auxiliary.

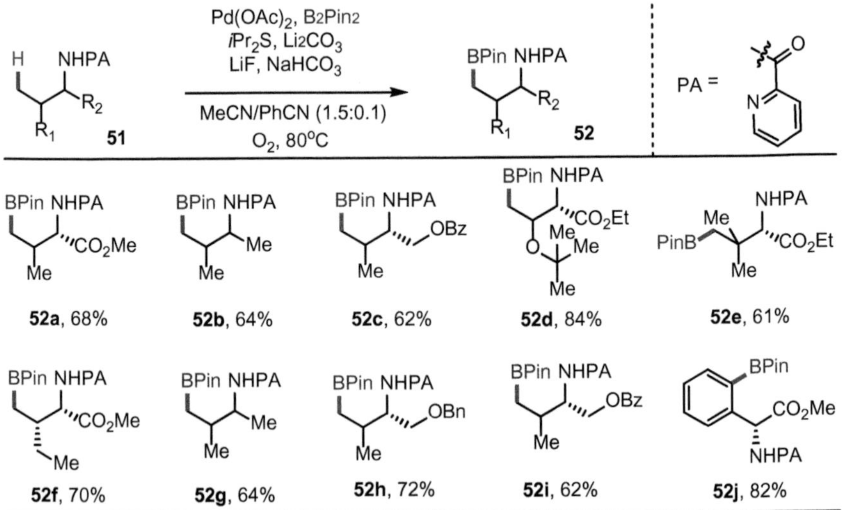

Scheme 9.19 Palladium-catalyzed γ-borylation using picolinamide auxiliary.

coupling partners (diastereomeric ratio up to 91:9) (Scheme 9.19).[23] A difficult task in this reaction was ensuring the stability of the carbon — boron bond in the oxidative alkaline reaction medium. Transmetalation in the presence of Pd(II) species may lead to the decomposition of the boronic esters. This was prevented by using Li_2CO_3 in the presence of additives like LiF. *i*-Pr_2S was used as the ligand that could stabilize the Pd(II) species and prevent its precipitation during the reaction. Moreover, use of molecular oxygen as the sole oxidant served to complete the catalytic cycle and prevented oxidation of the C—B bond in the product. A variety of amino acid, amino alcohol and alkyl amine derivatives were found to comply with the borylation transformation. Late stage modification of the products helped in the synthesis of important bioactive molecules.

9.2.6 Acetoxylation

Palladium-catalyzed C—H oxygenation reactions have attracted widespread scientific interest owing to their biosynthetic utility and transformational simplicity. Several groups like that of Sanford, Yu, and others, have contributed towards the development of such C—O bond formation protocols.[3b,24] Parallel to this, Chen and co-workers revisited their picolinamide-directed intramolecular C—H amination strategy to discover

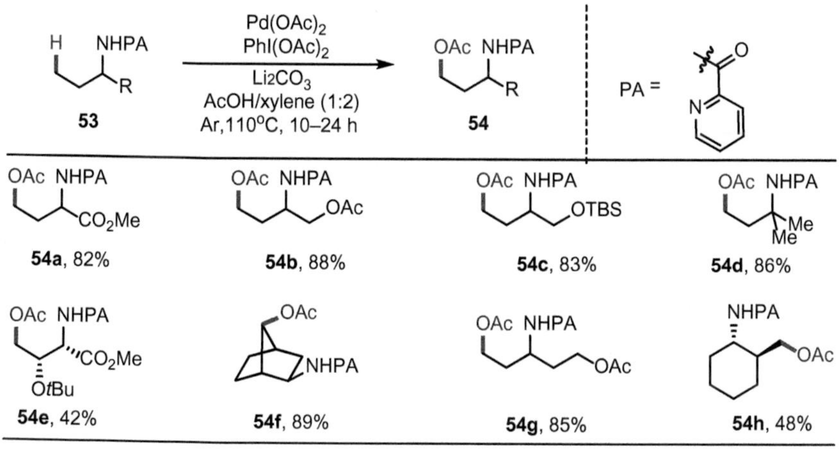

Scheme 9.20 Palladium-catalyzed intramolecular γ-acetoxylation using picolinamide auxiliary.

a novel aliphatic C—H oxygenation chemistry (Scheme 9.20).[25] Customization of the previous reaction conditions was made by introduction of a substoichiometric quantity of Li_2CO_3 and a solvent combination of AcOH/xylene (1:2) in the presence of $PhI(OAc)_2$ as the effective oxidant. This favored C—O reductive elimination from the in situ formed Pd(IV) intermediate to give acetoxylation products, thereby suppressing C—N reductive elimination and the corresponding C—H amination. Consequently, C—H acetoxylation of a wide array of N-alkylpicolinamides and *o*-toluidines was performed at the γ-position.

9.2.7 Cyanomethylation

Transition metal-catalyzed cross-coupling reactions with acetonitrile have attracted significant interest, owing to the possibility of obviating prefunctionalized substrates. It was presupposed that treatment of an arylpalladium (II) species with acetonitrile in the presence of a ligand could make the introduction of a cyanomethyl group into the substrate easier. Despite significant progress, direct cross-coupling of an sp^3 C—H bond with MeCN was not reported. Ge and co-workers recently reported a palladium-catalyzed highly regioselective cyanomethylation of aliphatic amides (Scheme 9.21).[26] This process involved a dehydrogenative cross-coupling reaction with 8-aminoquinoline as the directing group. The presence of copper salt was essential in the reaction medium

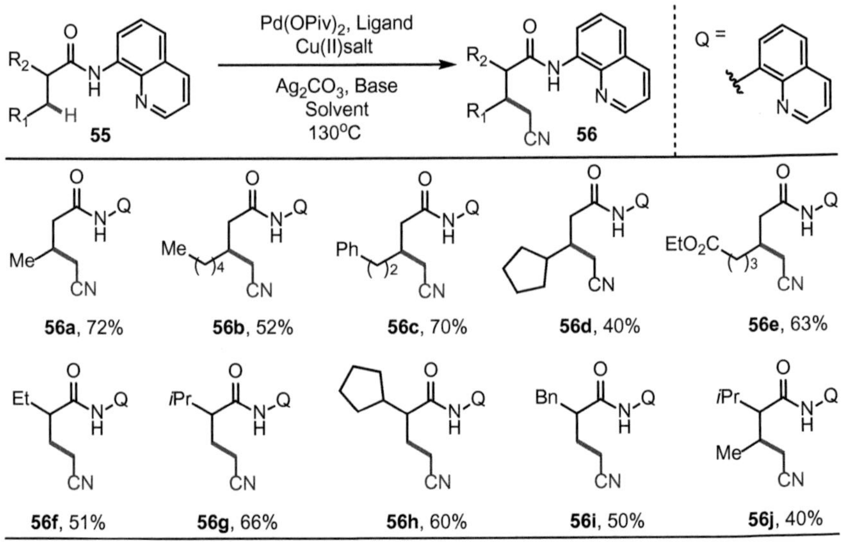

Scheme 9.21 Palladium-catalyzed cyanomethylation yielding γ-cyanated products using 8-aminoquinoline auxiliary.

because the facile copper promoted C—H activation of acetonitrile and the transmetalation of organocopper species onto an organopalladium (II) species. Substrate scope comprised of various linear aliphatic amides with exhibition of excellent functional group tolerance. However, there was a reactivity preference for the primary C—H bonds over the secondary methylene/benzylic C—H bonds. A primary isotope effect was obtained with respect to acetonitrile, thereby confirming sp^3 C—H cleavage of MeCN to be in the rate-determining step of the reaction.

9.2.8 Carbonylation

In 2015, the Zhao group first reported Pd-catalyzed γ-carbonylation of oxalyl amide-protected aliphatic amine with carbon monooxide to synthesize five-membered lactam (Scheme 9.22).[27] Benzylamine and allyl amine derivatives also worked very well in this reaction condition. Under the reaction condition, Pd(II) was reduced to Pd(0) and electron-deficient additive 3-(trifluoromethyl) benzoic acid stabilized palladium intermediate. Substrates with two β-substituents also gave excellent yields. Ester derivative, cyclopropanes were well tolerated in this optimized condition. The oxalyl amide directing group was easily removed using NaOH at 50°C to synthesize pyrrolidones. Under acidic

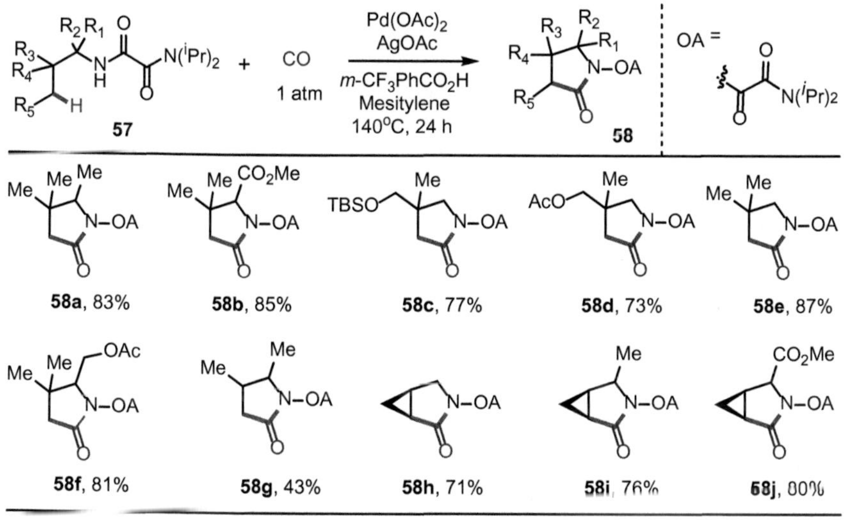

Scheme 9.22 Palladium-catalyzed γ-carbonylation using oxalylamide auxiliary.

hydrolysis these pyrrolidones can be easily transformed to unnatural γ-amino acids.

9.4 CONCLUSION

The development of transition metal-catalyzed chemical transformations has thus revolutionized the scenario of present day organic synthesis. Discovery of unique strategies have eased the formation of complex molecular frameworks from easily available smaller counterparts. A great variety of functionalizations have been attempted in modifying the unreactive sp^3 carbon cores in long chain acyclic and aliphatic systems, which itself is a living documentation of the scientific advances made in organometallic chemistry in the past few years. However, the veritable knowledge of the same is still a delineation of a small share of the boundless infinitum and a huge proportion of chemical miracles still remain to be mapped over the years.

REFERENCES

1. (a) Wencel-Delord, J.; Droge, T.; Liu, F.; Glorius, F. *Chem. Soc. Rev.* **2011,** *40*, 4740.
   (b) Mousseau, J. J.; Charette, A. B. *Acc. Chem. Res.* **2013,** *46*, 412.
   (c) Yang, L.; Huang, H. *Chem. Rev.* **2015,** *115*, 3468.

(d) Franzoni, I.; Mazet, C. *Org. Biomol. Chem.* **2014,** *12*, 233.
(e) Tobisu, M.; Chatani, N. *Science* **2014,** *343*, 850.
(f) Zhang, M.; Zhang, Y.; Jie, X.; Zhao, H.; Li, G.; Su, W. *Org. Chem. Front* **2014,** *1*, 843.
(g) Yang, G.; Butt, N.; Zhang, W. *Chinese J. Catal.* **2016,** *37*, 98.
(h) Dey, A.; Agasti, S.; Maiti, D. *Org. Biomol. Chem.* **2016,** *14*, 5440.

2. (a) Zaitsev, V. G.; Shabashov, D.; Daugulis, O. *J. Am. Chem. Soc.* **2005,** *127*, 13154.
(b) Shabashov, D.; Daugulis, O. *J. Am. Chem. Soc.* **2010,** *132*, 3965.
(c) Gutekunst, W. R.; Baran, P. S. *J. Am. Chem. Soc.* **2011,** *133*, 19076.
(d) Cochet, T.; Bellosta, V.; Roche, D.; Ortholand, J.-Y.; Greiner, A.; Cossy, J. *Chem. Commun.* **2012,** *48*, 10745.
(e) Larini, P.; Kefalidis, C. E.; Jazzar, R.; Renaudat, A.; Clot, E.; Baudoin, O. *Chem. Eur. J* **2012,** *18*, 1932.
(f) Petronijević, F. R.; Nappi, M.; MacMillan, D. W. C. *J. Am. Chem. Soc.* **2013,** *135*, 18323.
(g) Rit, R. K.; Yadav, M. R.; Sahoo, A. K. *Org. Lett.* **2014,** *16*, 968.
(h) Pedroni, J.; Boghi, M.; Saget, T.; Cramer, N. *Angew. Chem. Int. Ed.* **2014,** *53*, 9064.
(i) Yan, S.-B.; Zhang, S.; Duan, W.-L. *Org. Lett.* **2015,** *17*, 2458.
(j) Thompson, S. J.; Thach, D. Q.; Dong, G. *J. Am. Chem. Soc.* **2015,** *137*, 11586.
(k) Hu, J.; Lan, T.; Sun, Y.; Chen, H.; Yao, J.; Rao, Y. *Chem. Commun.* **2015,** *51*, 14929.
(l) Ozawa, J.; Tashiro, M.; Ni, J.; Oisaki, K.; Kanai, M. *Chem. Sci* **2016,** 7, 1904.
(m) Quinn, R. K.; Könst, Z. A.; Michalak, S. E.; Schmidt, Y.; Szklarski, A. R.; Flores, A. R.; Nam, S.; Horne, D. A.; Vanderwal, C. D.; Alexanian, E. J. *J. Am. Chem. Soc.* **2016,** *138*, 696.
(n) Nack, W. A.; Wang, B.; Wu, X.; Jiao, R.; He, G.; Chen, G. *Org. Chem. Front* **2016,** *3*, 561.

3. (a) Ano, Y.; Tobisu, M.; Chatani, N. *J. Am. Chem. Soc.* **2011,** *133*, 12984.
(b) Rit, R. K.; Yadav, M. R.; Sahoo, A. K. *Org. Lett.* **2012,** *14*, 3724.
(c) Chen, F.-J.; Zhao, S.; Hu, F.; Chen, K.; Zhang, Q.; Zhang, S.-Q.; Shi, B.-F. *Chem. Sci* **2013,** *4*, 4187.
(d) Chen, K.; Hu, F.; Zhang, S.-Q.; Shi, B.-F. *Chem. Sci* **2013,** *4*, 3906.
(e) Gou, Q.; Liu, G.; Liu, Z.-N.; Qin, J. *Chem. Eur. J* **2015,** *21*, 15491.
(f) Xiong, H.-Y.; Besset, T.; Cahard, D.; Pannecoucke, X. *J. Org. Chem.* **2015,** *80*, 4204.
(g) Rao, W.-H.; Zhan, B.-B.; Chen, K.; Ling, P.-X.; Zhang, Z.-Z.; Shi, B.-F. *Org. Lett.* **2015,** *17*, 3552.

4. (a) Rouquet, G.; Chatani, N. *Angew. Chem. Int. Ed.* **2013,** *52*, 11726.
(b) Shang, R.; Ilies, L.; Matsumoto, A.; Nakamura, E. *J. Am. Chem. Soc.* **2013,** *135*, 6030.
(c) Maity, S.; Agasti, S.; Earsad, A. M.; Hazra, A.; Maiti, D. *Chem. Eur. J* **2015,** *21*, 11320.
(d) Daugulis, O.; Roane, J.; Tran, L. D. *Acc. Chem. Res.* **2015,** *48*, 1053.

5. Nadres, E. T.; Daugulis, O. *J. Am. Chem. Soc.* **2012,** *134*, 7.
6. Reddy, B. V. S.; Reddy, L. R.; Corey, E. J. *Org. Lett.* **2006,** *8*, 3391.
7. Hyde, A. M.; Buchwald, S. L. *Angew. Chem. Int. Ed.* **2008,** *47*, 177.
8. Hyde, A. M.; Buchwald, S. L. *Org. Lett.* **2009,** *11*, 2663.
9. He, G.; Chen, G. *Angew. Chem. Int. Ed.* **2011,** *50*, 5192.
10. Werner, E. W.; Mei, T.-S.; Burckle, A. J.; Sigman, M. S. *Science* **2012,** *338*, 1455.
11. Fan, M.; Ma, D. *Angew. Chem. Int. Ed.* **2013,** *52*, 12152.
12. Rodriguez, N.; Romero-Revilla, J. A.; Fernandez-Ibanez, M. A.; Carretero, J. C. *Chem. Sci* **2013,** *4*, 175.
13. Franzoni, I.; Guenee, L.; Mazet, C. *Chem. Sci* **2013,** *4*, 2619.
14. Terao, Y.; Satoh, T.; Miura, M.; Nomura, M. *Tet. Lett* **1998,** *39*, 6203.
15. Mei, T.-S.; Patel, H. H.; Sigman, M. S. *Nature* **2014,** *508*, 340.

16. Li, S.; Chen, G.; Feng, C.-G.; Gong, W.; Yu, J.-Q. *J. Am. Chem. Soc.* **2014,** *136*, 5267.
17. Li, S.; Zhu, R.-Y.; Xiao, K.-J.; Yu, J.-Q. *Angew. Chem. Int. Ed.* **2016,** *55*, 4317.
18. He, G.; Zhang, S.-Y.; Nack, W. A.; Li, Q.; Chen, G. *Angew. Chem. Int. Ed.* **2013,** *52*, 11124.
19. Jiang, H.; He, J.; Liu, T.; Yu, J.-Q. *J. Am. Chem. Soc.* **2016,** *138*, 2055.
20. (a) Wolff, M. E. *Chem. Rev.* **1963,** *63*, 55.
    (b) Lu, H.; Zhang, X. P. *Chem. Soc. Rev.* **2011,** *40*, 1899.
    (c) Du Bois, J. *Org. Process Res. Dev.* **2011,** *15*, 758.
21. He, G.; Zhao, Y.; Zhang, S.; Lu, C.; Chen, G. *J. Am. Chem. Soc.* **2012,** *134*, 3.
22. Wang, C.; Chen, C.; Zhang, J.; Han, J.; Wang, Q.; Guo, K.; Liu, P.; Guan, M.; Yao, Y.; Zhao, Y. *Angew. Chem. Int. Ed.* **2014,** *53*, 9884.
23. Zhang, L.-S.; Chen, G.; Wang, X.; Guo, Q.-Y.; Zhang, X.-S.; Pan, F.; Chen, K.; Shi, Z.-J. *Angew. Chem. Int. Ed.* **2014,** *53*, 3899.
24. (a) Giri, R.; Liang, J.; Lei, J.-G.; Li, J.-J.; Wang, D.-H.; Chen, X.; Naggar, I. C.; Guo, C.; Foxman, B. M.; Yu, J.-Q. *Angew. Chem. Int. Ed.* **2005,** *44*, 7420.
    (b) Desai, L. V.; Hull, K. L.; Sanford, M. S. *J. Am. Chem. Soc.* **2004,** *126*, 9542.
25. Li, Q.; Zhang, S.-Y.; He, G.; Nack, W. A.; Chen, G. *Adv. Synth. Catal* **2014,** *356*, 1544.
26. Liu, Y.; Yang, K.; Ge, H. *Chem. Sci* **2016,** 7, 2804.
27. Wang, C.; Zhang, L.; Chen, C.; Han, J.; Yao, Y.; Zhao, Y. *Chem. Sci* **2015,** *6*, 4610.

CHAPTER 10

Palladacycles for Directed and Nondirected C—H Bond Functionalization of (Hetero)arenes

Vineeta Soni and Benudhar Punji
CSIR-National Chemical Laboratory (CSIR-NCL), Pune, Maharashtra, India

Contents

10.1 INTRODUCTION

Palladacycles are palladium compounds containing chelating ligand with at least one Pd—carbon bond. Palladacycles can be C-anionic four-electron donor (CE) or six-electron donor (ECE) types (E = donor group, Fig. 10.1). The former can exist as a mononuclear species or a bridged binuclear

Strategies for Palladium-Catalyzed Non-Directed and Directed C—H Bond Functionalization
DOI: http://dx.doi.org/10.1016/B978-0-12-805254-9.00010-4

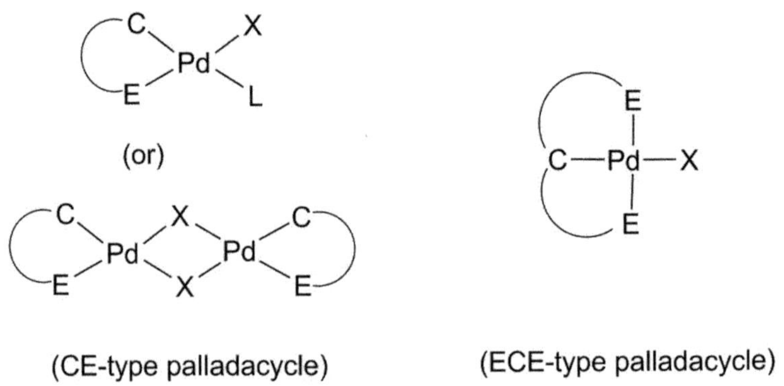

Figure 10.1 Types of palladacycles (E = N, O, P, S, etc.; X = Cl, Br, I, OAc, OTf, etc.).

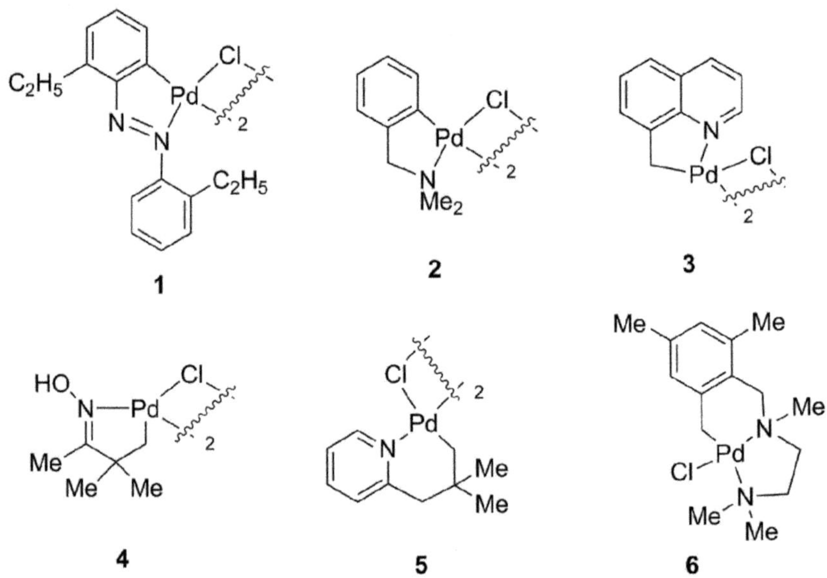

Figure 10.2 Examples of selected CE- and ECE-type palladacycles resulting from directed C—H activation.

complex, and the latter are generally referred to as pincer palladacycle complexes. Palladacycle complexes through the stoichiometric C—H activation process have been known since 1960 (Fig. 10.2).[1–8] Very often, palladacycles display unique chemical reactivities that are not feasible with other palladium complexes, and they show excellent catalytic activity compared to the traditional ligated palladium complexes. The unique feature of palladacycles is that they have high thermal stability with respect to decomposition, which has led to their exploitation as catalysts for many cross-coupling reactions. Though palladacycles were extensively explored in various traditional catalytic reactions,[9,10] their application as a catalyst in direct C—H bond functionalization is relatively less. In the first part of this chapter, we attempt to summarize the use of well-defined palladacycle precatalysts for C—H bond functionalization reactions. In addition to their application, mechanistic aspects of selected palladacycle-catalyzed C—H functionalization reactions are discussed. Because of the defined nature of the catalysts, the working mode of the catalysts is explored.

Direct C—H functionalization processes by in situ generated palladium catalysts have been extensively explored in last two decades. In these cases, the directing group strategy for selective C—H bond functionalization is employed, wherein the substrates containing a coordinating ligand bind to the palladium center and selectively deliver the catalyst to the proximal C—H bond. This coordination of the substrate and proximal C—H activation leads to the C—Pd bond formation that results in the generation of a palladacycle (Fig. 10.3).[11–17] The strong Pd—carbon σ-bond makes this palladacycle robust and allows catalytic reactions through higher oxidation states. In many cases, these palladacycles are intermediate for the C—H functionalization process, and they can be converted to new functional groups. The coordinating ligands on the substrate are generally Lewis-basic heteroatoms, such as pyridine, pyrimidine, benzo[*h*]quinoline, amines, imines, oxazolines, amides, ketones, esters, carboxylic acids, anilides, thioketones, etc. (Fig. 10.3). In these cases, substrates itself are the associate ligand for palladacycle intermediate. Hence, in the second part of this chapter, we describe the involvement of palladacycle intermediates in C—H bond functionalization, which were proposed with some experimental evidence, and/or are well-characterized. We have not included all other C—H functionalizations, wherein palladacycles have been assumed without much evidence. The second part of the chapter is organized on the basis of the type of palladacycle involved (N-, O-, S-containing palladacycle or bis-chelated palladacycle). In addition, we have also included the C—H functionalization of other substrate that proceeds via silanol, phosphoramidate, and phosphate-containing palladacycles.

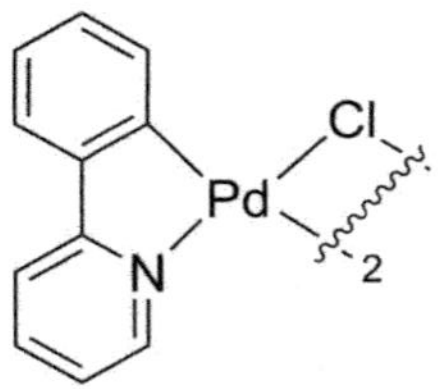

7

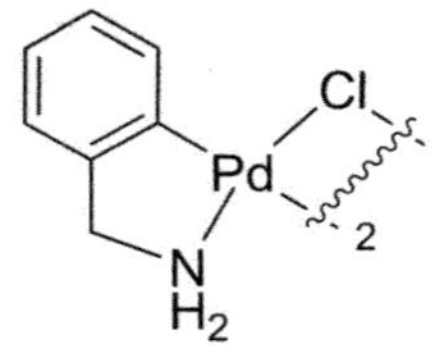

8

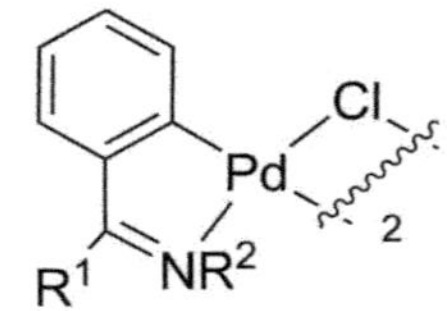

9

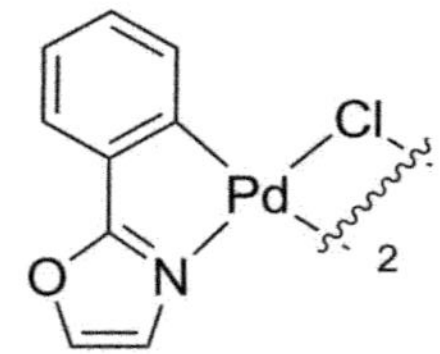

10

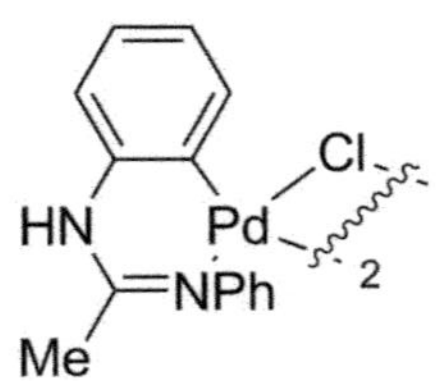

11

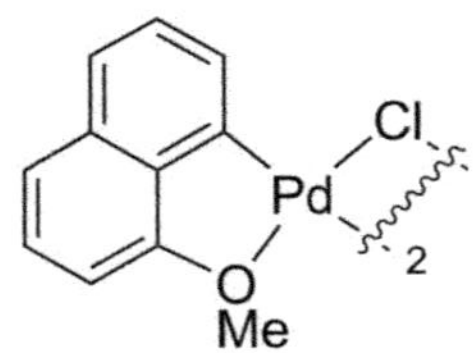

12

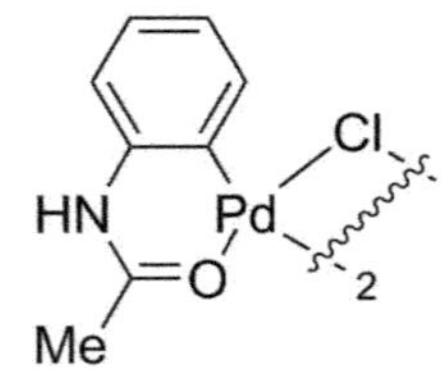

13

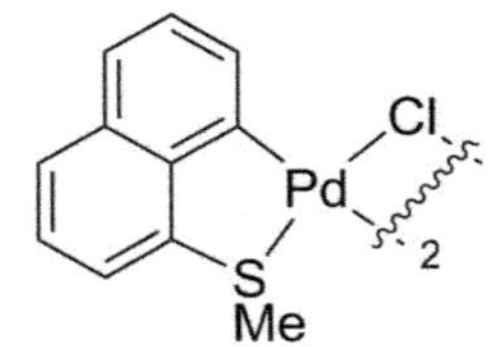

14

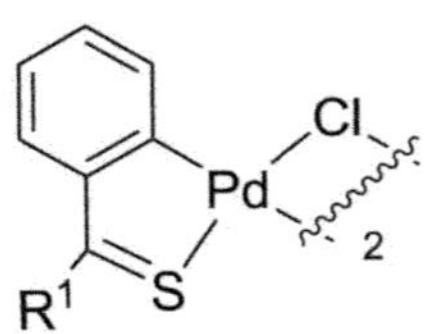

15

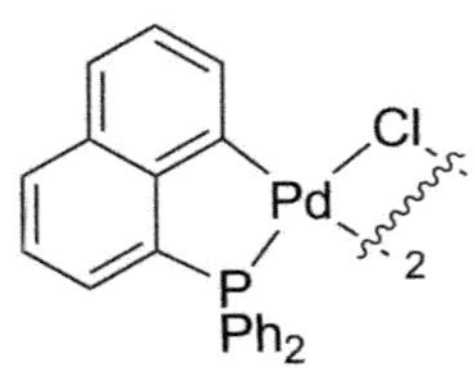

16

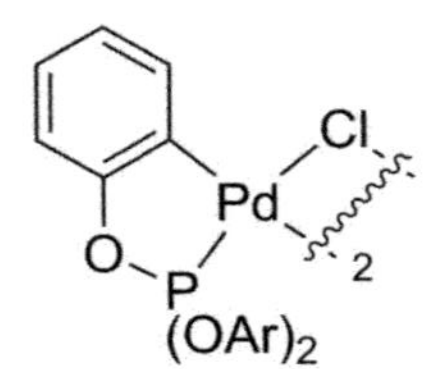

17

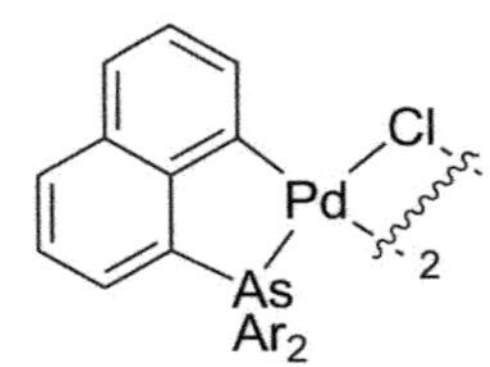

18

Figure 10.3 Examples of chloro-bridged binuclear palladacycles from directing group strategy C—H activation.

10.2 PALLADACYCLE PRECATALYST IN C—H FUNCTIONALIZATION

The stoichiometric synthesis of palladacycles via selective C—H bond activation has been known for decades. Palladacycles as catalysts came to prominence when Herrmann's palladacycle **19** was highlighted as a potential catalyst for cross-coupling reactions.[18] After this initial discovery, many palladacycles have been optimized and employed for various organic transformations.[9,10,19,20] Only recently, the application of palladacycles for direct C—H functionalization has been initiated. The well-known Herrmann—Beller palladacycle **19** was reported for the arylation of simple, unactivated arenes with various symmetrical diaryliodonium salts (Scheme 10.1).[21] Both the electron-donating and electron-withdrawing groups on iodonium salts reacted efficiently to produce the desired product in moderate-to-excellent yields. The *para-* and *meta-*substituted arenes were well tolerated, whereas the *ortho*-substitution was reported to be lower in yields. Interestingly, various important functional groups, such as halides, ester, and nitro groups, were tolerated in the aryl backbone and reacted with moderate-to-good activity. Key to the success of this reaction is the use of the Herrmann—Beller palladacycle precatalyst and trifluoroacetic acid, the latter of which enhances the electrophilicity of the palladium catalyst. The significant feature of this arylation is that it uses cheap, readily available, and recoverable arenes.

An oxime-based palladacycle **20** was shown to catalyze the direct arylation of benzothiazole with aryl iodides and bromides (Scheme 10.2).[22]

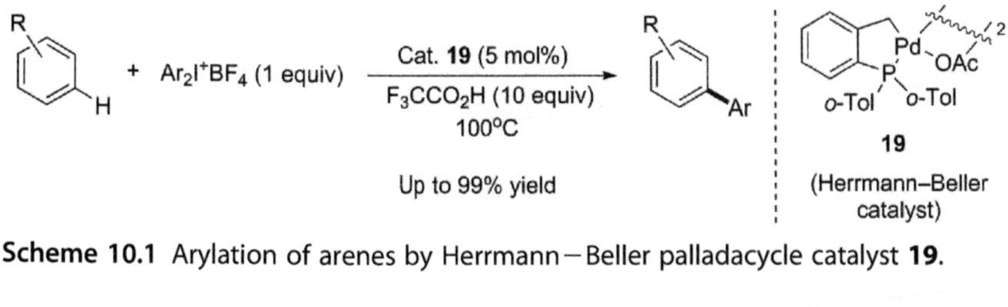

Scheme 10.1 Arylation of arenes by Herrmann—Beller palladacycle catalyst **19**.

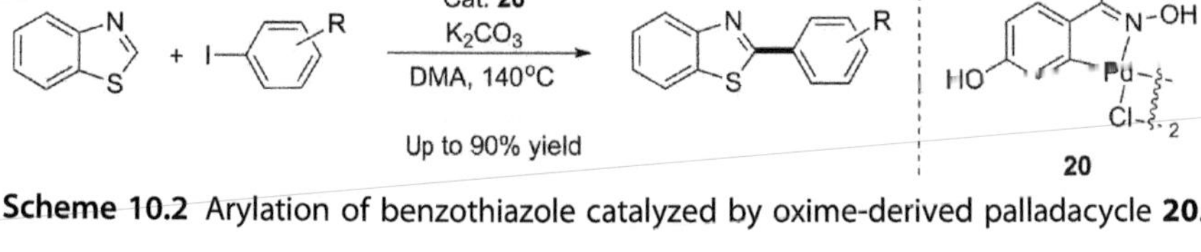

Scheme 10.2 Arylation of benzothiazole catalyzed by oxime-derived palladacycle **20**.

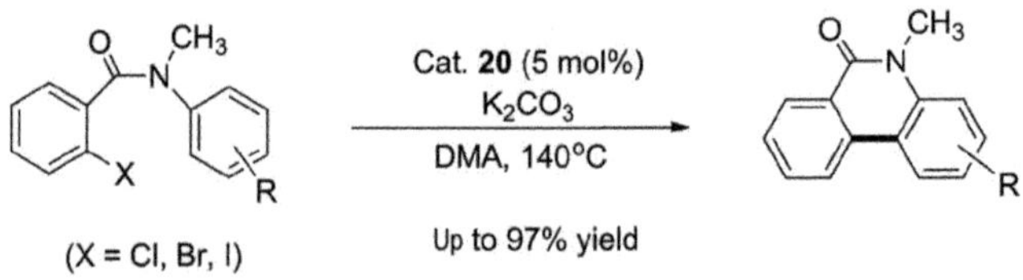

Scheme 10.3 Intramolecular arylation of haloamides catalyzed by palladacycle **20**.

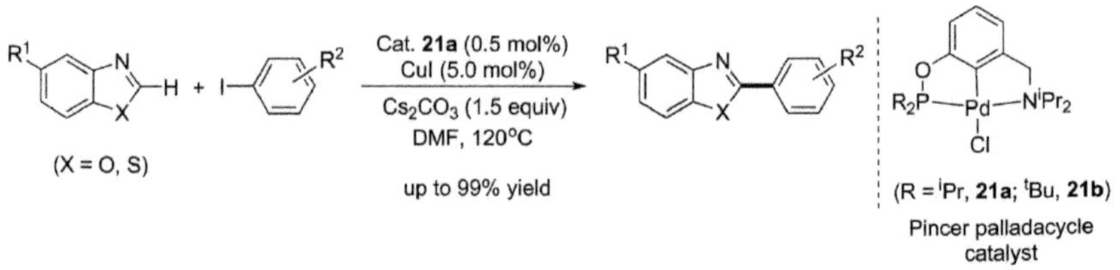

Scheme 10.4 Pincer palladacycle-catalyzed arylation of azoles with aryl iodides.

Generally, yields of the coupled products with aryl iodides were higher than those with aryl bromide electrophiles. Both the electron-donating and electron-withdrawing substituents on aryl halides exhibit similar reactivities. Interestingly, this phosphane-free catalyst **20** also catalyzes the intramolecular direct coupling of chloro-, bromo-, and iodoamides to furnish the phenanthridinone derivatives (Scheme 10.3).

Recently, Punji et al. developed a hybrid (POCN)Pd palladacycle catalyst system (**21a**,**b**) for the sp^2 C–H bond arylation of various substituted azoles with aryl iodides (Scheme 10.4).[23] This was the first pincer palladium system applied for the direct arylation of azoles. Employing the palladacycle **21a**, pharmaceutically important benzothiazoles, benzoxazoles, and 5-aryl oxazoles were coupled with diverse aryl iodides in the presence of CuI co-catalyst. These reactions were efficiently carried out at 120°C albeit using an expensive base, Cs_2CO_3. Interestingly, later on, a sterically less demanding pincer palladacycle **21c** was developed, which catalyzed the azoles arylation with aryl iodides using a relatively mild and inexpensive base, K_3PO_4.[24] The detailed mechanistic studies were performed for this reaction with catalyst **21c**, including kinetics and labeling experiments, controlled reactivity study, and DFT energy calculation (Scheme 10.5). A crucial palladacycle intermediate **22** was also isolated and structurally characterized. With all the mechanistic experimental studies, the authors have proposed a $Pd^{II}–Pd^{IV}–Pd^{II}$ catalytic pathway

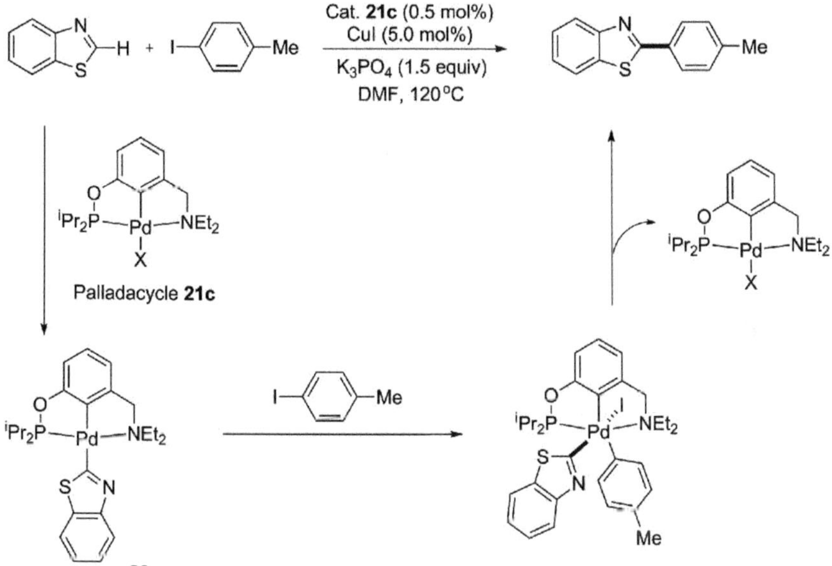

Scheme 10.5 Proposed pathway for palladacycle **21c**-catalyzed arylation of azoles.

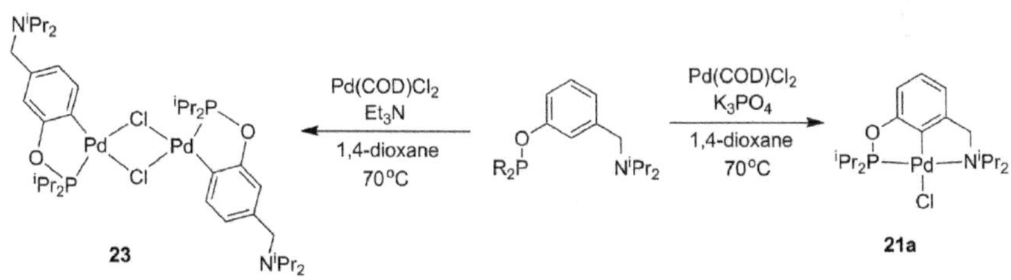

Scheme 10.6 Synthesis of mono- and binuclear palladacycles via regioselective C—H bond activation.

for this reaction with probable oxidative addition of Ar—I to the palladacycle species **22**.

The C—H bond arylation of azoles was also reported with the binuclear palladacycle catalyst **23** (Scheme 10.6).[25] Similar to the palladacycle catalyst **21c**, the binuclear palladacycle **23** catalyzed the arylation of various azoles with aryl iodides in the presence of mild base K_3PO_4 in DMF. It was shown that the mononuclear palladacycle **21a** was obtained exclusively in the presence of inorganic base K_3PO_4, whereas in the presence of coordinating base Et_3N, an exclusive formation of chloro-bridged binuclear palladacycle **23** was obtained (Scheme 10.6). Detailed mechanistic studies for this regioselective C—H palladation to achieve mono- or bi-nuclear palladacycle were conducted, which suggests the coordinating ability of the external base, sterics between the coordinated base and N-arm of the POCN ligand, as well as the electrophilicity of the palladium center were crucial to the regioselective palladations.

A few notable examples have been described where four-membered palladacycle catalysts were employed, or were the active catalysts in the C—H bond arylation of arenes. For example, Hartwig et al. reported that the arylation of pyridine *N*-oxide (PyO) with aryl halides, which was performed in the presence of $Pd(OAc)_2$ and tBu_3P, is catalyzed by palladacycle complex $[Pd(OAc)(tBu_2PCMe_2CH_2)]_2$ (**24**) (Scheme 10.7).[26] Initially, the same reaction was proposed to occur by the generation of $(tBu_3P)Pd(Ar)(OAc)$ and reaction of this with PyO.[27] However, with the detailed kinetic studies, Hartwig et al. established that the reaction of arylpalladium complex $(tBu_3P)Pd(Ar)(OAc)$ with PyO is catalyzed by palladacycle **24**.[26] It was suggested that the C—H bond of PyO is cleaved by the reaction with palladacycle complex **24**, whereas the other palladium fragment bearing tBu_3P ligand adds the aryl halide and forms the C—C coupled product (Scheme 10.7).

Herrmann—Beller palladacycle catalyst **19** is also reported for the cross-coupling of heteroarenes. For example, palladacycle **19** could be

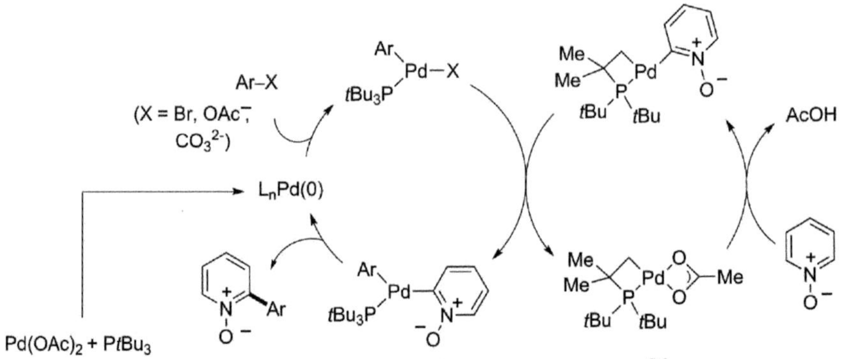

Scheme 10.7 Arylation of pyridine *N*-oxide by palladacycle **24**.

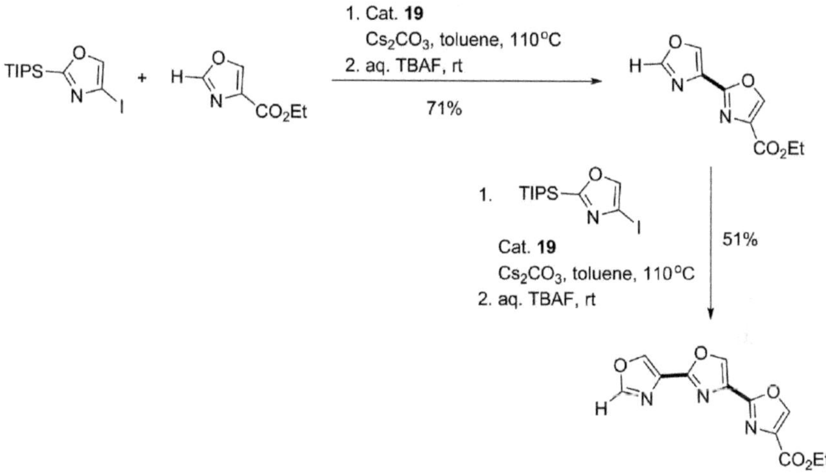

Scheme 10.8 Synthesis of bisoxazole and trisoxazole derivatives catalyzed by Herrmann–Beller palladacycle catalyst **19**.

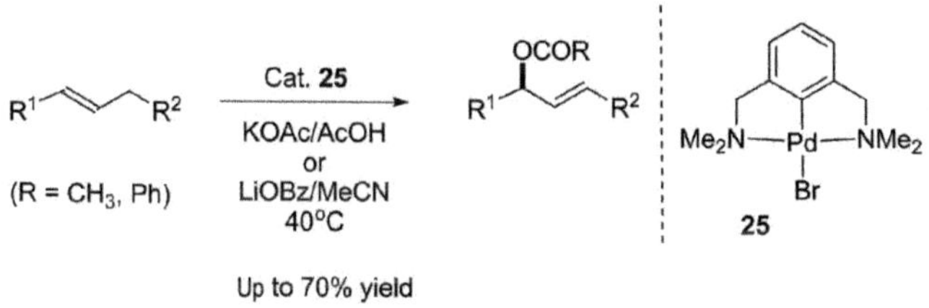

Scheme 10.9 Acetoxylation and benzoyloxylation of alkenes by palladacycle **25**.

employed in the synthesis of bisoxazole in very good yield (Scheme 10.8).[28] Upon deprotection of the triisopropylsilyl group from bisoxazole and further arylation in the presence of the same palladacycle catalyst, trisoxazole was afforded in moderate yield. By using catalyst **19**, the author has reported one of the quickest processes for the synthesis of trisoxazoles, which were found in Ulapualide A macrolide.

Acetoxylation and benzoyloxylation of sp^2 C—H bond were described by the pincer-based palladacycle catalyst **25** (Scheme 10.9).[29] Employing the palladacycle **25** as catalyst, various olefins could be used for the acetoxylation and benzoyloxylation with iodonium salts under mild conditions. Interestingly, the reaction can be performed under both acidic and nonacidic conditions with high regio- and stereoselectivity. Mechanistic

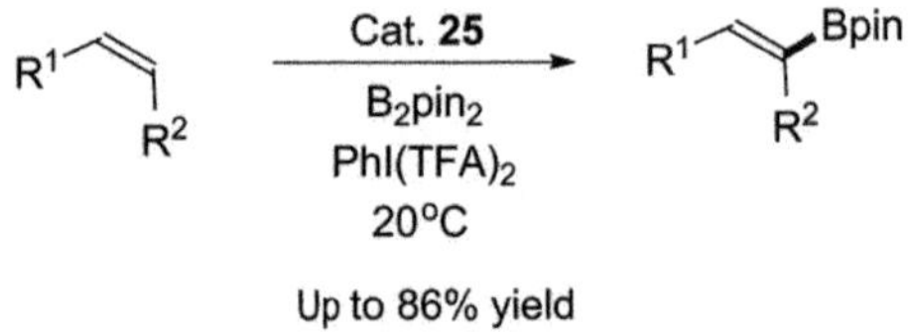

Scheme 10.10 Oxidative C—H borylation of alkenes by palladacycle **25.**

study shows that the reaction proceeds through (η^3-allyl)palladium intermediate and via the formation of a Pd(IV) species. Although the catalysis can be performed with simple $Pd(OAc)_2$, the same reaction by palladacycle **25** has been suggested through a Pd^{II}/Pd^{IV} pathway, which is complementary to the traditional Pd^0/Pd^{II} process. Szabo et al. also employed palladacycle **25** for the selective C—H borylation of alkenes (Scheme 10.10).[30] This reaction provided an easy access to pinacolboronates, which can successively be used for Suzuki—Miyaura coupling. The key steps involved in this reaction were the oxidative addition of the catalyst to a Pd^{IV} species, facile transmetalation with B_2pin_2, and subsequent reaction with olefins.

10.3 PALLADACYCLE INTERMEDIATE IN C—H FUNCTIONALIZATION

10.3.1 Pyridinyl-Based Palladacycle

Several breakthrough examples were reported in the last decade for the C—H bond functionalization of substrates bearing nitrogen functionality that proceeded through a nitrogen-containing palladacycle. Nitrogen functionality as a Lewis-base directing group has been extensively employed because of its strong coordination ability, which smoothly binds and directs the palladium into close proximity of a particular C—H bond. A variety of pyridine derived five-membered palladacycles served as intermediates, as well as catalyst, during pyridine-directed C(sp^2)—H and C(sp^3)—H functionalization. For example, the Sanford group has disclosed the intermediacy of acetate-bridged binuclear palladacycle **26** during the arylation of aromatics via the C—H activation process employing a hypervalent iodine(III) arylating derivative (Scheme 10.11).[31] The proposed mechanism involves chelation-assisted C—H activation to produce a palladacycle **26**, oxidation of palladacycle Pd(II) to Pd(IV) by iodonium reagent, Ph_2IBF_4, and reductive elimination to generate the arylated product. In addition to pyridine, other nitrogen-functionalities such as

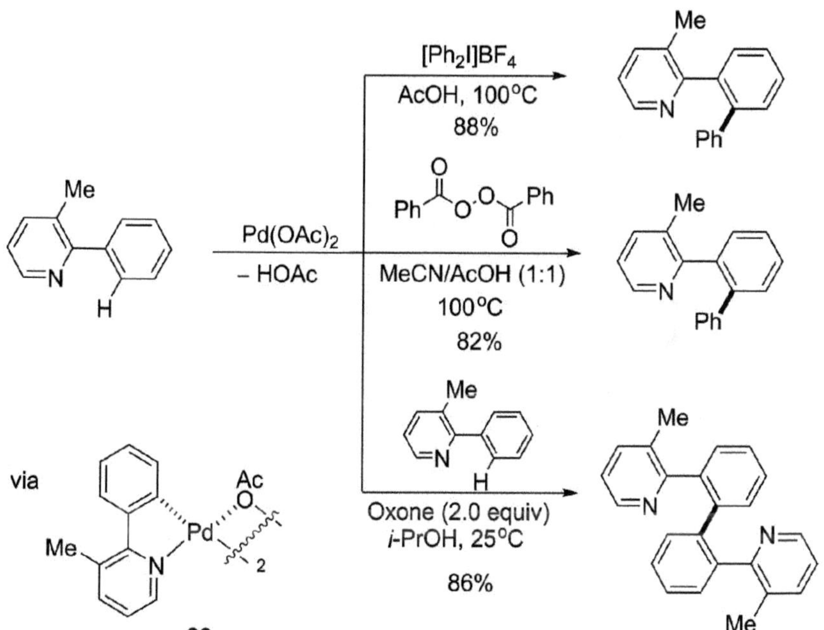

Scheme 10.11 Pyridine-coordinated palladacycle in arylation reaction of arenes.

quinoline, and oxazolidinone can be part of the palladacycle ring. Unsymmetrical aryl-substituted iodonium reagents $[Ar^1\text{-I-}Ar^2]BF_4$ could also be used to install different aryl groups.[32] Yu et al. reported the aryl acylperoxides as arylating reagents to effect the *ortho*-arylation of pyridine-, quinoline-, oxime ether-, and oxazoline-containing arenes through a five-membered palladacycle intermediate.[33] Interestingly, the arylation has also been described by the cross-couplings of two aromatics via a bis-chelate palladacycle intermediate upon double C—H activation in the presence of an external oxidant (Scheme 10.11).[34]

Shabashov and Daugulis described the arylation of the sp^3 C—H bond with aryl iodides using the external metal oxidant AgOAc, and the reactions are assumed to proceed through pyridine-, quinoline-, and pyrrozole-containing palladacycles (Scheme 10.12).[35] This process exhibits interesting functional group tolerability, such as bromide group is tolerated on the aryl iodide electrophile.

Pyridine-containing palladacycle **27** has been reported in the catalytic sp^2 C—H bond alkylation employing alkylboronic acids (Scheme 10.13).[36] In this system, Ag_2O was used as an oxidant and benzoquinoline was a promoter. Important functional groups, such as MeO, vinyl, and CF_3 were

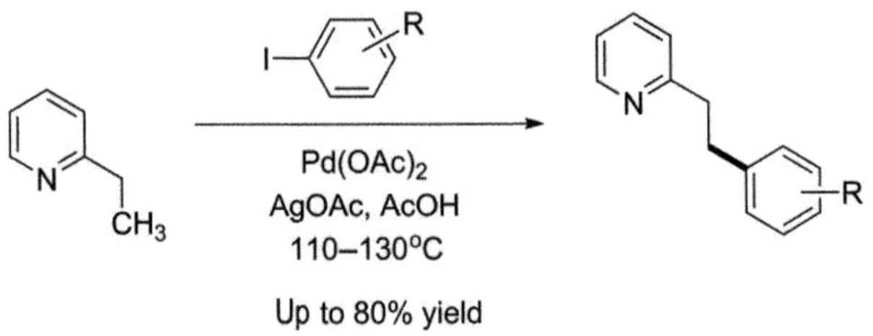

Scheme 10.12 C(sp^3)—H bond arylation with pyridine-directed activation.

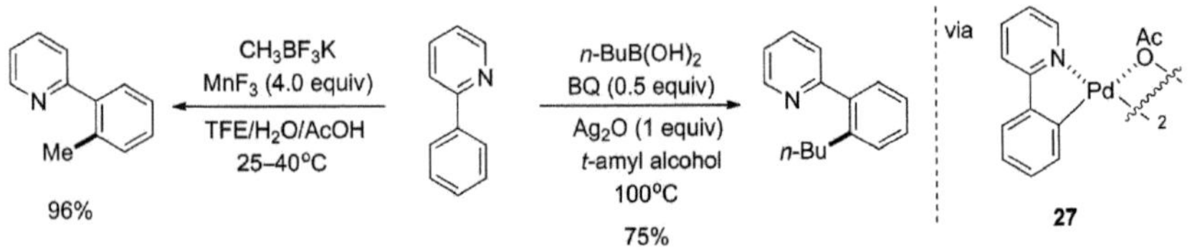

Scheme 10.13 Pyridine-based palladacycle in C—H alkylation.

well tolerated in the described conditions. Recently, Sanford et al. reported pyridine-directed alkylation using potassium alkyltrifluoroborates as an alkylating reagent and MnF_3 as oxidant.[37] These reactions proceeded under mild conditions in weakly acidic medium, and do not require an additional promoter, such as benzoquinoline.

Palladacycle **27** type intermediate has been proposed for the acylation of arenes via C—H bond cleavage (Scheme 10.14).[38] In this method, ketones were directly accessed from aldehydes through arene C—H activation. Interestingly, air was used as oxidant in this process and a series of functional groups, such as methoxy, fluoro, bromo, and COOMe were compatible under the reaction conditions. Similarly, Li et al. reported on the acylation of the sp^2 C—H bond with alcohol in the presence of peroxide oxidant.[39]

Yu et al. described oxidative ethoxycarbonylation of aromatic C—H bond with diethyl azodicarboxylate (DEAD) through the cyclopalladation of 2-phenyl pyridine and derivatives (Scheme 10.14).[40] Among the various oxidants, oxone was found to be efficient for this ethoxycarbonylation reaction. The palladacycle **27** is proposed to react with the ethoxyacyl radicals generated from the thermal decomposition of DEAD, to afford the product esters. In this process, benzo[*h*]quinoline, 8-methylquinoline, and oxime ether were also efficiently ethoxycarbonylated in good yields.

Cheng et al. independently synthesized the palladacycle **27** and employed it for the *ortho*-cyanation of the sp^2 C—H bond using CuCN

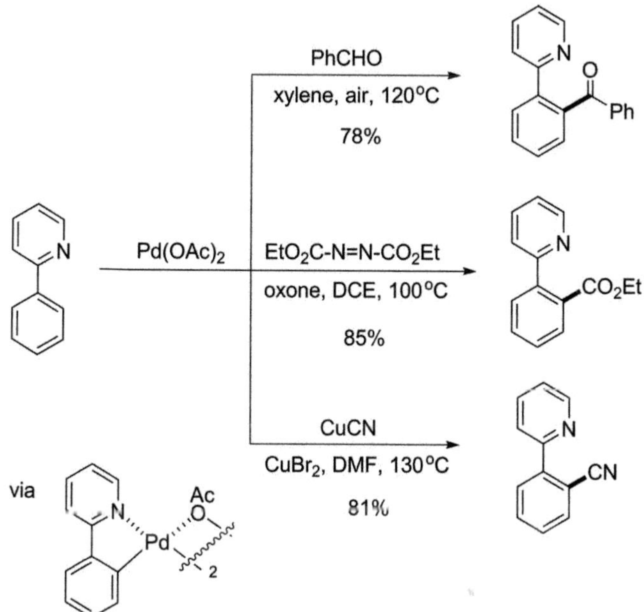

Scheme 10.14 Pyridine-based palladacycle in C—H acylation, ethxycarbonylation, and cyanation reactions.

as the CN-source (Scheme 10.14).[41] This approach represents a practical method for the synthesis of aromatic nitriles, and an alternative to the traditional Sandmeyer reaction. This reaction could also be performed in 10 mmol scale without much affect on the yield of product.

In addition to C—C bond formation by the C—H bond activation and functionalization through palladacycle catalysts and intermediates, the palladacycles have also been known to catalyze the C—O and C—halogen bond formations. The Sanford group described elegant work on a selective method for the oxidative functionalization of C—H bonds. For example, the palladacycle **27**, upon treatment with $PhI(OAc)_2$, afforded *ortho*-acetoxylated product (Scheme 10.15).[42,43] This method showed remarkably broad substrate scope and functional group tolerance. Pyrrolidinone as a directing group was also employed in the method to produce the desired acetoxylated products. Jiao et al. reported hydroxylation of the $C(sp^2)$—H bond via a palladacycle intermediate of 2-phenyl pyridines, employing *N*-hydroxyphthalimide (NHPI) co-catalyst and by dioxygen activation.[44] An initial palladacycle formation and NHPI-initiated radical process have been proposed for this reaction. The

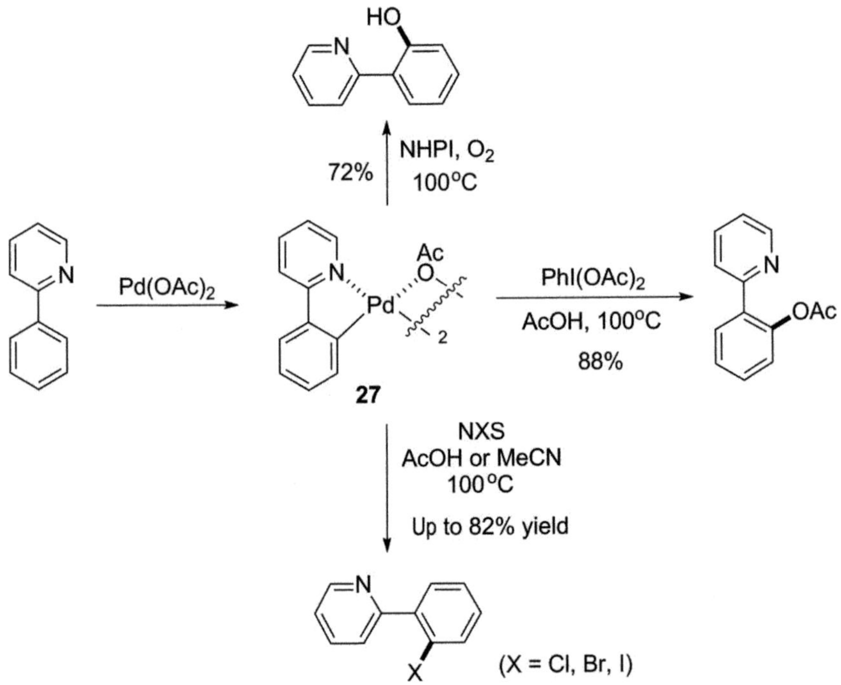

Scheme 10.15 Oxygenation and halogenation of C—H bond involving pyridine-based palladacycle.

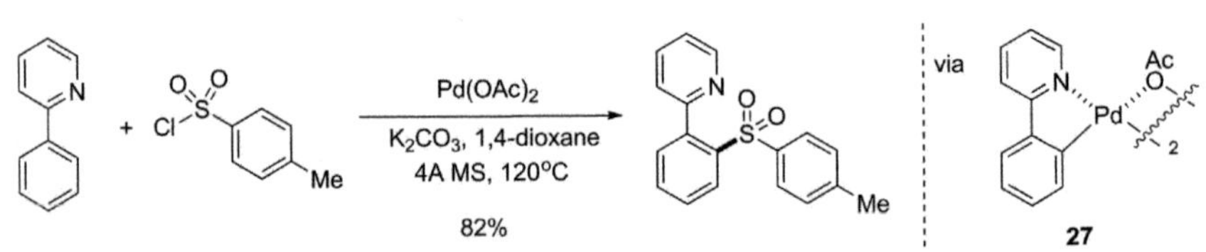

Scheme 10.16 Palladacycle **27** in C—H bond sulfonylation reaction.

palladium-catalyzed regioselective chlorination, bromination, and iodination of arene C—H bonds using *N*-halosuccinimides (NXS) as oxidants have been described by Sanford and coworkers.[45–47] These methods and product formations are complementary to those obtained via conventional electrophilic aromatic substitution reactions.

Dong et al. disclosed arene C—H bond sulfonylation using arylsulfonyl chlorides, which are useful synthetic intermediates (Scheme 10.16).[48] Palladacycle **27** was proposed to oxidize with sulfonyl chlorides, followed by reductive elimination to generate the product. This sulfonylation

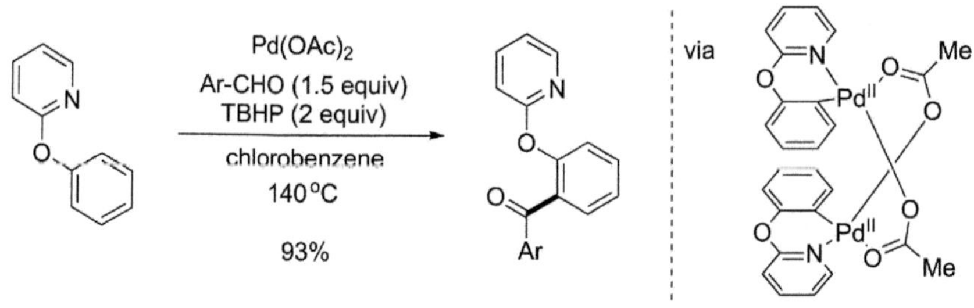

28

Scheme 10.17 Aroylation of 2-phenoxypyridine via six-membered palladacycle **28**.

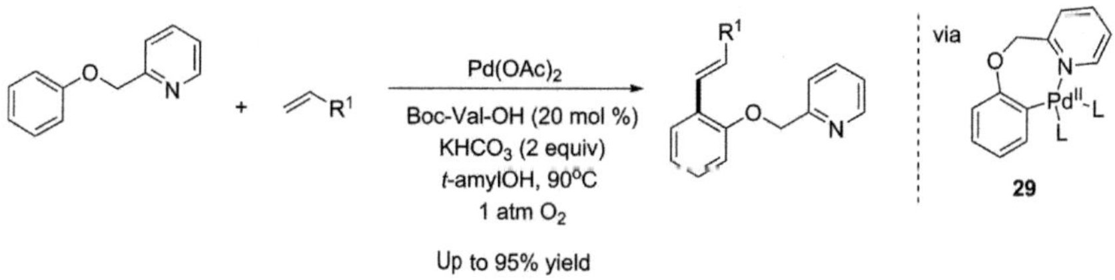

Scheme 10.18 Alkenylation of (pyridin-2-nyl methoxy)benzene via seven-membered palladacycle **29**.

condition was also employed for the substrates bearing pyrazole- and oxime-directing groups to produce the desired sulfones in good yields.

A pyridinyl-based six-membered palladacycle **28** as an intermediate has been described in various C—H functionalizations of 2-phenoxypyridine. Wu et al. reported *ortho*-aroylation with aldehyde using *tert*-butyl hydrogenperoxide as oxidant (Scheme 10.17).[49] The detailed mechanistic study reveals a radical path for this reaction, wherein the aroyl radical is generated from the reaction aldehyde with TBHP by one-electron abstraction. Similar acylation of 2-phenoxypyridine has also been described with benzylic alcohol in the presence of TBHP,[50] and with α-oxo carboxylic acids via decarboxylation.[51]

A seven-membered palladacycle **29** has been proposed during the sp^2 C—H olefination of 2-(phenoxymethyl)pyridine (Scheme 10.18). However, the six-membered and higher metallacycle intermediates are considered more unstable than the five-membered. Lan et al. have shown the efficient coupling of olefins with 2-(phenoxymethyl)pyridine through a seven-membered palladacycle.[52] This reaction is accelerated in the presence of an amino acid ligand (Boc-Val-OH). Importantly, a 2-pyridylmethyl directing group can be easily removed, affording *ortho*-alkenyl phenol. Similar, a seven-membered palladacycle has also been implicated for the alkenylation of *N*-(2-pyridyl)sulfonyl anilines and arylalkylamines with olefins.[53]

10.3.2 Quinolinyl-Based Palladacycle

Benzo[*h*]quinoline substrate is the general choice for most investigations into C—H functionalization, because it presents a single bond [C—H(10)] for directed C—H activation and does not bear any functional group. In fact, this substrate is well-known to undergo cyclopalladation under mild conditions.[3] Hull and Sanford reported the oxidative cross-coupling of benzo[*h*]quinoline with Ar—H substrates, which proceeded through palladacycle **30** (Scheme 10.19).[54] The oxidant benzoquinone promotes the C—H activation reaction at the Pd(II) center. For the substituted arenes, the reaction progressed with high selectivity for coupling at the less hindered position of Ar—H. This reaction was proposed to proceed via two discrete C—H activations, wherein the selectivities were controlled by proximity to a ligand (1st arene) and a steric environment around the second arene C—H bond. The same group has reported the acetoxylation of benzo[*h*]quinoline with $PhI(OAc)_2$ oxidant.[55] The isolated cyclopalladated complex **30** as a catalyst also provided high catalytic activity for this reaction. Interestingly, upon changing the solvent for this reaction from AcOH to alcohol (ROH), aryl ether products were formed in excellent yields. In situ reaction of the alcohol solvent with $PhI(OAc)_2$ to generate $PhI(OR)_2$ was suggested to be the key in these transformations.

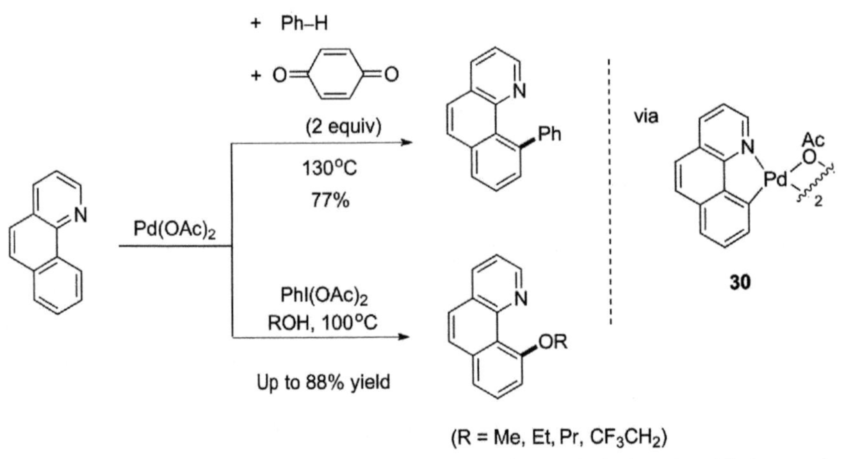

Scheme 10.19 Benzo[*h*]quinoline-based palladacycle in C—H bond oxidative arylation and oxygenation.

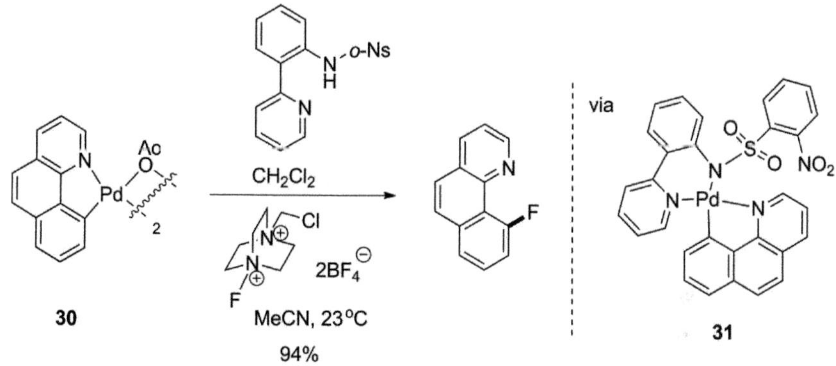

Scheme 10.20 Fluorination of benzo[*h*]quinoline via palladacycle **31**.

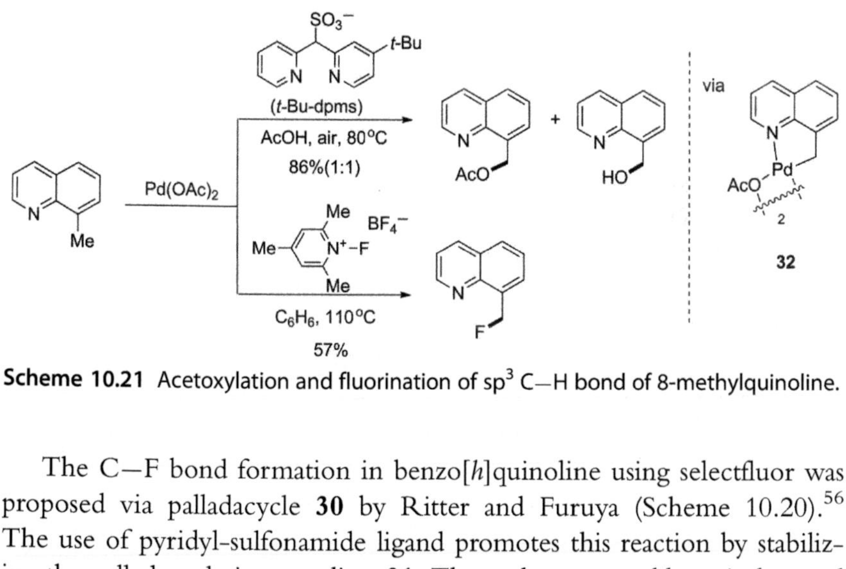

Scheme 10.21 Acetoxylation and fluorination of sp^3 C—H bond of 8-methylquinoline.

The C—F bond formation in benzo[*h*]quinoline using selectfluor was proposed via palladacycle **30** by Ritter and Furuya (Scheme 10.20).[56] The use of pyridyl-sulfonamide ligand promotes this reaction by stabilizing the palladacycle intermediate **31**. The authors were able to isolate and characterize many palladacycles, including **31** and a Pd(IV) intermediate. In addition, carbon-fluorine reductive elimination was described from a high-valent palladium fluoride complex.

The palladium-catalyzed sp^3 C—H bond oxidation of substituted 8-methylquinoline is reported through the formation of the five-membered palladacycle **32** employing air as the terminal oxidant (Scheme 10.21).[57] The use of anionic ligand *t*-Bu-dpms and acidic medium was very crucial for this reaction. Palladacycle **32** reacts with *t*-Bu-dpms ligand to afford the *t*-Bu-dpms chelated mononuclear palladacycle complex, which upon stirring in AcOH under air led to the formation of

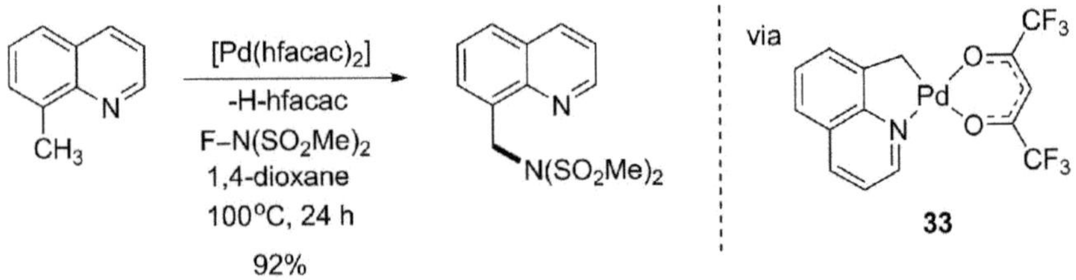

33

Scheme 10.22 Amidation of sp^3 C—H bond of 8-methylquinoline via palladacycle **33**.

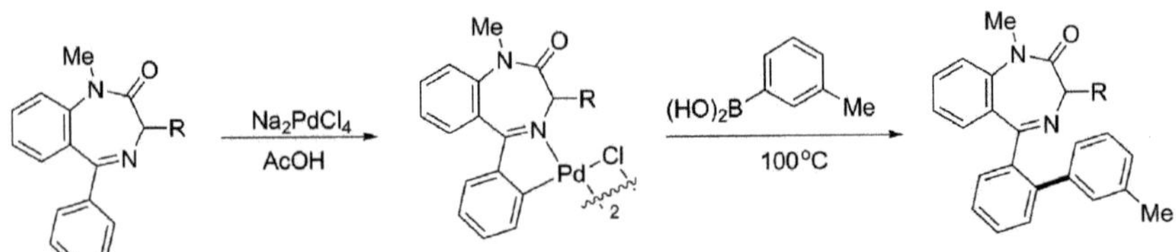

34

Scheme 10.23 Benzodiazepine-based palladacycle in C—H arylation.

8-quinolylmethyl acetate and 8-quinolylmethanol. When the same reaction was performed under O_2 in the presence of Ac_2O, exclusive formation of 8-quinolylmethyl acetate was observed.

Catalytic sp^3 C—H bond fluorination of 8-methylquinoline was demonstrated employing Selectfluor, *N*-fluoro-2,4,6-trimethylpyridinium tetrafluoroborate as an electrophilic fluorinating reagent (Scheme 10.21).[58] Microwave heating of this reaction decreased the reaction time, as well as minimizing the formation of undesired side products. Further, this process was applied for the *ortho*-fluorination of 2-arylpyridines using *N*-fluoropyridinium tetrafluoroborate as a fluorinating agent.

Muniz et al. reported the C(sp^3)—H bond amidation of 8-methylquinoline, which proceeds by the initial formation of the five-membered palladacycle compound **33** (Scheme 10.22).[59] This palladacycle compound was structurally characterized by X-ray diffraction study. The choice of *N*-fluorobis(phenylsulfonyl)imide (NFSI) electrophilic oxidant was very crucial for this oxidative amidation reaction.

Spencer et al. reported a series of benzodiazepine-derived palladacycles by sp^2 C—H bond activation (Scheme 10.23).[60] The dimeric palladacycles (**34**) were conveniently converted to a planar mononuclear palladacycle by treating them with monodentate ligands (PPh_3, Pyridine). Mononuclear palladacycles were structurally characterized. These complexes were stoichiometrically reacted with arylboronic acid to afford the *ortho*-arylated benzodiazepine.[61] Notably, this arylation reaction by the benzodiazepine-palladacycle catalyst was made catalytic by employing hypervalent iodonium salts as the arylating agent.

10.3.3 Amine-Based Palladacycle

Amine-based palladacycles were synthesized and their reactivity was studied by van Koten et al. in early 1990.[62,63] Recently, these palladacycles were demonstrated for C—H bond functionalizations. For example, Shi et al. showed highly regioselective olefination of substituted *N,N*-dimethylbenzylamine via a five-membered *N*-containing palladacycle **35** (Scheme 10.24).[64] The palladacycle **35** was independently synthesized and tested as an active catalyst for the olefination reaction.[2] The appropriate acidity of the reaction condition was key to the success of this reaction. The acid most likely enhances the electrophilicity of the metal center for facile electrophilic C—H activation. They have shown that the *ortho*-functionalized *N,N*-dimethylbenzylamines could be converted into 3-(2′-tolyl)propanoic acid derivatives under mild conditions.

Unsubstituted benzylamines could be *ortho*-arylated efficiently with aryl iodides in the presence of $Pd(OAc)_2$ in trifluoroacetic acid solvent, which most likely proceeded through the five-membered palladacycle intermediate (Scheme 10.25).[65] The reactions proceeded well with

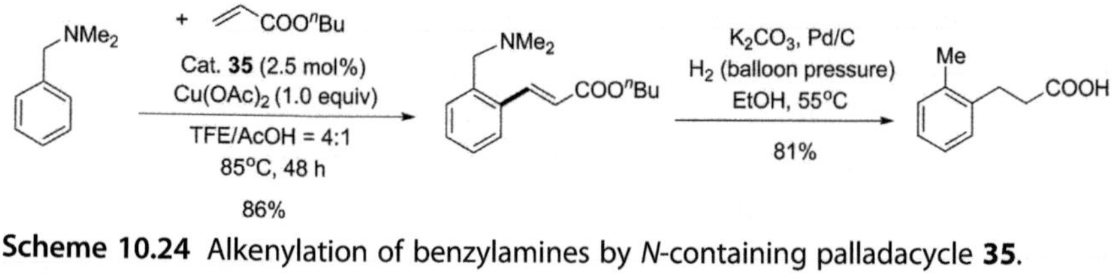

Scheme 10.24 Alkenylation of benzylamines by *N*-containing palladacycle **35**.

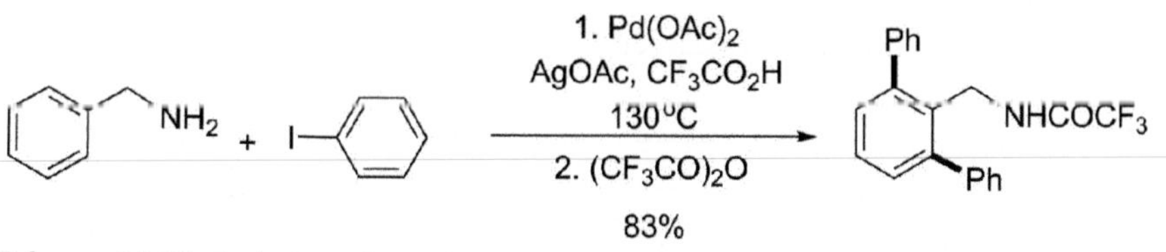

Scheme 10.25 Arylation of unsubstituted benzylamine with aryl iodides.

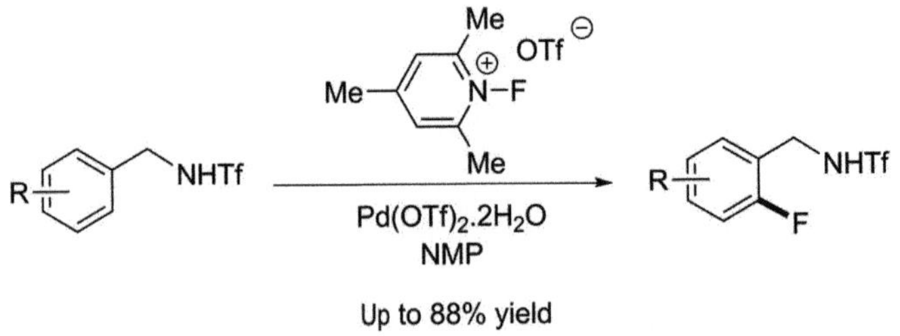

Scheme 10.26 Fluorination of triflamide-protected benzylamines.

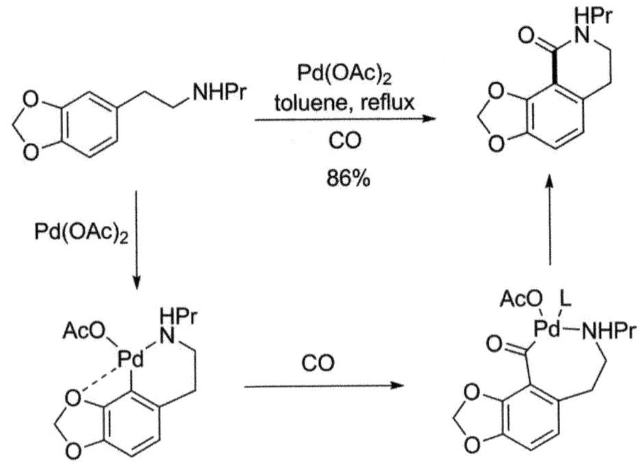

Scheme 10.27 Aromatic carbonylation of *N*-alkylphenylethylamines via six- and seven-membered palladacycles.

electron-rich benzylamines, whereas they were moderate with electron-poor substrates. Generally, for unsubstituted or 4-substituted benzylamines, 2,6-diarylation was observed. *N*-methylbenzylamines were also arylated efficiently under optimized reaction conditions. *Ortho*-fluorination of triflamide-protected benzylamine was reported by Yu et al. (Scheme 10.26).[66] The fluorinating reagent, *N*-fluoro-2,4,6-trimethylpyridinium triflate and 0.5 equiv. of NMP was crucial for this fluorination reaction. The authors have shown the utility of this process by displacing the triflamide directing group with a range of heteroatom and carbon nucleophiles.

The *N*-alkyl-ω-arylalkylamines were carbonylated to five- or six-membered benzolactam, which proceeded through six- and seven-membered palladacycle complexes **36** and **37** (Scheme 10.27).[67]

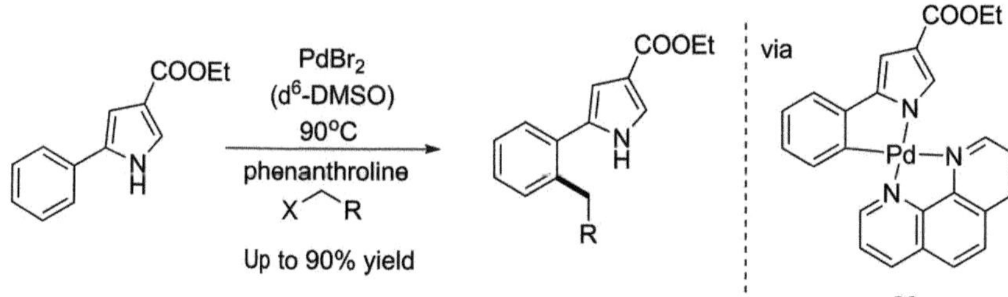

38

Scheme 10.28 2-Phenylpyrrole-based palladacycle for *ortho*-alkylation of arenes.

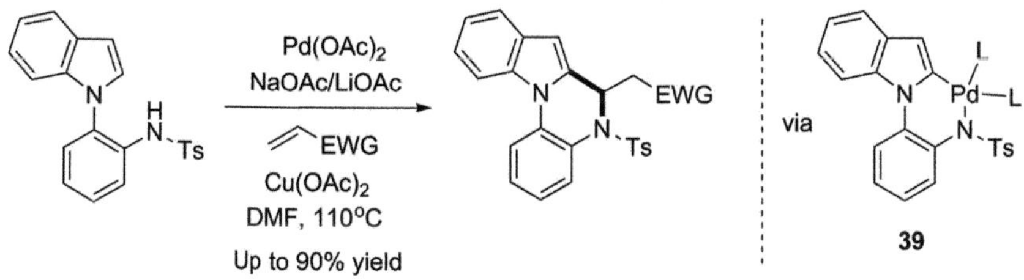

39

Scheme 10.29 Six-membered palladacycle during the synthesis of indolo[1,2-*a*] quinoxaline.

Interestingly, this catalytic system was phosphine-free and was carried out using one atmosphere of CO gas containing air. The reaction was proposed to start with *ortho*-palladation of *N*-alkyl-ω-arylalkylamine, and was followed by insertion of CO into the six-membered palladacycle **36**, leading to formation of the seven-membered acyl palladacycle **37**. The nucleophilic attack of amino group on the carbonyl group resulted in amide bond formation to generate benzolactam.

Recently, a phenanthroline stabilized palladacycle of 2-phenylpyrrole **38** has been synthesized in excellent yield and was structurally characterized (Scheme 10.28).[68] This palladacycle is likely intermediate for the methylation, alkylation, and benzylation of 2-phenylpyrrole. This report described the first pyrrole directed palladium-catalyzed $C(sp^2)$—H functionalization. Most importantly, this method tolerated various important functional groups, and the reaction was performed at mild reaction conditions.

The synthesis of indolo[1,2-*a*]quinoxaline derivatives through the six-membered palladacycle **39** has been described (Scheme 10.29).[69] Palladacycle **39** efficiently reacted with electron-deficient alkenes through the Fujiwara–Moritani process to generate the 2-alkenylated indole derivative, which by subsequent intramolecular cyclization via C—N bond formation produced indolo[1,2-*a*]quinoxaline derivatives. This reaction

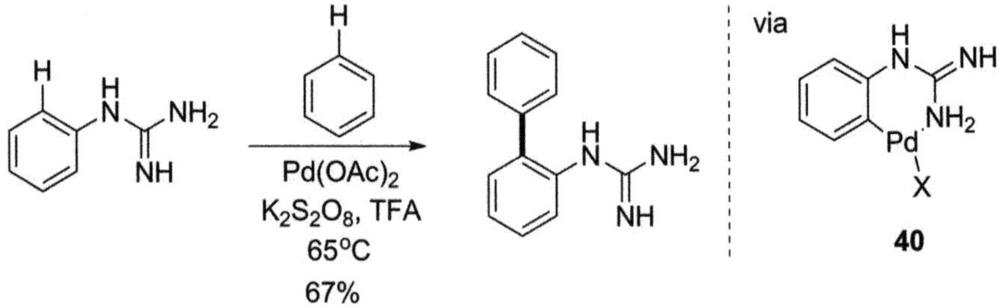

Scheme 10.30 Guanidine-based palladacycle during oxidative C—H arylation.

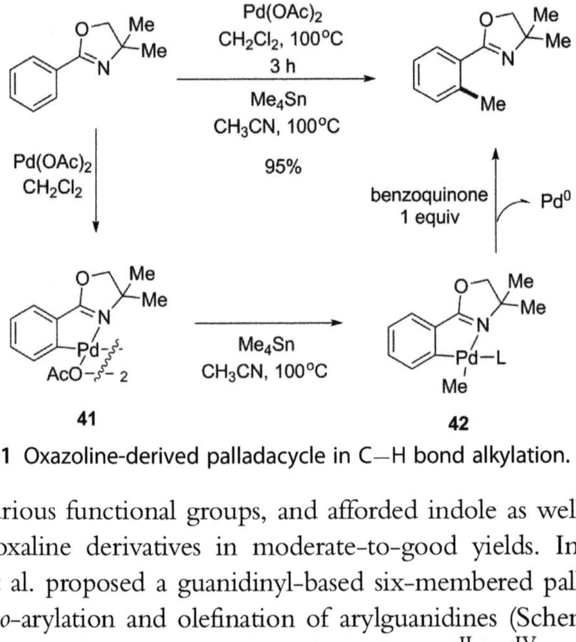

Scheme 10.31 Oxazoline-derived palladacycle in C—H bond alkylation.

tolerated various functional groups, and afforded indole as well as pyrrole-fused quinoxaline derivatives in moderate-to-good yields. In a different note, Yu et al. proposed a guanidinyl-based six-membered palladacycle **40** for the *ortho*-arylation and olefination of arylguanidines (Scheme 10.30).[70] The arylation reaction was proposed through a Pd^{II}/Pd^{IV} oxidative path, whereas the olefination reaction seems to follow a Pd^{0}/Pd^{II} pathway.

10.3.4 Azole and Oxime Ether-Based Palladacycle

Azoles and oxime-ethers are important nitrogen-containing functional groups that constitute part of a palladacycle during the directed C—H functionalization. Benzoxazole-directed C—H bond arylation with aryl iodides was described under oxidative conditions and proposed through a five-membered palladacycle.[71,72] This reaction was highly regioselective for *ortho*-arylation and could be performed with low catalyst loading. Yu et al. reported the first method for 2-oxazoline directed sp^2 C—H bond alkylation using organotin reagents (Scheme 10.31).[73] The reaction proceeded via a dimeric palladacycle complex **41**, which was independently synthesized to prove its authenticity.[74] The use of external oxidant, benzoquinone, made this alkylation reaction catalytic by promoting the reductive elimination step. Further, microwave irradiation to the reaction mixture drastically reduced the reaction time. In this alkylation reaction, the batch-wise addition of organotin reagent, to minimize undesired homo-coupling product, was the major limitation. Later, the same group reported the alkylation of sp^2 and sp^3 C—H bonds with alkylboronic acids via a similar palladacycle intermediate.[36] Another method for the C—H bond alkylation through a palladacycle has been described employing potassium organotrifluoroborates as the alkylating reagent and MnF_3 as an oxidizing agent under mild acidic conditions.[37]

In 2005, Yu et al. described the Pd-catalyzed oxazoline-directed acetoxylation of unactivated methyl groups using peroxyester MeCOOO*t*-Bu as oxidant and Ac_2O as a promoter.[75] The reaction proceeded through the formation of trinuclear palladacycle **43**, which was conveniently synthesized at room temperature from the reaction of oxazoline with $Pd(OAc)_2$ (Scheme 10.32). Among the two isomers (*syn* and *anti*) obtained, the *anti* isomer was structurally characterized. Palladacycle **43** reacted with five equivalents of TBHP followed by reduction with NH_2NH_2 afforded the hydroxylated product, whereas the same reaction in the presence of Ac_2O in place of NH_2NH_2 produced the acetate product. The catalytic diastereoselective acetoxylation of methyl groups was also achieved by this method using lauroyl or benzoyl peroxide as oxidant. The iodination of methyl groups of alkyl oxazoline has been described through a similar trinuclear palladacycle intermediate.[76] In this case, palladacycle **43** reacted with I_2 to afford iodinated product and PdI_2. The PdI_2 was converted to $Pd(OAc)_2$ by treatment with $PhI(OAc)_2$ in the presence of I_2, which was the active catalyst for this reaction. Sanford et al. shown oxazoline-directed sp^2 C—H halogenation through a six-membered palladacycle during their study of the Pd-catalyzed halogenation of other arenes.[46]

Further, Yu et al. showed the oxazoline-directed iodination of remote arene C—H bonds located four, five, or six bonds away from the directing group (Scheme 10.33).[77] These reactions were proceeded by employing $Pd(OAc)_2$ as a catalyst and iodoacetate (IOAc) as the oxidant, wherein reactions were proposed through six- or higher-membered palladacycles. An electrophilic palladation pathway has been proposed for the initial palladation of the C—H bond.

Oxime ether containing palladacycles by the stoichiometric C—H activation at the Pd(II) center have been well predicted.[78,79] Only

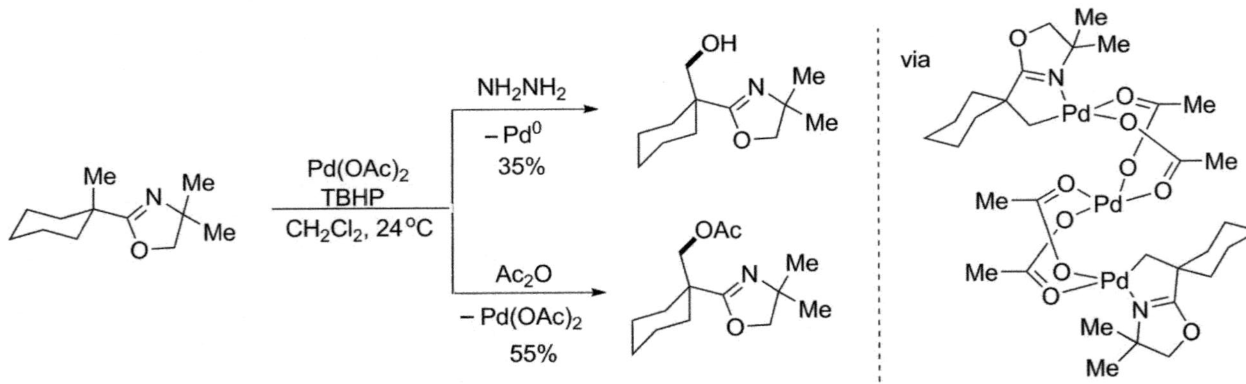

43

Scheme 10.32 Oxazoline-derived trinuclear palladacycle in C—H bond acetoxylation.

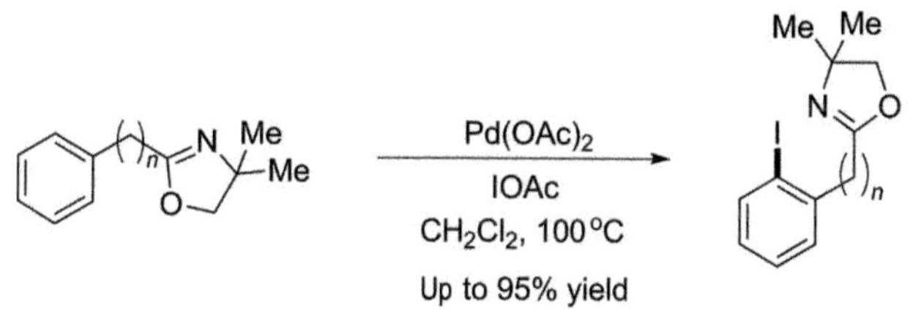

Scheme 10.33 Oxazoline-directed iodination through six- or higher membered palladacycle.

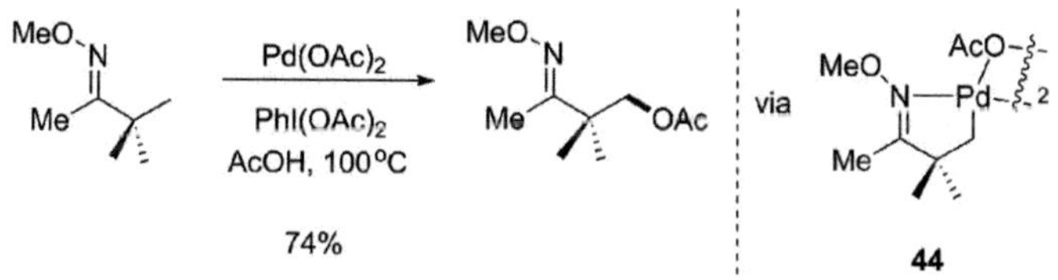

Scheme 10.34 Oxime ether based palladacycle in sp^3 C—H bond acetoxylation.

recently, the catalytic β-functionalization of oxime ether through the palladacycle has been reported. From the synthetic prospective, β-functionalization of oxime ether is important, as by this method the indirect functionalization of ketone is possible. Sanford et al. have shown that the sp^3 C—H bond in alkyl oxime ether can be catalytically oxygenated through palladacycle **44** with stoichiometric $PhI(OAc)_2$ as an oxidant (Scheme 10.34).[80] This reaction was highly selective for the oxidation of the primary carbon center rather than the secondary. Further, high reactivity of the β (in presence of α or γ) C—H was observed, which was due to the energetically favored five-membered palladacycle. Oxygenation of arene oxime ether has also been described (Scheme 10.35).[81] Interestingly, this reaction proceeded efficiently by the use of peroxide-based oxidants, such as oxone, $K_2S_2O_8$, etc. Oxone in acetic acid generally afforded the acetoxylated product, whereas the same reaction in methanol produced *ortho*-methoxylated arenes. Very recently, Jiao et al. showed the PPh_3 or DEAD ligand promoted Pd-catalyzed hydroxylation of arenes, which was directed by the oxime ether (Scheme 10.36).[82] This reaction was shown under neutral conditions and by employing the peroxide as the terminal oxidant. Further, with this reaction condition electron-deficient substrates were tolerated, and the reaction was highly selective for mono-hydroxylation.

Che et al. reported on the amidation of sp^2 and sp^3 C—H bonds, which proceeded through the initial formation of oxime ether containing

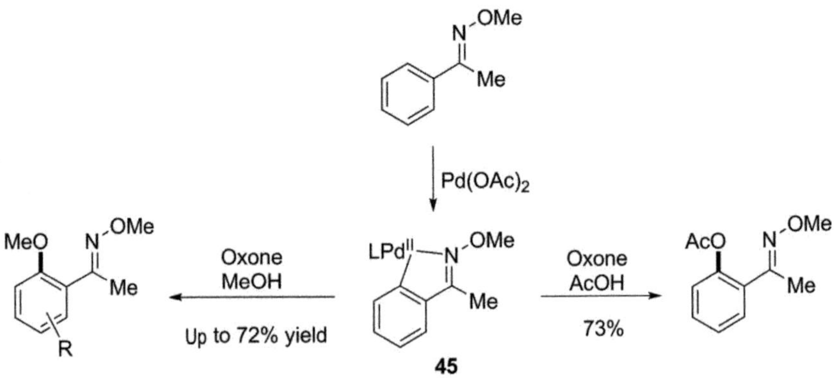

Scheme 10.35 Oxime ether-based palladacycle in sp^2 C–H acetoxylation using oxone as oxidant.

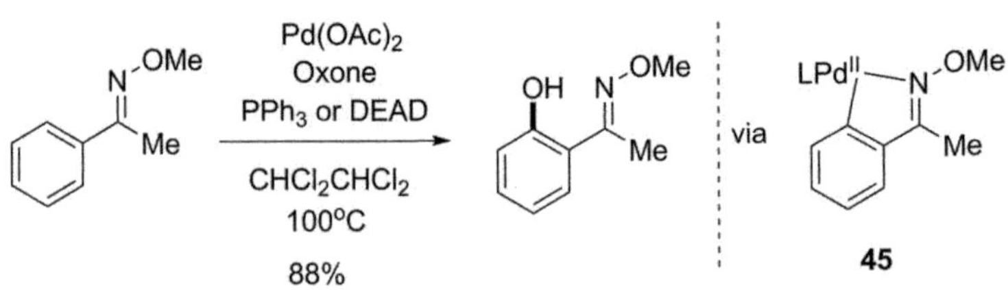

Scheme 10.36 Acetoxylation of arenes via oxime ether-based palladacycle under neutral conditions.

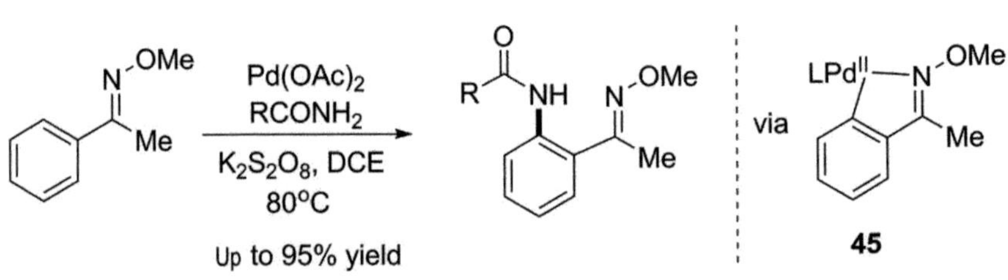

Scheme 10.37 Amidation of unactivated C–H bond through palladacycle **45**.

palladacycle **45** (Scheme 10.37).[83] An intermediacy of nitrene insertion to the Pd–C bond has been proposed after palladacycle formation. A variety of 1° amides, such as carbamates, acetamides, and sulfonamides were effectively employed for this Pd-catalyzed amidation reaction. The reaction was accomplished with very high regio- and chemoselectivity.

Cheng et al. described the Pd-catalyzed synthesis of 9-fluorenone from aromatic aldoxime ether and aryl iodides.[84] Benzaldehyde oxime ether reacted with $Pd(OAc)_2$ in CF_3COOH to generate a five-membered

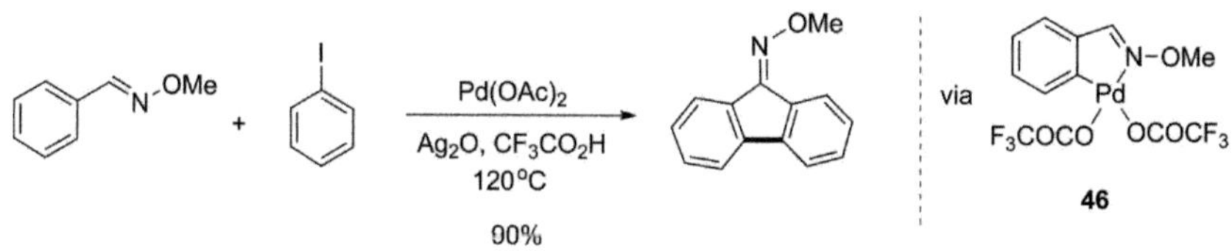

Scheme 10.38 Synthesis of fluorenone through the palladacycle of aromatic aldoxime ether.

palladacycle **46** (Scheme 10.38). This complex was fully characterized and its structure was elucidated by X-ray crystallography. The palladacycle **46** reacted with aryl iodides in the presence of Ag_2O as terminal oxidant to afford the *ortho*-arylated product, which upon oxidative Heck cyclization led to the formation of 9-fluorenone. This catalytic system was compatible with important functional groups, such as nitro, chloro, and fluoro substituents, and afforded the good yields of fluorenone derivatives. However, the reaction is very sensitive to steric effect and hence, the arylation and cyclization occurred at the less steric site of the substrates.

10.3.5 Amide-Based Palladacycle

Palladium-catalyzed ortho C—H functionalization of aryl and alkyl amides has been well demonstrated in last two decades. For example, *ortho*-arylation of *N*-substituted and unsubstituted benzamides with various aryl electrophiles were reported.[85–87] Similarly, arylation of benzamides with simple arenes in the presence of an external oxidant via double C—H activation was described.[88] The sp^3 C—H bond arylation of alkyl amides were also reported with various aryl coupling partners.[89–92] In addition, the alkylation of sp^3 and sp^2 C—H bonds of simple amides with various alkyl iodides was demonstrated.[93] All these functionalizations of amides were assumed to proceed through the five-membered *N*-coordinated ortho-palladacycle complexes, though the isolation and demonstration of amido-based palladacycle as intermediate has not been performed.

In 2011, Wang et al. reported the synthesis of phenanthridinones via palladium-catalyzed dual C—H activation, and the reaction was proposed to proceed through two different palladacycle intermediates (Scheme 10.39).[94] Initially, the arylation of *N*-methoxybenzamide with aryl iodide occurred through a palladacycle **47** type intermediate involving *N*-methoxybenzamide. The palladacycle intermediate reacted with PhI and Ag_2O in acetic acid to afford the phenanthridinone through C—N bond

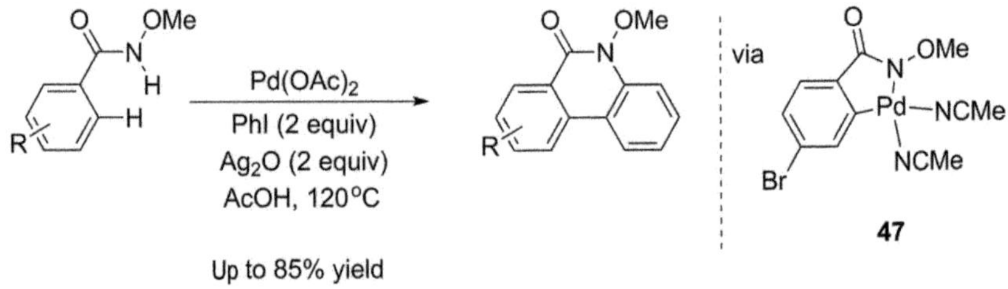

Scheme 10.39 Formation of phenanthridinones via palladacycle intermediate **47**.

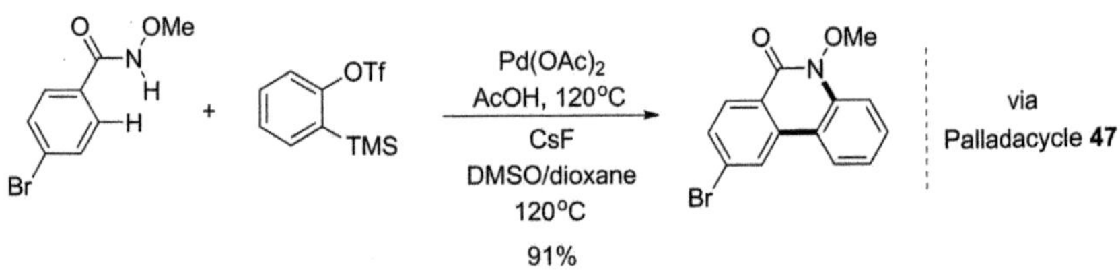

Scheme 10.40 Synthesis of phenanthridinone via palladacycle **47**.

formation involving a seven-membered palladacycle. The palladacycle **47** was independently synthesized from the reaction of benzamide with $Pd(OAc)_2$ in acetic acid, and was structurally characterized by growing the crystals in acetonitrile. Cheng and Karthikeyan also reported the formation of phenanthridinone via a similar five-membered palladacycle from the reaction of *N*-methoxybenzamide and simple arenes in the presence of $K_2S_2O_8$ as terminal oxidant in acidic conditions.[95] In both the methods, non-MeO-substituted benzamides did not afford satisfactory yields of the phenanthridinone product.

Palladium-catalyzed coupling of *N*-methoxyamides with strained alkynes and cyclooctynes through C—H/N—H activation, that proceeded via palladacycle **47** intermediate has been described by Xu et al. (Scheme 10.40).[96] A variety of phenanthridinones and isoquinolones were synthesized by this method. The compatibility of C—H activation and the reactive aryne intermediate was very crucial for this reaction. Palladacycle **47** was also synthesized and reacted with benzyne precursor under standard reaction conditions to obtain corresponding phenanthridinone, which suggested the intermediacy of **47** during this reaction.

In 2011, Wang et al. reported the synthesis isoindolinones via palladium-catalyzed C—H bond activation of *N*-methoxybenzamides, which was proposed through a five-membered (**48**) and a seven-membered palladacycle intermediate (Scheme 10.41).[97] This reaction begins with the *ortho* sp^2 C—H alkenylation of *N*-methoxybenzamide

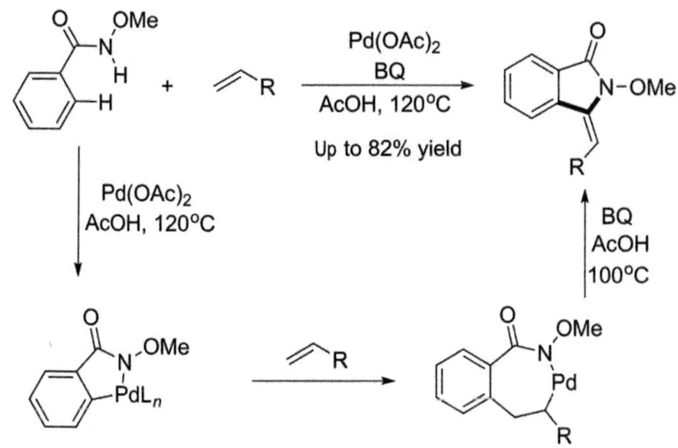

48

Scheme 10.41 Synthesis of isoindolinones through the palladacycle intermediates.

derivatives with different activated alkenes such as acrylate esters, followed by oxidative intramolecular amidation leading to the formation of isoindolinones. Overall, this reaction involved the cleavage of three C—H bonds and one N—H bond, as well as formation of one C—C and one C—N bond. In addition to acrylate esters, this reaction was extended to other olefins such as acrylamides and styrenes, leading to the isoindolinones in moderate-to-good yields. Notably, the isoindolinone products were exclusively obtained as *E*-isomers rather than *Z*-isomers.

In addition to the involvement of amide-based palladacycles in the C—C bond forming reaction, they were also well demonstrated as an intermediate in the oxidative C—O bond forming reactions through C—H bond activation. Wang et al. reported the alkoxylation of *N*-methoxybenzamide via direct sp^2 C—H bond activation (Scheme 10.42).[98] In this reaction the five-membered palladacycle **48**, formed by the *ortho* C—H activation of benzamide, undergoes oxidative functionalization in the presence of $K_2S_2O_8$ and ROH, leading to the formation of *ortho*-alkoxylated amides in moderate-to-good yields.

Similarly, Sahoo et al. showed the use of sulfoximines as a directing group for chemo and regioselective *ortho*-acetoxylation of arenes using $K_2S_2O_8$ in the solvent AcOH/$CHCl_3$.[99] This reaction was proposed to occur via five-membered palladacycle **49** (Fig. 10.4). During the acetoxylation, the stereointegrity of chiral sulfoximines was preserved. Interestingly, after acetoxylation the sulfoximine group can be easily deprotected, and can be used for further functionalization. Lu and Zhou reported

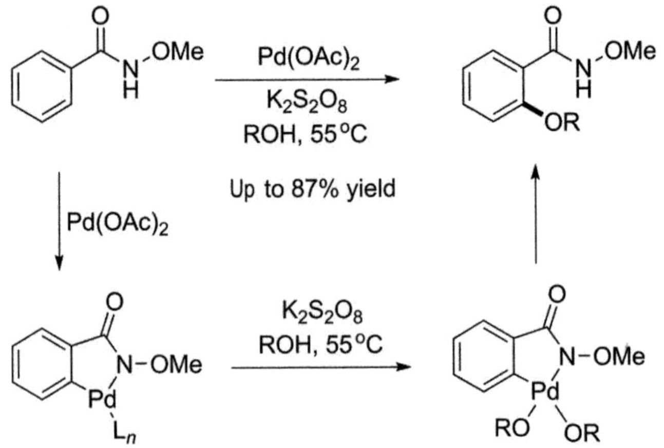

48

Scheme 10.42 Alkoxylation of benzamide through palladacycle intermediates.

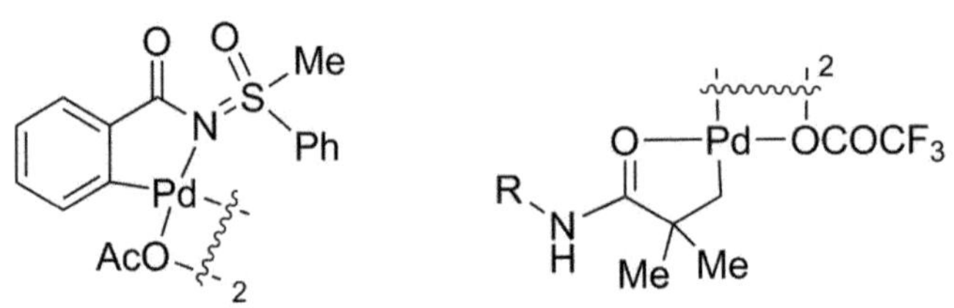

49 50

Figure 10.4 Palladacycles proposed in C—H bond oxygenation of amides.

β-acyloxylation of simple amide via sp^3 C—H bond activation through palladacycle **50** (Fig. 10.4).[100] The acetoxylation was conveniently performed using inorganic oxidant $K_2S_2O_8$ in the solvent CF_3COOH.

10.3.6 Ketone and Ester-Based Palladacycles

Ketone-derived palladacycles are also well predicted as intermediates during the sp^2 C—H bond functionalization of aryl ketone. In 2010, Cheng et al. reported the arylation of alkyl aryl ketones with aryl iodides, as well as the synthesis of phenanthrone from *sec*-alkyl aryl ketones, and both of these reactions proceeded through the palladacycle intermediates (**51a** and **51b**; Scheme 10.43).[101] Interestingly, when the alkyl group on alkyl aryl ketone is primary, only *ortho*-arylation products were obtained; whereas for secondary alkyl group (i.e., isopropyl or cyclic alkyl) the arylation reaction led to the enolate cyclization and formation of phenanthrone derivatives. This

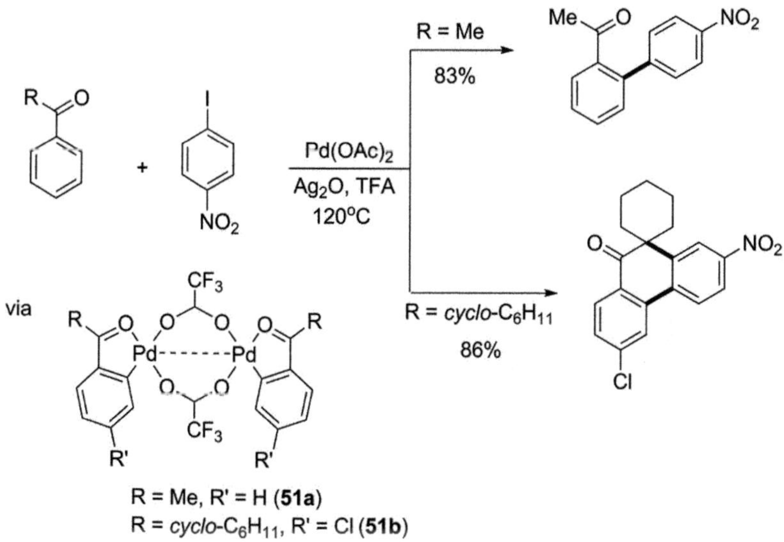

Scheme 10.43 Ketone-derived palladacycle in arylation and enolate cyclization of alkyl aryl ketones.

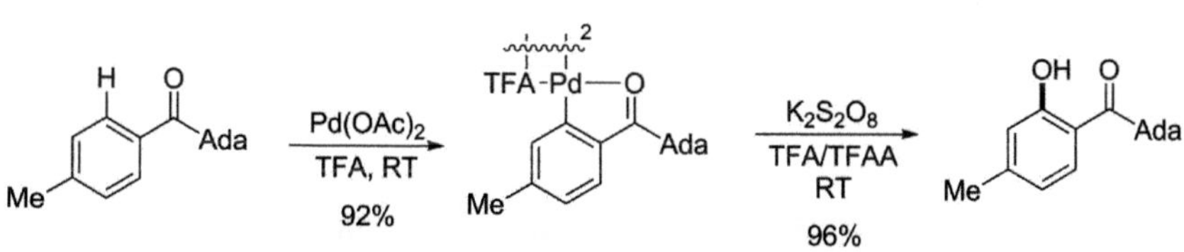

52

Scheme 10.44 Ketone-based palladacycle in oxygenation of adamantyl-ketones.

selective reactivity of the substrate was mainly due to the comparative stability of enolates during the catalytic reaction. The binuclear palladacycles **51a** and **51b** were independently isolated and the structure of **51b** was determined by single crystal X-ray diffraction study.

The C—H bond oxygenation of aromatic ketones, benzoates, acetanilides, etc., using palladium(II) acetate catalyst and $K_2S_2O_8$ as oxidant under strong acidic conditions was described by Rao et al. (Scheme 10.44).[102] The C—H bond cleavage occurred by an *ortho*-metalation process through weak coordination of carbonyl oxygen of aryl ketones, benzoates, etc., to the palladium center. The five-membered palladacycle **52** was isolated and characterized by X-ray crystallography. Stoichiometric reaction of palladacycle **52** with $K_2S_2O_8$ in the presence of TFA/TFAA produced the *ortho*-hydroxylated product, which confirmed the intermediacy of **52** during the

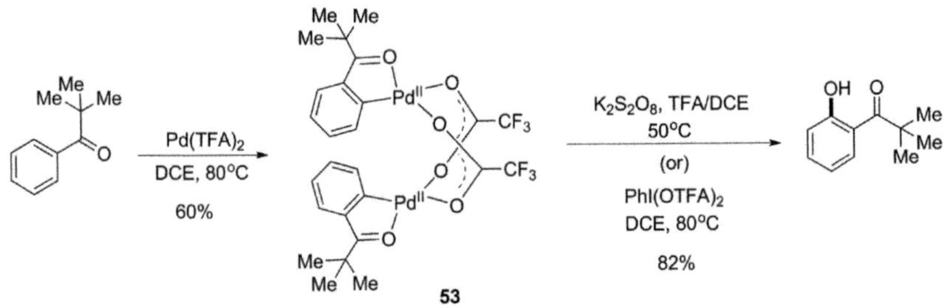

Scheme 10.45 Ketone-based palladacycle in oxygenation under acidic and neutral conditions.

hydroxylation reaction of aromatic ketones. The use of solvent TFA/TFAA was very crucial for this hydroxylation reaction. Broad substrate scope, excellent reactivity, and high yields were reported for this C—H oxygenation reaction. Around the same time, Dong et al. also reported on the hydroxylation of aromatic ketones via palladacycle intermediate **53** (Scheme 10.45).[103] The palladacycle **53** was isolated and structurally characterized. Interestingly, the authors have identified two different conditions for the oxygenation reaction: (1) neutral conditions: $PhI(OTFA)_2$, DEC, 80°C; and (2) acidic conditions: $K_2S_2O_8$, TFA/DCE, 50°C. With this methodology, they have provided an effective approach to access various *ortho*-acylphenol compounds from aryl ketones.

Liu et al. described the ortho C—H amidation of aromatic ketones with both sulfonamides and amides that progressed through the key palladacycle intermediates **54** and **55** (Scheme 10.46).[104] Both of these palladacycles were isolated and structurally characterized. The employment of electron-deficient palladium precursor, $Pd(OTf)_2$, was crucial for this amidation reaction. The use of HOTf was also important to tune the electrophilicity of the palladium catalyst. Under the optimized conditions, a variety of aryl ketones were coupled with numerous amides. The electron-rich ketones generally afforded higher yields than the electron-poor ones. Also, aryl alkyl ketones with tertiary alkyl group gave higher yields than the substrates with primary and secondary alkyl groups. This trend is due to the competition of sulfonamide with ketone for the coordination with Pd^{II}, hence an electron-rich ketone performs better because of its good coordinating ability.

10.3.7 Anilide-Based Palladacycle

Synthesis and reactivity of an anilide-based palladacycle has been known since early 1981. Horino and Inoue reported the preparation and

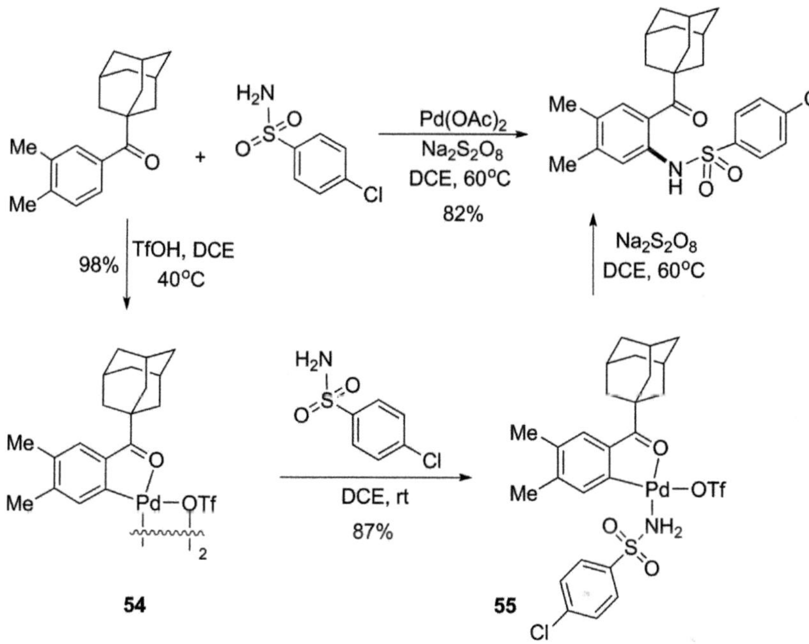

Scheme 10.46 Ketone-based palladacycle during C—H amidation of aromatic ketones.

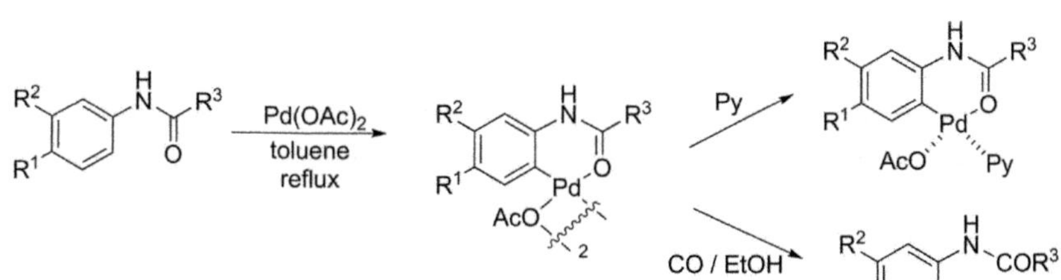

R^1 = H, Me, OMe, Cl, COMe, etc.
R^2 = H, Me, OMe, Cl
R^3 = Me, Et, ^{i}Pr, Ph

Scheme 10.47 Synthesis of anilide-based binuclear palladacycles and their reactivity.

characterization of a variety of binuclear palladacycle complexes (**56**) of acetanilide (Scheme 10.47).[15] The stoichiometric reaction of anilides with palladium(II) acetate in refluxing toluene afforded the greenish yellow palladacycle complexes, which were characterized by IR and NMR spectroscopy. They have also studied the stoichiometric reactivity of those palladacycles with monodentate ligands, olefins, CO/EtOH, and so on.

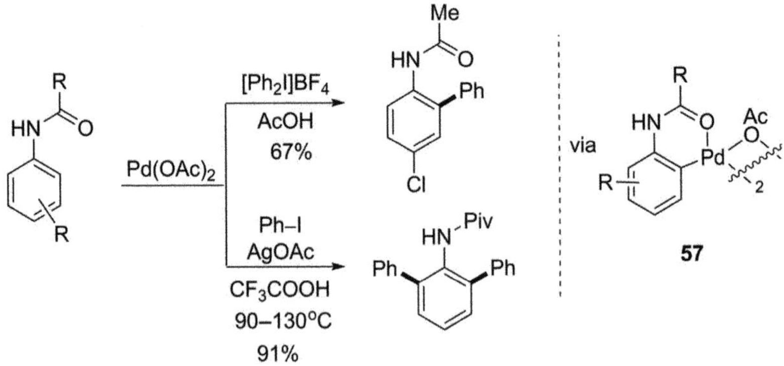

Scheme 10.48 Anilide-based palladacycle during the arylation reaction.

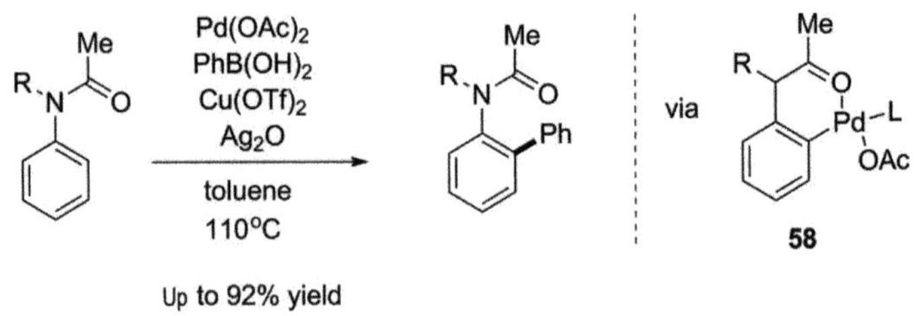

Scheme 10.49 Anilide-based palladacycle during the arylation reaction with aryl boronic acids.

In recent years, catalytic C—H bond functionalization of anilides that conquered through the six-membered palladacycle intermediates was reported. Many arylation reactions of anilides were described, wherein a palladacycle intermediate has been assumed or proposed after the preliminary investigations. Sanford et al. reported the *ortho* C—H bond arylation of anilides including other arenes, wherein the iodonium salt was used as an arylating agent under acidic conditions (Scheme 10.48).[31] Daugulis and Zaitsev described the arylation of the anilides with aryl iodides using AgOAc as a terminal oxidant (Scheme 10.48).[105]

Shi et al. reported on the arylation of anilides using various aryl boronic acids as the arylating reagent (Scheme 10.49).[106] The reaction was proposed through the six-membered anilide-based palladacycle **58**. In addition to the $Pd(OAc)_2$ catalyst, the reaction needed $Cu(OAc)_2$ as terminal oxidant and Ag_2O as an additive, and the arylation reactions were efficiently performed in toluene at 120°C. Numerous functional groups were tolerated under the catalytic conditions and afforded the arylated products in moderate-to-good yields.

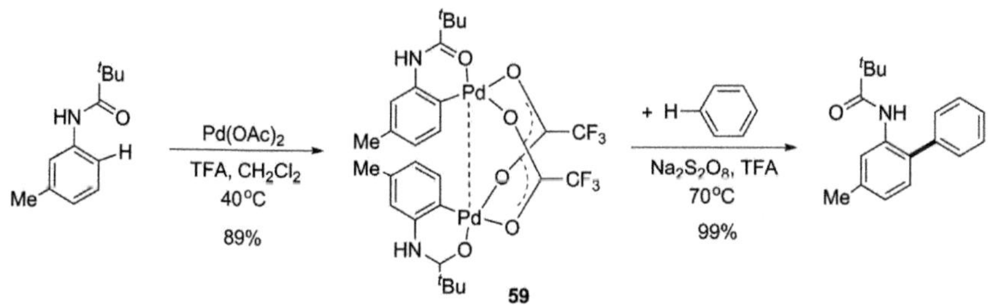

Scheme 10.50 Synthesis of anilide-based palladacycle and oxidative arylation reaction.

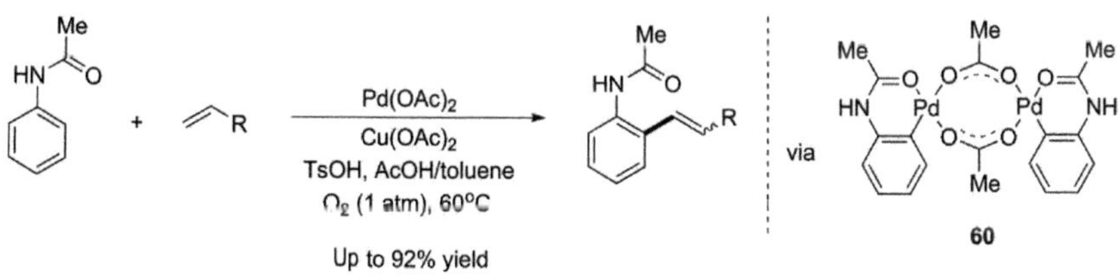

Scheme 10.51 Anilide-based palladacycle in alkenylation reaction.

The synthesis and characterization of a binuclear anilide-based palladacycle **59** was reported by Dong et al. during the *ortho*-arylation of anilides with arenes under oxidative conditions (Scheme 10.50).[107] The palladacycle **59** was conveniently synthesized from the reaction of anilide with $Pd(OAc)_2$ in the presence of TFA in dichloromethane, and was structurally characterized. The reaction of palladacycle **59** with benzene in the presence of TFA and oxidant $Na_2S_2O_8$ exclusively produced the *ortho*-phenylated anilide. This observation supported the intermediacy of palladacycle **59** during the oxidative arylation reaction. By this methodology, the authors have synthesized a number of arylated anilides using simple arenes as cross-coupling partners and $Na_2S_2O_8$ as an oxidant. This methodology was also applied to the intramolecular oxidative couplings of various anilides.

Previously, Horino and Inoue reported on the anilide-based palladacycles and studied their stoichiometric reactivity with olefin to produce the corresponding 2-[(acetylamino)phenyl]olefin.[15] Guo et al. made this reaction catalytic by the use of suitable reaction conditions (Scheme 10.51).[108] Hence, by using $Cu(OAc)_2$ as co-catalyst and molecular oxygen as oxidant, a number of anilides were ortho alkenylated with various alkenes through selective C—H activation. Generally, the electron-rich anilides gave high yields and electron-deficient olefins were efficiently coupled in the reaction.

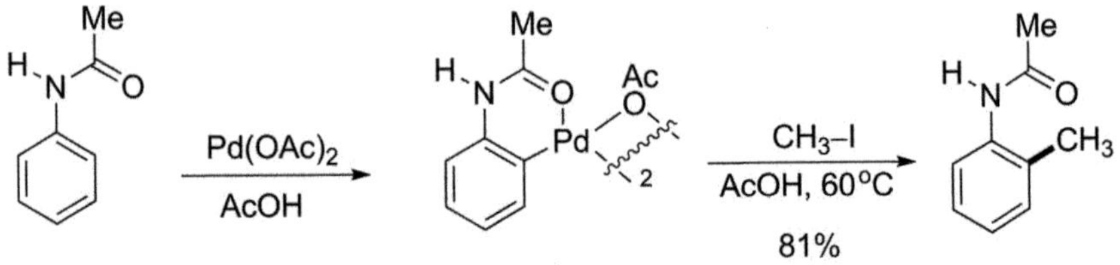

60

Scheme 10.52 Anilide-based palladacycle **60** in alkylation reaction.

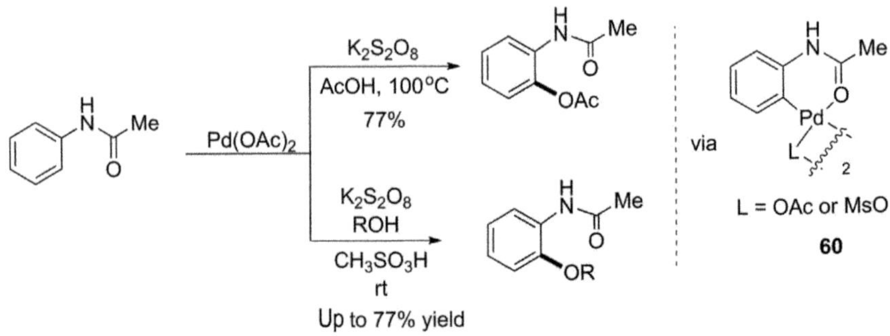

Scheme 10.53 Anilide-based palladacycle in acetoxylation and alkoxylation reactions.

In 1984, Tremont and Rahman reported the stoichiometric *ortho*-alkylation of acetanilides with alkyl halides using $Pd(OAc)_2$ (Scheme 10.52).[109] They have proposed the six-membered acetate-bridged binuclear palladacycle **60** as the initial intermediate for the reaction. This metalation reaction to produce a palladacycle was inhibited by donor solvents such as DMF, DMA, and CH_3CN, and was considered as the slow step of this alkylation reaction. The alkylation step to a palladacycle was proposed to be either electrophilic attack of MeI on Pd—C bond, or oxidative addition of MeI to Pd^{II} to generate Pd^{IV} complex, which was followed by the reductive elimination of the alkylated anilide product.

Palladium-catalyzed acetoxylation/oxygenation of arenes has been long known, but the oxidative oxygenation of anilides has recently been explored. Wang et al. reported *ortho*-acetoxylation of substituted anilides in the presence of palladium(II) acetate, inorganic oxidant $K_2S_2O_8$, and a mixture of AcOH/DCE solvents (Scheme 10.53).[110] This reaction is proposed through the acetate-bridged six-membered cyclo-palladated intermediate **60**, which formed upon chelation-assisted C—H activation. In the presence of $K_2S_2O_8$ and acetic acid, hexacoordinated Pd(IV) intermediate was assumed, followed by reductive elimination to produce the

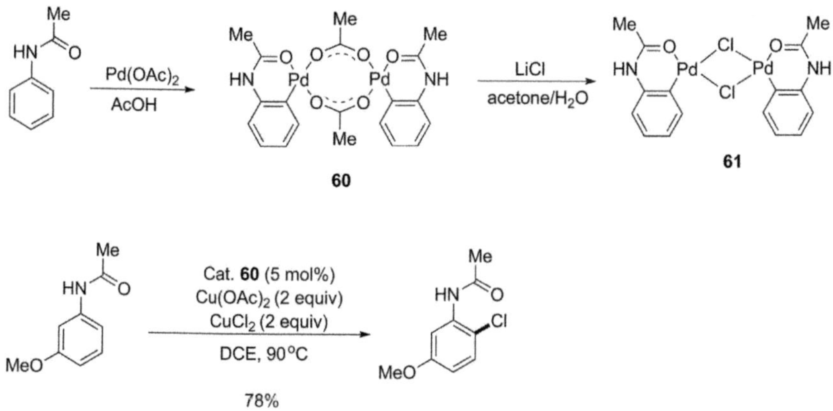

Scheme 10.54 Anilide-based palladacycle **60** in chlorination reaction.

acetoxylated anilides. This protocol was convenient, efficient, and applicable for various substituted anilides to furnish the corresponding *ortho*-acetoxylated anilides. The same protocol was extended to the *ortho*-alkoxylation of anilides using primary and secondary alcohols at ambient temperature (Scheme 10.53).[111] A catalytic amount of methanesulfonic acid was employed as an additive for this reaction, which promoted the reaction and afforded moderate-to-good yields of alkoxylated products.

Ortho C—H bond halogenation of anilides was described by Shi et al. using catalytic amounts of palladium(II) acetate and stoichiometric Cu $(OAc)_2$, $CuCl_2$ in dichloroethane solvent (Scheme 10.54).[112] This reaction proceeded through the binuclear palladacycle intermediate **60**. Interestingly, the palladacycle **60** was independently synthesized and shown as a catalyst precursor for the *ortho*-chlorination of anilides. It was observed that the pivalyl and acetyl groups can be used as an efficient directing group, whereas anilides with formyl, benzoyl, tosyl, and trifluoroacetyl were ineffective.

10.3.8 Carboxylic Acid-Based Palladacycle

The carboxylic acid-based five-membered oxapalladacycle **62** has been reported as intermediate during C—H functionalization of many carboxylic acid derivatives.[113] For example, in 2007, Daugulis et al. reported two methods for direct *ortho*-arylation of benzoic acids through oxapalladacycle intermediate **62** (Scheme 10.55).[114] The first method described the use of catalytic $Pd(OAc)_2$, stoichiometric AgOAc, and an aryl iodide coupling partner. This method tolerated chloride and bromide substitution and proposed

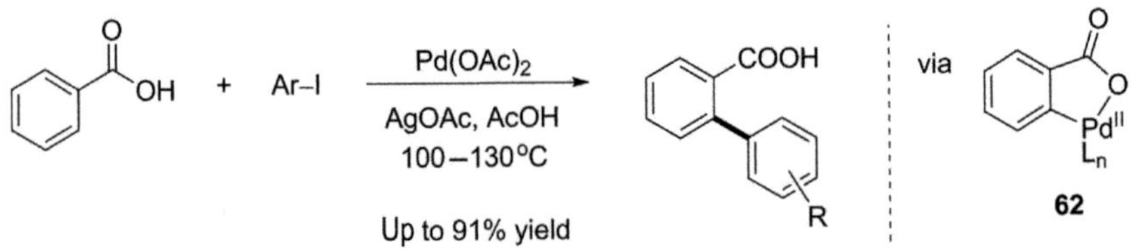

Scheme 10.55 Formation of *ortho*-arylated benzoic acid via oxapalladacycle intermediate.

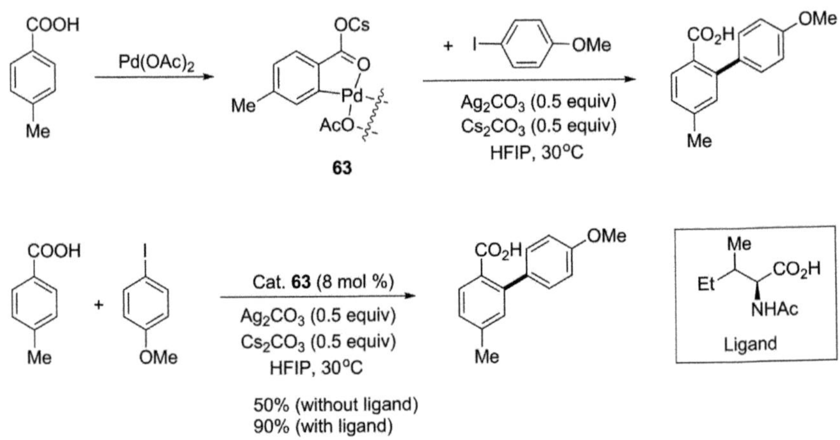

Scheme 10.56 Formation of oxapalladacycle and its reactivity.

proceeding through a Pd^{II}/Pd^{IV} coupling cycle. Both electron-rich and moderately electron-poor benzoic acids were reactive, and aryl iodides of any electronic properties were used. The second method involved the use of catalytic palladium in combination with *n*-butyl-di-1-adamantylphosphine, Cs_2CO_3 base, and an aryl chloride as a coupling partner. Benzoic acids of any electronic properties were reactive in this method.

Recently, the C—H bond arylation of benzoic acid with aryl iodides was reported under mild conditions via the formation of a five-membered oxapalladacycle **63** (Scheme 10.56).[115] This reaction proceeded through Pd-promoted ortho C—H bond cleavage of electron-deficient benzoic acid in the presence of HFIP solvent. A wide range of unactivated benzoic acids were shown to couple with different aryl iodides in good-to-excellent yields. The oxapalladacycle **63** was isolated and fully characterized, and shown to be an active catalyst in this coupling reaction. A monoprotected amino acid ligand proved beneficial for the C—H activation process, and for enhancing the catalyst lifetime in this reaction.

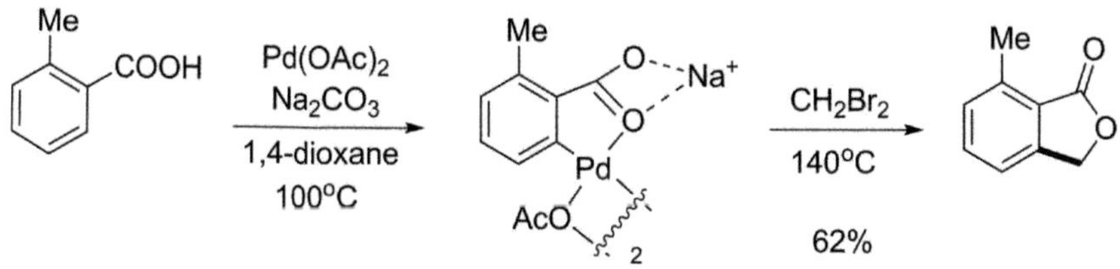

Scheme 10.57 Formation of benzolactone through palladacycle **64**.

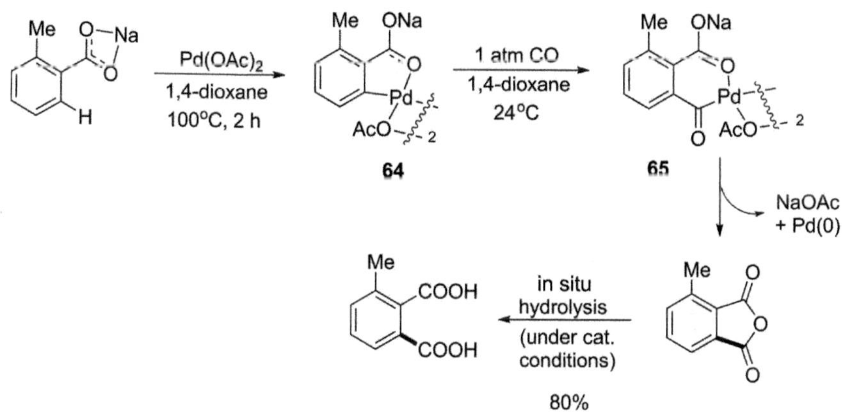

Scheme 10.58 Palladacycle intermediate **65** in carboxylation reaction.

Yu et al. described the methylation and arylation of sp^2 and sp^3 C—H bonds in carboxylic acid derivatives, and assumed the reaction through the five-membered oxapalladacycle intermediate **64**.[116] In this case, methyl boronic acid and phenylboronate were used as coupling partners for methylation and phenylation, respectively. The same group reported *ortho*-alkylation of benzoic acid with alkyl halides through the palladacycle **64**.[117] The authors have independently synthesized and fully characterized the palladacycle **64**, which upon treatment with CH_2Br_2 afforded the alkylative cyclized product benzolactone (Scheme 10.57).

Giri and Yu also showed the intermediacy of palladacycle **64** during the carboxylation reaction of aryl and vinyl carboxylic acids (Scheme 10.58).[118] The palladacycle **64** was converted into a tetrameric palladacycle by treating the same with PPh_3 in dichloromethane, which was structurally characterized by X-ray crystallography (not shown here). The palladacycle **64** was an active intermediate, which

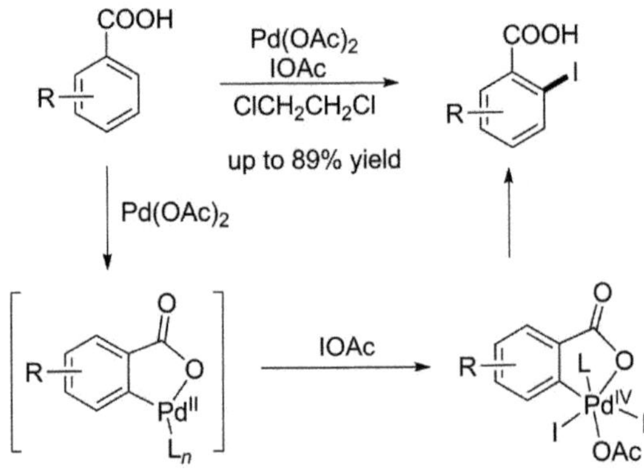

Scheme 10.59 *Ortho*-iodination of carboxylic acid via oxapalladacycle intermediates.

upon reaction with stoichiometric amounts of CO at room temperature afforded phthalic anhydride derivative quantitatively through palladacycle **65**. In the catalytic reaction, the anhydride hydrolyzes in the presence of base and trace amounts of water to produce the carboxylic acid derivatives.

The formation of oxapalladacycle has also been demonstrated during the *ortho*-iodination and *ortho*-bromination of arene carboxylic acids (Scheme 10.59).[119] Initially the five-membered palladacycle **66** is formed in the presence of base, and undergoes oxidative addition with IOAc to form a Pd(IV) species **67**. The reductive elimination of product regenerates the catalyst. The authors observed that the formation of an ion pair (carboxylic acid with IOAc/Bu_4NOAc) was crucial for this reaction. Further, it was unclear whether the Pd^{II} center initiates the C—H cleavage by coordinating with the lone pair of the carbonyl group or simply by the O-anion to form a Pd carboxylate.

The intermediacy of oxapalladacycle was proposed by Miura et al. during the oxidative cross-coupling of benzoic and naphthoic acids with alkenes (Scheme 10.60).[113] In this reaction, the regiochemistry of initial *ortho*-palladation and mode of the oxidative cyclization were sterically controlled, which decided the formation of phthalide or/and isocoumarin products. For example, the reaction of 1-naphthoic acid with styrene selectively produced the benzylidenephthalides, whereas the reaction of 2-naphthoic acid afforded isocoumarin as the sole product.

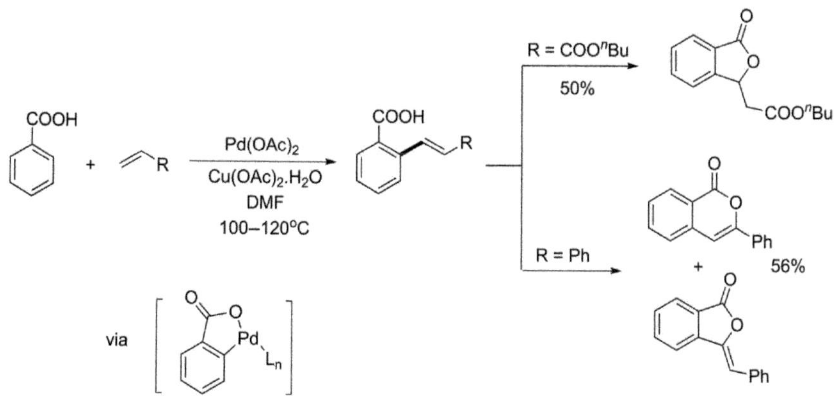

62

Scheme 10.60 Palladacycle during the oxidative coupling of benzoic acid with alkenes.

10.3.9 Sulfur-Based Palladacycle

A sulfur-containing palladacycle was described in thioamide directed C—H bond functionalizations. Murai et al. reported the arylation of thienyl thioamides which was catalyzed by $[Pd(phen)_2](PF_6)_2$ (Scheme 10.61).[120] They have isolated and fully characterized the mono-thioamide palladacycle **68** from the stoichiometric reaction of $[Pd(phen)_2](PF_6)_2$ and thienyl thioamides. Interestingly, the reactivity studies of mono-thioamide palladacycle ruled out the possibility of it being an active catalyst species. Instead, the reaction was proposed to proceed through an active bisthioamide palladacycle **69**, which was detected by ESI-MS analysis.

10.3.10 Bis-Chelated Palladacycle

Palladium-catalyzed regioselective functionalization of challenging C—H bonds has now been performed with the assistance of bidentate chelate auxiliaries, and these reactions were hypothesized to proceed through the bis-chelated palladacycle intermediate **70** (Scheme 10.62). It is believed that the bidentate directing group promotes the C—H bond activation via the formation of a stoichiometric amount of a stable palladacycle, and it confers new properties to the palladated complex. In that regard, the bidentate directing group was considered as an interesting alternative to a monodentate system which can be employed for the development of new catalytic strategies for C—H bond

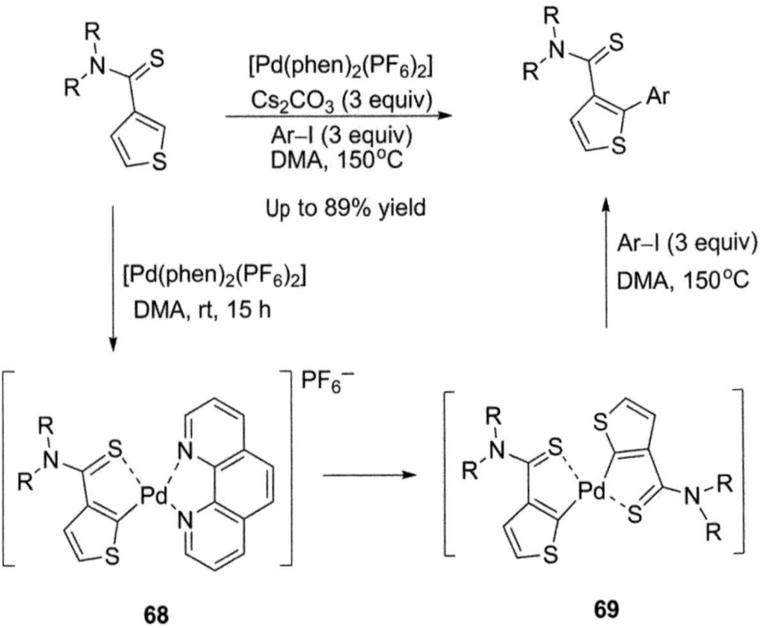

Scheme 10.61 Arylation of thioamide via palladacycle intermediates **68** and **69**.

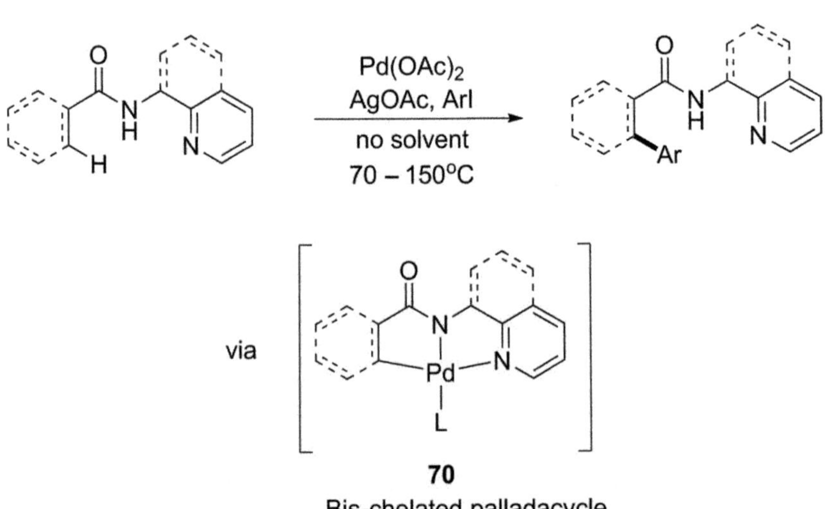

Scheme 10.62 Arylation of amides via a bis-chelated palladacycle intermediate.

functionalization. In 2005, Daugulis et al. reported for the first time bidentate chelate-assistance for the arylation of the sp^3 C—H bond with aryl iodides.[121] The arylation of unactivated sp^3 C—H bonds of 8-aminoquinoline amides and picolinamide derivatives were efficiently carried out. They have demonstrated that the directing group coordinates to the palladium center in a bidentate fashion during the formation of the palladacycle intermediate.

To gain mechanistic insight, Daugulis and Shabashov synthesized bis-chelated palladacycle complex **71** from the stoichiometric reaction of Pd$(OAc)_2$ and amide in pivalonitrile (Scheme 10.63).[122] This complex was obtained as a monomeric species and was stabilized by one molecule of pivalonitrile. The reaction of this complex with Br_2 afforded a stable Pd^{IV} complex **72**, which suggests the intermediacy of a Pd^{IV} species during the catalytic reaction. However, attempts to obtain a similar Pd^{IV} complex by the reaction of palladacycle **71** with aryl iodide were unsuccessful, and direct arylation product was obtained, suggesting that palladacycle **71** is a crucial intermediate during arylation reactions.

Following the initial finding of Daugulis, several reports for the functionalization of unactivated sp^2 and sp^3 C—H bonds of substrates bearing bidentate directing groups have been demonstrated through bis-chelated palladacycle intermediates. The arylation reaction of aliphatic amines,[123] non-natural α-protected amino acid derivatives,[124] primary and secondary sp^3 C—H bonds,[125–128] *β*-methyl C—H of alanine,[129] naphthylamides,[130] *N*-benzoyl α-amino ester derivatives,[131] and other substrates[132] were reported via a bis-chelated palladacycle intermediate using 8-aminoquinolinyl, picolinamide, or pyridine *N*-oxide as the directing group. Some of the proposed bis-chelated palladacycles during the arylation reactions are shown in Fig. 10.5.

Carretero et al. reported the *N*-(2-pyridyl)sulfonyl directed *γ*-arylation of sp^3 C—H bonds of *N*-(2-pyridyl)sulfonamide derivatives, and the reaction was proposed via a bis-chelated palladacycle intermediate (Scheme 10.64).[133] They have isolated a bimetallic palladacycle complex **79** during the arylation reaction, which highlighted the intermediacy of palladacycle during the reaction. The easy removal of the *N*-(2-pyridyl)sulfonyl directing group under reductive conditions (Zn powder at 60°C/NH_4Cl solution) afforded the free primary amine in excellent yields.

In addition to the arylation reaction of sp^2 and sp^3 C—H bonds, bis-chelated palladacycle as an intermediate has been proposed in other C—H functionalizations, such as alkenylation,[134–136] alkynylation,[134,137,138] alkylation,[132,139–142] alkoxylation/acetoxylation,[143–149] and amination[126,150,151] reactions. In most of the cases 8-aminoquinolinyl or picolinyl were reported as the directing group, and reactions were proposed via a Pd^{II}/Pd^{IV} pathway.

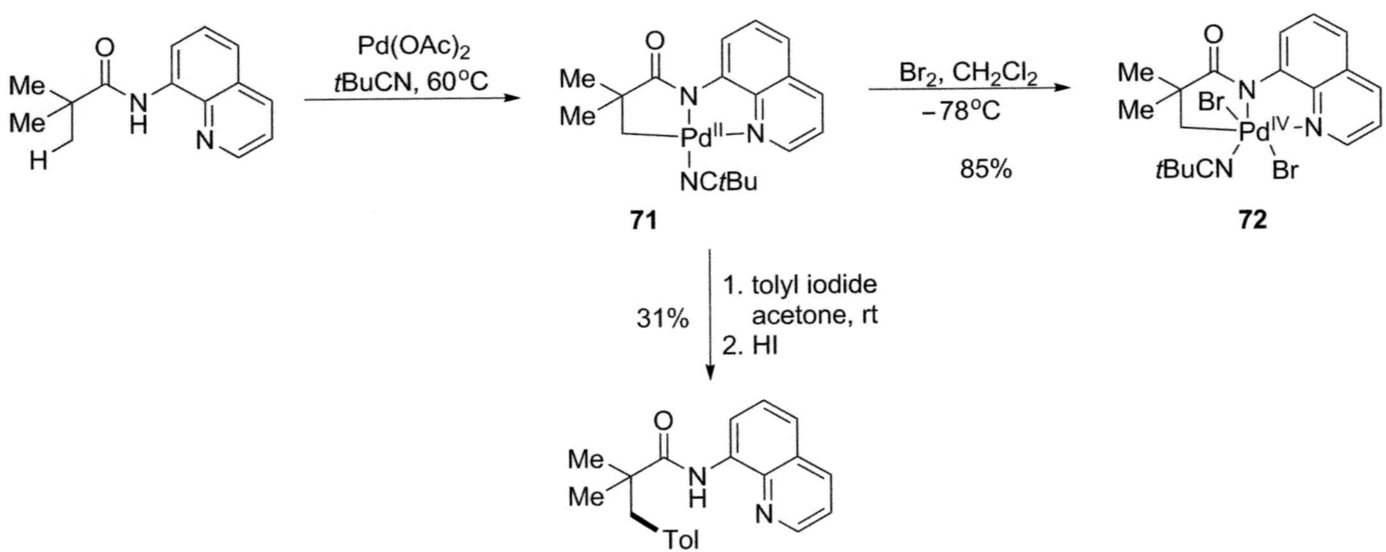

Scheme 10.63 Synthesis and reactivity of bis-chelated palladacycles **71** and **72**.

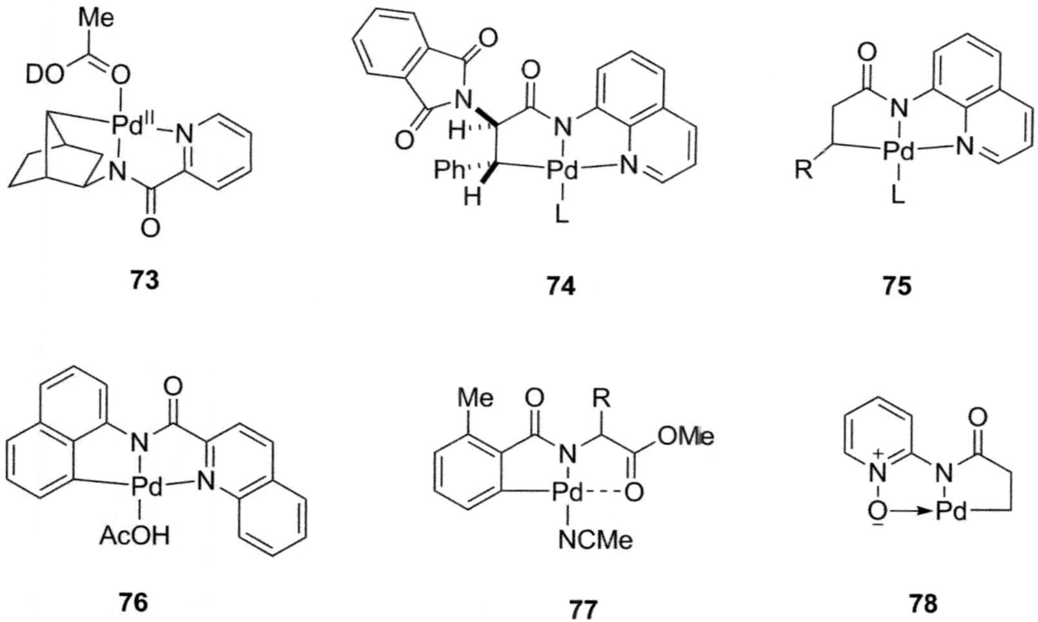

Figure 10.5 Some important bis-chelated palladacycles proposed during C—H functionalization.

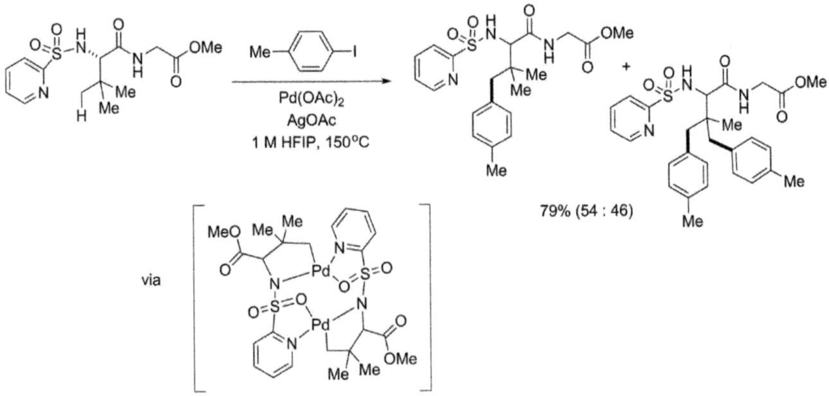

79

Scheme 10.64 Palladacycle intermediate **79** in γ-arylation of *N*-(2-pyridyl) sulfonamide.

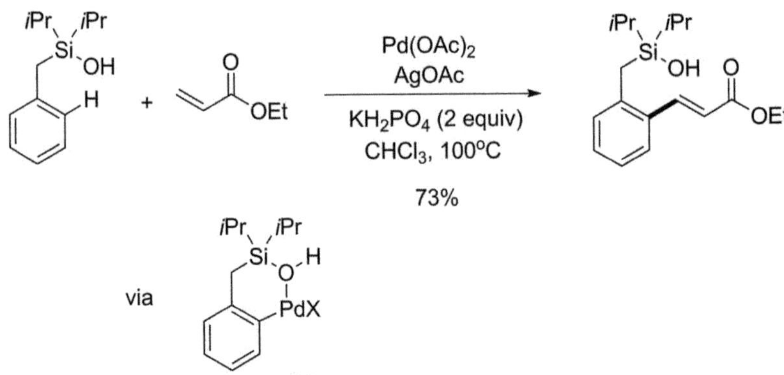

80

Scheme 10.65 Silicon-containing palladacycle in alkenylation reaction.

10.3.11 Other Palladacycles in C—H Functionalizations

Silicon and phosphorous-containing palladacycles are also reported as important intermediates during C—H functionalization reactions. In 2011, Ge and Wang reported palladium-catalyzed olefination of arenes containing silanol as a removable directing group (Scheme 10.65).[152] In this reaction, the C—H bond olefination of benzyldiisopropylsilanol derivatives occurred with the substituted olefins by using $Pd(OAc)_2$ catalyst and silver acetate as an oxidant in the presence of KH_2PO_4. Both the electron-donating and electron-withdrawing substituents on arenes were well tolerated to afford the alkenylated products in moderate-to-good yields. This process was proposed via the cyclopalladation of

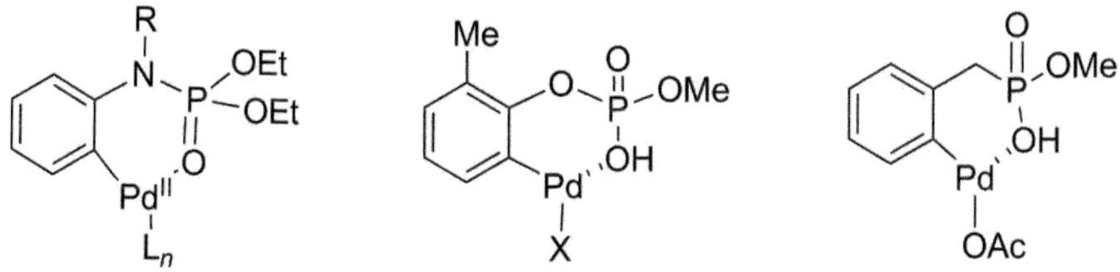

81 **82** **83**

Figure 10.6 Proposed palladacycles containing phosphoramidate, phosphates, and benzylic phosphonic monoester directing groups.

benzyldiisopropylsilanols with Pd(II) directed by a silanol group, which forms a six-membered palladacycle **80** (Scheme 10.65). The olefin coordination to palladacycle followed by 1,2-migratory insertion and subsequent β-hydride elimination afforded the desired product. Importantly, the silanol group can be easily removed from the product by treatment with tetrabutylammonium fluoride at room temperature. Gevorgyan et al. described the palladium-catalyzed silanol directed C—H alkenylation of phenol with activated olefins in the presence of AgOAc as the terminal oxidant.[153] The silanol group from the product can be conveniently removed by reaction with TBAF in THF to afford the alkenylated phenol. Although the authors have not proposed the reaction mechanism, the formation of silanol-containing palladacycle after the initial C—H activation was assumed as the intermediate for this reaction.

Phosphoramidate, phosphates, and benzylic phosphonic monoester groups were also employed as directing groups in the ortho C—H functionalization of arenes. In these cases the reactions were proposed through six-membered palladacycle intermediates (Fig. 10.6). After installing the phosphoramidate directing group, various useful arylation reactions were performed using $Pd(OAc)_2$ catalyst and aryl iodonium salts.[154] A stoichiometric amount of CuO was very crucial for this arylation reaction. Notably, the arylation reactions proceeded efficiently at room temperature and afforded very good-to-excellent yields of *ortho*-arylated phosphoramidate. *Ortho*-arylation of aryl phosphates and aryl hydrogen phosphate also occurred efficiently with diaryliodonium triflate in dichloroethane at 80—110°C.[155] The aryl hydrogen phosphate directing group could be used for the alkenylation of arenes.[156] In this reaction, AgOAc was used as the terminal oxidant to synthesize various alkenylated aryl hydrogen phosphates in excellent yields. Meng and Kim demonstrated the

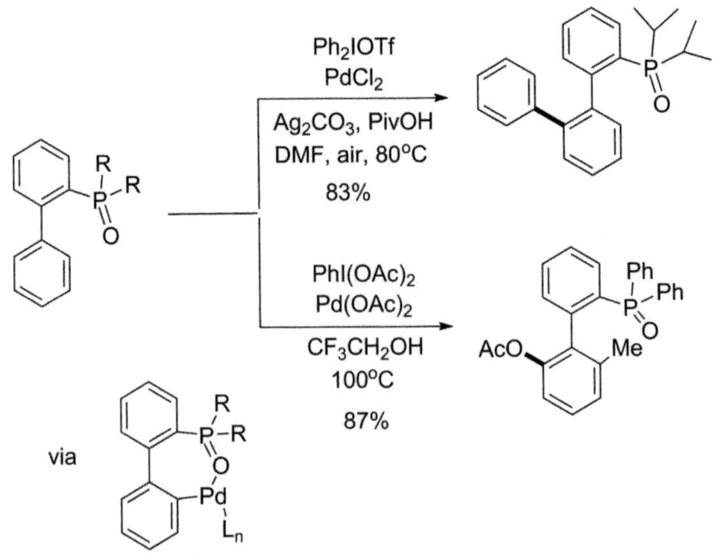

84

Scheme 10.66 Phosphoryl-containing palladacycle in arylation and acetoxylation reactions.

alkenylation of benzylic phosphonic monoester using $Pd(OAc)_2$ catalyst and AgOAc as the oxidant.[157]

Recently, Yang et al. described the $P(O)R_2$ directed arylation and acetoxylation of the sp^2 C—H bond, which proceeded through the phosphoryl-containing seven-membered palladacycle intermediate **84** (Scheme 10.66).[158] Arylation reactions were efficiently carried out with iodonium salt $Ph_2I(OTf)$ in the presence of $PdCl_2$ as catalyst, whereas the acetoxylation was performed with $Pd(OAc)_2$ catalyst using $PhI(OAc)_2$ as the acetoxylating reagent (Scheme 10.66).[159]

Catellani et al. have shown that C—H bond alkylation and alkenylation of arenes in the presence of norbornene proceeded via the formation of multiple palladacycles, such as **85** and **86** (Scheme 10.67).[160] The regioselective functionalization at the 1, 2, and 3 positions of arenes is most likely due to the different reactivity behavior of Pd^0, Pd^{II}, and Pd^{IV} species generated during the catalytic process. The synthesis of tri-substituted arenes by this method was very simple and straightforward, and the reaction was performed under mild reaction conditions. Notably, the substrate selectivity is very precise in this method, even in the presence of similar compounds, such as aromatic and aliphatic iodides or norbornene and olefins.

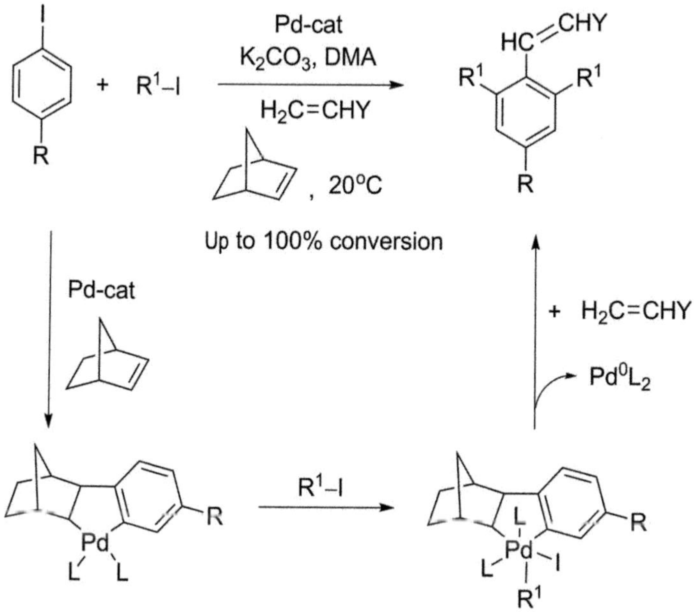

85

86

Scheme 10.67 Norbornene-mediated alkylation/alkenylation of arenes via intermediacy of palladacycles **85** and **86**.

10.4 CONCLUSION

Palladacycle catalysts have found enormous applications in catalytic organic transformations. Various functionalizations of the C—H bond on both directed and nondirected (hetero)arenes catalyzed by palladacycle catalysts have been discussed here. Palladacycle precatalysts have been used for both inter and intramolecular arylation of arenes and hetero arenes. In addition to this, C—H functionalization of directed arenes which proceeded through palladacycle intermediates has been discussed. Among the various directing functional groups, nitrogen-containing functional groups (such as pyridine, pyrimidine, quinoline, amine, azole, and oxime) were the best choices because of their better coordinating ability. These directing groups containing palladacycles have been involved in further functionalizations, such as arylation, alkylation, alkenylation, acylation, ethoxycarbonylation, cyanation, oxygenation, and halogenations. In many cases, palladacycles were isolated and characterized. Palladacycles bearing other groups like ketone and ester, anilide, carboxylic acid, and sulfur, as well as bis-chelated directing groups have been discussed for various functionalizations. Other than these groups, silanol, phosphoramidate, phosphates, and benzylic phosphonic monoester-based palladacycles were highlighted for arylation, alkylation, and acetoxylation reactions. Thus, palladium has been widely explored for many functionalizations using different directing groups that proceeded through palladacycle intermediates. The palladacycle formed in these reactions provides more stability to the system leading to facile C—H functionalization. Further, the well-defined nature of the palladacycle makes mechanistic understanding easier. Given the beneficial features of palladacycles for selective C—H bond functionalization, the development and application of more novel palladacycle precatalysts is expected.

ABBREVIATIONS

Ac	acetyl
acac	acetylacetonate
Ada	adamantyl
BQ	benzoquinoline
Boc	*tert*-butyloxycarbonyl
Val	valine (2-amino-3-methylbutyric acid)
Bu	butyl
CN	cyano
Cy	cyclohexyl
DCE	1,2-dichloroethane
DMA	*N,N*-dimethylacetamide
DEAD	diethyl azodicarboxylate
DMF	*N,N*-dimethylformamide
DMSO	dimethyl sulfoxide
Et	ethyl
HFIP	hexafluoroisopropanol
Me	methyl
NHPI	*N*-hydroxyphthalimide
NXS	*N*-halosuccinimides
NMP	1-methyl-2-pyrrolidone
NFSI	*N*-fluorobis(phenylsulfonyl)imide
Phen	phenanthroline
Piv	pivalate
Pr	propyl
Py	pyridinyl
Tf	trifluoromethanesulfonyl
TFA	trifluoroacetic acid
TFAA	trifluoroacetic anhydride
Ts	tosyl
TMS	trimethylsilyl

Tol	tolyl
TIPS	triisopropylsilyl
TBAF	tetra-*n*-butylammonium fluoride
TFE	trifluoroethanol
TBHP	*tert*-butylhydroperoxide
Dpms	di(2-pyridyl)methanesulfonate

REFERENCES

1. Cope, A. C.; Siekman, R. W. Formation of covalent bonds from platinum or palladium to carbon by direct substitution. *J. Am. Chem. Soc.* **1965,** *87*, 3272–3273.
2. Cope, A. C.; Friedrich, E. C. Electrophilic aromatic substitution reactions by platinum(II) and palladium(II) chlorides on *N,N*-dimethylbenzylamines. *J. Am. Chem. Soc.* **1968,** *90*, 909–913.
3. Hartwell, G. E.; Lawrence, R. V.; Smas, M. J. The formation of palladium(II)- and platinum(II)-carbon bonds by proton abstraction from benzo[*h*]quinoline and 8-methylquinoline. *J. Chem. Soc. Chem. Commun.* **1970**, 912.
4. Constable, A. G.; McDonald, W. S.; Sawkins, L. C.; Shaw, B. L. Palladation of dimethylhydrazones, oximes, and oxime *O*-allyl ethers: crystal structure of $[Pd_3(ON = CPrPh)_6]$. *J. Chem. Soc. Chem. Commun.* **1978,** 1061–1062.
5. Fuchita, Y.; Hiraki, K.; Uchiyama, T. Metallation of aliphatic carbon atoms. Part 1. Synthesis and characterization of the cyclopalladated complexes of 2-neopentylpyridine. *J. Chem. Soc., Dalton Trans.* **1983,** 897–899.
6. Baldwin, J. E.; Jones, R. H.; Najera, C.; Yus, M. Functionalisation of unactivated methyl groups through cyclopalladation reactions. *Tetrahedron* **1985,** *41*, 699–711.
7. Ryabov, A. D. Cyclopalladated complexes in organic synthesis. *Synthesis (Mass)* **1985,** 233–252.
8. Alsters, P. L.; Engel, P. F.; Hogerheide, M. P.; Copijn, M.; Spek, A. L.; van Koten, G. Rigid five- and six-membered *C,N,N'*-bound aryl-, benzyl-, and alkylorganopalladium complexes: sp^2 *vs.* sp^3 carbon-hydrogen activation during cyclopalladation and palladium(IV) intermediates in oxidative addition reactions with dihalogens and alkyl halides. *Organometallics* **1993,** *12*, 1831–1844.
9. Dupont, J.; Consorti, C. S.; Spencer, J. The potential of palladacycles: more than just precatalysts. *Chem. Rev.* **2005,** *105*, 2527–2572.
10. Szabó, K. J. Pincer complexes as catalysts in organic chemistry. *Top. Organomet. Chem.* **2013,** *40*, 203–242.
11. Kasahara, A. σ-Bonded 2-phenylpyridine palladium complex. *Bull. Chem. Soc. Jpn.* **1968,** *41*, 1272.
12. O'Sullivan, R. D.; Parkins, A. W. The synthesis of *N*-heterocycles using *ortho*-metallated primary benzylamine complexes of palladium(II) and platinum(II). *J. Chem. Soc. Chem. Commun.* **1984,** 1165–1166.
13. Girling, I. R.; Widdowson, D. A. Cyclopalladated imines in synthesis: the preparation of unsymmetrical stilbenes and 3-arylisoquinolones. *Tetrahedron Lett.* **1982,** *23*, 1957–1960.
14. Cameron, N. D.; Kilner, M. Six-membered *ortho*-metallated ring systems. *J. Chem. Soc. Chem. Commun.* **1975,** 687–688.
15. Horino, H.; Inoue, N. Ortho vinylation of aromatic amides via cyclopalladation complexes. *J. Org. Chem.* **1981,** *46*, 4416–4422.
16. Dupont, J.; Beydoun, N.; Pfeffer, M. Reactions of cyclopalladated compounds. Part 21. Various examples of sulphur-assisted intramolecular palladation of aryl and alkyl groups. *J. Chem. Soc. Dalton Trans.* **1989,** 1715–1720.

17. Alper, H. *Ortho*-Palladated and *ortho*-platinated complexes of diaryl thioketones. *J. Organomet. Chem.* **1973,** *61*, C62—C64.
18. Herrmann, W. A.; Brossmer, C.; Öfele, K.; Reisinger, C.-P.; Priermeier, T.; Beller, M.; Fischer, H. Palladacycles as structurally defined catalysts for the Heck olefination of chloro- and bromoarenes. *Angew. Chem. Int. Ed. Engl.* **1995,** *34*, 1844—1848.
19. van der Boom, M. E.; Milstein, D. Cyclometalated phosphine-based pincer complexes: mechanistic insight in catalysis, coordination, and bond activation. *Chem. Rev.* **2003,** *103*, 1759—1792.
20. Selander, N.; Szabó, K. Catalysis by palladium pincer complexes. *Chem. Rev.* **2011,** *111*, 2048—2076.
21. Storr, T. E.; Greaney, M. F. Palladium-catalyzed arylation of simple arenes with iodonium salts. *Org. Lett.* **2013,** *15*, 1410—1413.
22. Zhang, G.; Zhao, X.; Yan, Y.; Ding, C. Direct arylation under catalysis of an oxime-derived palladacycle: search for a phosphane-free method. *Eur. J. Org. Chem.* **2012,** 669—672.
23. Khake, S. M.; Soni, V.; Gonnade, R. G.; Punji, B. Design and development of POCN-pincer palladium catalysts for C—H bond arylation of azoles with aryl iodides. *Dalton. Trans.* **2014,** *43*, 16084—16096.
24. Khake, S. M.; Jagtap, R. A.; Dangat, Y. B.; Gonnade, R. G.; Vanka, K.; Punji, B. Mechanistic insights into pincer-ligated palladium-catalyzed arylation of azoles with aryl iodides: evidence of a Pd^{II}—Pd^{IV}—Pd^{II} pathway. *Organometallics* **2016,** *35*, 875—886.
25. Pandey, D. K.; Khake, S. M.; Gonnade, R. G.; Punji, B. Mono- and binuclear palladacycles via regioselective C—H bond activation: syntheses, mechanistic insights and catalytic activity in direct arylation of azoles. *RSC Adv.* **2015,** *5*, 81502—81514.
26. Tan, Y.; Barrios-Landeros, F.; Hartwig, J. F. Mechanistic studies on direct arylation of pyridine *N*-oxide: evidence for cooperative catalysis between two distinct palladium centers. *J. Am. Chem. Soc.* **2012,** *134*, 3683—3686.
27. Campeau, L.-C.; Rousseaux, S.; Fagnou, K. A solution to the 2-pyridyl organometallic cross-coupling problem: regioselective catalytic direct arylation of pyridine *N*-oxides. *J. Am. Chem. Soc.* **2005,** *127*, 18020—18021.
28. Flegeau, E. F.; Popkin, M. E.; Greaney, M. F. Direct arylation of oxazoles at C2. A concise approach to consecutively linked oxazoles. *Org. Lett.* **2008,** *10*, 2717—2720.
29. Pilarski, L. T.; Selander, N.; Böse, D.; Szabó, K. J. Catalytic allylic C—H acetoxylation and benzoyloxylation *via* suggested (η^3-allyl)palladium(IV) intermediates. *Org. Lett.* **2009,** *11*, 5518—5521.
30. Selander, N.; Willy, B.; Szabó, K. J. Selective C—H borylation of alkenes by palladium pincer complex catalyzed oxidative functionalization. *Angew. Chem., Int. Ed.* **2010,** *49*, 4051—4053.
31. Kalyani, D.; Deprez, N. R.; Desai, L. V.; Sanford, M. S. Oxidative C—H activation/C—C bond forming reactions: synthetic scope and mechanistic insights. *J. Am. Chem. Soc.* **2005,** *127*, 7330—7331.
32. Deprez, N. R.; Sanford, M. S. Synthetic and mechanistic studies of Pd-catalyzed C—H arylation with diaryliodonium salts: evidence for a bimetallic high oxidation state Pd intermediate. *J. Am. Chem. Soc.* **2009,** *131*, 11234—11241.
33. Yu, W.-Y.; Sit, W. N.; Zhou, Z.; Chan, A. S. C. Palladium-catalyzed decarboxylative arylation of C—H bonds by aryl acylperoxides. *Org. Lett.* **2009,** *11*, 3174—3177.
34. Hull, K. L.; Lanni, E. L.; Sanford, M. S. Highly regioselective catalytic oxidative coupling reactions: synthetic and mechanistic investigations. *J. Am. Chem. Soc.* **2006,** *128*, 14047—14049.
35. Shabashov, D.; Daugulis, O. Catalytic coupling of C—H and C—I bonds using pyridine as a directing group. *Org. Lett.* **2005,** 7, 3657—3659.

36. Chen, X.; Goodhue, C. E.; Yu, J.-Q. Palladium-catalyzed alkylation of sp^2 and sp^3 C—H bonds with methylboroxine and alkylboronic acids: two distinct C—H activation pathways. *J. Am. Chem. Soc.* **2006,** *128*, 12634—12635.
37. Neufeldt, S. R.; Seigerman, C. K.; Sanford, M. S. Mild palladium-catalyzed C—H alkylation using potassium alkyltrifluoroborates in combination with MnF_3. *Org. Lett.* **2013,** *15*, 2302—2305.
38. Jia, X.; Zhang, S.; Wang, W.; Luo, F.; Cheng, J. Palladium-catalyzed acylation of sp^2 C—H bond: direct access to ketones from aldehydes. *Org. Lett.* **2009,** *11*, 3120—3123.
39. Xiao, F.; Shuai, Q.; Zhao, F.; Baslé, O.; Deng, G.; Li, C.-J. Palladium-catalyzed oxidative sp^2 C—H bond acylation with alcohols. *Org. Lett.* **2011,** *13*, 1614—1617.
40. Yu, W.-Y.; Sit, W. N.; Lai, K.-M.; Zhou, Z.; Chan, A. S. C. Palladium-catalyzed oxidative ethoxycarbonylation of aromatic C—H bond with diethyl azodicarboxylate. *J. Am. Chem. Soc.* **2008,** *130*, 3304—3306.
41. Jia, X.; Yang, D.; Zhang, S.; Cheng, J. Chelation-assisted palladium-catalyzed direct cyanation of 2-arylpyridine C—H bonds. *Org. Lett.* **2009,** *11*, 4716—4719.
42. Kalyani, D.; Sanford, M. S. Regioselectivity in palladium-catalyzed C—H activation/oxygenation reactions. *Org. Lett.* **2005,** 7, 4149—4152.
43. Racowski, J. M.; Dick, A. R.; Sanford, M. S. Detailed study of C—O and C—C bond-forming reductive elimination from stable $C_2N_2O_2$ ligated palladium(IV) complexes. *J. Am. Chem. Soc.* **2009,** *131*, 10974—10983.
44. Yan, Y.; Feng, P.; Zheng, Q.-Z.; Liang, Y.-F.; Lu, J.-F.; Cui, Y.; Jiao, N. $PdCl_2$ and *N*-Hydroxyphthalimide co-catalyzed $C(sp^2)$—H hydroxylation by dioxygen activation. *Angew. Chem. Int. Ed.* **2013,** *52*, 5827—5831.
45. Kalyani, D.; Dick, A. R.; Anani, W. Q.; Sanford, M. S. A simple catalytic method for the regioselective halogenation of arenes. *Org. Lett.* **2006,** *8*, 2523—2526.
46. Kalyani, D.; Dick, A. R.; Anani, W. Q.; Sanford, M. S. Scope and selectivity in palladium-catalyzed directed C—H bond halogenation reactions. *Tetrahedron* **2006,** *62*, 11483—11498.
47. Stowers, K. J.; Sanford, M. S. Mechanistic comparison between Pd-catalyzed ligand-directed C—H chlorination and C—H acetoxylation. *Org. Lett.* **2009,** *11*, 4584—4587.
48. Zhao, X.; Dimitrijević, E.; Dong, V. M. Palladium-catalyzed C—H bond functionalization with arylsulfonyl chlorides. *J. Am. Chem. Soc.* **2009,** *131*, 3466—3467.
49. Chu, J.-H.; Chen, S.-T.; Chiang, M.-F.; Wu, M.-J. Palladium-catalyzed direct ortho aroylation of 2-phenoxypyridines with aldehydes and catalytic mechanism investigation. *Organometallics* **2015,** *34*, 953—966.
50. Kim, M.; Sharma, S.; Park, J.; Kim, M.; Choi, Y.; Jeon, Y.; Kwak, J. H.; Kim, I. S. Pd-catalyzed oxidative acylation of 2-phenoxypyridines with alcohols *via* C—H bond activation. *Tetrahedron* **2013,** *69*, 6552—6559.
51. Yao, J.; Feng, R.; Wu, Z.; Liu, Z.; Zhang, Y. Palladium-catalyzed decarboxylative coupling of α-oxocarboxylic acids with $C(sp^2)$—H of 2-aryloxypyridines. *Adv. Synth. Catal.* **2013,** *355*, 1517—1522.
52. Cong, X.; You, J.; Gao, G.; Lan, J. 2-Pyridylmethyl ether: a readily removable and efficient directing group for amino acid ligand accelerated *ortho*-C—H olefination of phenols. *Chem. Commun.* **2013,** *49*, 662—664.
53. García-Rubia, A.; Urones, B.; Gómez Arrayás, R.; Carretero, J. C. Pd^{II}-catalyzed C—H olefination of *N*-(2-pyridyl)sulfonyl anilines and arylalkylamines. *Angew. Chem. Int. Ed.* **2011,** *50*, 10927—10931.
54. Hull, K. L.; Sanford, M. S. Catalytic and highly regioselective cross-coupling of aromatic C—H substrates. *J. Am. Chem. Soc.* **2007,** *129*, 11904—11905.

55. Dick, A. R.; Hull, K. L.; Sanford, M. S. A highly selective catalytic method for the oxidative functionalization of C—H bonds. *J. Am. Chem. Soc.* **2004,** *126*, 2300—2301.
56. Furuya, T.; Ritter, T. Carbon—fluorine reductive elimination from a high-valent palladium fluoride. *J. Am. Chem. Soc.* **2008,** *130*, 10060—10061.
57. Zhang, J.; Khaskin, E.; Anderson, N. P.; Zavalij, P. Y.; Vedernikov, A. N. Catalytic aerobic oxidation of substituted 8-methylquinolines in Pd^{II}-2,6-pyridinedicarboxylic acid systems. *Chem. Commun.* **2008,** 3625—3627.
58. Hull, K. L.; Anani, W. Q.; Sanford, M. S. Palladium-catalyzed fluorination of carbon—hydrogen bonds. *J. Am. Chem. Soc.* **2006,** *128*, 7134—7135.
59. Iglesias, Á.; Álvarez, R.; deLera, Á. R.; Muñiz, K. Palladium-catalyzed intermolecular $C(sp^3)$—H amidation. *Angew. Chem. Int. Ed.* **2012,** *51*, 2225—2228.
60. Spencer, J.; Sharratt, D. P.; Dupont, J.; Monteiro, A. L.; Reis, V. I.; Stracke, M. P.; Rominger, F.; McDonald, I. M. Synthesis and evaluation of 5-phenyl-1*H*-1,4-benzodiazepin-2(3*H*)-one-based palladium complexes as precatalysts in C—C bond forming reactions. *Organometallics* **2005,** *24*, 5665—5672.
61. Spencer, J.; Chowdhry, B. Z.; Mallet, A. I.; Rathnam, R. P.; Adatia, T.; Bashall, A.; Rominger, F. C—H activations on a 1*H*-1,4-benzodiazepin-2(3*H*)-one template. *Tetrahedron* **2008,** *64*, 6082—6089.
62. Alsters, P. L.; Boersma, J.; van Koten, G. Reactivity of organopalladium compounds toward molybdenum peroxides: oxygen insertion *versus* carbon—chlorine and carbon—oxygen coupling. Evidence for palladium(IV) molybdate intermediates. *Organometallics* **1993,** *12*, 1629—1638.
63. Alsters, P. L.; Teunissen, H. T.; Boersma, J.; Spek, A. L.; van Koten, G. Oxygenation of cyclopalladated *N,N*-dimethylbenzylamine complexes by inorganic and organic peroxides: oxygen insertion into the palladium—carbon bond. *Organometallics* **1993,** *12*, 4691—4696.
64. Cai, G.; Fu, Y.; Li, Y.; Wan, X.; Shi, Z. Indirect ortho functionalization of substituted toluenes through ortho olefination of *N,N*-dimethylbenzylamines tuned by the acidity of reaction conditions. *J. Am. Chem. Soc.* **2007,** *129*, 7666—7673.
65. Lazareva, A.; Daugulis, O. Direct palladium-catalyzed *ortho*-arylation of benzylamines. *Org. Lett.* **2006,** *8*, 5211—5213.
66. Wang, X.; Mei, T.-S.; Yu, J.-Q. Versatile $Pd(OTf)_2 \cdot 2H_2O$-catalyzed *ortho*-fluorination using NMP as a promoter. *J. Am. Chem. Soc.* **2009,** *131*, 7520—7521.
67. Orito, K.; Horibata, A.; Nakamura, T.; Ushito, H.; Nagasaki, H.; Yuguchi, M.; Yamashita, S.; Tokuda, M. Preparation of benzolactams by $Pd(OAc)_2$-catalyzed direct aromatic carbonylation. *J. Am. Chem. Soc.* **2004,** *126*, 14342—14343.
68. Wiest, J. M.; Pöthig, A.; Bach, T. Pyrrole as a directing group: regioselective Pd(II)-catalyzed alkylation and benzylation at the benzene core of 2-phenylpyrroles. *Org. Lett.* **2016,** *18*, 852—855.
69. Wang, L.; Guo, W.; Zhang, X.-X.; Xia, X.-D.; Xiao, W.-J. Synthesis of indolo [1,2-a]quinoxalines *via* a Pd-catalyzed regioselective C—H olefination/cyclization sequence. *Org. Lett.* **2012,** *14*, 740—743.
70. Shao, J.; Chen, W.; Giulianotti, M. A.; Houghten, R. A.; Yu, Y. Palladium-catalyzed C—H functionalization using guanidine as a directing group: ortho arylation and olefination of arylguanidines. *Org. Lett.* **2012,** *14*, 5452—5455.
71. Yang, F.; Wu, Y.; Zhu, Z.; Zhang, J.; Li, Y. Direct *ortho*-arylation of 2-arylbenzoxazoles *via* C—H activation. *Tetrahedron* **2008,** *64*, 6782—6787.
72. Yang, F.; Wu, Y.; Li, Y.; Wang, B.; Zhang, J. Modified palladium-catalyzed regioselective *ortho*-arylation of sp^2 C—H bond substrates with a low catalyst loading. *Tetrahedron* **2009,** *65*, 914—919.

73. Chen, X.; Li, J.-J.; Hao, X.-S.; Goodhue, C. E.; Yu, J.-Q. Palladium-catalyzed alkylation of aryl C—H bonds with sp^3 organotin reagents using benzoquinone as a crucial promoter. *J. Am. Chem. Soc.* **2006,** *128*, 78—79.
74. Izumi, T.; Watabe, H.; Kasahara, A. The σ-bonded palladium(II) complexes of 2-aryl-4,4-dimethyl-2-oxazolines. *Bull. Chem. Soc. Jpn.* **1981,** *54*, 1711—1714.
75. Giri, R.; Liang, J.; Lei, J.-G.; Li, J.-J.; Wang, D.-H.; Chen, X.; Naggar, I. C.; Guo, C.; Foxman, B. M.; Yu, J.-Q. Pd-catalyzed stereoselective oxidation of methyl groups by inexpensive oxidants under mild conditions: a dual role for carboxylic anhydrides in catalytic C—H bond oxidation. *Angew. Chem. Int. Ed.* **2005,** *44*, 7420—7424.
76. Giri, R.; Chen, X.; Yu, J.-Q. Palladium-catalyzed asymmetric iodination of unactivated C—H bonds under mild conditions. *Angew. Chem. Int. Ed.* **2005,** *44*, 2112—2115.
77. Li, J.-J.; Giri, R.; Yu, J.-Q. Remote C—H bond functionalization reveals the distance-dependent isotope effect. *Tetrahedron* **2008,** *64*, 6979—6987.
78. Constable, A. G.; McDonald, W. S.; Sawkins, L. C.; Shaw, B. L. Transition—metal—carbon bonds. Part 45. Attempts to cyclopalladate some aliphatic oximes, *NN*-dimethylhydrazones, ketazines, and oxime *O*-allyl ethers. Crystal structures of $[Pd_2\{CH_2C(CH_3)_2C(=NOH)CH_3\}_2Cl_2]$ and $[Pd\{CH_2C(=NNMe_2)C(CH_3)_3\}(acac)]$. *J. Chem. Soc. Dalton Trans.* **1980,** 1992—2000.
79. Bezsoudnova, E. Y.; Ryabov, A. D. Water-soluble cyclopalladated aryl oxime: a potent "green" catalyst. *J. Organomet. Chem.* **2001,** *622*, 38—42.
80. Desai, L. V.; Hull, K. L.; Sanford, M. S. Palladium-catalyzed oxygenation of unactivated sp^3 C—H bonds. *J. Am. Chem. Soc.* **2004,** *126*, 9542—9543.
81. Desai, L. V.; Malik, H. A.; Sanford, M. S. Oxone as an inexpensive, safe, and environmentally benign oxidant for C—H bond oxygenation. *Org. Lett.* **2006,** *8*, 1141—1144.
82. Liang, Y.-F.; Wang, X.; Yuan, Y.; Liang, Y. M.; Li, X.; Jiao, N. Ligand-promoted Pd-catalyzed oxime ether directed C—H hydroxylation of arenes. *ACS Catal.* **2015,** *5*, 6148—6152.
83. Thu, H.-Y.; Yu, W.-Y.; Che, C.-M. Intermolecular amidation of unactivated sp^2 and sp^3 C—H bonds via palladium-catalyzed cascade C—H activation/nitrene insertion. *J. Am. Chem. Soc.* **2006,** *128*, 9048—9049.
84. Thirunavukkarasu, V. S.; Parthasarathy, K.; Cheng, C.-H. Synthesis of fluorenones from aromatic aldoxime ethers and aryl halides by palladium-catalyzed dual C—H activation and heck cyclization. *Angew. Chem. Int. Ed.* **2008,** *47*, 9462—9465.
85. Kametani, Y.; Satoh, T.; Miura, M.; Nomura, M. Regioselective arylation of benzanilides with aryl triflates or bromides under palladium catalysis. *Tetrahedron Lett.* **2000,** *41*, 2655—2658.
86. Shabashov, D.; Daugulis, O. *ortho*-Arylation of benzamides. *Org. Lett.* **2006,** *8*, 4947—4949.
87. Li, D.-D.; Yuan, T.-T.; Wang, G.-W. Palladium-catalyzed *ortho*-arylation of benzamides *via* direct sp^2 C—H bond activation. *J. Org. Chem.* **2012,** 77, 3341—3347.
88. Wang, X.; Leow, D.; Yu, J.-Q. Pd(II)-catalyzed *para*-selective C—H arylation of monosubstituted arenes. *J. Am. Chem. Soc.* **2011,** *133*, 13864—13867.
89. Wang, D.-H.; Wasa, M.; Giri, R.; Yu, J.-Q. Pd(II)-catalyzed cross-coupling of sp^3 C—H bonds with sp^2 and sp^3 boronic acids using air as the oxidant. *J. Am. Chem. Soc.* **2008,** *130*, 7190—7191.
90. Wasa, M.; Engle, K. M.; Yu, J.-Q. Pd(0)/PR_3-catalyzed intermolecular arylation of sp^3 C—H bonds. *J. Am. Chem. Soc.* **2009,** *131*, 9886—9887.
91. Wasa, M.; Yu, J.-Q. Amide-directed arylation of sp^3 C—H bonds using Pd(II) and Pd(0) catalysts. *Tetrahedron* **2010,** *66*, 4811—4815.

92. Xiao, K.-J.; Lin, D. W.; Miura, M.; Zhu, R.-Y.; Gong, W.; Wasa, M.; Yu, J.-Q. Palladium(II)-catalyzed enantioselective C(sp^3)—H activation using a chiral hydroxamic acid ligand. *J. Am. Chem. Soc.* **2014,** *136*, 8138—8142.
93. Zhu, R.-Y.; He, J.; Wang, X.-C.; Yu, J.-Q. Ligand-promoted alkylation of C(sp^3)—H and C(sp^2)—H bonds. *J. Am. Chem. Soc.* **2014,** *136*, 13194—13197.
94. Wang, G.-W.; Yuan, T.-T.; Li, D.-D. One-pot formation of C—C and C—N bonds through palladium-catalyzed dual C—H activation: synthesis of phenanthridinones. *Angew. Chem. Int. Ed.* **2011,** *50*, 1380—1383.
95. Karthikeyan, J.; Cheng, C.-H. Synthesis of phenanthridinones from *N*-methoxybenzamides and arenes by multiple palladium-catalyzed C—H activation steps at room temperature. *Angew. Chem. Int. Ed.* **2011,** *50*, 9880—9883.
96. Peng, X.; Wang, W.; Jiang, C.; Sun, D.; Xu, Z.; Tung, C.-H. Strain-promoted oxidative annulation of arynes and cyclooctynes with benzamides: palladium-catalyzed C—H/N—H activation for the synthesis of *N*-heterocycles. *Org. Lett.* **2014,** *16*, 5354—5357.
97. Li, D.-D.; Yuan, T.-T.; Wang, G.-W. Synthesis of isoindolinones *via* palladium-catalyzed C—H activation of *N*-methoxybenzamides. *Chem. Commun.* **2011,** *47*, 12789—12791.
98. Wang, G.-W.; Yuan, T.-T. Palladium-catalyzed alkoxylation of *N*-methoxybenzamides *via* direct sp^2 C—H bond activation. *J. Org. Chem.* **2010,** *75*, 476—479.
99. Yadav, M. R.; Rit, R. K.; Sahoo, A. K. Sulfoximines: a reusable directing group for chemo- and regioselective ortho C—H oxidation of arenes. *Chem. Eur. J.* **2012,** *18*, 5541—5545.
100. Zhou, L.; Lu, W. Palladium-catalyzed β-acyloxylation of simple amide *via* sp^3 C—H activation. *Org. Lett.* **2014,** *16*, 508—511.
101. Gandeepan, P.; Parthasarathy, K.; Cheng, C.-H. Synthesis of phenanthrone derivatives from sec-alkyl aryl ketones and aryl halides *via* a palladium-catalyzed dual C—H bond activation and enolate cyclization. *J. Am. Chem. Soc.* **2010,** *132*, 8569—8571.
102. Shan, G.; Yang, X.; Ma, L.; Rao, Y. Pd-catalyzed C—H oxygenation with TFA/TFAA: expedient access to oxygen-containing heterocycles and late-stage drug modification. *Angew. Chem. Int. Ed.* **2012,** *51*, 13070—13074.
103. Mo, F.; Trzepkowski, L. J.; Dong, G. Synthesis of *ortho*-acylphenols through the palladium-catalyzed ketone-directed hydroxylation of arenes. *Angew. Chem. Int. Ed.* **2012,** *51*, 13075—13079.
104. Xiao, B.; Gong, T.-J.; Xu, J.; Liu, Z.-J.; Liu, L. Palladium-catalyzed intermolecular directed C—H amidation of aromatic ketones. *J. Am. Chem. Soc.* **2011,** *133*, 1466—1474.
105. Daugulis, O.; Zaitsev, V. G. Anilide *ortho*-arylation by using C—H activation methodology. *Angew. Chem. Int. Ed.* **2005,** *44*, 4046—4048.
106. Shi, Z.; Li, B.; Wan, X.; Cheng, J.; Fang, Z.; Cao, B.; Qin, C.; Wang, Y. Suzuki—Miyaura coupling reaction by Pd^{II}-catalyzed aromatic C—H bond activation directed by an *N*-alkyl acetamino group. *Angew. Chem. Int. Ed.* **2007,** *46*, 5554—5558.
107. Yeung, C. S.; Zhao, X.; Borduas, N.; Dong, V. M. Pd-catalyzed *ortho*-arylation of phenylacetamides, benzamides, and anilides with simple arenes using sodium persulfate. *Chem. Sci.* **2010,** *1*, 331—336.
108. Wang, J.-R.; Yang, C.-T.; Liu, L.; Guo, Q.-X. Pd-catalyzed aerobic oxidative coupling of anilides with olefins through regioselective C—H bond activation. *Tetrahedron Lett.* **2007,** *48*, 5449—5453.
109. Tremont, S. J.; Rahman, H. *Ortho*-alkylation of acetanilides using alkyl halides and palladium acetate. *J. Am. Chem. Soc.* **1984,** *106*, 5759—5760.

110. Wang, G.-W.; Yuan, T.-T.; Wu, X.-L. Direct *ortho*-acetoxylation of anilides *via* palladium-catalyzed sp^2 C—H bond oxidative activation. *J. Org. Chem.* **2008,** *73*, 4717—4720.
111. Jiang, T.-S.; Wang, G.-W. Palladium-catalyzed *ortho*-alkoxylation of anilides *via* C—H activation. *J. Org. Chem.* **2012,** 77, 9504—9509.
112. Wan, X.; Ma, Z.; Li, B.; Zhang, K.; Cao, S.; Zhang, S.; Shi, Z. Highly selective C—H functionalization/halogenation of acetanilide. *J. Am. Chem. Soc.* **2006,** *128*, 7416—7417.
113. Miura, M.; Tsuda, T.; Satoh, T.; Pivsa-Art, S.; Nomura, M. Oxidative cross-coupling of *N*-(2-phenylphenyl)benzene-sulfonamides or benzoic and naphthoic acids with alkenes using a palladium copper catalyst system under air. *J. Org. Chem.* **1998,** *63*, 5211—5215.
114. Chiong, H. A.; Pham, Q.-N.; Daugulis, O. Two methods for direct *ortho*-arylation of benzoic acids. *J. Am. Chem. Soc.* **2007,** *129*, 9879—9884.
115. Zhu, C.; Zhang, Y.; Kan, J.; Zhao, H.; Su, W. Ambient-temperature *ortho* C—H arylation of benzoic acids with aryl iodides with ligand-supported palladium catalyst. *Org. Lett.* **2015,** *17*, 3418—3421.
116. Giri, R.; Maugel, N.; Li, J.-J.; Wang, D.-H.; Breazzano, S. P.; Saunders, L. B.; Yu, J.-Q. Palladium-catalyzed methylation and arylation of sp^2 and sp^3 C—H bonds in simple carboxylic acids. *J. Am. Chem. Soc.* **2007,** *129*, 3510—3511.
117. Zhang, Y.-H.; Shi, B. F.; Yu, J.-Q. Palladium(II)-catalyzed *ortho* alkylation of benzoic acids with alkyl aalides. *Angew. Chem. Int. Ed.* **2009,** *48*, 6097—6100.
118. Giri, R.; Yu, J.-Q. Synthesis of 1,2- and 1,3-dicarboxylic acids *via* Pd(II)-catalyzed carboxylation of aryl and vinyl C—H bonds. *J. Am. Chem. Soc.* **2008,** *130*, 14082—14083.
119. Mei, T.-S.; Giri, R.; Maugel, N.; Yu, J.-Q. Pd^{II}-catalyzed monoselective ortho halogenation of C—H bonds assisted by counter cations: a complementary method to directed ortholithiation. *Angew. Chem. Int. Ed.* **2008,** *47*, 5215—5219.
120. Yamauchi, T.; Shibahara, F.; Murai, T. Direct C—H bond arylation of thienyl thioamides catalyzed by Pd—phenanthroline complexes. *Org. Lett.* **2015,** *17*, 5392—5395.
121. Zaitsev, V. G.; Shabashov, D.; Daugulis, O. Highly regioselective arylation of sp^3 C—H bonds catalyzed by palladium acetate. *J. Am. Chem. Soc.* **2005,** *127*, 13154—13155.
122. Shabashov, D.; Daugulis, O. Auxiliary-assisted palladium-catalyzed arylation and alkylation of sp^2 and sp^3 carbon—hydrogen bonds. *J. Am. Chem. Soc.* **2010,** *132*, 3965—3972.
123. He, G.; Chen, G. A practical strategy for the structural diversification of aliphatic scaffolds through the palladium-catalyzed picolinamide-directed remote functionalization of unactivated $C(sp^3)$—H bonds. *Angew. Chem. Int. Ed.* **2011,** *50*, 5192—5196.
124. Tran, L. D.; Daugulis, O. Nonnatural amino acid synthesis by using carbon—hydrogen bond functionalization methodology. *Angew. Chem. Int. Ed.* **2012,** *51*, 5188—5191.
125. Feng, Y.; Wang, Y.; Landgraf, B.; Liu, S.; Chen, G. Facile benzo-ring construction *via* palladium-catalyzed functionalization of unactivated sp^3 C—H bonds under mild reaction conditions. *Org. Lett.* **2010,** *12*, 3414—3417.
126. Nadres, E. T.; Daugulis, O. Heterocycle synthesis *via* direct C—H/N—H coupling. *J. Am. Chem. Soc.* **2012,** *134*, 7—10.
127. Pan, F.; Shen, P.-X.; Zhang, L.-S.; Wang, X.-C.; Shi, Z.-J. Direct arylation of primary and secondary sp^3 C—H bonds with diarylhyperiodonium salts *via* Pd catalysis. *Org. Lett.* **2013,** *15*, 4758—4761.

128. Liu, J.; Xie, Y.; Zeng, W.; Lin, D.; Deng, Y.; Lu, X. Pd(II)-catalyzed pyridine *N*-oxides directed arylation of unactivated C(sp^3)—H bonds. *J. Org. Chem.* **2015,** *80*, 4618—4626.
129. Chen, K.; Zhang, S.-Q.; Xu, J.-W.; Hu, F.; Shi, B.-F. A general and practical palladium-catalyzed monoarylation of β-methyl C(sp^3)—H of alanine. *Chem. Commun.* **2014,** *50*, 13924—13927.
130. Huang, L.; Li, Q.; Wang, C.; Qi, C. Palladium(II)-catalyzed regioselective arylation of naphthylamides with aryl iodides utilizing a quinolinamide bidentate system. *J. Org. Chem.* **2013,** *78*, 3030—3038.
131. MisalCastro, L. C.; Chatani, N. Palladium(II)-catalyzed *ortho*-C—H arylation/alkylation of *N*-benzoyl α-amino ester derivatives. *Chem. Eur. J.* **2014,** *20*, 4548—4553.
132. Nadres, E. T.; Santos, G. I. F.; Shabashov, D.; Daugulis, O. Scope and limitations of auxiliary-assisted, palladium-catalyzed arylation and alkylation of sp^2 and sp^3 C—H bonds. *J. Org. Chem.* **2013,** *78*, 9689—9714.
133. Rodriguéz, N.; Romero-Revilla, J. A.; Fernández-Ibáñez, M. A.; Carretero, J. C. Palladium-catalyzed *N*-(2-pyridyl)sulfonyl-directed C(sp^3)—H γ-arylation of amino acid derivatives. *Chem. Sci.* **2013,** *4*, 175—179.
134. Zhao, Y.; He, G.; Nack, W. A.; Chen, G. Palladium-catalyzed alkenylation and alkynylation of *ortho*-C(sp^2)—H bonds of benzylamine picolinamides. *Org. Lett.* **2012,** *14*, 2948—2951.
135. Deb, A.; Bag, S.; Kancherla, R.; Maiti, D. Palladium-catalyzed aryl C—H olefination with unactivated, aliphatic alkenes. *J. Am. Chem. Soc.* **2014,** *136*, 13602—13605.
136. Wang, B.; Lu, C.; Zhang, S.-Y.; He, G.; Nack, W. A.; Chen, G. Palladium-catalyzed stereoretentive olefination of unactivated C(sp^3)—H bonds with vinyl iodides at room temperature: synthesis of β-vinyl α-amino acids. *Org. Lett.* **2014,** *16*, 6260—6263.
137. Ano, Y.; Tobisu, M.; Chatani, N. Palladium-catalyzed direct ethynylation of C(sp^3)—H bonds in aliphatic carboxylic acid derivatives. *J. Am. Chem. Soc.* **2011,** *133*, 12984—12986.
138. Ano, Y.; Tobisu, M.; Chatani, N. Palladium-catalyzed direct *ortho*-alkynylation of aromatic carboxylic acid derivatives. *Org. Lett.* **2012,** *14*, 354—357.
139. Zhao, Y.; Chen, G. Palladium-catalyzed alkylation of *ortho*-C(sp^2)—H bonds of benzylamide substrates with alkyl halides. *Org. Lett.* **2011,** *13*, 4850—4853.
140. Zhang, S.-Y.; He, G.; Nack, W. A.; Zhao, Y.; Li, Q.; Chen, G. Palladium-catalyzed picolinamide-directed alkylation of unactivated C(sp^3)—H bonds with alkyl iodides. *J. Am. Chem. Soc.* **2013,** *135*, 2124—2127.
141. Chen, K.; Shi, B.-F. Sulfonamide-promoted palladium(II)-catalyzed alkylation of unactivated methylene C(sp^3)—H bonds with alkyl iodides. *Angew. Chem. Int. Ed.* **2014,** *53*, 11950—11954.
142. Zhang, S.-Y.; Li, Q.; He, G.; Nack, W. A.; Chen, G. Pd-catalyzed monoselective ortho-C—H alkylation of *N*-quinolyl benzamides: evidence for stereoretentive coupling of secondary alkyl iodides. *J. Am. Chem. Soc.* **2015,** *137*, 531—539.
143. Reddy, B. V. S.; Reddy, L. R.; Corey, E. J. Novel acetoxylation and C—C coupling reactions at unactivated positions in α-amino acid derivatives. *Org. Lett.* **2006,** *8*, 3391—3394.
144. Gou, F.-R.; Wang, X.-C.; Huo, P.-F.; Bi, H.-P.; Guan, Z.-H.; Liang, Y.-M. Palladium-catalyzed aryl C—H bonds activation/acetoxylation utilizing a bidentate system. *Org. Lett.* **2009,** *11*, 5726—5729.
145. Zhang, S.-Y.; He, G.; Zhao, Y.; Wright, K.; Nack, W. A.; Chen, G. Efficient alkyl ether synthesis *via* palladium-catalyzed, picolinamide-directed alkoxylation of

unactivated C(sp^3)—H and C(sp^2)—H bonds at remote positions. *J. Am. Chem. Soc.* **2012,** *134*, 7313—7316.
146. Rit, R. K.; Yadav, M. R.; Sahoo, A. K. Pd(II)-catalyzed primary-C(sp^3)—H acyloxylation at room temperature. *Org. Lett.* **2012,** *14*, 3724—3727.
147. Ju, L.; Yao, J.; Wu, Z.; Liu, Z.; Zhang, Y. Palladium-catalyzed oxidative acetoxylation of benzylic C—H bond using bidentate auxiliary. *J. Org. Chem.* **2013,** *78*, 10821—10831.
148. Rit, R. K.; Yadav, M. R.; Sahoo, A. K. Pd(II)-catalyzed ortho-C—H oxidation of arylacetic acid derivatives: synthesis of benzofuranones. *Org. Lett.* **2014,** *16*, 968—971.
149. Lu, X.; Xiao, B.; Shang, R.; Liu, L. Synthesis of unnatural amino acids through palladium-catalyzed C(sp^3)—H functionalization. *Chin. Chem. Lett.* **2016,** *27*, 305—311.
150. He, G.; Zhao, Y.; Zhang, S.; Lu, C.; Chen, G. Highly efficient syntheses of azetidines, pyrrolidines, and indolines *via* palladium catalyzed intramolecular amination of C(sp^3)—H and C(sp^2)—H bonds at γ and δ positions. *J. Am. Chem. Soc.* **2012,** *134*, 3—6.
151. He, G.; Lu, C.; Zhao, Y.; Nack, W. A.; Chen, G. Improved protocol for indoline synthesis *via* palladium-catalyzed intramolecular C(sp^2)—H amination. *Org. Lett.* **2012,** *14*, 2944—2947.
152. Wang, C.; Ge, H. Silanol as a removable directing group for the Pd^{II}-catalyzed direct olefination of arenes. *Chem. Eur. J.* **2011,** *17*, 14371—14374.
153. Huang, C.; Chattopadhyay, B.; Gevorgyan, V. Silanol: a traceless directing group for Pd-catalyzed *o*-alkenylation of phenols. *J. Am. Chem. Soc.* **2011,** *133*, 12406—12409.
154. Chary, B. C.; Kim, S.; Park, Y.; Kim, J.; Lee, P. H. Palladium-catalyzed C—H arylation using phosphoramidate as a directing group at room temperature. *Org. Lett.* **2013,** *15*, 2692—2695.
155. Chan, L. Y.; Cheong, L.; Kim, S. Pd(II)-catalyzed *ortho*-arylation of aryl phosphates and aryl hydrogen phosphates with diaryliodonium triflates. *Org. Lett.* **2013,** *15*, 2186—2189.
156. Chan, L. Y.; Kim, S.; Ryu, T.; Lee, P. H. Palladium-catalyzed *ortho*-alkenylation of aryl hydrogen phosphates using a new mono-phosphoric acid directing group. *Chem. Commun.* **2013,** *49*, 4682—4684.
157. Meng, X.; Kim, S. Palladium(II)-catalyzed *ortho*-olefination of benzylic phosphonic monoesters. *Org. Lett.* **2013,** *15*, 1910—1913.
158. Hu, R.-B.; Zhang, H.; Zhang, X.-Y.; Yang, S.-D. Palladium-catalyzed $P(O)R_2$ directed C—H arylation to synthesize electron-rich polyaromatic monophosphorus ligands. *Chem. Commun.* **2014,** *50*, 2193—2195.
159. Zhang, H.; Hu, R.-B.; Zhang, X.-Y.; Li, S.-X.; Yang, S.-D. Palladium-catalyzed $R_2(O)P$ directed C(sp^2)—H acetoxylation. *Chem. Commun.* **2014,** *50*, 4686—4689.
160. Catellani, M.; Frignani, F.; Rangoni, A. A complex catalytic cycle leading to a regioselective synthesis of *o,o′*-disubstituted vinylarenes. *Angew. Chem., Int. Ed. Engl.* **1997,** *36*, 119—122.

CHAPTER 11

Mechanistic/Organometallic Aspects of Palladium-Catalyzed C—H Bond Functionalization

Vijay Gayakhe, Aniket Gholap and Anant R. Kapdi
Institute of Chemical Technology, Mumbai, Maharashtra, India

Contents

11.1 INTRODUCTION

Transition metal catalyzed C—H bond functionalization is an important tool in modern organic synthesis and catalysis and has become an attractive alternative[1] to traditional cross-coupling reactions because of its several advantages such as minimization of stoichiometric waste, high atom economy, high regioselectivity and excellent functional group tolerance, and non-requirement of pre-functionalization for one or both of the starting materials.[2]

During carbon—hydrogen bond functionalization (C—H functionalization) transition metal (Pd, Ru, Rh, or Ir) inserts into the strong C—H bond (90—105 kcal/mol) to form a new, weaker C—M bond (50—80 kcal/mol), followed by the generation of a new C—C bond. While tremendous development has taken place in C—H bond functionalization[1] still, many mechanisms for a variety of C—H activations are

Strategies for Palladium-Catalyzed Non-Directed and Directed C—H Bond Functionalization
DOI: http://dx.doi.org/10.1016/B978-0-12-805254-9.00011-6

unknown. Generally these have been categorized into four major types: (1) oxidative addition,[3] the earliest discovered pathway in which a metal center inserts into a carbon—hydrogen bond leading to the cleavage of the bond and in the process oxidizes the metal, producing an intermediate that can undergo reductive elimination to yield the organometallic reactive intermediate. This was later followed by: (2) electrophilic aromatic substitution[4] (S_EAr Mechanism) or electrophilic activation, which proceeds through the formation of organometallic reactive intermediate; and recently two more important pathways were disclosed as (3) concerted metalation deprotonation (CMD) mechanism[5] (σ-bond metathesis of the C—H bond) proceeding through a "four-centered" transition state in which bond breaks and forms in concerted fashion; and (4) Heck-type carbometalation[6] as shown in Scheme 11.1.

Organic chemists have strived for years to develop highly efficient methodologies in C—H bond functionalization[1,2] that can help to synthesize more complex structures, natural products[7] from simple and easily available precursors. However, there are still several major challenges in this area that have hindered its further growth such as selective activation, terminal C—H bond functionalization, development of highly active catalytic processes, and conversion of methane to methanol. Through the pioneering work done by Shilov,[8] Bercaw,[9] Bergman,[10] Crabtree,[11] Murai,[12] and Goldman,[13] C—H bond functionalization has blossomed over the past few years, but still a lot more needs to be done to understand this area further in terms of mechanistic aspects.[14] Detailed

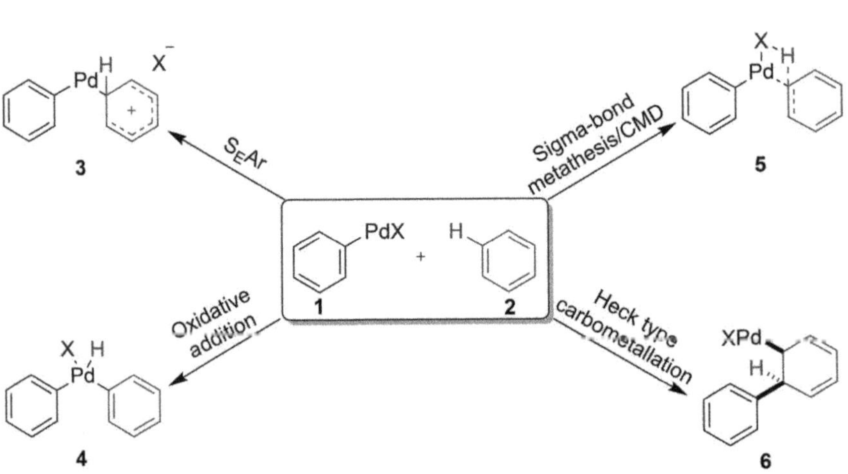

Scheme 11.1 Mechanistic pathways for C—H activation.

understanding of the mechanisms are important not only for the development of more efficient catalytic conditions, but also with a view of developing newer C—H bond functionalization strategies for tackling synthetic challenges.[15]

Palladium metal, in that sense, has a strong ability to insert into the inert C—H bond and thereby promote the synthesis of new compounds via direct arylation.[16] Therefore, in this chapter we are mainly going to focus on different aspects of mechanistic pathways that generally are known to be followed in palladium-catalyzed C—H activation,[17] giving more relevance to highlighting the indicators that could ease the identification of possible mechanisms in catalytic C—H bond functionalization processes.

11.2 EARLY FINDINGS

Earliest examples for transition metal catalyzed activation were performed on the activation of H—H bond and oxidation of alkanes, though many researchers have made great attempts to solve this problem in last few decades. This journey, however, started from the pioneering work by Halpern and co-workers,[18–20] who proposed initially the activation of H—H bond via electrophilic attack of the catalytically active Cu(II) species. Around the same time Vaska and co-workers[21] made a different attempt for H—H activation using an Iridium complex [IrCl(CO)(PPh_3)$_2$] **7** via oxidative addition (Scheme 11.2).

Chatt and co-workers[22] finally provided the first breakthrough in the area of C—H bond functionalization of arenes, activating naphthalene **9** using [Ru(0)(dmpe)$_2$] **10** catalyst via an oxidative addition type mechanism, which further helped to develop a highly efficient catalytic process for C—H bond functionalization of unactivated arenes[23] (Scheme 11.3).

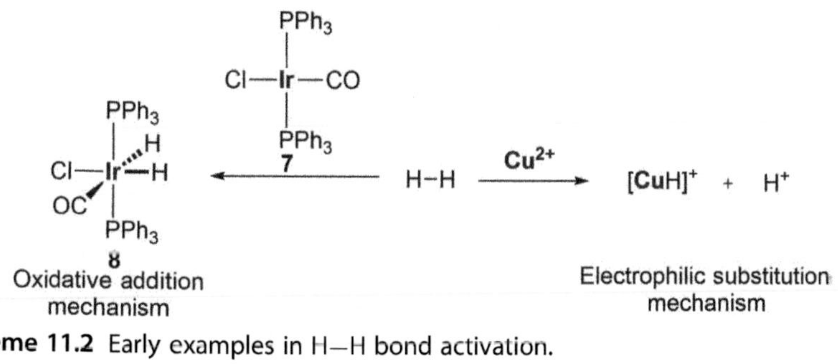

Scheme 11.2 Early examples in H—H bond activation.

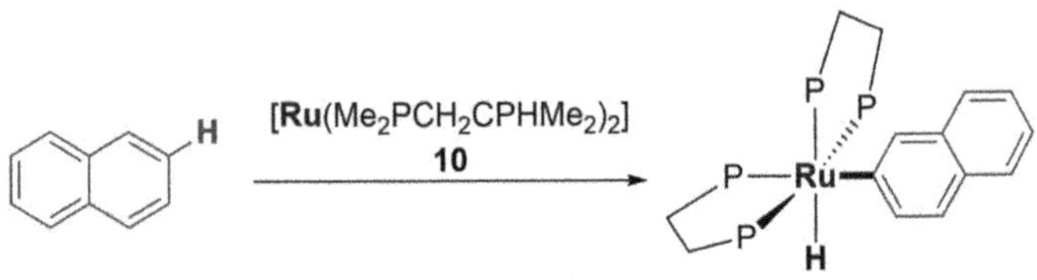

9 11

Scheme 11.3 An early example of C—H bond activation.

On the other hand, alkane oxidation was also a big hurdle in this area, which was solved by Shilov and co-workers in 1970, who had made a great attempt in the functionalization of sp^3 C—H bond of simple alkanes such as methane to C—O bonds using catalytically active Pt(II) salts along with stoichiometric amounts of Pt(IV) based oxidants.[24,25]

11.3 PALLADIUM-CATALYZED C—H ACTIVATION PATHWAYS

11.3.1 Electrophilic Aromatic Substitution (S_EAr Mechanism)

Regioselective C—H bond functionalization[26] has always been a challenge for synthetic chemists over the past few decades, with the earliest example for achieving regioselectivity reported by Catellani and co-workers.[27] The regioselectivity was achieved in the case of arenes **15** by the employment of norbornene **14** as a covalent linker exhibiting excellent selectivity that has helped to generate further development in this area of research. As a result of this, many other efficient methodologies have been reported in the past few years targeted towards tackling regioselectivity issues in complex molecules.[28] One of the key features of these transformations was the involvement of palladacyclic intermediate, whose presence has been taken as possible evidence for the electrophilic aromatic substitution pathway (Scheme 11.4).

Subsequently, a highly regioselective and efficient C-3 (hetero)arylation of indolizines[29] was reported by Gevorgyan and co-workers that was proposed to proceed through electrophilic aromatic substitution (Scheme 11.5). As a support for such a proposition, Gevorgyan and co-workers performed kinetic isotopic effect studies on deuterium labeled 2-carboethoxyindolizine suggesting the absence of kinetic isotope effect ($K_{H/D} = 1$). Mechanistic findings were also well supported by examples.

In a continuation of the above protocol, the same group also reported another example of an electrophilic aromatic substitution for the synthesis of substituted 1,2,3-triazoles via palladium-catalyzed C-5 arylation[30]

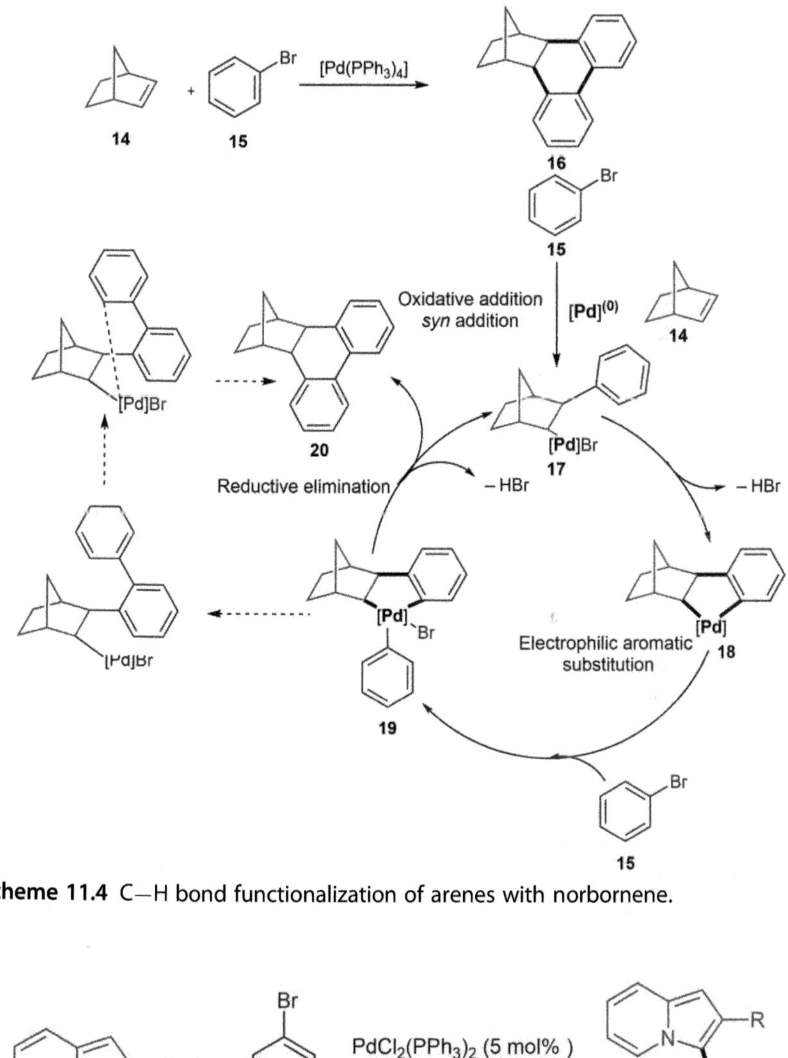

Scheme 11.4 C—H bond functionalization of arenes with norbornene.

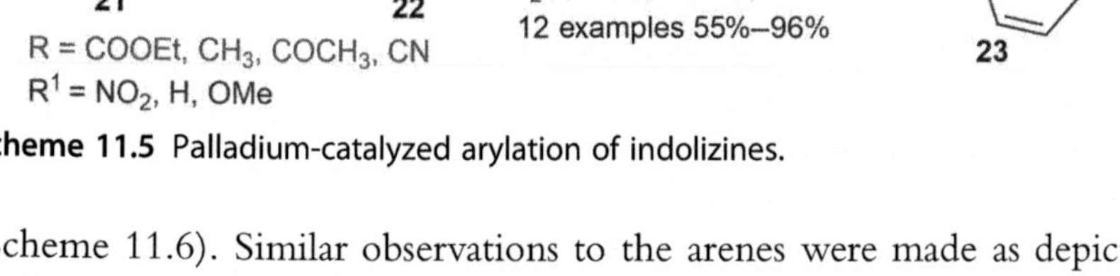

Scheme 11.5 Palladium-catalyzed arylation of indolizines.

(Scheme 11.6). Similar observations to the arenes were made as depicted in the above example, when kinetic isotopic studies performed on the deuterated triazole substrate **27** (Scheme 11.7) suggested the absence of any kinetic isotopic effect ($K_H/K_D = 1$), as shown in Scheme 11.7.

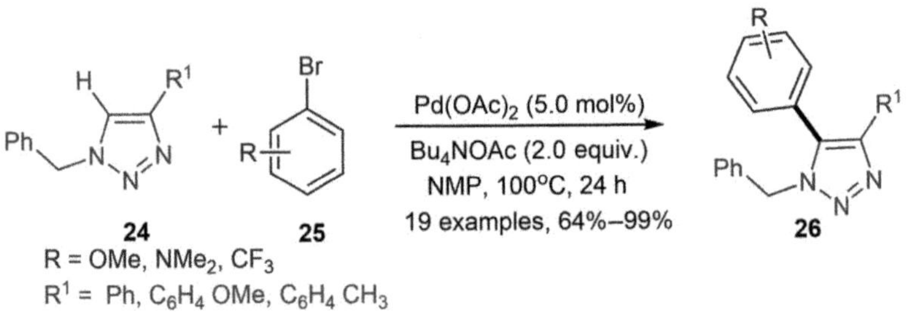

Scheme 11.6 Regioselective palladium-catalyzed arylation of 1,2,3-triazole.

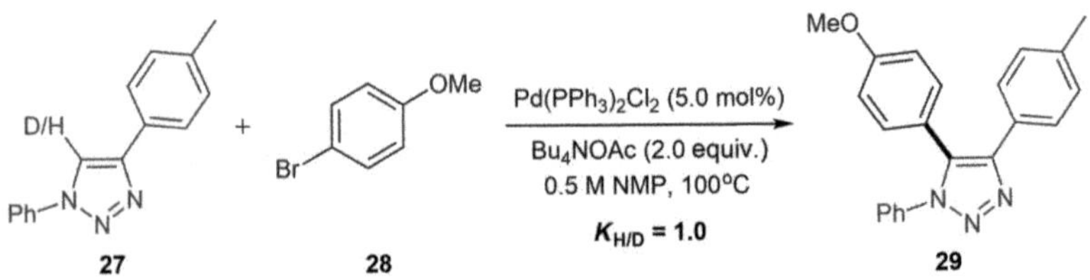

Scheme 11.7 Kinetic isotope effect studies.

Furthermore, DFT calculations of electrostatic potential charges at the C-4 and C-5 position of 1-methyl-1,2,3-triazole led to the observation that a substantial amount of negative charge (−0.45) developed at the C-5 position, and a positive charge (+0.13) at the C-4 position, thus supporting the electrophilic substitution pathway and explaining the reason for the high regioselectivity at the C-5 position.

Following on these results, Sames and co-workers[31] in 2005 reported a highly regioselective arylation of indoles using palladium-catalyzed C—H arylation (Scheme 11.8). The mechanistic proposal given on the basis of kinetic studies and Hammett plot showing a negative ρ value of −0.71 strongly supported the electrophilic palladation pathway. Kinetic isotopic effect studies performed at the C-2 and C-3 position of 1-methylindole **30** (Scheme 11.9) providing the value of $K_{H/D}=1.6$ at the C-3 position and $K_{H/D}=1.2$ was observed at the C-2 position of indole, also indicating the electrophilic attack at the C-3 position followed by 1,2 migration of palladium to furnish a C-2 arylated product. Arylation of (NH)-indole was carried out in the presence of an Mg salt, which was suggested to play a vital role, thus helping in deciding the regioselectivity at the C-2 position due to the steric effect of magnesium coordination sphere.

Meanwhile, one of the classical examples of electrophilic aromatic substitution was reported recently by Larock,[32] for the synthesis of substituted carbazoles and dibenzofurans. As a part of the synthetic transformation,

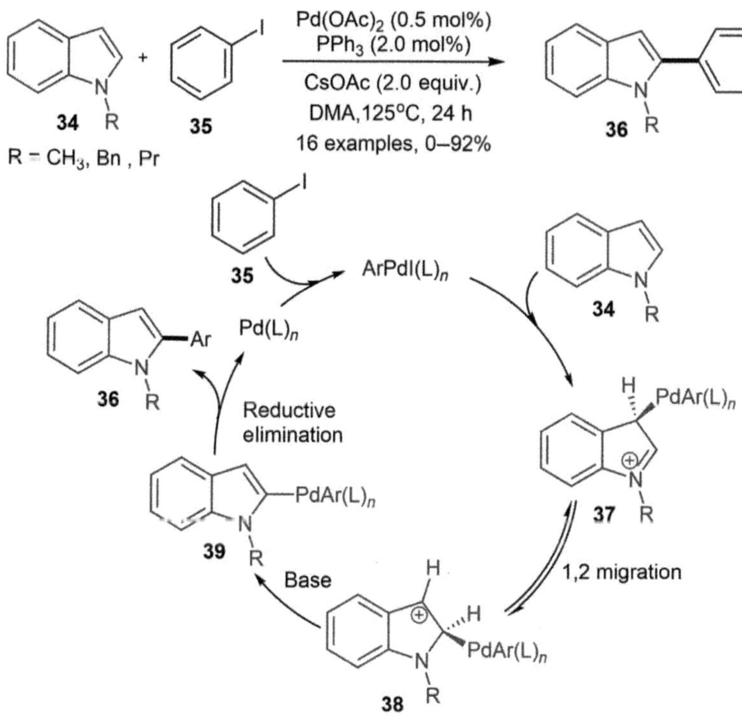

Scheme 11.8 Palladium-catalyzed C-2 arylation of indoles.

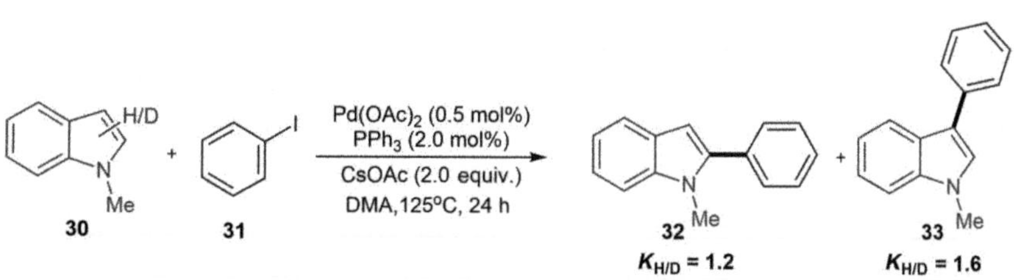

Scheme 11.9 Kinetic isotope effect studies at the C-2 and C-3 position of 1-methylindole.

oxidative addition of aryl iodide **40** to Pd(0) and subsequent alkyne carbopalladation leads to the generation of intermediate **44**. Further migration of palladium from vinylic to aryl position results in the formation of a palladacyclic intermediate **46**. This finally undergoes electrophilic aromatic substitution reaction and produces the desired cyclized product **42** (Scheme 11.10). The mechanistic proposal was supported by a deuterium labeling experiment (Scheme 11.11), where H/D exchange was analyzed

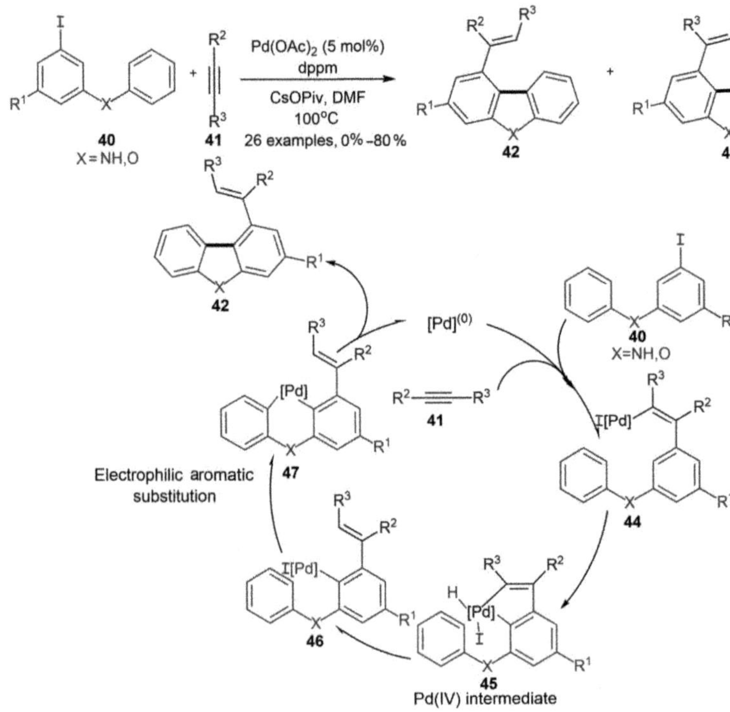

Scheme 11.10 Synthesis of substituted carbazoles and dibenzofurans via C—H bond functionalization.

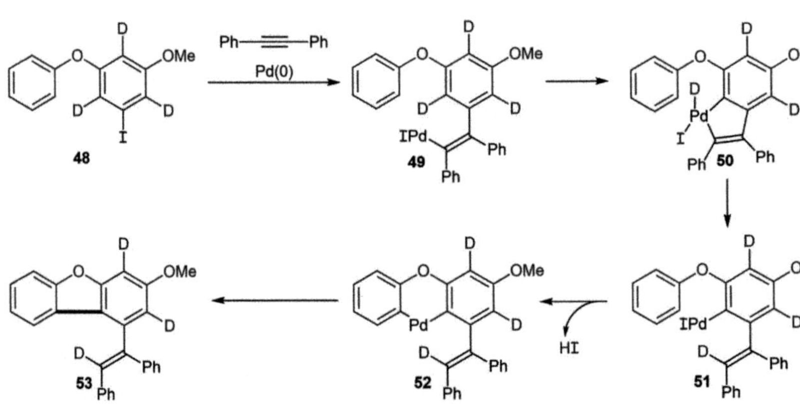

Scheme 11.11 Deuterium labeling experiments performed by Larock.

by ^{1}H NMR proving the possible existence of electrophilic aromatic substitution proceeding through the formation of a Pd(IV) intermediate **45**.

Another example of palladium-catalyzed regioselective coupling of (hetero)arenes[33] was recently reported involving the functionalization of furan-2-carbonyls with simple arenes (Scheme 11.12). The reaction was proposed to proceed through an electrophilic palladation pathway with the mechanistic evidence provided on the basis of kinetic isotopic effect studies performed on different substrates (Scheme 11.13).

When competition reaction of benzene **54** and benzene-d_6 **55** was carried out with furan-2-carbonyls **56** a primary KIE of 4.8 was observed (Scheme 11.13). However, there was no significant KIE ($K_{H/D}$=1.65) when the reaction was carried out between 2-acetyl furan **60** and C5-*d*-2-furfural **59** with benzene **54**, suggesting that C—H bond cleavage of furan-2-carbonyl was not the rate determining step. Instead, the activation of C-5 carbon of furan-2-carbonyl substrate proceeded via electrophilic palladation pathway.

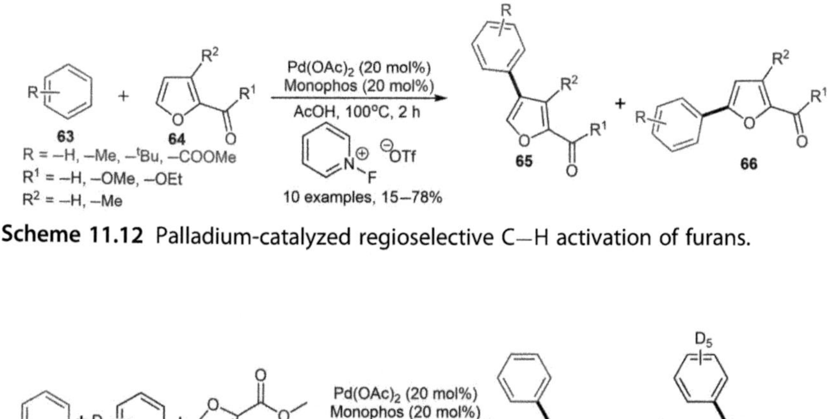

Scheme 11.12 Palladium-catalyzed regioselective C—H activation of furans.

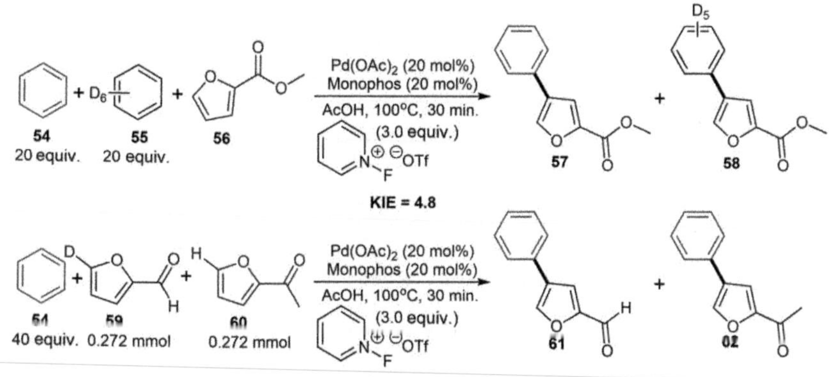

Scheme 11.13 Competition experiments for determination of KIE.

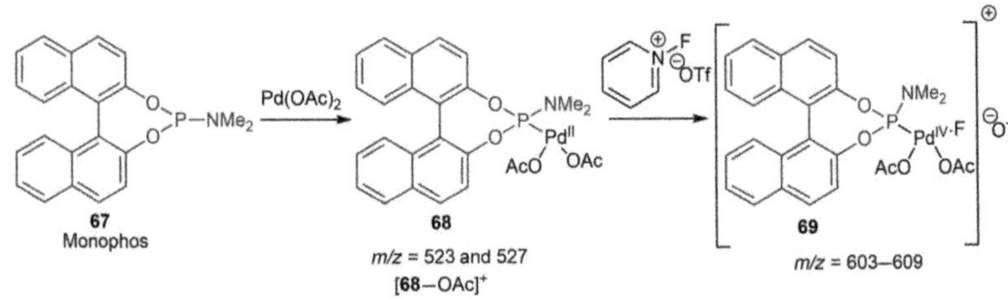

Scheme 11.14 Involvement of Pd(IV) species in catalytic cycle.

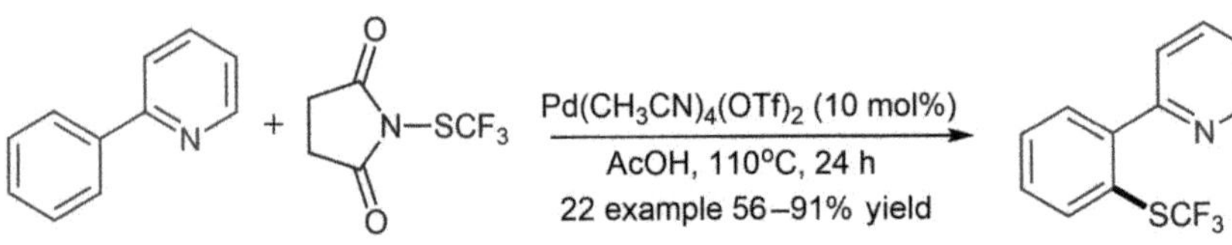

Scheme 11.15 Palladium-catalyzed trifluoromethylthiolation of arenes.

As a part of the studies, it was observed that the electronic factors on the arene affected the rate of reaction, and regioselectivity was mainly achieved due to the presence of F^+ oxidant. The possible presence of an active Pd(IV) intermediate was also proposed based on the analysis of the catalytic reaction by electrospray ionization mass spectrometry (Scheme 11.14).

The presence of the thiomethyl (-SCH_3) or trifluromethylthio (—SCF_3) structural motifs in organic compounds are of great interest due to their occurrence in agrochemicals and pharmaceutical intermediates.[34] The introduction of such functional groups[35] has been a synthetic challenge for several years that has now been made possible by the catalytic C—H bond functionalization. In one such example, Shen and co-workers[36] have disclosed an efficient method for trifluoromethylthiolation of 2-pyridyl benzenes using Pd(CH_3CN)($OTf)_2$ catalyst (Scheme 11.15).

The mechanism for the formation of the trifluoromethylated product was proposed to proceed through an electrophilic palladation pathway. In order to prove this assumption, the kinetic isotope effect was studied on different substrates to provide a value of intramolecular $K_{H/D}=1.3$ and intermolecular $K_{H/D}=1.2$ which ruled out the possibility of C—H activation as the rate limiting step, thus supporting an electrophilic substitution pathway (Scheme 11.16).

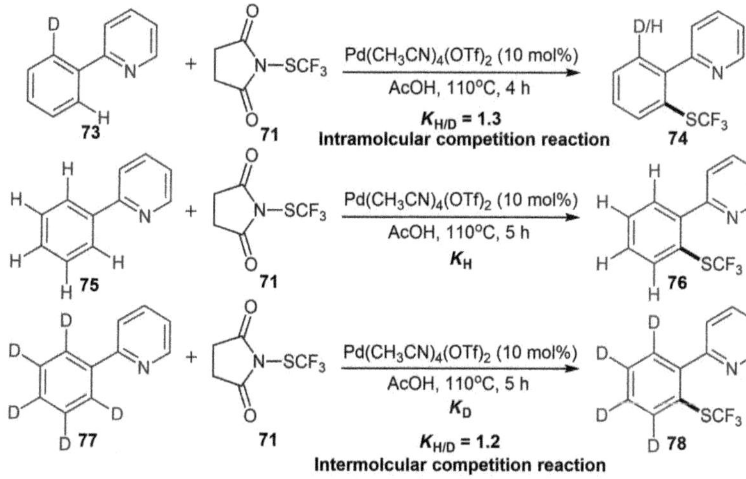

Scheme 11.16 Intra/intermolecular kinetic isotopic effect studies.

More recently, Yu and co-workers[37] have reported an excellent example of ligand-controlled palladium-catalyzed dehydrogenative coupling of arenes with complete para-selectivity following their previous report on para-selective C—H arylation of monosubstituted arenes[38] using F^+(NFSI) reagent as a bystanding oxidant (Scheme 11.17).

Both these protocols were found to proceed via an electrophilic palladation pathway that was supported by kinetic isotopic effect studies (KIE) carried out by intermolecular competition experiments between toluene **88** and toluene-d_8 **89** with *N*-(*m*-tolyl)pivalamide **87** that gave a value of $K_{H/D}=1$ (Scheme 11.18). Pyridine ligands used in the arylation reaction played a vital role in achieving para-selectivity by increasing electrophilicity and steric hindrance of Pd center allowing it to react at para-position selectively via electrophilic palladation pathway.

All these reports highlight the criteria that if achieved would allow the researchers to predict with certainty the mechanism followed in the catalytic C—H bond functionalization reaction to be the electrophilic aromatic substitution.[39–41] Kinetic isotopic effect studies have proved to be the most effective in determining the existence of such a pathway with values closer to $K_{H/D}=1$, suggesting the electrophilic palladation[42] to be the rate determining step. In some cases DFT[43] has also provided the prediction of possible centers undergoing electrophilic palladation.

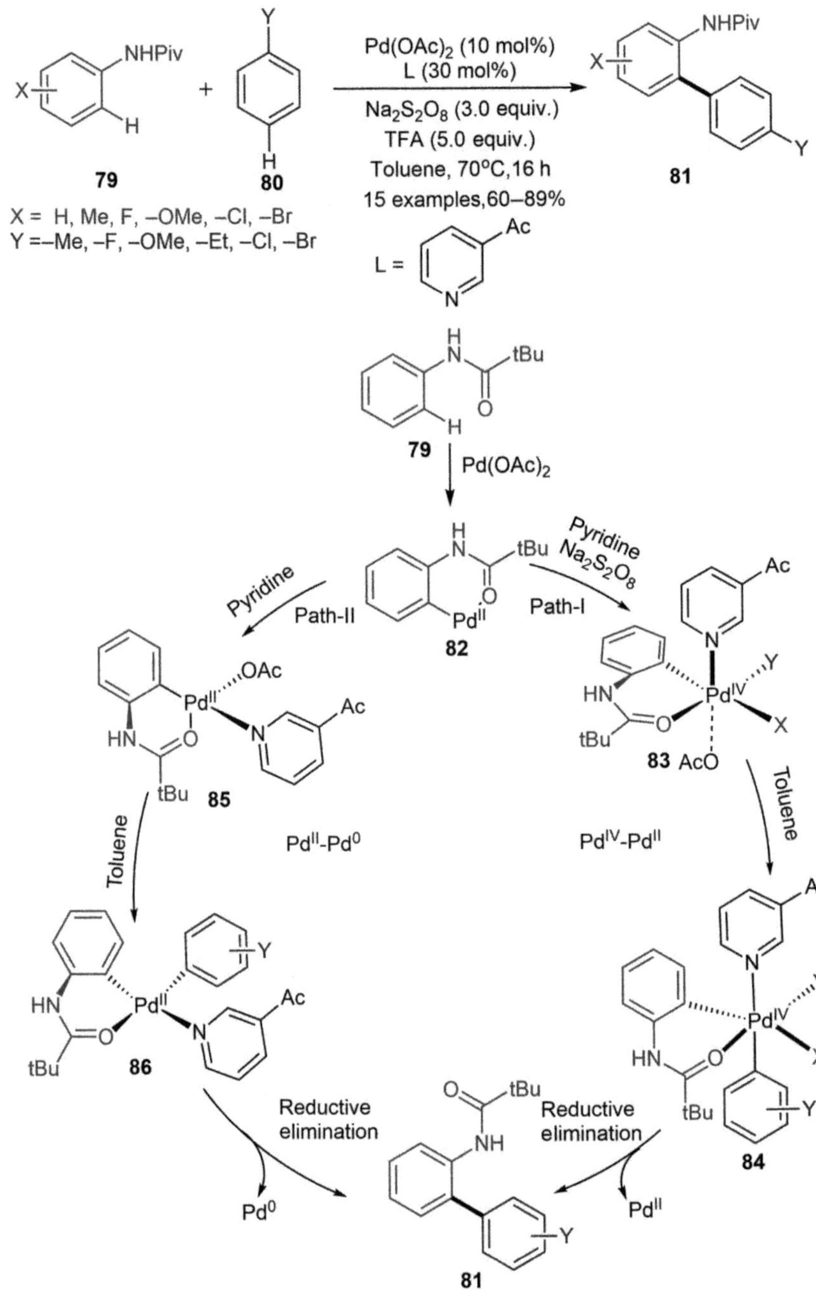

Scheme 11.17 Palladium-catalyzed dehydrogenative coupling of arenes.

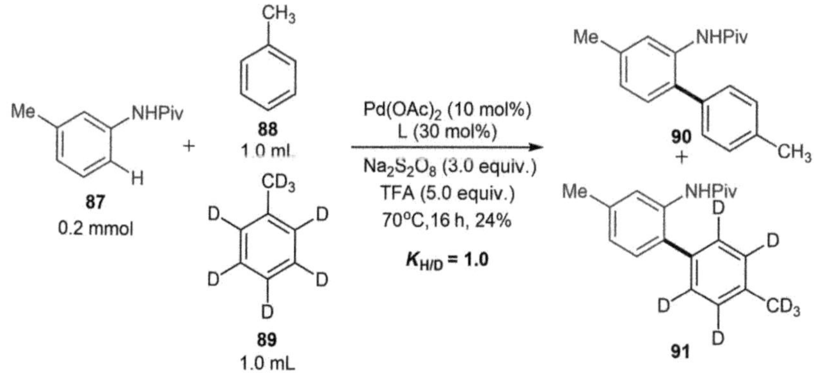

Scheme 11.18 Intermolecular kinetic isotopic effect studies.

11.3.2 Oxidative Addition Mechanism

Another important type of C H activation that has attracted greater interest in the last few years after electrophilic aromatic substitution is the palladium-catalyzed oxidative C—H activation,[44] in which both the coupling partners do not require prefunctionalization. The oxidative addition mechanism proceeds through the insertion of a metal center into the carbon—hydrogen bond, which cleaves the bond and oxidizes the metal, producing an intermediate which after reductive elimination gives rise to an organometallic reactive intermediate. Palladium-catalyzed oxidative C—H activation has therefore in recent years emerged as an exciting area that attracts greater synthetic interest. We have herein described the recent reports highlighting the mechanistic evidence provided for oxidative C—H activation.

In one of the earliest examples, Yu and co-workers[45] reported palladium-catalyzed selective C—H activation of BOC-protected *N*-methylamines using in situ generated IOAC as an efficient oxidant from $PhI(OAc)_2$[46] and I_2 (Scheme 11.19). Mechanistic evidence provided on the basis of various experimental studies and previous reports on the characterization of Pd(IV) intermediates[47] obtained by oxidizing Pd^{II} with MeI (X-ray),[48] $(PhCO_2)_2$ (^1H NMR),[49] and $PhI(CO_2Ph)$ (X-ray)[50] suggesting that a stoichiometric amount of IOAC promotes the oxidation of unstable Pd(II) complex **94** to Pd(IV) complex **95** and forms the acetoxylated product.

This was further supported by kinetic isotopic effect studies, showing higher values for both intramolecular $K_{H/D}=3.2$ and intermolecular $K_{H/D}=2.9$, suggesting that C—H bond cleavage was the rate determining step (Scheme 11.20).

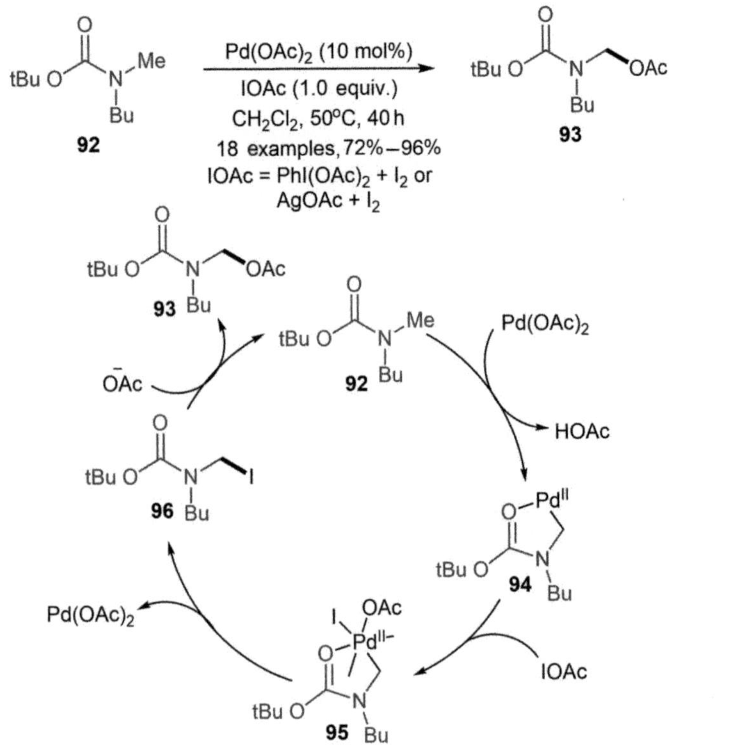

Scheme 11.19 Palladium-catalyzed acetoxylation of *N*-methylcarbamates.

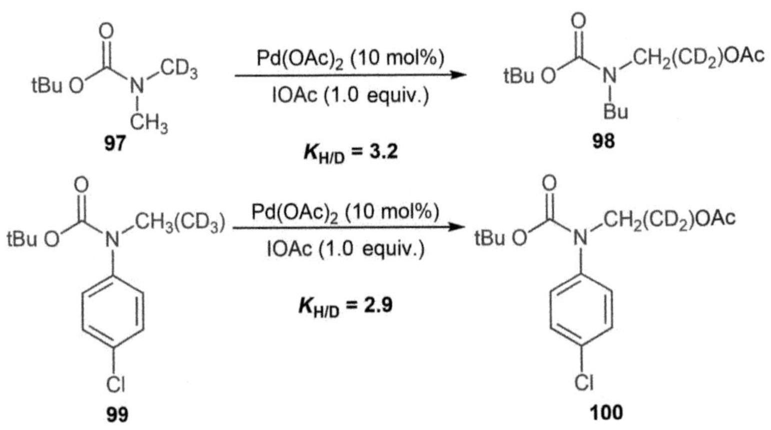

Scheme 11.20 Kinetic isotope effect studies.

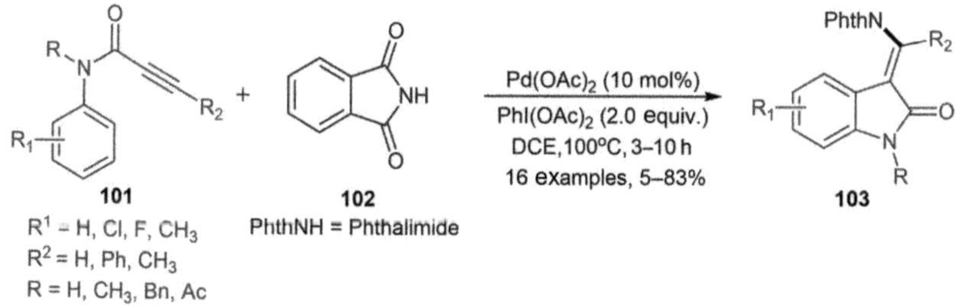

Scheme 11.21 Palladium-catalyzed C—H activation of *N*-phenylpropiolamides.

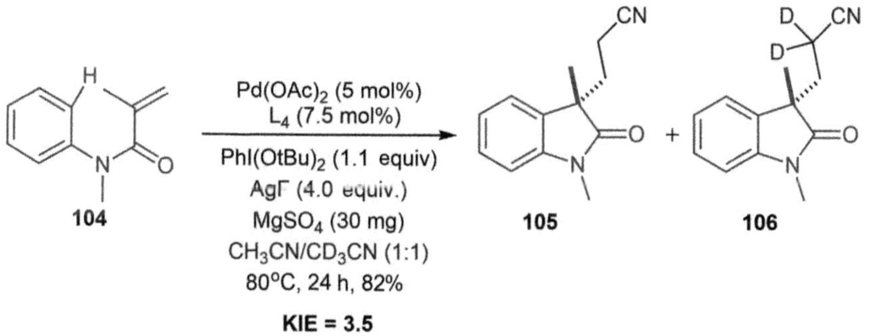

Scheme 11.22 Kinetic isotope effect studies with deuterated acetonitrile.

One of the classical examples of intermolecular aminopalladation of alkynes with phthalimide was reported for the first time by Li[51] in the presence of PhIOAc as an oxidant. The role of the oxidant is to oxidize Pd(II) intermediate formed by the insertion of Pd(II) in between nitrogen and alkyne into a Pd(IV) intermediate that further activates the ortho position of the arene C—H bond of *N*-aromatic ring to form a new C—C bond (Scheme 11.21).

Subsequently, Liu and co-workers[52] reported an unexpected alkylation of activated alkenes with acetonitrile via palladium-catalyzed oxidative C—H bond functionalization. To gain more insight into the mechanism of C—H activation, they performed deuterium KIE studies on the alkene substrate, which gives ($K_{H/D}=1$) for both inter- and intramolecular study. These results point more towards the likelihood of an electrophilic substitution mechanism, but when they carried out a reaction in a solvent mixture of CH_3CN:CD_3CN (1:1) a higher value of kinetic isotopic effect ($K_{H/D}=3.5$) was observed, indicating that C—H bond activation of acetonitrile is the rate determining step (Scheme 11.22).

Further experimental studies revealed that the reaction only proceeds in the presence of palladium catalyst, CH_3CN, AgF, and $PhI(OPiv)_2$. Therefore, a mechanistic pathway was disclosed on the basis of the performed study, as Pd(II) first coordinates with alkene and produces Pd(II) complex **111**, which was characterized by subjecting the reaction mixture to electrospray ionization mass spectral analysis showing a peak at $m/z = 438$ corresponding to Pd(II). CH_3CN then undergoes C—H bond activation in the presence of AgF and $PhI(OPiv)_2$ to generate Pd(IV) complex **112**, finally undergoing reductive elimination to afford the desired alkylated product **105** (Scheme 11.23).

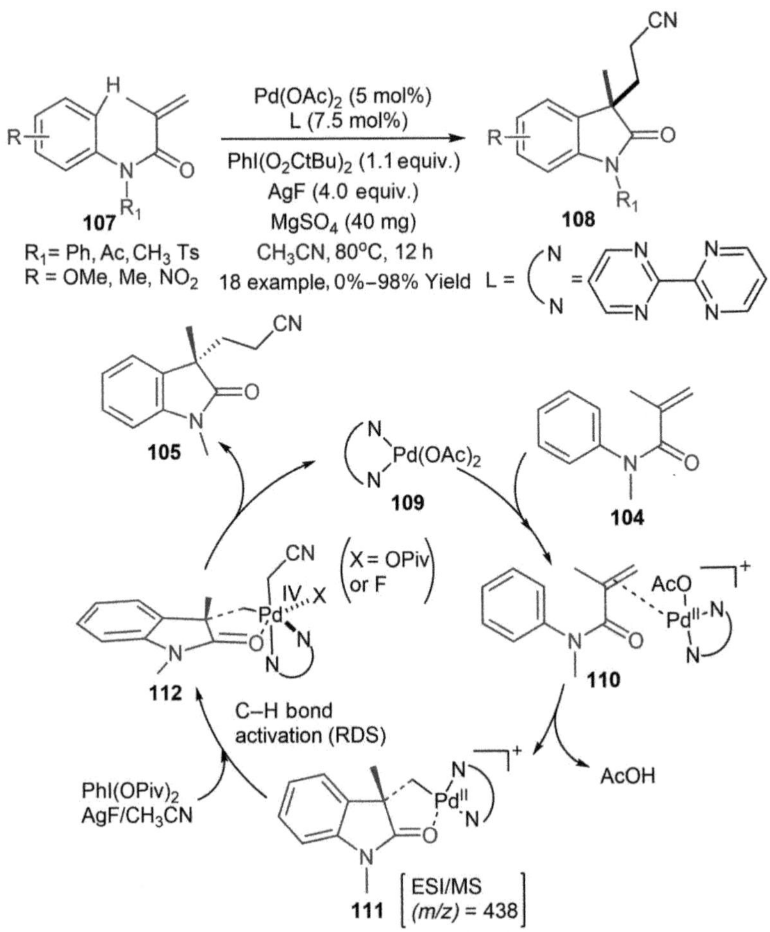

Scheme 11.23 Palladium-catalyzed oxidative C—H activation of alkene.

Palladium-catalyzed intramolecular C–H bond functionalization of coumaryl ethers[53] was disclosed recently by our group, giving easy access to benzofurocoumarins in good to excellent yields. The employment of Hermann–Beller palladacyclic complex[54] **115** was found to play a major role in affecting the transformation that was found to proceed via oxidative addition-type mechanism. The support for such an assumption was given on the basis of studies carried out using UV-Vis stopped flow analysis of the reaction mixture. Electrospray ionization mass spectral analysis of the catalytic reaction revealed the presence of the oxidative Pd(II) intermediate **117** at $m/z = 728 + H^+$ (Scheme 11.24).

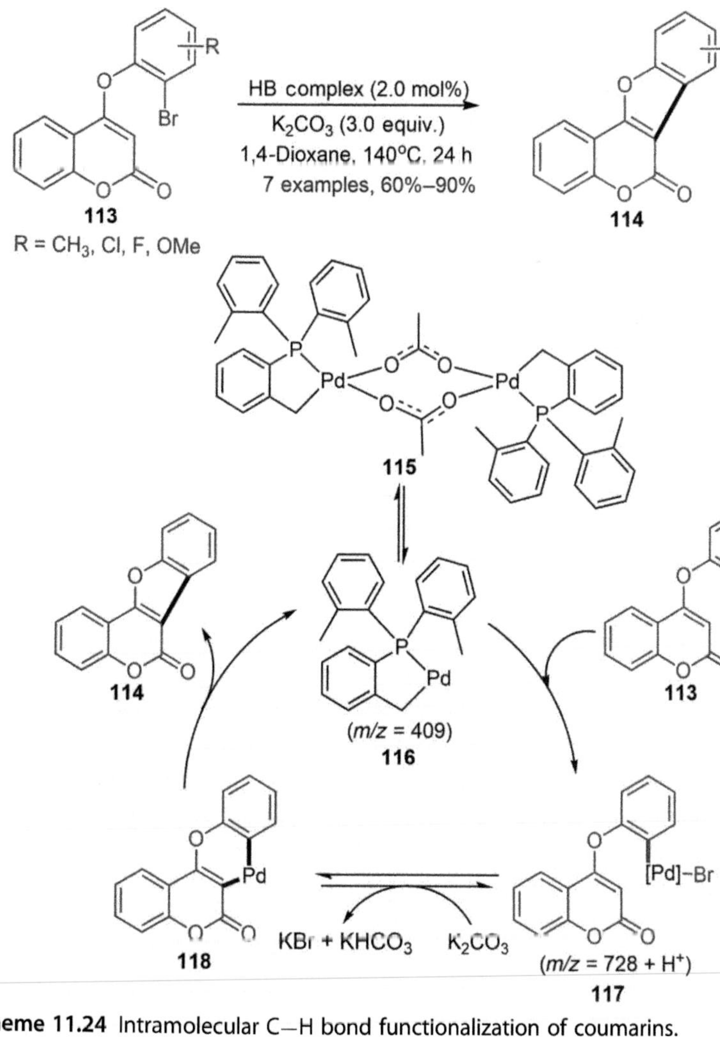

Scheme 11.24 Intramolecular C–H bond functionalization of coumarins.

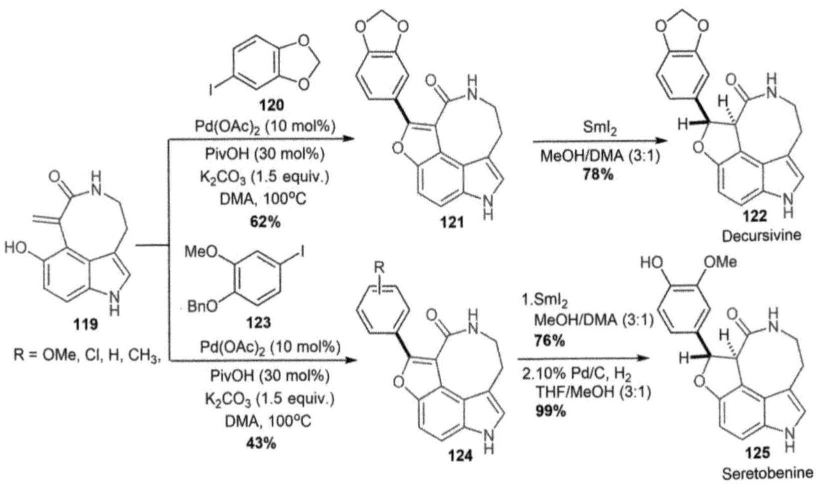

Scheme 11.25 Synthesis of decursivine and seretobenine via palladium-catalyzed oxidative C—H activation.

An efficient synthesis of aryl benzofurans via palladium-catalyzed oxidative C—H activation was reported by Jia and co-workers,[55] which allowed them to synthesize decursivine, serotobenine, and their analogues[56] without any requirement for a protection strategy (Scheme 11.25).

To describe the generality of this protocol, authors tried the synthesis of 2,3-diarylbenzofurans with the above-mentioned optimized conditions. Surprisingly, desired benzofuran **129** was obtained in only 5% yield, whereas Heck-type product **128** was obtained with 73% yield as the only Z isomer, which helped the authors to propose the reaction mechanism. As they carried out the reaction of substituted hydroxyl styrene **126** with 4-iodo anisole **127** at higher temperature (160°C) with the optimized conditions affording the olefinic product **128** (evidenced by NOESY) in 86% yield within 1 h, which on further treatment produced the desired product **129** in 82% yield, clearly suggesting that the olefin **128** is the intermediate formed in this tandem reaction (Scheme 11.26).

This was further supported by kinetic isotopic effect studies carried out on deuterated 2-hydroxy styryl derivative that gave the value of KIE ($K_{H/D}$=1.93), suggesting the C—H bond cleavage involvement in the rate determining step (Scheme 11.27).

Besides these examples there are several others that have been reported to proceed via an oxidative addition mechanism.[57–61] The key point to be remembered for the identification of this mechanism would be to

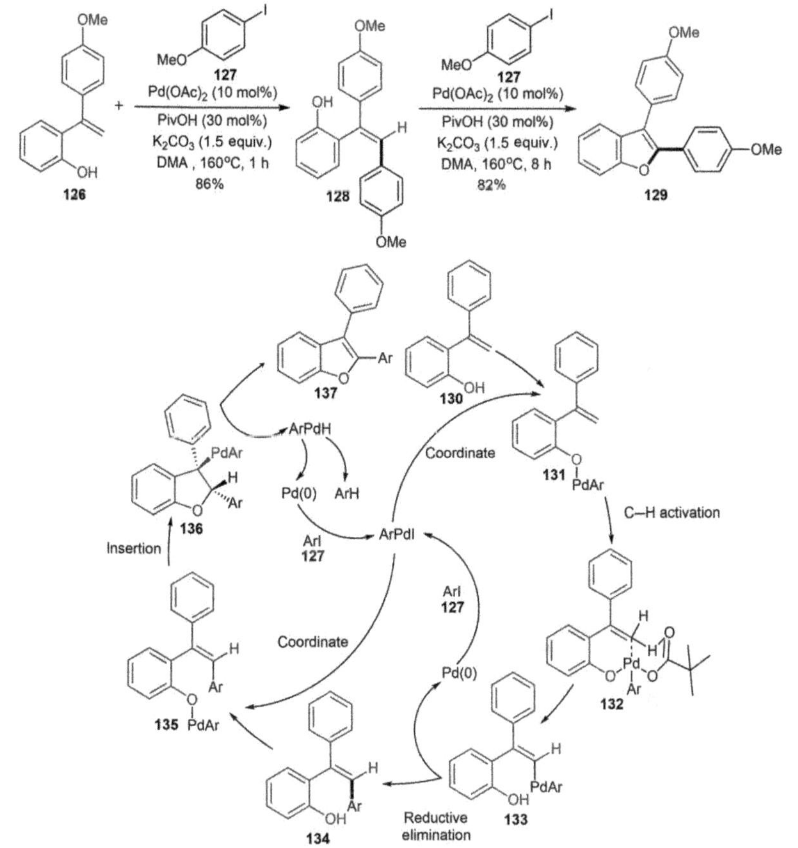

Scheme 11.26 Diarylbenzofuran synthesis via palladium-catalyzed oxidative C—H activation.

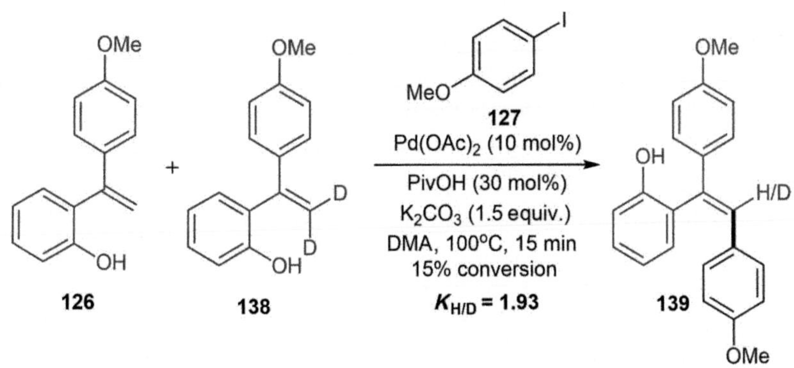

Scheme 11.27 Kinetic isotope effect.

perform kinetic isotopic effect studies, which commonly give higher values of K_H/K_D, a characteristic for oxidative addition. Electrospray ionization mass spectrometry has proved to be a useful technique for the identification of the oxidative addition intermediates, and could therefore serve as an important tool for mechanistic purposes.

11.3.3 Concerted Metalation Deprotonation Mechanism

Another important mechanistic pathway in C—H activation, which emerged recently after S_EAr and oxidative addition, is CMD[62–65] or σ-bond metathesis of the C—H bond generally proceeding via the formation of a four-membered transition state.

One of the classical examples in literature portraying the CMD pathway was put forward by Fagnou and co-workers[66] in 2006 for the highly efficient direct arylation of benzene with aryl bromides. The arylation was found to proceed smoothly in the presence of $Pd(OAc)_2$ and Davephos **150** as a phosphine ligand, along with a stoichiometric amount of pivalic acid **149** acting as a proton shuttle that was proposed to be the main component in C—H bond cleavage.

Experimental and computational studies performed on the catalytic reaction revealed that the biaryl palladium complex was formed by reacting phosphine-ligated aryl palladium pivalate with benzene via the CMD pathway. Depending on the interaction of pivalic acid **149** with palladium, two pathways were proposed, wherein pivalate anion reversibly interacts with palladium catalyst (pathway A) or it remains bound to the palladium metal throughout the catalytic cycle (pathway B). DFT analysis also supported the pivalate anion influence on C—H bond cleaving transition state, which decreases the transition state energy (24.9 kcal/mol) by 1.3 kcal/mol as compared to other bicarbonate anions (26.2 kcal/mol), clearly explaining the enhanced reactivity of palladium pivalate co-catalyst system in terms of energetic (Scheme 11.28).

From the above findings and the earlier reports of Fagnou on intermolecular direct arylation[67] of pentafluorobenzene with bromo toluene, it was clear that the electron deficient arenes also work well under stoichiometric amounts of organometallic reagents and carbonate as the base. Thus, the reactivity of arene is more dependent on the C—H acidity rather than the nucleophilicity of arenes. This was later supported by mechanistic work done for palladium-catalyzed direct arylation of arenes by Echavarren and co-workers[68,69] involving the significant role of carbonate or related ligand in proton abstraction, and the reaction was

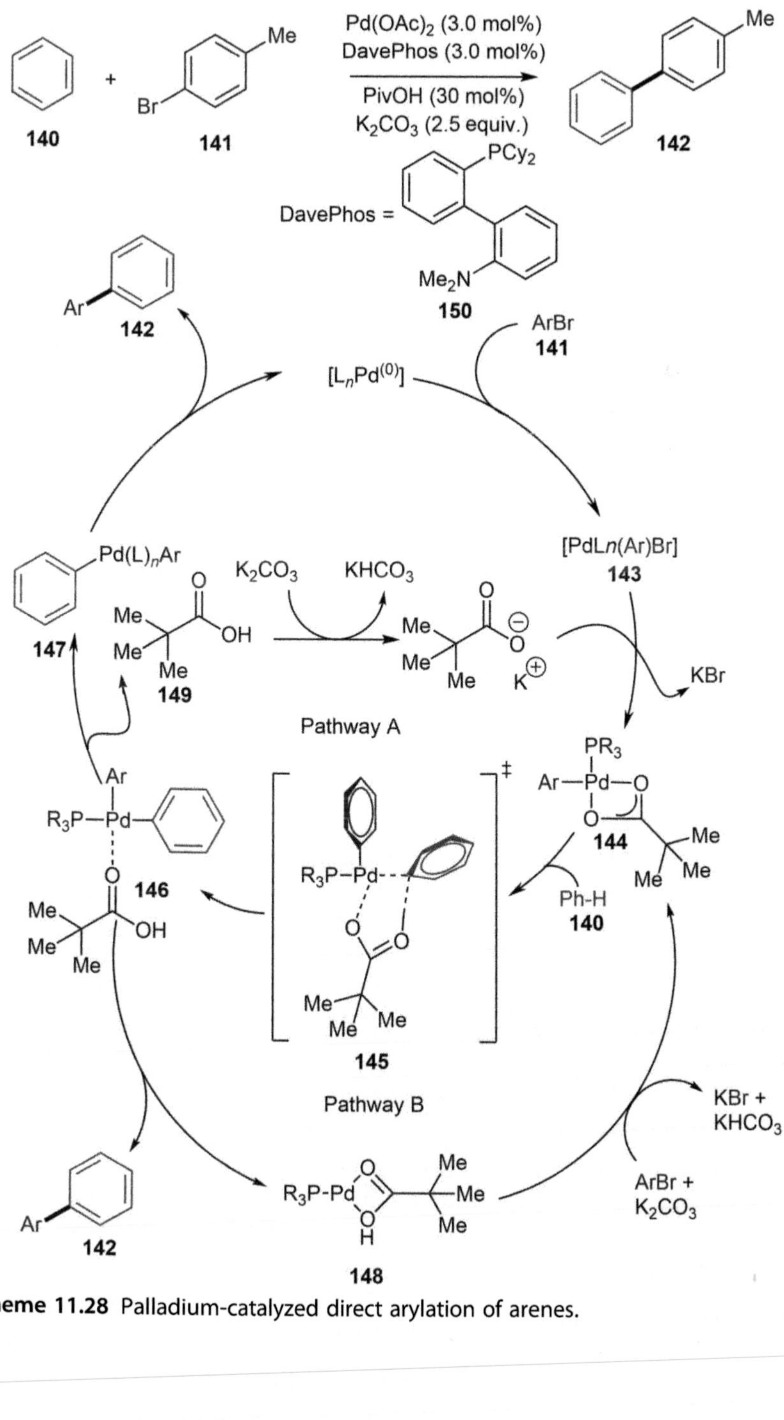

Scheme 11.28 Palladium-catalyzed direct arylation of arenes.

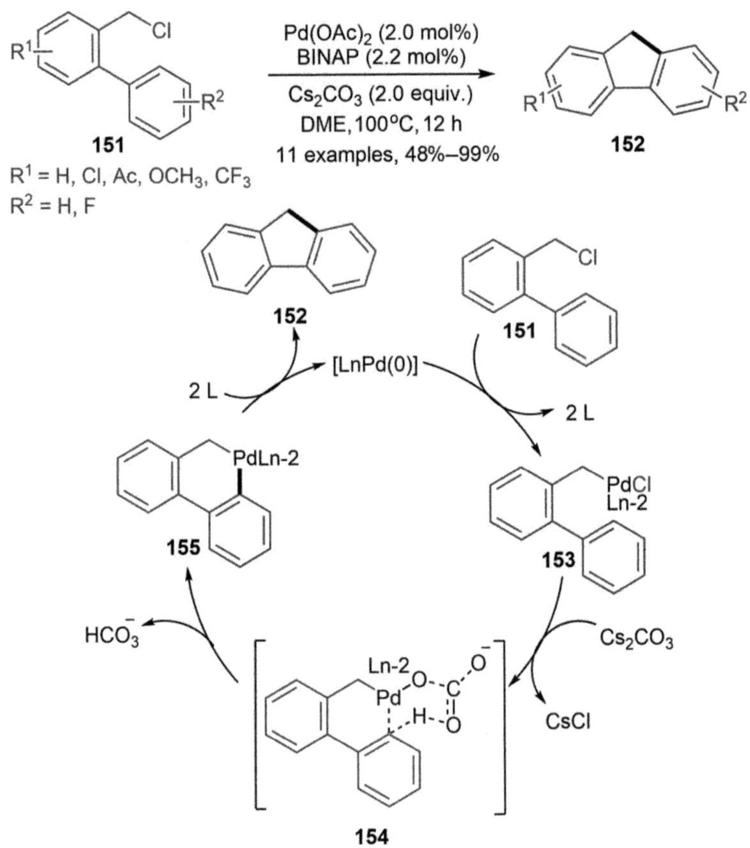

Scheme 11.29 Synthesis of fluorenes via C–H bond activation.

therefore proposed to proceed through a CMD pathway. These findings overcome the limitations of a S_EAR mechanism, where nucleophilicity of arene[70] was important and therefore, the CMD pathway could help the development of more efficient protocols for the arylation of a variety of unactivated arenes for the synthesis of biaryls and other complex molecules via direct arylation[71]

In 2009, Chang and co-workers reported a palladium-catalyzed highly efficient and milder protocol for the synthesis of various derivatives of fluorenes[72] (Scheme 11.29). A mechanistic proposal given on the basis of Hammett studies showed a positive ρ value of 0.42, indicating partial negative charge buildup during the rate limiting transition state. Further, kinetic isotope effect studies (Scheme 11.30) also showed intermolecular (KIE) $K_{H/D}=2.1$, thus ruling out the possibility of an electrophilic palladation pathway. The authors therefore proposed the reaction

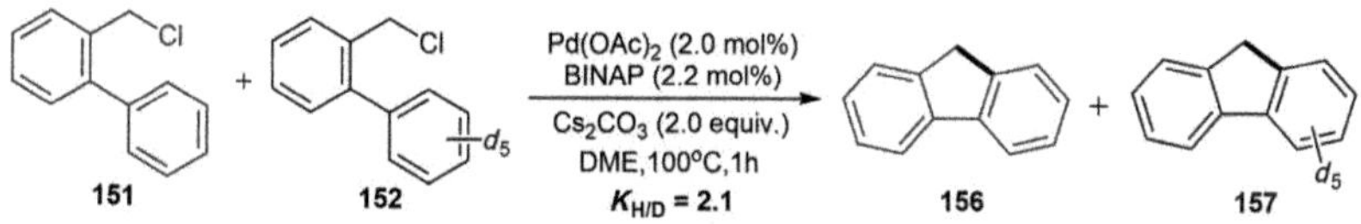

Scheme 11.30 Intermolecular KIE experiment.

to proceed via a base-assisted protonative metalation pathway through the formation of a palladacyclic intermediate **155** that finally undergoes reductive elimination to yield the desired fluorene **152.**

On a similar note in 2010, Fairlamb and co-workers reported palladium-catalyzed regioselective C—H bond functionalization of 2-pyrones **158**[73] that provide researchers with an easy access to tricyclic pyrone compounds. Based on the various stoichiometric studies and experimental data, two pathways were depicted as shown in Scheme 11.31, wherein the CMD pathway seems to be the most probable one.

Another example on the regioselective arylation of *N*-alkylated indoles[74] via palladium-catalyzed dual C—H bond functionalization came out during the same time and was proposed to be following the CMD pathway. Interesting findings were put forth by the authors revealing the *N*-alkylated indoles not getting arylated under the highly acidic conditions that are most likely for oxidative arylation of benzofurans,[75] resulting in rapid decomposition. But when the mixture of PivOH with a slight excess of AgOAc was used, the reaction was found to work smoothly without any decomposition. The mechanistic proposal given on the basis of experimental and computational studies involved the concerted metalation deprotonation pathway. Also, when the kinetic isotopic effect studies were performed on arenes, a large KIE of >3 was obtained, indicating that the C—H bond cleavage of arenes was the rate determining step (Scheme 11.32).

More recently, Yu and co-workers disclosed an efficient intramolecular base-assisted direct arylation of phenyl acetic acid with aryl boron reagent in the presence of Ac-Ile-OH as amino acid based ligand and Ag_2CO_3 as an oxidant.[76] To gain insight into the mechanism which was proposed to operate via carboxylate ion assisted[77] concerted metalation deprotonation, competition experiments were performed between electron rich and electron poor phenyl acetic acid substrates. Different rates of reactions were obtained for both electron rich and electron poor substrates, as in the presence of ligand, electron-poor substrates gave a higher initial rate (k_a/k_b=2.17), and in the absence of ligand, electron-rich substrates gave lower initial rates (k_a/k_b=0.48), as shown in Scheme 11.33. Therefore, on

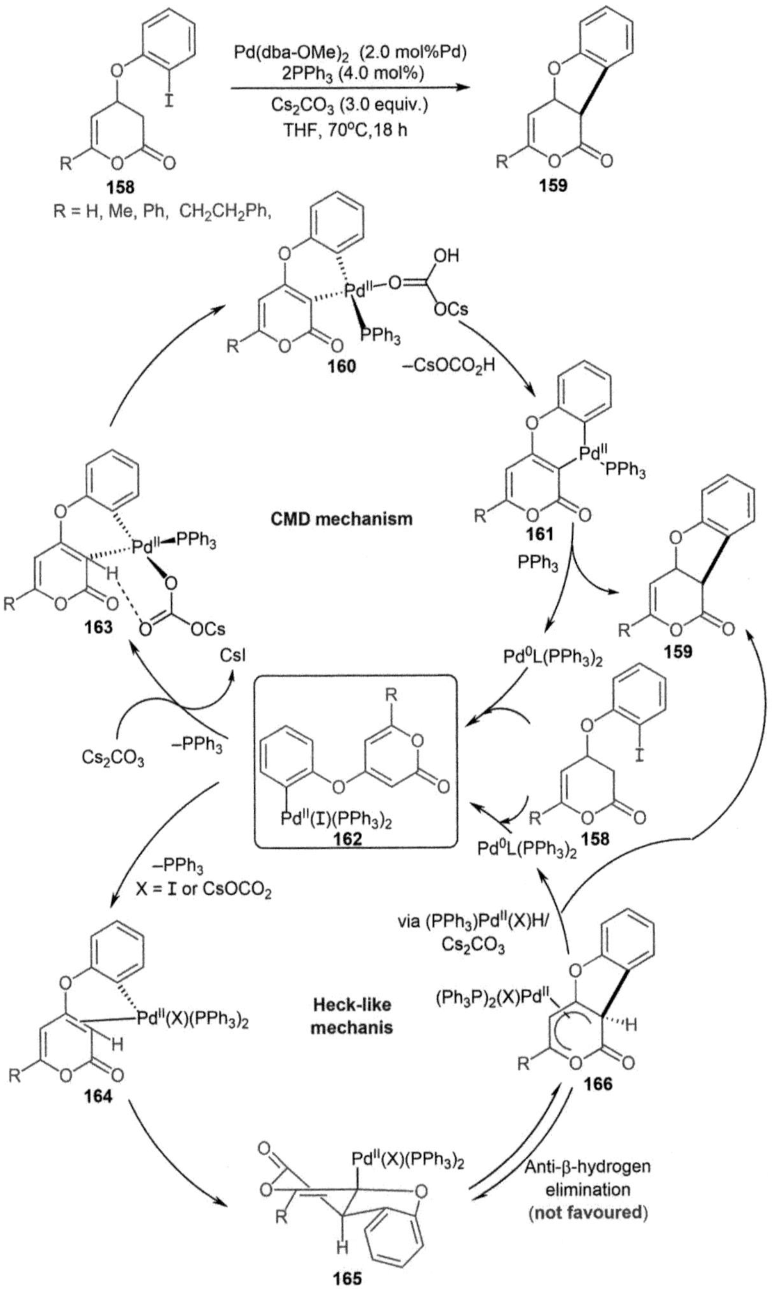

Scheme 11.31 Palladium-catalyzed regioselective C-3 arylation of 2-pyrones via C—H bond functionalization.

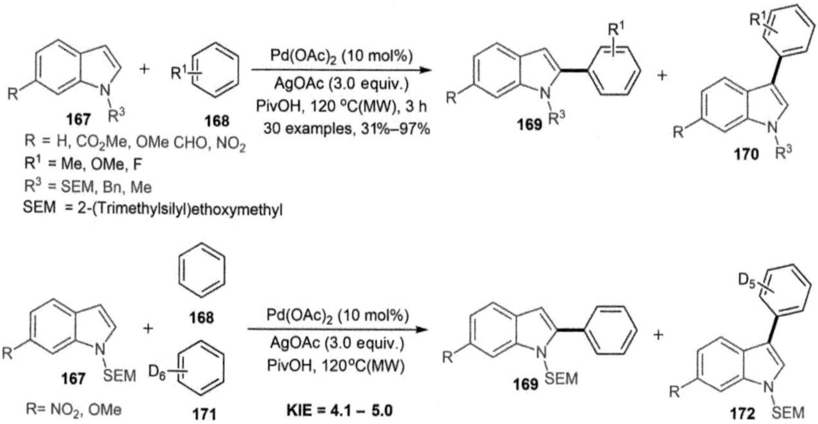

Scheme 11.32 Regioselective arylation of indoles via palladium-catalyzed C—H activation.

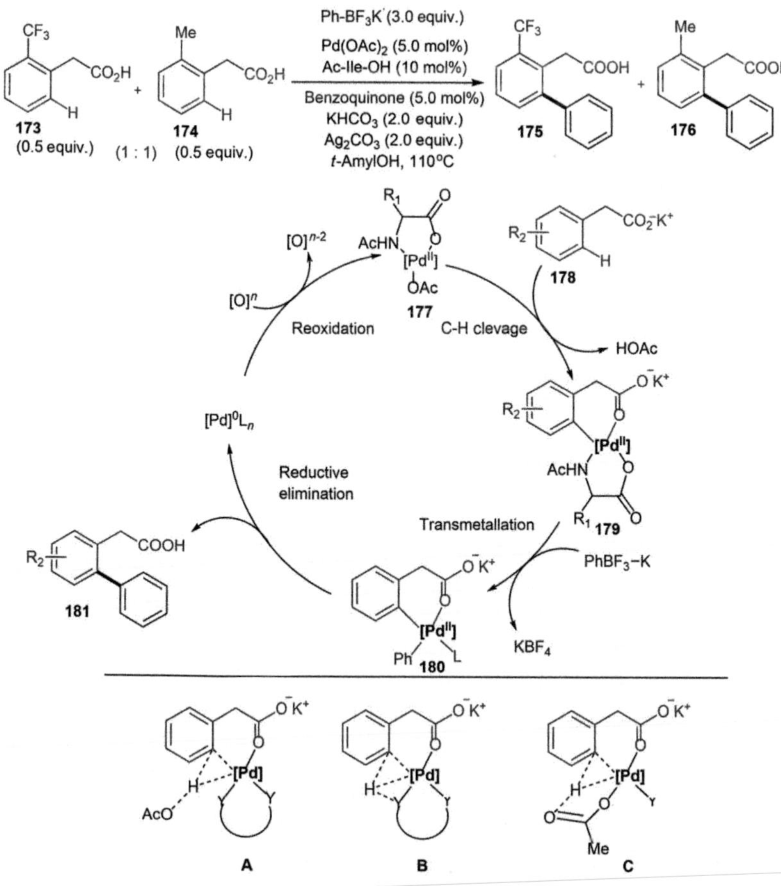

Scheme 11.33 Carboxylate-assisted direct arylation of phenylacetic acid.

the basis of these findings it was proposed that C—H bond cleavage proceeds via a concerted metalation deprotonation pathway in the presence of Ac-Ile-OH ligand and in the absence of ligand, it follows an electrophilic palladation pathway. Subsequent transmetalation with organoboron reagent produces diarylated Pd(II) species **180**, which after reductive elimination yield the desired arylated phenyl acetic acid product **181**.

In 2014, Zeng and co-workers disclosed a protocol for arylation of unactivated C—H bonds with aryl bromides to produce new arylated products via palladium-catalyzed intermolecular C—H activation.[78] The directing group 8-aminoquinolinyl[79] used in the reaction, along with K_2CO_3 base, plays an important role in deciding the regiospecificity at the β-position of carboxamides. The mechanistic proposal given on the basis of DFT studies performed on the reaction of **182** with bromobenzene suggested that the C—H activation proceeds via the concerted metalation deprotonation pathway. The 8-aminoquinolinyl group on the substrate acts as an L- and X-type ligand to bind with Pd(II) by proton abstraction, further followed by β-C—H bond cleavage via concerted metalation deprotonation (Scheme 11.34).

Many other examples[80–82] besides those mentioned above have been reported in the literature that are proposed to follow the concerted metalation deprotonation pathway.

11.3.4 Heck-Type Carbometalation Mechanism

The Heck reaction has become one of the most fundamental transition metal catalyzed reactions for C—C bond formation over the past few decades for the synthesis of a variety of complex organic molecules, despite the limitation of activation steps for the formation of an aryl or vinyl halide followed by the Pd(0) catalyzed coupling of both reaction partners. A Heck-type carbometalation or oxidative Heck reaction would help to resolve the need for preactivation.[6]

Fujiwara and co-workers disclosed one of the earliest examples of direct arylation of olefins with arene[83] via palladium-catalyzed oxidative Heck reaction after their previous reports on the synthesis of stilbene derivatives by direct substitution of arenes on styrene palladium chloride complex.[84] After these initial findings, several reports have come out showing a Heck-type carbometalation mechanism, with those giving mechanistic evidence described below.

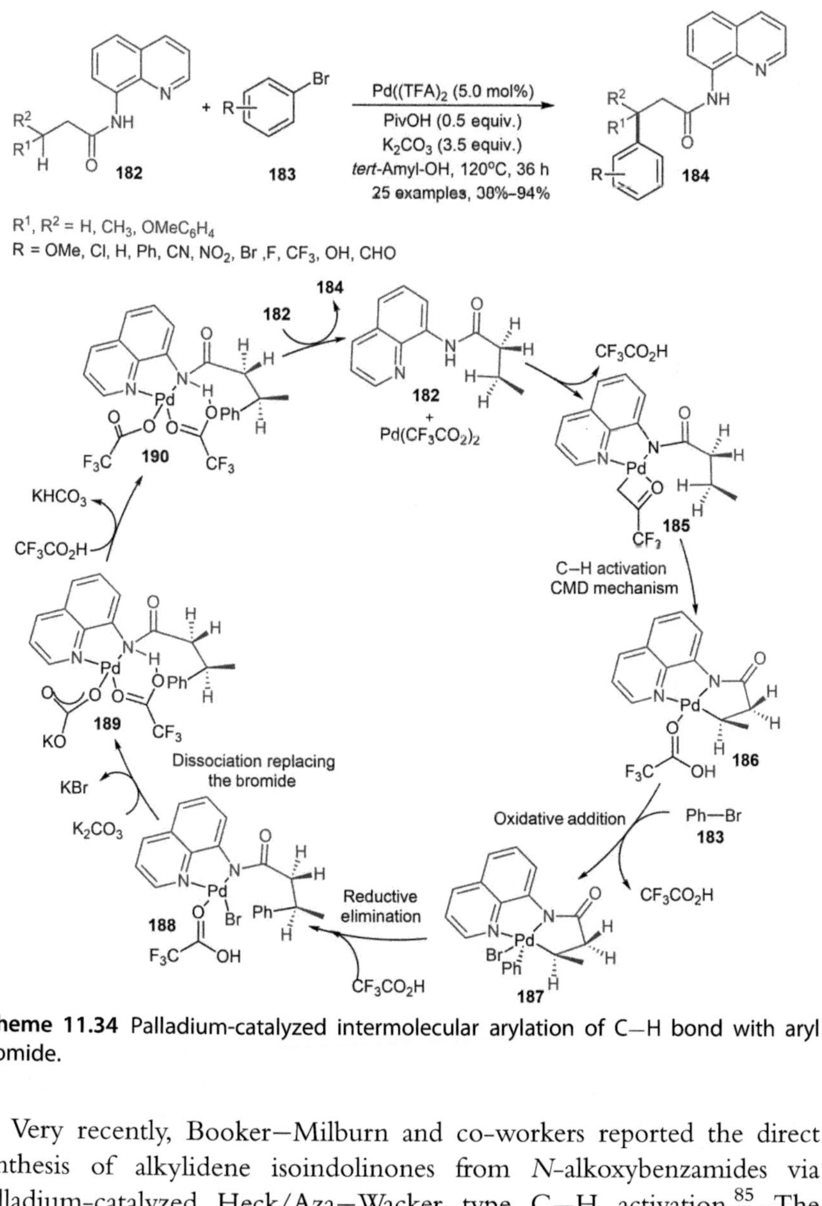

Scheme 11.34 Palladium-catalyzed intermolecular arylation of C–H bond with aryl bromide.

Very recently, Booker–Milburn and co-workers reported the direct synthesis of alkylidene isoindolinones from *N*-alkoxybenzamides via palladium-catalyzed Heck/Aza–Wacker type C–H activation.[85] The reaction was found to proceed in the presence of benzoquinone and dioxygen as an oxidant, giving exceptional E-selectivity. A mechanistic proposal was given on the basis of preliminary experiments, in which palladacyclic intermediate **194** is formed via insertion of Pd(II) in between the C–H bond that further undergoes Heck-type addition to produce another intermediate **195**, finally undergoing β-hydride elimination to generate the desired product **197**. In another Pd(II) catalytic cycle, this intermediate **197** follows Aza—Wacker sequence to produce cyclized product **199**, which after C—C bond rotation and β-hydride elimination generates the desired alkylidine isoindolinone **193** (Scheme 11.35).

Another example of Heck-type C—H functionalization was disclosed by Fu and co-workers as an extensive report on the mechanistic finding for the ligand-controlled regioselective arylation of thiophenes,[86] previously reported by Itami and co-workers.[87] Based on extensive computational studies, it was concluded that the use of 2,2′-bipyridyl ligand in arylation of thiophenes gives rise to an α-arylated (C-5) product and proceeds via a

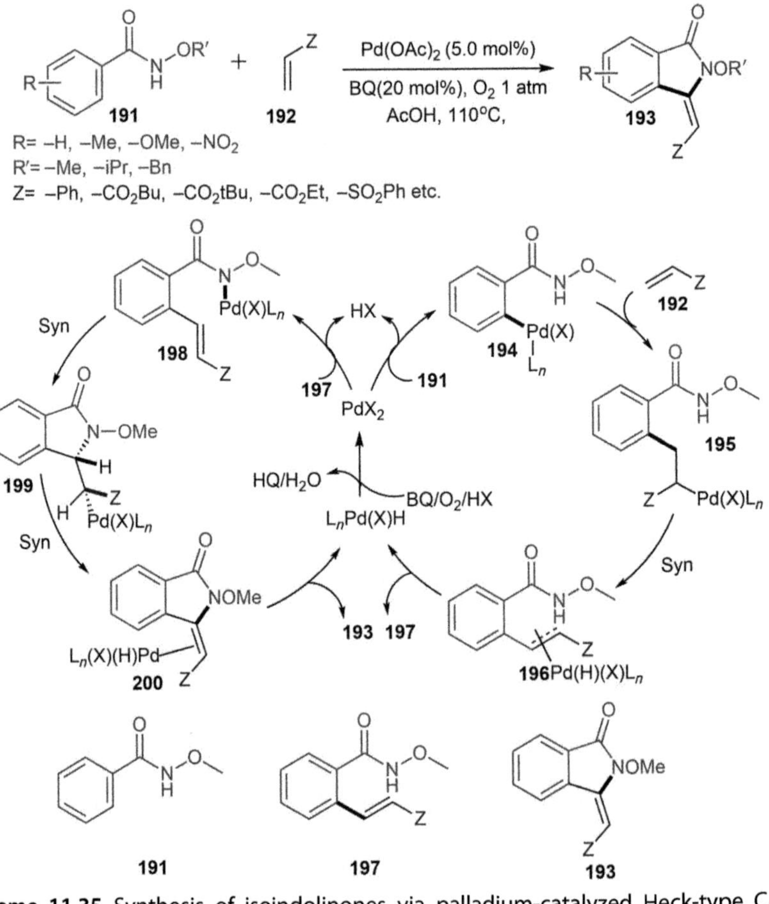

Scheme 11.35 Synthesis of isoindolinones via palladium-catalyzed Heck-type C—H activation.

metalation/deprotonation pathway that is energetically favored. The use of the bulky fluorinated phosphine ligand $P[OCH(CF_3)_2]_3$ resulted in β-arylation (C-4) of thiophene, proceeding via a Heck-type arylation which is a kinetically favored pathway. Theoretical evidence also supported the Heck-type arylation that could proceed through an anti-β-hydride elimination process (Scheme 11.36).

Subsequently, a Fujiwara–Moritani arylation protocol with the product obtained via a β-hydride elimination process was disclosed by Jana and co-workers.[88] An efficient synthesis of 2-aryl indoles and indolines was achieved through palladium-catalyzed C–H activation at room temperature (Scheme 11.37). Based on the control experiments, a mechanistic cycle was proposed to pass through cascade C–H insertion/1,2-carboamination/β-hydride elimination to produce indoles and indolines via C–H insertion and 1,2 carboamination passing through a π-benzyl Pd stabilization.

Another example giving direct access to indoles[89] and substituted 2-quinolinones was disclosed recently by Maiti and co-workers[90] (Scheme 11.38) after reacting diphenyl amines **208** with substituted cinnamic acids **209** under optimized conditions. A mechanistic proposal was given on the basis of competition experiments and deuterium labeling studies (Scheme 11.39) giving higher values of intramolecular $K_{H/D}=4.5$ and intermolecular $K_{H/D}=4.7$, indicating C–H bond

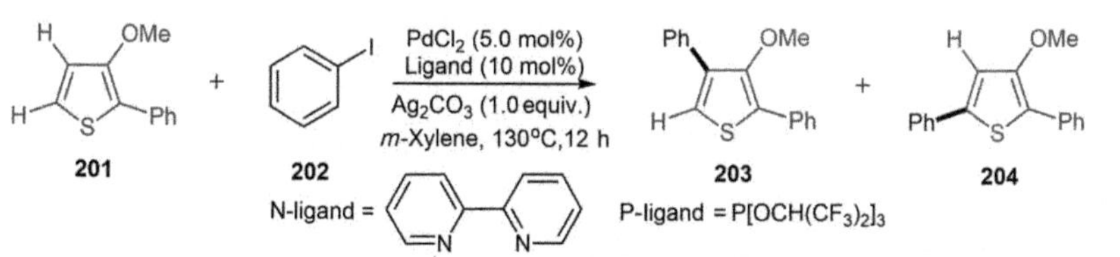

Scheme 11.36 Regioselective arylation of thiophenes via Pd-catalyzed C–H activation.

Scheme 11.37 Palladium-catalyzed synthesis of indoles.

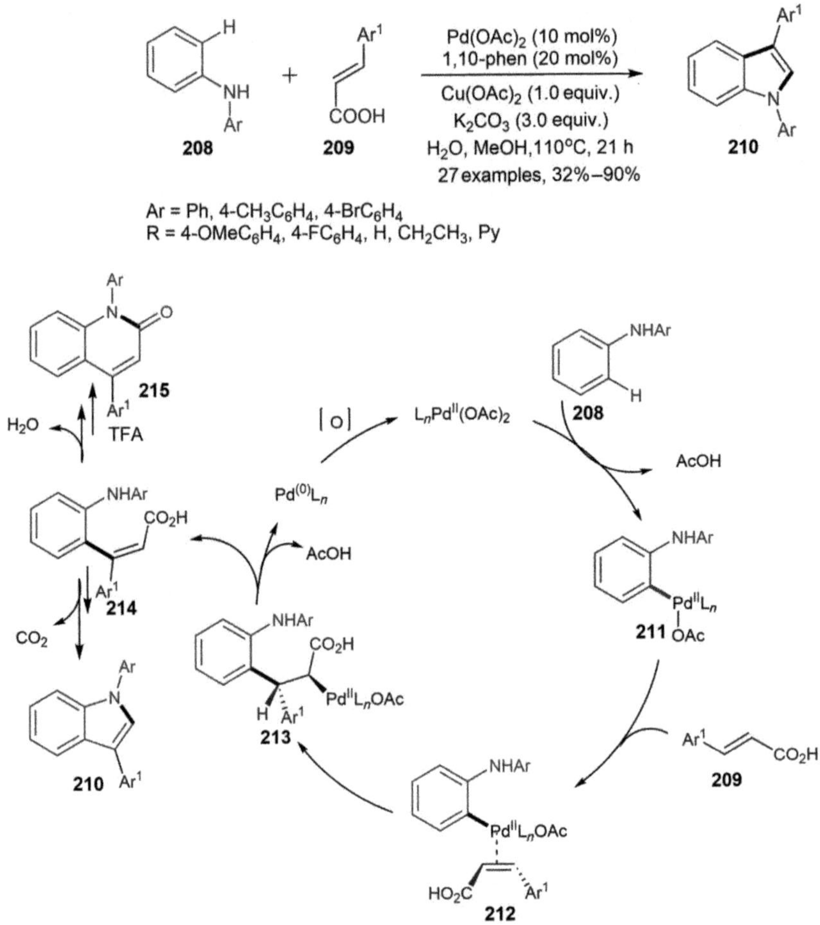

Scheme 11.38 Synthesis of indoles via palladium-catalyzed C—H activation.

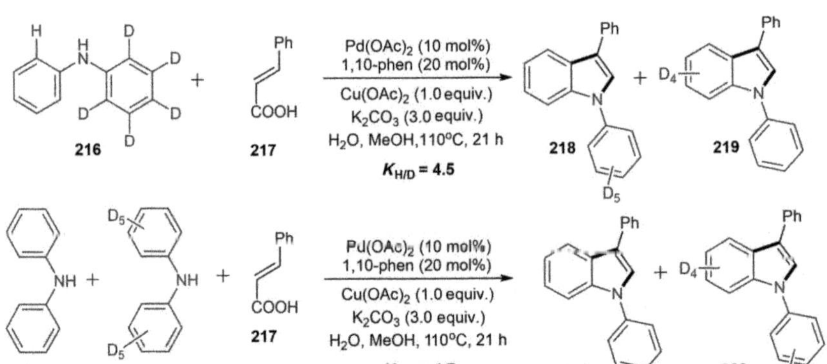

Scheme 11.39 Deuterium labeling experiments.

cleavage involvement in the rate-determining step. The mechanistic cycle proceeds via the formation of an intermediate **211** by ortho-palladation of diarylamine **208**, which further interacts with α,β-unsaturated acid **209** leading to the formation of **212**. Subsequent formation of ortho-olefinated intermediate **214** takes place via β-migratory insertion followed by β-hydride elimination. Further, **214** undergoes decarboxylation in the absence of TFA following β-hydride elimination furnishing the desired indole **210**.

We have highlighted herein some of the representative examples following Heck-type carbometalation. Besides this there are several other examples reported in the literature which proceed via Heck-type C—H activation[91–94] and concerted S_E3[95] processes.

11.4 CONCLUSION

In this chapter we have described the various mechanistic pathways that are commonly applicable for palladium-catalyzed C—H activation. These mechanism have exhibited certain characteristics that have been highlighted as a part of the chapter to assist the readers to predict with greater certainty the type of mechanism operating in their catalytic processes.

ACKNOWLEDGMENTS

A.R.K. would like to thank the Department of Science and Technology for the DST Inspire Faculty Award (IFA12-CH-22). The authors also would like to thank University Grants Commission, India for UGC-JRF and UGC-SAP fellowships to A.G. and V.G. The Alexander von Humboldt Foundation is also thanked by the authors for an equipment grant.

LIST OF ABBREVIATIONS

CMD	Concerted Metalation Deprotonation
DFT	Density Functional Theory
Dmpe	Dimethylphosphinoethane
Dppm	Bis(diphenylphosphino)methane
KIE	Kinetic isotopic effect
PivOH	Pivalic acid
S_E3	Electrophilic Substitution Trimolecular
S_EAr	Electrophilic Aromatic Substitution
TEMPO	2,2,6,6-Tetramethylpyridinium oxide
UV-Vis	Ultra-violet and Visible Spectroscopy

REFERENCES

1. For reviews on C—H bond functionalizations, see (a) Hartwig, J. F. *Chem. Soc. Rev.* **2011,** *40*, 1992. (b) Gutekunst, W. R.; Baran, P. S. *Chem. Soc. Rev.* **2011,** *40*, 1976. (c) Mkhalid, I. A. I.; Barnard, J. H.; Marder, T. B.; Murphy, J. M.; Hartwig, J. F. *Chem. Rev.* **2010,** *110*, 890. (d) Willis, M. C. *Chem. Rev.* **2010,** *110*, 725. (e) Ackermann, L.; Potukuchi, H. K. *Org. Biomol. Chem.* **2010,** *8*, 4503. (f) Fagnou, K. *Top. Curr. Chem.* **2010,** *292*, 35. (g) Lyons, T. W.; Sanford, M. S. *Chem. Rev.* **2010,** *110*, 1147. (h) Dudnik, A. S.; Gevorgyan, V. *Angew. Chem. Int. Ed.* **2010,** *49*, 2096. (i) Ackermann, L.; Vicente, R.; Kapdi, A. R. *Angew. Chem. Int. Ed.* **2009,** *48*, 9792. (j) Miura, M.; Satoh, T. In *In Modern Arylation Methods;* Ackermann, L., Ed.; Wiley-VCH: Weinheim, 2009. (k) Kakiuchi, F.; Kochi, T. *Synthesis* **2008,** 3013. (l) Alberico, D.; Scott, M. E.; Lautens, M. *Chem. Rev.* **2007,** *107*, 174. (m) Stuart, D. R.; Fagnou, K. *Aldrichimica Acta* **2007,** *40*, 35. (n) Seregin, I. V.; Gevorgyan, V. *Chem. Soc. Rev.* **2007,** *36*, 1173. (o) Satoh, T.; Miura, M. *Chem. Lett.* **2007,** *36*, 200. (p) Fairlamb, I. J. S. *Chem. Soc. Rev.* **2007,** *36*, 1036. (q) Godula, K.; Sames, D. *Science* **2006,** *312*, 67. (r) Dyker, G. *Angew. Chem. Int. Ed.* **1999,** *38*, 1698. (s) Trost, B. M. *Science* **1991,** *254*, 1471.
2. Some recent reviews of C—H bond activation by organometallic complexes: (a) Sen, A. *Acc. Chem. Res.* **1998,** *31*, 550. (b) Guari, Y. S.; Etiennne, S.; Chaudret, B. *Eur. J. Inorg. Chem.* **1999,** 1047. (c) Jones, W. D. *Science* **1942,** *2000*, 287. (d) Ritleng, V.; Sirlin, C.; Pfeffer, M. *Chem. Rev.* **2002,** *102*, 1731. (e) Dyker, G. *Handbook of the C—H Transformation;* Wiley-VCH: Weinheim, 2005.
3. (a) Bedford, R. B. *Chem. Commun.* **2003,** 1787. (b) Tang, S. Y.; Yu, H. Z.; You, W. L.; Guo, Q. X. *Chin. J. Chem. Phys.* **2013,** *26*, 415.
4. Gretz, E.; Oliver, T. F.; Sen, A. *J. Am. Chem. Soc.* **1987,** *109*, 26.
5. Gorelsky, S. I.; Lapointe, D.; Fagnou, K. *J. Am. Chem. Soc.* **2008,** *130*, 10848.
6. Grimster, N. P.; Gauntlett, C.; Godfrey, C. R. A.; Gaunt, M. J. *Angew. Chem. Int. Ed.* **2005,** *44*, 3125.
7. McMurray, L.; O'Hara, F.; Gaunt, M. J. *Chem. Soc. Rev.* **2011,** *40*, 1885.
8. Shilov, A. E.; Shteinman, A. A. *Coord. Chem. Rev.* **1977,** *24*, 97.
9. (a) Labinger, J. A.; Herring, A. M.; Lyon, D. K.; Luinstra, G. A.; Bercaw, J. A. *Organometallics* **1993,** *12*, 895. (b) Luinstra, G. A.; Labinger, J. A.; Bercaw, J. A. *J. Am. Chem. Soc.* **1993,** *115*, 3004. (c) Luinstra, G. A.; Wang, L.; Stahl, S. S.; Labinger, J. A.; Bercaw, J. A. *Organometallics* **1994,** *13*, 755. (d) Labinger, J. A.; Bercaw, J. E. *Nature* **2000,** *417*, 507.
10. (a) Colby, D. A.; Bergman, R. G.; Ellman, J. A. *Chem. Rev.* **2010,** *110*, 624. (b) Lewis, J. C.; Bergman, R. G.; Ellman, J. A. *Acc. Chem. Res.* **2008,** *41*, 1013. (c) Bergman, R. G. *Nature* **2007,** *446*, 391.
11. (a) Crabtree, R. H. *Chem. Rev.* **2010,** *110*, 575. (b) Crabtree, R. H. *J. Chem. Soc. Dalton Trans.* **2001,** *17*, 2437. (c) Crabtree, R. H. *J. Organomet. Chem.* **2004,** *689*, 4083.
12. Kakiuchi, F.; Murai, S. *Acc. Chem. Res.* **2002,** *35*, 826.
13. Goldman, A. S.; Goldberg, K. I. *Activation and Functionalisation of C—H Bonds. ACS Symposium Series*, Vol. 885. American Chemical Society: Washington, DC, 2004.
14. Labinger, J. A.; Bercaw, J. E. *J. Organomet. Chem.* **2015,** *793*, 47.
15. (a) Kapdi, A. R. Organometallic aspects of C—H bond activation/functionalisation. In *Organometallic Chemistry: Volume 38 Specialist Periodical Reports;* Fairlamb, I. J. S., Ed.; RSC: Cambridge, 2012; p 48. (b) Li, H.; Li, B.-J.; Shi, Z.-J. *Catal. Sci. Technol.* **2011,** *1*, 191.
16. (a) Ames, D. E.; Bull, D. *Tetrahedron* **1982,** *38*, 383. (b) Akita, Y.; Inoue, A.; Yamamoto, K.; Ohta, A.; Kurihara, T.; Shimizu, M. *Heterocycles* **1985,** *23*, 2327. (c) Shaikh, T. M.; Hong, F. E. *J. Organomet. Chem.* **2016,** *801*, 139.

17. Engelin, C. J.; Fristrup, P. *Molecules* **2011,** *16*, 951.
18. (a) Halpern, J.; Peters, E. *J. Chem. Phys.* **1955,** *23*, 605. (b) Peters, E.; Halpern, J. *J. Phys. Chem.* **1955,** *59*, 793.
19. (a) Webster, A. H.; Halpern, J. *J. Phys. Chem.* **1956,** *60*, 280. (b) Halpern, J. *J. Phys. Chem.* **1959,** *63*, 398.
20. Halpern, J. *Discuss. Faraday Soc.* **1968,** *46*, 7.
21. (a) Vaska, L.; DiLuzio, J. W. *J. Am. Chem. Soc.* **1962,** *84*, 679. (b) Vaska, L. *Acc. Chem. Res.* **1968,** *1*, 335. (c) Vaska, L.; Werneke, M. F. *Ann. N. Y. Acad. Sci.* **1971,** *172*, 546.
22. Chatt, J.; Davidson, J. M. *J. Chem. Soc.* **1965,** 843.
23. Tang, S.-Y.; Yu, H.-Z.; You, W.-L.; Guo, Q.-X. *Chin. J. Chem. Phys.* **2013,** *26*, 415–423.
24. (a) Goldshleger, N. F.; Eskova, V. V.; Shilov, A. E.; Shteinman, A. A. *Zh. Fiz. Khim.* **1972,** *46*, 1353. (b) Kushch, L. A.; Lavrushko, V. V.; Misharin, Y. S.; Moravskii, A. P.; Shilov, A. E. *Nouv. J. Chim.* **1983,** 7, 729.
25. For references of Shilov system using metals other than Platinum see: For Palladium-(a) Gretz, E.; Oliver, T. F.; Sen, A. *J. Am. Chem. Soc.* **1987,** *109*, 8109. (b) Kao, L. C.; Hutson, A. C.; Sen, A. *J. Am. Chem. Soc.* **1991,** *113*, 700.For Mercury-Periana, R. A.; Taube, D. J.; Evitt, E. R.; Loffler, D. G.; Wentrcek, P. R.; Voss, G.; Matsuda, T. *Science* **1993,** *259*, 3.
26. Regioselective palladium-catalyzed C–H bond functionalization see:(a) Choy, P. Y., Lau, C. K.; Kwong, F. K. *J. Org. Chem.* **2011,** *76*, 80.
27. (a) Catellani, M.; Chiusoli, G. P. *J. Organomet. Chem.* **1982,** *239*, 35. (b) Catellani, M.; Fagnola, M. C. *Angew. Chem. Int. Ed. Engl.* **1994,** *33*, 2421. (c) Catellani, M.; Frignani, F.; Rangnoni, A. *Angew. Chem. Int. Ed. Engl.* **1997,** *36*, 119.For recent reviews on Catellani reactions, see(d)Catellani, M.; Motti, E.; Della, N. *Acc. Chem. Res.* **2008,** *41*, 1512.(e)Catellani, M. *Top. Organomet. Chem.* **2005,** *14*, 21.(f)Faccini, F.; Motti, E.; Catellani, M. *J. Am. Chem. Soc.* **2004,** *126*, 78.(g)Della, N.; Maestri, C. G.; Catellani, M. *Chem. Eur. J.* **2009,** *15*, 7850.(h)Catellani, M.; Chiusoli, G. P. *J. Organomet. Chem.* **1985,** *286*, C13.
28. (a) Catellani, M.; Motti, E.; Paterlini, L.; Bocelli, G.; Righi, L. *J. Organomet. Chem.* **1999,** *580*, 191. (b) Catellani, M.; Cugini, F.; Bocelli, G. *J. Organomet. Chem.* **1999,** *584*, 63.
29. Park, C. H.; Ryabova, V.; Seregin, I. V.; Sromek, A. W.; Gevorgyan, V. *Org. Lett.* **2004,** *6*, 1159.
30. Chuprakov, S.; Chernyak, N.; Dudnik, A. D.; Gevorgyan, V. *Org. Lett.* **2007,** *9*, 2333.
31. Lane, B. S.; Brown, M. A.; Sames, D. *J. Am. Chem. Soc.* **2005,** *127*, 8050.
32. Zhao, J.; Larock, R. C. *J. Org. Chem.* **2006,** *71*, 5340.
33. Juwaini, N. A. B.; Ng, J. K. P.; Seayad, J. *ACS Catal* **2012,** *2*, 1787.
34. (a) Leo, A.; Hansch, C.; Elkins, D. *Chem. Rev.* **1971,** *71*, 525. (b) Hansch, C.; Leo, A.; Taft, R. W. *Chem. Rev.* **1991,** *91*, 165.
35. (a) Boiko, V. N. *Beilstein J. Org. Chem.* **2010,** *6*, 880. (b) Tlili, A.; Billard, T. *Angew. Chem. Int. Ed.* **2013,** *52*, 6818. (c) Liang, T.; Neumann, C. N.; Ritter, T. *Angew. Chem. Int. Ed.* **2013,** *52*, 8214.
36. Xu, C.; Shen, Q. *Org. Lett.* **2014,** *16*, 2046.
37. Xu, H.; Shang, M.; Dai, H.-X.; Yu, J.-Q. *Org. Lett.* **2015,** *17*, 3830.
38. Wang, X.; Leow, D.; Yu, J.-Q. *J. Am. Chem. Soc.* **2011,** *133*, 13864.
39. Shen, P. X.; Wang, X. C.; Wang, P.; Zhu, R. Y.; Yu, J. Q. *J. Am. Chem. Soc.* **2015,** *137*, 11574.
40. Zhao, Y., Xie, Y.; Xia, C.; Huanga, H. *Adv. Synth. Catal.* **2014,** *356*, 2471.
41. Tjutrins, J.; Arndtsen, B. A. *J. Am. Chem. Soc.* **2015,** *137*, 12050.

42. Bunescu, A.; Piou, T.; Wang, Q.; Zhu, J. *Org. Lett.* **2015,** *17*, 334.
43. Gu, H.; Qiu, Z.; Zhang, Z.; Li, J.; Yan, B. *Dalton Trans.* **2015,** *44*, 9839.
44. Examples of Palladium-catalyzed oxidative C–H activation: (a) Xie, P.; Xie, Y.; Qian, B.; Zhou, H.; Xia, C.; Huang, H. *J. Am. Chem. Soc.* **2012,** *134*, 9902. (b) Wang, L.; Huang, J.; Peng, S.; Liu, H.; Jiang, X. *Angew. Chem. Int. Ed.* **2013,** *52*, 1768. (c) Mazuela, J.; Bannerjee, D.; Backvall, J. *J. Am. Chem. Soc.* **2015,** *137*, 9559.
45. Wang, D.; Hao, X.; Wu, D.; Yu, J. Q. *Org. Lett.* **2006,** *15*, 3387.
46. For the direct involvement of PhI(OAc)2 in C–H bond oxidation see:(a) Yoneyama, T.; Crabtree, R. H. *J. Mol. Catal. A: Chem.* **1996,** *108*, 35. (b) Dick, A. R.; Hull, K. L.; Sanford, M. S. *J. Am. Chem. Soc.* **2004,** *126*, 2300.
47. For more details on Pd(IV) catalysis see: (a) Sehnal, P.; Taylor, R. J. K.; Fairlamb, I. J. S. *Chem. Rev.* **2010,** *110*, 824. (b) Hickman, A. J.; Sanford, M. S. *Nature* **2012,** *484*, 177. (c) Topczewski, J. J.; Sanford, M. S. *Chem. Sci.* **2015,** *6*, 70.
48. Byers, P. K.; Canty, A. J.; Skelton, B. W.; White, A. H. *J. Chem. Soc. Chem. Commun.* **1986,** 1722.
49. Canty, A. J.; Denney, M. C.; Skelton, B. W.; White, A. H. *Organometallics* **2004,** *23*, 1122.
50. Dick, A. R.; Kampf, J. W.; Sanford, M. S. *J. Am. Chem. Soc.* **2005,** *127*, 12790.
51. Tang, S.; Peng, P.; Pi, S.-F.; Liang, Y.; Wang, N.-X.; Li, J.-H. *Org. Lett.* **2008,** *10*, 1179.
52. Wu, T.; Mu, X.; Liu, G. *Angew. Chem. Int. Ed.* **2011,** *50*, 12578.
53. Kapdi, A. R.; Karbelkar, A.; Naik, M.; Pednekar, S.; Fisher, C.; Schulzke, C.; Tromp, M. *RSC. Adv.*, 3; 2013, 20905.
54. For reviews on cyclometallated palladacyclic complexes see: (a) Niu, J.-L.; Hao, X.-Q.; Gong, J.-F.; Song, M.-P. *Dalton Trans.* **2011,** 5135. (b) Zhao, Q.; Huang, C.; Li, F. *Chem. Soc. Rev.* **2011,** *40*, 2508. (c) Selander, N.; Szabó, K. J. *Chem. Rev.* **2011,** *111*, 2048. (d) Alonso, D. A.; Nájera, C. *Chem. Soc. Rev.* **2010,** *39*, 2891. (e) Chi, Y.; Chou, P.-T. *Chem. Soc. Rev.* **2010,** *39*, 638. (f) Albrecht, M. *Chem. Rev.* **2010,** *110*, 576. (g) Baranoff, E.; Yum, J.-H.; Graetzel, M.; Nazeeruddin, M. K. *J. Organomet. Chem.* **2009,** *694*, 2661. (h) Djukic, J.-P.; Sortais, J.-B.; Barloy, L.; Pfeffer, M. *Eur. J. Inorg. Chem.* **2009,** 817. (i) Dupont, J.; Pfeffer, M., Eds. *Palladacycles. Synthesis, Characterization and Applications;* Wiley-VCH: Weinheim, 2008. (j) Djukic, J.-P.; Hijazi, A.; Flack, H. D.; Bernardinelli, G. *Chem. Soc. Rev.* **2008,** *37*, 406. (k) Moral-Morales, D.; Jensen, C. M., Eds. *The Chemistry of Pincer Compounds;* Elsevier: Amsterdam, 2007. (l) Mohra, F.; Privér, S. H.; Bhargava, S. K.; Bennett, M. A. *Coord. Chem. Rev.* **2006,** *250*, 1851. (m) Dupont, J.; Consorti, C. S.; Spencer, J. *Chem. Rev.* **2005,** *105*, 2527. (n) Pérez, M. A.; Quijada, R.; Ortega-Jiménez, F.; Alvarez-Toledano, C. *J. Mol. Catal. A: Chem.* **2005,** *226*, 291. (o) Beletskaya, I. P.; Cheprakov, A. V. *J. Organomet. Chem.* **2004,** *689*, 4055. (p) Bedford, R. B. *Chem. Commun.* **2003,** 1787. (q) van der Boom, M. E.; Milstein, D. *Chem. Rev.* **2003,** *103*, 1759. (r) Dupont, J.; Pfeffer, M.; Spencer, J. *Eur. J. Inorg. Chem.* **2001,** 1917. (s) Herrman, W. A.; Böhm, V. P. W.; Reisenger, C.-P. *J. Organomet. Chem.* **1999,** *576*, 23.
55. Guo, L.; Zhang, F.; Hu, W.; Li, L.; Jia, Y. *Chem. Commun.* **2014,** *50*, 3299.
56. (a) Qin, H.; Xu, Z.; Cui, Y.; Jia, Y. *Angew. Chem. Int. Ed.* **2011,** *50*, 4447. (b) Hu, W.; Qin, H.; Cui, Y.; Jia, Y. *Chem. Eur. J.* **2013,** *19*, 3139.
57. (a) Reinartz, S.; White, P. S.; Brookhart, M.; Templeton, J. L. *J. Am. Chem. Soc.* **2001,** *123*, 12724. (b) Norris, C. M.; Reinartz, S.; White, P. S.; Templeton, J. L. *Organometallics* **2002,** *21*, 5649. (c) Meiere, S. H.; Brooks, B. C.; Gunnoe, T. B.; Sabat, M.; Harman, W. D. *Organometallics* **2001,** *20*, 1038. (d) Brown, S. N.; Myers, A. W.; Fulton, J. R.; Mayer, J. M. *Organometallics* **1998,** *17*, 3364.

58. (a) Boller, T. M.; Murphy, J. M.; Hapke, M.; Ishiyama, T.; Miyaura, N.; Hartwig, J. F. *J. Am. Chem. Soc.* **2005,** *127*, 14263. (b) Salem, B.; Delort, E.; Klotz, P.; Suffert, J. *Org. Lett.* **2003,** *5*, 2307. (c) Bour, C.; Suffert, J. *Org. Lett.* **2005,** 7, 653. (d) Mota, A. J.; Dedieu, A.; Bour, C.; Suffert, J. *J. Am. Chem. Soc.* **2005,** *127*, 7171.
59. (a) Larsson, J. M.; Zahao, T. S. N.; Szabo, K. J. *Org. Lett.* **2011,** *13*, 1888. (b) Racowski, J. M.; Ball, N. D.; Sanford, M. S. *J. Am. Chem. Soc.* **2011,** *133*, 18022. (c) Nix, J. L.; Hao, X. Q.; Gong, J. F.; Song, M. P. *Dalton. Trans.* **2011,** *40*, 5135. (d) Zhou, W.; Li, H.; Wang, L. *Org. Lett.* **2012,** *14*, 4594.
60. (a) Coombs, J. R.; Haeffner, F.; Kliman, L. T.; Morken, J. P. *J. Am. Chem. Soc.* **2013,** *135*, 11222. (b) Chen, M.; Ren, Z. H.; Wang, Y. Y.; Guan, Z. H. *Angew. Chem. Int. Ed.* **2013,** *52*, 14196. (c) Zhai, W. Q.; Peng, R. F.; Jin, B.; Wang, G. W. *Org. Lett.* **2014,** *16*, 1638. (d) Zhang, J. Z.; Bellomo, A.; Trongsiriwat, N.; Jia, T.; Carroll, P.; Dreher, S. D.; Tudge, M. T.; Yin, H.; Robinson, J. R.; Schelter, E. J.; Walsh, P. J. *J. Am. Chem. Soc.* **2014,** *136*, 6267.
61. (a) Larry, M. W.; Thiel, W. *J. Org. Chem.* **2014,** *79*, 12136. (b) Furukawa, S.; Endo, M.; Komatsu, T. *ACS. Catal.* **2014,** *4*, 3533. (c) Samala, S.; Pallavi, P.; Kumar, R.; Arigela, R. K.; Singh, G.; Ampapathi, R. S.; Amulya, P.; Datta, S.; Patra, A.; Kundu, B. *Chem. Eur. J* **2014,** *20*, 14344. (d) Sun, W.; Wang, M.; Zhang, Y.; Wang, L. *Org. Lett.* **2015,** *17*, 426.
62. Gorelsky, S. I.; Lapointe, D.; Fagnou, K. *J. Am. Chem. Soc.* **2008,** *130*, 10848.
63. For previous proposals of a CMD mechanism, see(a) Sokolov, V. I.; Troitskaya, L. L., Reutov, O. A. *J. Organomet. Chem.* **1979,** *182*, 537. (b) Davies, D. L.; Donald, S. M. A.; Macgregor, S. A. *J. Am. Chem. Soc.* **2005,** *127*, 13754. (c) Tenn, W. J., III; Young, K. J. H.; Bhalla, G.; Oxgaard, J.; Goddard, W. A., III; Periana, R. A. *J. Am. Chem. Soc.* **2005,** *127*, 14172. (d) for a review, see Ryabov, A. D. *Chem. Rev.* **1990,** *90*, 403.
64. Oxgaard, J.; Tenn, W. J., III; Nielsen, R. J.; Periana, R. A.; Goddard, W. A., III *Organometallics* **2007,** *26*, 1565.
65. Boutadla, Y.; Davies, D. L.; Macgregor, S. A.; Poblador-Bahamonde, A. I. *Dalton Trans.* **2009,** 5820.
66. Lafrance, M.; Fagnou, K. *J. Am. Chem. Soc.* **2006,** *128*, 16496.
67. Lafrance, M.; Rowley, C. N.; Woo, T. K.; Fagnou, K. K. *J. Am. Chem. Soc.* **2006,** *128*, 8754.
68. (a) Garcia-Cuadrado, D.; Braga, A. A. C.; Maseras, F.; Echavarren, A. M. *J. Am. Chem. Soc.* **2006,** *128*, 1066. (b) García-Cuadrado, D.; de Mendoza, P.; Braga, A. A. C.; Maseras, F.; Echavarren, A. M. *J. Am. Chem. Soc.* **2007,** *129*, 6880.
69. Pascual, S.; de Mendoza, P.; Braga, A. A. C.; Maseras, F.; Echavarren, A. M. *Tetrahedron* **2008,** *64*, 6021.
70. (a) Lane, B. S.; Sames, D. *Org. Lett.* **2004,** *6*, 2897. (b) Li, W. J.; Nelson, D. P.; Jensen, M. S.; Hoerrner, R. S.; Javadi, G. J.; Cai, D.; Larsen, R. D. *Org. Lett.* **2003,** *5*, 4835. (c) Mori, A.; Sekiguchi, A.; Masui, K.; Shimada, T.; Horie, M.; Osakada, K.; Kawamoto, M.; Ikeda, T. *J. Am. Chem. Soc.* **2003,** *125*, 1700. (d) Wang, X.; Lane, B. S.; Sames, D. *J. Am. Chem. Soc.* **2005,** *127*, 4996.
71. Thveau, L.; Verrier, C.; Lassalas, P.; Martin, T.; Dupas, G.; Querolle, O.; Hijfte, L. V.; Marsais, F.; Hoarau, C. *Chem. Eur. J.* **2011,** *17*, 14450.
72. Hwang, S. J.; Kim, H. J.; Chang, S. *Org. Lett.* **2009,** *11*, 20.
73. Burns, M. J.; Thatcher, R. J.; Taylor, R. J. K.; Fairlamb, I. J. S. *Dalton Trans.* **2010,** *39*, 10391.
74. Potavathri, S.; Pereira, K. C.; Gorelsky, S. I.; Pike, A.; LeBris, A. P.; DeBoef, B. *J. Am. Chem. Soc.* **2010,** *132*, 14676
75. Dwight, T. A.; Rue, N. R.; Charyk, D.; Josselyn, R.; DeBoef, B. *Org. Lett.* **2007,** *9*, 3137.

76. Engle, K. M.; Thuy-Boun, P. S.; Dang, M.; Yu, J. Q. *J. Am. Chem. Soc.* **2011,** *133*, 18183.
77. Ackermann, L. *Chem. Rev.* **2011,** *111*, 1315.
78. Wei, Y.; Tang, H.; Cong, X.; Rao, B.; Wu, C.; Zeng, X. *Org. Lett.* **2014,** *16*, 2248.
79. For selected reviews on the use of 8-aminoquinolinyl as directing groups in C–H activation reactions, see:(a) Rouquet, G.; Chatani, N. *Angew. Chem. Int. Ed.* **2013,** *52*, 11726. (b) Corbet, M.; De Campo, F. *Angew. Chem. Int. Ed.* **2013,** *52*, 9896.
80. (a) Johnson, L. K.; Huff, R. L.; White, P. S.; Brookhart, M. *J. Am. Chem. Soc.* **2000,** *122*, 6686. (b) Kurzeev, S. A.; Kazankov, G. M.; Ryabov, A. M. *Inorg. Chim. Acta* **2002,** *340*, 192. (c) Hennessy, E. J.; Buchwald, S. L. *J. Am. Chem. Soc.* **2003,** *125*, 12085. (d) Mota, A. J.; Dedieu, A.; Bour, C.; Suffert, J. *J. Am. Chem. Soc.* **2005,** *127*, 7171.(e)Davies, D. L.; Donald, S. M. A.; Macgregor, S. A. *J. Am. Chem. Soc.* **2005,** *127*, 13754.(f)Chaumontet, M.; Piccardi, R.; Audic, N.; Hitce, J.; Peglion, J. L.; Clot, E.; Baudoin, O. *J. Am. Chem. Soc* **2008,** *130*, 15157.
81. (a) Cook, A. K.; Sanford, M. S. *J. Am. Chem. Soc.* **2015,** *137*, 3109. (b) Liu, H.; Shi, G.; Pan, S.; Jiang, Y.; Zhang, Y. *Org. Lett.* **2013,** *15*, 4098. (c) Maleckis, A.; Kampf, J. W.; Sanford, M. S. *J. Am. Chem. Soc.* **2013,** *135*, 6618.
82. (a) Haines, B. E.; Musaev, D. G. *ACS Catal.* **2015,** *5*, 830. (b) Duan, Y.; Li, L.; Chen, M. W.; Yu, C.-B.; Fan, H.-J.; Zhou, Y.-G. *J. Am. Chem. Soc.* **2014,** *136*, 7688. (c) Musaev, D. G.; Kaledin, A.; Shi, B. F.; Yu, J. Q. *J. Am. Chem. Soc.* **2012,** *134*, 1690.
83. Fujiwara, Y.; Moritani, I.; Danno, S.; Asano, R.; Teranishi, S. *J. Am. Chem. Soc.* **1969,** *91*, 7166.
84. Moritani, I.; Fujiwara, Y. *Tetrahedron Lett.* **1967,** *8*, 1119.
85. Wrigglesworth, J. W.; Cox, B.; Lloyd-Jones, G. C.; Booker-Milburn, K. I. *Org. Lett.* **2011,** *13*, 19.
86. Tang, S.-Y.; Guo, Q.-X.; Fu, Y. *Chem. Eur. J.* **2011,** *17*, 13866.
87. (a) Yanagisawa, S.; Ueda, K.; Sekizawa, H.; Itami, K. *J. Am. Chem. Soc.* **2009,** *131*, 14622. (b) Ueda, K.; Yanagisawa, S.; Yamaguchi, J.; Itami, K. *Angew. Chem. Int. Ed.* **2010,** *49*, 8946.
88. Manna, M. K.; Hossian, A.; Jana, R. *Org. Lett.* **2015,** *17*, 672.
89. For selected examples for synthesis of indoles see:(a) Fischer, E.; Jourdan, F. *Ber. Dtsch. Chem. Ges.* **1883,** *16*, 2241. (b) Robinson, B. *Chem. Rev.* **1969,** *69*, 227. (c) Rutherford, J. L.; Rainka, M. P.; Buchwald, S. L. *J. Am. Chem. Soc.* **2002,** *124*, 15168. (d) Shen, M.; Leslie, B. E.; Driver, T. G. *Angew. Chem. Int. Ed.* **2008,** *47*, 5056.(e)Sharma, U.; Kancherla, R.; Naveen, T.; Agasti, S.; Maiti, D. *Angew. Chem. Int. Ed.* **2014,** *53*, 11895.(f)DeAngelis, A.; Wang, D.-H.; Buchwald, S. L. *Angew. Chem. Int. Ed.* **2013,** *52*, 3434.
90. Kancherla, R.; Naveen, T.; Maiti, D. *Chem. Eur. J.* **2015,** *21*, 1.
91. (a) Hansen, A. L.; Ebran, E. P.; Ahlquist, P.; Norrby, P. O.; Skrydstrup, T. *Angew. Chem. Int. Ed.* **2006,** *45*, 3349. (b) Ebran, J. P.; Hansen, A. L.; Gogsig, T. M.; Skrydstrup, T. *J. Am. Chem. Soc.* **2007,** *129*, 6931. (c) Lyons, T. W.; Sanford, M. S. *Tetrahedron* **2009,** *65*, 3211.
92. (a) Liu, L.; Yuan, H.; Fu, T.; Wang, T.; Gao, X.; Zeng, Z.; Zhu, J.; Zhao, Y. *J. Org. Chem.* **2014,** *79*, 80. (b) Hilton, M. J.; Xu, L. P.; Norrby, P. O.; Wu, Y. D.; Wiest, O.; Sigman, M. S. *J. Org. Chem.* **2014,** *79*, 11841. (c) Ardolino, M. J.; Morken, J. P. *J. Am. Chem. Soc.* **2014,** *136*, 7092.
93. (a) Gordillo, A.; Ortuno, M. A.; Lopez-Mardomingo, C.; Lledos, A.; Ujaque, G.; Jesus, E. D. *J. Am. Chem. Soc.* **2013,** *135*, 13749. (b) Agasti, S.; Sharma, U.; Naveen, T.; Maiti, D. *Chem. Commun.* **2015,** *51*, 5375. (c) Sharma, U.; Kancherla, R.; Naveen, T.; Agasti, S.; Maiti, D. *Angew. Chem. Int. Ed.* **2014,** *53*, 11895.
94. Agasti, S.; Maity, S.; Szabo, K. J.; Maiti, D. *Adv. Synth. Catal.* **2015,** *357*, 2331.
95. Zollinger, H. *Adv. Phys. Org. Chem* **1964,** *2*, 162.

CHAPTER 12

Recent Developments in Palladium-Catalyzed Natural Product Synthesis via C—H Activation

Neetipalli Thrimurtulu, Aniruddha Dey, Debabrata Maiti and Chandra M.R. Volla
Indian Institute of Technology Bombay, Mumbai, Maharashtra, India

Contents

12.1 INTRODUCTION

The central dogma of synthetic organic chemistry at the molecular level revolves around the formation of complex targets from simpler counterparts. As advances in this field unfolded over the years, the importance of chemical transformations which proceeded with uncompromised selectivity were duly realized. Contemporary synthetic chemists have recognized

Strategies for Palladium-Catalyzed Non-Directed and Directed C—H Bond Functionalization
DOI: http://dx.doi.org/10.1016/B978-0-12-805254-9.00012-8

the suitability of C—H functionalization as an effective tool in ensuring selective modifications within a complex molecular framework.[1–5] "Site selective C—H bond functionalization" subsequently became a tangible perception in retrosynthetic analysis. This spans over the benign protocols which offer a convenient and effective route to perform functionalization at selective sites within an organic core, besides providing notable levels of chemo, regioselectivity, and stereoselectivity in various cases. Most often, these are guided by the presence of a metal that steers the reactivity and the subsequent success of the protocol. Nevertheless, metal free reaction conditions have also been developed in parallel to metal-catalyzed conditions that serve to be equally efficient.

Palladium has been of elite mention in the field of metal-catalyzed organic transformations. As rich as its contribution had been in classical organometallic chemistry, it has equally dominated the emerging frontiers in organic catalytic transformations. Over the years, a tremendous pace of development was achieved through palladium-catalyzed methodologies which have certainly been the primary interest for innumerable research groups around the globe.[6–8] Cost-effectivness, easy procurement, and high catalytic competence made palladium chemistry among the most famous reactions recently discovered, the likes of which include the noble prize winning Heck, Suzuki and Negishi reactions to name a few. Following these milestones, the disclosure of palladium-catalyzed C—C, C—O, and C—N bond formation reactions have dominated the literature in the last decade, thus enriching it on a relentless scale. Importantly, the success of these protocols relied mostly upon the uncompromised functional selectivity that was obtained with palladium, and this attracted huge attention from synthetic chemists. Consequently, the notion of incorporating this superlative property into useful productivity achieved widespread recognition. Achieving the synthesis of complex targets was thus simplified by breaking up the synthetic outline into smaller fragments which defined metal-catalyzed methodological conversions. The art of retrosynthesis dwells upon the devising of a most feasible and economical route that can mimic the formation of naturally abundant products. Palladium-catalyzed reactions with their unique properties of easy catalyst loading, with exceptional tolerance towards functional groups, could be duly effected in devising retrosynthetic disconnections. Moreover, it is the availability of the variegated oxidation states of palladium, e.g., Pd(0), Pd(II), and Pd(IV), each with their innate chemical properties, that made it suitable for its smooth role in the catalytic cycle. Besides this, the last decade saw the unprecedented emergence of palladium-catalyzed C—H activation reactions. This was understood so that the importance of the C—H bond transformation was now coupled with a definite regioselectivity. Few of the earlier chapters in this book have already discussed the lucrative prospects of palladium-mediated C—H activation. What remains yet to be covered now are ideal applications of palladium-catalyzed reactions to simplify the art of total synthesis of natural products from simple precursors.

The goal of total synthesis has always been to achieve the target in the fewest possible steps. Adoption of metal-catalyzed transformations helped in this task, since multiple steps could now be superseded by a chemical reaction in a single step. In this chapter we intend to discuss and drive out the most exciting examples of total synthesis of natural products, interspersed by palladium-catalyzed methodologies. The chapter revolves around some elegant examples from the past literature, and weaves a composite discussion about the development of a strategic synthetic pathway to prepare a complex molecule. The chapter encompasses the unique challenges which synthetic chemists came upon while formulating the protocols, and how they ended up utilizing C—H activation as an important tool to eradicate these challenges. We present a glimpse of palladium-catalyzed sp^2 and sp^3 C—H functionalization reactions as part of the synthetic process in formation of large complex molecules and natural products.

12.2 TOTAL SYNTHESIS OF ASPIDOSPERMIDINE

The aspidosperma family has been nominated to be one of the largest groups among indole alkaloids. This indole alkaloid has attracted tremendous attention from the synthetic community since the 1960s, due to its attractive constitutional framework and the presence of multiple stereogenic centers. Thorsten Bach and co-workers initiated the synthesis of aspidospermidine[9] using a method which involves a selective 2-alkylation of indole in the presence of a 10 mol% palladium catalyst and norbornene. Bach developed the α-alkylation of indole derivative **3**, from indole and alkyl bromide in the presence of palladium and norbornene at 70°C. The 2-alkylated product **3** was obtained in 65% yield. Treatment of **3** with lithiumhexamethyldisilazide (LiHMDS) and quenching the generated lactam enolate with allylbromide afforded α-allyl-δ-lactam **4** in a high yield. A hydroboration – oxidation sequence of compound **4** gave alcohol **5**, which was oxidized with Dess-Martin Periodinane (DMP). The aldehyde intermediate **6** was transformed to the key intermediate aminoalcohol **8** from **7** by a three-step sequence, as shown in Scheme 12.1. The key intermediate aminoalcohol **8** was treated with MsCl and then with *t*-BuOK to promote an intramolecular cyclization reaction. The final step of the synthesis was accomplished by reducing imine with $NaBH_4$, affording aspidospermidine in nine steps in an overall yield of 15% (Scheme 12.1).

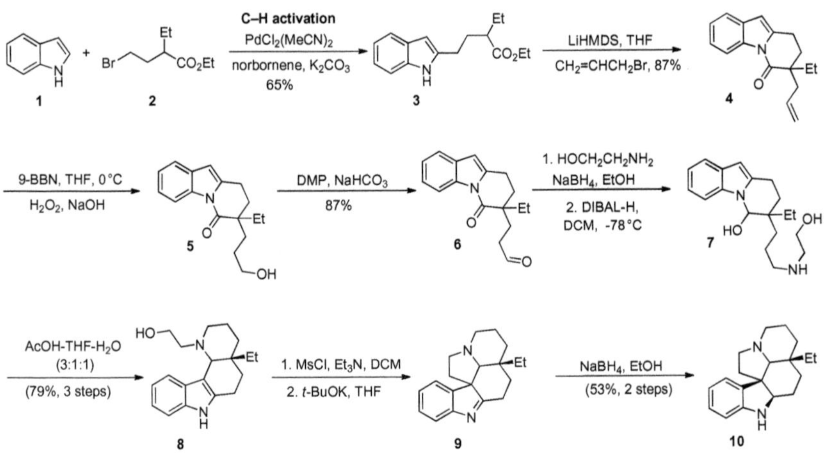

Scheme 12.1 Synthesis of aspidospermidine.

12.2.1 Total Synthesis of Goniomitine

Total synthesis of goniomitine was planned by 2-alkylation of TBS-tryptophol **11** using alkyliodide **12**.[9] It was found that the reaction under standard conditions was rather sluggish. The reaction was optimized with **12** in the presence of $PdCl_2$ and norbornene under air. Treatment of **13** with LiHMDS and quenching the generated lactam enolate with allyl bromide afforded α-allyl-δ-lactam **14** in 87% yield. A hydroboration–oxidation sequence of compound **14** gave alcohol **15**. Mitsunobu reaction of alcohol **15** with diphenylphosphorylazide (DPPA) gave azide **16**. Treatment of **16** with $LiAlH_4$ was followed by cyclization under mild acidic conditions, as well as deprotection of the silyl ether. The total synthesis of goniomitine was accomplished in seven steps from tryptophol **11** and iodide **12** with an overall yield of 35% with high diastereoselectivity (Scheme 12.2).

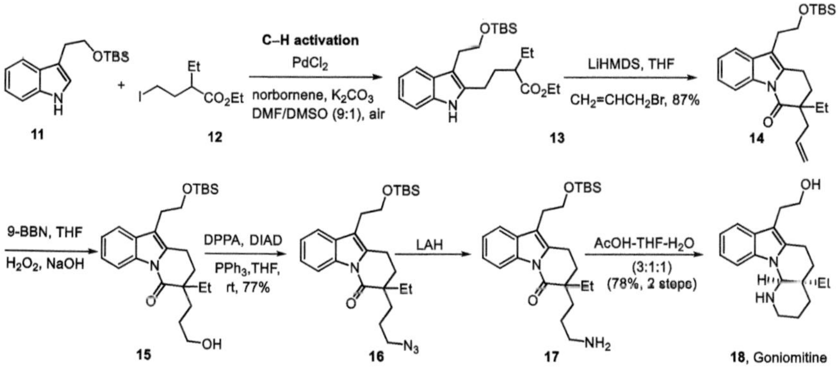

Scheme 12.2 Synthesis of goniomitine.

12.3 TOTAL SYNTHESIS OF (+) -LITHOSPERMIC ACID

Yu and co-workers demonstrated the synthesis of (+) -lithospermic acid[10] via a late stage intermolecular C—H olefination method. Yu started the synthesis from O-eugenol and arrived at **20** in three steps. Reacting **20** with pyrrolidinyl (*S*)-lactamide under Mitsunobu conditions led to the formation of **21** with an appended lactamide type chiral auxiliary in 83% yield. Removal of the benzyl group gave the corresponding phenol in quantitative yield, which was then protected with 3,4-dimethoxybenzyl bromide to provide the cyclization precursor **22** in 86% yield. Treatment of **22** with diazo transfer reagent in the presence of 1,8-Diazobicyclo[5.4.0]undec-7-ene (DBU) at room temperature afforded diazo compound **23**, which was immediately treated with 0.5 mol% of Davies' catalyst $Rh_2(S\text{-}DOSP)_4$. This provided the *trans*-dihydrobenzofuran core in good yield and diastereoselectivity (85%, 8:1). Basic hydrolysis conditions afforded the carboxylic acid **24** in 86% yield. Olefination of **24** with acrylate coupling partner **25** in the presence of 5 mol% of $Pd(OAc)_2$, 5 mol% of Ac-Ile-OH, and 2 equiv. of $KHCO_3$ in a solution of *t*-amyl alcohol at 85°C afforded the desired product **26** in 93% isolated yield. The final step of the synthesis was accomplished by methyl deprotection with TMSI-quinoline (Scheme 12.3).

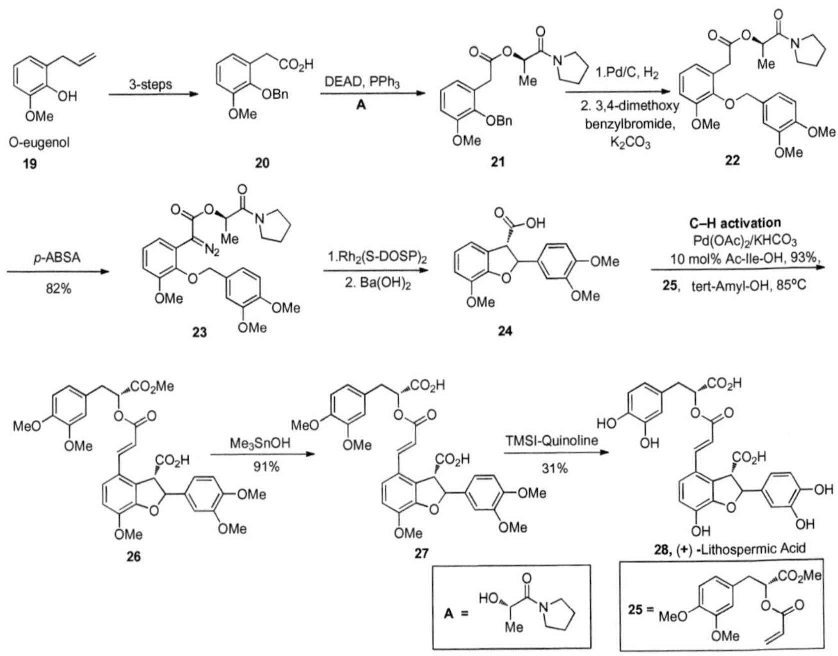

Scheme 12.3 Synthesis of (+)-lithospermic acid via a late stage intermolecular C—H olefination.

12.4 TOTAL SYNTHESIS OF PIPERCYCLOBUTANAMIDE A

The isolation of pipercyclobutanamide A from *Piper nigrum* was performed by Fujiwara, and its selective inhibition towards cytochrome P450 2D6 was later demonstrated. Owing to the presence of four different substituents on the cyclobutane ring, the synthesis of this natural product has remained elusive. A concise synthesis of the proposed structure of pipercyclobutanamide A[11] was reported by Baran and co-workers in 2012. Baran started the synthesis with methylcoumalate as a precursor which underwent a photochemical 4π-electrocyclization at reduced temperature to give the photopyrone **30**. This unstable intermediate was immediately hydrogenated and coupled to 8-aminoquinoline in the presence of 1-Ethyl-3-(3-dimethylaminopropyl)carbodiimide (EDC) to give the desired C—H olefination precursor **32** in 54% overall yield. Following this, reaction of cyclobutane **32** with 1-iodo-3,4-methylenedioxybenzene in the presence of palladium acetate and pivalic acid in *t*-BuOH at high concentration led to the formation of **33**. Monoarylated **33** was directly subjected to the C—H olefination reaction with styrenyl iodide in the presence of $Pd(OAc)_2$ and 1.5 equivalent of AgOAc in toluene to produce the all-*cis* cyclobutane **34** (Scheme 12.4). The required stereochemistry was obtained from **34** by using sodium methoxide. Ester hydrolysis and cleavage of the directing group was done with 1N NaOH and DIBAL-H to give the aldehyde **35**. The crude aldehyde **35** was coupled with piperidine to give the amide **36** in 45% yield upon isolation after three steps. Finally, treatment of aldehyde **36** with Ando phosphonate **C** in the presence of KO*t*-Bu afforded the pipercyclobutanamide A **37** in 80% yield.

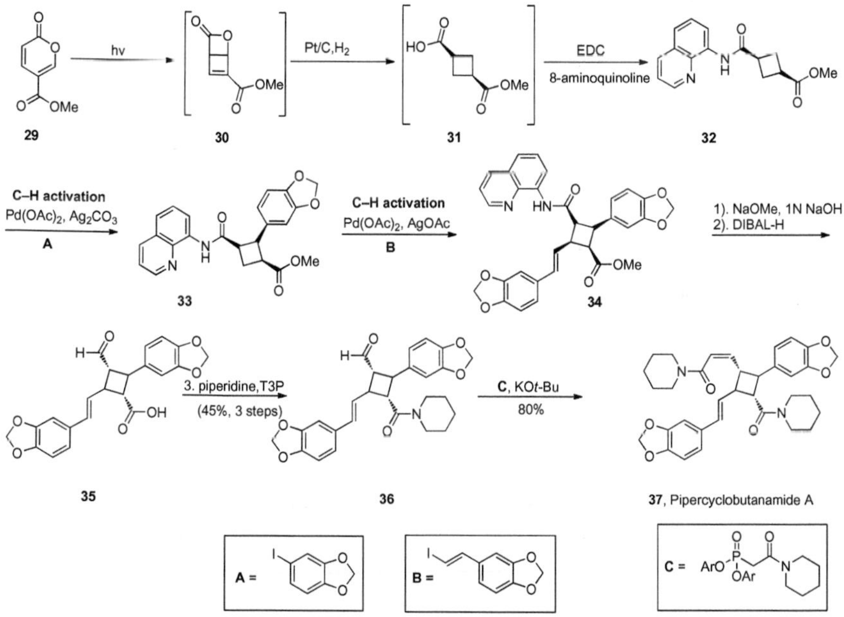

Scheme 12.4 Synthesis of pipercyclobutanamide A.

12.5 TOTAL SYNTHESIS OF PIPERARBORENINE B AND D

Baran and co-workers reported palladium-mediated C—H activation on a cyclobutane ring.[12] Similar to the synthesis of pipercyclobutanamide A, methyl cumulate undergoes photocyclization to an intermediate photopyrone which was further reduced with Pt/C to provide the desired carboxylic acid. An EDC coupling with 2-aminothioanisole provided the C—H arylated substrate. The selective C—H cross-coupling of **38** with

3,4,5-trimethoxyiodobenzene provided the challenging intermediate **39** in 52% yield (Scheme 12.5).

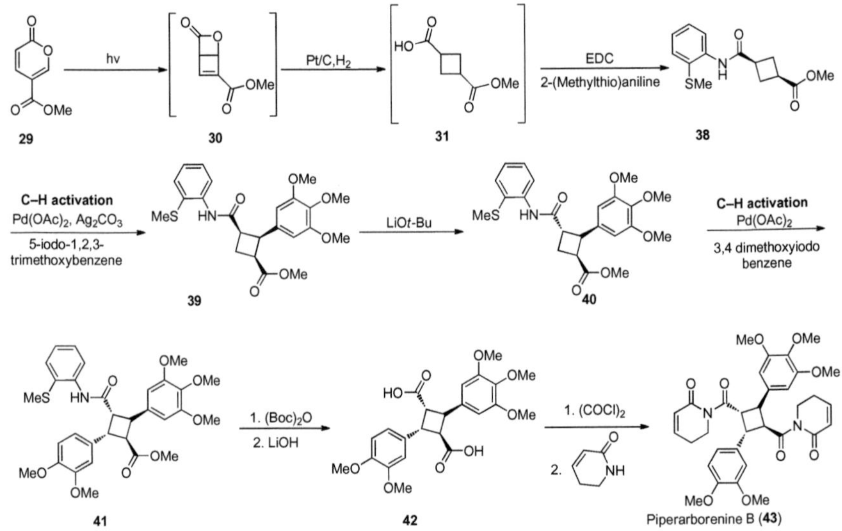

Scheme 12.5 Synthesis of piperarborenine B.

Treatment of **39** with LiO*t*-Bu in a nonpolar solvent selectively epimerized the C-1 stereocenter to give cyclobutane **40** in 79% yield. Subsequently, **40** underwent a second C—H arylation reaction with 3,4-dimethoxyiodobenzene to provide the corresponding amide **41** in 46% yield. Finally, cleavage of the directing group was carried out with lithium hydroxide. Fortunately, the methyl ester also cleaved under these conditions but the carefully constructed stereotetramer remained intact. After conversion of both carboxylic acids to corresponding acid chlorides and heating with dihydropyridone in the presence of molecular sieves, piperarborenine B **43** was obtained in 77% yield.

In order to prepare the stereoisomeric piperaborenine D (**48**), epimer **44** was subjected to a second round of C—H arylation with 3,4-dimethyliodobenzene to provide **45** in 81% yield. Refluxing **45** with sodium hydroxide in ethanol resulted in the formation of bis-carboxylic acid **45** in 86% yield. Conversion to the bis-acid chloride and coupling with dihydropyridone resulted in the proposed structure of piperarborenine D **48** in 78% yield (Scheme 12.6).

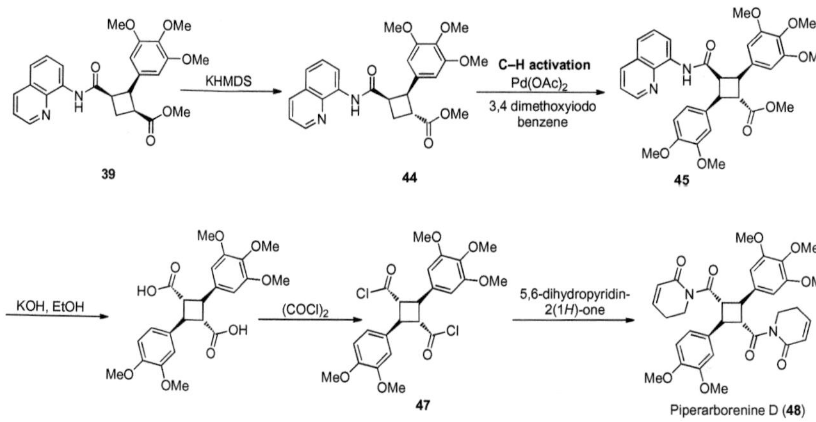

Scheme 12.6 Synthesis of piperarborenine D.

12.6 TOTAL SYNTHESIS OF PODOPHYLLATOXIN

Maimone and co-workers developed a synthetic route for the therapeutically important natural product podophyllatoxin[13] from 6-bromopiperonal. Epoxidation of 6-bromopiperonal followed by benzocyclobutenol formation and subsequent silylation could be afforded in a three-step procedure. Thermal cycloaddition of **49** with fumarate-derived amide afforded C—H activation precursor **50** via an o-quinodimethane intermediate. Attempted coupling of **50** with 3,4,5-trimethoxyiodobenzene under catalytic C—H arylation protocols afforded β-lactam **51** as the major product, along with substantial amounts of recovered starting material (Scheme 12.7).

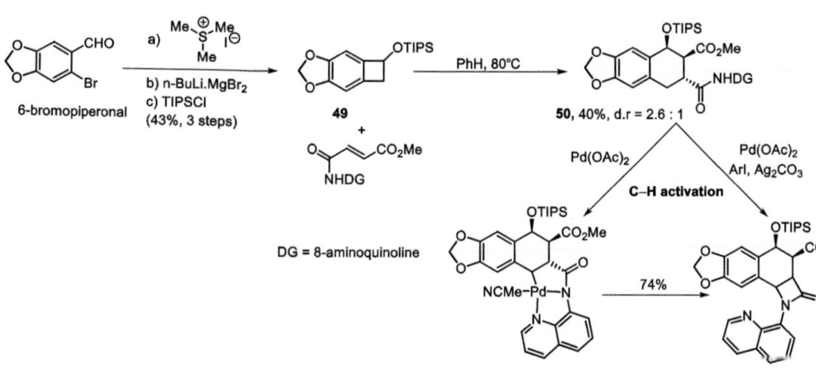

Scheme 12.7 β-lactam synthesis.

The Maimone group synthesized the conformationally distinct substrate **52** by selective ester reduction/desilylation/ketalization starting with **50** (Scheme 12.8). When **52** was heated in the presence of excess 3,4,5-trimethoxyiodobenzene and K_2CO_3, arylated product **53** was formed as the major product (50%) along with a small amount of β-lactam (10% yield). Subsequently, they began to examine catalytic conditions for the C—H arylation reaction.

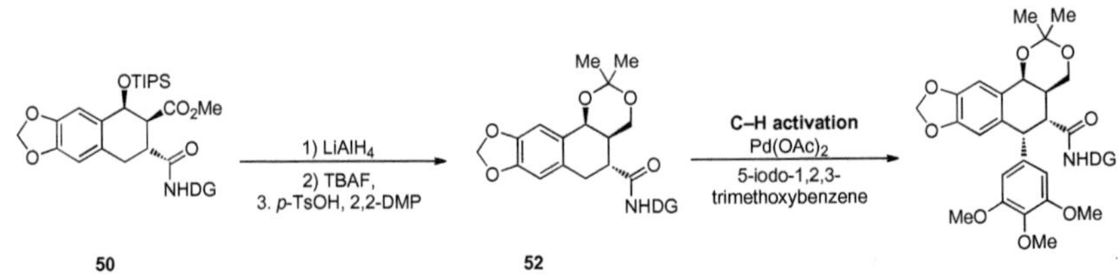

50

52

53, 50% + 10% (β-lacatam)

Scheme 12.8 Aminoquinoline directed C—H arylation.

2-Methylthioaniline-containing amide **55** was prepared in one step from monomethyl fumarate. Addition of superhydride directly to this mixture led to efficient in situ ester reduction. Ketalization produced precursor **56** in 41% yield as a single diastereomer. After adjusting the concentration and reaction time, they were able to achieve a synthetically useful yield of 58% for the arylated product using 2 equivalents of aryliodide and 15 mol % of $Pd(OAc)_2$. Finally, stirring the product in a mixture of TFA/THF/H_2O at room temperature directly afforded podophyllotoxin **57** in 43% as well as C-4 epi-**57** in 33% yield (Scheme 12.9).

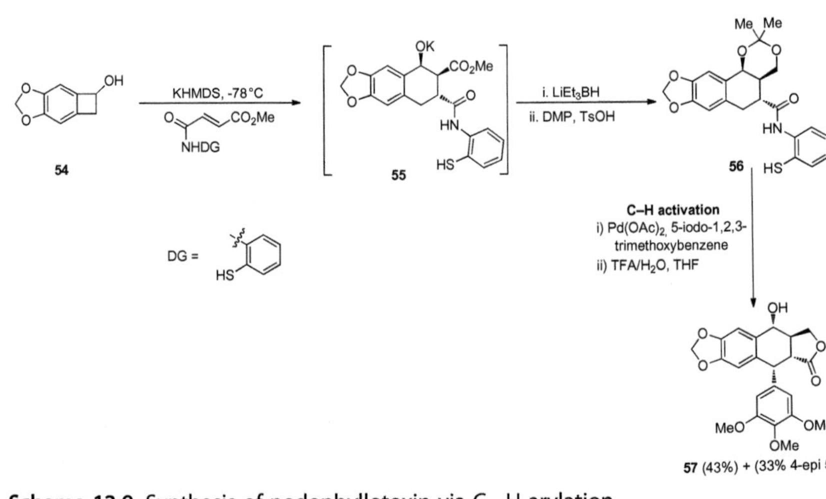

54

55

56

57 (43%) + (33% 4-epi **57**)

Scheme 12.9 Synthesis of podophyllotoxin via C—H arylation.

12.7 TOTAL SYNTHESIS OF CLAVICIPITIC ACID

Jia and co-workers chalked out the synthesis of clavicipitic acid[14] using tryptophan methyl ester **58**. Under the optimized reaction conditions, compound **58** and 2-methyl-3-buten-2-ol reacted smoothly, generating the allylalcohol **59** in 17% yield and the cyclized product **60** in 14% yield. Surprisingly, Pd-mediated cyclization conditions could not convert alcohol **59** to cyclized product **60**. However AgOAc was shown to be an effective catalyst in this cyclization. Treatment of compound **60** with *t*-BuOK in DMSO at room temperature afforded enamine **61** in 80% yield. After several experiments, enamine **61** was reduced with Mg in dry methanol to give *cis*-ester **62** in 36% yield and *trans-ester* **63** in 44% yield. Finally, alkaline hydrolysis of **62** and **63** isomers furnished (+) *cis*-clavicipitic acid **64** and (−)*trans*-clavicipitic acid **65** (Scheme 12.10).

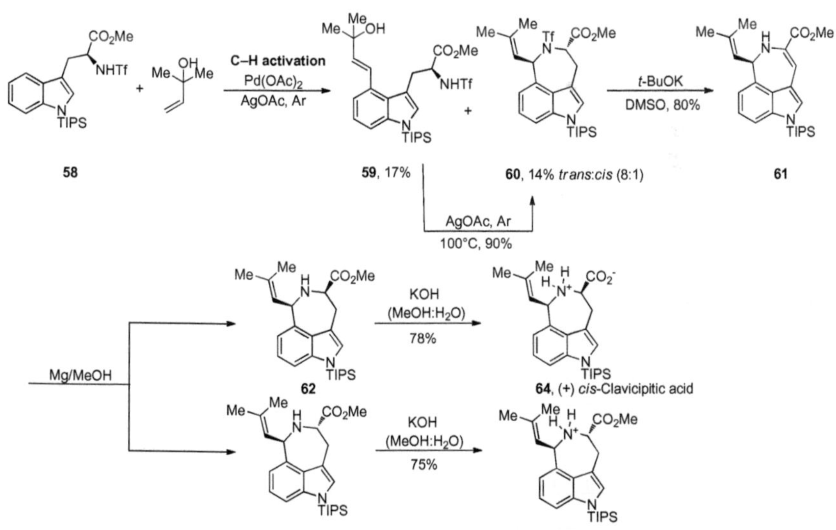

Scheme 12.10 Synthesis of clavicipitic acid.

12.8 TOTAL SYNTHESIS OF HIBISPEPTIN A

Chen and co-workers explored a series of bidentate carboxamide auxiliary groups for the palladium-catalyzed arylation of the γ − methyl group of isoleucine to construct the Ile-Hpa (Isoleucine-4-hydroxy-phenacylamine) pseudo-dipeptide motif in hibispeptins.[15] 4-Hydroxy acetophenone **66** was monoiodinated with I_2. An azido group was then installed at the α-position of the carbonyl group via $CuBr_2$ mediated α-bromination followed by S_N^2 substitution with NaN_3. Staudinger reduction of the N_3 group, Boc protection, and protection of the phenolic —OH with BnBr gave mono-boc phenacylamine in good yield. Treatment of mono-boc phenacylamine with excess $(Boc)_2O$/DMAP gave tri-Boc protected compound **68** in 75% yield (Scheme 12.11).

With the optimized conditions, arylation of pyridylmethylamine (PR)- Ile **69** with tri-Boc protected aryliodide **68** proceeded smoothly to give the desired γ-C—H arylation product **70** in 47% yield.

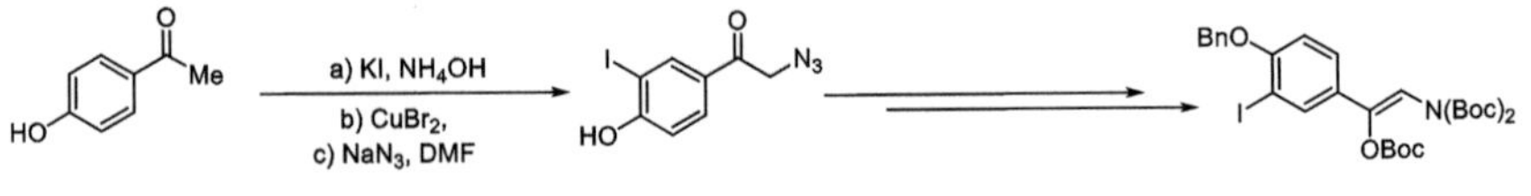

Scheme 12.11 Aryliodide synthesis.

Deprotection of TIPS in **70** with HF/pyridine, oxazolidone formation with triphosgene, and subsequent cleavage of the acyl oxazolidone with $LiOH/H_2O$ gave the key Ile-Hpa residue **71** in good yield and excellent purity (Scheme 12.12). The NPhth protecting group of Ile-Hpa **71** was swapped with a Cbz group by treatment with $NH_2NH_2.H_2O$ and CbzCl. Saponification followed by amide coupling with H-Pro-Me gave tripeptide **72** in excellent yield. The Boc group of tripeptide **72** was removed with HCl and then coupled with dipeptide using EDCI/HOBt to give pentapeptide **73**. Hydrolysis of methyl ester of **73** with LiOH and Boc deprotection with HCl gave the macrolactoanization precursor. The desired macrolactonization at the Pro/Leu site was mediated by pentafluorophenyl diphenylphosphinate (FDPP) proceeded smoothly to give cyclic pentapeptide **74**. The Cbz and Bn groups of **74** were then deprotected by Pd-catalyzed hydrogenolysis. Subsequent coupling with pyroglutamic acid using EDCI/HOBt gave hibispeptin A **75** in moderate yield (Scheme 12.13).

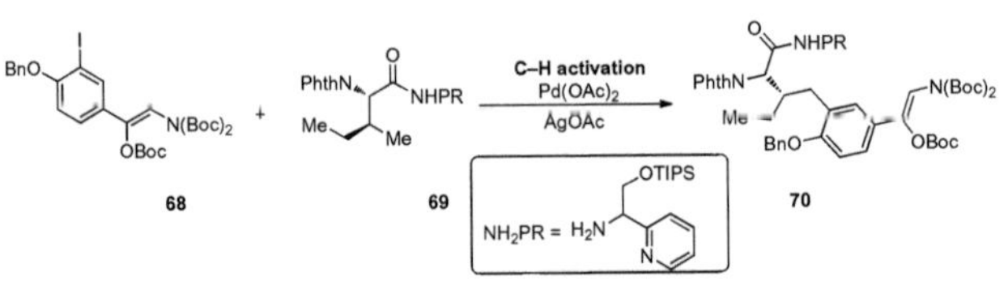

Scheme 12.12 γ-C—H arylation.

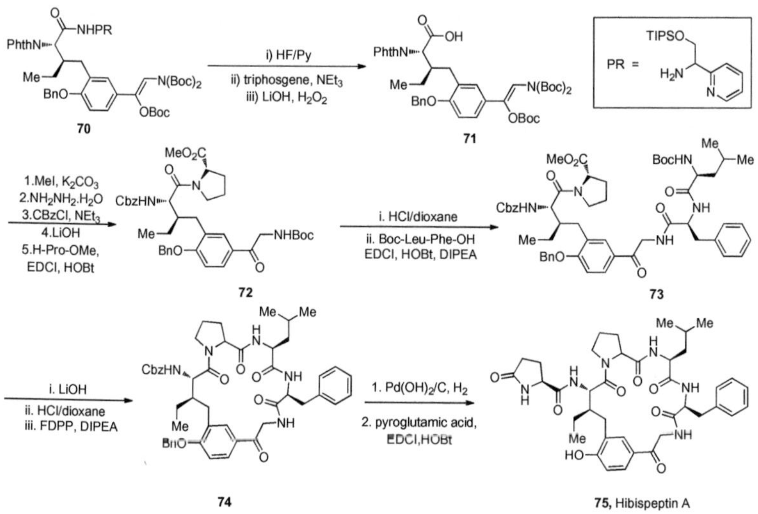

Scheme 12.13 Synthesis of hibispeptin A.

12.9 DIVERGENT SYNTHESIS OF AERUGINOSINS (SP^3 C—H ALKENYLATION AND ARYLATION)

Baudoin and co-workers reported a general and scalable access to the aeruginosin family of natural products using sp^3 C—H activation reactions.[16] Synthesis of the C—H activation precursor was performed from L-alanilol and *tert*-Butyldiphenylsilyl (TBDPS)-protected dibromo-cyclohexenol which provided a 1:1 mixture of two diastereomers. The amine group of the desired diastereoisomer was then protected as trifluoroacetamide, leading to the C—H activation precursor. The key Pd-catalyzed intramolecular C—H alkenylation was then examined. Optimization of palladium sources and ligands allowed the yield to be increased to 71% in gram scale (Scheme 12.14).

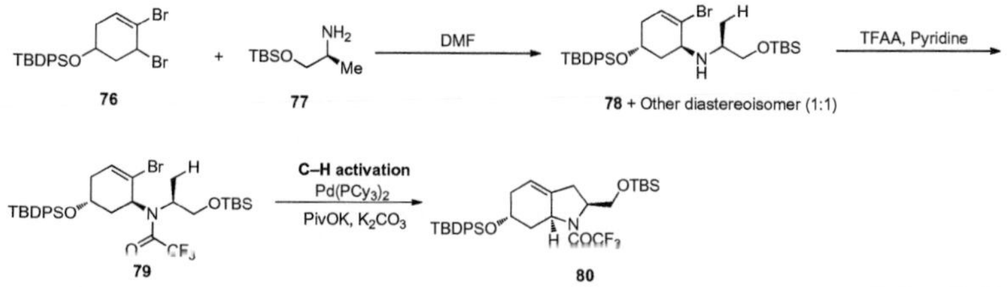

Scheme 12.14 Pd-catalyzed intramoecular C—H alkenylation.

Diastereoselective hydrogenation on the convex face of the cyclic product **80** was achieved by Rh-catalyzed hydrogenation. The *cis*-configuration of the corresponding product was treated with 0.5% HCl in MeOH which provided the alcohol **81**. This primary alcohol was then oxidized under Jones conditions. The corresponding acid was esterified and the trifluoroacetyl group was removed under reductive conditions to give amino ester **82** in good yield (Scheme 12.15).

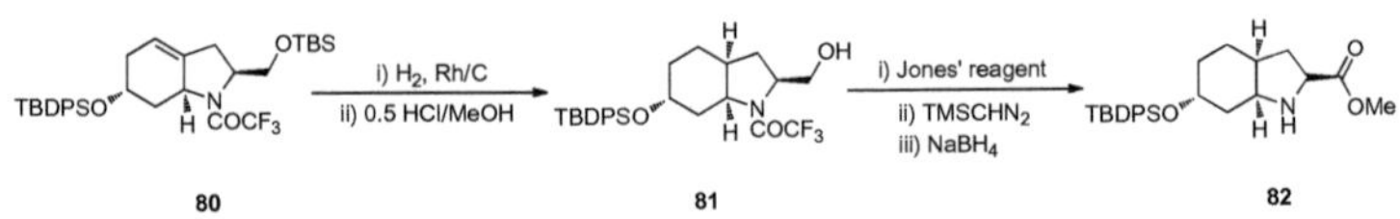

Scheme 12.15 Synthesis of amino ester fragment.

Benzyl protected iodophenol was prepared and submitted to the optimized C–H activation conditions. The corresponding arylated product **84** was isolated in 78% yield without detectable - racemization. Hydrolysis of the amide afforded the acid **85**, which on subsequent coupling with D-leucine methyl ester gave rise to the southern part of aerugniosins **86** (Scheme 12.16).

The ester part of fragment **86** was hydrolyzed and the corresponding acid was subjected to peptide coupling with **82**. The resulting product **87** was obtained in 76% yield. Methyl ester **87** was then hydrolyzed and coupled with L-argol fragment. Finally, acid mediated cleavage of the TBS group on the argol fragment and hydrogenolysis of benzyl and Cbz groups afforded aeruginosin 298 A **89** (Scheme 12.17).

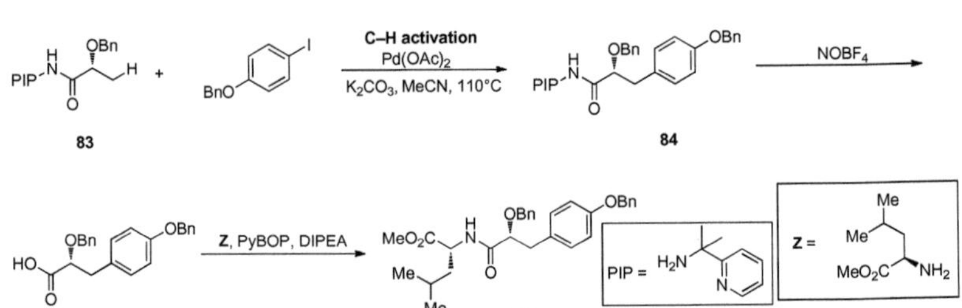

Scheme 12.16 Synthesis of southern part aerugniosins.

12.10 TOTAL SYNTHESIS OF (+) LINOXEPIN

The first application of the Catellani reaction in enantioselective, protection group free natural product synthesis was reported by Lautens

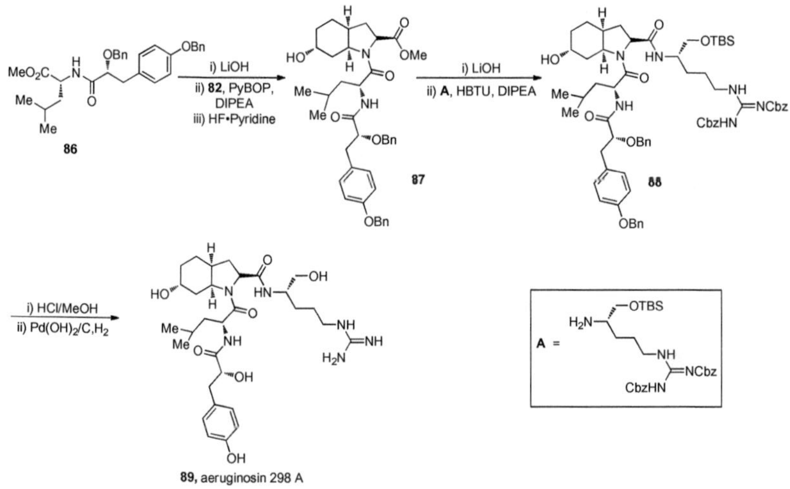

Scheme 12.17 Synthesis of aeruginosin 298 A.

asymmetric synthesis of (+) linoxepin.[17] *Ortho*-alkylation of aryl iodide **92** with enantio-enriched iodolactone **90** followed by olefination with *t*-butylacrylate **91** afforded the key intermediate **93** in 89% yield. The key intermediate **93** was oxidatively cleaved under Lemieux–Johnson conditions followed by $TiCl_4$ mediated intramolecular aldol condensation afforded **94**. Palladium-catalyzed intramolecular Mizoroki–Heck conditions produced (+) linoxepin (Scheme 12.18).

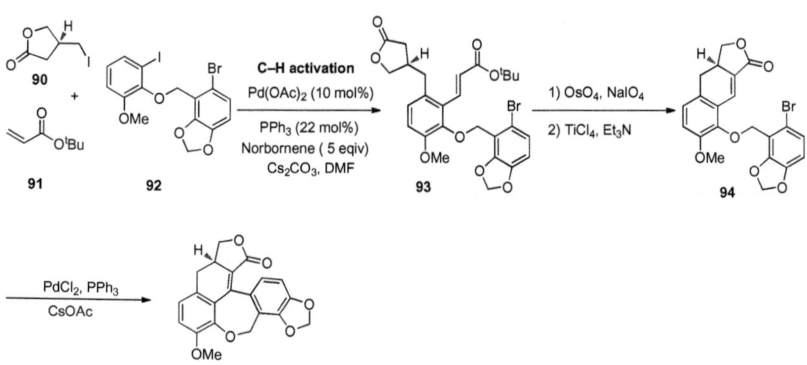

Scheme 12.18 Synthesis of (+) linoxepin *via* Catellani reaction of aryliodide.

12.11 RAPID TOTAL SYNTHESIS OF RHAZINAL

Gu and co-workers reported the palladium-catalyzed chemoselective Catellani reaction of iodopyrroles for the efficient synthesis of rhazinal.[18]

In these cases, the remote ester group on the iodopyrrole **95** allows the tandem Catellani-type *ortho*-arylation/intramolecular Heck reaction under palladium chloride conditions with 2-nitrobromobenzene **96**. This afforded the cyclized key core fragment **97**. With the core fragment in hand, completion of the natural product rhazinal **98** was realized in three additional steps (Scheme 12.19).

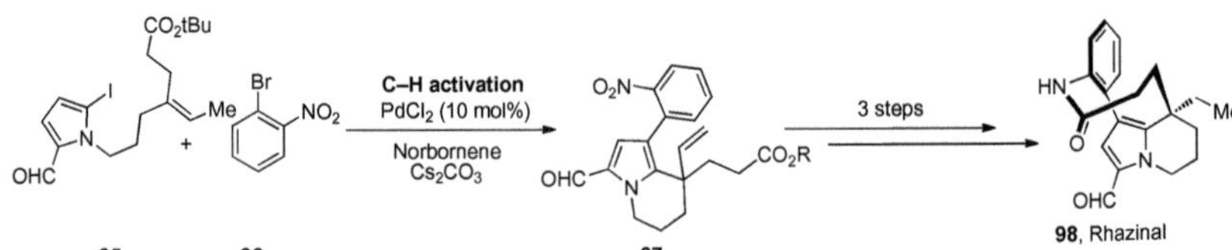

Scheme 12.19 Synthesis of rhazinal via chemoselective Catellani reaction of iodopyrroles.

12.12 TOTAL SYNTHESIS OF GAMENDAZOLE

Guillaumet and co-workers reported a three-step synthesis for gamendazole via palladium-catalyzed oxidative direct alkenylation of indazoles.[19] Regioselective C-3-alkenylation of indazole **99** with methyl acrylate yielded the olefinated compound **100**. Finally, ester hydrolysis with LiOH.H_2O in a mixture of THF and MeOH gave gamendazole **101** in 91% yield (Scheme 12.20)

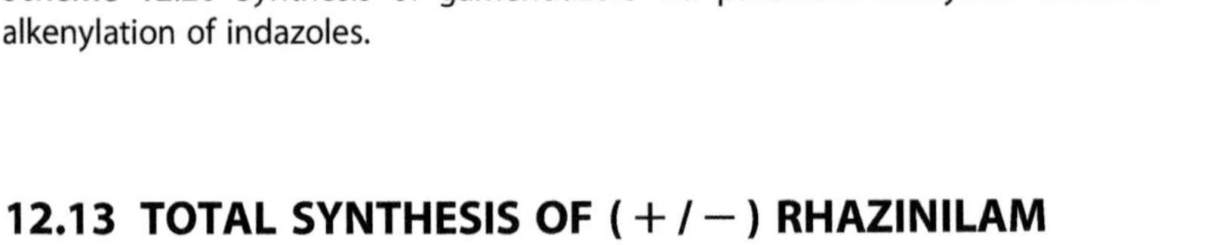

Scheme 12.20 Synthesis of gamendazole via palladium-catalyzed oxidative direct alkenylation of indazoles.

12.13 TOTAL SYNTHESIS OF (+ / −) RHAZINILAM

Yao[20] and co-workers reported the synthesis of (+ / −) rhazinilam via selective C-5-intramolecular alkenylation of pyrrole derivaties. The synthesis of (+ / −) rhazinilam was from phenyl acrylate with TOSMIC through Van Lausen reaction which provided the corresponding substituted pyrrole **102**. This was treated with iodoalkene **103** to give the alkylated pyrrole **104** with a full carbon skeleton. Regioselective intramolecular palladium-catalyzed C—H activation provided the architecture of rhazinilam **105** in 56% yield in 4:7 ratio. The relative configuration was confirmed by X-ray crystallography. Hydrogenation of the core structure followed by decarboxylation, hydrogenation, and macrolactoanization fulfilled the total synthesis of rhazinilam **107** in four steps with a 56% yield (Scheme 12.21).

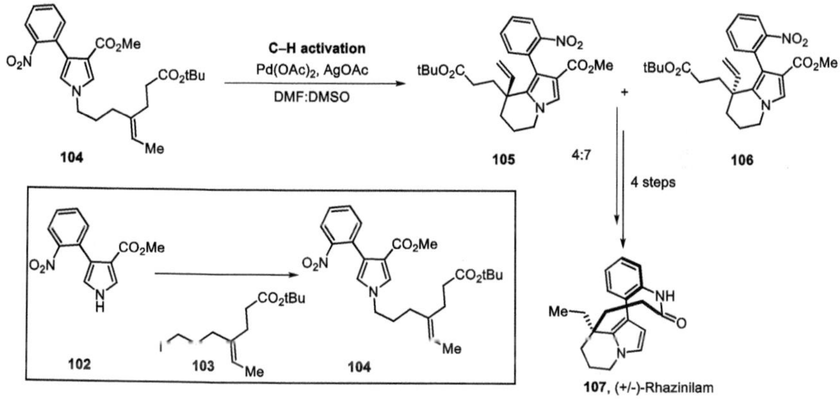

Scheme 12.21 Synthesis of (+/−) rhazinilam *via* selective C5-intramolecular alkenylation of pyrrole derivatives.

12.14 CONCLUSION

Total synthesis of organic compounds has often been eulogised as a divine art in chemical science. It has been an example of a primordial institution since the discovery of science itself but never termed as banal. Application of total synthesis has been shown to span a wide variety of transformations which are of immense significance in the pharmaceutical, materials, and polymer industries. Nevertheless, the foundation of this pristine institution is defined by the numerous synthetic challenges which chemists often encounter while synthesizing complex products. As evident from the exemplars discussed in this chapter, adoption of C—H activation has proven to be a successful venture in spite of these challenges which often include unprecedented site selectivity. This ensured simplified reaction routes, and made them atom- and step-economical, beside being cost-effective. Although significant developments have been made, a plethora of discoveries still await revelation. Strong perception of organic reactions coupled with sound chemical intuition are worthy benefactors in this regard. A united cultivation of scientific research, favorable exchange of erudite ideas, and a diligent commitment towards science are the prime elements which delineate the future developments for this branch of organic synthesis.

REFERENCES

1. Corbet, M.; De Campo, F. *Angew. Chem. Int. Ed.* **2013,** *52*, 9896.
2. Chen, Z.; Wang, B.; Zhang, J.; Yu, W.; Liu, Z.; Zhang, Y. *Org. Chem. Front.* **2015,** *2*, 1107.
3. Yamaguchi, J.; Amaike, K.; Itami, K. Transition Metal-Catalyzed Heterocycle Synthesis via C—H Activation*;* Wiley-VCH Verlag GmbH & Co. KgaA: Weinheim, 2016505.
4. Gutekunst, W. R.; Baran, P. S. *Chem. Soc. Rev.* **2011,** *40*, 1976.
5. Davies, H. M. L.; Morton, D. *J. Org. Chem.* **2016,** *81*, 343.
6. Lyons, T. W.; Sanford, M. S. *Chem. Rev.* **2010,** *110*, 1147.
7. Nakamura, I.; Yamamoto, Y. *Chem. Rev.* **2004,** *104*, 2127.
8. Chen, X.; Engle, K. M.; Wang, D.-H.; Yu, J.-Q. *Angew. Chem. In. Ed.* **2009,** *48*, 5094.
9. Jiao, L.; Herdtweck, E.; Bach, T. *J. Am. Chem. Soc.* **2012,** *134*, 14563.
10. Wang, D.-H.; Yu, J.-Q. *J. Am. Chem. Soc.* **2011,** *133*, 5767.
11. Gutekunst, W. R.; Gianatassio, R.; Baran, P. S. *Angew. Chem. Int. Ed.* **2012,** *51*, 7507.
12. Gutekunst, W. R.; Baran, P. S. *J. Am. Chem. Soc.* **2011,** *133*, 19076.
13. Ting, C. P.; Maimone, T. J. *Angew. Chem. Int. Ed.* **2014,** *53*, 3115.
14. Liu, Q.; Li, Q.; Ma, Y.; Jia, Y. *Org. Lett.* **2013,** *15*, 4528.
15. He, G.; Zhang, S.-Y.; Nack, W. A.; Pearson, R.; Rabb-Lynch, J.; Chen, G. *Org. Lett.* **2014,** *16*, 6488.
16. Dailler, D.; Danoun, G.; Ourri, B.; Baudoin, O. *Chem. Eur. J.* **2015,** *21*, 9370.
17. Weinstabl, H.; Suhartono, M.; Qureshi, Z.; Lautens, M. *Angew. Chem. Int. Ed.* **2013,** *52*, 5305.
18. Sui, X.; Zhu, R.; Li, G.; Ma, X.; Gu, Z. *J. Am. Chem. Soc.* **2013,** *135*, 9318.
19. Naas, M.; El Kazzouli, S.; Essassi, E. M.; Bousmina, M.; Guillaumet, G. *Org. Lett.* **2015,** *17*, 4320.
20. Su, Y.; Zhou, H.; Chen, J.; Xu, J.; Wu, X.; Lin, A.; Yao, H. *Org. Lett.* **2014,** *16*, 4884.

INDEX

Note: Page numbers followed by "*f*" and "*t*" refer to figures and tables, respectively.

D

K

L

M

Q

R

S

T

V